W9-BAA-258

"Information through Innovation"

Systems Analysis, Design, and Implementation

John G. Burch
University of Nevada, Reno

boyd & fraser

bf

boyd & fraser publishing company

TO MATTHEW GILLAND BURCH

A tough fighter

Publisher: Thomas Walker
Acquisitions Editor: James H. Edwards
Production Editor: Barbara Worth
Interior Design, Composition, and Project Management: Spectrum Publisher Services, Inc.
Cover Art: "Cross Country," © Nancy DeYoung and Alon Gallery
Cover Design: LMY Studio, Inc.
Cover Art Research: HeeMoon Chang

boyd & fraser

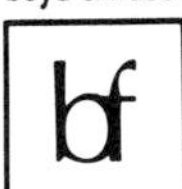

© 1992 boyd & fraser publishing company
A Division of South-Western Publishing Company
Boston, MA 02116

Manufactured in the United States of America

Library of Congress Cataloging-in-Publication Data
Burch, John G.
Systems analysis, design, and implementation / John Burch.
p. cm.
Includes bibliographical references and index.
ISBN 0-87835-818-8
1. System analysis. 2. System design. I. Title.
QA76.9.S88B87 1992
004.4'1—dc20 91-33841
CIP

1 2 3 4 5 6 7 8 9 D 4 3 2 1

Brief Contents

Contents

Preface

In today's competitive world, one of the most valuable resources is a well-designed, user-oriented information system. Such information systems can improve productivity by automating labor-intensive tasks, eliminating nonvalue-added processes, and coordinating disparate activities. Information systems have the potential to enhance product and service differentiation by giving customers easier access to products and services, improved quality and quicker response, more tracking and status information, and a broader range of products and services. Strategically, information systems can help managers combat competitors, innovate, reduce conflict, and adapt to head-spinning changes in the marketplace.

ABOUT THIS BOOK

This book is intended to be used as the textbook for a course in computer information systems development. This book assumes a reasonable understanding of computer concepts, terminology, and programming. Additionally, this book should prove beneficial for people in industry and others who are seeking an understanding of information systems development.

Although the vast majority of people who will take a course in computer information systems development will become systems professionals, others will also benefit from this book. Anyone who plans to be a user of computer information systems, such as marketing, accounting, and production managers, auditors, and various operations personnel, needs to gain a better understanding of how computer information systems are developed. Indeed, user participation is critical in developing successful information systems. User participation is normally much more effective when users have an understanding of how computer information systems are supposed to be developed.

This book can be used as the mainline textbook for lecture-based, case-based, or project-based courses. If the course is offered as a two-term sequence, often used in project-based courses, Parts I and II are applicable in the first term and Parts III and IV are applicable in the second term. Other instructors, however, may use Parts I, II, and III for the first term, and Part IV for the second term. If the course is single-term, often found in lecture-based and some case-based courses, the entire book is applicable. The book works well with supplementary material, such as computer-aided systems and software engineering (CASE) tools.

DISTINGUISHING FEATURES

This book has been tested in several courses using lecture-based, case-based, and project-based approaches. It works well with all approaches. An overall distinguishing feature coming from these classroom tests was that student evaluations of the book were excellent. A pervasive comment was: "I like the scope of the book and its real-world flavor." Other specific distinguishing features follow.

Real-World Cases

This book includes an optimum balance between concepts and techniques. Difficult-to-understand concepts and techniques are introduced by real-world cases in which such concepts and techniques are applied. Real-world cases aid understanding and motivate students to study the material. In addition, an ongoing case is integrated throughout the book beginning in Chapter 3.

User Participation

A major emphasis of this book is user participation. Computer information systems development is normally much more successful when users become involved to help develop "their" system. Systems work, indeed, is much more than methodologies, modeling tools, and techniques. Students need to gain an understanding of the user or people side of systems development.

Systems Development Life Cycle (SDLC)

The SDLC presented in this book includes, in addition to systems analysis, general systems design, detailed systems design, and implementation phases, systems planning and systems evaluation and selection phases. More and more, systems professionals are emphasizing the need for systems planning, and managers are increasingly demanding that systems be stringently evaluated both qualitatively and financially before major funds are committed for their implementation.

Joint Application Development (JAD)

JAD sessions (or workshops) are ideal ways to foster user participation. JAD is explained and applied in several places in the book.

Rapid Application Development (RAD)

RAD is a technique that gets systems developed and operating quickly. Elements of RAD include JAD, specialists with advanced tools (SWAT) teams, CASE tools, and prototyping.

Structure-Oriented and Object-Oriented Approaches

Although this book focuses more on the classical structure-oriented approach, the object-oriented approach is introduced and compared to the structure-oriented approach in Chapter 6 and applied in Chapter 16.

Systems Design Components

The book divides computer information systems into six design components: output, input, process, controls, database, and technology platform, which includes telecommunication networks and computers. Each systems design component is treated in detail, enabling students to understand how to design each in accordance with user requirements.

Systems Maintenance

A pressing problem in many companies today is information systems that are difficult and costly to maintain. How to build computer information systems that are easier and less costly to maintain is stressed throughout the book. Moreover, Chapter 20 is devoted to how to manage and perform systems maintenance efficiently and effectively.

Pedagogy

The book gets the student off to a fast start. The first three chapters set the foundation and the remaining 17 chapters deal with topics and applications in the same sequence as the SDLC. As stated earlier, this book is filled with cases to augment understanding. Each chapter starts with learning objectives and an introduction. Where appropriate, figures are presented to illustrate and clarify text material. After the principal body of the chapter is presented, a review of chapter learning objectives is given along with a summary checklist. Key terms are included at the end of each chapter. A glossary of key terms is included at the end of the book. Assignment material includes review questions, chapter-specific problems that are based on chapter material and require an exact solution, and think-tank problems that expand student perspectives and require a feasible response or approach, not necessarily an exact solution. As stated earlier, this book is flexible enough to support a lecture-based, case-based, or project-based course.

ANCILLARIES

Comprehensive support packages accompany *Systems Analysis, Design, and Implementation*. These ancillary materials are available to instructors upon request from boyd & fraser.

Instructor's Manual

Material in the Instructor's Manual follows the organization of the book. Chapters of the Instructor's Manual include:

- Solutions to assignment material
- Transparency masters

CASE Package and Tutorial

A CASE package is provided to enable students to create designs on the computer. The CASE tutorial walks students through a series of lessons that demonstrate the use of CASE for systems analysis and design.

Test Item File

A test bank includes over 2000 items based on true/false and multiple choice formats. Answers are provided. MicroSWAT III, a microcomputer-based testing program, is also available.

Study Guide and Casebook

This ancillary is for the student to gain further insight into this book's concepts and techniques. It includes:

- Self-test and answers
- Study tips
- Problem-solving tips
- Demonstration cases using methodologies, modeling tools, and techniques selected from this book

ORGANIZATION OF THIS BOOK

This book contains a table of contents, 20 chapters, and a glossary. It is organized the same way that computer information systems are developed. With the exception of Part I, topics of this book unfold as the development of systems unfolds.

The book is organized into four parts. Part I, Chapters 1 through 3, provides an introduction to systems development. Chapter 1 introduces the systems development life cycle (SDLC), prototyping, and joint application development (JAD). Chapter 2 presents the key features of the structured approach to developing systems and shows how to apply supporting modeling tools. Chapter 3 introduces ways to manage systems projects and discusses full-featured computer-aided systems and software engineering (CASE) technologies. After completing Part I, students have acquired an understanding of methodologies, modeling tools, and techniques necessary to develop successful information systems. With this background, students are ready to pursue systems development.

There are two broad categories of systems development: front-end phases and back-end phases. Part II treats the front-end phases. Chapter 4 is the first phase of the systems development life cycle (SDLC), and it determines those systems projects that are feasible and worthwhile for development. Chapter 5 is the second front-end phase, and it establishes systems scope and specifies user requirements. Chapter 6 is the third front-end phase, and it creates conceptual design alternatives that meet user requirements. Chapter 7 is the last front-end phase. Its purpose is to evaluate all conceptual designs with the objective of selecting the optimum one.

After successfully completing the front-end phases, students are ready to begin the back-end phases. Part III covers detailed systems design, the first back-end phase. The purpose of this phase is to convert the selected conceptual design into a functional design. To do so requires the development of detailed specifications for six systems design components: output, input, process, database, controls, and technology platform.

Chapter 8 shows how to design output, both in form and substance, that is acceptable to users. Chapter 9 treats ways to capture input. Chapter 10 introduces ways to convert input to output. Chapter 11 covers how stored data are managed and organized to support the system. Chapter 12 includes a treatment of techniques used to protect the information system from a number of hazards and abuses. Chapter 13 shows how telecommunications can be used to establish a network backbone for the system. Chapter 14 discusses ways that computer technology can be configured to meet systems design objectives. Thus Chapters 13 and 14 establish ways to design the technology platform for the system under development.

The results of detailed systems design is a systems "blueprint." This blueprint is used for systems implementation, the final back-end phase of the SDLC and the subject of Part IV. Chapter 15 introduces the software development life cycle to be used in those situations that call for in-house software development. Chapter 16 discusses tools used to convert systems design to software design. Chapter 17 describes programming languages and the way the languages are used to code software. Chapter 18 covers tools used to test software. Chapter 19 pulls together and explains all the tasks necessary to implement the total system and convert it from development to operations. Chapter 20 covers those tasks that must be performed after the system becomes operative.

The job-order cost system (JOCS) case runs parallel with Chapters 4 through 20. The purpose of this case is to show students how selected methodologies, modeling tools, and techniques are applied in a real-world situation.

A glossary is included at the end of the book. It defines key terms used throughout the book.

TO THE STUDENT

Following are some items that will help you come to grips with the book material.

Learning Objectives and Chapter Review of Learning Objectives

Learning objectives listed at the beginning of each chapter indicate what you will achieve by studying and applying chapter material. At the end of the main body of the chapter, a review of chapter learning objectives is presented, which provides a nice closure to the body of the chapter.

Summary Checklist

Immediately following the review of chapter learning objectives is a summary checklist. This section provides a list of how to apply the methodologies, tools, and techniques presented in the chapter.

Key Terms and Glossary

Significant words are introduced in the text in boldface where they are defined and used in context. At the end of each chapter is a list of key terms appearing in that chapter. At the end of the book is a glossary that defines key terms.

Embedded Cases

Wherever hard-to-understand methodologies, modeling tools, and techniques are presented, we have an accompanying specific case to show how they are applied in a real-world setting. Such cases not only show you how methodologies, modeling tools, and techniques are applied, but also show you why.

Running Case

The job-order cost system (JOCS) is a running case that applies selected methodologies, modeling tools, and techniques from each chapter, starting in Chapter 3 and ending in Chapter 20. Its purpose is to demonstrate how certain methodologies, modeling tools, and techniques are applied and how a system is developed from start to finish. Should your instructor assign a project, JOCS, as well as the embedded cases, will provide you with guidance in performing your project.

Assignment Material

The assignment material consists of three elements: review questions, chapter-specific problems, and think-tank problems. The review questions require you to recall important chapter material. The chapter-specific problems require you to apply what you have learned in the chapter, and, on occasion, from previous chapters to derive an exact solution. The think-tank problems require you to go beyond the bounds of chapter material, make your own assumptions, and create a feasible approach. Think-tank problems mirror real-world situations, which do not always lend themselves to exact solutions.

Study Guide and Casebook

A Study Guide and Casebook accompanies the book. It is designed to help you understand the book material better and to gain additional working knowledge about how to develop computer information systems. It contains self-tests with answers, problem-solving tips, and demonstration cases.

A Final Note

We wish you success in your study of the exciting, important, and rapidly changing field of computer information systems development. We hope you find it as interesting, challenging, and rewarding as we do.

ACKNOWLEDGMENTS

I would like to acknowledge several people for their contributions in the development of this book.

I am grateful to students in my systems analysis courses, who used preliminary versions of this book. I appreciate not only their many helpful suggestions but also their patience and support.

I am also indebted to Jeff Carleton, both for his diligence in preparing the figures and for his work on several cases and problems and the development of ancillary material. His creativity and hard work have made this a better book.

Thanks are also due to Dana Edberg for her work on JOCS.

My thanks to the following reviewers for their comments and suggestions during the preparation of this book:

Robert Clark
Buffalo State College

Mohammed Dadashzadeh
Wichita State University

Constanza Hagmann
Kansas State University

Cary Hughes
Middle Tennessee State University

Karen-Ann Kievit
Loyola Marymount University

David Yen
Miami University

The efforts of the staff of boyd & fraser deserve high praise. My thanks to James H. Edwards, senior acquisitions editor, for his vision and faith in this book; Tom Walker, vice president and publisher, for his encouragement and support; Barbara Worth, production editor, and Becky Herrington, director of production, for their attention to detail and skills in the book's production; and Abigail Grissom and Arthur Weisbach for their developmental editing. My thanks to Rosanne Coit, editorial assistant, for her thoughtfulness and conscientious work.

My deepest thanks, however, go to my wife, Glenda, who gave her full support throughout the project. She performed most of the hard work, and without her, it is doubtful that this book would ever have been written.

John G. Burch

Part I
Introduction to Systems Development

The purpose of Part I is to describe the methods, tools, and techniques used to develop information systems. Chapter 1 introduces the systems development life cycle (SDLC) and prototyping.

Chapter 2 presents the structured approach that applies an engineering discipline to systems analysis and design and to software development. Chapter 2 covers data flow diagrams (DFD), data dictionary, entity relationship diagrams (ERD), state transition diagrams (STD), structure charts, structured program flowcharts, process specification tools, Warnier–Orr diagrams (WOD), and Jackson diagrams.

Chapter 3 explains systems development management. It also covers computer-aided systems and software engineering (CASE), which helps to automate systems development and maintain systems.

Chapter 1
Systems Development Methodologies

WHAT WILL YOU LEARN IN THIS CHAPTER?

After studying this chapter, you should be able to:

1 Cite causes of systems development failure, and list ways to overcome these causes.
2 Define the systems development life cycle (SDLC) and discuss each of its phases.
3 Explain prototyping and its use in developing systems.
4 Relate the information engineering methodology (IEM) to the SDLC of this chapter.
5 Describe joint application development (JAD) and discuss the role of users in systems development.

INTRODUCTION

This chapter introduces systems development work. Chapters 2 and 3 present modeling tools, project management techniques, and systems development technologies. These three chapters set the foundation for the rest of the book. Chapters 4 through 20 present additional concepts and techniques and apply the material presented in the first three chapters. Real-world cases show how this material is used for building information systems. As part of the casework, the book introduces Peerless, Incorporated, a manufacturing firm that is developing a job-order costing system known as JOCS. The Peerless story is a running case, portions of which appear at the end of each chapter, beginning with Chapter 4 and ending with Chapter 20. The Peerless case demonstrates how methodologies, modeling tools, and techniques are used to develop JOCS for Peerless.

PEOPLE AND SYSTEMS DESIGN

Systems analysts and systems designers are the **SYSTEMS PROFESSIONALS** who build information systems. The people who employ those systems are generally referred to as **USERS** (also end users). They are the managers and workers who interact with the information system and depend on it to perform their job.

All kinds of users are served by information systems. A manufacturing firm consists of engineers, factory workers, plant managers, logistics personnel, and so forth, all of whom require different things from information systems. A bank employs loan officers, tellers, and managers who must depend on the bank's information systems. Similarly, nurses, hospital administrators, doctors, and support personnel are relying more and more on hospital information systems. Some of these users, such as nurses or bank tellers, interact directly with systems through terminals. Others benefit from information generated by someone else, such as the bank officer who receives a list of delinquent accounts prepared by loan officers using a system. As the user/system interface becomes friendlier and easier to use, more and more users will interact directly with information systems.

Any change in information systems will force users to change their behavior. The users may resist change. Systems professionals can get users involved in systems development. Thereby they can develop information systems that perform as users want them to perform, rather than users who perform as the information system wants them to perform. If users are involved in systems development, the information systems can become their systems, and they will perceive that no one will be shoving the systems down their throats.

Another payoff from involving users in systems development is that systems professionals gain from it a better understanding of the organization's operations and business functions. Users know what works and what does not work as far as their jobs are concerned, and they have better insight as to how any information system should work for them. Indeed, users and systems professionals working together, both bringing their expertise to systems development, will almost always ensure success. We therefore stress user involvement throughout this book.

Systems professionals must have a strong ability to move between user and technical environments. This book is written for those who aspire to become systems professionals, working in both worlds, like Larry Riggins in the Rutherford LSI sample case.

An Organized Approach to Systems Development at Rutherford LSI

Rutherford LSI is a medium-sized (approximately 1100 employees) producer of high-quality semiconductor chips for use in a variety of commercial applications. Some of these large-scale integrated (LSI) electronic circuits have been used in computers, automobiles, and home appliances. The company was formed in the late 1960s primarily as a military contractor, using the then-new semiconductor integration technology to produce lightweight guidance and control systems for jet fighter aircraft and submarines.

These military products were produced in small numbers, but due to their complexity and the high level of reliability required, they were lucrative for the company. Unfortunately, in the early 1970s the company suffered several technical setbacks and its products gained a reputation for poor quality. In 1973, Rutherford lost several of its largest contracts and filed for bankruptcy.

As part of its corporate reorganization plan, strategic management at Rutherford decided to phase out all remaining military product lines over a period of two years and to begin producing electronic circuits for the commercial and consumer markets. These new products did not require the extreme reliability of the military circuits, and management felt the company could compete using its adequate, though somewhat outdated technological base. The real challenge would be found in converting Rutherford's low-volume production operation over to the high-volume, mass production facility required to compete in the company's new target markets.

By the late 1970s it was apparent that Rutherford LSI had been successful in its reorganization. The company had even solved its earlier technological problems and had updated its production methods significantly. Now it was feeling a new pressure, that of increasing foreign presence in the semiconductor industry. Robert McCallister, then Rutherford's chief executive officer (CEO), decided that the company needed to implement a computer information system in the production areas and management offices. In a memo to his managers, McCallister stated: "The use of computers, linked throughout the company, will streamline our operations by providing timely information to managers. This is precisely what we need to combat our low-cost competitors in the Far East." He then instructed the managers to determine how many terminals would be needed in each department and where each should be located. Meanwhile, a special room was set up to house the mainframe and related offices. Rutherford then sought bids from computer and software vendors, and by late 1979 installation of the new system was completed.

Rutherford LSI survived the 1980s, certainly not as a result of its information system, but by introducing innovative new products in advance of its foreign competition. Everyone agrees that the information system installed back in 1979 is a real hindrance to productivity. The current CEO, Linda Jameson, is very much aware that a lack of planning was the reason for the failure of the present information system. She knows that proper development of a replacement system will be a long, involved process, but will be necessary if a new system is to be successful. For this reason, she has mandated the replacement of this system by the end of the next fiscal year, with development work beginning immediately.

Next, a systems development team is created, including a new hire, Larry Riggins. Larry's experience in systems engineering at his former job results in the unanimous decision to make him team leader. Soon, everyone on the team has heard of the "systems development life cycle (SDLC)," the organizational approach Larry is using to coordinate the team's efforts. Later, they learn about and make use of other techniques, such as "prototyping" and "joint application development (JAD)."

This chapter is concerned with describing the SDLC and some of these associated systems analysis and design techniques. Larry Riggins' system development team can be assured that by making use of these methodologies, the information system ultimately implemented at Rutherford LSI will be a success.

THE CHALLENGE TO DEVELOP BETTER INFORMATION SYSTEMS

Why should we be concerned about learning the methodologies, modeling tools, and techniques used for developing information systems? Why should we involve users in systems development? Why don't we just start building information systems, let things evolve, and solve problems as we come to them? Or why don't we simply acquire computers and software and then permit people to use them as they see fit? The history of developing information systems in a haphazard, unplanned, and unstructured fashion testifies to the importance of taking an organized approach instead.

What Makes a System Unsuccessful?

Causes often cited for the lack of success in developing viable information systems include the following:

- Systems developed that did not support business strategies and objectives
- Poor systems planning and inadequate project management
- Failure to define or understand user requirements and get users involved in systems development
- Negligence in estimating costs and benefits of the systems project
- Creation of a myriad of design defects and errors
- Acquisition of computers and software that no one needs or knows how to use
- Installation of incompatible or inadequate technology
- Negligence in implementing adequate controls
- Development of unstructured, unmaintainable software
- Inadequate implementation tasks

Problems caused by poorly developed systems can range from loss of life to loss of assets, to loss of customers and revenue, to management's making wrong decisions based on inaccurate or untimely information, to wasted time and decreased productivity.

What Makes a System Successful?

Elimination of the causes of systems failure lie in the application of methodologies, modeling tools, and techniques to design information systems development and build information systems that not only meet the needs of users but also are delivered on time and within budget.

Information systems professionals have worked for years to develop, teach, and use workable methodologies, modeling tools, and techniques that will reduce, if not eliminate, these problems. The general intent is to transform the development of systems from an informal, sloppy art to a structured, engineered, and managed approach that involves the user throughout the process. Nowadays, with the systems development life cycle (SDLC) and prototyping methodologies, modeling tools, project management techniques, and **COMPUTER-AIDED SYSTEMS AND SOFTWARE ENGINEERING (CASE)** systems (covered in Chapter 3), the transformation is closer to reality than ever before. Systems professionals now have ways to perform successful systems development by:

- Stressing user involvement in systems development
- Implementing systems planning and using project management techniques
- Developing alternative systems designs for evaluation before making major commitments to final design, technology, and software development
- Designing all systems design components functionally
- Using the detailed functional design as a guide for software design, coding, and testing

- Preparing clear, complete, and current documentation
- Using a coordinated, planned approach to systems implementation
- Performing postimplementation reviews
- Designing for and performing systems maintenance

SDLC

The **SYSTEM DEVELOPMENT LIFE CYCLE (SDLC)**, portrayed in Figure 1.1, consists of six phases:

1 Systems planning

2 Systems analysis

3 General (or conceptual) systems design

4 Systems evaluation and selection

5 Detailed (or functional) systems design

6 Systems implementation

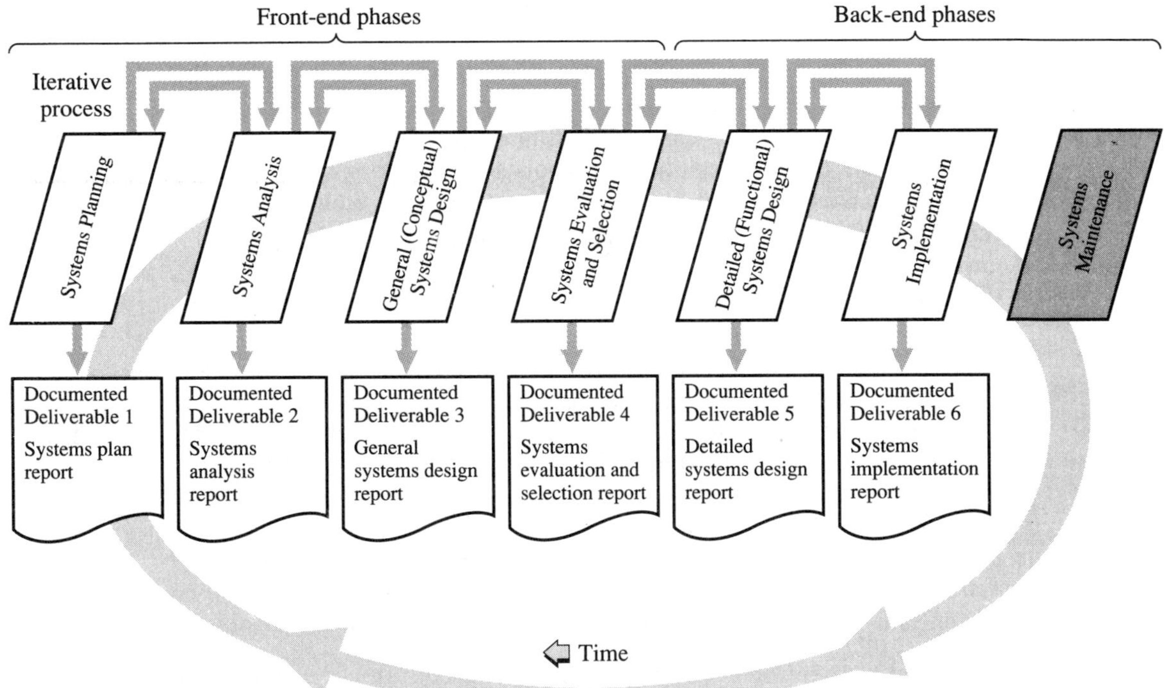

Figure 1.1
The systems development life cycle (SDLC).

The first four phases are sometimes called the **FRONT-END PHASES**; the last two development phases and systems maintenance are often termed the **BACK-END PHASES**. After the new system is developed and converted to operations, it goes into a systems maintenance phase that lasts several years, sometimes 10 to 20 years, or longer. When it becomes no longer efficient and effective to maintain the system, it is discontinued and a new system is developed to take its place. The SDLC begins all over again.

Figure 1.1 depicts key features of SDLC. Systems development is project-oriented. It requires time to complete; thus a time flow is shown. Also, the time flow represents iteration within SDLC and total systems life cycle. Once the system is converted to operations, it has to be maintained over the system's life; when the old systems die, the SDLC is repeated. The figure shows that systems development is not necessarily sequential, but iterative (or recursive) in nature. The slanted phases indicate that one phase may start before the preceding phase is finished.

The DOCUMENTED DELIVERABLES provide documentation of the system. They also aid in showing progress of the system under development. Further, they are used as a base for forward planning.

The Front-End Phases of Systems Development

The front-end phases of systems development are:

- Systems planning
- Systems analysis
- General (conceptual) systems design
- Systems evaluation and selection

The front-end phases represent the conceptual aspects of systems development. They must be user-driven. (The back-end phases consist of the functional aspects of systems development; they are chiefly designer- and technologist-driven.) One of the main reasons one conducts the front-end phases is to explore new systems concepts and to determine exactly what users need before designing a system in detail. To do otherwise causes backtracking and costly delays.

Each phase of the SDLC produces a documented deliverable that discloses the results of work conducted during that phase. The documented deliverable also indicates that a milestone has been reached.

Documented deliverables provide users and managers with information on how the system is being developed. Users and managers get the opportunity to request changes or sign-off on the documented deliverable if they agree with what has been developed to that point. Continuation of the systems project into the next phase depends on review and sign-off by these people. Each documented deliverable becomes a logical extension of prior reports. This way, documented deliverables provide a clear trail from planning through implementation.

The documented deliverables for the front-end phases are written for users of the system. An outline of these documented deliverables is presented in Figure 1.2. Material in Part II will help you prepare the content of these deliverables. The following material, however, will give you an introduction to how each front-end phase is conducted.

Figure 1.2
Outline of front-end documented deliverables.

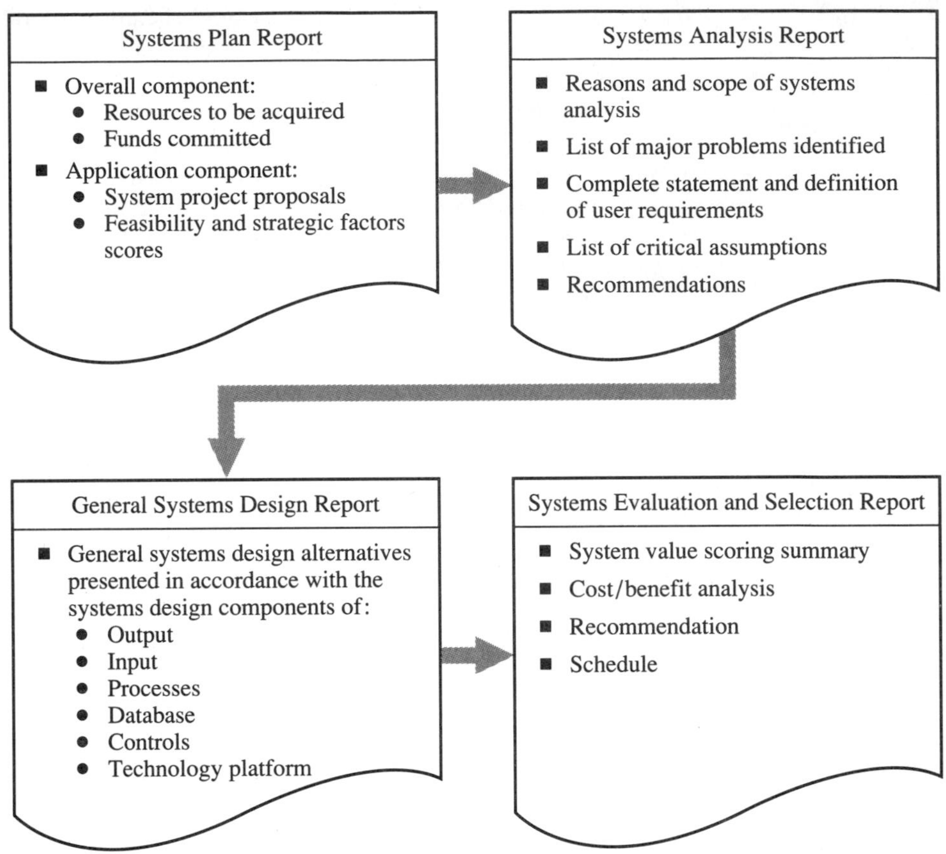

Systems Plan Report
■ Overall component:
• Resources to be acquired
• Funds committed
■ Application component:
• System project proposals
• Feasibility and strategic factors scores
Systems Analysis Report
■ Reasons and scope of systems analysis
■ List of major problems identified
■ Complete statement and definition of user requirements
■ List of critical assumptions
■ Recommendations
General Systems Design Report
■ General systems design alternatives presented in accordance with the systems design components of:
• Output
• Input
• Processes
• Database
• Controls
• Technology platform
Systems Evaluation and Selection Report
■ System value scoring summary
■ Cost/benefit analysis
■ Recommendation
■ Schedule

The Systems Planning Phase

The SYSTEMS PLANNING phase establishes a broad strategic framework and clear vision of the new information system that will satisfy the users' information needs. Systems planning sessions are conducted. They involve senior management, users, and systems professionals to find how the system can support the organization's business plan. Proposed systems projects are evaluated and prioritized. Those with high priorities are selected for development. New resources are planned for, and funds are committed to support systems development.

During the systems planning phase, FEASIBILITY FACTORS, which refer to the likelihood of the information system being successfully developed and used, and STRATEGIC FACTORS, which relate to the information system's support of business goals, are weighted for each proposed project. The resulting scores are evaluated to determine which systems project receives the highest priority.

TELOS Feasibility Factors A proposed system must be feasible; that is, it must meet these criteria:

- TECHNICAL FEASIBILITY shows if the proposed system can be developed and implemented using existing technology or if new technology is needed.

- **ECONOMIC FEASIBILITY** shows if adequate funds are available to support the estimated cost of the proposed system.
- **LEGAL FEASIBILITY** shows if there is conflict between the system under consideration and the company's ability to discharge its legal obligations.
- **OPERATIONAL FEASIBILITY** shows if existing procedures and personnel skills are sufficient to operate the proposed system or if added procedures and skills will have to be acquired.
- **SCHEDULE FEASIBILITY** means that the proposed system must become operative within an acceptable time frame.

Together, these feasibility factors are referred to by the acronym **TELOS**. We will apply the TELOS feasibility factors in Chapters 4 and 7.

PDM Strategic Factors Besides being feasible, a proposed information system project must also support strategic factors. The critical strategic factors include the following:

- **PRODUCTIVITY** measures the amount of outputs produced by inputs. Its purpose is to reduce or eliminate nonvalue-added costs. It can be measured by ratios, such as total weekly labor costs compared to number of units produced during the week or amounts of raw material input during the week compared to the number of finished goods produced during the week.
- **DIFFERENTIATION** measures how well an enterprise can offer a product or service that is significantly unlike in kind and character from its competitors' products and services. Differentiation can be achieved through increased quality, variety, special handling, quicker service, lower cost, price, and so on.
- **MANAGEMENT** shows how well the information system provides information to aid managers in planning, controlling, and decision making. Providing a report about production efficiency every day rather than once a month or converting reams of computer printouts to one meaningful graph may improve the management strategic factor.

Together, the strategic factors are referred to by the acronym **PDM**. We will apply the PDM strategic factors in Chapters 4 and 7. Information systems that improve PDM significantly can become some of the most valuable strategic resources for those enterprises.

The **SYSTEMS PLAN REPORT** is the chief documented deliverable produced by the systems planning phase. It covers resources and funds required to develop and operate the new information system, all the systems project proposals that will make up the total information system, and the TELOS feasibility factors and PDM strategic factors scores for each systems project proposal. A complete Systems Plan Report will be prepared in Chapter 4.

The Systems Analysis Phase

The Systems Plan Report produced in the systems planning phase provides a base to form a systems project team and begin systems analysis. During **SYSTEMS**

ANALYSIS, the systems project team gains a clearer understanding of the reasons for developing a new system. The scope of systems analysis is defined. Systems professionals interview prospective users and work with the users to ferret out problems and define user requirements. Some aspects of the system under development may not be fully known at this point, so critical assumptions are made to permit continuation of the SDLC. At the conclusion of systems analysis, the SYSTEMS ANALYSIS REPORT is prepared; it contains findings and recommendations. If agreement is achieved, the systems project team is ready to embark on the general systems design phase. Should there be disagreement, the systems project team must perform additional analysis until all participants are in agreement.

The General Systems Design Phase

The GENERAL SYSTEMS DESIGN phase creates conceptual design alternatives for user review. These are an extension of user requirements. Conceptual design alternatives allow managers and users to select the design best suited to their needs.

The essence of general systems design is to describe, in fairly broad, high-level terms, how each systems design component of output, input, processes, controls, database, and technology platform will be designed. A description of these systems design components is the main content of the GENERAL SYSTEMS DESIGN REPORT.

The Systems Evaluation and Selection Phase

The end of the general systems design phase provides a major checkpoint for an investment decision. It is therefore in the SYSTEMS EVALUATION AND SELECTION phase that the qualitative value of the system and cost/benefit of proceeding with the systems project is carefully assessed and disclosed in the SYSTEMS EVALUATION AND SELECTION REPORT. The back-end phases of detailed systems design and systems implementation are expensive and time-consuming. If none of the conceptual design alternatives produced in the general systems design phase prove to be justifiable, then all should be scrapped. Normally, several should prove justifiable, and the one with the highest rating is selected for the back-end work.

The Back-End Phases of Systems Development

Completion of the front-end phases provides a solid foundation on which to implement the conceptual systems design. This work involves two back-end phases, which are:

- Detailed (functional) systems design
- Systems implementation

An outline of the documented deliverables for the back-end phases is shown in Figure 1.3. Material presented in Parts III and IV will help you prepare the contents for these documented deliverables.

The Detailed Systems Design Phase

The DETAILED SYSTEMS DESIGN phase provides specifications for the conceptual design. In it, all components are designed and described in detail.

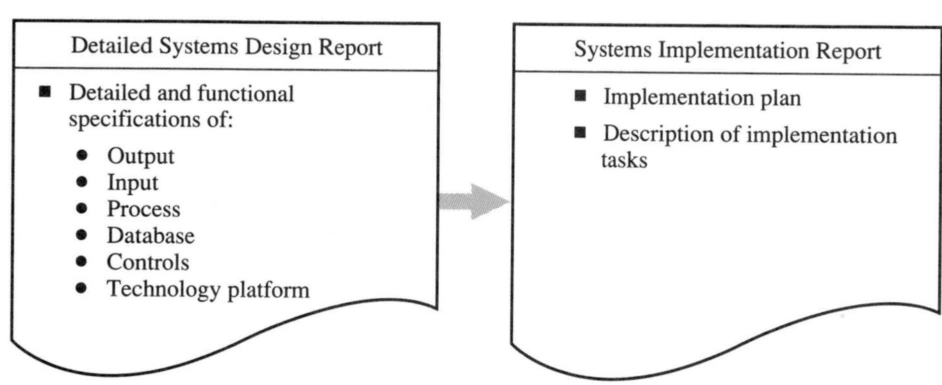

Figure 1.3 Outline of back-end documented deliverables.

OUTPUT layouts are designed for all screens, special forms, and printed reports. All outputs are reviewed and approved by users and documented. All INPUTS are specified and formats, both screen and paper forms, are also reviewed and approved by users and documented. Based on output and input designs, specific PROCESSES are designed to convert the inputs to outputs. Transactions may be captured and entered online or batched. Various models are developed to transform data into information. Procedures are written to guide users and operations personnel on how to work with the system. A DATABASE is designed to allow storage and ready access to interim data. CONTROLS that are needed to protect the new system from various threats and errors are specified. In some systems projects, a new or different TECHNOLOGY PLATFORM is required calling for the design and acquisition of various computers, peripherals, and telecommunication networks.

If the systems project is fairly sizable and complex, the detailed systems design phase normally requires a number of different technical specialists on the systems project team. Depending on the application, these specialists may include the following:

- Forms designers
- Database specialists, computer auditors, security, and control personnel
- Expert system and decision support system specialists
- Computer and software engineers
- Telecommunication technicians

At the conclusion of the detailed systems design process, a DETAILED SYSTEMS DESIGN REPORT is compiled. It may contain thousands of documents with all the specifications for each systems design component integrated into a whole. This report represents a complete guide for software designing, coding, and testing; equipment installation; training; and other implementation tasks.

Although a number of people have reviewed and approved each systems design component, a SYSTEMS DESIGN WALKTHROUGH (i.e., complete review) of the Detailed Systems Design Report should be conducted by both users and management personnel and, possibly, by other systems professionals not involved in the design process (for peer review). The purpose of a systems design walkthrough is to catch errors and design flaws before implementation begins. If errors, design flaws, or omissions can be found prior to implementing the system, valuable resources will be saved and costly mistakes will be avoided. For large systems projects that span many departments or divisions, several individual systems design walkthroughs may be required by different user groups. After all systems design walkthroughs are completed, changes are made and users and managers sign-off on the Detailed Systems Design Report.

The Systems Implementation Phase

Now the system is ready for construction and installation; that is, the **SYSTEMS IMPLEMENTATION** phase. A number of tasks must be coordinated and conducted.

The **SYSTEMS IMPLEMENTATION REPORT** is composed of two segments. The first is the systems implementation plan, typically in the form of a **GANTT CHART** or **PROGRAM AND EVALUATION REVIEW TECHNIQUE (PERT) CHART**. (Gantt and PERT charts are project scheduling and management techniques that are introduced in Chapter 3 and used in several other places in the book.) The second segment of the Systems Implementation Report describes those tasks necessary to perform systems implementation, including:

- Software development
- Site preparation
- Equipment installation
- Testing
- Training
- Documentation preparation
- Conversion
- Postimplementation review

Software development is a major construction and implementation task in most new systems. In fact, in some systems projects this task has its own development methodology, called the **SOFTWARE DEVELOPMENT LIFE CYCLE (SWDLC)**, which is presented in the last part of the book. The SWDLC includes three phases:

- Software designing
- Software coding
- Software testing

After detailed systems design is performed, it will require, as part of its support, software—either programs acquired from software vendors or programs developed in-house. If the software is developed in-house, detailed systems design is decomposed to produce **SOFTWARE DESIGN**. The second phase of SWDLC, which converts the software design into program code, is **SOFTWARE CODING**. The third and final phase of SWDLC is **SOFTWARE TESTING**, which observes, examines, and evaluates the structure, quality, and functionality of the software.

If new technology is recommended, site preparation and equipment acquisition are started as soon as the detailed systems design phase is completed and accepted. Management may want to prepare the computer site or sites and install computers, telecommunication networks, and systems software well before the detailed systems design phase is complete so that the technology platform will be in place for other tasks, such as software development, testing, training, and conversion. The installation of a technology platform can require much time and can delay implementation of a new system if it is not started early in the systems project.

Four types of documentation are finished during implementation:

- Systems documentation
- Software documentation
- Operations documentation
- User documentation

SYSTEMS DOCUMENTATION comes from the detailed systems design phase. (Its documented deliverable describes functional design features of the new system.) **SOFTWARE DOCUMENTATION** describes the structure of the software program, related input and output, and test cases and results. This documentation is important for software maintenance. **OPERATIONS DOCUMENTATION** gives computer operators and operation managers a technical description of systems operations and security and control procedures. **USER DOCUMENTATION** ensures that different users receive the documentation they need to perform their tasks. All of this documentation must reflect the training materials in form and content. The training tasks must be coordinated with the documentation task, because training materials amplify documentation with examples and step-by-step instructions. (The remaining chapters, especially Chapter 17, will help you prepare these four types of documentation.)

To cut over to the new system, hardware, software, and database conversion must be performed. There are four basic conversion methods: direct, parallel, phase-in, and pilot. Each method is defined and described in detail in Chapter 19.

The final systems implementation task is sometimes called a **POSTIMPLEMENTATION REVIEW**. This process generates its own report, which serves as an addendum to the Systems Implementation Report. Such a review is conducted at any time from a few weeks to six months after systems conversion. A set of interviews is conducted during the postimplementation review to discuss users' reactions to the new system now in operation. These interviews and observations can uncover enhancements that can be made during systems maintenance.

MAINTAINING THE SYSTEM AFTER IT IS IMPLEMENTED

Although **SYSTEMS MAINTENANCE** is often the longest and costliest phase of the systems life cycle, it is not a SDLC phase. Systems maintenance is stressed in this book because the success of SDLC phases is to a great extent dependent on how well systems maintenance can be and is performed after the system converts from development to operations. Chapter 20 is devoted to systems maintenance.

Some causes of systems maintenance problems are that:

- Systems design did not meet user requirements and organizational needs.
- Inflexible and defect-laden design did not allow systems maintenance to be performed, or forced it to be performed at a high cost.

The first cause can be overcome by following SDLC and generating documented deliverables. The second cause can be eliminated by using modeling tools and techniques that support SDLC.

USING PROTOTYPING TO DESIGN A WORKING SYSTEMS MODEL

PROTOTYPING is the production of a working model of a system or subsystem. The prototyping process is like storyboarding. Storyboarding is a technique from the theatrical world used to describe a series of panels on which is tacked a set of small rough drawings of the important changes of a scene and its action. Prototyping is an analysis and design technique that allows users to take part in defining requirements and shaping what the system will do to meet those requirements. In many cases, prototyping helps define user requirements that could not have been defined during systems analysis. Depending on the system, the number of prototypes can range from several to hundreds.

Prototypes can be hand-drawn or created by an application program, such as Apple's MacDraw or MacPaint, American Intelliware's Storyboarder, Microsoft's PowerPoint, and IBM's PC Storyboard. Computer-aided systems and software engineering (CASE) technologies, discussed in Chapter 3, provide prototyping. Also, the **FOURTH-GENERATION LANGUAGES (4GLs)**, discussed in later chapters, can serve as prototyping tools. Such application programs can easily be installed on a microcomputer. By working with the users, a series of prototypes can be viewed and revised interactively.

Why Prototyping Is Used

A prototype may be used as a basis for a more refined prototype; used as a pilot and then later thrown away; used in conjunction with the systems development life cycle (SDLC) to develop difficult-to-define user requirements and to expedite the SDLC process; or sometimes used as is.

An example of the prototyping process is presented in Figure 1.4. The systems professional works with a user or users to find user requirements. A screen prototype is developed; it is based on what is thought to be a full

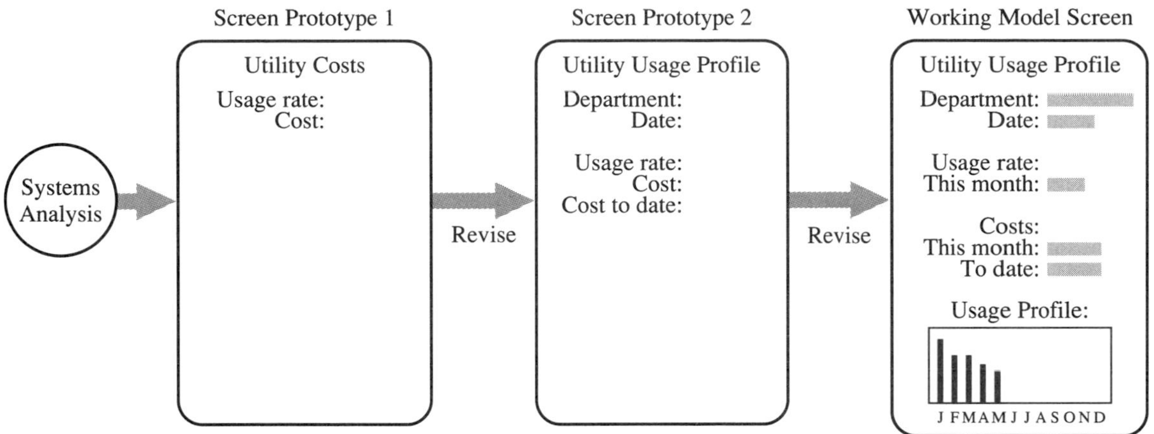

Figure 1.4
The prototyping process.

definition of user requirements. The systems professional and user evaluate the prototype to decide if it meets user requirements and to determine if it can be improved. In our example, the user was not satisfied with Screen Prototype 1, because it did not provide enough identification and useful data. So, following user comments, this prototype was revised to produce Screen Prototype 2, from which a working model was created with which both the systems professional and user are satisfied.

The results of closer working relationships between the users and systems professional narrows the gap between what users say or think they want from the system and what they actually get. The users are brought directly into the process so that the application becomes their project.

Prototyping is best used to develop systems that are poorly defined. Prototyping is also appropriate for unique small systems applications. In almost any case, prototypes enhance visualization and communications.

In many instances, what users ask for is not what they want, and what they want is not really what they need. Prototyping helps solve this double dilemma. If users don't like a prototype, another prototype is built, and so on until consensus is reached. Applied in this manner, prototyping really amounts to building systems by learning and discovery.

How Prototyping Works in Conjunction with SDLC

SDLC is much longer and broader in scope than the prototyping life cycle. Prototyping is a small-scale view of SDLC, except that prototyping creates a working physical model of proposed designs, where SDLC creates conceptual and general systems design strategies. How prototyping is used in conjunction with SDLC is shown in Figure 1.5.

Prototyping can help during systems development by clarifying user requirements and reviewing alternative design models. As a design tool, prototyping is used only to arrive at a functional definition of what the user wants from the new system. After the prototype has been developed to the users' satisfaction, it is used as a reference for the remainder of the systems project.

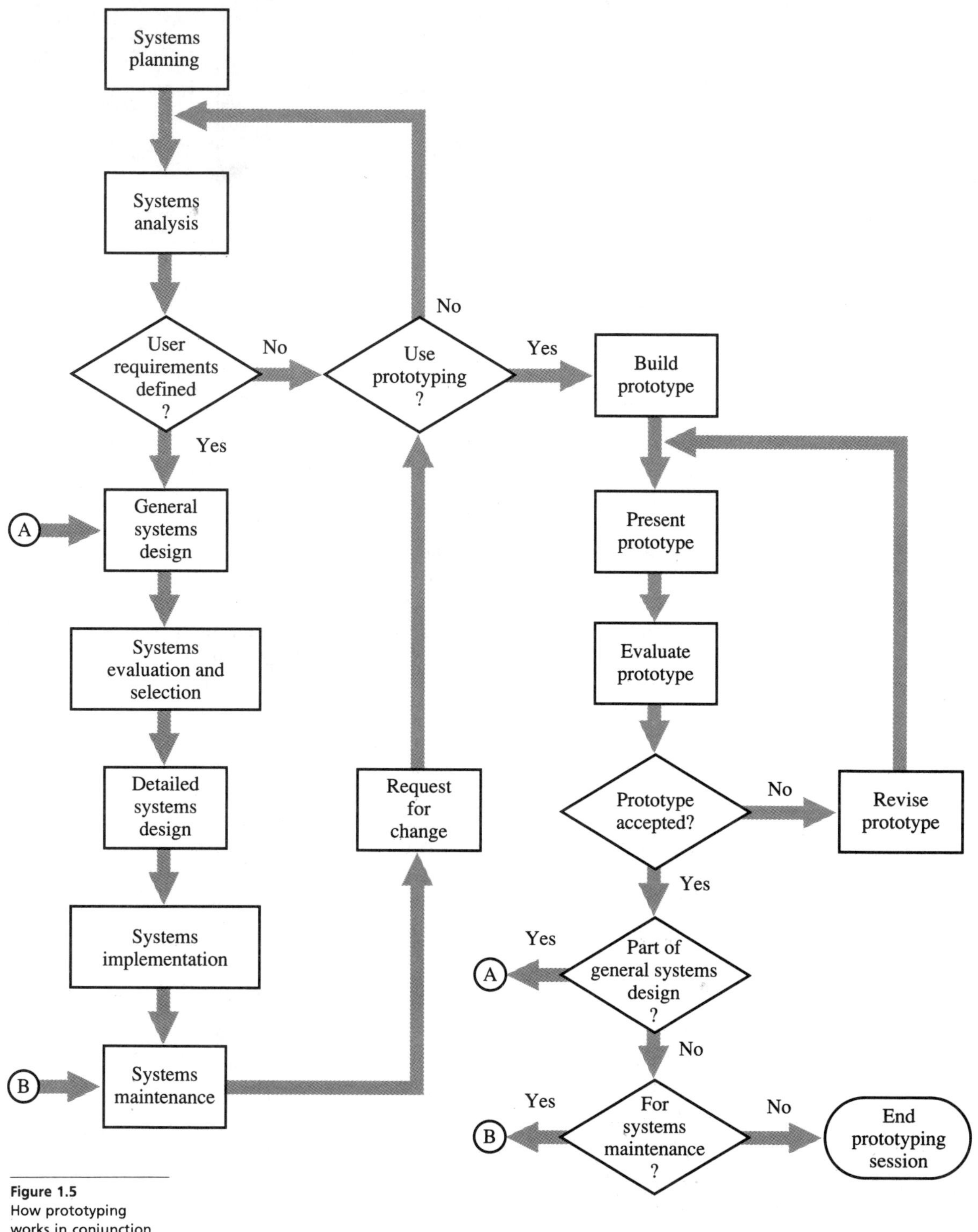

Figure 1.5
How prototyping works in conjunction with the SDLC.

System Characteristics	Methodology	
	Prototyping	SDLC
User requirements	User has difficulty in defining requirements	User requirements are generally well defined
Input, output, and transactions	Low in volume	High in volume
Database	Small number of records and elements within records	Large number of records and elements within records
Controls	Basic editing controls	Extensive system of controls, including sophisticated security controls
Technology	Usually a standalone computer with a "private" database	Usually a widespread multiuser computer system, often interconnected by an enterprisewide telecommunication network
Type of application	Parochial and ad hoc such as special reports, decision support systems (DSS), or executive information systems (EIS)	Distributed throughout the organization for general functions such as transaction processing, inventory control, production scheduling, and accounting
Number of users	Few	Many

Figure 1.6
Comparison of prototyping and the SDLC based on system characteristics.

See Figure 1.6 for characteristics of prototyping and SDLC. This figure suggests various applications of both methodologies.

THE INFORMATION ENGINEERING METHODOLOGY

Different organizations use different systems development methodologies that may have more or fewer steps and be called by different names. However, they all have phases similar to the methodology covered in this book. For example, the **INFORMATION ENGINEERING LIFE CYCLE (IELC)** is divided into the following phases:

- Information strategy planning
- Business analysis
- Business systems design
- Technical design
- Construction
- Transition
- Production

Another systems development methodology is the **INFORMATION ENGINEERING METHODOLOGY (IEM)**. It includes four phases:

- Systems planning
- Systems analysis
- Systems design
- Systems construction and implementation

These phases are presented in Figure 1.7.

The Ideal Systems Development Methodology

To eliminate those causes of unsuccessful systems development alluded to earlier in this chapter, a systems development methodology, with its supporting modeling tools and techniques, should:

- Interact with and encourage user involvement throughout systems development
- Reduce time and cost of developing systems
- Provide rigor and discipline to the systems development process
- Improve systems quality
- Produce complete and accurate design specifications and documentation
- Produce a system that meets user requirements and is easy to maintain

JOINT APPLICATION DEVELOPMENT: A TECHNIQUE FOR INVOLVING USERS IN SYSTEMS DEVELOPMENT

JOINT APPLICATION DEVELOPMENT (JAD) is a technique developed by James Martin and Associates for the express purpose of involving users in systems development. JAD sessions can be used in all phases of any systems development methodol-

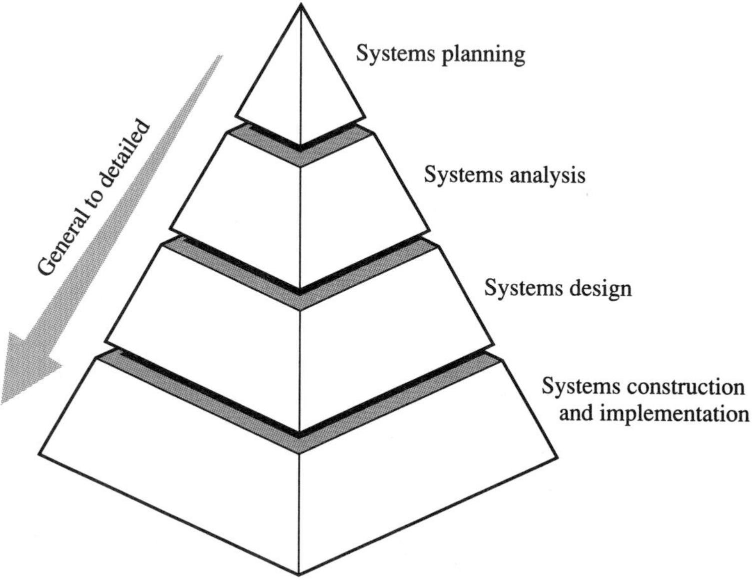

Figure 1.7
A view of the information engineering methodology (IEM).

ogy, but they are particularly helpful during systems planning, systems analysis, and general (or conceptual) systems design.

A Typical JAD Layout

A typical JAD layout is shown in Figure 1.8. The JAD room is a separate room configured specifically for JAD sessions. Participants sit at a U-shaped table. They are end users and systems professionals. Separate tables are provided for the facilitator, scribe, and observers. White boards are used to capture dialogue. Flip charts may also be used, and flip-chart paper may be hung on walls. A workstation containing computer-aided systems and software engineering (CASE) is used by the scribe to capture user specifications and to display prototypes on the screen. The JAD room does not contain telephones, beepers,

Figure 1.8
Layout of a JAD room.

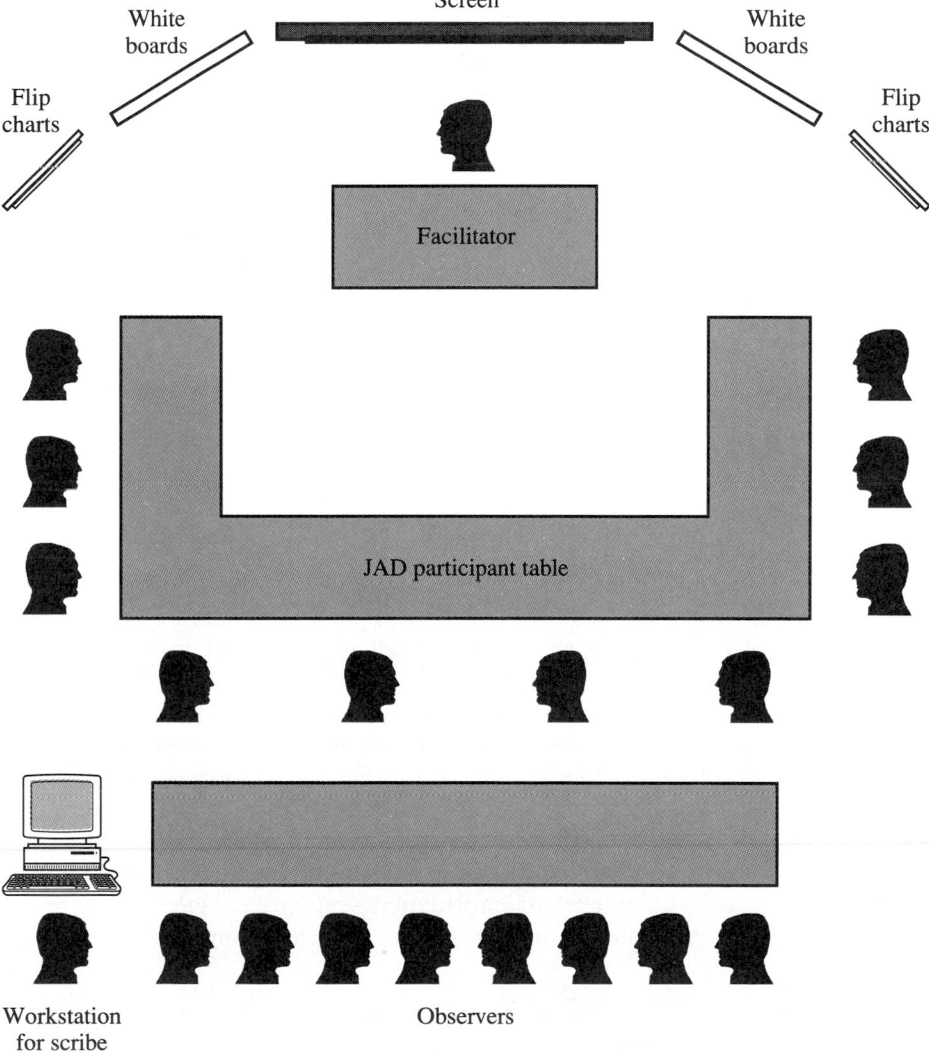

or other means of communicating with the outside world. The objectives are to get the participants away from day-to-day business operations and to focus on the JAD session. The outcome is agreement between users and systems professionals as to what the general information system requirements are.

None of the participants can pull rank, no matter what their authority is outside the JAD room. All egos are left behind. The facilitator must have a special ability to encourage communication among participants. He or she may not be a systems expert, but must be a person who understands group dynamics who can remain independent and objective.

The scribe's function is to record findings of the JAD session. Some of the observers may be other employees in the enterprise. Still others may be invited specialists, such as database administrators, telecommunication technicians, or lawyers who may be called on to resolve technical, legal, or business issues.

User Involvement: The Key to Successful Systems Development

It's a Matter of Working Together

During John Glenn's first ride into space, several heat-shield tiles came loose from his spacecraft. No one in mission control was quite sure if the problem would cause Glenn's orbiter to burn up upon reentry into the earth's atmosphere.

While the engineers at mission control scurried around trying to figure out what to do about the problem, a fellow astronaut suggested they level with Glenn about the predicament. "He's a pilot," the astronaut argued. "He needs to know the condition of his craft."

Perhaps Glenn wouldn't have known all the combinations and permutations of what could happen if those heat shields blew off, but he was the pilot and his life was on the line. He needed to be part of the process and help steer the spacecraft during the reentry. He couldn't do everything, but he needed to be a part of the decision. He needed to help weigh the risks.

John Glenn accepted the challenge when he was informed about the situation, and he accepted the challenge of his damaged spacecraft. He did his best to steer the spacecraft and returned to earth a hero.[1]

These days, systems professionals need to take a tip from the astronaut in mission control. Systems professionals need to treat managers and employees, the users of information systems, as mission control treated John Glenn and involve them in the systems development process. Users can't always deal with all the technical details, but they can and should be in a position to steer. It's their company, their job, and their careers that are on the line.

Users should be informed about choices and options so they can make the right decisions. Indeed, information system decisions are as strategic and important as many other business decisions, and a partnering approach will nearly always result in the best decisions being made.

Systems professionals should communicate with users in plain English. A large number of information system decisions are technology-dependent and technology-oriented. Normally, users don't understand much about informa-

---

[1] Excerpted from Cheryl Currid, "IS Staffers Need to Share Technology Wheel," *PC Week*, June 24, 1991, p. 71.

tion technology and its ramifications. On the other hand, users do have time to understand how this information technology will help them do their jobs better.[2]

JAD sessions make possible strong user involvement in systems development. On the whole, user involvement is healthy. It should reduce some of the mistrust and conflict between users and systems professionals that may have existed in the past. Ideally, the system becomes the users' system. When users have a psychological ownership of the system, they are motivated to work with it and strive for its success.

Most people resist change. Users are no exception. If, however, users are involved in systems development, and if they understand their options, they are more likely to make a strong commitment to change when it is needed.

Because users know more about what is needed to do certain tasks, user-oriented design results in better solutions. User interfaces are defined according to users' criteria. The resulting design is meaningful to users and easy to use.

Another reason for involving users in systems development is to make it possible for systems professionals to understand what users must do to perform their work, and for users to understand what it takes to develop a multiuser system. For example, a user who works with a microcomputer and spreadsheet for a personal, ad hoc application may not understand why it takes so long to develop a system that is integrated to serve a large number of users. Users may have little appreciation for following a systems development methodology, for preparing documentation, and for developing systems that are easy to maintain. However, through jointly participating in systems development, users soon learn what tasks must be performed and what constraints exist in building multiuser systems.

In the past, after the system was developed, a great deal of time and effort had to be spent in training users how to work with the system. Today, the amount of time and effort spent on training can be much less because users gain more knowledge about the system while it is under development. Also, in some instances, some users involved in systems development will become superusers who are highly skilled in how the system works. These superusers can effectively train other users.

As the role of users changes, so does the role of systems professionals. As users become more knowledgeable in systems and information technology, systems professionals will, more and more, play consulting, researching, and educating roles. Systems professionals will not take sole charge of systems projects, but will serve as guides and educators for users in a joint effort. In this way, users are treated as customers who actively participate in designing what they will be using to perform their tasks. Such an approach provides opportunities for increasing systems development productivity, systems quality, and user acceptance, which are the final goals of users and systems professionals alike.

---

[2] *Ibid.*

REVIEW OF CHAPTER LEARNING OBJECTIVES

The major goals of this chapter were to enable each student to achieve five learning objectives:

Learning objective 1:
Cite causes of systems development failure, and list ways to overcome these causes.

Systems development failure includes single causes or combinations of causes, including:

- Lack of matching systems development with business strategies and objectives
- Failure to determine user requirements and get users involved in systems development
- Deficient quality evaluation and cost/benefit analysis
- Inclusion of design defects and errors
- Unplanned acquisition of computers and software and installation of incompatible technology
- Development of unmaintainable systems
- Poorly planned and performed implementation tasks

To overcome these causes requires systems professionals to use various methodologies, tools, and techniques. These are:

- The systems development life cycle (SDLC)
- Prototyping
- Modeling tools
- Project management techniques
- Computer-aided systems and software engineering (CASE) technologies
- Joint application development (JAD)
- User involvement

Learning objective 2:
Define the systems development life cycle and discuss each of its phases.

SDLC is a logical, engineered process for developing systems from the planning stage through implementation. SDLC is made up of six development phases. The systems planning, systems analysis, general systems design, and systems evaluation and selection phases are the first four phases, referred to as the front-end phases because they are performed first. They are very user-oriented. The back-end phases are composed of detailed systems design and systems implementation. They are conducted after the front-end phases, and with the exception of tasks performed in the latter part of implementation, these phases are technically oriented. Systems maintenance is performed after the system is implemented and in operation. How well systems maintenance can be conducted is directly dependent on the quality of work performed during systems development.

Each SDLC phase generates documented deliverables that disclose the results of specific tasks performed in each phase. The planning phase produces the Systems Plan Report, which includes requests for systems proposals, TELOS feasibility, and PDM strategic-factor priority grades for each proposal. The Systems Plan Report triggers the analysis phase. The analysis phase generates a Systems Analysis Report, which defines the systems scope, user requirements, and problems with the present system. The general design phase produces the General Systems Design Report, which describes in broad and conceptual terms the system design components of several design alternatives. The evaluation and selection phase creates the Systems Evaluation and Selection Report, which discloses the qualitative value and cost-to-benefit ratio of each general systems design alternative. The best alternative is selected for detailed design, the next phase, which generates the Detailed Systems Design Report. This documented deliverable describes the functions of each systems design component of the chosen system in technical and elementary detail. The result of this work is a blueprint for software development and other implementation tasks. Implementation is the last phase of SDLC. Its documented deliverable, the Systems Implementation Report, is made up of two segments. The first segment is a plan (or schedule) of all implementation tasks. The second segment describes each task performed. After the new system is converted to operations, a postimplementation review is conducted to determine if the new system is operating satisfactorily. Much of the work done during this review will aid and provide input for maintenance.

Learning objective 3:
Explain prototyping and its use in developing systems.

Like a mockup of an engineered product, the prototype methodology enables designers and users to examine and use alternative designs. To a great extent, prototyping is used for the development of a specific, parochial application for one or a few users. It is especially useful in applications where users cannot specify information needs in the abstract, but only after they have seen a concrete representation of the design. Many design versions may be examined before one is selected as a working model.

With a prototype, users have the opportunity to get the feel of a working system, understand its strengths and weaknesses, and request improvements and extensions. This try-before-you-buy approach also works well with SDLC during the analysis and design phases, or during maintenance work.

Learning objective 4:
Relate the information engineering methodology (IEM) to the SDLC of this chapter.

IEM signifies a disciplined way to develop systems. Both IEM and this chapter's SDLC contain a systems planning phase. IEM, however, does not contain a separate systems evaluation and selection phase, nor does it divide its systems design phase into general (conceptual) and detailed (functional) design. The systems implementation of SDLC is referred to as systems construction and implementation in the IEM. This chapter's SDLC and IEM use similar or the same techniques and tools throughout systems development and maintenance.

Learning objective 5:
Describe joint application development (JAD) and discuss the role of users in systems development.

JAD is a technique for getting users and systems professionals to work together in systems development. A JAD facilitator augments interaction between users and systems professionals. A JAD scribe displays and records results of these interactions. Specialists may be called into JAD sessions to answer questions that cannot be answered by users or systems professionals.

In the past, users were not given much say in systems development. They had little sense of ownership in the system with which they had to work. Their special requirements were often not taken into consideration. Often, such an approach led to systems failure.

Today, enlightened systems professionals work with users in developing systems that will give users a sense of ownership and a vested interest in the system's success. In some enterprises, systems professionals act as consultants, researchers, educators, and communicators of standards and guidelines for systems development. Users act as participating customers for the system under development. In effect, the systems that are developed in this manner are owned by the users.

SYSTEMS DEVELOPMENT CHECKLIST

Following is a checklist of the system professional's responsibilities.

1 Be aware of the history of information systems development. This history is invaluable as a source of examples of both successful and unsuccessful systems, and the approaches and techniques that were used.

2 Model your efforts after successful systems and avoid the pitfalls encountered in unsuccessful systems.

3 Know the causes of unsuccessful systems development presented early in this chapter, and ask yourself at every step of your work: "Have I acknowledged and adequately worked to eliminate these causes?"

4 Incorporate the systems development life cycle at an early stage. The SDLC must have the full support and resources for all of its phases.

5 Get users involved in systems development and stress that the system under development is their system.

6 Complete the front-end phases of the SDLC. The sequence of phases is not cast in stone, because the process of systems development is frequently an iterative one. Additionally, some phases may be performed in parallel.

7 Document each SDLC phase clearly and completely. The documented deliverables are not just progress reports for management. They also become reference documents, which, if properly written, can help the company get the most out of its information system.

8 Complete the back-end phases of the SDLC. These may be performed in parallel and, possibly, iteratively. Good documentation is vital here if the system is to be operated and maintained effectively.

9 Do not ignore systems maintenance. Throughout the SDLC, always bear in mind the maintainability of the ultimate design.

10 Make use of prototyping in your work. This is a great way to involve users in the development of the system and to ascertain their requirements. Incorporate this user feedback into your analysis and design work, always keeping the user as the principal focus.

11 Consider the use of information engineering methodology as a possible alternative to this book's SDLC. This methodology is widely used and, depending on the experience and preferences of your systems development colleagues, may be more appropriate. In either case, the key phase of the book's SDLC and IEM that differentiates them from some of the earlier systems development methodologies is the systems planning phase. Today, most practitioners and academics emphasize this phase.

12 If IEM is used, include an evaluation and selection analysis phase after design, especially the application of a rigorous cost/benefit analysis of the system under development.

13 Consider the use of joint application development sessions, especially in the systems planning and analysis phases, and as a technique to involve users jointly with systems professionals in systems development.

KEY TERMS

Back-end phases

Computer-aided systems and software engineering (CASE)

Controls

Database

Detailed systems design

Detailed Systems Design Report

Differentiation

Documented deliverables

Economic feasibility

Feasibility factors

Front-end phases

Fourth-generation languages (4GLs)

Gantt chart

General systems design

General Systems Design Report

Information engineering life cycle (IELC)

Information engineering methodology (IEM)

Inputs

Joint application development (JAD)

Legal feasibility

Management

Operational feasibility

Operations documentation

Output

PDM strategic factors

Postimplementation review

Processes

Productivity

Program and evaluation review technique (PERT) chart

Prototyping

Schedule feasibility

Software coding

Software design

Software development life cycle (SWDLC)

Software documentation

Software testing

Strategic factors

Systems analysis

Systems Analysis Report

Systems design walkthrough

Systems development life cycle (SDLC)

Systems documentation

Systems evaluation and selection

Systems Evaluation and Selection Report

Systems implementation

Systems Implementation Report

Systems maintenance

Systems Plan Report

Systems planning

Systems professionals

Technical feasibility

Technology platform

TELOS feasibility factors

User documentation

Users

REVIEW QUESTIONS

1.1 Why should aspiring systems professionals be interested in learning about systems development methodologies, tools, and techniques?

1.2 Cite the causes for the lack of success in developing information systems.

1.3 List and discuss ways to develop successful information systems.

1.4 Define the systems development life cycle (SDLC).

1.5 List and discuss the phases and documented deliverables of SDLC.

1.6 Why is the systems maintenance phase emphasized in SDLC?

1.7 List, define, and give the purpose of documented deliverables produced by each phase of SDLC.

1.8 What are the front-end phases? Why are they called front-end phases?

1.9 What are the back-end phases? Why are they called back-end phases? Is maintenance a back-end phase? If so, why is it referred to as a back-end phase? If not, why not?

1.10 Explain a sign-off, its purpose, and how it is used in conjunction with documented deliverables.

1.11 List and explain the five feasibility factors. What acronym do they form? What role do they play in systems planning? Give an example of each feasibility factor.

1.12 List and explain the three strategic factors. What acronym do they form?

1.13 Why is it a good idea not to rush through the front-end phases? If the front-end phases are skipped, what problems will usually occur?

1.14 Why is it important for all participants to review thoroughly all documented deliverables before accepting or signing-off on them?

1.15 What are the four types of documentation prepared and finalized during the systems implementation phase? What is the purpose of each type?

1.16 Define prototyping and explain how it is used in developing systems. How is it used in conjunction with the SDLC? List and discuss systems characteristics that are especially applicable to prototyping.

1.17 List and briefly describe phases of the information engineering methodology (IEM). Compare and contrast IEM with this chapter's SDLC.

1.18 Define joint application development (JAD) and explain its purpose. During what phases of the SDLC is JAD most appropriate?

CHAPTER-SPECIFIC PROBLEMS

These problems require exact responses based directly on concepts and techniques presented in the text.

1.19 You are asked to develop an information system for Smallville Bell, a local telephone company. Among other things, this system will print customer billing statements each month.

Required: Create a prototype billing statement for Smallville Bell. Then, with the help of another student or friend as a user, evaluate this prototype. Create a second prototype billing statement based on this user feedback.

1.20 The Wavemaster Sailboat Company recently completed the development of a new information system in its production facilities, and it is very pleased with the results. Now that it realizes the merits of SDLC, it would like to incorporate a similar methodology in the development of a new 30-foot sailboat.

Required: Create a development methodology like the six-phase SDLC presented in this chapter for the new sailboat. Include two or three specific tasks to be completed as part of each phase. (*Note:* The development of an information system or a sailboat is project-oriented.)

1.21 Review the following random statements:

________________ We will be able to beat the competition by providing our customers with online tracking of their orders.

________________ The objective of our new system is to provide executives with trends and changes in the market's demographics so that they can make better strategic decisions.

_______________ At the present time, management is not willing to commit funds for the development of an online order-entry system.

_______________ The particular medical data that will be stored in the new database are extremely sensitive. Unauthorized access and dissemination of such data could subject the hospital to an array of lawsuits.

_______________ We can't wait six months for the application. We need it by the first of next month.

_______________ What you are presenting to us is too complex and sophisticated for our employees. We will need something much simpler.

_______________ The system will connect our suppliers and customers to our inventory management system. It will eliminate the need for data-entry clerks; reduce all paperwork, such as purchase orders, invoices, bills of lading, checks, and so forth; and decrease the inventory replenishment time by 60 percent.

Required: In the blank next to each statement, place the letter that best describes one of the TELOS feasibility factors or one of the PDM strategic factors. For example, if the statement implies an increase in productivity (P), place the letter P in the blank next to the statement.

THINK-TANK PROBLEMS

These problems call for a feasible approach rather than a precise solution. Although the problems are based on chapter material, extra reading and creativity may be required to develop workable solutions.

1.22 Northwestern Cycles, whose product is a large adult tricycle marketed to the elderly, has recently experienced significant expansion of its operations. Approximately two years ago, the development of an information system for the company was begun by an outside firm. However, due to economic conditions at the time, the development was discontinued suddenly and never resumed. The outside systems development firm made use of the information engineering methodology (IEM), and had completed the systems planning and systems analysis phases. The management of Northwestern has on file the documented deliverables it received after each of these two phases was completed. Northwestern Cycles has now decided to proceed with the development of the information system. The company has asked you to head the development team, and has provided you with the documented deliverables it kept on file from two years ago. Your experience, and that of the rest of your team, is with the SDLC methodology, and you decide to use the SDLC for this project, rather than the IEM.

Required: How do the phases completed two years ago by the outside firm using the IEM correspond to the phases of the SDLC your development team has decided to use? Since the first two phases of the IEM were completed and you have been provided with the documented deliverables, will your team merely pick up where the outside firm left off? Why or why not? How do the "causes of unsuccessful systems development" defined in this chapter relate to your decision? Would your decision have been different had the company not expanded significantly in the last two years?

1.23 Top-Flight Company is a small ski equipment manufacturer. Pat Norman, CEO, has just completed a mission statement for Top-Flight. A key part of the mission statement is: "To become the low-cost producer of ski equipment and deliver orders to our distributors in 24 hours, or less. Within six months, we will convert our traditional method of manufacturing to just-in-time (JIT) manufacturing. The development of a new information system will replace our present batch processing system. We expect to leverage information technology to support JIT manufacturing and give us a competitive advantage." You have been hired as the project manager in charge of systems development. You have been told, in no uncertain terms, that systems work at Top-Flight has a history of missed deadlines and lack of user involvement. Pat Norman wants to know what you will do to change the way that systems are developed at Top-Flight in the future.

Required: Prepare a brief report for Pat Norman outlining how you plan to develop systems at Top-Flight and overcome past problems.

SUGGESTED READING

Andriole, Stephen J. *Storyboard Prototyping*. Wellesley, MA: QED Information Sciences, 1989.

Burch, John, and Gary Grudnitski. *Information Systems: Theory and Practice,* 5th ed. New York: John Wiley, 1989.

Currid, Cheryl. "IS Staffers Need to Share Technology Wheel." *PC Week,* June 24, 1991.

Ernst & Young. "Information Engineering Methodology: Strategic CASE Implementation." *CASExpo,* Santa Clara, Calif., Fall 1990.

Frantzen, Trond, and Ken McEvoy. *A Game Plan for Systems Development*. Englewood Cliffs, N.J.: Yourdon Press, A Prentice-Hall Company, 1988.

Heck, Michael. "Mission: Made Possible." *Infoworld,* September 25, 1989.

Inmon, W. H. *Information Engineering for the Practitioner*. Englewood Cliffs, N.J.: Prentice-Hall, 1988.

McCusker, Tom. "Why Business Analysts Are Indispensable to IS." *Datamation,* January 15, 1990.

Meyer, Gary. "The Best Laid Plans." *Information Strategy: The Executive's Journal,* Winter 1990.

Mimno, Pieter. "Rapid Prototyping: How to Build Applications Fast Through Prototyping." *CASExpo,* Santa Clara, Calif., Fall 1990.

Stewart, Gary. "CASE . . . A Practical Perspective." *Computing,* June 1990.

Stock, Michael. "Systems Analyst Updated." *Digital News,* May 14, 1990.

Wolman, Rebekah. "Managing Technical Professionals." *Information Center,* March 1990.

Yourdon, Edward. *Managing the Systems Life Cycle,* 2nd ed. Englewood Cliffs, N.J.: Yourdon Press, A Prentice-Hall Company, 1988.

Chapter 2

Taking a Structured Approach and Using Modeling Tools

WHAT WILL YOU LEARN IN THIS CHAPTER?

After studying this chapter, you should be able to:

1 Discuss the structured approach and list its key features.
2 Explain the types of modeling tools and the roles they play in developing systems.
3 State the purpose of a data flow diagram (DFD) and describe the steps necessary to prepare a DFD.
4 Explain the construction of a data dictionary and discuss its purpose.
5 Cite the purpose of an entity relationship diagram (ERD) and list the steps necessary to prepare an ERD.
6 Describe a state transition diagram (STD) and explain how it is applied.
7 Describe the purpose of a structure chart and explain the steps required to create a structure chart.
8 Describe the structured program flowchart and discuss its usage.
9 Explain how process specification tools relate to the data flow diagram and cite the purpose of each of the four process specification tools.
10 Explain the purpose of a Warnier–Orr diagram (WOD) and describe the steps necessary to develop a WOD.
11 State the purpose of a Jackson diagram and list the steps required to develop a Jackson diagram.

INTRODUCTION

Chapter 1 introduced the systems development life cycle (SDLC) and prototyping methodologies. This background provides a basis for introduction to the structured approach and supporting modeling tools. The structured approach is a disciplined way to develop systems; the modeling tools provide the means for applying the structured approach.

TAKING A STRUCTURED APPROACH TO DEVELOPING SYSTEMS

The **STRUCTURED APPROACH** is used in systems analysis, systems design, and software development. It is a disciplined, engineered approach that employs discrete phases as defined by the SDLC. The goal of the structured approach is to have at the end of the systems project an information system that meets user requirements, is on time and within budget, and is easy to work with, under-

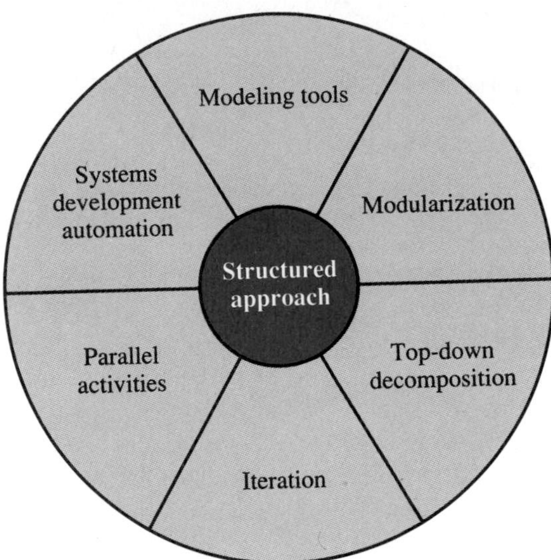

Figure 2.1
The structured approach and its features.

stand, and maintain. Major features that support the structured approach, as depicted in Figure 2.1, include the following:

- Modeling tools
- Modularization
- Top-down decomposition
- Iteration
- Parallel activities
- Systems development automation

What Are Modeling Tools?

MODELING TOOLS, the major subject of this chapter, are used to model and describe various systems, subsystems, and software designs on paper or screen for review and evaluation by both systems professionals and users. One value of modeling tools is that a picture is worth a thousand words. A key task of modeling tools is to break a system down into manageable parts and to communicate to all viewers the system's conceptual and functional characteristics. In this way, a model serves as a representation of a system on paper or video for review and evaluation.

Designing in Modules

MODULARIZATION is a process that divides a system into independently operable modules. The technique of dividing functions that together achieve a larger objective is used in a number of industries. The components of cars, televisions, airplanes, and houses, and the interfaces between these components, are so well defined and engineered that they can be manufactured by different companies or divisions and assembled efficiently into a workable unit. Later, these finished products can also be maintained and repaired easily.

The common thread running through all of these products is modularization, because all the products are composed of standardized modules used together to form a whole. This modularization technique can also be applied to building information systems. Each module becomes a contiguous, bounded group of system functions or software code having a unique name by which it can be referenced as a unit.

Additional benefits are derived by dividing modules again and again until each module contains one well-defined function. The objective is reduced complexity, increased simplicity, and improved maintainability.

Some say that doubling the size of a system can quadruple the time and effort necessary to deal with it. Indeed, complexity grows much faster than size. For example, a 500-line program is much more complex and difficult to maintain than ten 50-line modules.

Modularization is, indeed, a key technique of the structured approach. Many benefits accrue from its use. In fact, benefits of the structured approach rely on modularization. These benefits are:

- The total system or software program is made simpler, because it can be understood, designed, coded, tested, debugged, and changed module by module.
- Errors are reduced, because people are working on simple, well-understood pieces of the system rather than complex structures.
- More accurate cost estimates are possible.
- Testing is easier, because software is tested module by module and errors can be readily isolated.
- Turnover of project team members has less impact on the project, because no single person is responsible for the entire project.
- Optimization efforts can be applied to critical areas; this increases overall efficiency.
- Certain modules that execute common functions can be used by other projects; this substantially reduces development costs.

Working from the Top Down

Problem solving is more difficult when all aspects of the problem are considered at the same time. It is easier if the problem can be solved piece by piece, level by level. The main idea behind the top-down technique, combined with modeling tools and modularization, is that all problems can be solved and all systems can be dealt with by decomposing them to their elementary levels. **TOP-DOWN DECOMPOSITION** entails identifying major high-level user requirements and systems functions and breaking them down more and more until function-specific modules can be designed.

Using Iteration

Only the simplest of systems can proceed step by step in one pass from analysis to design to implementation. **ITERATION**, supported by SDLC and prototyping, helps systems professionals to go back to earlier phases and tasks to make small improvements and respond to better definitions of user requirements. But too many iterations reach a point of diminishing returns, where the project team must move on to the next task. Moreover, excessive iterations may indicate that work in earlier phases was inadequate. For instance, the systems analysis phase must be conducted thoroughly to provide a sound basis for the general systems design phase.

Using Parallel Activities

PARALLEL ACTIVITIES enable some overlapping of systems analysis, design, and implementation tasks. Because of SDLC, prototyping, modularization, and top-down decomposition, more than one person can develop different parts of the system at the same time and thus speed up the systems design and software development tasks. For example, one group may be installing a telecommunication network while another group is developing a database, or one programmer may be coding module A while another is coding module B.

Automating Systems Development

SYSTEMS DEVELOPMENT AUTOMATION is the application of computer hardware and software to systems development. In recent years, automated systems development packages have become available. Generally, these packages are known as **COMPUTER-AIDED SYSTEMS AND SOFTWARE ENGINEERING (CASE)**. They are systems that help build systems. With CASE, systems professionals can sit at their workstations with a kit of computerized tools and build systems and automatically generate executable software code. More about CASE technology is included in Chapter 3.

ROLES AND TYPES OF MODELING TOOLS

Sound analysis and design rely on models of a system. Modeling tools are used to develop these models of existing or planned systems.

Roles of Modeling Tools

Specifically, modeling tools play three roles in systems development:

- Communication
- Experimentation
- Prediction

Communication

Modeling tools help to systematize the often abstract concepts and difficult-to-define user requirements that surface in the early phases of SDLC. One of the major reasons for creating models is to facilitate communication with the user.

Heavy user involvement throughout the systems analysis and general systems design phases increases the project team members' understanding of the system, and that understanding can then be fed back to the user for confirmation. In this way, misunderstandings arc corrected quickly without affecting deadlines.

Experimentation

An inherent aspect of systems development is trial and error through iteration. Models enable people to see how a system could work as they attempt to discover the optimal design of that system. These tools provide conceptual models of a system before functional (or physical) design and implementation begin. The conceptual design, or designs, can be configured in a form that is easily understood by end users. Design errors that may not be apparent to systems professionals can be pointed out and corrected by users who are more familiar with the business operations for which the system is being developed. These models allow users to understand and evaluate the system early in the development process. It is not uncommon for a model to undergo three or more iterations before the participants are finally satisfied with the results.

Prediction

Models foretell how a system will work. A model that is finally selected for implementation is a complete and coherent representation on paper or video screen of what the new system will be like and how it will work. To achieve the goals of senior management, systems professionals will need to develop information systems considerably more complex than traditional systems. As new information systems become the backbone of the company's competitive strategy, systems professionals will be placed in extremely visible, high-risk, high-reward situations. The ability to predict how a system will work from a model before massive funds are spent to implement a real system can help systems professionals reduce their risk and increase their reward.

As you continue reading this chapter, keep in mind the three roles of a model: communication, experimentation, and prediction. Also, begin to evaluate the effectiveness of each modeling tool on the basis of these roles. As you examine a tool, ask yourself some of the following questions:

- Is this modeling tool easy to understand?
- Does this tool function as an aid to clear thinking?
- Would this tool help communicate the design of a current system to people? Could it be used to communicate the design of a new system?
- Would this modeling tool assist in experimentation with new ideas and encourage good systems design?
- Could this tool be used to help predict the actual functioning of a new system?
- Does this tool help add structure to the systems analysis and design phases?

These questions will assist you to evaluate and classify each modeling tool. It's important that analysts be able to analyze their own work, and their own tools, as well as information systems.

Types of Modeling Tools

More than twenty modeling tools of the trade exist. Most of these modeling tools are based on the structured approach. A large number of the tools work practically the same way, their essential difference being more cosmetic than substantive. Some are obscure. Others resemble fancy solutions looking for problems. In this chapter we present those that have fairly wide acceptance. We use them for specific applications throughout the book. These are:

- Data flow diagram (DFD)
- Data dictionary
- Entity relationship diagram (ERD)
- State transition diagram (STD)
- Structure chart
- Structured program flowchart
- Process specification tools
- Warnier–Orr diagram (WOD)
- Jackson diagram

No one systems professional will use all these tools during a given systems project. Rather, each individual will have a favorite set of tools as part of his or her personal modeling tool kit.

CREATING A DATA FLOW DIAGRAM

The purpose of a **DATA FLOW DIAGRAM (DFD)** is to show the processes that data undergo in a system. But before we begin, we first present the set of symbols used to prepare DFDs.

What Are DFD Symbols?

Tom DeMarco, Edward Yourdon, Chris Gane, and Trish Sarson are credited with development of the data flow diagram. They differ, however, on DFD symbols as depicted in Figure 2.2. All those cited use symbols to represent the elements discussed below.

Source and Sink

A rectangle (Yourdon and DeMarco) and a square (Gane and Sarson) indicate the net inputs to the system and net outputs from the system. **SOURCES**, which can be persons, organizations, departments, or other systems, are the net inputs to the system being modeled. Sometimes sources are called **ORIGINS**. **SINKS**,

Figure 2.2
DFD main symbols.

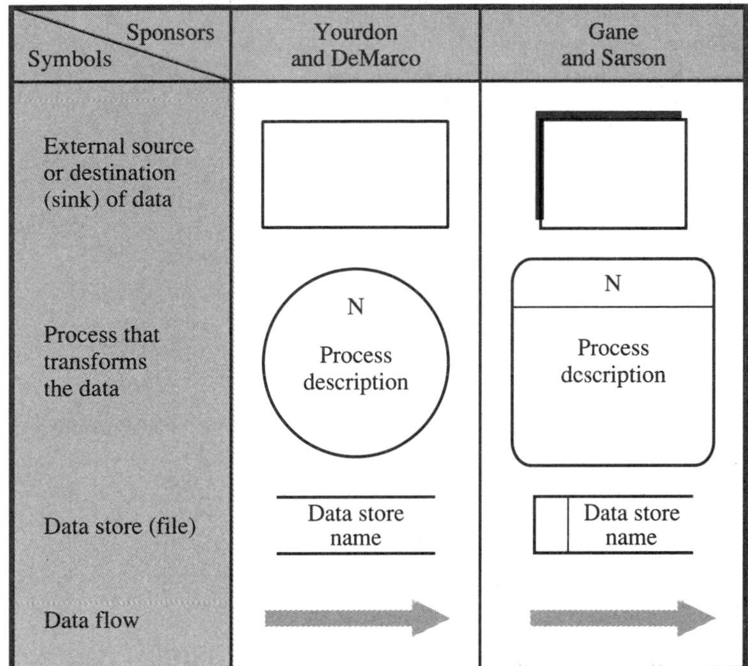

which can also be persons, organizations, departments, or other systems, are the net outputs of the system being modeled. Sinks are also referred to as **DESTINATIONS**. Sources and sinks define the boundaries of the system being modeled. Labels of the sources and sinks should be descriptive, such as Customer or Credit Department.

To avoid crossing data flow lines on a DFD, sources and sinks may be duplicated. Normally, sources and sinks should be located on the perimeters of the page or screen. This placement is consistent with their definition as systems boundaries.

Process

A **PROCESS** is represented by a circle (Yourdon and DeMarco), also called a bubble or transform, and a rounded rectangle (Gane and Sarson). Processes transform inputs into outputs. The number of the process (N) and a brief description identify processes. The details of the processes are not shown. A process may be manual or automated. For example, sorting mail might be a manual process, and sorting bank drafts might be automated.

Process names consist of a verb and an object or object clause. Process names selected depend on the level of detail required on the DFD. General process names may be: Check Credit, Order Entry, or Update Accounts Receivable. Processes of detailed DFDs should be given more specific names. Examples are: Update Salary Codes For Department B, or Identify Customer Accounts Over Sixty Days Past Due. Furthermore, names that identify the processor, such as person, location, machine, or computer, are helpful. For example, if the Personnel Manager updates salaries, the process may be named as follows:

1.1
Personnel manager: Update salary codes for department B

All processes must have both inputs and outputs. A process that shows inputs but no outputs is called a black hole, because data enter the process and disappear. A process with output but no input is creating something from nothing, which is a miracle.

Data Store

A **DATA STORE** is a file of any kind: paper, magnetic, or optical. A data store is represented by either two parallel lines or an open-ended rectangle. Each data store is referenced by its name, such as Customer, Inventory, or Payroll. Names should not specify storage media. The small square in the Gane and Sarson data store symbol may also contain letters or numbers for additional identification and reference. To avoid crossing data flow lines, it is permissible to duplicate data stores in the diagram.

Only processes may connect to data stores. Only processes can use or update data in the data stores. A data flow from a data store means that the process uses the data. A data flow to a data store means that the process updates (i.e., adds, deletes, or modifies) the data in the data store. If this is the case, use separate data flows for the use and the update.

Data Flow

A **DATA FLOW** is displayed by a line and an arrow. A data flow represents the transfer of data among data stores, sources or sinks, and processes. Occurrences of a data flow must contain data, and all data flows either initiate a process or result from a process. For example, imagine a customer placing a purchase order with a company. The customer is a source transferring the purchase order, which is a data flow, to the process of order entry in the company. Figure 2.3 displays the DFD for this example. Each data flow should have a noun clause next to the flow line describing the data that are being transferred.

Steps in Preparing DFDs

During systems analysis, DFDs are useful for modeling the scope of the systems project and for analyzing and modeling study facts gathered during analysis. DFDs are used for portraying the overview of the entire system under development to depicting the detailed processing of a single transaction. Struc-

Figure 2.3
DFD showing how a customer order is handled.

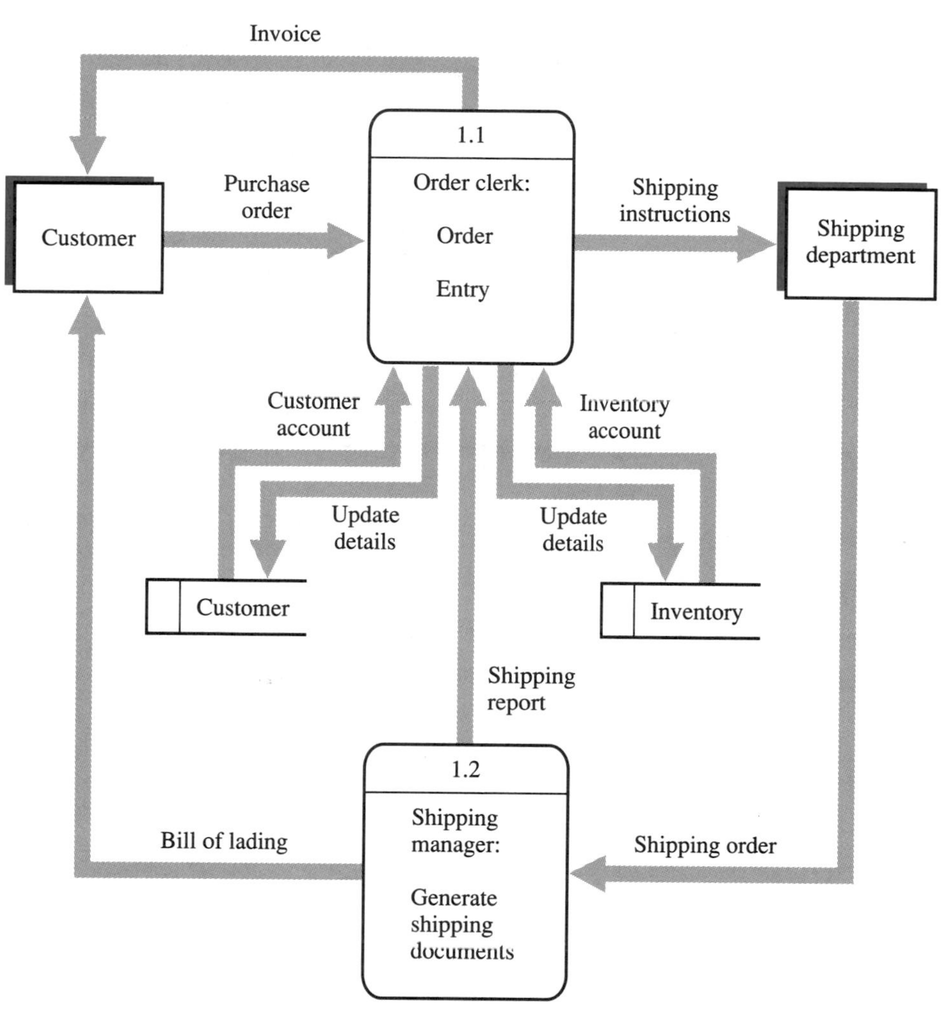

tured top-down decomposition is employed to start at a context level and decompose the system little by little until the system is defined at its most elementary level. The entire collection of the resulting DFDs is termed a leveled set. The steps used to produce this leveled set are depicted in Figure 2.4 and explained as follows:

Step 1. Draw a Context-Level Diagram. The **CONTEXT-LEVEL DIAGRAM** shows the main sources, sinks, processes, and scope of the system under development. Details are deferred until lower-level DFDs are drawn.

A context-level diagram for an item processing system at a commercial bank is shown at the top of Figure 2.4. Items in a bank are such instruments as drafts or checks, deposits, and cash slips. The purpose of the context-level diagram is to conceptualize the general sources and sinks of data. In this example, the customer and clearinghouse are sources of bank drafts (bank

Context-Level DFD

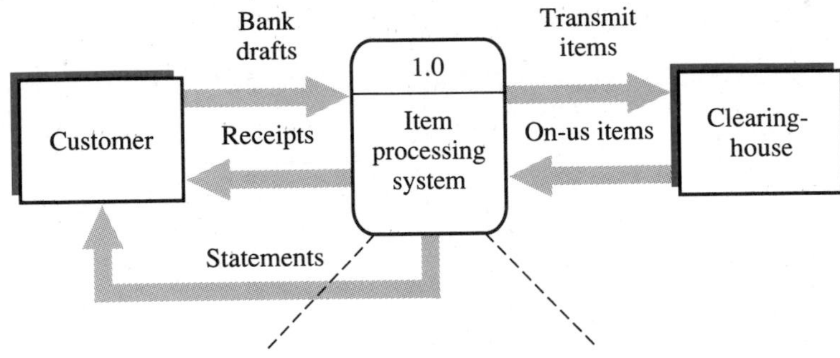

Bank drafts
1.0
Item processing system
Transmit items
Customer
Receipts
On-us items
Clearing-house
Statements

Second-Level DFD

Bank drafts
1.1
Accept transaction
Batched draft
1.2
Proof transaction
Proofed batches
Customer
Receipts
Clearing-house
Customer account
On-us items
Account details
Transmit items
1.3
Sort batch
Sorted on-us items
On-us items
On-us items details
1.4
Generate statement
Statement
Invalid item
Exception items
Customer

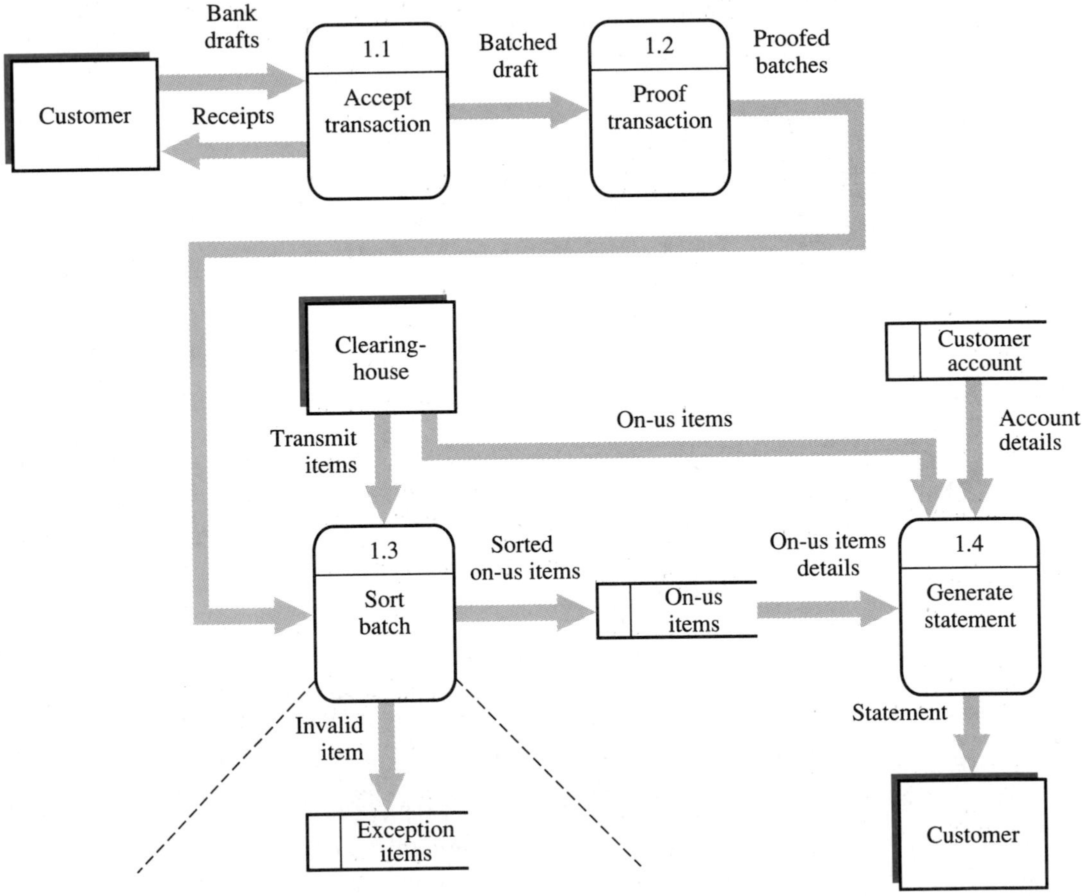

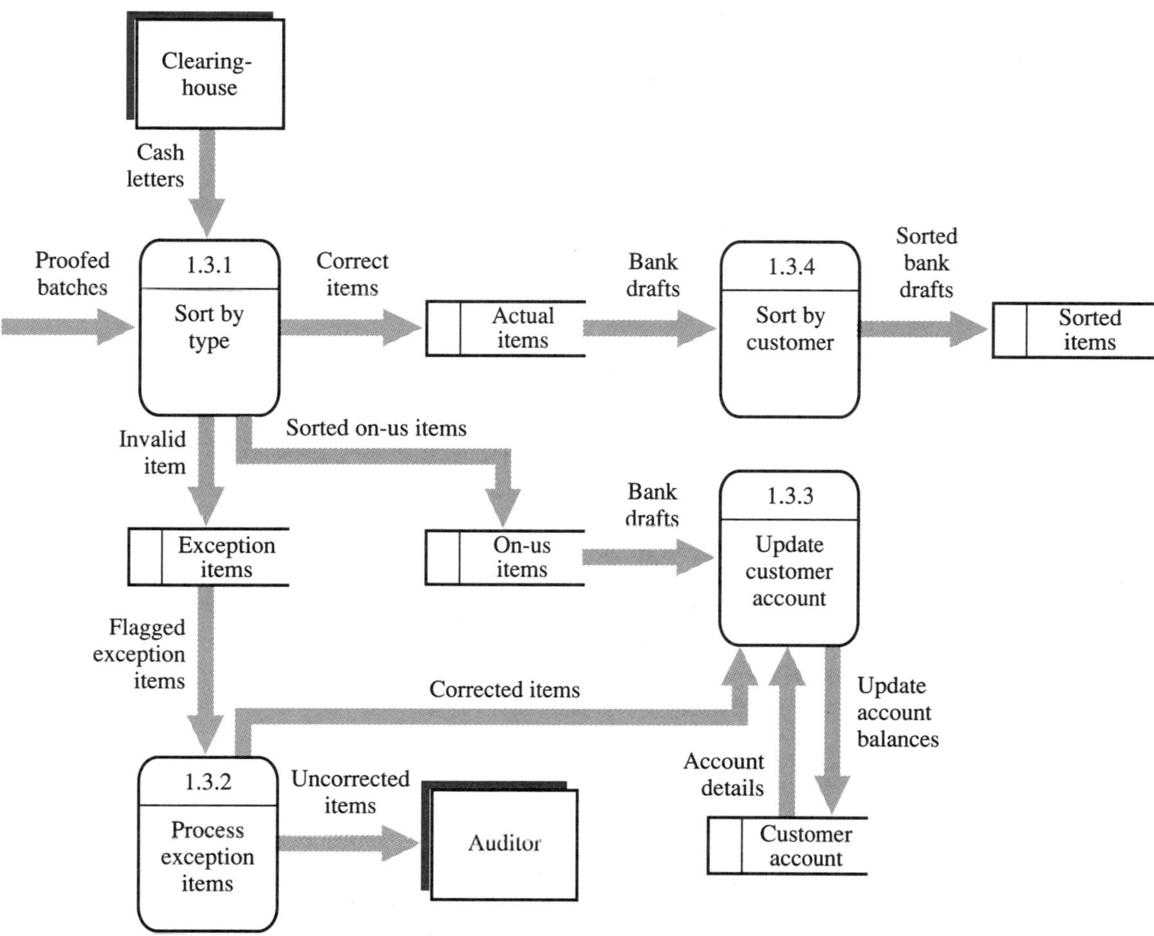

Figure 2.4 (continued)

transactions such as checks and deposits) to the item processing system. The customer and clearinghouse are represented by sinks. The auditor who receives uncorrected items is a sink. What one auditor does with these items becomes part of another system.

Step 2. Decompose the Context-Level Diagram. The context-level diagram is exploded or decomposed into a second-level DFD. The 1.0 item processing system process is decomposed into:

- 1.1 Accept transaction
- 1.2 Proof transaction
- 1.3 Sort batch
- 1.4 Generate statement

There may be more processes that represent decompositions of processes 1.1, 1.2, and 1.4, but these will be detailed on other DFDs.

All the data flows on the context-level diagram have been carried down to the second-level diagram. This balances the diagrams. Balancing ensures consistency between diagram levels. Notice that new data flows are added.

Step 3. Decompose to an Elementary Level. Process 1.3, sort batch, is decomposed to an elementary-level DFD. This level of diagram explains in even greater detail the actual processes that are occurring to sort a batch of bank drafts. Processes 1.1, 1.2, and 1.4 may even need more leveling until enough detail is described for the systems analyst and users to understand the system fully. Leveling occurs until consensus and complete understanding is achieved among analyst and users.

One must be sure to label and reference all sources, sinks, data flows, data stores, and processes. Numbering each process allows the analyst to trace the hierarchy of processes being completed. Analysts periodically conduct a review of DFDs with users and project team members to determine whether the diagram accurately represents the system under development. If necessary, go back and make changes to the diagram. The DFD is a tool to create a model. Don't let the tool become the goal; instead, use the tool to create an understandable and meaningful model.

USING A DATA DICTIONARY FOR SYSTEMS ANALYSIS

The data elements in the DFD can easily number in the thousands. A formal means of cataloging these elements is useful considering that each element has its own internal composition and structure. The **DATA DICTIONARY** provides the organization necessary to make sense of a mountain of data.

A data dictionary is like any other dictionary, in that it defines each data element name. The data dictionary contains definitions for all data elements in the system being modeled, no more and no less.

As an example of a typical data dictionary entry, consider the following definition for the data element VENDOR-ORDER:

```
VENDOR-ORDER = VENDOR-NAME +
               ACCOUNT-NUMBER +
               VENDOR-ADDRESS +
               [SALES-REP | "CATALOG ORDER"] +
               (ORDER-PRIORITY) +
               {ITEM-ORDER}
```

This description tells us, with no supporting explanation, that a VENDOR-ORDER consists of a VENDOR-NAME, together with an ACCOUNT NUMBER and VENDOR-ADDRESS, along with either the name of the vendor's SALES-REPresentative for this order or the literal text string "CATALOG ORDER," plus an optional code to indicate the ORDER-PRIORITY level, and finally zero or more instances of an ITEM-ORDER.

Thus the notation used in a data dictionary definition is all that must be known to understand a data element fully. The symbols and their meanings are:

Symbols	Typical Notation	Meaning
+	x = a + b	x consists of a and b
[|]	x = [a | b]	x consists of either a or b
( )	x = a + (b)	x consists of a and an optional b
{ }	x = {a}	x consists of zero or more occurrences of a

Many of the component parts of the data element VENDOR-ORDER are in themselves data elements, and they should also be defined in the data dictionary. A definition for VENDOR-ADDRESS, for instance, might be

```
VENDOR-ADDRESS = [STREET-ADDRESS | PO-BOX] +
                 CITY +
                 STATE +
                 ZIPCODE
```

Data elements should continue to be defined down to the lowest appropriate level of detail. For instance, in the definition of VENDOR-ORDER, the data element ORDER-PRIORITY is included. Most likely, ORDER-PRIORITY is simply assigned a number, maybe between 1 and 5, and is not composed of any lower-level data elements. Hence ORDER-PRIORITY is the lowest level of detail that we would define in the data dictionary. Data elements at the lowest level might require units of measure, such as dollars, miles, or gallons, to define them completely.

CREATING AN ENTITY RELATIONSHIP DIAGRAM

A good use of an **ENTITY RELATIONSHIP DIAGRAM (ERD)** is to model data stores in a DFD, independent of processing performed with those data stores. An ERD depicts data at rest—a file, for instance. ERDs can also be used for other modeling tasks. For example, we will use an ERD for enterprisewide modeling in Chapter 4, and we will use a modified ERD for modeling object-oriented software designs in Chapter 16. But for now, we will use an entity relationship diagram (ERD) in its traditional role of modeling data stores.

What Are ERD Symbols?

Symbols used to develop entity relationship diagrams (ERDs) are shown in Figure 2.5. The symbols depict entities and relationships. An **ENTITY** is a set of persons, places, or things, all of which have a common name, a common definition, and a common set of properties or attributes. A **RELATIONSHIP** shows how entities interact and work together. For example, a user might describe the following problem: Customers occasionally mail us checks without identifying their account numbers, or they visit one of our branches to make a deposit but don't know their account numbers. In this example, Customers, Checks, Ac-

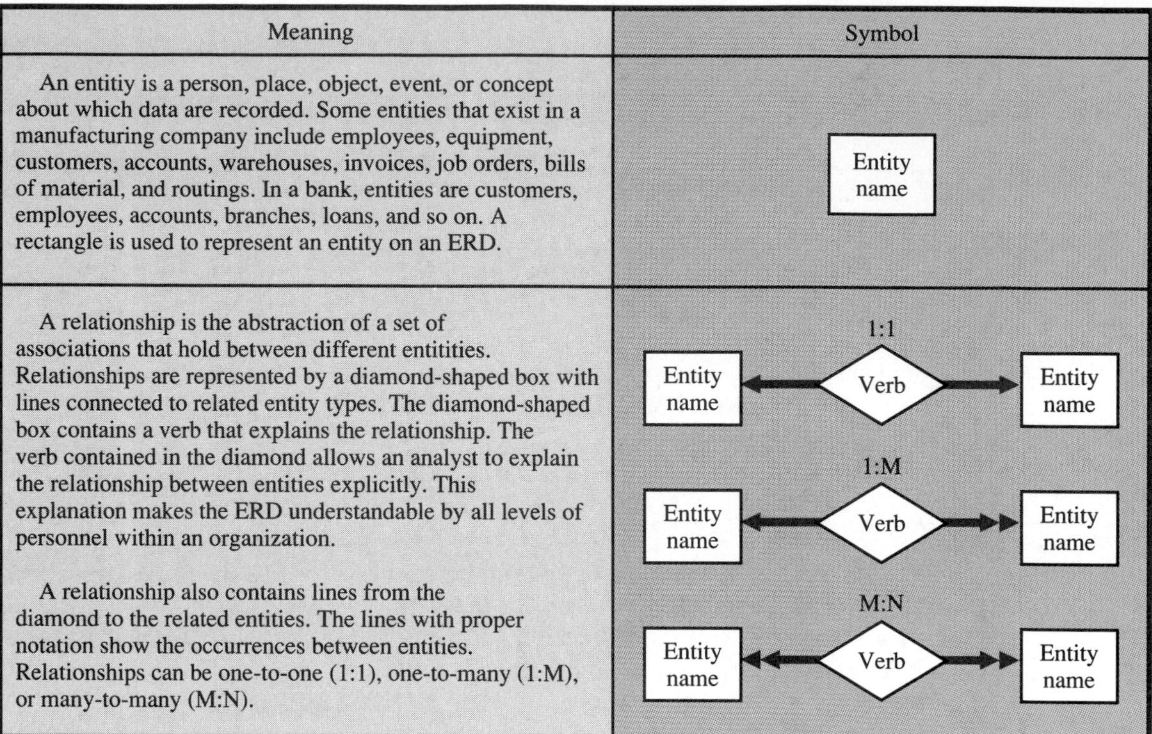
Meaning
Symbol
An entitiy is a person, place, object, event, or concept about which data are recorded. Some entities that exist in a manufacturing company include employees, equipment, customers, accounts, warehouses, invoices, job orders, bills of material, and routings. In a bank, entities are customers, employees, accounts, branches, loans, and so on. A rectangle is used to represent an entity on an ERD.
Entity name
A relationship is the abstraction of a set of associations that hold between different entitities. Relationships are represented by a diamond-shaped box with lines connected to related entity types. The diamond-shaped box contains a verb that explains the relationship. The verb contained in the diamond allows an analyst to explain the relationship between entities explicitly. This explanation makes the ERD understandable by all levels of personnel within an organization.
A relationship also contains lines from the diamond to the related entities. The lines with proper notation show the occurrences between entities. Relationships can be one-to-one (1:1), one-to-many (1:M), or many-to-many (M:N).
1:1
Entity name
Verb
Entity name
1:M
Entity name
Verb
Entity name
M:N
Entity name
Verb
Entity name

Figure 2.5
ERD symbols and their meanings.

count, and Deposit are entities. A relationship exists between customers and their accounts. The account number is an attribute that helps define both customers and account entities.

Steps in Preparing ERDs

If properly labeled, entities (nouns) and relationships (verbs) read like simple sentences, such as: Customer (entity) buys from (relationship) salesperson (entity). An entity noun may include a modifier, such as Desk Salesperson, Retail Customer, or Pending Order. The steps required to create ERDs are presented next. These steps explain the ERD illustrated in Figure 2.6.

Step 1. Identify Entities. This step involves identifying and labeling those entities that comprise the system under development, which is a university reporting system. These entities are: Departments, Faculty, Students, Majors, Courses, Sections and Grades.

Step 2. Indicate Relationships Between Entities. Entities and their relationships connected by lines make up the ERD. For example, Faculty are Assigned To Sections.

Step 3. Define Keys for Each Entity. Keys are data elements that uniquely identify each entity. Examples of keys are Department-Number, Course-Number, Section-Number, Social-Security-Number (SSN), and so on.

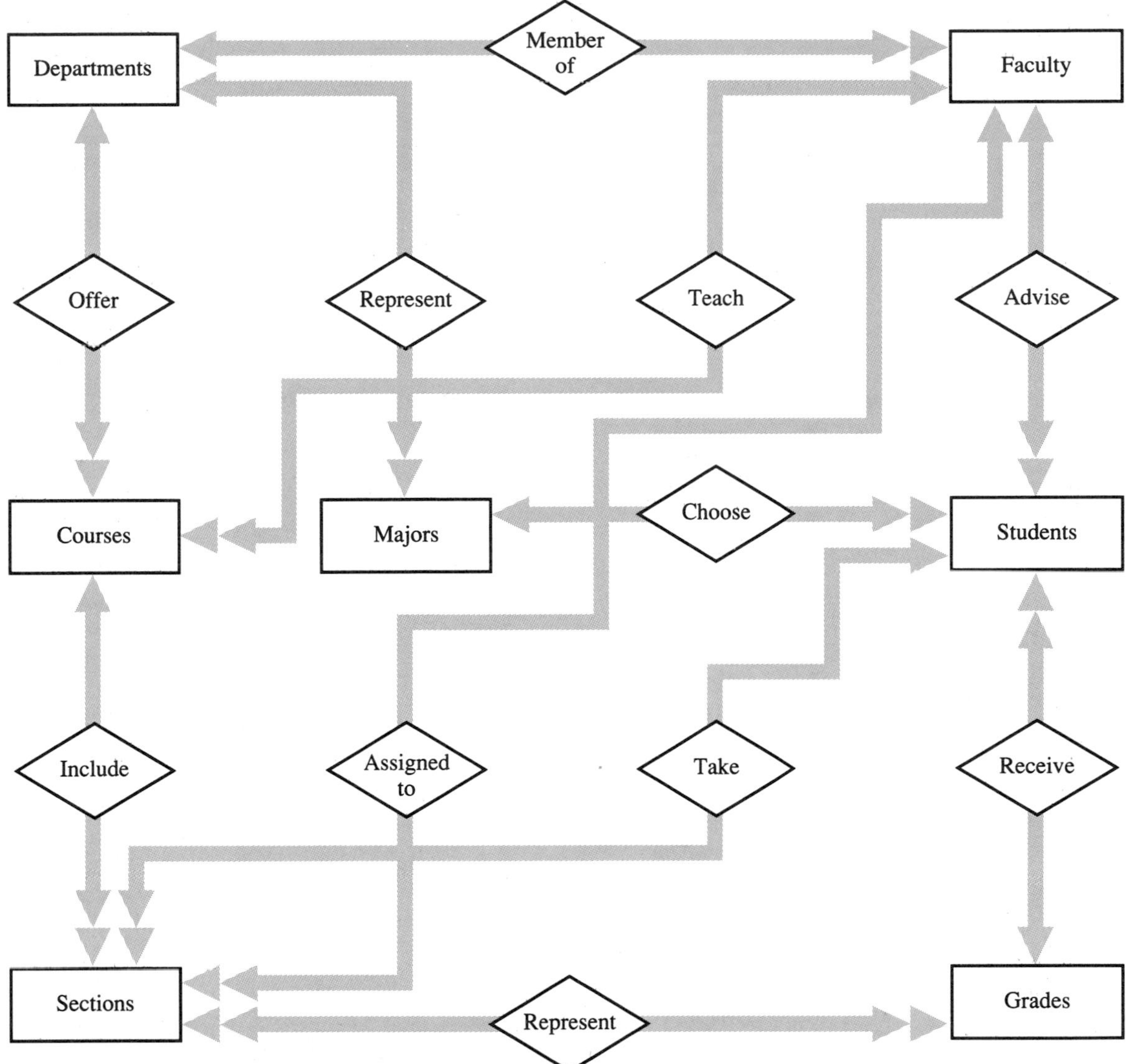

Figure 2.6
ERD of a university reporting system.

Step 4. Define and Map Data Elements for Each Entity. Data elements, also called data attributes, represent the data that define the entities. The results of defining and mapping data elements is a data model for the ERD disclosed in Figure 2.7. Each data structure of the data model contains the name of the entity, the underlined key or keys that uniquely identify each entity, and other data elements.

Step 5. Normalize the Data Model. A data model must be flexible and stable, with limited data redundancy. Normalization achieves this objective. Normalization is performed to prepare data for implementation. It is treated in Chapter 11, which presents database design.

Figure 2.7
Data elements for each entity.

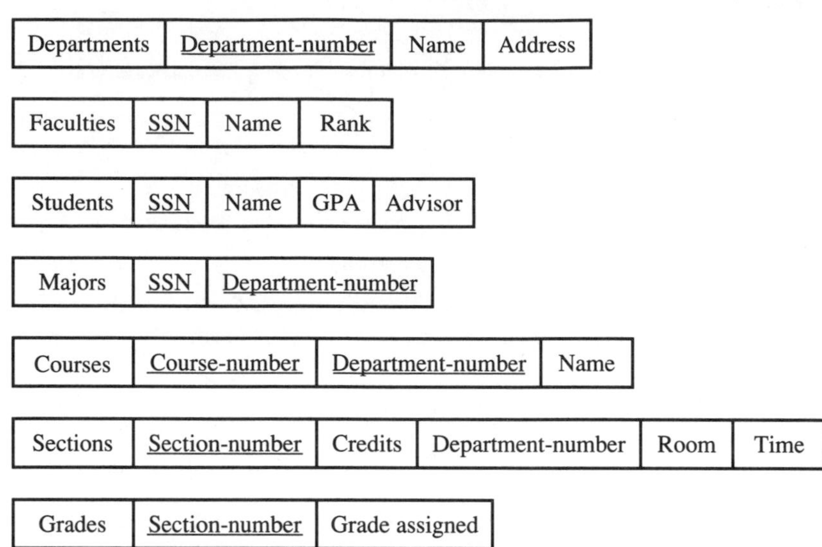

CREATING A STATE TRANSITION DIAGRAM

The modeling tools discussed thus far do not adequately represent the temporal, or time-dependent, aspects of a system. For real-time systems, a modeling tool which represents these temporal aspects is necessary. Such a tool is the **STATE TRANSITION DIAGRAM (STD)**. STDs are used to model the sequencing of many real-time systems, from online data-entry systems to temperature control systems in factories to mission programs for the Space Shuttle. No matter what its application, the construction of a STD requires two steps:

Step 1. Identify all possible finite states for the system.

Step 2. Identify activities that cause a transition from one finite state to another.

Figure 2.8 shows several states in the order which they occur, for an online real-time raw material requisition system. This system uses screen menus and data entry dialogue interfaces to control the transfer of raw materials from inventory to the production floor of a manufacturing operation. States are identified by determining the equilibrium of a system when no outside condition interferes. For instance, our online real-time system will remain waiting for raw material information until a user physically enters a raw material name at the computer terminal.

Figure 2.9 is the complete STD, showing the six states previously identified and the transitions between these states. Transitions, represented by arrows, are composed of conditions (C) which initiate them, and actions (A) that result in a new state. In this example, the transition from waiting for requisitioning department information to waiting for raw material information is initiated by the entry of either a department name or a department number. Either of these

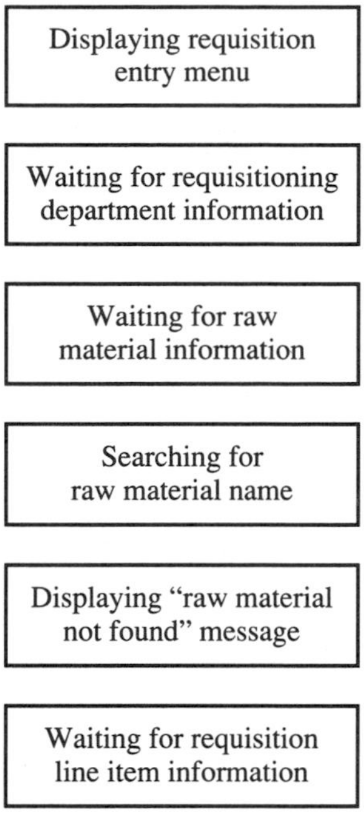

Figure 2.8
Some possible states for an online real-time raw material requisition system.

Displaying requisition entry menu
Waiting for requisitioning department information
Waiting for raw material information
Searching for raw material name
Displaying "raw material not found" message
Waiting for requisition line item information

conditions causes the system to transition to a new state where it awaits the entry of a raw material name.

The states of a STD must be mutually exclusive. In our example, the system cannot be displaying two different menus at the same time. Transitions, on the other hand, are not always paired up one-to-one with states. In the example, the entry of either a department name or a department number results in transition to the same new state. However, if a raw material name is not found, different responses from the user result in transition to different states.

CREATING A STRUCTURE CHART

Structure charts are typically used in conjunction with DFDs. The DFDs model the system, and the structure chart models the software, coded and tested to support the system.

A **STRUCTURE CHART** represents a hierarchy of software program modules, including documentation of interfaces between the modules. The structure chart acts not only as the design blueprint for software coding but also as a guide to software testing. It guides the coding of modules, the order in which the modules will be built and tested, and the assignment of coders to specific modules. It also serves as external documentation for software maintenance later on after the system has been implemented.

Structure Chart Control Rules

A structure chart is a control chart much like an organization chart showing who reports to whom. Only one module is at the top of the structure chart.

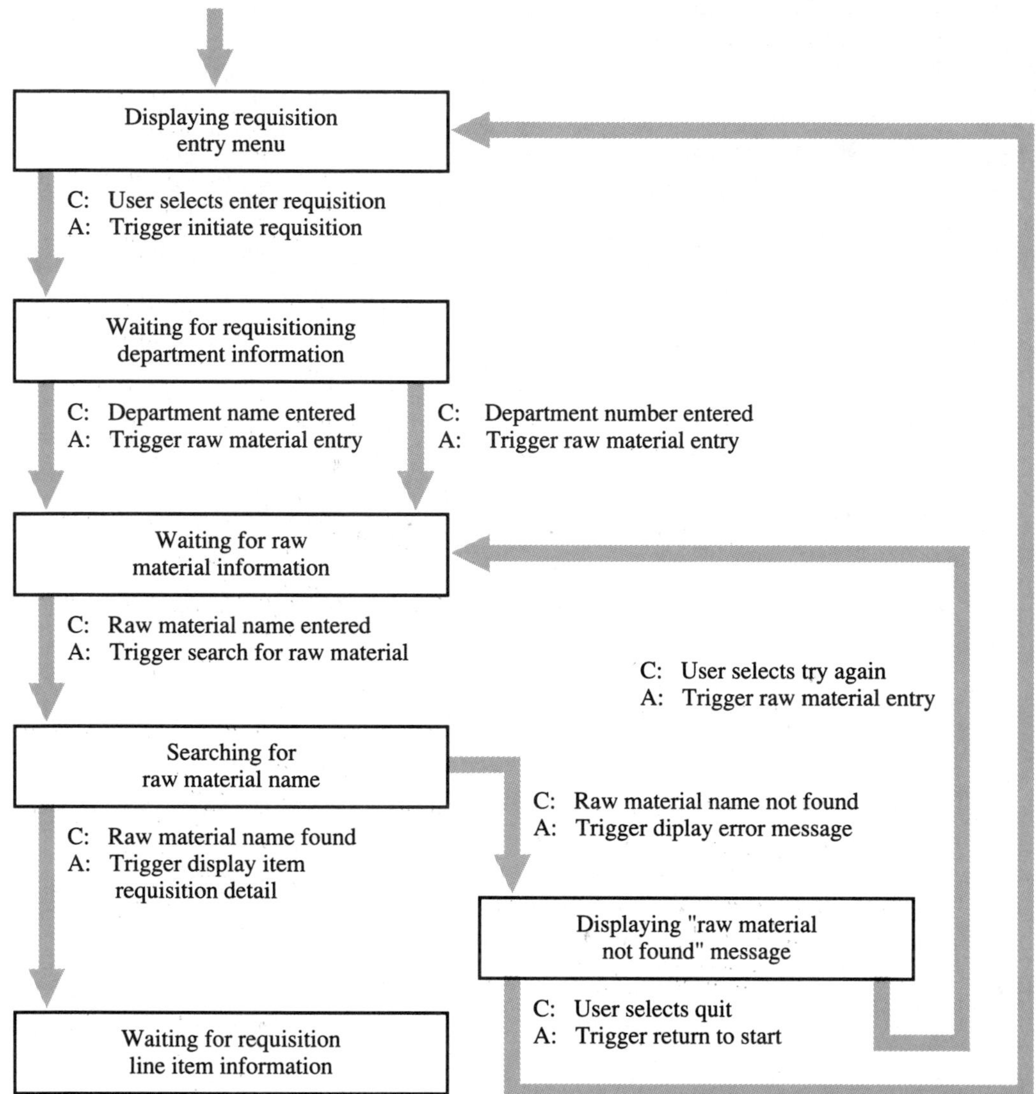

Figure 2.9
A state transition diagram for an online real-time raw material requisition system.

Called the root or executive module, it is where control starts. From the root module, control is passed down the structure chart level by level to other modules. Control is always returned to the invoking or calling module. When the program completes executing, control returns to the root module.

One control relationship at most exists between any two modules. Module A can call module B, but module B cannot also call module A. A module cannot call itself. Both module A and module B can invoke module C, called a common module.

When control is transferred between two modules, data are normally transferred as well. Data may be transferred in either direction between modules.

What Are Structure Chart Symbols?

Symbols that comprise structure charts are displayed in Figure 2.10. The basic symbols of a structure chart are rectangles used to designate modules. Arrows indicate connections between modules. Short arrows with circular tails are couples that show data and control elements communicated between modules.

Figure 2.10
Structure chart symbols and their meaning.

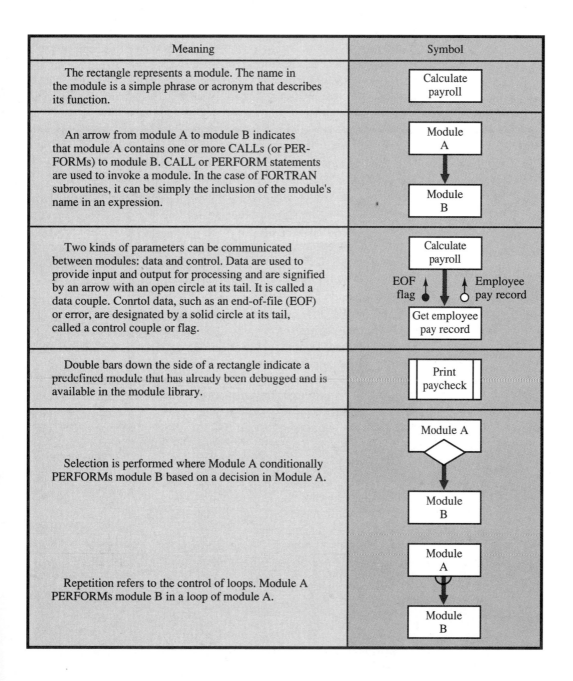

Meaning	Symbol
The rectangle represents a module. The name in the module is a simple phrase or acronym that describes its function.	Calculate payroll
An arrow from module A to module B indicates that module A contains one or more CALLs (or PERFORMs) to module B. CALL or PERFORM statements are used to invoke a module. In the case of FORTRAN subroutines, it can be simply the inclusion of the module's name in an expression.	Module A → Module B
Two kinds of parameters can be communicated between modules: data and control. Data are used to provide input and output for processing and are signified by an arrow with an open circle at its tail. It is called a data couple. Conrtol data, such as an end-of-file (EOF) or error, are designated by a solid circle at its tail, called a control couple or flag.	Calculate payroll; EOF flag; Employee pay record; Get employee pay record
Double bars down the side of a rectangle indicate a predefined module that has already been debugged and is available in the module library.	Print paycheck
Selection is performed where Module A conditionally PERFORMs module B based on a decision in Module A.	Module A; Module B
Repetition refers to the control of loops. Module A PERFORMs module B in a loop of module A.	Module A; Module B

A diamond and semicircle are used to indicate decisions and loops, respectively.

Steps in Preparing Structure Charts

Structure charts are derived from a leveled set of DFDs. A completed structure chart is the software design that will be coded and tested for implementation. Essentially, a structure chart provides a detailed view of a low-level process on a DFD. The process symbol at the top of Figure 2.11, for example, is part of a job costing system modeled by a DFD. It requires the calculation of the payroll as one of the requirements of the job costing system. The structure chart decomposes or explodes this elementary process to depict the design of the software required to implement the process. In addition, it produces payroll checks. The steps necessary to develop a structure chart are:

Step 1. Identify the Central Transform. The central transform is the portion of the DFD that contains the essential functions of the system. In our example, 4.3.3 calculate payroll is the central transform.

Step 2. Produce a First-Cut Structure Chart. Identify the executive or root module. In our example, the root module is Calculate Payroll. This module serves as the "traffic cop" for the structure chart hierarchy of modules. Then add the read and write modules. In our example, these modules are in the module library ready for reuse on a number of applications dealing with employees. Therefore, these modules will not require coding.

Step 3. Add Modules and Show Data Flows. Modules that perform the central functions are added. For example, these modules are typically computation and error-handling (not shown here for simplicity) modules. Each module should represent one self-contained function and be labeled with a descriptive name. Creating modules is discussed in more detail later in the book.

Add data up and data down symbols that depict the flow of data between superior and subordinate modules. The arrows with an open circle show data passing between modules. The arrow with a filled circle shows control data passing between modules. Control data generally represent flags or end-of-file (EOF) indicators. Their values cause the receiving module to perform or not perform specific functions.

Step 4. Write Module Details. The structure chart depicts the overall design of the software program to be coded and tested by showing all the modules and their interrelationships. Details of the modules are specified in an algorithmic language such as pseudocode or structured English. Decision trees and decision tables may also be used. These techniques are discussed later in this chapter and throughout the book. A brief example of module details for the Get Rate And Hours module is:

```
Get Rate and Hours.
    Get Employee-Pay-Rate
         and Hours-Worked
         From Employee-File
         Using Employee-ID of
         Employee-Record.
```

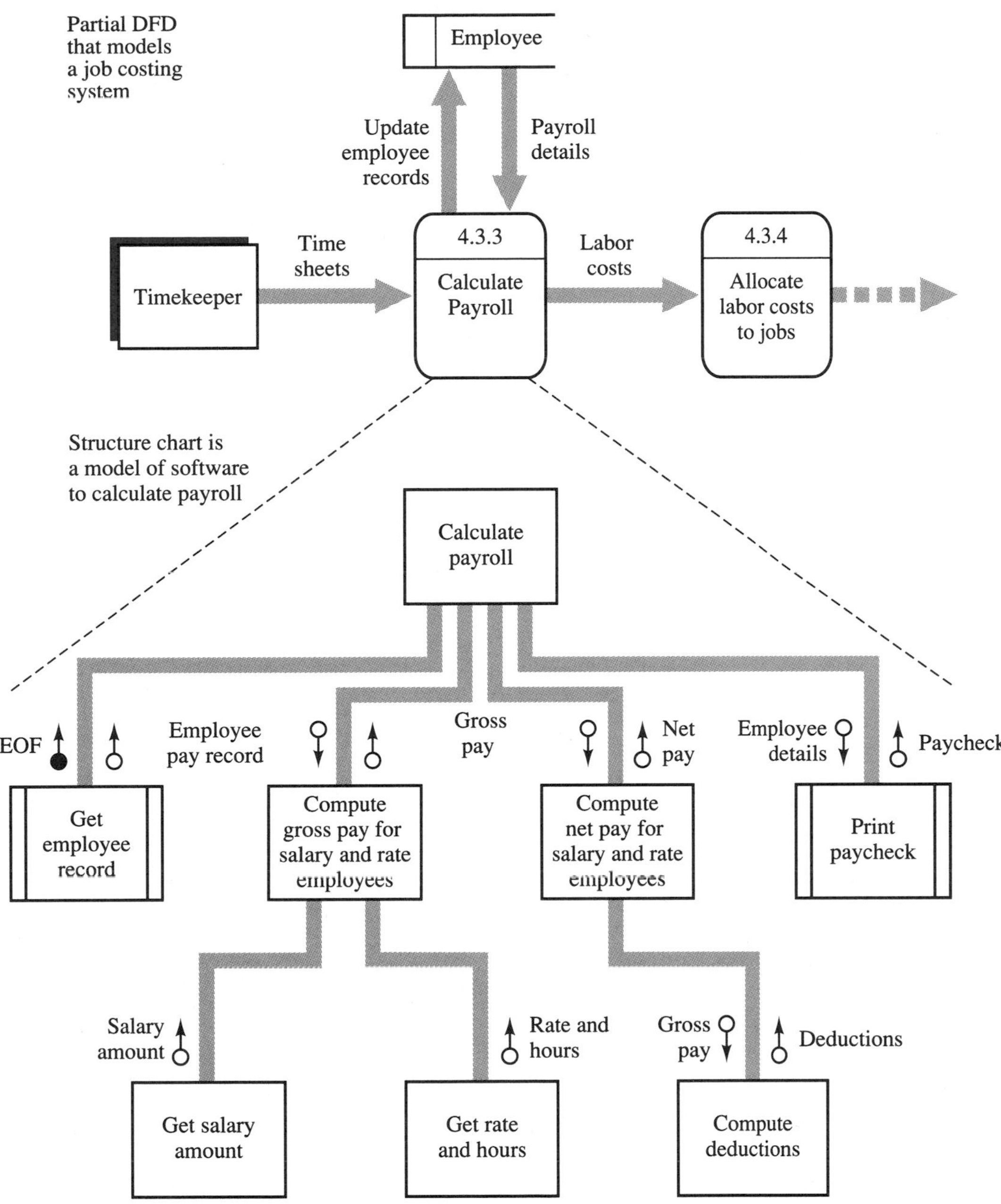

Figure 2.11 Structure chart prepared from an elementary-level DFD.

Step 5. Conduct Software Design Walkthrough. After the module details are specified, a **SOFTWARE DESIGN WALKTHROUGH** (i.e., review and evaluation) is conducted to see if the structure chart design works. Usually, the people conducting the software design walkthrough are different from the people who created the structure chart design. For example, the systems analysts who created the DFD may not be the ones who developed the structure chart. In this case, the systems analysts would make excellent software design walkthrough reviewers to ensure that the structure chart design correctly implements the requirements of the DFD. Walkthroughs are discussed further later in the book.

CREATING A STRUCTURED PROGRAM FLOWCHART

Several types of flowcharts are used in information systems, the most common being the **STRUCTURED PROGRAM FLOWCHART**. Flowcharts were one of the earliest modeling tools. They were designed to help standardize the process of systems analysis and design.

A structured program flowchart is an excellent tool to describe the software logic. The style of a structured program flowchart that we describe here is particularly applicable to the structured approach for developing software. Structured programs are designed on the basis of three **STANDARD CONTROL CONSTRUCTS** (also called **CONTROL STRUCTURES**).

- A sequence of instructions.
- A selection of instructions based on some decision criteria in the form of an IF-THEN-ELSE construct.
- A repetition of instructions based on DO WHILE or DO UNTIL.

These standard control constructs must each show a single entry and single exit. (Structured code reads from top to bottom without backward references.) This makes the code easy to test and maintain. The structured program flowchart illustrated in Figure 2.12 represents the standard control constructs.

In a program, the sequence control construct is found in such simple statements as PRINT, READ, WRITE, or in algebraic operations such as C = A + B. Such a construct is characterized by sequential execution by the computer, where each operation is completed before the next is begun.

The selection control construct is the decision-making aspect of a structured program. According to some condition, the computer will execute one series of operations to the exclusion of one or several others. For instance, many programs include a command to quit or end the execution of the program. If the user issues this command, a confirmation response, such as "are you sure?" may be required before the program will actually end. At this point in the program structure, a selection control construct of the form IF-THEN-ELSE is used which tests for the user's response of "Y" (an answer of "yes" to the confirmation). If this test result is positive, a series of operations to end the program will be executed by the computer. Otherwise, a different series of operations to cancel the initial quit command will be executed. The more interactive a program is with its user, the more selection control constructs will be found in its structure.

The repetition control construct causes the computer to perform a series of operations more than once. This repetition may continue as long as a particular condition exists (DO WHILE), or as long as the condition does not exist (DO

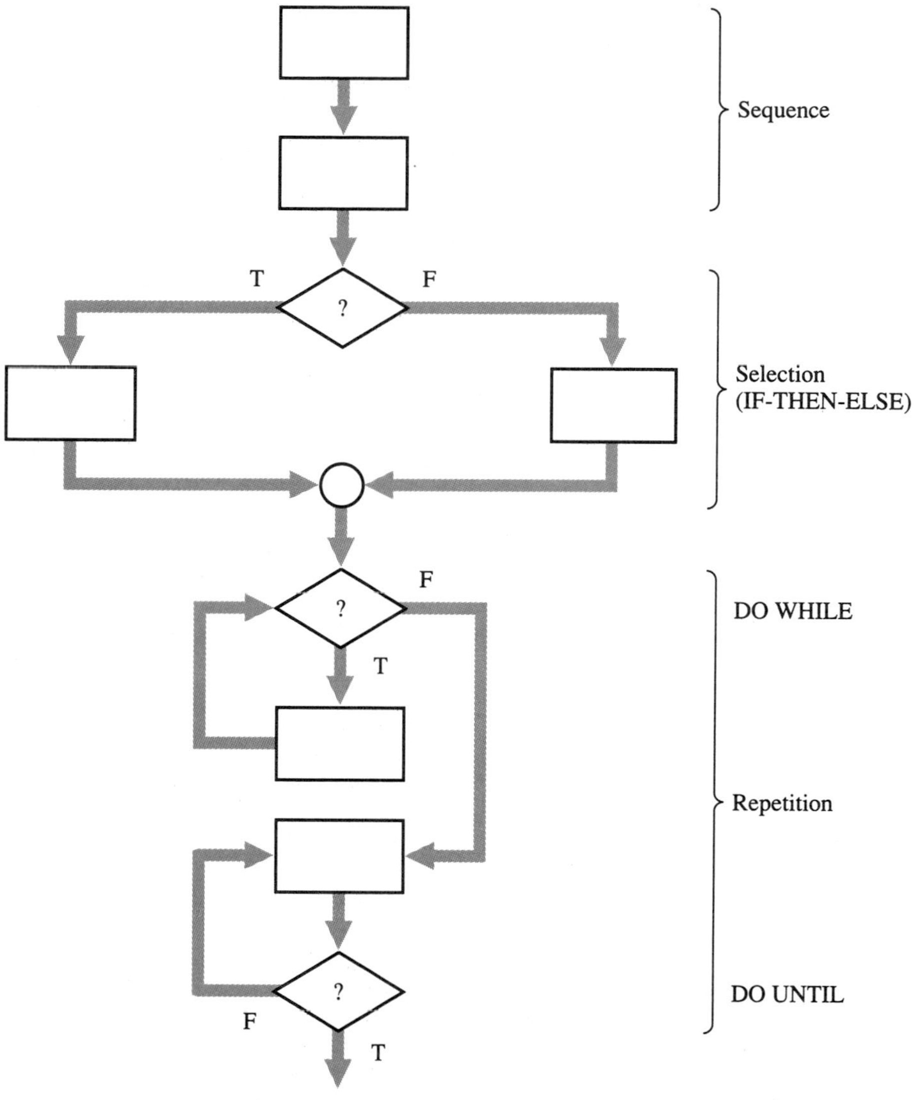

Figure 2.12
Use of a structured program flowchart to illustrate the sequence, selection, and repetition constructs.

UNTIL). The repetition construct is the foundation on which the computer's labor- and time-saving abilities are built.

WHAT ARE PROCESS SPECIFICATION TOOLS?

In our discussion of the data flow diagram (DFD), we saw that a process in the context-level DFD could be exploded in a second-level DFD to show more detail. This explosion can be repeated down to a detailed-level DFD, which consists of processes at an elementary level. **PROCESS SPECIFICATION TOOLS** provide further description of elementary-level processes. Process specification tools include:

- Structured English
- Decision tables
- Decision trees
- Equations

Structured English

Typical English narrative describing specifications can contain many ambiguities. For this reason, a subset of the English language called **STRUCTURED ENGLISH** is used. It consists of a few small groups of words that clearly describe the three standard control constructs of sequence, selection, and repetition.

Figure 2.13 shows some of the keywords used in structured English description of the three control constructs. Additional keywords are used for logic. While some of these keywords may appear as syntax in many computer languages, structured English is always language-independent.

Keywords used to describe the selection construct can be of the standard IF-THEN-ELSE variety, where a selection is made between two paths. They may also be of the IF-ELSEIF-ELSEIF-ELSE (or SELECT-WHEN-WHEN) variety, where a selection is made from more than two options. This multiple-option type of selection construct is sometimes called a case construct.

Figure 2.14 is a structured English listing for a routine called VALIDATE-ITEM. Note how no knowledge of the syntax of a programming language is necessary to make sense of the routine. This is the strength of structured English. It is understandable to the technical and nontechnical person alike.

Figure 2.13 Typical structured English keywords.

Construct	Structured English Keywords
Sequence	No keywords are used to show sequence. Frequently the sequence is preceded by a title. The end of the sequence is often indicated by the word EXIT.
Selection	IF...THEN...ELSE...ENDIF IF...ELSEIF...ELSEIF...ELSE...ENDIF SELECT...WHEN...WHEN...ENDSELECT
Repetition	DO WHILE, REPEAT WHILE, LOOP WHILE DO UNTIL, REPEAT UNTIL, LOOP UNTIL FOR ALL, FOR EACH...WHERE END, ENDDO, ENDREPEAT, ENDLOOP, ENDFOR

Logic Keywords	
AND, OR	GE (greater than or equal)
GT (greater than)	LE (less than or equal)
LT (less than)	

Figure 2.14
A structured English listing for the routine VALIDATE_ITEM.

```
VALIDATE_ITEM

*   VALIDATE GENERAL FORMAT.
    CHECK GENERAL FORMAT.
    IF ERRORS
        WRITE ERROR-MESSAGE
    ENDIF.

*   VALIDATE SPECIAL FORMAT.
    IF NEW ITEM
        CHECK NAME AND TITLE
        CHECK FOR NUMERIC BOX
        CHECK ADDRESS
        CHECK FOR PAYMENT
        IF ERROR
            SET INVALID FLAG
        ELSE
            SET VALID FLAG
        ENDIF
    ENDIF.
    IF CANCEL
        SET CANCEL INDICATOR
    ENDIF.
    IF INVALID FLAG
        WRITE ERROR-MESSAGE
    ENDIF.
```

Comments within the structured English listing are indicated by asterisks. Titles of sequences in this listing are "CHECK ADDRESS" and "SET VALID FLAG," while selection is shown by numerous IF-ELSE-ENDIF keywords.

Decision Tables

The logical control of an information system, as in the IF-THEN selection control construct, is easily organized using a **DECISION TABLE.** Figure 2.15 illustrates just such a table. It might be used in the design of a newspaper subscription system.

The upper half of the table includes all "if" conditions that affect the decision. Some of these conditions may be mutually exclusive, while others may occur simultaneously. For instance, the two conditions "Renewal" and "Cancellation" are mutually exclusive; they cannot occur simultaneously. It is possible, however, for the conditions "Credit check OK" and "Renewal" to coexist. The number of possible combinations of the conditions is represented by the number of columns in the decision table.

The lower half of the table includes all "then" actions that might be required as a result of the decision. To use the table, identify the conditions that are present and locate the matching column in the upper half of the table. Then follow that column down to the lower half of the table, where the required actions, numbered in sequence, are found. Sometimes, if the sequence of the actions is unimportant, these numbers may be replaced by ×'s or other marks.

Figure 2.15
Decision table for a newspaper subscription system.

Conditions				
Credit check OK	NO	YES	YES	YES
New subscription	—	YES	NO	NO
Renewal	—	NO	YES	NO
Cancellation	—	NO	NO	YES
Actions				
Print error message	1			
Create record		1		
Print bill		2	1	
Change expiration date			2	
Indicate deletion				1
Issue refund				2

Decision tables are most useful where many actions are required for any given decision. For decisions that result in only one or two actions, a simpler tool, the decision tree, is called for.

Decision Trees

Figure 2.16 shows a typical **DECISION TREE**, based on the same decision logic we used in our decision table example. This simple example lends itself well to the tree format. It is easily read from left to right. The reader's eye follows pathways, choosing the correct path at each junction based on a single set of mutually exclusive conditions.

Sometimes decision tables are huge matrices, with hundreds of conditions and many possible actions. Such complicated decision logic is best left in table form, because it does not convert easily to a decision tree format. For our simple example, however, pick a set of conditions and follow the decision logic in both the decision table and the decision tree. You'll find that each results in the same set of required actions.

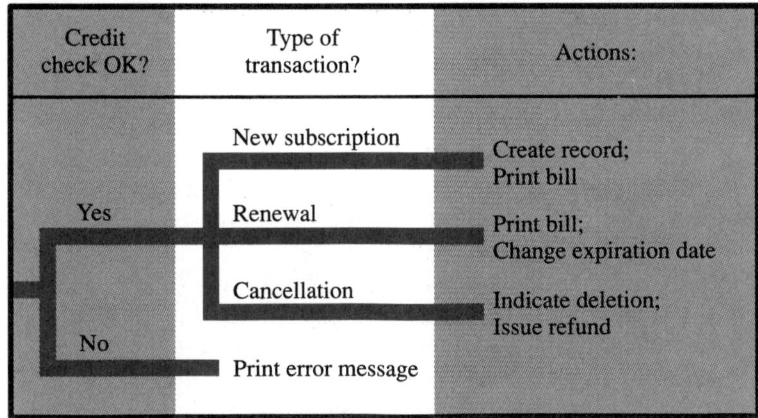

Figure 2.16
Decision tree for a newspaper subscription system.

Equations

While many processes can be specified by the keywords of a structured English listing or the logic of a decision table or tree, sometimes a process is too complex for these tools. **EQUATIONS** provide a fourth way of specifying a process that perhaps involves much mathematical computation.

It would be difficult and cumbersome to describe a process that computes, say, the present value of an annuity, using words alone. An equation, on the other hand, succinctly and completely specifies the process.

Even a simple process that takes a retail purchase total and adds sales tax to give a final total payable by the customer is most clearly expressed by an equation:

```
PAYTOTAL = PURCHASETOTAL * (1 + TAXRATE)
```

The equation format has the added advantage of easy coding by programming people during the back-end phases of SDLC.

CREATING A WARNIER–ORR DIAGRAM

The **WARNIER–ORR DIAGRAM (WOD)** modeling tool uses brackets (or braces) to show the hierarchical decomposition of processes or data. This decomposition can represent a high-level overview of software structure or detailed software logic.

What Are WOD Symbols?

The WOD is named after its developers: Jean-Dominique Warnier and Kenneth T. Orr. The main symbol in the WOD is the brace. Sequence, selection, and repetition are the three standard control constructs that represent various data and process structures. All the symbols used in WOD are presented in Figure 2.17.

Figure 2.17
WOD symbols.

Meaning	Symbol
Shows decomposition and hierarchy (A consists of B and B consists of C)	A { B { C
Exclusive OR (A or B but not both)	⊕
Inclusive OR (A or B or both)	+
Arithmetic operators	⊞ ⊟ ⧄ ⧆
Negation	———
Repetition one to N times (DO UNTIL)	(1,N)
Repetition zero to N times (DO WHILE)	(0,N)
Repetition occurs N times	(N)

Figure 2.18
Venn diagram showing logical relationships.

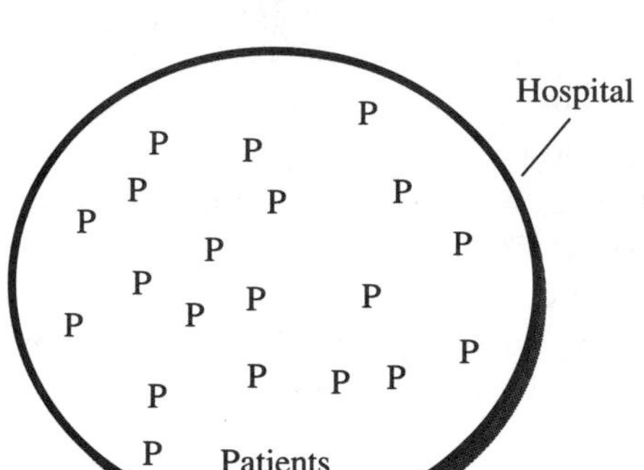

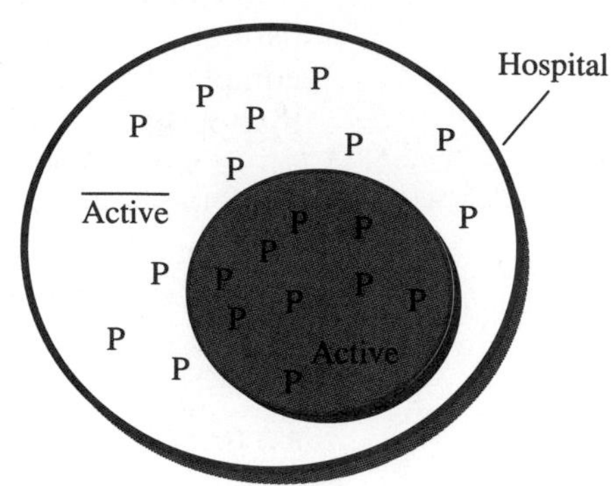

Steps in Preparing WODs

Assume that you work as a systems professional for a hospital. You have been assigned to develop a system to print active patient bills. Further assume that you have a file that contains data on each patient in the hospital. To model the system, employ the following steps.[1]

Step 1. Prepare a Logical View. You can use Venn diagrams to show the relationship of all patients. At the left of Figure 2.18 is a Venn diagram that shows this relationship. This hospital set is made up of patients. Only active patients shown at the right of the Venn diagram are to be billed. The small circle defines this subset. All patients outside the small circle are nonactive. They are not to receive a bill.

Step 2. Map the Solution into a WOD. Figure 2.19 represents the mapping of a solution into a WOD. The patient file contains P records. Each will be read until an end-of-file (EOF) indicator terminates the process. The WOD solves the hospital billing problem by computing and billing all active patients and skipping all inactive patients. This is a simple problem and a simple solution. More complex systems can be analyzed and designed in a similar manner.

CREATING A JACKSON DIAGRAM

In developing a **JACKSON DIAGRAM**, input and output data of a system are used to create the program structure. The creator of Jackson diagrams is Michael Jackson, a noted systems authority.

What Are Jackson Diagram Symbols?

Jackson diagrams are based on the ability to represent data and software with standard control constructs specified by the symbols in Figure 2.20. These symbols are combined in an infinite number of ways to form full data and software structures.

[1] Kenneth T. Orr, *Structured Systems Development* (New York: Yourdon Press, 1977), pp. 63–71.

Figure 2.19
Hospital billing solution.

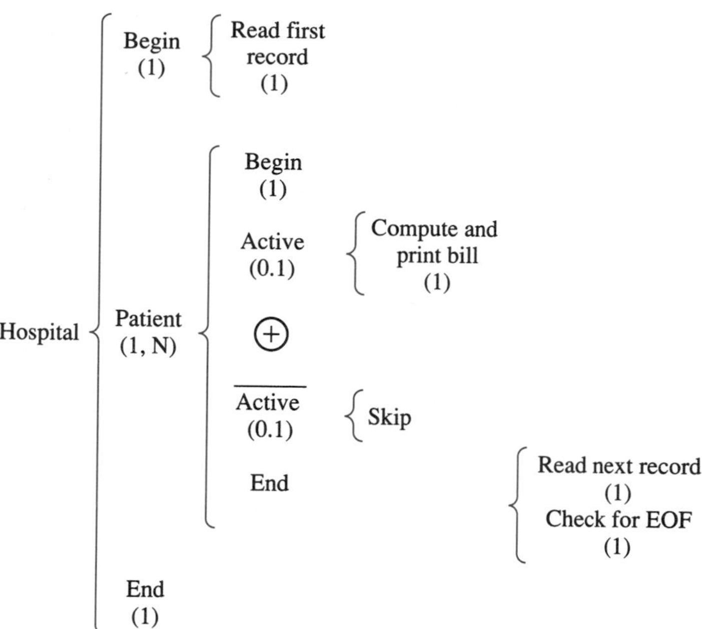

Figure 2.20
Jackson diagram symbols and their meaning.

Meaning	Symbol
The sequence construct shows that A is internally composed of B, C, and D and is read from left to right.	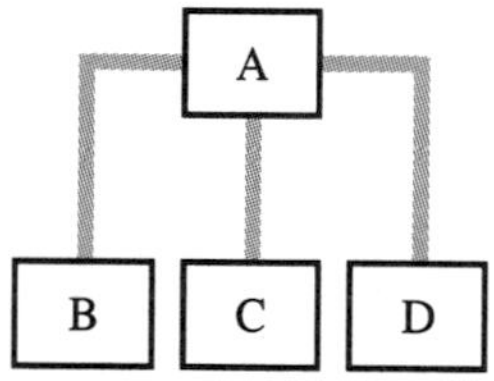 A → B, C, D
The selection construct indicates that A is composed of B, C, or D and only one of these in any single case. The small circles in the boxes for B, C, and D indicate the parts of the selection component.	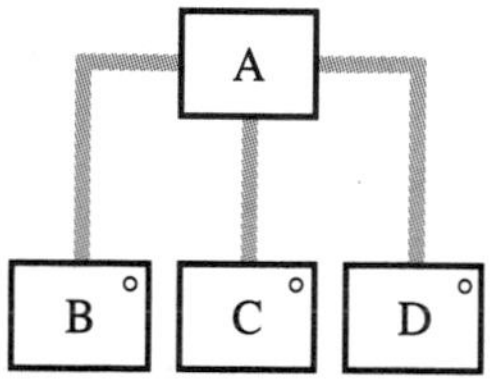 A → B°, C°, D°
The repetition construct shows that A is composed of zero or more Cs. It is denoted by an asterisk in C.	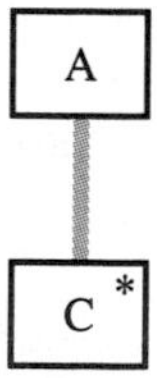 A → C*

Figure 2.21
Input and output and their correspondences.

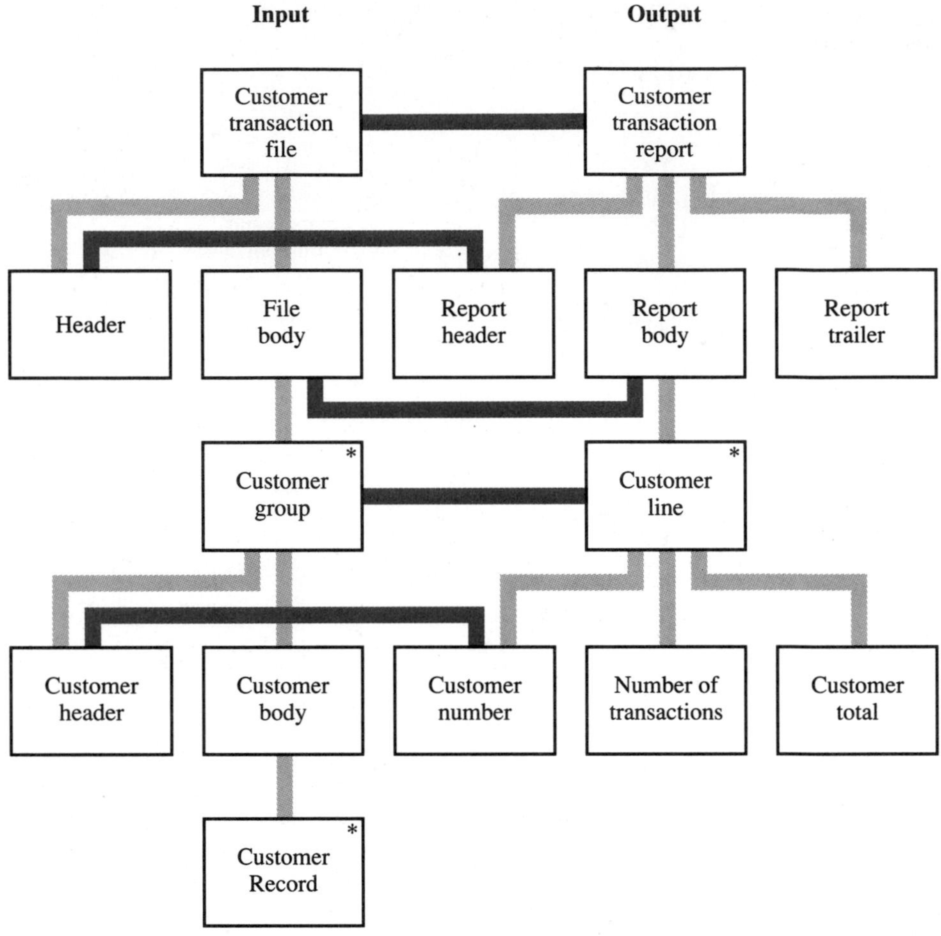

Steps in Preparing Jackson Diagrams

Our objective is to generate a simple customer transaction report, which represents output. The input is a sorted customer transaction file.[2] There are four steps needed to do this.

Step 1. Define Input and Output Data Structures. The input is a hierarchical structure of the Customer Transaction File. The output to be produced is a hierarchical structure of the Customer Transaction Report. These input and output hierarchies are shown in Figure 2.21.

Step 2. Determine the Program Correspondences Between Data Structures. The correspondences indicated by bold lines show a consume–produce relationship. For example, the input header is used to produce the report header.

[2] David King, *Creating Effective Software: Computer Program Design Using the Jackson Methodology* (Englewood Cliffs, N.J.: Yourdon Press, A Prentice-Hall Company, 1988), pp. 57–66.

Step 3. Develop the Software Structure. The software structure is modeled in hierarchical form similar to a structure chart or a WOD, except that a WOD lies on its side. This software structure is depicted in Figure 2.22.This structure shows how input is converted to output.

Step 4. Write Structured Text. Each module in the software structure diagram is translated into **STRUCTURED TEXT**, presented in Figure 2.23. Structured text is a type of pseudocode or structured English. All three are linguistic tools that are used to describe software logic.

REVIEW OF CHAPTER LEARNING OBJECTIVES

The major goals of this chapter were to enable each student to achieve eleven learning objectives.

Figure 2.22
Software structure.

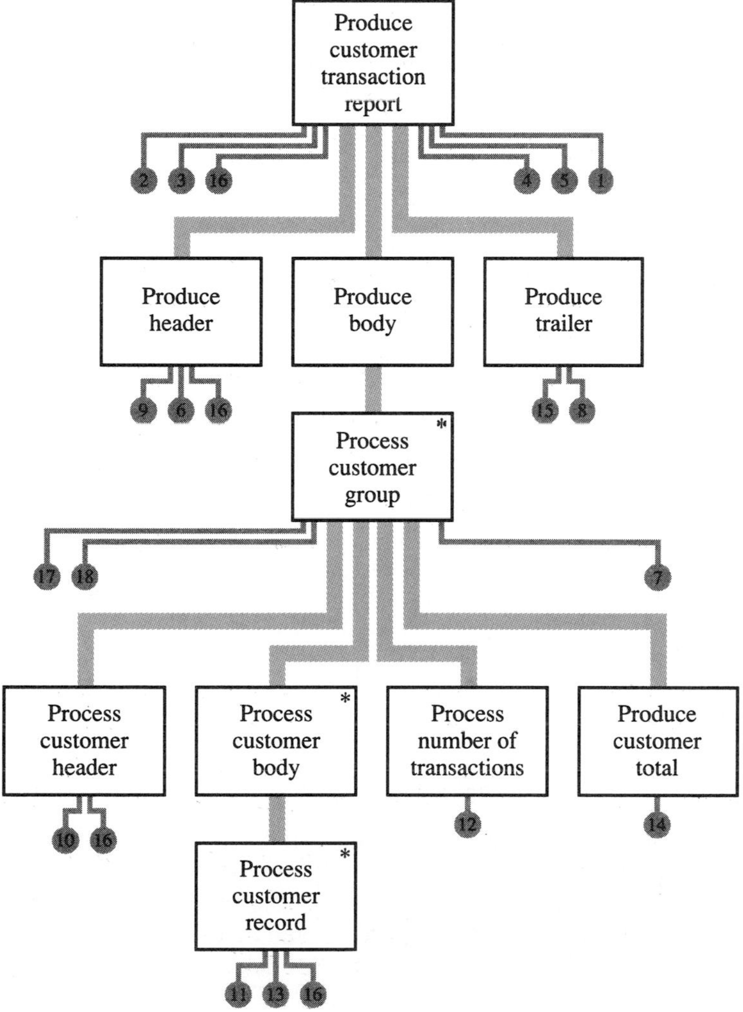

Figure 2.23
Structured text.

1 STOP
2 OPEN INPUT
3 OPEN OUTPUT
4 CLOSE INPUT
5 CLOSE OUTPUT
6 WRITE REPORT HEADER
7 WRITE CUSTOMER LINE
8 WRITE REPORT TRAILER
9 MOVE REPORT HEADER TO PRINT LINE
10 MOVE CUSTOMER NUMBER TO PRINT LINE
11 ADD 1 TO TRANSACTION
12 MOVE TRANSACTION TO PRINT LINE
13 ADD AMOUNT TO TOTAL
14 MOVE TOTAL TO PRINT LINE
15 MOVE REPORT TRAILER TO PRINT LINE
16 READ INPUT
17 SET TRANSACTION TO ZERO
18 SET TOTAL TO ZERO

Learning objective 1:
Discuss the structured approach and list its key features.

The structured approach is a disciplined, engineered way to develop systems and software. Major features that support this approach are:

- Modeling tools
- Modularization
- Top-down decomposition
- Iteration
- Parallel activities
- Systems development automation

Learning objective 2:
Explain the types of modeling tools and the roles they play in developing systems.

Creating models greatly facilitates the development of systems and software. Modeling tools used to create these models play three key roles. The model serves as a communication medium between users and systems professionals. Both users and systems professionals experiment with the model to gain a clearer view of its functionality. Models can also be used to predict how the system will operate in the future.

Learning objective 3:
State the purpose of a data flow diagram (DFD) and describe the steps necessary to prepare a DFD.

The purpose of a DFD is to show the processes that convert input into output at the systems level. At the beginning of analysis, the systems analyst draws a context level diagram to provide a broad model and scope of the system under development. The context-level diagram is then decomposed to a lower-level diagram, showing more details. This step is repeated until a full set of leveled DFDs is prepared, thus modeling the system under development in detail.

Learning objective 4:
Explain the construction of a data dictionary and discuss its purpose.

The data dictionary is composed of all data elements found in the system. Each data element has a corresponding definition, which shows how the element is made up of more elementary data elements. For the most elementary data elements, the definition consists of a range of possible values and perhaps a unit of measure. The purpose of a data dictionary is to define and catalog potentially thousands of data elements.

Learning objective 5:
Cite the purpose of an entity relationship diagram (ERD) and list the steps necessary to prepare an ERD.

An ERD creates a data model of the system under development. Steps necessary to create an ERD are:

1 Identify entities.

2 Indicate relationships between entities.

3 Define keys for each entity.

4 Define and map data elements for each entity.

5 Normalize the data model.

Learning objective 6:
Describe a state transition diagram (STD) and explain how it is applied.

A STD shows finite states in which a system can exist and the transitions which result in new finite states. These diagrams are useful in expressing the time-dependent aspects of a system. Hence they are most often used in modeling online real-time systems.

Learning objective 7:
Describe the purpose of a structure chart and explain the steps required to create a structure chart.

A structure chart is used for software design. The resulting design is used to communicate with both software design walkthrough reviewers and the programmers (or coders) who will code the software program.

Steps required to create a structure chart are:

1 Identify the central transform.

2 Produce a first-cut structure chart.

3 Add modules and show data flows.

4 Write module details.

5 Conduct software design walkthroughs.

After the structure chart design is accepted, it is ready to be used by programmers to code, and testers to test, the software program.

Learning objective 8:
Describe the structured program flowchart and discuss its usage.

The structured program flowchart models the standard control constructs of sequence, selection, and repetition. Structured program flowcharts are widely used because they help to standardize the program development process.

Learning objective 9:
Explain how process specification tools relate to the data flow diagram and cite the purpose of each of the four process specification tools.

Process specification tools are used to provide complete descriptions of the processes found in the elementary level data flow diagrams. Structured English is a collection of simple keywords that convey the three standard control constructs. Decision tables show complex, logical relationships based on numerous conditions and actions. Decision trees are similar, except that they show more clearly simple logic structures with few required actions. Equations succinctly convey processes which involve mathematical computation.

Learning objective 10:
Explain the purpose of a Warnier–Orr diagram (WOD) and describe the steps necessary to develop a WOD.

A WOD is another structured modeling tool that aids systems analysis and software design. Steps required to develop a WOD are:

1 Prepare a logical view.

2 Diagram the solution.

Learning objective 11:
State the purpose of a Jackson diagram and list the steps required to develop a Jackson diagram.

The Jackson diagram is an input- and output-driven structured modeling tool. Its primary purpose is to model a software program structure from the data structures. Steps required to develop a Jackson diagram are:

1 Define input and output data structures.

2 Determine the program correspondences between data structures.

3 Develop the software structure.

4 Write structured text.

STRUCTURED APPROACH AND MODELING TOOLS CHECKLIST

Following is a checklist of the structured approach and modeling tools used by systems professionals.

1 To develop systems following the structured approach, employ modeling tools, modularization, top-down decomposition, iteration, parallel activities, and systems development automation.

2 Use modeling tools to communicate with users, experiment with alternative designs, and predict how a systems design will work.

3 For modeling the flow of data and the processes that the data undergo, use a data flow diagram.

4 For defining all the data elements in the system being modeled, use a data dictionary.

5 For modeling the data stores of a data flow diagram, or creating enterprisewide models, or preparing object-oriented software designs, use an entity relationship diagram.

6 For modeling time-dependent processes of a system, use a STD.

7 For converting elementary-level processes of the data flow diagram into a hierarchy of software modules, including descriptions of the interfaces between the modules, use a structure chart.

8 For modeling the standard control constructs of sequence, selection, and repetition of a structured software design, use a structured program flowchart.

9 For specifying processes rigorously and precisely, in a way that can be understood by systems professionals and end users alike, use structured English.

10 For specifying, in tabular form, what actions to take when complex combinations of conditions exist in a decision-making situation, use a decision table.

11 For specifying, in a logical manner, a sequence of decisions in which each branch of the tree represents a decision rule and a single action, use a decision tree.

12 For specifying business, mathematical, and statistical models of processes, use an equation.

13 For modeling software standard control constructs of sequence, selection, and repetition, use a Warnier–Orr diagram as an alternative tool to a structure chart, structured program flowchart, structured English, decision table, and decision tree. The Warnier–Orr development approach is to start with output and work backward through the system until you finally arrive at the processing steps that are required, the database you need, and finally the inputs you have to collect to produce the outputs.

14 For modeling the software standard control constructs of sequence, selection, and repetition, build input and output structures, and specify processes, use a Jackson diagram as an alternative tool to a structure chart, structured program flowchart, structured English, decision table, decision tree, and Warnier–Orr diagram.

KEY TERMS

Computer-aided systems and software engineering (CASE)

Context-level diagram

Data dictionary

Data flow

Data flow diagram (DFD)

Data store

Decision table

Decision tree

Entity

Entity relationship diagram (ERD)

Equations

Iteration

Jackson diagram

Modeling tools

Modularization

Parallel activities

Process

Process specification tools

Relationship

Sinks (destinations)

Software design walkthrough

Sources (origins)

Standard control constructs (control structures)

State transition diagram (STD)

Structure chart

Structured approach

Structured English

Structured program flowchart

Structured text

Systems development automation

Top-down decomposition

Warnier–Orr diagram (WOD)

REVIEW QUESTIONS

2.1 Define the structured approach. List and describe the key features of the structured approach.

2.2 What is the goal of the structured approach?

2.3 List and briefly describe major features that support the structured approach.

2.4 List and briefly describe the three roles played by modeling tools.

2.5 What is the purpose of a data flow diagram (DFD)? Describe the steps required to develop a DFD.

2.6 Describe the DFD symbols sponsored by Yourdon and DeMarco and Gane and Sarson.

2.7 What is a leveled set?

2.8 What is the purpose of a context-level DFD?

2.9 What is a detailed-level DFD?

2.10 What is a data dictionary, and what is it used for?

2.11 What is the purpose of an entity relationship diagram (ERD)? Describe the steps necessary to create an ERD.

2.12 For what purpose are entities, relationships, and data elements (or attributes) used in ERDs?

2.13 What is a state transition diagram? What is its purpose?

2.14 Structure charts normally work in conjunction with what other modeling tool?

2.15 What is the purpose of a structure chart? Describe the steps required to develop a structure chart.

2.16 What is a structured program flowchart used for?

2.17 Why are structured English, decision tables, decision trees, and equations called process specification tools?

2.18 How is structured English used in conjunction with structure charts or structured program flowcharts?

2.19 What are decision tables used for?

2.20 What are decision trees used for?

2.21 What is the purpose of equations?

2.22 What is the purpose of a Warnier–Orr diagram? What steps are required to create a WOD?

2.23 What is the purpose of a Jackson diagram? What steps are necessary to develop a Jackson diagram?

CHAPTER-SPECIFIC PROBLEMS

These problems require exact responses based directly on concepts and techniques presented in the text.

2.24 Following is a list of systems development objectives:

_______________ Go back to previous phases for creating more precise definitions.

________________ Present models and describe a system at progressive levels of detail.

________________ Use CASE technologies.

________________ Use linguistics and graphical diagrams to communicate with end users and systems professionals, experiment with alternative designs, and predict how systems will perform.

________________ Divide a design into well-defined, function-specific units.

________________ Work on a number of phases and activities simultaneously.

Required: Insert the name of the structured approach feature that best meets each systems development objective.

2.25 A customer sends a customer order to a validate order process. An invalid order from the validate order process is sent to a process not defined, and a valid order from the validate order process is sent to a fill from inventory process, which queries and updates an inventory data store. The fill from inventory process sends a process order to a ship order process. The ship order process accesses rates from a shipping rate data store.

Required: Prepare a data flow diagram that models the system described above.

2.26 Truck drivers for ACE trucking company are supposed to keep their trucks in good repair. If a truck driver detects a malfunction, he or she is required to request repairs from the company shop immediately. At the shop, a repair order is prepared and a mechanic is assigned according to availability status from the mechanic master file. The assigned mechanic diagnoses the problem, orders parts, and performs the repairs. Parts information is stored on a parts master file. Parts come from parts inventory. After repairs are made, the mechanic submits charges to accounting and releases the truck to the truck driver. Accounting sends charges to the appropriate division and notifies the truck driver of the amount of charges made to his truck.

Required: Prepare two data flow diagrams that describe the foregoing processes. The first DFD should be a context-level diagram. The second DFD should decompose the request repair process of the context diagram into its appropriate components.

2.27 A data element CUSTOMER-ORDER is shown on a data flow diagram. CUSTOMER-ORDER is composed of a CUSTOMER-NAME, together with an ACCOUNT-NUMBER, WHOLESALE-CUSTOMER, or RETAIL-CUSTOMER, together with SALESPERSON, together with ITEM-NUMBER.

Required: Prepare a data dictionary entry describing the data element.

2.28 Customers purchase air compressors made to order. Each customer is assigned a specific account. Purchase orders are prepared by customers. Each purchase order must contain an account designation. The plant manager reviews purchase orders and releases a work order to start work in process. After a compressor is completed, it is transferred to finished goods and inspected by the plant manager. The plant manager releases the compressor for shipment to the customer.

Required: Construct an entity relationship diagram that describes the foregoing customer order process. The ERD should contain eight entities.

2.29 There are three data stores on a DFD: departments, employees, and projects. However, the DFD tells us nothing about their relationships. Based on further analysis, you have defined the relationships as: employees belong to departments, employees are assigned to projects, and departments initiate projects.

Required: Prepare an entity relationship diagram that models the system above.

2.30 Create a state transition diagram for a combined furnace and air-conditioning control system. This system has three states: (1) the furnace is on while the air conditioner is off; (2) the air conditioner is on while the furnace is off; (3) neither the air conditioner nor the furnace is on. Model the transitions between these states based on the temperature of the room. The temperature settings are as follows:

Unit	Temperature at Which Unit Turns On	Temperature at Which Unit Turns Off
Air conditioner	**84 degrees**	**78 degrees**
Furnace	**65 degrees**	**72 degrees**

2.31 Part of processing a customer order involves accessing the customer credit record via the customer number and verifying the charge by calculating the check digit, checking the credit limit, and issuing an authorization code. The results of this processing are formatted and displayed on a screen.

Required: Draw a structure chart that describes this process.

2.32 Review the following instructions:

```
IF C1
      IF C2
            PERFORM P1
      ELSE
            NEXT SEQUENCE
ELSE
      IF C3
            PERFORM P2.
```

Required: Draw a structured program flowchart that represents these instructions.

2.33 Review the following instructions:

```
IF C1
     PERFORM P1
     IF C2
          NEXT SEQUENCE
     ELSE
          PERFORM P2
          PERFORM P3
ELSE
     IF C3
          PERFORM P4
     ELSE
          PERFORM P5.
```

Required: Draw a structured program flowchart that represents these instructions.

2.34 Fits Rite Clothing Stores is creating a new customer check acceptance policy for use in its discount stores. A simple means of implementing this policy at the cash register is needed. As an analyst at Fits Rite, you have suggested that a decision table be placed at each cash register for the cashiers to follow.

Required: Create a decision table for the check acceptance policy. The actions specified by this table will be either to accept the check or refuse it. The decision rules will be based on a check acceptance policy of your own making, but will be dependent on the following conditions:

- Check amount is less than $100.
- Customer has a driver's license.
- Customer has a check guarantee card.
- Check is a two-party check.
- Customer is a "preferred customer."

2.35 The following routine in structured English illustrates the standard control constructs of sequence and selection.

```
GET CUSTOMER INFO
GET ADDRESS RECORD
IF NO SUCH ADDRESS
        THEN
          DISPLAY ''NO SUCH ADDRESS''
        ELSE
          EDIT CUSTOMER INFO
          IF ERROR IN CUSTOMER INFO
        THEN
```

```
                DISPLAY ERROR
            ELSE
                IF CUSTOMER IS PREFERRED
                    THEN
                        PRINT ''PREFERRED CUSTOMER''
                    ELSE
                        PRINT ''STANDARD CUSTOMER''
                ENDIF
            UPDATE ADDRESS RECORD
            SAVE RECORD TO BACKUP FILE
        ENDIF
ENDIF
```

Required: Transform this listing to a decision tree, where selection alternatives are represented by the branches of the tree, and sequence is represented by actions.

2.36 The information system of Supra Company processes one payroll file each week. One-to-N employee records are contained within the payroll file. Each record contains employee name, address, social security number, pay rate, and employee code. The employee code designates the employee as either a salaried or an hourly employee.

Required: Construct a Warnier–Orr diagram and Jackson diagram to describe this.

2.37 Review the following Warnier–Orr diagram.

```
              ⎧ Header
              ⎪                 ⎧ Payment
              ⎪                 ⎪ (0,1)
Accounts      ⎪ Customer        ⎪
Receivable    ⎨ Records         ⎨ +
File(1)       ⎪ (1,N)           ⎪
              ⎪                 ⎪ Invoice
              ⎪                 ⎩ (0,1)
              ⎩ Trailer
```

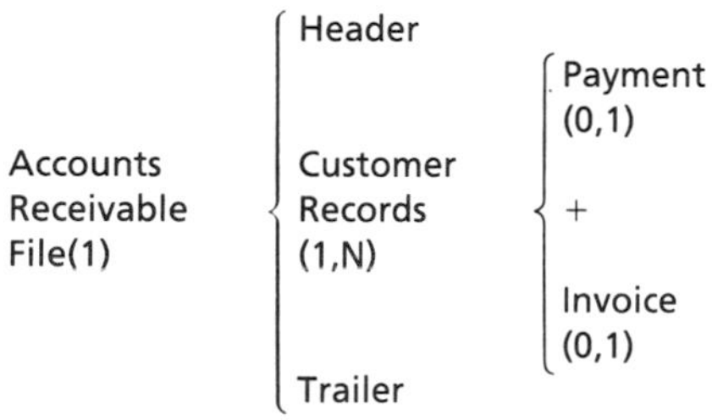

Required: Convert this Warnier–Orr diagram into a Jackson diagram.

2.38 The input to an edit module is transaction records. Edited transaction records represent output. The function of the edit module is to ensure that the customer-ID is valid. If the transaction record customer-ID is invalid, it is rejected and a new record is requested from the calling module. If the record is valid, it is passed back to the calling module for processing.

Required: Using a Jackson diagram, draw the input and output structures and indicate the correspondences. Then write the structured text to describe the edit module.

2.39 The following Warnier–Orr diagram describes the processing of an accounts receivable file:

```
      ⎧ B   ⎧ F   ⎧ J
      ⎪ C   ⎨ G   ⎨
  A   ⎨     ⎩ H   ⎩ K
      ⎪ D
      ⎩ E   { I
```

Required: Construct an equivalent Jackson diagram.

2.40 Review the following systems development situations:

________________ End user requires a narrative description of a process.

________________ The process is time-dependent.

________________ A broad view of how data flow through the system and the processes these data undergo is to be modeled.

________________ A number of actions are to be taken based on a complex combination of conditions.

________________ A hierarchical structure of software modules is to be modeled from an elementary-level data flow diagram.

________________ Elements of data are to be defined.

________________ Data stores of a data flow diagram are to be modeled.

________________ The standard control constructs are to be defined in a flowchart manner.

________________ The standard control constructs are to be defined along with input and output correspondences and structured narrative (or text).

Required: In the space provided, insert the name of the tool you believe to be most appropriate for the foregoing systems development situations.

THINK-TANK PROBLEMS

These problems call for a feasible approach rather than a precise solution. Although the problems are based on chapter material, extra reading and creativity may be required to develop workable solutions.

2.41 Success Seminars, Inc., is developing an information system to process registrations for its small business leadership seminars. You have been hired as a systems analyst to participate in the development of this system. In an effort to clarify and communicate some of the time-dependent aspects of this system, you choose to model the registration process using a state transition diagram.

Required: Create a state transition diagram that models the registration process. This diagram should identify all the finite states of the process, from the time a customer's application is submitted until the seminar is over and the registration for the customer has been achieved. Allow for the fact that there is a limited number of seats available, that the customer may cancel his registration, and that the customer may reapply for a different section of the seminar.

2.42 The Pitout Company manufactures machines that remove the pits from peaches for peaches that will be canned. The company is considering integrating its manufacturing and capacity planning systems. Currently, manufacturing is controlled through a computerized material requirements planning (MRP) system. Capacity planning is performed by the combined expertise of a group of analysts.

The manufacturing process involves five departments: Marketing, Engineering, Production Control, Manufacturing, and Purchasing. The Engineering Department is responsible for preparing bills of materials. A bill of materials is similar to a recipe; it lists all the parts that are necessary to make a peach pitter. For example, it takes ten general parts to make a peach pitter. To make one of the ten general parts requires another five parts. The Engineering Department makes sure that all bills of materials are current in the computer system. Each individual inventoried part necessary to put together a peach pitter is also in the computer system. If the part must be manufactured, the manufacturing operations (called a routing) are stored in the computer separate from the part. If the part must be purchased, the price and lead time are included with the part. The quantity on hand is, of course, maintained for each part.

The manufacturing process is started by a forecast from the Marketing Department. For example, the Marketing Department might forecast that 20 peach pitters will be sold during the next quarter. This forecast is entered into the computer. The computer then produces a listing, based on the bill of materials, showing which parts must be purchased and which parts must be manufactured. This listing is called the Material Requirements Planning Master Report (MRPMR). Purchase and manufacture dates based on part lead times are included on the listing. The MRPMR for purchase parts is used by the Purchasing Department to begin the purchasing process.

The MRPMR for manufactured parts is used by the Production Control Department to schedule the production of manufactured parts. The Production Control Department issues work orders based on the MRPMR to communicate with the Manufacturing Department to actually produce the peach pitters. The company wants to make sure that it has enough capacity, in terms of people and equipment, to produce its peach pitters. If Pitout waits until a forecast is made and an MRPMR is generated, there isn't enough time to buy new manufacturing equipment and hire new people to facilitate the manufacturing process. An industrial engineer, working with a financial analyst and a marketing analyst, examines past peach-pitter production and long-range marketing to plan future capacity requirements. The analysts use computer simulations to assist them in their decision making, but there is no link between the MRP system and the capacity planning system.

Required: You have been hired as a consultant by Pitout to analyze the feasibility of combining the MRP and capacity planning functions. Which tool would you use to model the system described above? Would you be more concerned with selecting a tool for communication or experimentation? Model the system described above with the tool you have selected.

SUGGESTED READING

Aktas, A. Ziya. *Structured Analysis and Design of Information Systems*. Englewood Cliffs, N.J.: A Reston Book, Prentice-Hall, 1988.

Burch, John, and Gary Grudnitski. *Information Systems: Theory and Practice,* 5th ed. New York: John Wiley, 1989.

Hackathorn, John C., and Jahangir Karimi. "A Framework for Comparing Information Engineering Methods." *MIS Quarterly,* Vol. 12, No. 2, June 1988.

Inmon, W. H. *Information Engineering for the Practitioner*. Englewood Cliffs, N.J.: Yourdon Press, A Prentice-Hall Company, 1988.

King, David. *Creating Effective Software: Computer Program Designing Using the Jackson Methodology*. Englewood Cliffs, N.J.: Yourdon Press, a Prentice-Hall Company, 1988.

Klein, H. K., and R. A. Hirscheim. "A Comparative Framework of Data Modeling Paradigms and Approaches." *Computer Journal,* Vol. 30, No. 1, 1987.

Leeson, Marjorie. *Systems Analysis and Design*. Chicago: Science Research Associates, 1985.

McIntyre, Scott C., and Lexis F. Higgins. "Object-Oriented Systems Analysis and Design: Methodology and Application." *Journal of Management Information Systems,* Vol. 5, No. 1, Summer 1988.

Martin, James, and Carma McClure. *Diagramming Techniques for Analysts and Programmers*. Englewood Cliffs, N.J.: Prentice-Hall, 1985.

Merlyn, Vaughan, and Greg Boone. "Sorting Out the Tangle of Tool Types." *Computerworld,* Vol. 23, No. 13, March 27, 1989.

Necco, Charles. "Evaluating Methods of Systems Development: A Management Survey." *Journal of Information Systems Management,* Vol. 6, No. 1, Winter 1989.

Orr, Kenneth T. *Structured Systems Development.* New York: Yourdon Press, 1977.

Page-Jones, Meilir. *The Practical Guide to Structured Systems Design,* 2nd ed. Englewood Cliffs, N.J.: Yourdon Press, A Prentice-Hall Company, 1988.

Stevens, Wayne P. *Using Structured Design.* New York: John Wiley, 1981.

Yourdon, Edward. *Managing the System Life Cycle,* 2nd ed. Englewood Cliffs, N.J.: Yourdon Press, A Prentice-Hall Company, 1988.

Chapter 3

Systems Development Management and Automation

WHAT WILL YOU LEARN IN THIS CHAPTER?

After studying this chapter, you should be able to:

1 Discuss project management functions.
2 Explain project management techniques.
3 Describe computer-aided systems and software engineering (CASE) packages, which provide systems development and maintenance automation technologies.

INTRODUCTION

In Chapter 1, we presented the systems development life cycle (SDLC) and prototyping as guides and ways to develop information systems. The structured approach and selected modeling tools, covered in Chapter 2, add an engineering style to systems development. This chapter presents systems management and automation functions and techniques. Put them all together and you have a powerful set of methodologies, modeling tools, and techniques to aid in the development of successful information systems. To gain an insight into the importance of systems development management, review the sample case.

Systems Development Management at Desert Heavy Equipment

As a subsidiary of a national heavy equipment leasing corporation, Desert Heavy Equipment Company (DHEC) had some special challenges facing it when the company decided to develop its new information system. DHEC's recent growth had placed it in a position where it needed an integrated system to manage its construction lease contracts, provide time scheduling for its heavy machinery (backhoes, earth movers, and tractor-trailers), and keep maintenance records for this equipment. It was also desired that the accounting function be integrated into the system as well. The general manager of DHEC, Michael Reed, had introduced the idea of such a system at an employee meeting early last year, and it met with overwhelming positive response. Later that month, Michael approved a plan to create a systems development team, headed by Deborah Hainsworth, the company's leasing director.

Since funding for the new information system's development would come from its national parent corporation, DHEC had to do some preliminary planning before funding for the project as a whole was firmly in place. Unfortunately, the parent corporation had been unreceptive to similar projects in the past, for it viewed DHEC as a very small component of its operations. Thus DHEC was bargaining from a position of very low priority in the eyes of the national parent corporation, and its preliminary systems planning and organization would have to be first rate if it were to make a favorable impression and secure the needed funding.

With DHEC's growth over the previous two years, Mr. Reed and his directors had decided to develop a general business plan, which outlined the company's goals, objectives, and strategies as an organization. With this business plan in place, Deborah knew that the first step in the systems development process would be to establish a systems plan that would, like the more general business plan, give direction in terms of goals, objectives, and strategies for the new information system. More would be required, however, if the support of DHEC's parent corporation was to be secured. This would require a more detailed, specific plan of just how the systems development was to proceed.

Working together, Mr. Reed and Ms. Hainsworth developed a systems project plan that broke down the overall information system into more manageable subprojects, then into phases and individual tasks as part of the systems development life cycle. Deborah used a project management software package on her office microcomputer to schedule each specific task, then created a Gantt chart to show when each task and phase would be started and completed. She then used the software to perform calculations of most likely times for each project task, and created a program evaluation and review technique (PERT) diagram which allowed her to visualize the task sequence and critical path. The software even helped her to rearrange the tasks and eliminate some of the slack time in her original schedule. The systems development proposal DHEC would submit to its national parent was really beginning to take shape!

Ms. Hainsworth then set about organizing the systems development project team members. This involved hours of consultation with the company's maintenance and operations directors, along with other members of the team, to determine the individual strengths of each person. Some people were cut out to perform people-type work or were good leaders, while others preferred to work on their own in a more technical environment. Looking at the project's changing needs as its development proceeded through the life cycle, as well as the specific tasks involved, she attempted to place this multitude of personalities in positions where they would each contribute the most. She was then able to create an organizational chart for the information system's development.

All this work paid off. DHEC's national parent was so impressed by the management and planning of the system thus far that it approved full funding, and decided to make the system development project at DHEC a model for similar projects at its other subsidiaries. With the support of senior management, and the end-user involvement this project had from its inception, smooth progress was virtually assured.

WHAT IS PROJECT MANAGEMENT?

Everything that we have discussed to this point, and everything that we will discuss throughout the remainder of this book, is project oriented. A systems project is an undertaking with a defined beginning and end that produces specific documented deliverables and an implemented system. It can and should be managed like any other project.

In the next part of this book, we begin the systems development life cycle. To make sure that the SDLC is applied correctly and works as it should, project management must be involved. Essentially, project management is a matter of employing the following four management functions:

- Planning
- Organizing
- Controlling
- Leading

These are explained in detail in Figure 3.1. We will now explore these management functions and learn how they are applied to systems development.

Planning

In systems work, there are two levels of planning. Systems planning, discussed extensively in Chapter 4, gives senior managers, users, and information systems personnel an overview of various systems projects that will demand resources over a long period of time. High-level planning establishes what should be done. It also sets a budget for the total cost for all planned systems projects. In addition to development costs, this budget includes costs of other resources, such as computers, telecommunication, new sites, and so on. Low-level planning is systems project planning, which involves setting a plan for the development of each systems project. The two most common systems project problems are budget overruns and missed schedules. Therefore, quality systems project planning requires accurate schedules and budgets, the key elements of systems project planning.

Estimating How Long It Will Take

Scheduling also entails not only how long it will take to perform a phase or task but also at what time a particular resource will be needed. Project planning must consider the fact that the quantity and type of resources required will vary during the SDLC. Usually, resource requirements start at a low level during analysis, general design, and evaluation and selection, build to a plateau during detailed design and implementation, and taper off toward the end of implementation. The type of skills needed during the SDLC will also vary. This affects staffing decisions, which are discussed in a later section.

Estimating What It Will Cost

Scheduling works hand in glove with budgeting. Where budgeting is cost-based, scheduling is time-based. Cost is normally a function of time. For example, if programmers are charged to a project at a rate of $20 per hour, and it is estimated to require 3000 hours to code a program, the salary budget for coding is $60,000.

A systems project budget is normally set before systems development gets under way. As the SDLC evolves and more becomes known about the systems project under development, the original systems project budget may be ad-

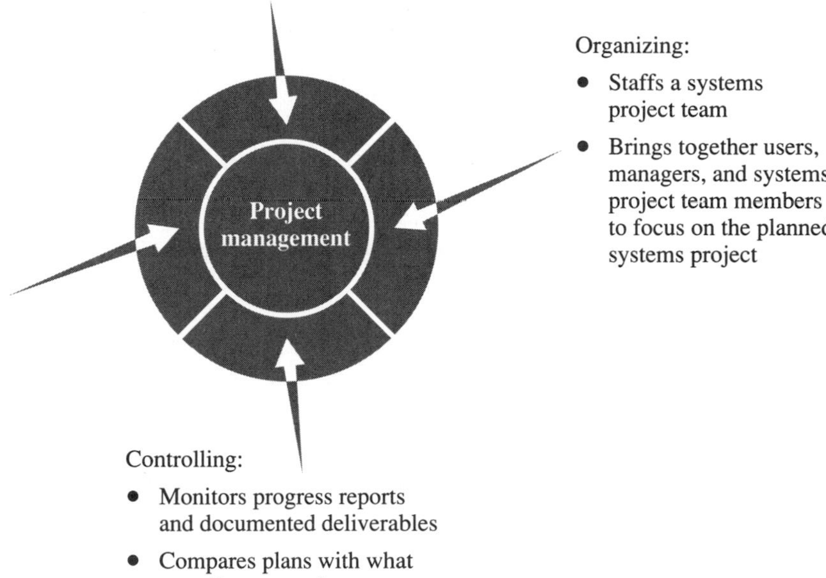

• Guides and induces people to perform at their maximum abilities

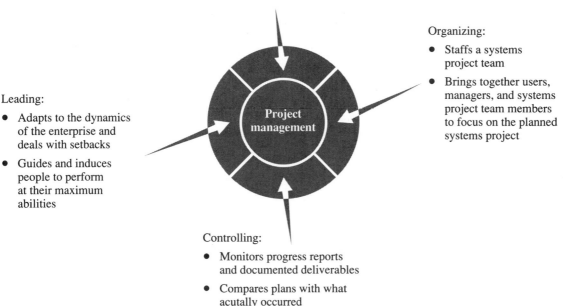

Figure 3.1
Project management functions.

justed up or down. Some people refer to this process as continuous budgeting, which reflects the dynamics of the enterprise and forces managers to rethink systems progress and needs at all times.

Organizing

Figure 3.2 presents an organization chart of a type found in large information systems. At the top of this organization chart is the **CHIEF INFORMATION OFFICER (CIO)**, who is in charge of the total information system. The CIO is usually a senior executive with complete knowledge of the enterprise's business and information system needs, and a strategic vision and plan for the information system. The **STEERING COMMITTEE**, made up of senior managers, performs an oversight function and gives broad, strategic guidance to systems development. The **PROJECT DIRECTOR** coordinates and manages all systems projects. The **PROJECT MANAGER** (also called **PROJECT LEADER**) is responsible for one systems project and one systems project team. Systems personnel who possess specific skills are team members. In very small companies, many of the tasks performed by a team of systems professionals are performed by only one person, usually referred to as a **PROGRAMMER/ANALYST**. (Such a person has to be fairly familiar with most of the material presented in this book.) Organization support includes office space, telephones, CASE technology, and secretarial support.

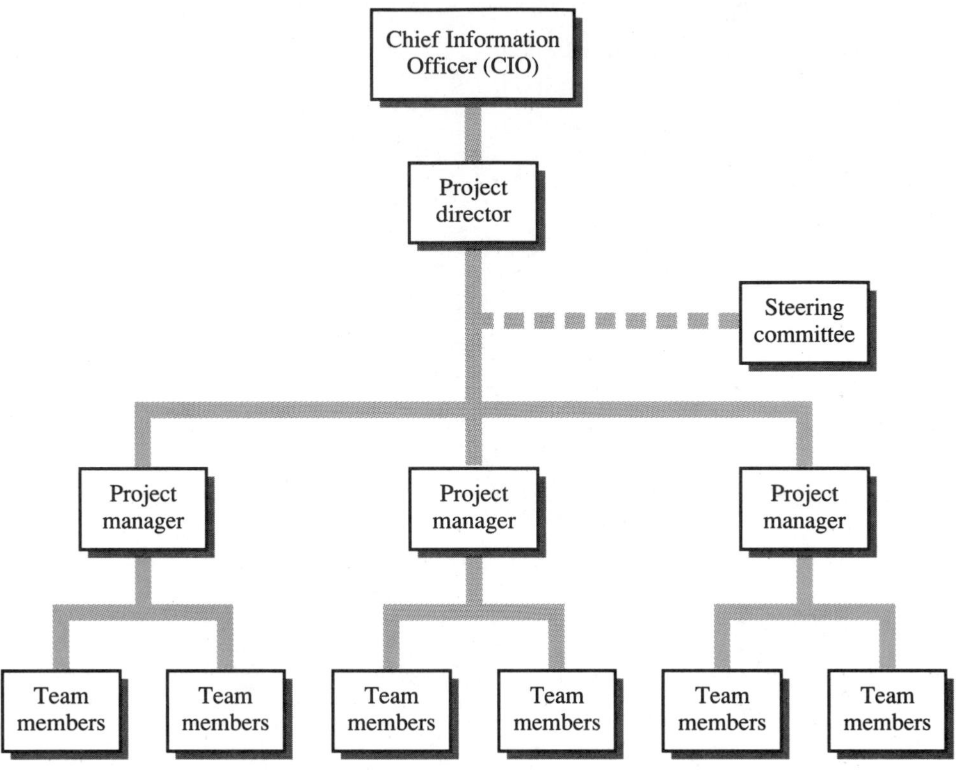

Figure 3.2
Organization for system projects.

Organizing the Work for the SDLC

During the phases of SDLC, different skills and focuses are required. This often requires that the project team organization, staff, and size shift from phase to phase to reflect these changes as described next.

SDLC Phase	Project Team
Systems planning	The steering committee and various other persons play a dominant role in this phase. The project team is formed after a proposed systems project is cleared for development.
Systems analysis	Business systems analysts who are especially knowledgeable in business operations and who also possess systems analysis skills are the main team members.
General systems design	Business systems analysts with some technical support people are needed here.
Systems evaluation and selection	The same skills as above with possible accounting support are needed here.
Detailed systems design	A host of systems and technical designers are needed here.
Systems implementation	A number of systems analysts, programmers, and special technicians are needed in this phase.

To get a better handle on a large systems project, a project manager may treat each front-end SDLC phase as a separate subproject. When development evolves to the detailed design phase, the systems project may be divided into a subproject for each systems design component. A team leader will be assigned to each subproject to manage a team of people who are trained for that component. For instance, a small project team will be assigned to design output in the form of graphs and tabular reports both on paper and video screen; another team will design forms and various other inputs; another team will design processes. Another team will work on logical and physical design of databases; another team made up of security and control people will design controls; and telecommunications and computer technicians who know how to design, configure, and acquire a technology platform will do those tasks. Because all of these systems components must evolve into an integrated whole, a great deal of coordination and communication are required among all subproject teams.

During implementation, project leaders may also be assigned to manage special teams. If customized software is a major task of implementation, a project leader who has experience in this area will be assigned to manage a team of software designers, coders, and testers.

Organizing the Work for People- and Technical-Oriented Phases

Not only do project managers coordinate and manage team leaders and members; they also work with users. Any systems project not only involves a varying mix of skills throughout the SDLC, but also a fluctuating mix of people-oriented and technical-oriented phases. For example, look at Figure 3.3, which depicts a **V CURVE**.

The graph takes on a V shape because the front-end phases, especially systems planning and systems analysis, involve users more intensely than the

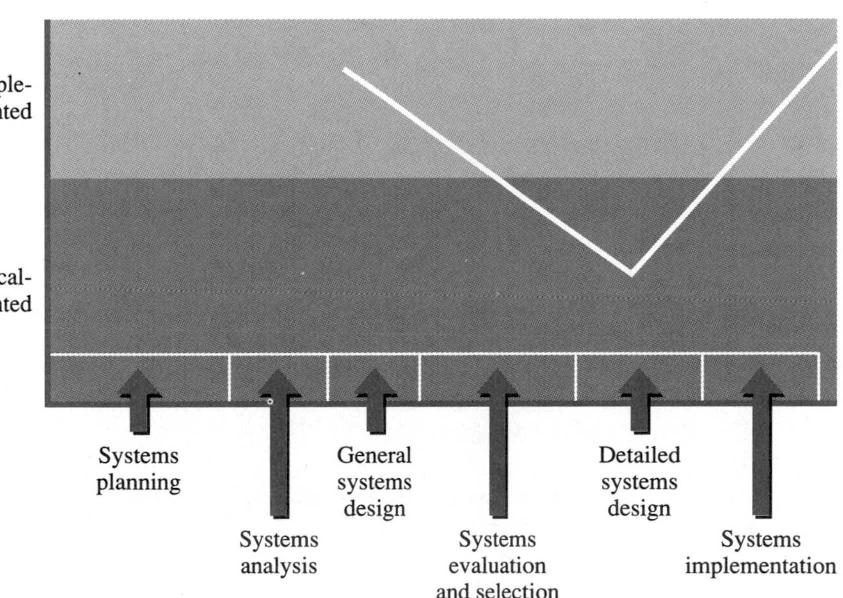

Figure 3.3 People-oriented and technical-oriented phases of SDLC (Adapted from F. Collins, "The V-Curve: A Road Map for Avoiding People-Problems in System Change," *Journal of Systems Management*, February 1983.)

back-end phase of detailed design. During systems planning and analysis, users and systems professionals work with each other in defining user requirements as well as identifying existing system capabilities and problems. With this knowledge, project team members develop conceptual designs and establish a basis for selecting the optimal design alternative. Detailed systems design is the engineering of the system. In it, functional design components are specified at their elementary level. During this phase, people-oriented tasks occur less frequently, and technical-oriented tasks occur more frequently.

After the programming, site preparation, and equipment installation tasks of systems implementation are completed, people-oriented tasks such as user training are conducted. The system is presented to users for operations. At this point, interaction between systems development personnel and users is often intense. In fact, users are the judge and jury of acceptance of the new system.

Probably no other single undertaking causes change more in an enterprise than the implementation of a new information system. The resistance of corporate culture to change is well known. It can be passive or active. Passive resistance to change involves withholding support or permitting failure when intervention could bring success. Active resistance involves taking specific action to ensure that the system fails. The pattern of resistance to change is the same as the SDLC V curve. That is, resistance to change will be more intense during the front-end phases of systems planning and analysis, less intense during detailed design, and intense again during the late part of implementation. Thus the V curve not only shows the pattern of resistance to change but also enables project managers to plan to prevent or reduce resistance to change. The more involved users become in systems development, the less they resist change brought about by the new information system. Some of the things that a project manager can do to reduce resistance to change are discussed below.

Enlist End Users as Team Members Involving end users on the project team helps establish a strong liaison with the user group. Moreover, users with a business perspective and systems professionals with an information systems perspective work together to figure out what they want right from the start and establish the scope and content of the proposed system. During team meetings, nontechnical end users and information systems staff meet on common ground to gather information and hammer out systems solutions that meet the needs of everyone throughout the enterprise.

Encourage Involvement and Support from Senior Management The systems planning phase is key to getting senior managers involved in the systems development process. If senior management, through the oversight of a steering committee, supports a project, the systems project team will receive much more cooperation from others in the organization, and the development process will go more smoothly.

Keep Every Key Participant Informed on the Progress of Systems Development This aspect is handled quite well by periodic team meetings and the preparation and reporting of progress reports and the documented deliverables.

Allay Fears of Job Security and Status If people will be displaced by the new system, candor is recommended as well as a method to help displaced people find other jobs or be retrained.

Controlling

Project control is based on reporting of documented deliverables and progress reports. Actual work achieved and expenditure of resources are compared with planned work and budgets. If, at some point in time, actual work achieved is more than expected and resources expended are in line with or below what was planned, continuation of such performance is encouraged and supported. If, on the other hand, actual work achieved is less than expected and resources expended more than planned, then corrective action is taken.

Feedback begins with documented deliverables, progress reports, and formal meetings. The agendas for these meetings include:

- Presentation and discussion of progress to date
- Budget status to date
- Planned tasks for the next time period
- Any areas of concern or unresolved issues

Preparing Weekly Progress Reports

Interim progress reports are prepared during a phase and disclose progress on specific tasks. A **WEEKLY PROGRESS REPORT** gives the project manager some idea of progress and the related expenditures. A weekly progress report should include:

- Tasks that should have been started and completed in a given week but were not, and an estimation as to when they will be started and completed
- Tasks completed or started in a given week
- Explanation for variances
- Unresolved problems encountered

These interim reports can be in writing or made orally at weekly systems project meetings.

Preparing Monthly Progress Reports

If a phase extends over several months, the project manager should prepare a monthly progress report for senior management and the steering committee. In large companies with many projects being developed at the same time, a progress report will be channeled to a project director, who will in turn summarize all project progress reports and present them to senior management and the steering committee. In smaller organizations, a project manager may present his or her report directly to top management and the steering committee.

To a great extent, the **MONTHLY PROGRESS REPORT** is a summary of the weekly progress reports prepared by systems project team members or systems project leaders. The monthly progress report, therefore, includes:

- Actual progress versus what was planned
- A new estimate of cost to complete the project for comparison with budgeted costs
- A comparison of the new project completion date with the scheduled one
- If time and cost variances exist, the reasons for them should be given. (If the variances are unfavorable, the systems project leader should explain how these variances will be corrected and how the systems project will be put back on track.)

Leading

In people-intensive work such as systems development, the solution to a number of problems, such as schedule slippage and cost overruns, is often found in leading the project team properly. Indeed, in seeking to increase productivity and develop high-quality systems on schedule and within budget, project managers must handle both technical and people issues. In most systems projects, people issues far outweigh technical issues.

What Makes a Good Leader?

Leadership by the systems project manager (leader) depends on influencing systems project team members to direct their efforts toward specific tasks. Six important traits of system project-manager leadership are:

- Emotional maturity and ability to handle crises and conflicts
- Ability to get along well with others
- Self-assurance and self-control
- Ability to motivate people
- Ability to delegate responsibility
- Loyalty to and ability to support team members

Leadership Style

Effective leadership style depends on the situation. In systems project teams that include a mix of systems professionals, a democratic systems project leader may prove to be superior, because such a team usually favors this style of leadership.

A democratic systems project leader fosters high achievement. When systems project team members know what they are supposed to be doing, and they have the freedom to achieve it, they become goal-oriented, and they derive satisfaction from getting things done. If a crisis develops, however—for instance, schedule slippage—a systems project leader may have to change temporarily to an authoritarian leadership style until the crisis is corrected. At the other extreme of the leadership spectrum, the laissez-faire leadership style can be very effective when used with highly motivated, highly skilled systems project team members. These people are most productive when they are left alone to do their work. They do not need either direction or encouragement from the systems project leader.

PROJECT MANAGEMENT TECHNIQUES

Project management techniques, which include the Gantt chart and the program evaluation and review technique (PERT) chart, help control timing and reduce the cost of systems projects. This section provides an overview of these two techniques and a discussion of how they are used in project management.

Gantt Chart

The **GANTT CHART**, named after its developer Henry Gantt, is a bar chart that shows phases or tasks on the left side of the chart and units of time (e.g., days, weeks, and months) across the top (also sometimes shown at the bottom of the chart) (see Figure 3.4). On this chart, the clear bars are the times originally scheduled for the SDLC phases of a project. The actual time spent on each phase to date is indicated by a colored bar. The letter "C" at the end of a colored bar means that the phase is complete.

The Gantt chart is used to compare planned performance against actual performance to determine whether the project is ahead of, behind, or on schedule. An important advantage of the Gantt chart is its simplicity. People can comprehend a schedule for the complete systems project easily. The Gantt chart, however, fails to show relationships among specific tasks. This is a shortcoming not found with a program evaluation and review technique (PERT) chart. This is why the Gantt chart is often used to schedule a complete systems project by phases of SDLC, while the PERT chart is used to schedule a

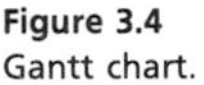

Figure 3.4
Gantt chart.

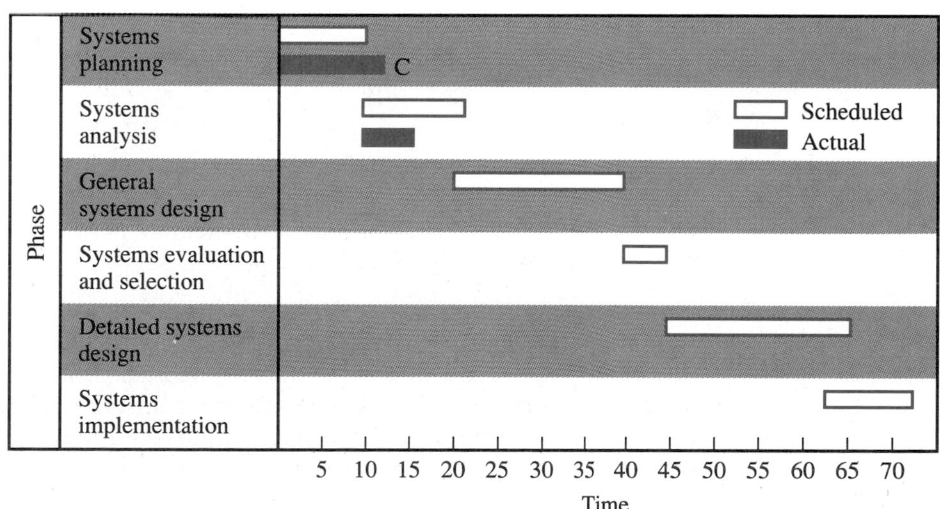

multiplicity of specific tasks. In other instances, Gantt charts are used to schedule tasks of front-end phases, and PERT is used to schedule tasks of back-end phases. The reason for this use of Gantt and PERT is that front-end tasks are fewer in number and sequential in nature, whereas the back-end phases contain a large number of varied tasks (some of which are performed simultaneously) that require different skills which must be coordinated to keep the total system on track and schedule.

Program Evaluation and Review Technique Chart

A **PROGRAM EVALUATION AND REVIEW TECHNIQUE (PERT) CHART** is used to estimate, schedule, and control a network of interdependent tasks shown by arrows and nodes or circles. The arrows represent project tasks that require time and resources. The nodes represent milestone points or events, showing the completion of one or more tasks and the initiation of one or more subsequent tasks. In this way, the PERT chart indicates which tasks must be done before others and which tasks can be performed simultaneously. The PERT chart is used to determine the minimum time needed to complete a project or a phase, based on the chart's critical, or longest-time, path.

The following step-by-step procedure can be used as a model for applying PERT. This example illustrates PERT applied to the detailed systems design phase. Later in this book, when we cover systems implementation, we will present additional features of PERT.

Step 1. Identify Tasks. The general tasks included in the detailed systems design phases include the design of these components:

- Output
- Input
- Processes
- Database
- Controls
- Network
- Computer

There are many subtasks too numerous to mention here. We present these later when we cover detailed systems design.

Step 2. Determine the Proper Sequence of Tasks. Because output design is the controlling factor of all the other design tasks, it should be completed first. Then input, processes, database, and controls can be designed pretty much together to produce the specified output. Next, the technology platform that supports the systems design components can be designed in accordance with a telecommunication network and computer architecture. The sequence of all these tasks and their sequence codes are specified as follows:

Sequence Code	Task
1.1	Output design
2.1	Input design
2.2	Processes design
2.3	Database design
2.4	Controls design
3.1	Network design
3.2	Computer design

Step 3. Estimate the Time Required to Perform Each Task. Time estimates can be made in hours, days, weeks, months, quarters, or years. Expected time (te) to complete a task is computed as follows:

$$te = \frac{o + 4ml + p}{6}$$

where o equals the most optimistic time, ml the most likely time, and p the most pessimistic time. Although the statistical meaning of this formula is beyond the scope of this book, you can see that the most likely time should be weighted more heavily than the optimistic and pessimistic time estimates. Let us assume that experienced systems professionals have used the preceding formula to estimate the expected time to complete each detailed systems design task in our example and the results are:

Sequence Code	Task	Estimated Time (days)
1.1	Output design	20
2.1	Input design	15
2.2	Processes design	30
2.3	Database design	35
2.4	Controls design	25
3.1	Network design	40
3.2	Computer design	30

Step 4. Prepare a Time-Scaled Chart of Tasks and Events and the Critical Path. The following is a simplified version of a PERT chart, showing the critical path (CP), that is, the longest sequence. Later in this book we present a

more complex, conventional version of a PERT chart. For now, the PERT chart displayed in the box provides a simple illustration of how PERT works and how it can be used in systems development management.

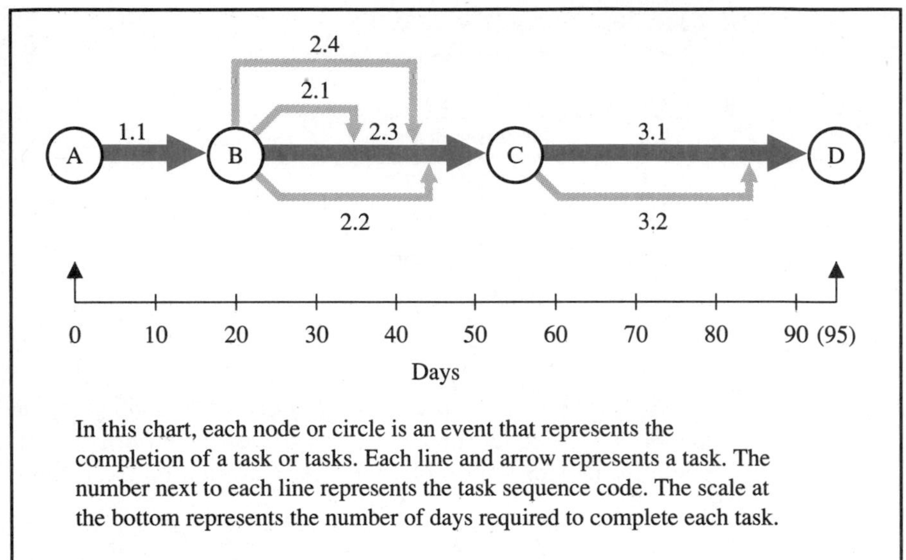

In this chart, each node or circle is an event that represents the completion of a task or tasks. Each line and arrow represents a task. The number next to each line represents the task sequence code. The scale at the bottom represents the number of days required to complete each task.

The time scale in the PERT chart shows that the detailed systems design phase will take 95 days. Tasks 1.1, 2.3, and 3.1 make up the critical path, which dictates total time to complete this phase. This means that slack time exists in the other tasks. Therefore, the project manager may be able to shift personnel and other resources from these tasks to the ones on the critical path to meet or beat target dates.

PERT suggests how tasks depend on each other. It shows the sequence in which tasks must be performed, and it provides an explicit basis for evaluating progress during SDLC. When a task is begun or finished, this is an event or milestone. A PERT chart, therefore, shows when each event should be reached and in what order, so that the next set of tasks can begin.

Project Management Software

More than one hundred project management software packages are available. They can be very helpful in coordinating projects and tasks within projects that involve many people.

These packages provide interactive Gantt and PERT displays. Some provide what-if scenarios. A project manager can monitor a schedule in almost any way.

Some software is for microcomputers; other programs require larger computers. All the software packages perform the routine, time-consuming aspects of project management and progress reporting. Main features include:

- Developing and displaying Gantt and PERT charts
- Tracking project status

- Performing all necessary calculations
- Preparing various status reports on tasks in process, tasks completed, planned tasks remaining, cost and time budgets, expenditures, and variances

COMPUTER-AIDED SYSTEMS AND SOFTWARE ENGINEERING SYSTEMS

COMPUTER-AIDED SYSTEMS AND SOFTWARE ENGINEERING (CASE) systems automate the methodologies, modeling tools, and techniques discussed to this point. CASE's primary goal is to increase the productivity of systems professionals and improve the quality of systems produced. Coupled with systems development productivity and quality is the CASE promise to make software maintenance process efficient and productive. Figure 3.5 presents a model of a full-featured CASE system. The model is full-featured because it assists in automating tasks in SDLC's front-end and back-end phases, and it also facilitates software maintenance once the system is converted to operations. Each feature is explained below. (*Note:* The CASE presented in Figure 3.5 is an ideal system. No CASE vendor offers all these features in one package.)

Setting Up a CASE Workstation

A CASE workstation, also called a workbench, is typically a microcomputer, with a mouse and laser printer that interacts with the CASE software. In some companies, these workstations are linked to form a systems-project team network. Each team member is assigned a workstation. A central processor, generally a minicomputer, acts as a central coordinator and server to the CASE features and also a central data file manager, or repository.

Central Repository

The **CENTRAL REPOSITORY** is the nucleus of the CASE system. It gathers systems specifications from all workstations and automatically checks their completeness and consistency. Also, each workstation has a local repository with its own portion of the total systems specifications.

Think of the central repository as a database management system that is capable of storing, coordinating, and producing the following items:

- Models derived from modeling tools
- Project management elements
- Documented deliverables
- Screen prototypes and report designs
- Software code produced by an automatic code generator
- Module and object libraries of reusable code
- Reverse engineering, reengineering, and restructuring features

Any of these features can be accessed instantly via workstations. The central repository also serves as a database (sometimes called a project dictionary or

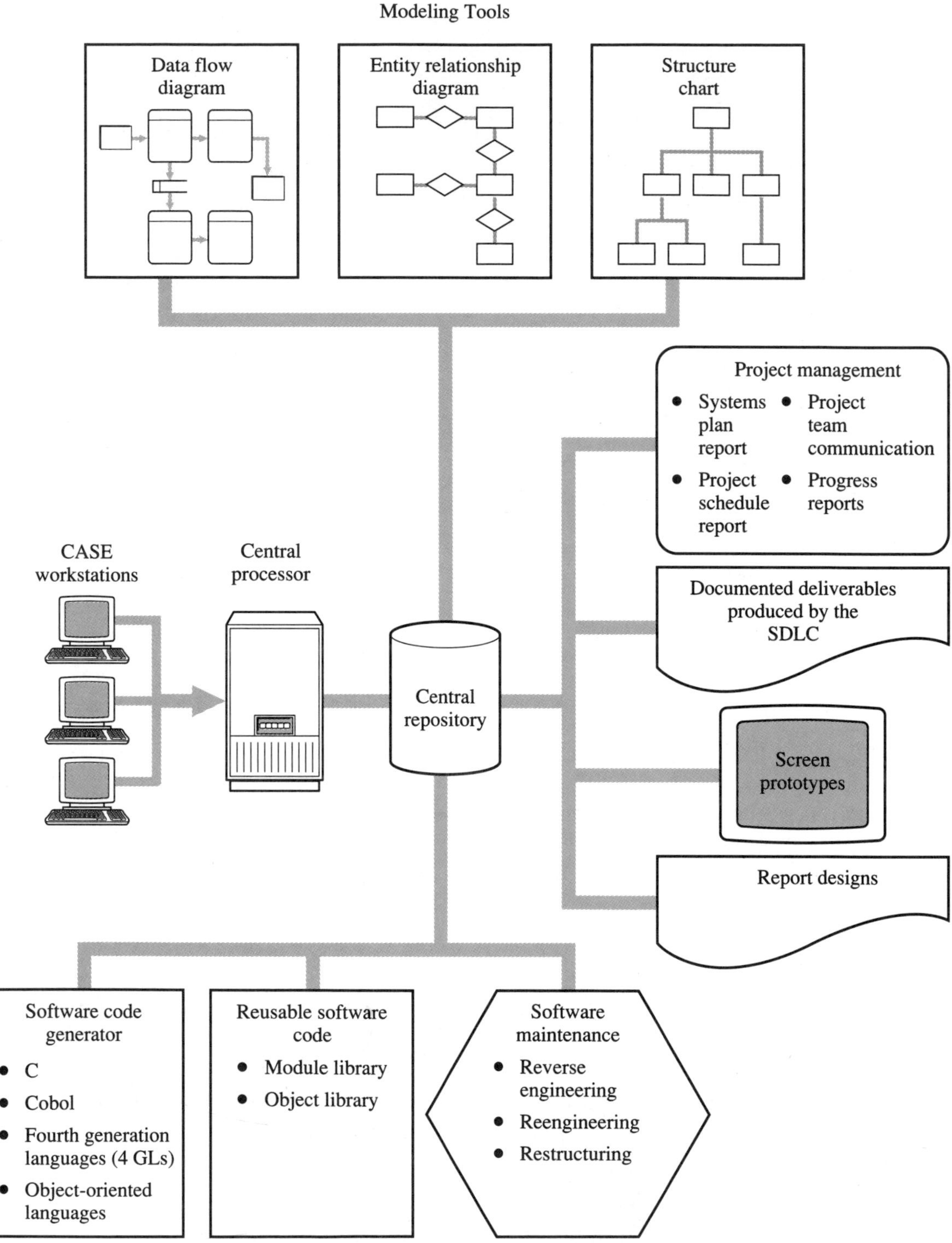
Modeling Tools
Data flow diagram
Entity relationship diagram
Structure chart
Project management
Systems plan report
Project team communication
Project schedule report
Progress reports
CASE workstations
Central processor
Documented deliverables produced by the SDLC
Central repository
Screen prototypes
Report designs
Software code generator
C
Cobol
Fourth generation languages (4 GLs)
Object-oriented languages
Reusable software code
Module library
Object library
Software maintenance
Reverse engineering
Reengineering
Restructuring

knowledge base) of legacy systems (i.e., existing systems), hardware, and systems maintenance histories.

Using Modeling Tools

Figure 3.6 is an example of a series of leveled DFDs created from the front-end CASE features. This figure also displays the structure charts, entity relationship diagrams (ERDs), screen designs, report designs, equations, and other components that assist systems professionals with back-end phases.

Models help end users understand systems people, and they help systems people understand business operations more clearly. Using a CASE system, you can create and link professional models without the use of paper, pencil, erasers, and templates. These models can be modified to reflect corrections and changes suggested by users. You don't have to start over. Also, CASE systems provide tools that can check your models for mechanical errors, consistency errors (between diagrams), and completeness errors.

Creating Reports and Providing Communications to Help Manage the Systems Project

Most systems projects require a team of people rather than only one individual. Managing a team of people requires that a set of techniques be incorporated into a CASE system. The CASE project management feature includes items discussed in the following subsections.

Systems Plan Report

The **SYSTEMS PLAN REPORT** is the documented deliverable produced by the systems planning phase, the subject of Chapter 4. It contains the resources budgeted over the next several years and the systems projects that have been accepted for development.

Project Schedule Report

The **PROJECT SCHEDULE REPORT** pertains to each separate systems project under development. It identifies the project. It presents committed resources, Gantt and PERT charts, and masters of team members assigned to the project with their responsibilities.

Project Team Communication

The CASE system makes it easy for team members to contact each other through electronic mail and messaging systems. Also, CASE performs scheduling functions automatically through computerized calendars and meeting planners.

Progress Reports

Progress reports help to keep the systems project on track. Deviations from what was planned are published for all team members.

Using CASE to Support the SDLC

The CASE system supports all phases of SDLC and produces documented deliverables for each phase. Subcomponents of CASE assist with both front-

Figure 3.6
Structured model of CASE for front-end and back-end phases.

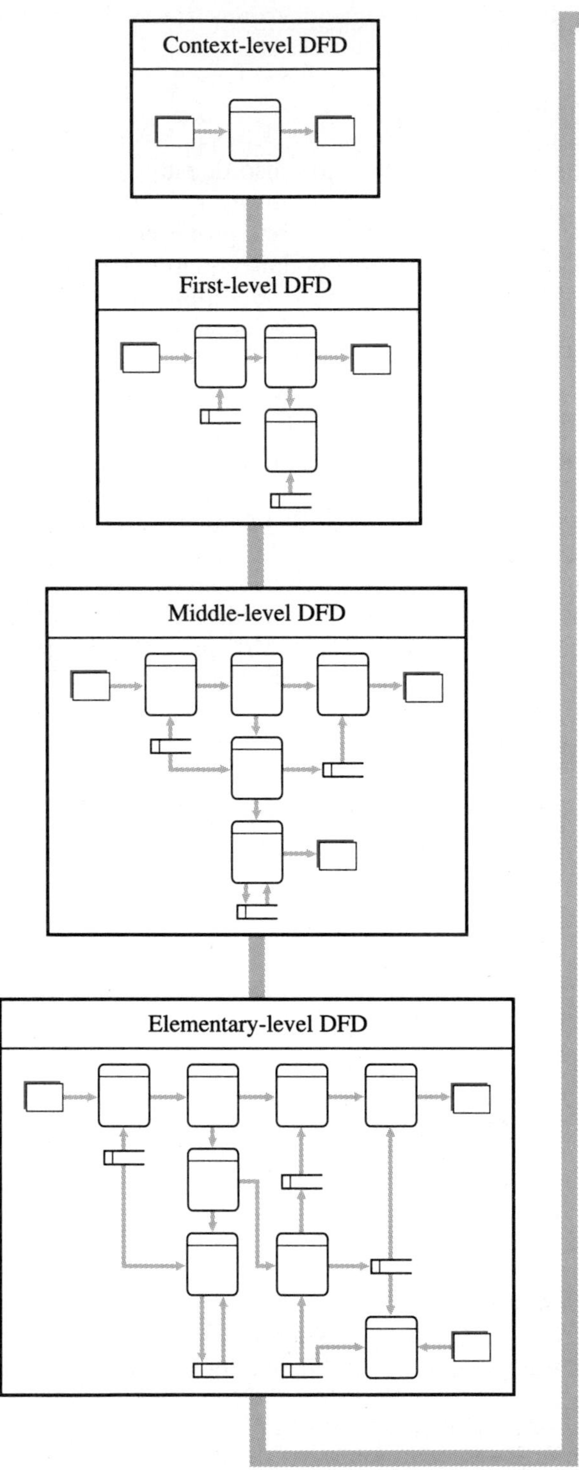

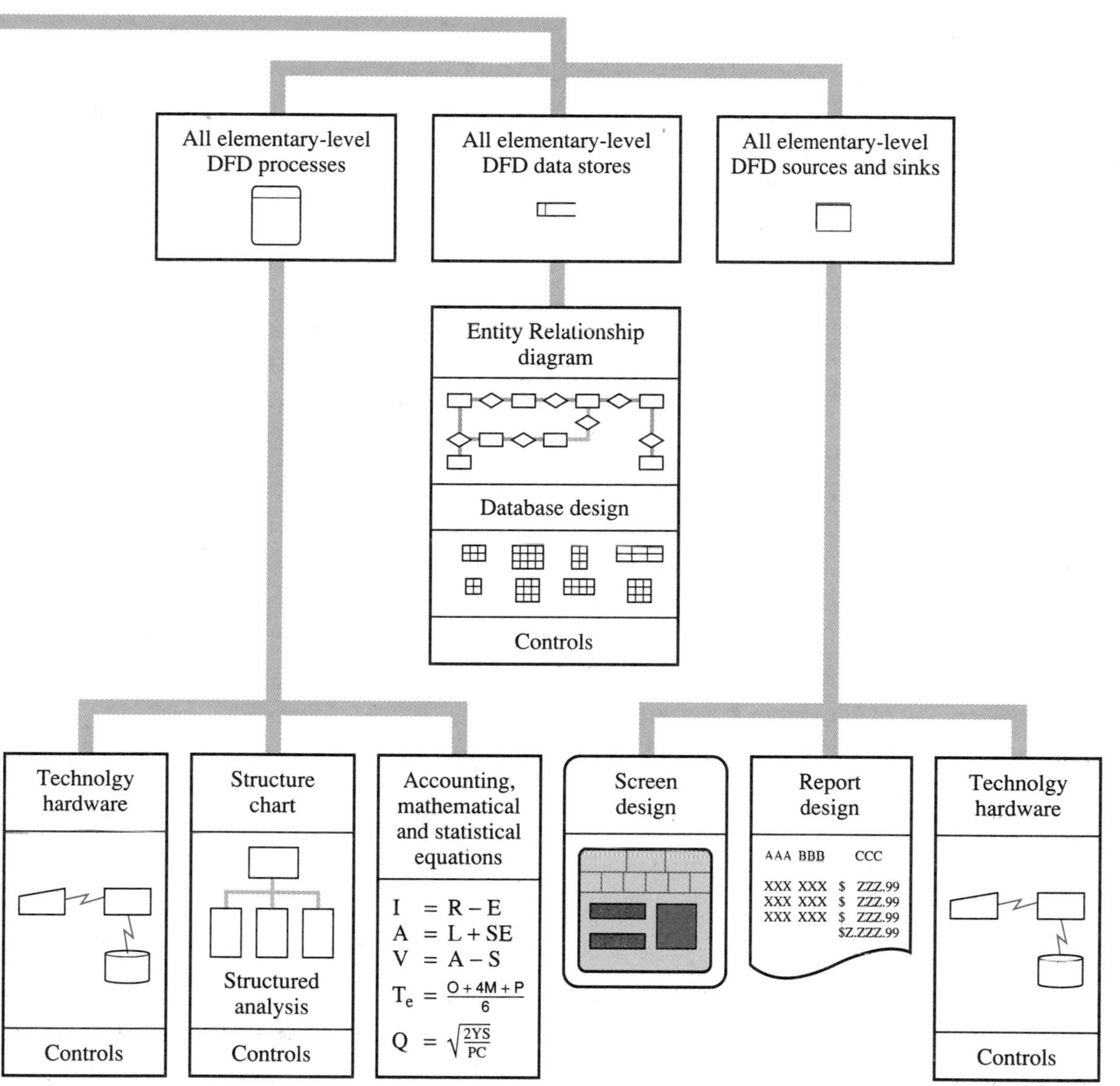

All elementary-level DFD processes
All elementary-level DFD data stores
All elementary-level DFD sources and sinks
Entity Relationship diagram
Database design
Controls
Technolgy hardware
Controls
Structure chart
Structured analysis
Controls
Accounting, mathematical and statistical equations
I = R – E
A = L + SE
V = A – S
Te = (O + 4M + P)/6
Q = √(2YS/PC)
Screen design
Report design
AAA BBB CCC
XXX XXX $ ZZZ.99
XXX XXX $ ZZZ.99
XXX XXX $ ZZZ.99
$Z.ZZZ.99
Technolgy hardware
Controls

end and back-end phases. (An enterprise that acquires a CASE system should have SDLC in place already.)

Prototyping with CASE

CASE systems enable online design of prototypes such as reports and screens, sometimes called screen painting. The ability to generate screen and report designs for presentation to, and feedback from, the potential users is a powerful feature of CASE in helping ensure that user requirements are being met precisely.

Automatically Generating Software Code

When the systems design is decomposed to the software design level of a structure chart and structured English, or Warnier–Orr diagram (WOD) or Jackson diagram, a CASE **AUTOMATIC CODE GENERATOR** transforms the software design into software code using languages such as C, COBOL, fourth-generation languages (4GLs), or object-oriented languages. This feature does away with the need for hand coding. This increases the productivity of customized software development.

Reusable Software Code

Many systems require the same functions. These functions can be designed, coded, and tested once and then reused many times. These reusable modules or objects are stored in a library for immediate access and implementation.

Maintaining the Software

Software maintenance of programs that are unstructured and undocumented is difficult and often unsuccessful because to do it requires a form of mental **REVERSE ENGINEERING**. Maintenance programmers must attempt to find other people's ideas in the software code. They must extract enough of the original design from the spaghetti-like, undocumented code to make the maintenance change request. Because maintenance programmers are usually unable to perform this task, change requests are often returned as infeasible. CASE provides the technology to reverse-engineer the old software, thus abstracting meaningful design specifications that can be used by maintenance programmers to perform their maintenance tasks.

REENGINEERING is the examination and changing of a system to reconstitute it in both form and functionality, along with its subsequent reimplementation. Reengineering, for example, uses reverse engineering and modeling tools to refurbish an old software program. This CASE feature changes the old software program's functionality and restructures and documents the code. This may include modifications with respect to new user requirements not met by the original system.

RESTRUCTURING is a form of reengineering, except this feature does not change the old software program's functionality. It merely restructures code into the standard control constructs: sequence, selection, and repetition. It also documents this code. Restructuring is also often used as a form of preventive maintenance to improve the physical state of the subject system with respect to some standard, such as standard variable names, standard language, and standard data models.

These features improve the maintainability of old software. Many old software programs were developed without following the structured approach. Without reverse engineering, reengineering, and restructuring, maintenance of this old software can be time-consuming, expensive, or, in some instances, impossible.

Developing a Plan for Implementing CASE Technology

Earlier, we defined users as managers and employees who interact with the information system and depend on it to perform their tasks. With CASE technology, systems professionals become users themselves, because they interact and work with a system. That is, systems professionals use CASE systems to perform their own tasks. Like other users, systems professionals normally have a strong resistance to change, especially when learning to use new tools. As a result, CIOs and project directors must ensure that the proper training, technical support, and commitment are put in place before beginning CASE implementation. Here are some guidelines that will help in successfully implementing CASE:

- Create a sense of mission.
- Treat the implementation of CASE systems like the implementation of all other information systems. After all, systems professionals are supposed to be the experts when it comes to developing and implementing information systems for users. They should be able to do the same for themselves.
- Make sure that SDLC is in place.
- Train systems people in the use and application of a set of modeling tools and the use of prototyping.
- Start small and build slowly.
- Consider using a well-defined systems project for CASE's first pilot project.
- Train a highly motivated systems project team to use CASE on the pilot project. Later this team can be used to train other systems personnel.
- Emphasize improvement in quality.

REVIEW OF CHAPTER LEARNING OBJECTIVES

The major goals of this chapter were to enable each student to achieve three important learning objectives. We will now summarize the responses to these learning objectives.

Learning objective 1:
Discuss project management functions.

Planning, organizing, controlling, and leading are the four project management functions. The planning function states what should be done, estimates how long it will take, and projects what it will cost. The organizing function staffs the systems project team and brings together team members, users, and managers to achieve the systems project plan. The controlling function monitors progress reports and documented deliverables. It compares plans with what actually occurred. Swift action is taken to correct significant variances. The leading function guides and motivates team members to do their best.

Learning objective 2:
Explain project management techniques.

Two popular project management techniques are Gantt and PERT. They are used to estimate, schedule, and control the systems project. Gantt is often used to provide the SDLC schedule and then to compare planned performance with actual performance. PERT is appropriate for estimating, scheduling, and controlling multiple interdependent tasks.

Learning objective 3:
Describe computer-aided systems and software engineering (CASE) packages, which provide systems development and maintenance automation technologies.

A full-featured CASE system includes features that aid and help automate both front-end and back-end phases. These features include:

- Workstations
- Central and local repositories
- Modeling tools
- Project management features
- SDLC and prototyping support
- Automatic code generators
- Libraries of reusable code
- Reverse engineering, reengineering, and restructuring facilities

SYSTEMS DEVELOPMENT MANAGEMENT CHECKLIST

Following is a checklist on how to manage systems development projects effectively and how to make use of automation technology to streamline the development process.

1 Establish a macrolevel systems plan to determine the enterprise's information system needs and strategies, and to set a budget for systems projects.

2 Establish a microlevel systems project plan to set schedule and budget requirements for specific systems projects within the systems plan.

3 Create an organization chart for the enterprise's information systems projects, clearly showing responsibility for individual systems projects and perhaps, if the projects are large, for specific phases of these projects.

These areas of responsibility should be assigned according to the specific skills required by the project phases, which may be people-oriented or technical-oriented.

4 Enlist the participation of people from all parts of the organization in each systems development project. Involving senior managers in a project will create an atmosphere of top-down support and cooperation, greatly smoothing the development process. At the other end of the spectrum, bringing end users into the process, perhaps as team members, will assure that the system meets the needs of those who will actually use it.

5 Use weekly and monthly progress reports, in addition to the systems development life cycle's documented deliverables, to inform organization members of actual progress, updated cost estimates, and estimated completion dates.

6 Involve technology to assist with various project management functions. A microcomputer project management software package can perform complex critical path calculations, create Gantt charts and PERT diagrams, and keep track of a project's status.

7 Use a CASE package to assist and automate the entire systems development process. Use of a CASE system does not replace fundamental business and systems planning, nor should it be used prior to the establishment of a structured systems development methodology such as SDLC. A CASE system does, however, provide a useful collection of tools, in an integrated package, which can assist with front-end and back-end tasks.

KEY TERMS

Automatic code generator

Central repository

Chief information officer (CIO)

Computer-aided systems and software engineering (CASE)

Gantt chart

Monthly progress report

Program evaluation and review technique (PERT) chart

Programmer/analyst

Project director

Project manager (project leader)

Project Schedule Report

Reengineering

Restructuring

Reverse engineering

Steering committee

Systems Plan Report

V curve

Weekly progress report

REVIEW QUESTIONS

3.1 List and briefly discuss the four project management functions.

3.2 What are the two levels of systems planning? Explain each.

3.3 What are typically the two most recognized systems project problems?

3.4 What are the two key elements of systems project planning?

3.5 Name the main players in large information systems and state their responsibilities.

3.6 List each SDLC phase and give a brief profile of the kind of systems professional needed for each phase.

3.7 Explain why SDLC phases are sometimes divided into subprojects. Give examples.

3.8 Describe the V curve and its relationship to the SDLC and people-oriented and technical-oriented phases.

3.9 Explain how the corporate culture's resistance to change can be reduced.

3.10 Explain how the controlling function is effected.

3.11 Explain the purpose of weekly and monthly progress reports.

3.12 List the six important traits of leadership.

3.13 Name and briefly describe the three leadership styles.

3.14 If team members are highly motivated and extremely competent, what kind of leadership style is generally the most effective?

3.15 Name and briefly describe two techniques that help estimate, schedule, and monitor the progress of the SDLC.

3.16 Describe how the Gantt chart is used in systems development.

3.17 Describe how the PERT chart is used in systems development.

3.18 List and briefly describe the steps necessary to prepare a PERT chart.

3.19 What is the critical path?

3.20 What's the primary goal of CASE?

3.21 Describe a full-featured CASE system.

3.22 Describe the central repository and its purpose.

3.23 List and briefly describe each feature of a full-featured CASE system.

3.24 Usually, CASE systems are thought of more as systems development or forward engineering technologies. Explain why CASE systems should also be thought of as powerful systems maintenance technologies.

3.25 Define reverse engineering, reengineering, and restructuring. Explain how each is used to aid software maintenance.

CHAPTER-SPECIFIC PROBLEMS

These problems require exact responses based directly on concepts and techniques presented in the text.

3.26 A new inventory control systems project is approved for development. Systems analysis is estimated to take 15 days, general systems design 20 days, systems evaluation and selection 5 days, detailed systems design 40 days, and systems implementation 30 days. None of the phases overlap. Five days have already been expended on systems analysis.

Required: Draw a Gantt chart for the systems project.

3.27 A subproject is composed of the following coded tasks and their estimated times to complete:

Sequence Code	Estimated Time (days)
1.1	5
1.2	10
2.1	10
2.2	15
2.3	5
3.1	5

Required: Prepare a time-scaled PERT chart of tasks and events, and specify the critical path with a bold line.

3.28 Review the following situations:

- The subject system is an old system whose design is unknown; that is, it is a black box. Its functionality is to be kept intact, but a design is necessary to reveal the system's modules and their interrelationships. Once representations of the system's modules are fully developed, some of the modules may be selected as candidates for reuse in the development of new systems.
- The subject system requires renovation and reclamation. It is to contain new form and functionality.
- The subject system's external behavior (i.e., functionality and semantics) is to be preserved. The task is to involve a code-to-code transform that recasts a software program from one that is unstructured (i.e., spaghetti-like) and undocumented to a software program that is structured and documented.

Required: Review the software maintenance section of the chapter and specify what software maintenance function should be performed: reverse engineering, restructuring, reengineering, or a combination of two or more.

THINK-TANK PROBLEMS

These problems call for a feasible approach rather than a precise solution. Although the problems are based on chapter material, extra reading and creativity may be required to develop workable solutions.

3.29 You have recently been hired by Atlas Manufacturing to overhaul completely its information system and its approaches to systems and software development. In the past, systems were developed in a haphazard fashion with little planning, direction, or discipline. Users have complained that little, if any, of their requirements were ever met. The old systems people tried to develop designs in narrative form. These narratives were voluminous and often ambiguous. In many instances, systems projects were started but never completed. Some systems people would work on a systems project for a year or so, then abandon it without any evidence of ever having delivered anything. Of those few software programs developed, none were structured and documented. Each, however, requires a great deal of maintenance.

Required: Write a detailed or authoritative report for top management at Atlas, explaining what needs to be done to develop successful information systems.

3.30 You have been hired to analyze a simple payroll system at a construction company. Imagine that you are looking at the set of programs used to generate paychecks for employees. Employees are either hourly or salaried at this company. The payroll system gets employee records from an employee master file and determines whether an employee is hourly or salaried. If an employee is hourly, the system must get the number of hours worked from the time master file. The system then calculates gross pay and deductions to determine net pay. The result of the net pay calculation is printed on a paycheck for a given employee.

Required: Create a structure chart with pencil and paper to organize the facts about the payroll system. How long did it take you to prepare this chart? Imagine that you are familiar with CASE systems described in this chapter. How long do you estimate it would have taken to prepare it using CASE technology?

You have just learned that the construction company also generates paychecks for contract workers. Contract workers are hired to complete special assignments or when the company needs additional help. Contract workers have no deductions taken from their paychecks. Modify your structure chart to include payroll processing for contract workers. How long did it take you to modify your chart? What are the advantages of a CASE package if you have to change the structure chart several times?

3.31 Acme Manufacturing automated its information system approximately 15 years ago. The company now has over 2,000 programs, performing the following functions: accounting, manufacturing control and requirements planning, and engineering support. Acme has an MIS staff of 30 employees. Fifteen of these people are programmer/analysts. Management at Acme is concerned about the application backlog (it's about two years) and is contemplating hiring more people and purchasing a CASE package for systems maintenance and new systems development. You have been hired to advise management on this acquisition. While at Acme, you discover that many of Acme's existing applications have been developed without the use of a disciplined methodology. The systems are not integrated, and there is a great deal of redundant data stored on the computer. Currently, each employee in MIS is given complete design and programming freedom when developing computerized applications.

Required: Prepare a report to management recommending what must be done before acquiring a CASE package. Include in your report a plan to implement the CASE package. Make some suggestions about how to ensure a successful implementation for the package.

3.32 First Fidelity Bank automated its demand deposit accounting system 20 years ago. The system contains 55 COBOL programs and ten programs written in IBM Assembler. The programs require a great deal of maintenance caused by changes in government regulations and computer equipment updates. Since the programs were written such a long time ago, each time a new piece of computer equipment is added to the system, at least a few of the programs must be modified to make use of the new hardware. Every program in the system has been modified many times by many different programmers. There is very little program documentation and no design documentation for the system. The programs also contain defects. The defects are not blatant errors that show up each month; instead, they appear sporadically, resulting from unusual data input. For instance, management discovered last month that the program for money market accounts calculated interest wrong for balances exactly equaling $4500. Management was not sure how long the program had been calculating interest incorrectly, because there was no record of updates to that program. Last week, a programmer found out that at least 20 programs in the system will not allow a processing date beyond 1999 to be input. There was no allowance for a "2" in the first position of the year of a date.

Management at First Fidelity Bank has been faithfully reading computer journals. It has read about CASE technology. According to the ads, it should be easy to apply reengineering concepts to the demand deposit accounting system to effect the necessary changes and make it work correctly. You have been hired by the management at First Fidelity to evaluate the purchase of a CASE package.

Required: Prepare a report for management concerning the feasibility of using CASE technology on the demand deposit accounting system. When preparing your report, make sure that you think about the cost-effectiveness of the current system. Do you think the bank should reengineer this system or perhaps consider systems euthanasia?

3.33 Following is a random list of statements that pertain to Ace Company.

_______________ The company has a legacy of old, unstructured, undocumented systems.

________________ The company does not have SDLC in place.

________________ The company is in the truck leasing business.

________________ The company's systems people believe systems and software development cannot be engineered. Therefore, none of them are trained in modeling tools.

________________ The company plans to add 50 trucks to its fleet.

________________ Ninety percent of the company's total life-cycle costs are devoted to software maintenance; this means that new systems development is almost at a standstill.

________________ Top management wants to overhaul its systems development area completely and install a modern approach to developing in-house systems.

________________ The company's home office is located in Chicago.

________________ In the future top management wants to develop systems that will be easily maintained and that will incur significantly reduced maintenance costs.

Required: Insert in each blank, ''R'' for relevant or ''IR'' for irrelevant regarding the decision to acquire CASE technology for Ace Company. Do you recommend that Ace acquire a CASE system? Explain why. What changes should Ace make before it decides to acquire a CASE system?

3.34 Given that a college education can be regarded as a systems project (even to the point of having a ''maintenance'' phase following graduation), what tasks would you associate with the creation of a ''business plan,'' a ''systems plan,'' and a ''systems project plan'' as illustrated in this chapter? If a CASE package could be developed to assist you with the development of your college education ''system,'' what specific tools would you include? Create a Gantt chart and a PERT diagram using most likely times for the completion of tasks in this system.

SUGGESTED READING

Chikofsky, Elliot J., and James H. Cross. ''Reverse Engineering and Design Recovery: A Taxonomy.'' *IEEE Software,* Vol. 7, No. 1, January 1990.

Collins, F. ''The V Curve: A Road Map for Avoiding People-Problems in System Change.'' *Journal of Systems Management,* February 1983.

Doke, E. Reed. ''Application Development Productivity Tools: A Survey of Importance and Popularity.'' *Journal of Information Systems Management,* Vol. 6, No. 1, Winter 1989.

Gibson, Michael L. "A Guide to Selecting CASE Tools." *Datamation,* July 1, 1988.

Hazzah, Ali. "Making Ends Meet: Repository Manager." *Software Magazine,* December 1989.

Hodge, Bartow. "CASE Tools Increase Productivity." *Information Executive,* Spring 1989.

Leavitt, Don. "Team Techniques in System Development." *Datamation,* November 15, 1987.

Martin, James. "RAD Techniques Are a Must for Retooling the IS Factory." *PC Week,* December 15, 1989/January 1, 1990.

Oman, Paul W. "CASE Analysis and Design Tools." *IEEE Software,* Vol. 7, No. 3, May 1990.

Pallatto, John. "CG1 Systems to Launch CASE Collections for LANs." *PC Week,* June 11, 1990.

Percy, Tony. "What CASE Can't Do Yet." *Computerworld,* Vol. 22, No. 25, June 20, 1988.

Pope, Stephen T., Adele Goldberg, and L. Peter Deutsch. "Object-Oriented Approaches to the Software Lifecycle Using the Smalltalk-80 System as a CASE Toolkit." *IEEE Software,* 1987.

Shlaer, Sally, and Stephen J. Mellor. *Object-Oriented Systems Analysis*. Englewood Cliffs, N.J.: Prentice-Hall, 1988.

Smith, L. Murphy. "Using the Microcomputer for Project Management." *Journal of Accounting and EDP,* Summer 1989.

Snyders, Jan. "The CASE of the Artful Dodgers." *Infosystems,* Vol. 35, No. 3, March 1988.

Wasserman, Anthony I., Peter A. Pircher, and Robert J. Muller. "An Object-Oriented Structured Design Method for Code Generation." *ACM Sigsoft Software Engineering Notes,* Vol. 14, No. 1, January 1989.

Williamson, Mickey. "Toward the Holy Grail of True CASE Integration." *Digital News,* May 14, 1990.

Yourdon, Edward. *Managing the System Life Cycle*. Englewood Cliffs, N.J.: Yourdon Press, A Prentice-Hall Company, 1988.

Part II

The Front-End Phases of the Systems Development Life Cycle

The purpose of Part II is to present each front-end phase of the systems development life cycle. Chapter 4, which treats systems planning, represents the initial phase of SDLC. During this phase, systems project proposals are graded and prioritized on the basis of technical, economic, legal, operational, and schedule (TELOS) feasibility factors and productivity, differentiation, and management (PDM) strategic factors. Those systems proposals that receive high grades are candidates for systems analysis. The Systems Plan Report is the documented deliverable produced by the systems planning phase. It includes all the systems project proposals that have been approved for development. To help you relate to and understand the chapter material better, we provide you with the beginning of the JOCS case at the end of the chapter.

Chapter 5 treats systems analysis. It presents reasons for performing systems analysis and discusses systems scope. It shows how to gather study facts using various study-fact gathering techniques, and how to analyze these study facts. Ways to prepare and present the Systems Analysis Report are discussed, along with the outcome of systems analysis. At the end of this chapter, the JOCS case continues by applying selected material from the chapter.

The subject of Chapter 6 is general (conceptual) systems design, which explains the analysis-to-design transition. It introduces three systems categories: global-, group-, and local-based. Object- and structure-oriented design approaches are compared. The documented deliverable produced by this phase is the General Systems Design Report. Material from this chapter is applied in the JOCS case study.

After general systems design alternatives have been created, it is time to evaluate and select the optimum one according to both qualitative and quantitative measurements. Therefore, the subject of Chapter 7 is systems evaluation and selection. The result is a Systems Evaluation and Selection Report. Material from this chapter is also used in the JOCS case.

Chapter 4
Systems Planning

WHAT WILL YOU LEARN IN THIS CHAPTER?

After studying this chapter, you should be able to:

1 Explain the reasons for systems planning, describe the systems planning participants, define the components of the Systems Plan Report, and relate the systems planning phase to the systems analysis phase.
2 Define and describe in detail the technical, economic, legal, operational, and schedule (TELOS) feasibility factors and the productivity, differentiation, and management (PDM) strategic factors and their relationship to systems planning.
3 Discuss the three approaches to developing systems project proposals.
4 Describe a method of evaluating and prioritizing systems project proposals.
5 State how systems project planning and reporting are conducted.

INTRODUCTION

Not too many years ago, the notion that information systems could support tactical and strategic business objectives of organizations would probably have struck many senior managers as ludicrous. Management viewed information systems as number crunchers—nothing more than transaction processors. Now, many senior managers view information systems as a key resource in enabling companies to gain a competitive advantage. More and more companies are conducting systems planning to determine ways that information systems can support **BUSINESS PLANS**, which spell out the business mission, goals and objectives, and strategies of the total enterprise.

Systems planning is not an exercise to be performed once every few years, but an ongoing activity. It is an integral part of the chief information officer's job. A significant part of the CIO's job is to exploit information system opportunities, effect the organization's business objectives, and position the company advantageously in the competitive environment. These aims are achieved through systems planning, which is discussed in the following sections.

Before we move into the details of systems planning, the Mammoth Machinery case gives insight into why companies should conduct systems planning. The case also outlines some steps that should be taken to begin the planning process, specifically the development of a business model and an information systems model. An **ENTERPRISEWIDE MODEL** describes all the major entities of the enterprise and their relationships. This model can be built with an entity relationship diagram (ERD). A GENERALIZED INFORMATION SYSTEMS MODEL illustrates a very broad technology platform that is sufficient to support the enterprisewide model, made up of computers, storage media, and a telecommunication network. Neither the enterprisewide model nor the generalized information systems model shows specifics; together they do give a broad vision and provide enough details for budgeting and inclusion in the Systems Plan Report. Specifics of both models are defined in more detail as the SDLC unfolds. Also, an important point made in the Mammoth Machinery sample case is how information systems can support business plans.

Systems Planning at Mammoth Machinery

Mammoth Machinery Company is an enterprise that manufactures large earth-moving equipment. Over the years, Mammoth's sales have declined because of stiff competition from Japanese companies. Moreover, Mammoth's manufacturing process is no longer state-of-the-art. Assembly lines are sprawled throughout the plant, with huge work-in-process (WIP) inventories. Some of the WIP products are defective. They have to be reworked. It takes Mammoth's plant 20 days' lead time to build and ship a customer order. The competition can do it in four days.

Mammoth's manufacturing process is the batch production process common in many manufacturing plants. In it, workers spend an inordinate amount of time on nonvalue-added tasks, such as moving items from one point to another, inspecting and reworking defective items, waiting for production equipment to be maintained or repaired, and waiting for WIP to move from one process to another.

Tom Nolan, plant manager, looked at all the extra materials handling, WIP inventories, wasted floor space, defective products, and excess lead time and called a meeting of all managers, including the chief executive officer, to formulate a business plan that would change the way that Mammoth manufactures its products and conducts its business.

From this strategic meeting came a concerted feeling that, in the words of one manager: "There's got to be a better way. Otherwise, we will not be able to compete in a worldwide market and continue as a viable company." A group of executives and their staff were charged with the responsibility of developing a new business plan.

Jane Thompson, CIO, was put in charge of forming a systems planning team. This systems planning team would be monitored by a steering committee composed of executives. Some of these executives, including Jane, are also on the business planning committee.

Jane believes that besides a well-formulated business plan, another key ingredient for ensuring a successful systems planning process is total user involvement and participation. Therefore, Jane scheduled a joint application development session to be conducted soon after completion of the business plan. She believes that the JAD session will generate a number of specific systems proposals that will be aligned with the new business plan and, consequently, will facilitate and support the business plan. From the JAD session will come the major components of the information systems plan that will be, to a great extent, a systems plan developed by users. Thus the systems plan will receive full endorsement, cooperation, and participation from users—a critical factor for ensuring success of systems development.

Several months later, a new business plan was completed. The major component of the business plan is a simplified production process that includes automated flexible work cells, just-in-time (JIT) manufacturing, total quality control (TQC), robotics, and automated materials handling equipment. The new business plan also stresses integration of entities throughout the enterprise. As CEO Sam Matheson pointed out: "What we need is integration from the executive suite to the factory floor."

After reviewing the business plan, Jane recommended the preparation of an enterprisewide model to show the entities of Mammoth and how they must be related and integrated to support the new production process spelled out in the business plan. She asked Ted Fischer, a systems analyst, to prepare this enterprisewide model using an entity relationship diagram. To prepare this enterprisewide model required that Ted spend a great deal of time working with users in the JAD session.

The final enterprisewide model for Mammoth is displayed in Figure 4.1. This model reflects how Mammoth will be integrated in the future to support the new business plan; it is not a model of how Mammoth conducts its business now.

After reviewing the enterprisewide model, the users in the JAD session asked Jane to prepare a general model of how the new information system will integrate with the new production process. The general information systems model prepared by Jane and her staff is shown in Figure 4.2. Jane cautioned users that the enterprisewide model and the general information systems model provide only a broad vision of the kind of information system that will eventually be developed. She emphasized that a great deal of systems development work will be required before the new information system is implemented.

The enterprisewide model reflects a computer-integrated manufacturing (CIM) environment. Such integration permits instant communication from financial planning and market analysis to material and capacity requirements planning so that valuable time is not lost in reacting to the dynamics of a manufacturing environment. Moreover, an integrated system provides management with information about manufacturing operations in terms of tracking and scheduling, product costing, and performance measurements.

The financial accounting LAN performs business transaction processing (e.g., sales and purchase orders) and financial analysis for budgeting and capital investments. All financial reports and documents, some of which may be confidential, are controlled by the mainframe. Functions performed by the management accounting LAN include:

- Product costing
- Materials, work-in-process, and finished goods control
- Plant and equipment control and depreciation accounting
- Performance measurement reporting

The management accounting LAN also transmits product cost information to the financial accounting LAN.

The engineering LAN performs:

- Master scheduling
- Material requirements planning
- Capacity requirements planning
- Bill of materials

The engineering LAN also receives budget data and sales forecasts from corporate headquarters for scheduling and planning functions. The factory floor LAN controls manufacturing operations. It also collects manufacturing data via bar-code scanners that are transmitted to both the management accounting LAN and the engineering LAN.

Keep in mind that the generalized information systems model serves as a skeleton outline of the target information system. It is a starting point for systems development. The details of the final detailed systems design will be defined precisely from systems proposals and user requirements.

What Jane and her systems staff did during and after the JAD session is explained in the following chapter material. Note that the chapter material does not relate to a specific organization. The purpose of the chapter material is to describe a means by which systems planning can be performed in all organizations.

SYSTEMS PLANNING OVERVIEW

Systems planning attempts to identify the strategic importance of information systems within organizations. Its objective is to look for opportunities to exploit information technology and develop systems projects that support business objectives. In fact, the beginning of any systems planning process should have its roots firmly planted in business planning, because systems planning is performed to spring from and support objectives outlined in the business plan. Thus the systems planning process provides a natural and congruent linkage to the organization's goals and those individual systems projects that support these goals.

In this section we discuss reasons for conducting systems planning, systems planning participants, and components of the systems plan. Then we show how the systems plan initiates the systems analysis phase.

Why Plan Systems?

Systems planning is a process that carefully develops a Systems Plan Report to employ the information system resources in a manner that aligns with and supports the business goals and operations of the organization. In this way, the systems plan is connected to the business plan. It helps to shape the competitive strategy of the enterprise.

Another reason to perform systems planning is to avoid a host of penalties that are suffered by the company when inadequate or no systems planning is performed. Some of these penalties are:

- Missed opportunities to take advantage of productivity, differentiation, and management, or PDM strategic factors
- Exorbitant prices paid for hardware, software, and telecommunication networks because they were ordered too late
- Rushed work toward the end of the project, causing the project team not to spend enough time on critical implementation tasks, such as site preparation, testing, training, and conversion

Figure 4.1
Mammoth's enterprise-wide model.

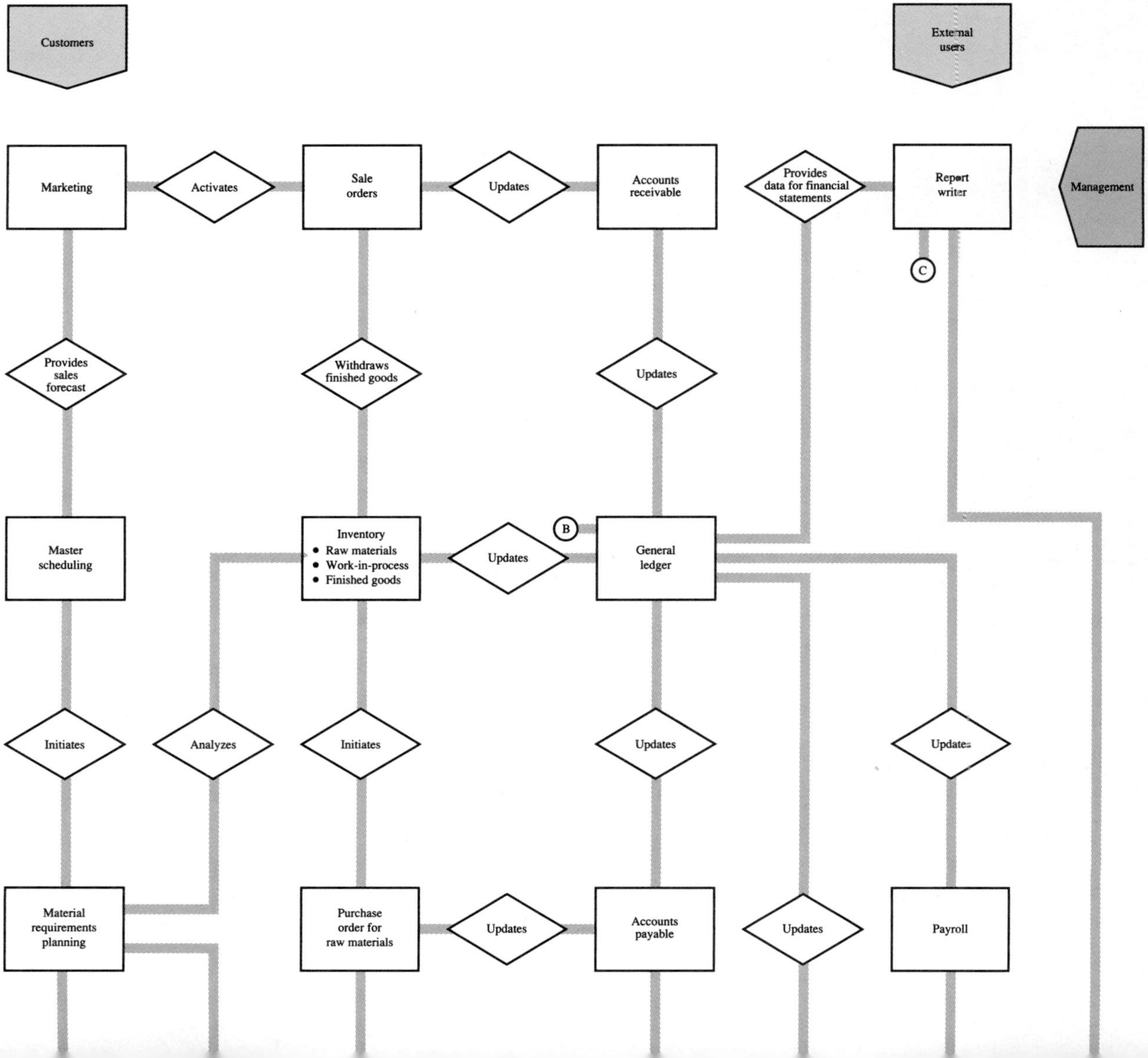

Customers
External users
Marketing
Activates
Sale orders
Updates
Accounts receivable
Provides data for financial statements
Report writer
Management
C
Provides sales forecast
Withdraws finished goods
Updates
Master scheduling
Inventory
• Raw materials
• Work-in-process
• Finished goods
Updates
B
General ledger
Initiates
Analyzes
Initiates
Updates
Updates
Material requirements planning
Purchase order for raw materials
Updates
Accounts payable
Updates
Payroll

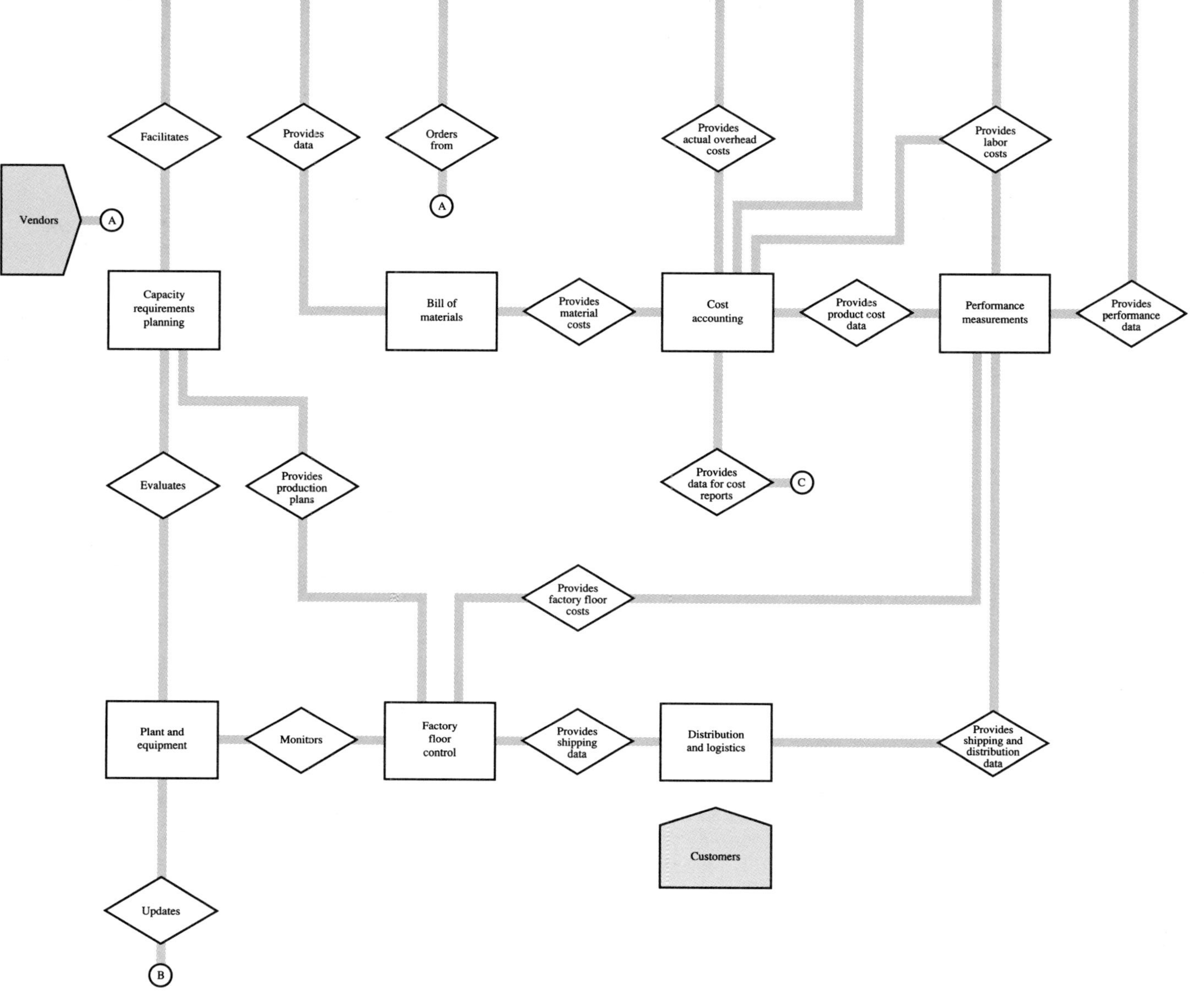

Facilitates
Provides data
Orders from
Provides actual overhead costs
Provides labor costs
Vendors
A
A
Capacity requirements planning
Bill of materials
Provides material costs
Cost accounting
Provides product cost data
Performance measurements
Provides performance data
Evaluates
Provides production plans
Provides data for cost reports
C
Provides factory floor costs
Plant and equipment
Monitors
Factory floor control
Provides shipping data
Distribution and logistics
Provides shipping and distribution data
Customers
Updates
B

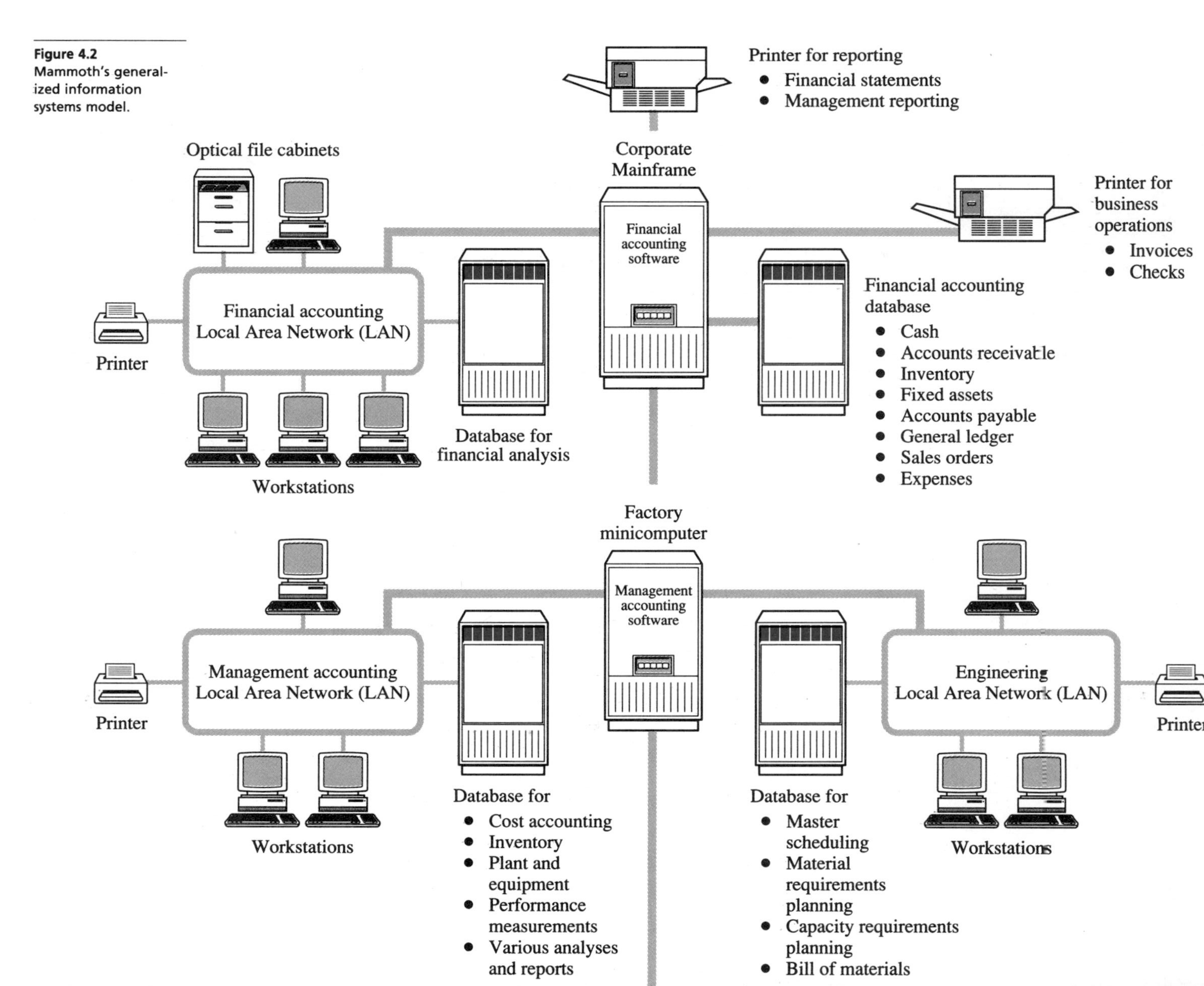
Printer for reporting
Financial statements
Management reporting
Corporate Mainframe
Financial accounting software
Printer for business operations
Invoices
Checks
Optical file cabinets
Printer
Financial accounting Local Area Network (LAN)
Workstations
Database for financial analysis
Financial accounting database
Cash
Accounts receivable
Inventory
Fixed assets
Accounts payable
General ledger
Sales orders
Expenses
Factory minicomputer
Management accounting software
Printer
Management accounting Local Area Network (LAN)
Workstations
Database for
Cost accounting
Inventory
Plant and equipment
Performance measurements
Various analyses and reports
Database for
Master scheduling
Material requirements planning
Capacity requirements planning
Bill of materials
Engineering Local Area Network (LAN)
Printer
Workstations

Figure 4.2
Mammoth's generalized information systems model.

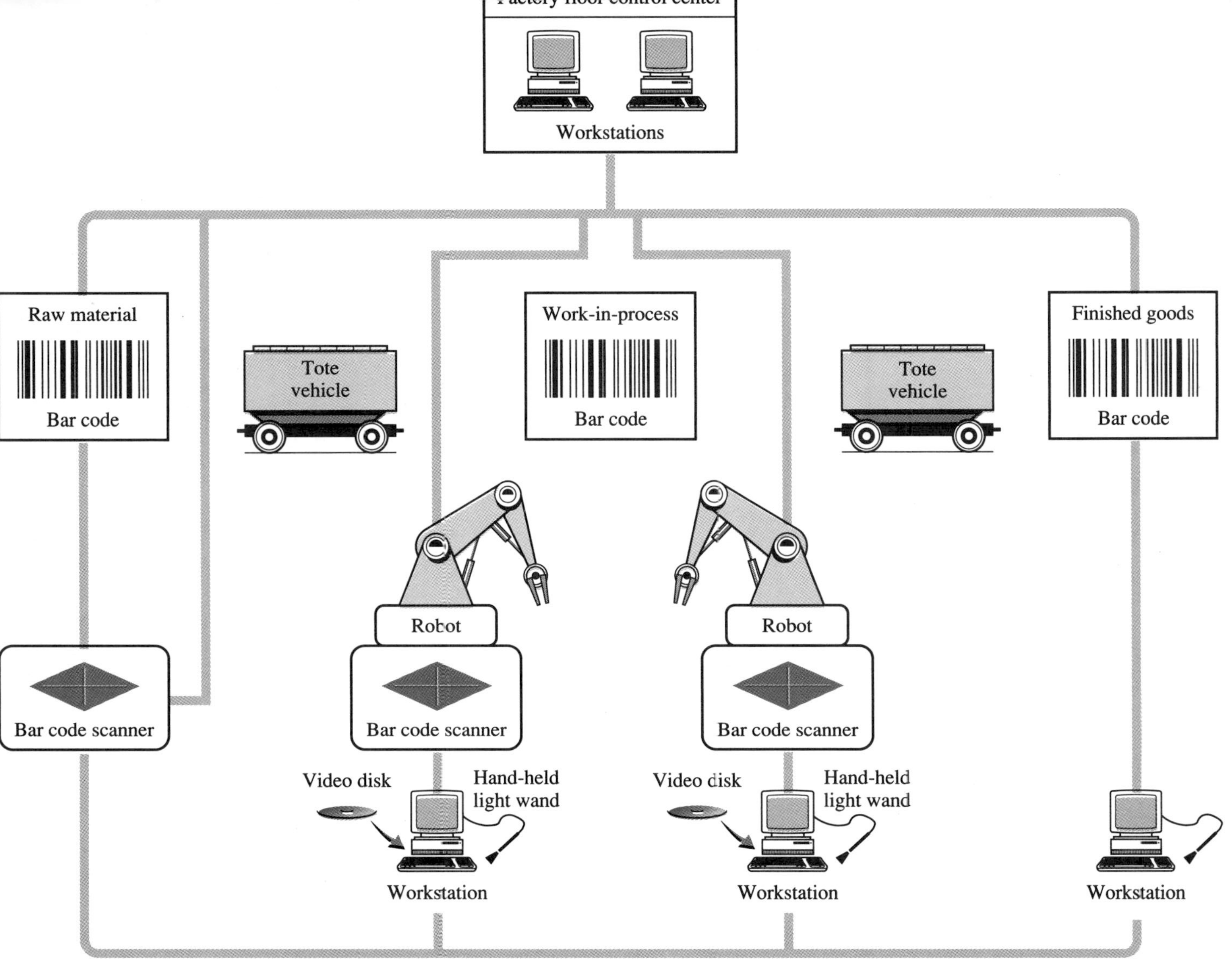
Workstations
Raw material
Bar code
Tote
vehicle
Work-in-process
Bar code
Tote
vehicle
Finished goods
Bar code
Robot
Robot
Bar code scanner
Bar code scanner
Bar code scanner
Video disk
Hand-held
light wand
Workstation
Video disk
Hand-held
light wand
Workstation
Workstation
Factory floor
Local Area Network (LAN)

- Disruption of users and their work and company operations during peak production because of poor scheduling
- Loss of credibility with users because of missed target dates (if the project manager is not careful during development, this problem can occur even with good systems planning)

Who Plans Systems?

In some organizations, a systems planning team is established under the direction of the CIO in concert with a steering committee. The systems planning team should consist of people who have a unique blend of management, business operations, and systems background coupled with an in-depth knowledge of the company and its business plan. In addition to this core planning team, staff members with specialized expertise or skills can be used for consultation on an as-needed basis. Indeed, a systems planning team that draws on multiple perspectives and varied experiences within the enterprise, perhaps supplemented by external consultants with firsthand industry and technical knowledge, is well positioned to create a highly useful systems plan.

Although the development of a systems plan is a shared responsibility between user groups and systems professionals, there should be a strong user-led process, because the major emphasis is development of systems projects that meet user requirements and support the business objectives of the company. Moreover, systems planning is an extremely people-oriented process. (Remember the V graph in Chapter 3?) Much of the work required throughout systems planning depends on open lines of communication among all participants and interpersonal skills rather than technical skills.

In most organizations, a steering committee plays a major role in systems planning. The steering committee, composed of senior managers from major segments of the organization, guides the planning process and prioritizes systems project proposals. Later the steering committee provides oversight and monitoring of the development of systems projects included in the systems plan.

A steering committee is not involved in day-to-day work, nor does it make technical decisions. A steering committee is an advisory group empowered to make top-level decisions. Ideally, the steering committee should include the CIO, CEO, chief financial officer (CFO), and those senior executives who represent other user groups. Conceptually, the steering committee is the linkage between business goals and the information system helping to meet these goals. Specifically, some of the things a steering committee might do are:

- Resolve territorial or political conflicts arising from development of a new system.
- Determine **SYSTEMS PROJECT PROPOSAL** priorities.
- Approve the systems plan and budgets.
- Review project progress once the development process begins.
- At specific checkpoints, determine whether a particular project should be continued or abandoned.

What Are the Components of the Systems Plan Report?
The **SYSTEMS PLAN REPORT** has two major components:

- The overall component
- The application component

The Overall Component
The overall component deals with resources that will be acquired over a three- to five-year time frame. Such resources include new personnel, hardware, software, telecommunications equipment, new computer sites, and security devices. These resources are stated in very general terms, because during the systems planning phase, it is not known precisely what kinds of resources will be required. For example, a list of resources may contain the following:

- Six systems analysts, one database administrator, and four programmers
- Two mainframes with appropriate operating system and utilities
- Twenty workstations
- Three laser printers
- A database management system for home office corporate data
- A telecommunication network to integrate all entities and functions of the enterprise
- An **UNINTERRUPTIBLE POWER SUPPLY (UPS)** that safeguards computers from power losses and power surges

This list of resources is broad enough to allow systems professionals to select optimal resources, considering such variables as education, price, financing terms, performance benchmarks, size, type, model, vendor, and the like. As systems development unfolds modifications will have to be made to the original list of resources. For example, two superminicomputers may be acquired rather than two mainframes.

Coupled with a list of resources is a budget for each item. Any restrictions or time constraints on funds are also disclosed. The list of resources required, along with when and how much money is allocated to their acquisition, is organized in a **RESOURCE REQUIREMENTS MATRIX**, shown in Figure 4.3. Years 1995 and 1996 are omittted. Few new resources are required during these years, because changes in information systems resources typically follow a step function. In contrast, the growth in resource requirements tends to be steady and continuous. This relationship is shown in Figure 4.4.

This graph illustrates how technology upgrades, including software, mainframes, auxiliary storage, channels, peripherals, telecommunication networks, and systems personnel must be made to handle growth and satisfy capacity requirements over time as spelled out in the systems plan. At the points where the dashed line is about to intersect with the load line, maximum capacity has been reached. Upgrades are necessary prior to this point to prevent the degradation of the information system and its service to users.

Year	Month	Resources							
		Computer with Operating Systems and Utilities	Disk Storage	Tape Storage	Workstation and Remote Terminals	Other Packages and Peripherals	Telecommunications Equipment	Personnel	Estimated Cost (1000s)
1994	Jan.						Install fiber optic backbone	Communications technician	$250
	Feb.							Three systems analysts	150
	Mar.						Install communications processor	Two programmers	370
	Apr.							One database administrator	80
	May	Two mainframes	Two disk drives	Two tape drives		One database management system			390
	Jun.					One UPS			40
	Jul.				Ten workstations	Two laser printers			100
	Aug.								
	Sep.								
	Oct.								
	Nov.								
	Dec.								
	⋮	⋮	⋮	⋮	⋮	⋮	⋮	⋮	
1997	Jan.							Two programmers	60
	Feb.								
	Mar.		Two disk drives		Ten workstations	One laser printer		Three systems analysts	250
	Apr.								
	May								
	Jun.								
	Jul.								
	Aug.								
	Sep.								
	Oct.								
	Nov.								
	Dec.								
									$1,690

Figure 4.3
Resource Requirements Matrix.

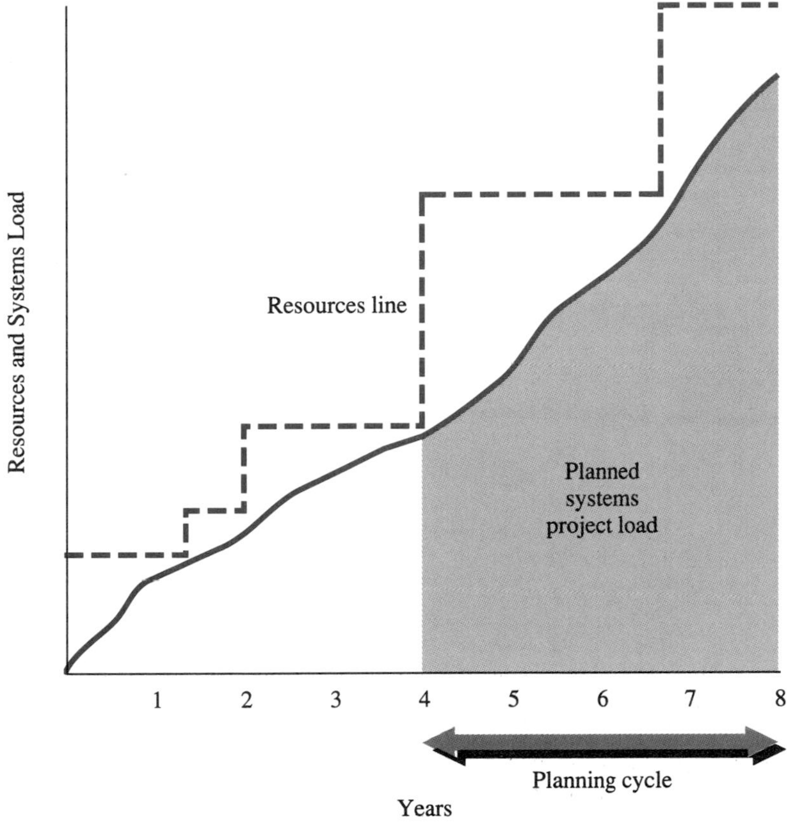

Figure 4.4
Load-line graph and resource upgrades.

Top management begins to allocate additional funds to the information systems budget when the graph suggests an impending overload. Such allocations of funds may be necessary several times during the three to five years typically covered by the systems plan. In the example, two large infusions of funds are required in years 4 (1994) and 7 (1997) to support the systems projects during that period.

The Application Component

The application component, the main subject of this chapter, contains an approved portfolio of systems project proposals. This component, to a great extent, dictates what's included in the overall component. That is, the amount and kind of resources to be acquired are a direct function of systems projects to be developed. (Some systems projects require investment in new technology and personnel, while others leverage existing information system resources.)

A final version of the completed Systems Plan Report, the first documented deliverable of SDLC, is prepared and issued to the steering committee, users, and members of the planning team. A structure for this report is shown in Figure 4.5. The key person to receive the Systems Plan Report is the CIO, who uses it to assign specific projects to project managers. The project managers in

Figure 4.5
Systems Plan Report.

SYSTEMS PLAN REPORT

Overall Component*

Resources to Be Acquired	Funds Committed
Item 1	$ 99,999.99
Item 2	99,999.99
.	
.	
.	
Item n	99,999.99
Total budget	$999,999.99

**Note:* A Resource Requirements Matrix could also be used here.

Application Component:

Systems Project Proposals Portfolio Statement

	Application	Feasibility Factors	Feasibility Factors Score	Strategic Factors	Strategic Factors Score
Ⓐ					
Ⓑ					
.					
.					
.					
Ⓩ					

(Note: This statement is fully explained later in this chapter.)

turn prepare systems project plans, select and organize their team members, and begin systems development.

Normally, the CIO or a project director will want general project schedules from each project manager showing how long it will take to complete the systems project. A Gantt chart is an ideal technique for preparing such a schedule. Figure 4.6 shows how project managers would present their schedules using a Gantt chart.

How Is Systems Planning Related to Systems Analysis?

Systems planning is clearly linked to the systems analysis phase. Both deal with defining user requirements. The primary differences in the two phases are scope and level of detail. The systems plan must define all accepted systems project proposals that will require resources, so it will have a wide scope. Very broad requirements are given enough definition to develop an information sys-

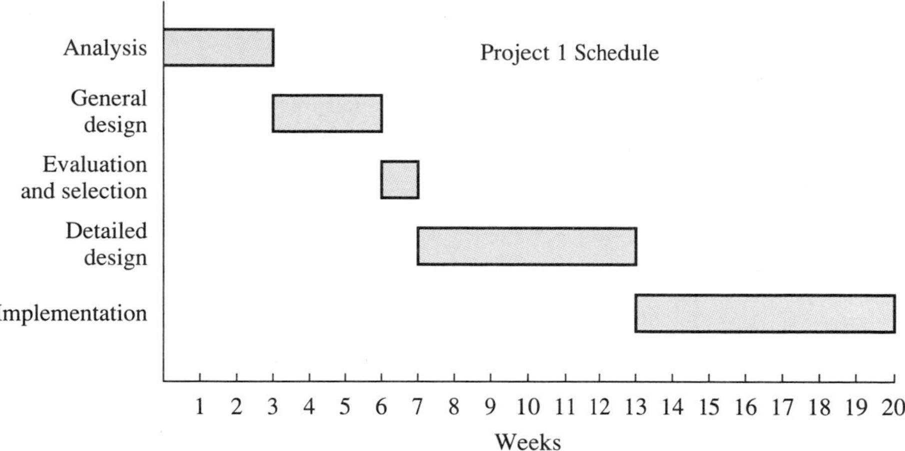

tems budget that entails systems development, personnel, hardware, software, and telecommunication costs over a period of time, usually a year. From a level-of-detail standpoint, the systems plan defines systems projects at a very high and general level. More specifics about user requirements are gathered during the systems analysis phase.

The individual systems projects approved and included in the systems plan become the point of departure for the systems analysis phase and the beginning of project assignments and the project management process. Thus some systems professionals view systems planning as the initial phase of SDLC; others call it the zero phase. Whatever it's called, systems planning should always precede systems analysis.

The relationship between systems planning and systems analysis is shown in Figure 4.7. Assume that three systems projects have been approved and are included in the systems plan. Each project manager will initiate the systems analysis phase for his or her own SDLC.

REVIEWING FEASIBILITY AND STRATEGIC FACTORS

For a systems project proposal to become a candidate for systems development, it must achieve high feasibility and strategic factors scores. Technical, economic, legal, operational, and schedule (TELOS) feasibility factors and productivity, differentiation, and management (PDM) strategic factors are two key elements used in evaluating the potential of systems project proposals made during the systems planning process.

TELOS Feasibility Factors

The five feasibility factors represented by the TELOS acronym must be highly rated before a systems project proposal is accepted for development. These factors are:

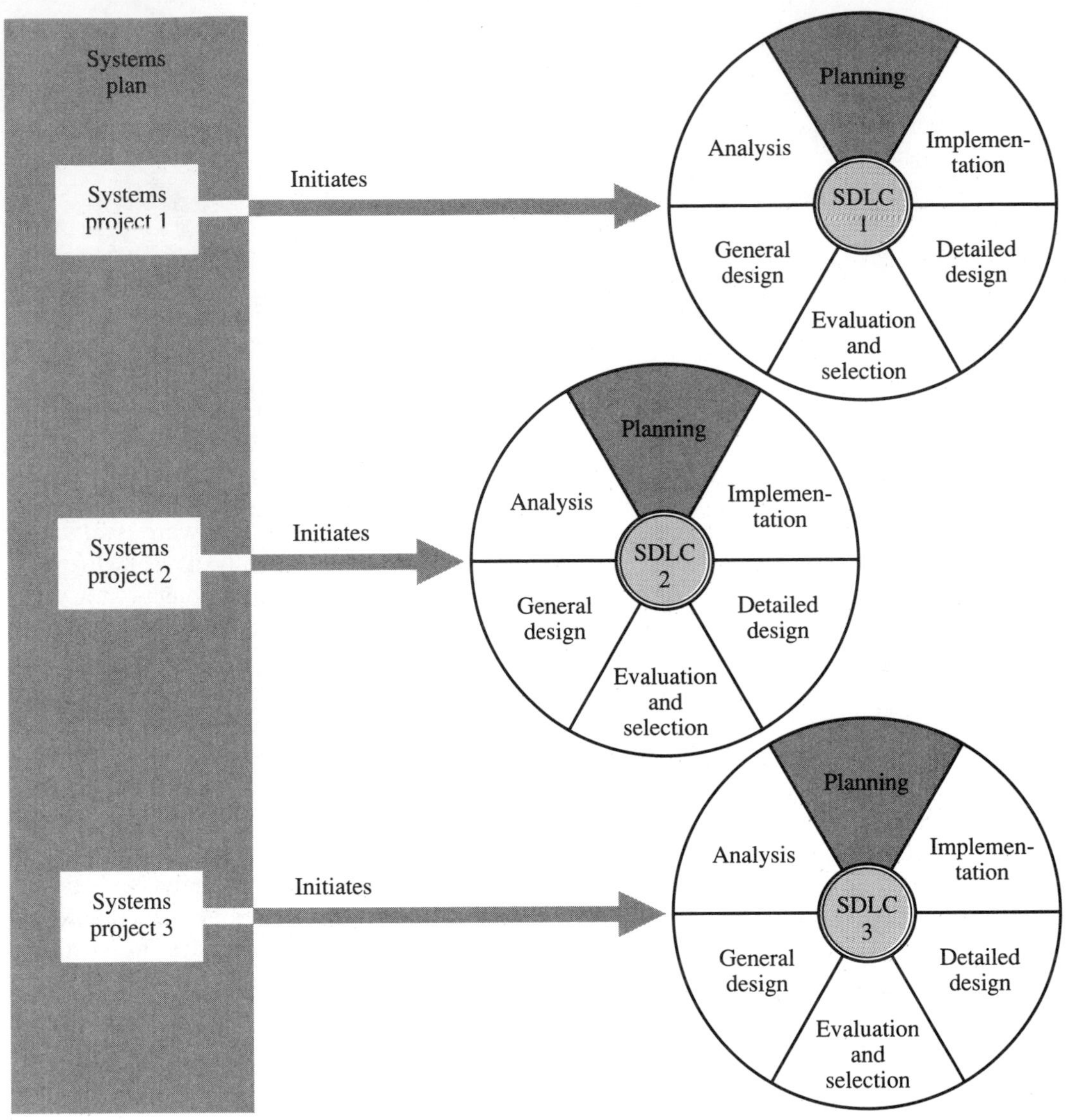

Figure 4.7
Linkage between systems planning and systems analysis phases of SDLC.

- Technical feasibility
- Economic feasibility
- Legal feasibility
- Operational feasibility
- Schedule feasibility

Technical Feasibility

TECHNICAL FEASIBILITY is based on the availability of existing technology or the acquisition of additional technology that supports the proposed systems project. For example, a systems project for Energy Exploration Company that plans to handle online data transmission between the home office and exploration camps located in the Amazon jungle may not be feasible, because of the company's existing technology. However, if the company decided to implement a telecommunication network using technology new to it, such as **VERY SMALL APERTURE TERMINALS (VSAT)**, the project suddenly becomes quite feasible from a technical viewpoint.

Assessment of technical feasibility also includes the technical expertise that the company has now and will have in the foreseeable future. Generally, technology that exists in the marketplace is far ahead of people's ability to apply it effectively. So in most cases, technical feasibility from the viewpoint of availability is a nonissue if the company is willing to acquire it. The ability to apply and use it may, however, be very much at issue. Indeed, because technology is the physical supporting platform for the other information system design components, the level of access to technology and its use will have a significant impact not only on the proposed project's technical feasibility but also on how much the proposed system would have to be modified to make it technically feasible.

Economic Feasibility

The **ECONOMIC FEASIBILITY** factor raises a basic question: Will the company commit sufficient funds to develop and implement a systems project, given the requirement of other capital projects within the organization? If so, what is the level of financial commitment? The level of design and scope are directly related to economic support.

Legal Feasibility

LEGAL FEASIBILITY pertains to how well the systems project proposal complies with the law. (An example of a law that requires a strong system of controls is the **FOREIGN CORRUPT PRACTICES ACT OF 1977**.) Many regulations and statutes deal with invasion of privacy and confidentiality; systems professionals are wise to consider the legal ramifications of the system that they develop. Sensitive medical databases must be secure, with stringent access controls. The transmission of financial data dealing with mergers and acquisitions must be tightly controlled. Many regulatory bodies have regulations dealing with retention of records and the level of security required. Failure to comply with these regulations can cause criminal and civil suits to be brought against the company and even the systems professionals who developed the system. The **INTERNAL REVENUE SERVICE (IRS)** and the **SECURITIES AND EXCHANGE COMMISSION (SEC)** may require compliance. These government agencies, among others, have a great deal to say about:

- What information is reported
- How information is reported
- To whom information is reported
- How information should be safeguarded from unauthorized access
- How long information should be stored

Operational Feasibility

OPERATIONAL FEASIBILITY relates to the efficacy and functionality of the systems project proposal. Will the systems project proposal be based on the company's existing procedures and personnel? If not, can enough skills be acquired, people trained, and procedural changes made to make the system operational? If the answer is no, the proposed project will have to be rejected or the design modified so that it will be operational within existing conditions.

Schedule Feasibility

SCHEDULE FEASIBILITY determines if the systems project proposal will meet a proposed timetable. This feasibility factor simply means that the system must become operational by a target date. If not, the design or the target date must be changed.

PDM Strategic Factors

PDM strategic factors are the things that companies do to gain or maintain a competitive advantage. These are the things that must go well to ensure the enterprise's success. Information systems are being used by innovative companies to support or shape one or more of these PDM strategic factors:

- Productivity
- Differentiation
- Management

Productivity

The **PRODUCTIVITY** strategic factor means that the systems project proposal can increase the effectiveness and efficiency of operations. High productivity reduces expenses and generates cost advantages over the competition. Nonvalue-added costs, such as move, wait, storage, and rework costs, cause a drag on productivity. Eliminating these nonvalue-added costs increases productivity and reduces expenses. Reducing expenses is sometimes referred to as improving the "middle line." Increased sales revenue, the "top line," less decreased expenses, the "middle line," generates greater net income, which is the "bottom line."

Companies pursuing a productivity strategy use information systems to reduce costs by improving labor efficiency and output or by improving use of other resources, such as machinery and inventory. Aspects of productivity require efficiency-based facilities for both knowledge and operation workers; stringent cost reduction, especially in overhead; and avoidance of marginal customer accounts, just to name a few. No doubt, systems projects can be developed that monitor and control efficiency-based machines and provide timesaving applications, reduce middle management and bureaucratic layers, and identify marginal customers.

Differentiation

The **DIFFERENTIATION** strategic factor means that the systems project proposal has the potential of making the enterprise's products and services different from its competition. Many companies, including American Airlines, Caterpillar, McKesson, and Toyota, have information systems that help differentiate their products and services. This is a very strong strategy that influences customer decisions to use one product or service over another. The main results of successful differentiation are increased sales revenue, which is referred to by some accountants as improving the top line. Federal Express uses a computer and telecommunication technology platform to track the exact location of a package in transit, thus improving its service to its customers.

Companies seeking a differentiation strategy use information systems to add unique features or to contribute to quality. A distribution company responds to customers' orders within 24 hours or less by using devices installed in the customers' systems that are in turn connected to the company's system. Another example of product and service differentiation is an automaker that provides online diagnostic and service information to its distributors, who in turn provide this service to their customers.

Management

The MANAGEMENT strategic factor means that the systems project proposal is able to help managers perform their planning, controlling, and decision-making tasks better. The great nemesis of management is uncertainty. Quality information produced by the information system reduces uncertainty, providing a means to enhance management performance. Moreover, a number of opportunities exist to develop systems projects that perform low-level decision-making tasks, thus enabling management to spend more time on high-level issues.

Companies searching for ways to improve the management function can use:

- Information systems to implement basic accounting, budgeting, and cost control systems
- Expert systems
- Decision support systems
- Executive information systems that gather and process economic, demographic, and political data for determining trends

Procter & Gamble has been able to increase its marketing prowess through use of information systems. From the P&G system, marketing managers can access consumer information dealing with age, income, education levels, mobility, and taste to define consumer groups and develop products tailored to their needs. Information for these managers answers such questions as:

- Who buys?
- How do they buy?
- When do they buy?
- Why do they buy?

If their information shows that a market segment for a particular product is large enough, a special product for that segment is developed.

An example illustrates how an enterprise may leverage all PDM strategic factors at the same time: Ace Heating Oil Company maintains a database that records each customer's consumption pattern, the date of the last refill of the customer's storage tank, and the weather conditions since the last refill. Using this information, the company can forecast heating oil usage and decide when to make the next delivery. Such an information system provides Ace with a way to enhance all three strategic factors. The information system makes Ace more productive; differentiates its service from competitors who don't have a similar system; and provides managers with information that will help them in deciding when to replenish Ace's storage tanks and dispatch delivery trucks.

DEVELOPING SYSTEMS PROJECT PROPOSALS

The primary trigger for developing systems project proposals is the business plan. Based on the business plan's objectives, users working in concert with the planning team generate systems project proposals that offer PDM strategic factors. They in turn support the business plan.

Other means of triggering proposals are to convert problems into opportunities and to seek ways to take advantage of new information technology. All of these triggering mechanisms working together provide a fountainhead for systems project ideas. We will start this section by showing how a problem is converted to an opportunity.

Turning a Problem into an Opportunity

Often, user groups view a particular situation as a problem. The converse of a problem, however, is an opportunity. Some planners have found it advisable to change problem statements into opportunity statements. Opportunity statements become the bases for systems proposals. For example, a problem statement is: "Inventory is out of control. It is causing poor customer service and obsolescence." An opportunity statement that could serve as a systems project proposal would be: "Develop an inventory control system that includes sales forecasts, economic order quantities, and reorder points that will reduce total investment in inventory while at the same time increasing customer service." After several meetings, proposals begin to materialize and are identified and documented in a **SYSTEMS PROJECT PROPOSAL FORM**, as depicted in Figure 4.8. This form is used for all systems project proposals regardless of the way they are generated.

The Systems Project Proposal Form has a twofold purpose. In Part 1 it enables users to formulate concisely a systems project proposal. Parts 2 and 3 are used to enter TELOS feasibility factors and PDM strategic factors scores, respectively. As the figure shows, one of the managers, Jane Hanifan, has submitted her proposal for an inventory control system to help in managing inventory and servicing customers. The information systems staff, with preliminary analysis, determines that this application will require a software package and moderate systems work and resources. (For now, ignore the TELOS feasibility factors and PDM strategic factors. We discuss their role in scoring systems project proposals in the next major section.)

Figure 4.8
Systems Project Proposal Form.

SYSTEMS PROJECT PROPOSAL FORM

Part 1 (to be completed by requester)

Date submitted: DD/MM/YY

Request for: ☐ Modification of system ☐ Redesign of system ☒ New system

Submitted by: Jane Hanifin Department: Purchasing

Nature of request: Develop an inventory control system

Reason for request: To achieve a more effective means of managing inventory and servicing customers.

Part 2 (to be completed by the CIO and staff)

Systems work required: ☐ Minor ☒ Moderate ☐ Major

Implementation will require: ☒ Software ☐ Hardware ☐ Personnel

New resources required: ☐ Minor ☒ Moderate ☐ Major

Feasibility factors score: T 9 E 7 L 9 O 8 S 7 Score: 8.0

Part 3 (to be completed by the steering committee or user groups)

Strategic factors score: P 10 D 4 M 7 Score: 7.0

Priority assigned: High Starting date: DD/MM/YY

☒ Approved ☐ Tentatively approved ☐ Rejected

Using the Business Plan to Create Systems Project Proposals

One or a combination of the PDM strategic factors make up the means for connecting the business plan and its objectives to those of the systems plan. Figure 4.9 demonstrates this relationship and linkage.

The business plan is produced by senior management to give a sense of direction and mission for the company as a whole. Everyone in the organization knows what business they are in; what their purposes are; how they plan to

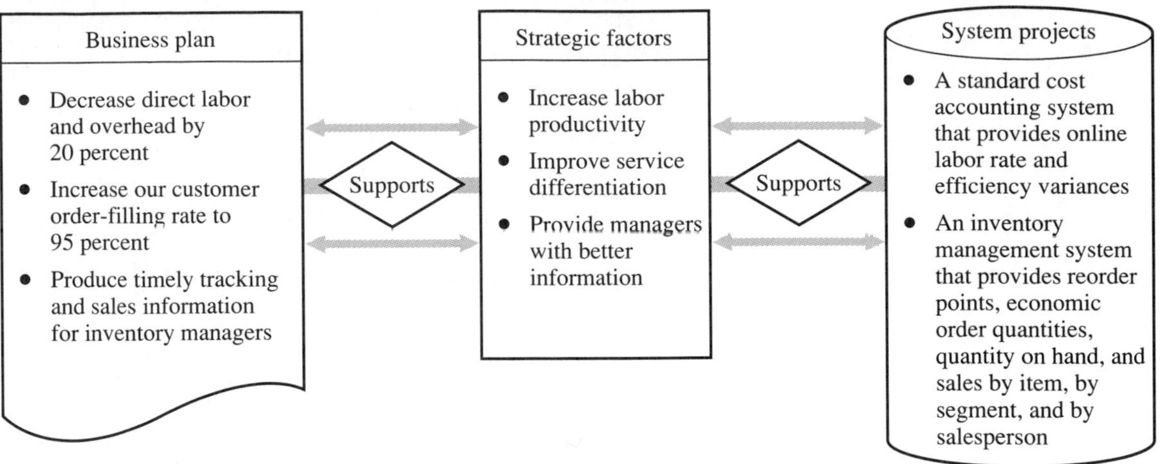
Business plan
Decrease direct labor and overhead by 20 percent
Increase our customer order-filling rate to 95 percent
Produce timely tracking and sales information for inventory managers
Supports
Strategic factors
Increase labor productivity
Improve service differentiation
Provide managers with better information
Supports
System projects
A standard cost accounting system that provides online labor rate and efficiency variances
An inventory management system that provides reorder points, economic order quantities, quantity on hand, and sales by item, by segment, and by salesperson

Figure 4.9
Relationship and linkage of proposed systems projects with strategic factors and the business plan.

conduct business; who their main customers are; and what their products and services are. The plan also includes some very specific objectives as indicated in our example. The strategic factors articulate with the business plan. In turn, the systems project proposals help produce and support the strategic factors. It is during the systems planning process that the linkage is made between the company's business plan objectives and systems project proposals.

Using Information Technology to Create Systems Project Proposals

Today, knowledgeable managers, especially those who work for world-class manufacturing enterprises, do not permit their companies to fall behind or become trapped in technological obsolescence. They recognize the value of information technology, such as:

- Telecommunications, computers, advanced software packages, and multifunctional workstations
- Transmission of combinations of voice, text, number, and graphics via digital telecommunication media
- Innovative applications such as computer-integrated manufacturing, teleshopping, electronic mail, and teleconferencing

Proactive managers are also beginning to implement information technology such as **ELECTRONIC DATA INTERCHANGE (EDI)**, to decrease paper-flow bottlenecks and keying errors by automating business transactions, thereby increasing productivity and reducing the time necessary to complete a transaction. EDI can change the way an enterprise conducts its business by forming a network to link its plants, suppliers, and dealers.

To be sure, telecommunications is an area that provides a number of strategic opportunities for companies that choose to plan strategically and invest properly. Building a telecommunication network (also called a backbone) for a company is analogous to building a highway system. The company may not know exactly all the kinds of data (traffic) that the telecommunication highway will carry, but the point is that the highway is there ready for use for anyone who needs it, and new opportunities are discovered that leverage the PDM strategic factors. Telecommunication networks enable companies to:

- Enter new markets
- Offer new products and services
- Deliver old products and services in new ways
- Support more efficient and effective manufacturing operations
- Provide more timely information to managers
- Coordinate operations throughout the enterprise

In effect, telecommunication networks are enablers of many new applications. They make technically feasible many systems proposals that, without networks, would not be feasible.

PRIORITIZING SYSTEMS PROJECT PROPOSALS

After a set of systems project proposals has been reduced to a manageable number by the process discussed in the preceding section, systems professionals and user groups, represented by a steering committee, grade each systems project proposal and rank it in order of priority. Parts 2 and 3 of the Systems Project Proposal Form in Figure 4.8 is used to determine and document both TELOS feasibility factors and PDM strategic factors grades.

Computing TELOS Feasibility Factors Scores

The TELOS feasibility factors are each weighted from zero to 10 by the CIO and other systems professionals. A TELOS feasibility factors score (sum of the weights divided by 5) of 10 means that a systems project proposal is totally feasible. The example proposal is feasible, with a TELOS feasibility factors score of 8.0.

But a systems project proposal may be quite feasible although the same proposal may not score very well from a strategic factors viewpoint, or vice versa. In any case, the next step is to document a PDM strategic factors score normally assigned by the steering committee.

Computing PDM Strategic Factors Scores

If the systems project proposal is feasible, the steering committee reviews it to determine how well it is aligned with and how well it supports the PDM strategic factors. Bear in mind that during the proposal development process, proposers of systems projects were not constrained in any way; otherwise, ideas may have been stifled. But not all ideas (i.e., systems project proposals) are necessarily good ideas. Therefore, all proposals should be scrutinized by the steering committee to determine which proposals promise the greatest potential in supporting the PDM strategic factors. Each strategic factor is given a weight of 10 or less. The sum of weights divided by 3 equals the PDM strategic factors score. In our example, the PDM strategic factors score is 7.0 out of a possible 10. Before Hanifan's proposal is approved or rejected, however, it must be prioritized and compared with other systems project proposals.

The Prioritization Process

Each systems project proposal is measured and prioritized by its TELOS feasibility factors and PDM strategic factors scores. This prioritization process is performed using a priority grid presented in Figure 4.10. Systems project proposals, according to their TELOS and PDM scores, can fall into one of four categories. Those in the low-potential category, with TELOS feasibility factors and PDM strategic factors scores of 2 or less, are not worthy of further consideration and are thereby rejected. Those in the moderate-potential category, with TELOS feasibility factors and PDM strategic factors scores between 2 and 5, are worthy of systems development, but not at the present time. Systems project proposals in the high-potential category, with TELOS feasibility factors and PDM strategic factors scores between 5 and 8, are worthy of systems

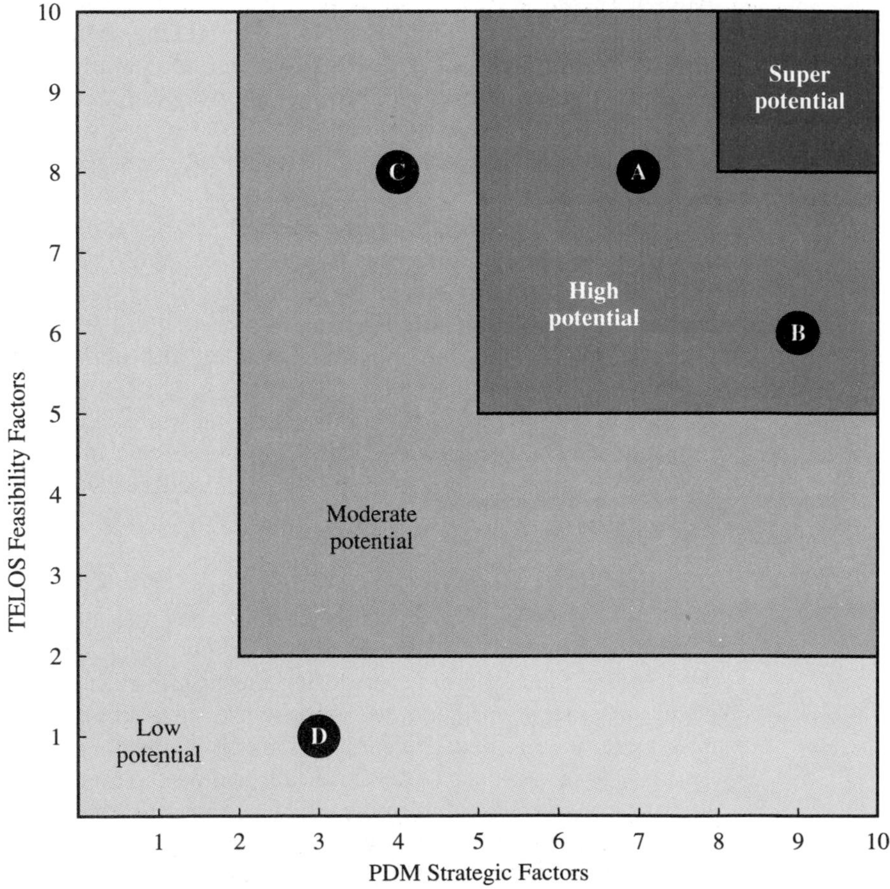

Figure 4.10
Priority grid.

development beginning this fiscal period. Finally, those systems project proposals that receive TELOS feasibility factors and PDM strategic factors scores of greater than 8 are worthy of systems development to begin immediately. These are considered blockbuster proposals. Note that Hanifan's inventory control system, project A, with TELOS feasibility factors score of 8.0 and a PDM strategic factors score of 7.0, falls into the high-potential category.

Now assume that other systems projects were submitted and graded in the same manner as project A. Project B, for example, is an electronic data interchange (EDI) proposal with very strong PDM strategic factors, and its TELOS feasibility factors score is fairly strong. It also falls into the high-potential category along with project A. Project C is an online payroll timekeeping system. Its TELOS feasibility factors score is strong, but its PDM strategic factors position is ordinary. It, therefore, falls into the moderate-potential category. Project D is a cellular-based message system using digital transmission. It has weak TELOS feasibility factors and somewhat weak PDM strategic factors, and therefore falls into the low-potential category and is eliminated from further consideration at this point. It may, however, be considered later.

The prioritizing process reduces, if not eliminates, the squeaky wheel syndrome, a well-known affliction of many organizations, which results in an inappropriate and unfair distribution of grease. Too much lubrication can result in a mess, and too little causes friction and dysfunction. The steering committee, by carefully assigning PDM strategic factors scores, helps ensure that the approved systems project proposals are goal-congruent for the entire organization, not solely for a single person or department.

Systems Project Proposals Portfolio

The key component of the systems plan is the portfolio of approved and tentatively approved systems project proposals. In our case, the three systems project proposals selected for systems development are projects A, B, and C, with A and B receiving the highest priority. Formally, they are entered into a **SYSTEMS PROJECT PROPOSALS PORTFOLIO STATEMENT**, presented in Figure 4.11. This is the statement used to assign specific projects to individual project managers.

Figure 4.11 Systems Project Proposals Portfolio Statement.

SYSTEMS PROJECT PROPOSALS PORTFOLIO STATEMENT

	Application	Feasibility Factors					TELOS Feasibility Factors Score	Strategic Factors			PDM Strategic Factors Score
		T	E	L	O	S		P	D	M	
Ⓐ	Inventory control system	9	7	9	8	7	8.0	10	4	7	7.0
Ⓑ	Electronic data interchange (EDI)	6	6	6	5	7	6.0	10	10	7	9.0
Ⓒ	Online payroll timekeeping	9	7	6	9	9	8.0	6	1	5	4.0
Ⓓ	Cellular-based digital messaging	1	1	1	1	1	1.0	4	1	4	3.0

The Systems Project Proposals Portfolio Statement also takes systems professionals out of any political crossfire. A consensus of participants validates the top systems project proposals and builds support and commitment for them throughout the organization. This commitment makes the project team's job a lot easier, because potential users will be inclined to cooperate fully. Users realize that the information system is part of the fabric of the organization with top management and budgetary support. This is especially true in organizations that have a CIO who participates with top management in setting the strategic direction of the enterprise.

PLANNING AND REPORTING ON THE SYSTEMS PROJECT

Either the CIO or project director assigns specific projects from the Systems Project Proposals Portfolio Statement to systems project managers (leaders). Systems project managers (leaders), in turn, start planning for development of systems projects under their control by using the project management techniques discussed in Chapter 3.

Creating a Project Schedule

The key document on which to base a project plan is the **PROJECT SCHEDULE REPORT**, shown in Figure 4.12. As the project evolves, progress reports are compared with the Project Schedule Report to determine how well the project is proceeding. This report represents each project manager's summary of estimating, assigning, and scheduling of the systems projects assigned to them.

The top of the Project Schedule Report contains standard project identification data. The next part summarizes the resources needed. These items should equal those amounts set in the systems budget. The amounts in the report are divided into three sections:

- Personnel costs for developing the system
- Programming costs
- Technology platform costs such as computers and telecommunications

The next part is a Gantt chart that gives a schedule for all SDLC phases. The last part is a breakdown of each phase into its respective tasks. Estimated start and completion times are given for each task. These estimates will be used to prepare a PERT chart. People who are responsible for each task are also listed.

Copies of the first three parts of the Project Schedule Reports will be collected and organized by the CIO (or project director) and steering committee. This information will be used to help these strategic-level managers monitor and control overall systems development progress. As estimates are revised, new reports will be prepared. Previous ones will be kept on file for reference. (A complete file will give an excellent history of systems projects.) A complete review of all documentation will disclose the validity of estimates,

Project Schedule Report

Project Identification

Project name: Project number:

Prepared by: Date:

Priority: Completion date:

Resource Summary

Systems development		Programming		Technology	
Estimate	Actual	Estimate	Actual	Estimate	Actual
$	$	$	$	$	$

Gantt Chart for Total Project by Phases

Analysis

General design

Evaluation and selection

Detailed design

Implementation

Time

Task Schedule

Task	Workdays to complete	Start		Complete		Responsibility
		Estimated	Actual	Estimated	Actual	

Figure 4.12 The Project Schedule Report.

showing where slippages occurred, and why. In these reviews, valuable lessons are learned. A complete file serves as a basis for planning and scheduling future systems projects.

Reporting Project Progress

The ability to evaluate systems project progress against a Project Schedule Report is at the heart of project management. Project progress evaluation, however, depends on the collection, measuring, and reporting of timely and accurate information at various checkpoints.

A checkpoint is a significant place or milestone in systems development where progress is assessed. To the project manager, checkpoints are both task- and phase-related. For example, the completion date for the interviewing task may be set for August 5, and the completion date for the systems analysis phase may be set for September 18.

August 5 is an important checkpoint or milestone for interviewing, and September 18 is a major checkpoint for systems analysis, but interim checkpoints are often set weekly and monthly to call for progress reports. These reports contain measures of progress for each phase and task under development. One of the goals of project management is to spot potential trouble areas before they reach crisis proportions. If a task or phase begins to slip badly, more frequent checkpoints are warranted.

Analyzing Project Progress

Progress reports contain estimates of the percent complete for each task and phase of the systems project under development. These percent-complete figures should be compared to the Project Schedule Report to determine which phases or tasks within phases are on, ahead of, or behind schedule. Each behind-schedule condition must be analyzed to determine whether the slippage was due to a one-time problem or to a problem that may occur again. From this information, management can determine what corrective action should be taken.

Revising the Project Schedule

Should the progress report differ significantly from the Project Schedule Report in which corrective action is warranted, all schedules and budgets must be revised on the basis of progress to date. Before schedules can be revised, new estimates must be made for the effort required for the remaining phases and tasks.

The new estimates for effort and resources to be expended should take into account only the figures for the work completed so far. The original estimates should be ignored. For example, assume that a particular task was originally estimated to require 100 workdays to complete. After 20 workdays of effort, the analyst reports that it is 10 percent completed. Assuming that the remaining 90 percent of the work will proceed at the same rate, the total project will require:

$$\text{Total workdays to complete project at present rate} = \frac{20 \text{ workdays}}{0.10 \text{ amount completed}} = 200 \text{ workdays}$$

The most reliable estimate for the remaining work is 180 workdays (200 workdays required to complete project at present rate − 20 workdays of effort already expended).

New schedules and budgets must be generated based on current estimates. Automated project management techniques discussed in Chapter 3 make these changes fairly easy. For example, new expected times, slack values, and the new critical path for a PERT chart can be completed automatically by a software package.

REVIEW OF CHAPTER LEARNING OBJECTIVES

The major goals of this chapter were to enable you to achieve five important learning objectives. We now summarize the responses to these learning objectives.

Learning objective 1:
Explain the reasons for systems planning, describe the systems planning participants, define the components of the Systems Plan Report, and relate the systems planning phase to the systems analysis phase.

Without systems planning, a company is prone to develop systems that are not congruent with goals of the organization that it is supposed to serve. Insufficient methods of setting priorities for proposed systems projects, inappropriate allocation of resources, and unrealistic (or unavailable) schedules contribute to this problem. Companies that don't perform systems planning can usually be recognized by their inability to meet schedules or budget targets, lack of project status awareness by managers or users, and frequent duplication of effort. Further, a truism in management is: If you don't plan a project, you will never be able to control it.

The main participants in the systems planning process are the:

- Steering committee
- Planning team
- Systems group

The steering committee's main function in systems planning is to oversee the planning process and assign PDM strategic factors scores to each systems project proposal made by the planning team. The planning team is made up of people representing all user groups. These participants should be experienced in business operations and knowledgeable about the company's business plan. Their major responsibility is to generate systems project proposals that they believe will support the business plan. A group of systems professionals assigns feasibility factors scores to each proposal. In some companies, the chief information officer (CIO) may serve on the steering committee. In other companies, the steering committee is completely independent from the information system and its management.

The Systems Plan Report is composed of two major components:

- Overall component
- Application component

The main ingredient in the overall component is a general list of systems resources to be acquired over the next several years. Combined with this list are amounts budgeted for each item. The application component is composed primarily of the Systems Project Proposals Portfolio Statement. This statement contains all the systems project proposals that have been accepted for development. To a great extent, the budget in the overall component is a function of the portfolio statement. This statement is also used by the CIO or project director to assign specific projects to individual systems project managers (leaders), who in turn prepare for the next phase, systems analysis.

Learning objective 2:
Define and describe in detail the TELOS feasibility factors and the PDM strategic factors and their relationship to systems planning.

TELOS feasibility factors and PDM strategic factors are the key parameters used to evaluate and score systems project proposals. Technical feasibility means that the systems project can work on available technology, or new technology that will be acquired. Economic feasibility means that resources will be available for its development. Legal feasibility means that the systems project will be in compliance with all laws and regulations. Operational feasibility means that the systems project will have the quality of being functional or operative in the environment (or conditions) that exists in the company today or will exist when the systems project is implemented. Schedule feasibility means that the completion date of the systems project fits other company time constraints and time frames.

Productivity, differentiation, and management represent the PDM strategic factors. Productivity means that the systems project proposal has the potential of increasing efficiency and decreasing costs. Differentiation means that the systems project will distinguish the company's product or service from its competitors. Management means that the new systems project will make well-defined low-level decisions and produce quality and usable information to improve various managers' performance.

Learning objective 3:
Discuss the three approaches to developing systems project proposals.

The three ways to develop systems project proposals are:

- Convert a problem into an opportunity
- Use business plan objectives
- Exploit new technology

All three approaches work together as triggering mechanisms to generate ideas for systems project proposals. The controlling approach, however, is always the business plan, because all systems project proposals must be congruent with the company's business plan.

Some kind of form is used, such as the Systems Project Proposal Form, to present proposals. The form shows:

- Date submitted
- Proposer
- Department
- Nature of request
- Whether the request will require modification of the existing system (if there is one), redesign of the existing system, or a new system
- TELOS feasibility factors scores
- PDM strategic factors scores
- Whether the system is approved, tentatively approved, or rejected (if approved, a starting date is also assigned)

Learning objective 4:
Describe a method of evaluating and prioritizing systems project proposals.

After generating a number of systems project proposals, the next critical step is to evaluate their feasibility of becoming successful. Normally, a group of systems professionals, working in a consulting capacity for the steering committee, assigns TELOS feasibility factors scores. This score indicates the possibility of the systems project proposal being developed successfully from a systems viewpoint.

Those proposals with relatively good feasibility scores are then graded on their capability of supporting the PDM strategic factors. The steering committee, possibly assisted by people with detailed knowledge of and experience with the specific business operations, assigns PDM strategic factors scores based on how well they think the systems proposal will support these factors.

For a graphic summary of the TELOS feasibility factors and PDM strategic factors scores and the priority of each systems project proposal, a priority grid is used. This priority grid divides proposals into four groups:

- Low potential
- Moderate potential
- High potential
- Super potential (or blockbuster)

Those proposals that are finally accepted for development are listed in a Systems Project Proposals Portfolio Statement, the key element of the Systems Plan Report. This statement contains the application name and scores of each accepted systems proposal. It is used by the CIO or project director to assign specific projects to systems project managers.

Learning objective 5:
State how systems project planning and reporting are conducted.

Once the systems project managers are assigned their systems projects, they immediately begin project planning and organizing. A key document of their project plans is the Project Schedule Report, which identifies the systems project and its estimated completion date; a summary of resources committed to the project; a Gantt chart that provides an overview schedule of the entire project broken down by phases of the SDLC; and a detailed schedule showing tasks, workdays to complete, start and completion times, and team members assigned to specific tasks. Usually, a systems project manager will develop a PERT chart from these data to show task sequence, interdependencies, and the project's critical path.

SYSTEMS PLANNING CHECKLIST

Following is a checklist on how to conduct systems planning.

1 Develop a business plan, or at least define key business goals and objectives. The business plan, strategies, and objectives should be developed by strategic-level executives.

2 Organize a steering committee, systems planning team, and a joint application development session.

3 Build an enterprisewide model using an entity relationship diagram that reflects how the organization's entities and their relationships should be formulated to support the business plan. It may be helpful in some situations to build an enterprisewide model of the present organization before business and systems planning begins to serve as a springboard for such planning activities. However, the final enterprisewide model should represent how the enterprise wants to conduct its business in the future.

4 Prepare a generalized information systems model using the enterprisewide model as a guide. The generalized information systems model represents a broad view of the technology platform with heavy emphasis on the network and nodes (e.g., computers, printers, and secondary storage devices) connected to the network. Also, a general description of departments (e.g., accounting, engineering, and manufacturing) is presented. However, a specific technology platform cannot be defined until after systems analysis and general systems design phases have been conducted.

5 Prepare a Resource Requirements Matrix and an overall component, especially a budget for the Systems Plan Report. Preparation of a Resource Requirements Matrix is based on the generalized information systems model.

6. Schedule a JAD session to generate a Systems Project Proposals Portfolio Statement. Also, use Systems Project Proposals Forms to document each systems project proposal.
7. Create Gantt charts and PERT diagrams to indicate the estimated schedules of specific projects.
8. Use TELOS feasibility factors and PDM strategic factors to grade each systems project proposal and include the results in a priority grid.
9. From the priority grid, select high-priority systems project proposals for immediate development and place them in a Systems Project Proposals Portfolio Statement.
10. Prepare a Project Schedule Report for each systems project proposal in the Systems Project Proposals Portfolio Statement. Assign to systems project managers, organize systems project teams, and begin the systems analysis phase of the SDLC.

Note: The preceding systems planning steps should be conducted for a systems project of any size. Even small systems project development does not preclude the need for systems planning, although the breadth of planning may be much narrower for small systems projects. With the completion of systems planning steps, systems professionals and users have reached a high level of consensus and cooperation, which will help start systems analysis on a sound footing. Now, systems professionals and users have a strong mandate as to what they will be working on and trying to develop, and a clear vision of the enterprise's ultimate objectives in congruence with and supportive of the enterprise's business plan.

KEY TERMS

Business plan

Differentiation

Economic feasibility

Electronic data interchange (EDI)

Enterprisewide model

Foreign Corrupt Practices Act of 1977

Generalized information systems model

Internal Revenue Service (IRS)

Legal feasibility

Management

Operational feasibility

Productivity

Project Schedule Report

Resource Requirements Matrix

Schedule feasibility

Securities and Exchange Commission (SEC)

Systems Plan Report

Systems Project Proposal Form

Systems project proposals

Systems Project Proposals Portfolio Statement

Technical feasibility

Uninterruptible power supply (UPS)

Very small aperture terminals (VSAT)

REVIEW QUESTIONS

4.1 Summarize the reasons for systems planning. List the penalties suffered by companies not conducting systems planning.

4.2 Explain why systems planning is an integral part of the CIO's job.

4.3 Explain how systems planning relates to the business plan and describe an enterprisewide model and a generalized information systems model. What are the purposes of these models?

4.4 Who are the main participants in the systems planning process? Describe their responsibilities.

4.5 Name and describe the chief components of a Systems Plan Report.

4.6 Explain how systems planning relates to systems analysis.

4.7 Briefly define the TELOS feasibility factors and the PDM strategic factors, and discuss their role in systems planning.

4.8 During systems planning, it is recommended that three approaches be used to generate systems project proposals. Describe and give an example of each approach.

4.9 List and describe the purpose of each element that makes up the Systems Project Proposal Form.

4.10 Once systems project proposals have been generated, describe how they are evaluated and prioritized.

4.11 Describe the purpose of a priority grid and Systems Project Proposals Portfolio Statement.

4.12 What is a Project Schedule Report? What is it used for?

4.13 How is project progress reporting handled? What are checkpoints?

4.14 What is the purpose of progress analysis?

4.15 Explain how revisions of schedules should be made.

CHAPTER-SPECIFIC PROBLEMS

These problems require exact responses based directly on concepts and techniques presented in the text.

4.16 You have been assigned to prepare an enterprisewide model for Pelican Supply Company. Following are notes taken at a JAD session: A customer places an order. The order is checked against inventory to determine if the ordered item is in stock. If the item is in stock, the item is shipped by the shipping department, and the accounting department sends an invoice to the customer. If the item is not in stock, it is backordered and the marketing department sends the customer a backorder notice.

Required: Using an entity relationship diagram, prepare a model of the preceding notes.

4.17 Sierra Medical Center is a new large hospital and clinic complex located in Pleasant Town, which is in the center of the state. Senior management plans to develop a remote access clinical system (RACS) for physicians who practice in outlying rural communities. RACS would permit physicians to gain access to a comprehensive diagnostic database. The RACS system project proposal has a TELOS rating of 9, 8, 6, 8, and 9, respectively, and a PDM rating of 8, 10, and 9, respectively. RACS will require a network with dual mainframes, hard disks, magnetic tape for backup, and a printer. The dual mainframes serve as the central hub in the star network. Located in the northern, eastern, western, and southern sections of the state are four rural communities with small clinics that will contain workstations and printers and will be connected to the hub in Pleasant Town by leased telephone lines.

Required: Compute RACS TELOS feasibility factors and PDM strategic factors scores and determine its priority category.

THINK-TANK PROBLEMS

These problems call for a feasible approach rather than a precise solution. Although the problems are based on chapter material, extra reading and creativity may be required to develop workable solutions.

4.18 Senior management at Big Wheel Trucking Company, along with the newly hired chief information officer, have concluded that Big Wheel's essential business objective is to provide overnight delivery of processed meat to any point in the contiguous states. After further consideration and a great deal of brainstorming, senior management states that strategies necessary to accomplish the business objective include running trucks around the clock with scheduled maintenance and monitoring the whereabouts and status of each truck at all times. The business plan necessary to implement the strategies includes hiring and training professional long-haul drivers, and buying a fleet of heavy-duty tractor rigs and refrigerated vans.

Required: Develop two specific systems project proposals that you believe have high or super potential in supporting the strategies and business plan of Big Wheel. Develop one of your systems proposals to be the deployment of a satellite-based truck tracking system that promises strong productivity, differentiation of services, and management factors. For example, service is obviously a key to success in transportation. Trucks have to be on time. The new system will help ensure that trucks are at their destination on time and deliver on time. The truck tracking system will provide hourly position reports and status messages, such as a truck entering a weighing station or truck terminal. Truckers can also quickly check on new assignments and be driving toward their next destination without having to sit and wait for a phone call. This systems proposal seems also to have high feasibility factors. In preparing your Systems Project Proposal Forms, make any other assumptions that you deem necessary. Also, grade both proposals and prepare a priority grid and Systems Project Proposals Portfolio Statement.

4.19 Direct computer-to-computer electronic data interchange linkage among all trading partners, as shown in Figure 4.13, takes advantage of electronic communication speed and accuracy. The cost for such a connection is justified by high volumes of data and speed of delivery. Moreover, a clearinghouse component, like a full-service bank, can significantly reduce the traditional bookkeeping functions for all participants. Clearinghouses with expertise in EDI can pay freight bills and invoices, audit these payments, protect against duplicate payments, and reformat data and transmit them directly to subscribers' computers for reconciliation and management reports.

The obvious benefits of EDI are in the areas of productivity, timeliness and accuracy of information that differentiates services, and quicker reporting to management. Productivity is a key benefit. For example, it is inefficient planning to have a computerized accounting system spit out paper documents that are mailed to another company that rekeys these documents into another computerized accounting system. EDI reduces paper consumption and shuffling, cuts down on error rates, and enables companies to use their personnel far more wisely. With an EDI-based system, costs are low and accuracy is high. Simply put, EDI is a more efficient method of handling business transactions.

Figure 4.13 Computer-to-computer EDI.

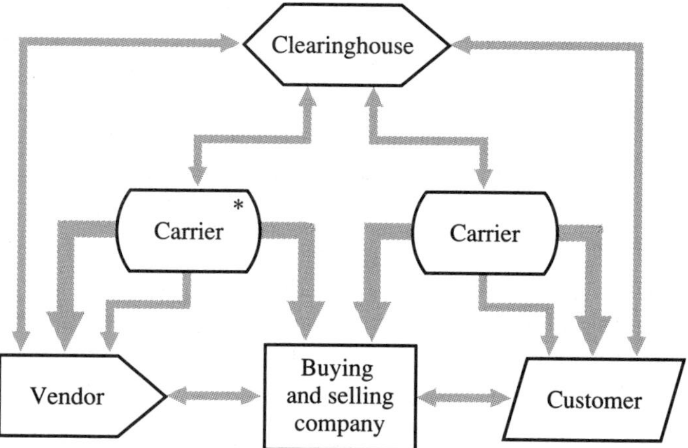

*A freight company that physically transports the goods.

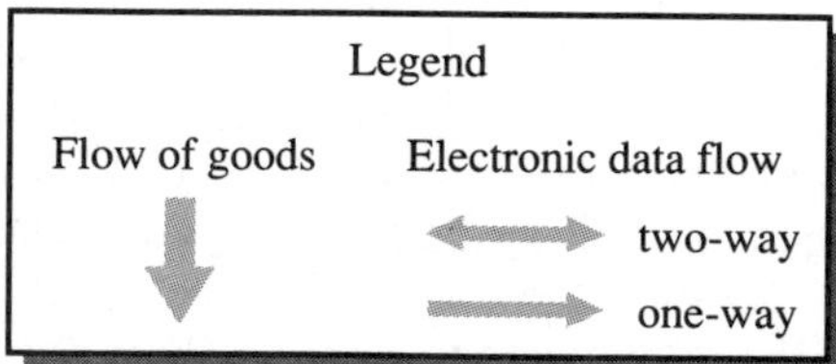

Required: Make any assumptions you deem necessary in completing the following assignments: Prepare a Systems Project Proposal form that includes the EDI application described in this problem, and prepare a Resource Requirements Matrix. Assign weights to the TELOS and PDM factors, compute their scores, and give the reason for the weight given each factor. Prepare a Systems Plan Report. Then prepare a Project Schedule Report. Include in these reports resources that you believe necessary to support the EDI application and their costs. Also, in the task schedule of the Project Schedule Report, list only three tasks. Again, make any assumptions that you deem necessary.

SUGGESTED READING

Below, Patrick J., George L. Morrisey, and Betty L. Acomb. *The Executive Guide to Strategic Planning.* San Francisco: Jossey-Bass Publishers, 1987.

Burch, John G. "Analysts of the World Unite!" *Information Strategy: The Executive's Journal,* Fall 1987.

Emery, James C. *Management Information Systems: The Critical Strategic Resource.* New York: Oxford University Press, 1987.

Gray, Paul, William R. King, Ephraim R. McLean, and Hugh J. Watson. *The Management of Information Systems.* Chicago: The Dryden Press, 1989.

Hicks, Jr., James O. and Wayne E. Leininger. *Accounting Information Systems,* 2nd ed. St. Paul, Minn.: West Publishing Company, 1987.

Karlof, Bengt. *Business Strategy in Practice.* New York: John Wiley, 1987.

Long, Larry E. *Design and Strategy for Corporate Information Services: MIS Long-Range Planning.* Englewood Cliffs, N.J.: Prentice-Hall, 1982.

Marchand, Donald A., and Forest W. Horton, Jr. *Infotrends: Profiting from Your Information Resources.* New York: John Wiley, 1986.

Moad, Jeff. "DuPont Seeks Global Communications Reach." *Datamation,* January 15, 1988.

Mula, Rose. "Trains, Planes and Automobiles: Automation Boom Drives Need for MIS." *Computerworld,* March 7, 1988.

Porter, Michael. *Competitive Advantage: Creating and Sustaining Superior Performance.* New York: Free Press, 1985.

Rifkin, Glenn, and Mitch Betts. "Strategic Systems Plans Gone Awry." *Computerworld,* March 14, 1988.

Synott, William R. *The Information Weapon.* New York: John Wiley, 1987.

Umbaugh, Robert E., ed. *The Handbook of MIS Management.* Pennsauken, N.J.: Auerbach Publishers, 1985.

Wiseman, Charles. *Strategic Information Systems.* Homewood, Ill.: Irwin,1988.

Introduction to JOCS: System Planning at Peerless, Inc.

Peerless, Inc. is a manufacturing company that produces hydraulic lifting equipment for deep-sea vessels. Peerless manufactures a product called the "hydronautical lifter" that is used to lift large ships for maintenance and cleaning when they are in dry dock. The company sells its equipment internationally; its customers include governmental entities such as military installations, deep-sea fishing conglomerates, oceanographic research institutes, and commercial shipping lines. Peerless management proudly states: "In ten years, every deep-sea port in the world will contain at least one hydronautical lifter."

The company has been in operation for seven years. It is considered to be a medium-sized manufacturing concern. It currently has total assets of $65 million, fixed assets of $46 million, a 25 percent market share, and gross revenues last year were $48 million. The company employs approximately 260 employees in one manufacturing location. All engineering, marketing, management, accounting, and manufacturing are done in the one location.

Hydronautical lifters are made to order based on the specifications for a given customer. The lifters usually take three months to a year to produce and range in price from $150,000 to $4.5 million for a single machine. Engineers designing the lifters, and manufacturing personnel producing the lifters, work closely with customers to make sure that the machines are created in accordance with the customer's specifications. A lifter that will not fit a given port, or lift the size ship in question, would be worthless to a Peerless customer.

Peerless is functioning in a market with some competition; there are four other manufacturers producing a machine that competes with the hydronautical lifter. Even though Peerless has only been in operation for seven years, the company has been able to compete very successfully with these other companies because Peerless management has stressed the importance of customer service and attention. Employees of Peerless strive to design and build high-quality lifters and still deliver them on time to the customers. This process requires coordination between all the departments at Peerless.

It is critical for Peerless employees to work closely with their customers to produce a machine that can be designed, engineered, and manufactured as quickly as possible. The business plan for Peerless states that "the number one strategic objective is to achieve a 35 percent market share within the next five years." However, management at Peerless is unwilling to sacrifice current profitability to attempt to buy market share. Management feels that by providing excellent service, a good product, and a competitive price they should be able to meet their objective while maintaining above-industry-average profitability ratios.

Kyle Bartwell, CEO of Peerless, is just beginning to understand the importance of integrating information systems into company operations. When the company began operations, a computer was purchased to assist in standard transaction-based accounting operations, including payroll, accounts payable, accounts receivable, and general ledger. Software was purchased to handle these functions. The hardware configuration consisted of a business-oriented multiuser computer capable of running 45 terminals and approximately 15 concurrent programs, two data processing quality printers, and 10 workstations. Kyle Bartwell did not see any need to employ a staff of information systems professionals. He simply hired a couple of data-entry people and gave responsibility for the system to his chief financial officer, Burt Flanders.

Burt Flanders directed the purchase of a manufacturing software package at that time. It is a process-costing manufacturing package based on a standard flow-type manufacturing system, where items are manufactured and placed in inventory before being sold to a customer. The system was purchased because a major computer vendor said it was a good package and would help Peerless management keep track of the manufacturing process. Unfortunately, the computer vendor did not understand that Peerless performs manufacturing, purchasing, and costing on a job basis. As a result, management has been trying to adapt the input data, as well as the reports and display screens generated by the software, to the needs of a job-oriented manufacturing operation.

Since the initial purchase of computing resources, many individual managers have purchased microcomputers, software, and other types of computers out of their departmental operating budgets. After seeing computers in the engineering, marketing, finance, purchasing, and customer service departments, Kyle Bartwell performed a survey to find out how much money was really being used to support computing at Peerless. He discovered that approximately 17 percent of gross revenue was being dedicated to computing, but less than a third of that figure was controlled directly by Burt Flanders.

Kyle Bartwell decided that it was time to form a steering committee to help consolidate and integrate current computer resources into the existing business plan. The members of the steering committee are shown in Figure 4.14. The first order of business for the committee was to recommend hiring a CIO to coordinate the company's information systems. Mary Stockland, an MBA with seven years' experience in information systems, was hired as CIO for Peerless and added to the steering committee depicted in Figure 4.14. Mary Stockland was given a month to acquaint herself with the systems in place at Peerless and to discuss new systems with her co-workers. After that time, a meeting of the steering committee was scheduled to take place to discuss the most urgent systems needs for the company.

While in the meeting of the steering committee, Mary Stockland says that she is satisfied with the operation of the current accounting system, but the manufacturing system is not meeting the needs of the company. When Kyle Bartwell asks for her opinion about an overall systems plan for Peerless, she responds: "I would like to be able to say we need two new minicomputers, 25 workstations, an Ethernet backbone, and four printers. Instead, I have

Figure 4.14
Steering committee members.

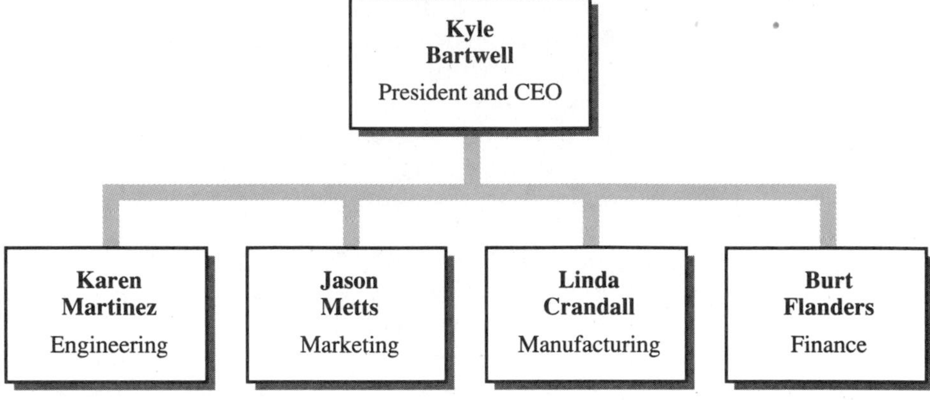

to tell you that we need to identify the kinds of applications we want to develop for Peerless. As a committee, we have to decide on the direction that our systems will take to support our organization. Before we discuss any hardware, we must prioritize our system needs." She identifies six new systems that are high priority for Peerless: a job costing system (JOCS); a marketing information system (MARKS); an engineering support system (ESS); a financial decision support system (FIDS); an EDI system for frequent vendor transactions (EDIS); and a companywide office automation system (COFAS). Mary Stockland feels that all six systems are critical installations to be completed within the next five years. However, it is her job to identify the systems that she feels must be completed within the next year. She provides systems applications proposals to the steering committee for only three of the systems above: COFAS, JOCS, and FIDS. Karen Martinez, director of engineering, is angry that ESS has not been rated in the top three systems. Mary Stockland attempts to explain her reasoning:

> We need to install systems immediately that will tie together our manufacturing, marketing, engineering, and financial operations. Right now, each department is running its own independent information systems department because each department has been unhappy with the information available. We can't continue to run separate information systems departments.

Karen Martinez agrees with this statement, but she says:

> Why can't we start with the engineering system? Engineering is the beginning of all our products.

At this point, Jason Metts jumps in and says:

> Engineering doesn't get to design any products until an order is placed. Marketing starts everything. Without an order, the rest of you don't do anything.

Mary Stockland tries to forestall further discussions about which department is most important by saying:

All departments are necessary for selling a good-quality product and making a profit. However, in computer information systems, some systems must be installed before other systems will work. We have to be able to control our day-to-day operations before we can expand and manage the rest of our company. We have to be able to manage before we can provide the data necessary for complex decision support systems. All six systems will be installed. But the first three I have identified have to be in place before we have the data to install the other three systems.

Mary Stockland distributes the three systems project proposals shown in Figure 4.15a to 4.15c to the steering committee. These proposals include Parts 1 and 2 shown in this chapter. Part 3 is completed during the steering committee meeting. Each is discussed in detail during the committee meeting. The office automation project, COFAS, receives a lower TELOS score than the other two projects, because of technological and economic limitations. COFAS would require a large hardware investment and would also require that many

Figure 4.15
(a) Systems Project Proposal for COFAS.

SYSTEMS PROJECT PROPOSAL

Part 1 (to be completed by requester)

Date submitted: 04/12/91

Request for: ☐ Modification of system ☐ Redesign of system ☒ New system

Submitted by: Mary Stockland Department: I.S.

Nature of request: Analyze and implement a companywide office automation system.

Reason for request: To create an office environment that will support easy intercommunication and shared access to office data.

Part 2 (to be completed by the CIO and staff)

Systems work required: ☒ Minor ☐ Moderate ☐ Major

Implementation required: ☒ Software ☒ Hardware ☒ Personnel

New resources required: ☐ Minor ☐ Moderate ☒ Major

Feasibility factors score: T 7 E 7 L 9 O 8 S 8 Score: 7.8

Figure 4.15
(b) Systems Project Proposal for JOCS.

SYSTEMS PROJECT PROPOSAL

Part 1 (to be completed by requester)

Date submitted: 04/12/91

Request for:
- [ ] Modification of system
- [ ] Redesign of system
- [X] New system

Submitted by: Mary Stockland   Department: I.S.

Nature of request: Develop a job order costing system.

Reason for request: To achieve a more effective means of controlling manufacturing costs and increasing productivity.

Part 2 (to be completed by the CIO and staff)

Systems work required: [ ] Minor [X] Moderate [ ] Major

Implementation required: [X] Software [X] Hardware [X] Personnel

New resources required: [ ] Minor [X] Moderate [ ] Major

Feasibility factors score: T 9 E 9 L 8 O 9 S 8   Score: 8.6

different computers be interconnected. JOCS and FIDS, on the other hand, could be installed with existing hardware technology and minimal additional equipment.

Mary Stockland, however, feels strongly that JOCS will not only provide information for day-to-day control of manufacturing operations, but will also provide the data necessary for the rest of the proposed application systems. She feels that JOCS is the most appropriate starting point for systems development since data from JOCS could be used as input to FIDS.

As a result, the JOCS project has the highest rating in strategic factors of the three systems. The three proposals are summarized in the Systems Project Proposals Portfolio Statement shown in Figure 4.16. JOCS would benefit all three strategic factors, while the other two projects are aimed primarily at increasing productivity and providing better management control. JOCS will

Figure 4.15
(c) Systems Project Proposal for FIDS.

SYSTEMS PROJECT PROPOSAL

Part 1 (to be completed by requester)

Date submitted: 04/12/91

Request for: ☐ Modification of system ☐ Redesign of system ☒ New system

Submitted by: Mary Stockland   Department: I.S.

Nature of request: Develop a financial information decision support system.

Reason for request: To achieve a more effective means of evaluating capital budgeting decisions and increasing the return received for invested funds.

Part 2 (to be completed by the CIO and staff)

Systems work required: ☐ Minor ☐ Moderate ☒ Major

Implementation required: ☒ Software ☒ Hardware ☒ Personnel

New resources required: ☐ Minor ☒ Moderate ☐ Major

Feasibility factors score: T 9 E 9 L 8 O 7 S 7 Score: 8.0

Figure 4.16
Systems Project Proposals Portfolio Statement for the three application projects for Peerless, Inc.

SYSTEMS PROJECT PROPOSALS PORTFOLIO STATEMENT

Application	Feasibility Factors					Score	Strategic Factors			Score
	T	E	L	O	S		P	D	M	
1. COFAS	7	7	9	8	8	7.8	8	2	7	5.7
2. JOCS	9	9	8	9	8	8.6	9	8	10	9.0
3. FIDS	9	9	8	7	7	8.0	8	5	9	7.3

achieve those two goals. It may also aid in product differentiation. Depending on the final design for the JOCS project, the project has the potential to:

1 Provide standards for products, thus enabling engineers to communicate from a much improved position of knowledge when discussing product design with customers and manufacturing personnel.

2 Keep accurate records for production, allowing manufacturing personnel to provide correct delivery dates for customers.

3 Increase product control helping management to charge the most competitive price for a product without lowering the bottom-line profits.

Mary Stockland's thoughtful analysis of the three projects convinces the other members of the steering committee to vote to forward the JOCS project to the next step of the systems development process. JOCS is the project selected to be scrutinized in depth during the next step of systems analysis.

Chapter 5

Systems Analysis

WHAT WILL YOU LEARN IN THIS CHAPTER?

After studying this chapter, you should be able to:

1 Relate the systems analysis phase to other phases of SDLC and discuss why the application of modern methodologies, modeling tools, and techniques helps to reduce consumption of system resources and increases the likelihood of successful systems development.
2 State the reasons for defining the systems scope and list three sources of study facts.
3 Describe the techniques used to gather study facts, especially those used to fill in gaps after JAD and prototyping sessions have been conducted.
4 Discuss the preparation and presentation of the Systems Analysis Report.
5 List and discuss alternative outcomes of the systems analysis phase.

INTRODUCTION

For an overview of where we are relative to SDLC and subsequent chapters, look at Figure 5.1. The subject of this chapter is systems analysis, the second phase of SDLC.

Systems analysis provides the foundation for the subsequent SDLC phases. The systems analysis phase investigates and begins to answer many questions about the new system. Major tasks of systems analysis include the following:

- Establishing the systems scope
- Gathering study facts
- Analyzing study facts
- Communicating findings via a **SYSTEMS ANALYSIS REPORT**

STUDY FACTS are pieces of information that reveal realities, situations, and relationships that warrant careful analysis and modeling. Study facts come from several sources which we discuss throughout the chapter.

Remember that the SDLC does not normally progress in linear fashion; that is, one complete phase leads to the next phase, and so on. Often, it is necessary to go back to a previous phase if continuing to the next phase would result in improper analysis, design, or implementation. It is certainly common

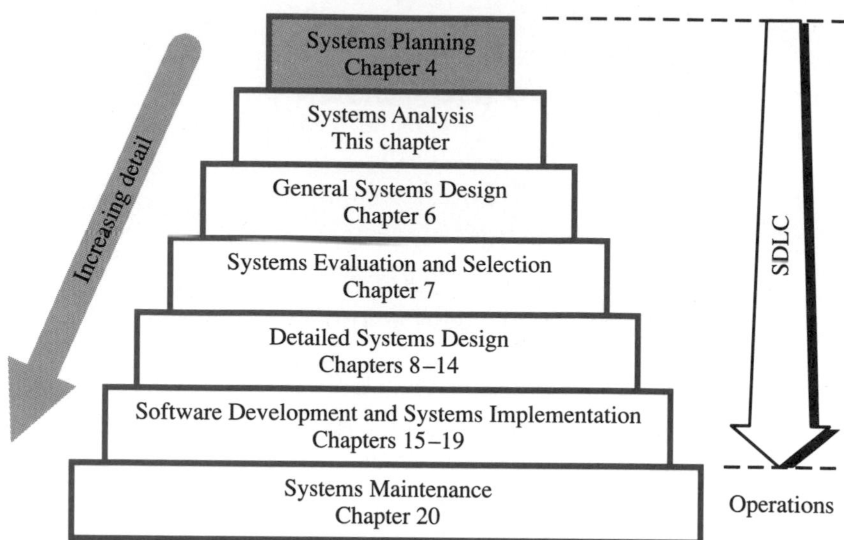

Figure 5.1
SDLC phases and their related chapters in this book.

for systems professionals to return to the analysis phase several times before completing the general systems design phase, because these two phases are highly interrelated. This recursive process of iterating and revising is normal in most systems projects.

WHAT IS THE RELATIONSHIP BETWEEN SYSTEMS ANALYSIS AND SYSTEMS DEVELOPMENT?

After systems planning, systems professionals begin development for specific systems projects. Therefore, now is a good time to put everything that supports development of systems projects into perspective.

The Dangers of Using Outmoded Methods to Develop Systems

Consumption of systems resources often resembles an iceberg, as portrayed in Figure 5.2. The visible portion is systems development, and the much larger submerged portion is systems maintenance. In some companies, systems maintenance may be consuming as much as 80 to 90 percent of systems resources. This unfortunate situation means that only 10 to 20 percent of systems resources can be used for systems development.

One of the major reasons for this **ICEBERG EFFECT** is the use of outmoded methodologies, modeling tools, and techniques. Outmoded methods typically include an undefined or poorly defined systems development methodology; old-fashioned flowchart templates, and employment of pencils, erasers, scissors, tape, and large sheets of paper. Using such methods often means that systems analysts spend weeks or months constructing flowcharts that are at best difficult to follow.

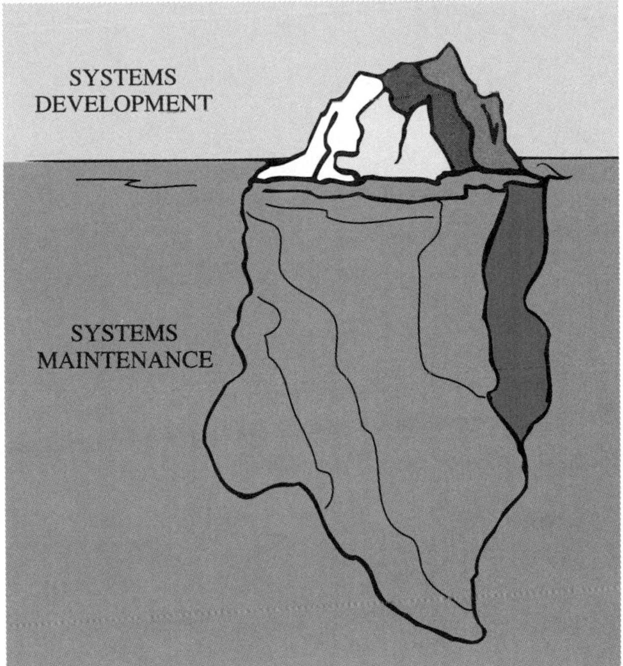

Figure 5.2
The systems "iceberg."

All too often, using outmoded methods produces systems that are:

- Difficult to maintain
- Unusable without major upgrades made after implementation
- Unreliable
- Difficult to extend to meet expanding user requirements

Systems people have long been searching for ways to respond to end-user requirements and develop systems effectively and efficiently. Modern techniques that represent the results of this search are presented in Figure 5.3.

Using Modern Methods to Reduce the Iceberg Effect

Figure 5.4 illustrates the consumption of systems resources using outmoded methods compared to consumption of systems resources using modern techniques. As the figure shows, modern methodologies, tools, and techniques require fewer systems resources for development and maintenance. For example, JAD promotes user participation, fostering better and more rapid definition of user requirements. CASE technologies automate use of modeling tools and software coding, and once the system is converted to operations, systems maintenance is easier to perform.

It is unlikely that any one company will use all of the modeling tools presented in Figure 5.3. For example, company A may use structured English

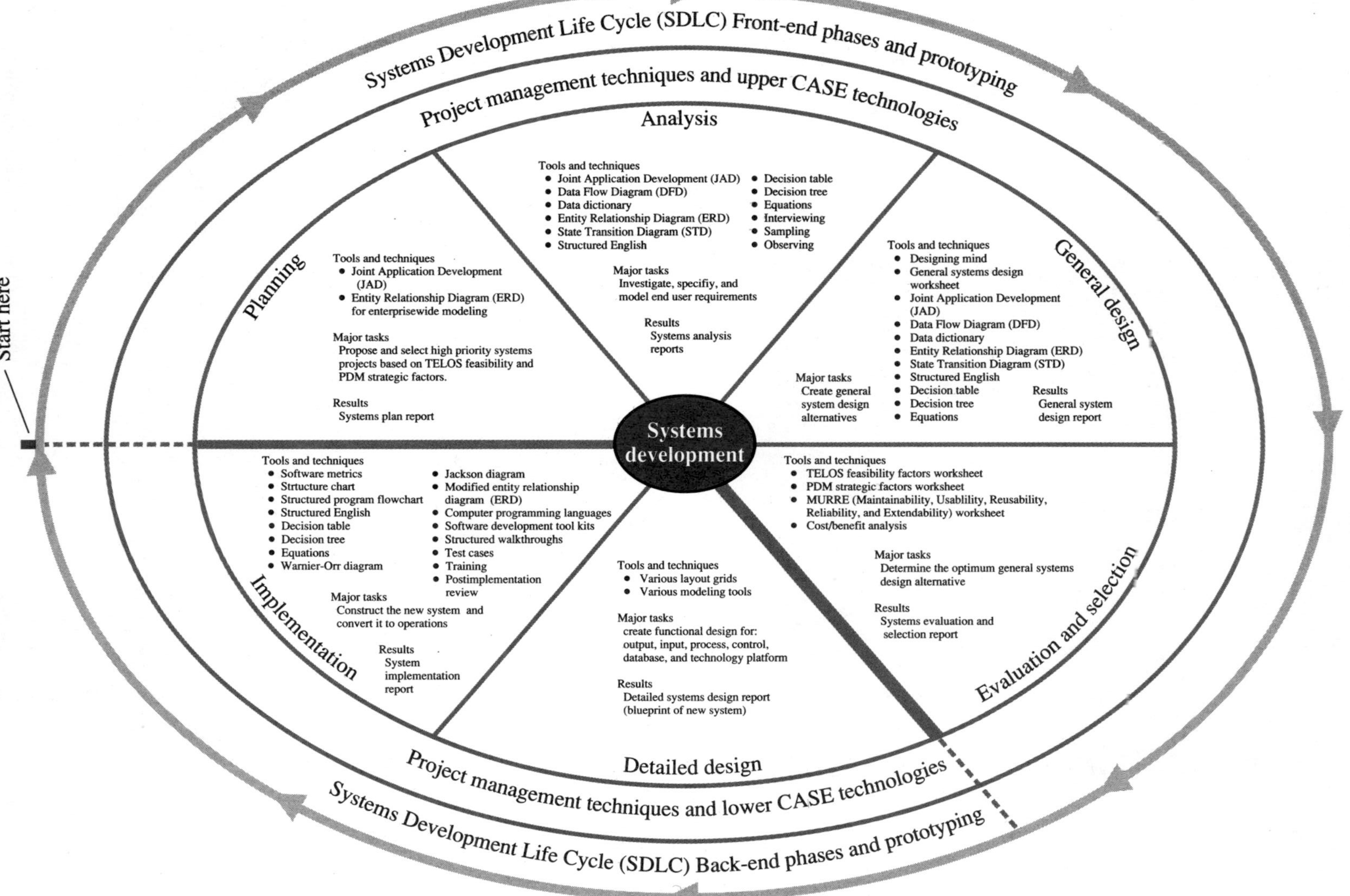

Systems Development Life Cycle (SDLC) Front-end phases and prototyping
Project management techniques and upper CASE technologies
Analysis
Tools and techniques
• Joint Application Development (JAD)
• Data Flow Diagram (DFD)
• Data dictionary
• Entity Relationship Diagram (ERD)
• State Transition Diagram (STD)
• Structured English
• Decision table
• Decision tree
• Equations
• Interviewing
• Sampling
• Observing
Major tasks
Investigate, specifiy, and model end user requirements
Results
Systems analysis reports
Planning
Tools and techniques
• Joint Application Development (JAD)
• Entity Relationship Diagram (ERD) for enterprisewide modeling
Major tasks
Propose and select high priority systems projects based on TELOS feasibility and PDM strategic factors.
Results
Systems plan report
General design
Tools and techniques
• Designing mind
• General systems design worksheet
• Joint Application Development (JAD)
• Data Flow Diagram (DFD)
• Data dictionary
• Entity Relationship Diagram (ERD)
• State Transition Diagram (STD)
• Structured English
• Decision table
• Decision tree
• Equations
Major tasks
Create general system design alternatives
Results
General system design report
Start here
Systems development
Implementation
Tools and techniques
• Software metrics
• Strtucture chart
• Structured program flowchart
• Structured English
• Decision table
• Decision tree
• Equations
• Warnier-Orr diagram
• Jackson diagram
• Modified entity relationship diagram (ERD)
• Computer programming languages
• Software development tool kits
• Structured walkthroughs
• Test cases
• Training
• Postimplementation review
Major tasks
Construct the new system and convert it to operations
Results
System implementation report
Detailed design
Tools and techniques
• Various layout grids
• Various modeling tools
Major tasks
create functional design for: output, input, process, control, database, and technology platform
Results
Detailed systems design report (blueprint of new system)
Evaluation and selection
Tools and techniques
• TELOS feasibility factors worksheet
• PDM strategic factors worksheet
• MURRE (Maintainability, Usablility, Reusability, Reliability, and Extendability) worksheet
• Cost/benefit analysis
Major tasks
Determine the optimum general systems design alternative
Results
Systems evaluation and selection report
Project management techniques and lower CASE technologies
Systems Development Life Cycle (SDLC) Back-end phases and prototyping

Figure 5.3 Methodologies, modeling tools, and techniques for systems development (*Source:* John Burch and Fritz Grupe, "Methods and Tools for Upgrading Systems Development," *Information Executive,* Summer 1991, p. 51.)

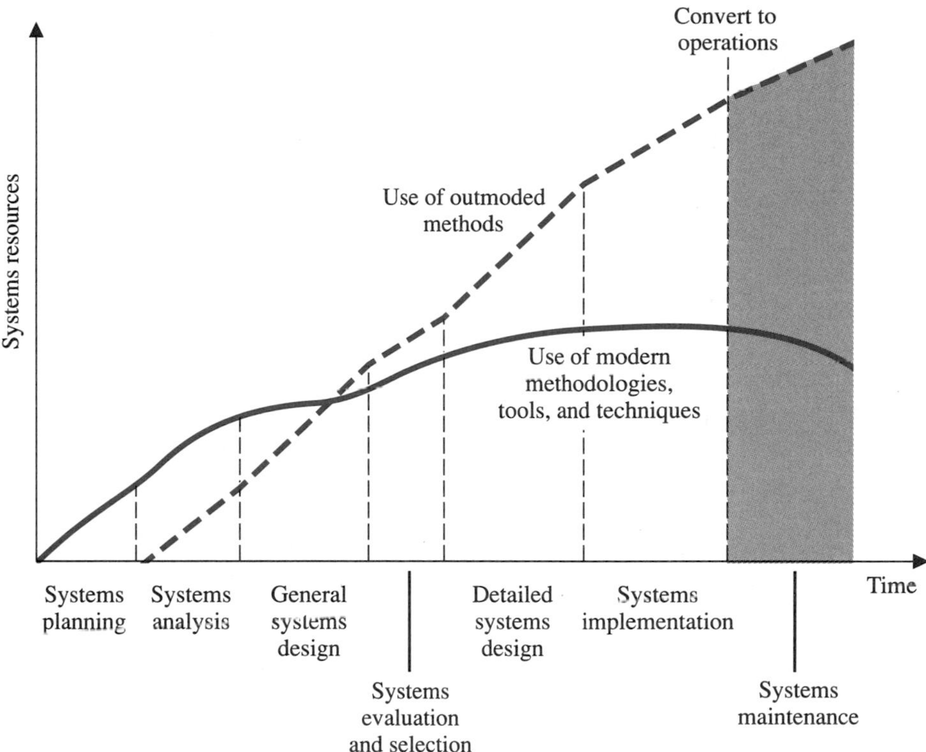

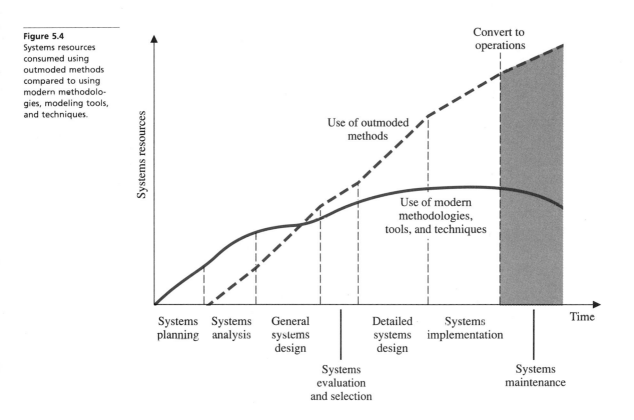

Figure 5.4
Systems resources consumed using outmoded methods compared to using modern methodologies, modeling tools, and techniques.

instead of a decision table. Company B may use a structured program flowchart for software design and documentation, whereas company C may use a structure chart. Some companies, on the other hand, may try to use a purported all-in-one tool, such as a Jackson diagram or a Warnier–Orr diagram. Also, some companies employ teams of **SPECIALISTS WITH ADVANCED TOOLS (SWAT) TEAMS**. SWAT teams are composed of highly motivated systems professionals who are skilled in most, if not all, of the modern methodologies, modeling tools, and techniques. Such teams develop high-quality systems in a fraction of the time it takes the average systems project team. So the combination of modern methodologies, modeling tools, and techniques in the hands of highly motivated, skilled systems professionals can increase systems development productivity, systems quality, and reduce, if not eliminate, the iceberg effect.

Applying Modeling Tools and Techniques in a Sample Case

The chief objective of the systems analysis phase is to gather study facts and analyze them in order to define user requirements. JAD is an excellent technique to get users involved in systems development and to gather study facts and define user requirements. Modeling tools, such as data flow diagrams and decision trees, help model these requirements. The Dillinson's case reveals how JAD and modeling tools are used.

Systems Analysis at a Major Department Store Chain: Dillinson Discovers JAD

Dillinson's Department Stores is a national chain of 43 retail units, most located within major shopping malls and centers. The stores vary in age, the oldest being the original store in Chicago, which was built in the mid-1920s, while the newest is now under construction in San Jose, California. Over the past two decades, information systems have developed in an almost piecemeal fashion throughout the chain. There are as many different data and reporting standards among these systems as there are different stores, and the task of tying all the systems together has been considered virtually impossible. In fact, all reporting from the individual units to Dillinson's head office in Chicago is still done on paper via overnight courier, much as for the last 50 years.

Jim Montecucco, CIO at Dillinson's head office, feels the time is right to standardize the chain's information system across the board. Rather than attempting to connect each unit's aging and nonstandard system into the head office's reporting network, he envisions developing one standard, state-of-the-art system and implementing it throughout the chain. Jim and his systems managers have already completed a good deal of systems planning, including the development of an enterprisewide model and a generalized information systems model. They're still at a general, conceptual level of the systems development life cycle, but Jim and his managers, Brenda Clemens and Clarence Jankowski, feel they're ready to move on to the systems analysis phase. Here, they'll take a closer look at the individual information needs of the individual retail units as well as those of the head office.

Brenda scheduled a meeting for the following Monday with herself, Jim, Clarence, and Stephen Wells, Dillinson's CEO, in attendance. Brenda knows that Stephen had been director of Dillinson's Atlanta store before coming to the head office and had been instrumental in developing that store's information system some 15 years ago. She worried that he might bring some preconceived notions to the meeting about how the current systems project should proceed, even though Atlanta's system was viewed as one of the least successful of those in the chain. While she is firmly committed to the systems development life cycle (as are Jim and Clarence), Mr. Wells may require convincing. In particular, Brenda is very interested in trying some innovative methods and tools that she's been reading about in the systems journals, and she's going to need his support.

Last Monday the four met in a small room adjoining Mr. Wells' office. Stephen was quiet at first, listening as the rest of the group discussed the results of the systems planning phase and presented him with a copy of the Systems Plan Report. Then the discussion turned to the problems the group was facing as it moved into the systems analysis phase, in particular the difficulty of satisfying the information needs, with one integrated system, of 43 stores that had heretofore been served by as many unique systems. Still wondering what was going through Stephen's mind, Brenda said: "I've been reading about a technique called JAD, or joint application development, which seems ideal for us in this project." Clarence piped up. "Yeah, I've heard of that. Doesn't JAD make use of CASE technology?"

Finally, Stephen spoke. "Brenda, what is this JAD you keep referring to? And what on earth is CASE? You know, I did that system down in Atlanta a few years back, but I sure don't recall any of these terms you're using." This was what Brenda was afraid of, but she continued. "Well, joint application development is a kind of workshop, a means of gathering systems user expertise and capturing design specifications in this phase of systems development. I think you can see the tremendous task we face here,

Mr. Wells. Bringing knowledgeable users in from each of our retail units for a JAD session would seem to be the best way to get a handle on things." Before Stephen could reply, Clarence announced, "Not only that, but CASE, computer-aided systems engineering, can be used during a JAD session to capture and simulate options the JAD participants come up with! The participants can discuss their requirements, a few ideas can be tossed around and modeled on the CASE system, then everyone gets to see a prototype firsthand and evaluate the idea. Not only that, but the right CASE system will even write the software code once a particular idea is agreed upon!"

Stephen said, "It seems as though things have come a long way since the project in Atlanta. I can see where some planning and analysis could have created a much better system back then, but I can see that this new project's in good hands. I agree with you, Brenda, that a JAD workshop should be part of your systems analysis efforts. And Jim, see to it that Clarence selects the CASE system you'll be using. Anybody that's as excited about something as he is must know what's going on."

Brenda breathed a sigh of relief. Her fears had been unwarranted, and it looked as though there was full support for this project from the highest levels of Dillinson's management. Everything she'd ever read about systems development stressed the importance of that.

That afternoon, Brenda and Jim made travel arrangements for information systems users from all the Dillinson's retail stores, including managers, lead salesclerks, and computing people. Jim was careful to select participants who could communicate well, who had good business knowledge, and who had some authority. A handful of store directors also chose to make the trip to Chicago, and Clarence managed to secure consulting services from three local systems professionals. They would all converge in three weeks for a JAD workshop at the Great Lakes Resort and Conference Center just outside the city. Brenda envisioned herself, Clarence, and Jim, along with the end users from each store, as the JAD participants. The store directors, along with the consultants, would be observers. She planned to retain one consultant as the JAD facilitator. Clarence suggested that one of his technicians, Darrell Laananen, should serve as the JAD scribe due to his CASE expertise. Brenda then created a layout for the JAD room which would accommodate everyone, as shown in Figure 5.5.

Three weeks later, the group met at the resort for the JAD workshop. Jim addressed the small crowd seated in the conference room. "Many of you here today are probably wondering just what this JAD session is all about and what we hope to accomplish with it. Brenda here has been instrumental in bringing us all together, and we're hoping to give this technique our best shot." Jim was interrupted by Dick Logan, director of Dillinson's Sioux City, Iowa, store. "Jim, forgive me for being blunt, but why are we wasting our time with all this arm waving hocus-pocus? Aren't you and your people in charge of developing the new system? Why can't you just build the system, then show us all how to use it once you get it done?"

Brenda turned to Jim, "May I handle this one? Dick, this workshop is designed to provide us with alternatives. We have some idea of the information needs at each of our stores, but our first order of business today will be to confirm with each of you what your needs are. From there, we can begin to generate some alternative designs and see how they might work." Dick seemed unconvinced, so Clarence interjected. "Think of the process of systems development as being akin to the construction of a custom home. The architect can do some preliminary work, like looking at the site the home will be built on and determining about how many square feet it will be, before he or she ever meets the customer. Well, we've gone about that far with this information system now that the systems planning phase is complete. The architect would now sit down with the client and discuss every aspect of the home's design, to determine just what the cus-

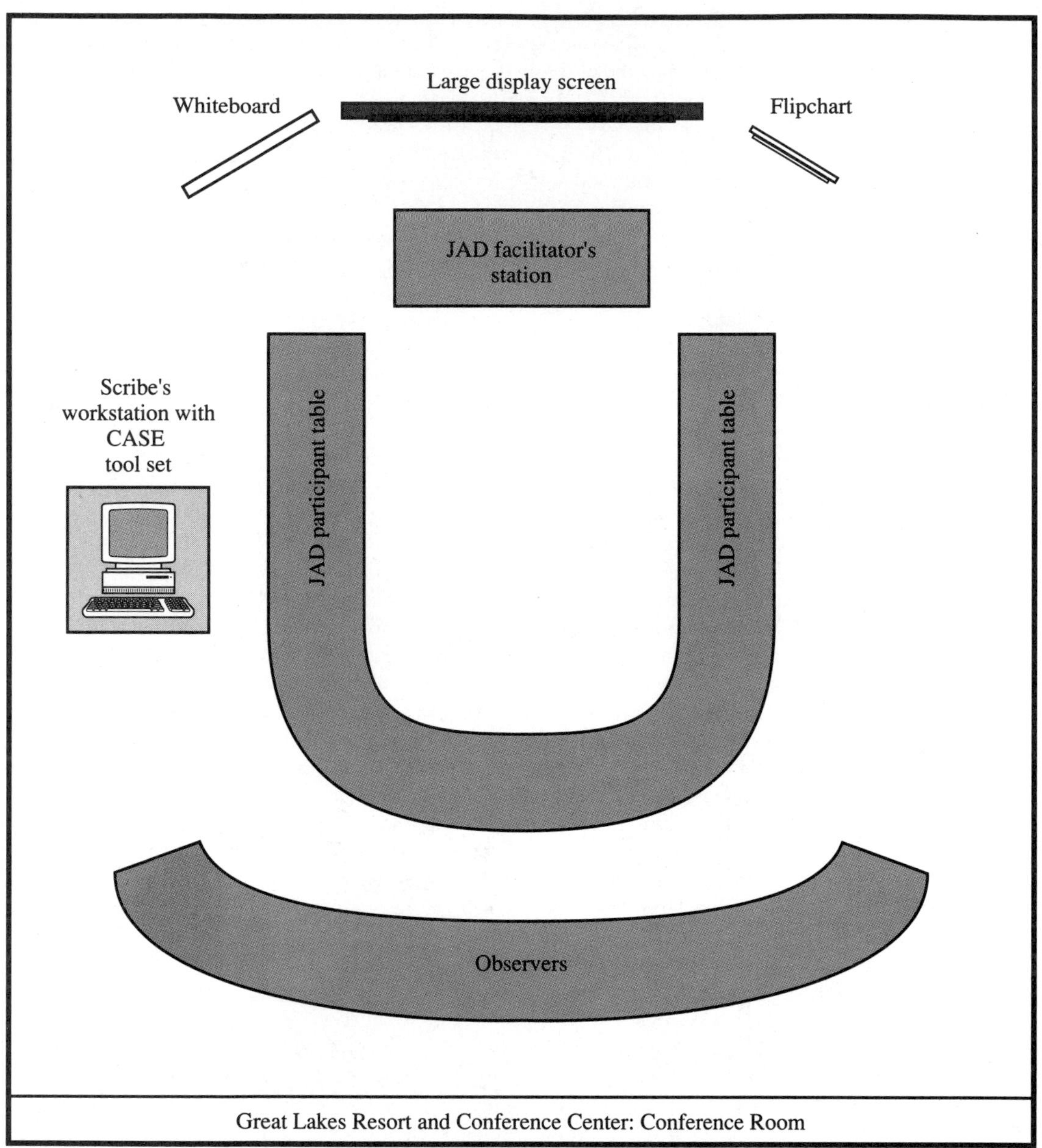

Figure 5.5
JAD workshop room layout for Dillinson's Department Stores.

tomer wants. Then the architect would come back with design alternatives and models to show the client. We're fortunate here in that we'll be able to show you alternative designs almost instantaneously. Darrell will be building models for us on this CASE system as we speak, so we'll be able to see right away how an alternative might look and work."

Jim spoke: "Correct me, Brenda, if I'm wrong, but this JAD session will result in numerous design alternatives from which we'll be able to select an optimal configuration. This is just the front end. After the workshop is over, my team still faces many months of detail work to bring our new system to the point where it can be implemented. I suppose, going back to the architect, all that detail work is analogous to the creation of blueprints for the custom home. Only then can construction begin!"

Dick Logan nodded his head, apparently in understanding, and it seemed as if everyone in the room was now clear on the purpose of the JAD workshop. Over the next two days, the JAD sessions went smoothly. A particular success was the full-featured CASE tool set that Clarence had selected, and Darrell was adept at bringing the numerous prototypes and models to life on the large display screen. One of the first models to be created using this dynamic presentation method was a data flow diagram. The systems consultant who served as the JAD facilitator really helped the group put the DFD together, while the CASE system made it easy to build the context-level DFD on-screen for all to see. From there, Darrell was able to decompose this context-level diagram into second and third levels of detail, as shown in Figure 5.6.

Later, the focus of the group turned to user interfaces. The information systems end users from the various stores were really able to help in this regard, because they'd all been using their own systems and knew what they liked and didn't like about each of them. One user interface the group explored was a sales-entry system used by clerks out on the sales floors. The sales-entry system is a time-dependent one, where the system waits for input from the clerk, then moves on to additional screens and waits for other information.

Agnes Goldstein from the Newark, New Jersey store had nothing but complaints regarding the sales-entry system her store had been using. "Our system is so confusing.

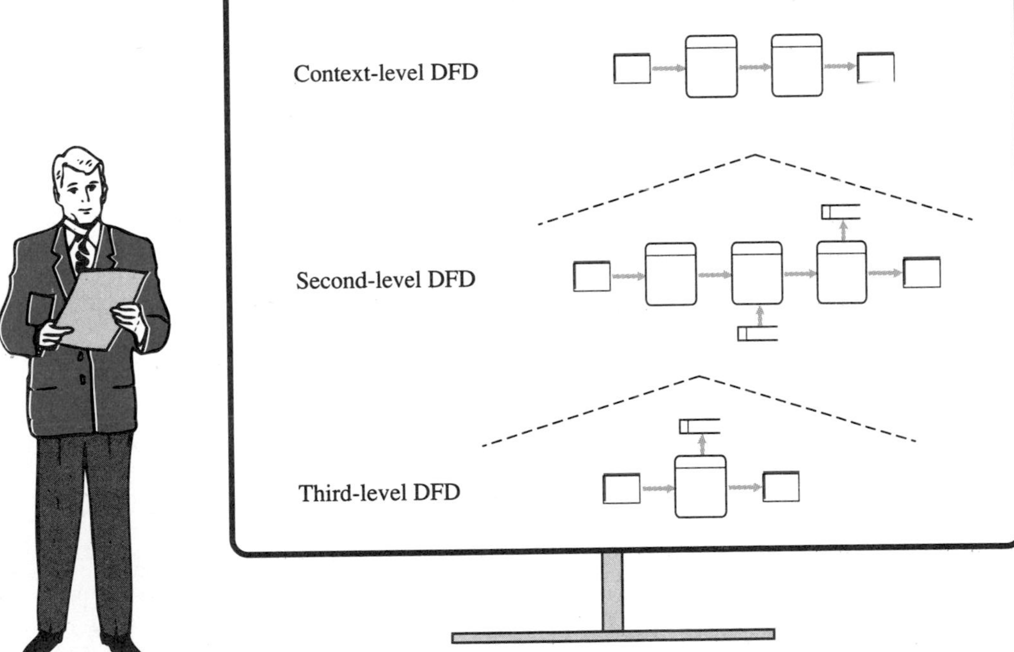

Figure 5.6
Data flow diagram generated by a full-featured CASE package at Dillinson's JAD workshop.

It gives me instructions when I first log-on to the terminal, but then the screen goes blank and I'm expected to remember in what order everything gets entered. Sometimes the system's so slow to respond that I forget which items I've entered and which still need to be keyed in."

Using Agnes' suggestions and those from other JAD participants, Darrell used the CASE tool set to put together a state transition diagram for the sales-entry system. Now everyone could see how the system would respond to a salesclerk's input in a time-dependent manner, while Jim and his people were able to see how they could synchronize sales-entry events and keep the system from getting bogged down when it got busy on the sales floor. The STD the group developed is shown in Figure 5.7.

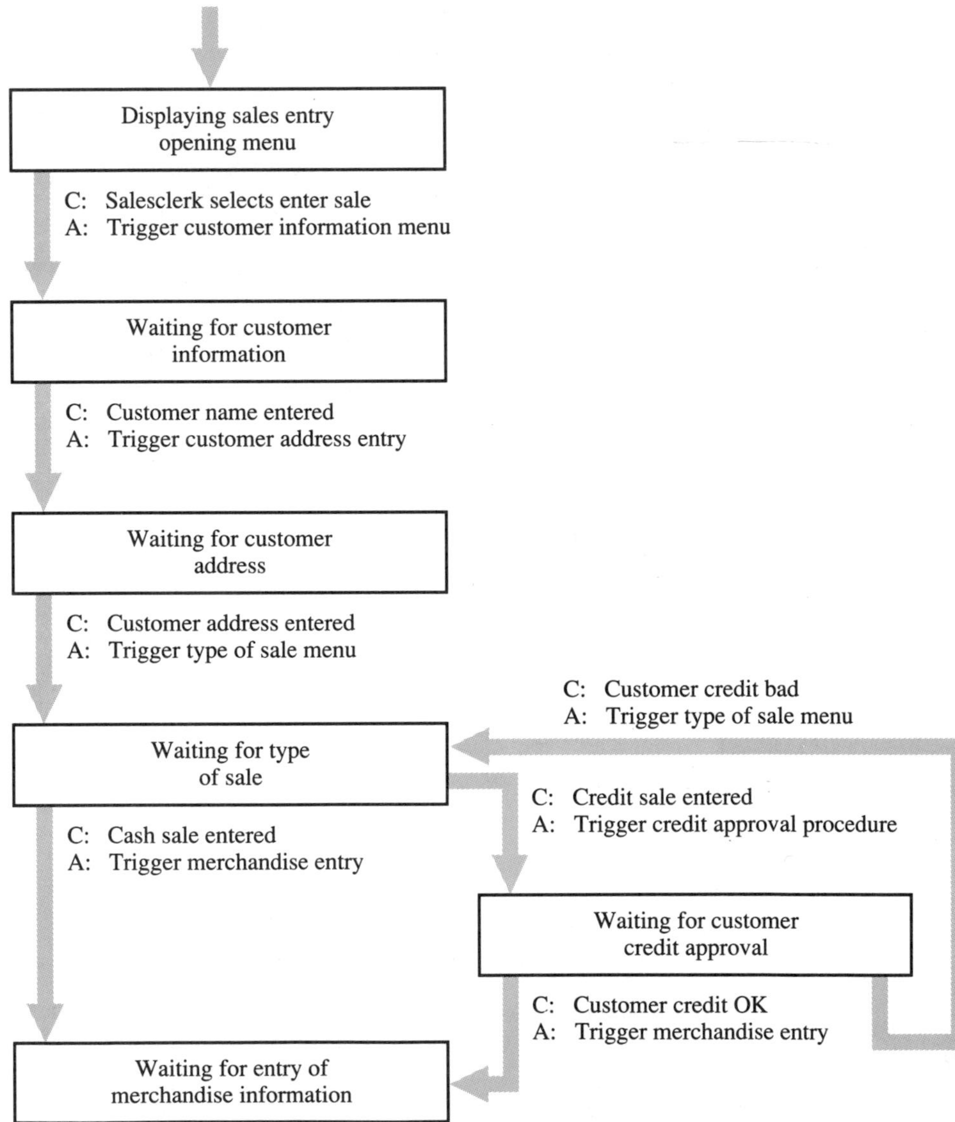

Figure 5.7
State transition diagram for a sales-entry system at Dillinson's Department Stores.

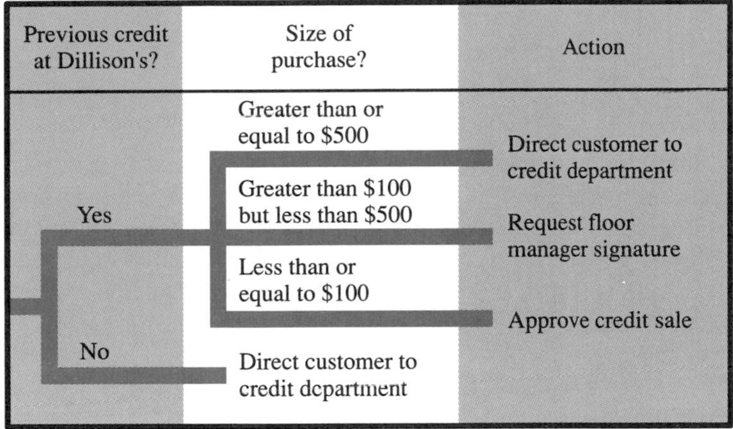

Figure 5.8 Decision tree for Dillinson's Department Stores customer credit approval.

One problem encountered at the workshop as the group discussed the sales-entry system was a difficulty in specifying the decision processes used. Brenda remarked, "I know we're still working at a general analysis level and that we're more concerned with the 'what' than with the 'how' of the sales-entry system. But a decision tree might be just what we need to clarify the decision logic of this system. You know, some of our older units aren't currently using an automated sales-entry system, and I think the folks here from those stores are having trouble understanding what we're proposing. Darrell, can you generate for us a decision tree showing the logic our prototype sales-entry system would use?"

Darrell did. Within a few minutes a decision tree appeared on the large screen at the front of the room, as shown in Figure 5.8. Agnes took one look at the diagram and exclaimed, "Oh! Now I understand what Clarence was talking about earlier with charge versus cash purchases. I didn't realize the system would automatically check for charge authorization, then base the decision on that." Brenda was surprised to find how well the decision tree conveyed this context-level process, because she'd never used this type of diagram for analysis before—only for more detailed design work. She noted, "You know Jim, I guess there's more of a blur between analysis and design than I thought!"

The remainder of the JAD workshop also went smoothly. Some of the participants even stayed on for a couple of days to enjoy tennis or boating at the resort, but all went away feeling that JAD had solved a common problem in systems development—the lack of effective communication between the end users of a system and the systems professionals who develop it. Jim and his team were especially pleased, for the new system now had support at all levels of Dillinson's organization, from the CEO to the salesclerks.

DEFINING THE SCOPE OF THE NEW SYSTEM AND GATHERING THE INFORMATION THAT AFFECTS IT

To perform their work, systems analysts need to determine the **SYSTEMS SCOPE** of the new system and gain a great deal of information, also termed study facts, about a number of things concerning it. Three sources of study facts are:

- The current system
- Other internal sources
- External sources

What Should the New System Include?

The activities and events comprising systems analysis are for the most part directed toward answering the question, "What is the new system to include?" In many cases this question can be more accurately phrased as: "What more is the current system to include?" In answering these general questions, the analyst must address the following specific questions:

- What information is needed?
- Who needs it?
- When is it needed?
- In what form is it required?
- Where does it originate?
- When and how can it be collected?

Moreover, the scope of systems analysis can vary widely in terms of duration, complexity, and expense. Consequently, the scope must be defined somewhat arbitrarily at times to meet constraints such as time and cost. The primary problem for both the novice analyst and the skilled systems professional is the need to convert (unconsciously) an instruction such as: "I want a daily purchasing report" into "Develop a new purchasing and inventory management system." A context-level data flow diagram (DFD) is an excellent tool for defining the scope of systems analysis.

In practice, an analyst who fails to define the scope of systems analysis properly either fails to achieve objectives or achieves them at a great loss of both time and money. It must be understood, however, that unduly restrictive constraints on the scope of the analysis limit the potential solutions and recommendations that result from the analysis. As a rule, the first definition of scope, as well as any given objectives and constraints, are subject to redefinition at a later date, depending on findings during analysis.

It is imperative to get a time commitment from users. Users are typically very enthusiastic to participate until they see how much time it actually takes away from their jobs.

What Are the Pros and Cons of Analyzing the Current System?

An analyst is not always provided with an opportunity to develop a system from scratch. In many cases a system or subsystem exists that serves the organization. As a result, the analyst is confronted with the following types of decisions:

- What role does the current system have with respect to the new system?
- Is the current system any good? Good enough?

- Should I analyze and model the current system?
- If so, what subsystems in the current system should I analyze?

Advantages

Several advantages exist for analyzing the current system. Studying and modeling the current system provides an opportunity to determine whether that system is satisfactory, is in need of minor repair, requires a major overhaul, or should be replaced. Analyzing and modeling the current system can also provide the analyst with an immediate source of design ideas to help the analyst to identify the resources available for the new system. When the new system is implemented, the analyst is responsible for having identified what tasks and activities will be needed to phase out the current system and begin operating the new system. To identify conversion requirements, the analyst must know not only what activities will be performed, but also what activities were performed. Studying and modeling the current system gives the analyst the answers.

Disadvantages

Some authorities do not approve of analyzing and modeling the current system. Doing so introduces a problem known as the "current physical tar pit."[1] Systems analysts begin analyzing and modeling the current system, but like the unlucky dinosaurs that wandered into the LaBrea Tar Pits, soon find themselves trapped in a modeling morass of the current system instead of modeling the system that is to replace it. Managers typically do not look favorably on such an unrewarding undertaking and usually cancel systems projects that spend too much time in the current physical tar pits.

There are, indeed, several disadvantages connected to analyzing the current system. In situations where the new system is unique or totally different from the current system, studying it may be misleading. At worst, the current system is probably irrelevant, and analyzing it may be a waste of time and money. As one manager asked, "Why analyze and model a system that's going to be thrown away?" In such situations it makes more sense to identify what is required from the new system than to rehash what's wrong with the current system. Also, an analysis of an existing system may result in unnecessary barriers or artificial constraints being included in the design of the new system. For example, in the current system, in a given department, there may be a document flow and a series of actions taken with that document. The analyst can become so involved with improving those actions that the involvement of the department and the value of the document are left unquestioned. The more familiar an analyst becomes with a system, the more likely it is that some perspective or objectivity concerning it will be lost.

Gathering Information from the People Who Will Use the System

The single most important source of study facts is the people who are going to use the new system. End users range from novice to skilled with respect to

[1] Warren Keuffel, "House of Structure," *Unix Review,* Vol. 9, No. 2, February 1991, p. 36.

their technical knowledge, sometimes called syntactic knowledge. But regarding knowledge about their business and job tasks (semantic knowledge), they are the experts. To develop a workable system, systems analysts gather semantic data (sometimes referred to as macrospecs) from end users. This semantic data will guide the systems analysts during analysis and design. Later, analysts and designers will discuss technical data with systems technicians, such as programmers, to convey detailed systems specifications, sometimes referred to as microspecs or minispecs. Thus systems analysts must communicate with end users on one side who have semantic knowledge and, on the other side, with systems technicians who have syntactic knowledge.

A secondary source of study facts for the analyst lies in the existing paperwork or documents within the organization. The documents in most organizations can be classified as anything which describes how the organization is structured; what the organization is or has been doing; and what the organization plans to do. Figure 5.9 shows a partial list, by types, of some organization documents that are useful to analysts while conducting analysis.

Gathering Information from Sources Outside the User Organization

Reviewing other information systems outside the organization can reveal practical ideas and techniques. Many industries form groups and sponsor seminars that freely exchange information systems experiences and recommend better ways of doing things.

If systems analysts are not familiar with the company's business and operations, they can often relate them to other, more familiar systems projects on which they have worked. Significant differences in systems applications across organizations are actually rare. Developing a network for a hospital complex with remote clinics may be similar to developing a network for a hotel chain.

Figure 5.9
Various types of documents available to the analyst in an organization from which study facts may be obtained pertaining to systems analysis.

Documents Describing How the Organization Is Organized	Documents Describing What the Organization Does	Documents Describing What the Organization Plans to Do
Policy statements	Financial statements	Mission statement
Methods and procedure manuals	Performance reports	Business plan
Organization charts	Staff studies	Budgets
Job descriptions	Historical reports	Schedules
Performance standards	Transaction files (including purchase orders, customer orders, invoices, expense records, customer correspondence)	Forecasts
Chart of accounts	Legal papers (including copyrights, patents, franchises, trademarks, judgments)	Corporate minutes
		Government regulations

An inventory management system works pretty much the same whether one is referring to items on shelves, houses on streets, or rooms in a building. To realtors, the houses and property they have for sale are their inventory. To hostlers, rooms in hotels are their inventory. If analysts relate differences and similarities of a system they do know to one that they do not know very well, they will increase their ability to figure out what is really required.

TECHNIQUES FOR GATHERING ADDITIONAL INFORMATION

JAD sessions are a wellspring of study facts, as are the sources just presented. However, after using these sources, some gaps may still exist. Three techniques help fill these gaps:

- Interviewing
- Sampling
- Observing

Interviewing

INTERVIEWING is probably the most common and effective technique used during systems development.

What is Interviewing?

An interview is an exchange of information between an interviewer (i.e., the systems analyst) and interviewee (i.e., the end user). It is planned, and it has a specific purpose. It consists of asking and answering questions. As used in systems analysis, it is a feedback mechanism and a prime way to gather study facts and fill in the gaps.

What Types of Questions Should Be Asked?

Two basic types of questions exist: open-ended and closed-ended.

Open-Ended Questions These questions are neutral and nonrestrictive. They permit interviewees freedom in answering questions, and they encourage them to disclose information previously unknown to the interviewer. The nonrestrictive nature of open-ended questions means, however, that the direction and progress of the interview may be controlled by the interviewee's answers rather than the interviewer's questions. Also, responses to open-ended questions may result in the disclosure of irrelevant information. An example of an open-ended question is: "What are your feelings about changing from paper forms to electronic forms?"

Closed-Ended Questions These questions, on the other hand, are specific and provide the interviewer more control over the direction and progress of the interview. Closed-ended questions, however, are limiting in that they usually achieve only what they ask and therefore prevent interviewees the freedom to open up and reveal relevant information that is unanticipated by the interviewer.

The greatest drawback to using closed-ended questions is their tendency to be loaded or leading. The best way to guard against asking loaded or leading closed-ended questions is to avoid either/or questions that make assumptions, and statements with question endings, such as "could you?" and "isn't it?" An example of an either/or question is: "Should we use a PC-based network or a departmental computer?" Maybe the interviewee doesn't agree with either and wants to present other alternatives. An example of a question that makes an assumption is: "When was the last time that you turned in your report late?" Maybe the interviewee has never turned in a late report. An example of a statement with a question ending is: "You agree with this report format, don't you?" The interviewer is trying to lead the interviewee, and clearly only one response is acceptable. The interviewee must take pains to ensure that closed-ended questions are not loaded or leading and are therefore neutral.

Questions can be divided into two categories: primary and secondary. Both can be open- or closed-ended.

Primary Questions Primary questions address a specific topic. They are planned. Like all interview questions, they should be neutral.

Secondary Questions Secondary questions are follow-up or probing questions designed to elicit more information than was given in response to a primary question. Whereas primary questions are planned, secondary questions are unplanned. But they are always linked to primary questions. The proper ad-libbing of secondary questions can help you pinpoint and draw specific study facts that you are seeking. An example of a secondary question is: "Please bear with me, but I do not quite understand how you propose to handle that operation; would you please give me more detail?"

Using Prompts to Improve the Effectiveness of the Interview

There are three nonquestioning prompts that the interviewer can use to improve further the effectiveness of an interview. First, the interviewer can make encouraging statements, such as: "Please continue." Second, the interviewer can vary the response by making noncommittal statements, such as: "I see" or "Mmm." Finally, the interviewer can simply remain silent. As odd as it may seem, silence can be a strong prompt if used with appropriate body language to signal the interviewee to continue. The interviewer can look expectant or nod

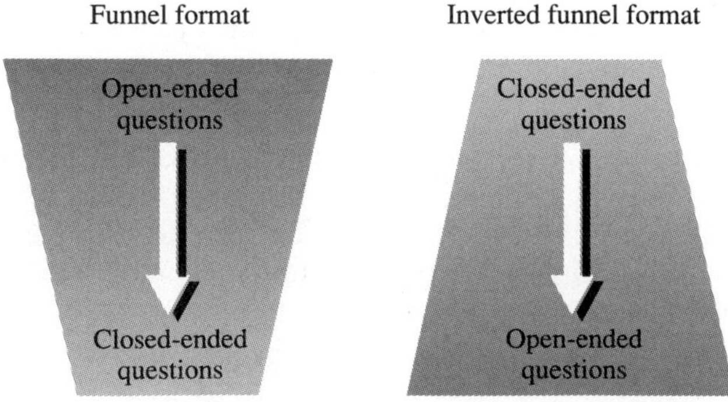

Figure 5.10
Interview sequencing formats.

Figure 5.11
Interview using the funnel format.

INTERVIEW AGENDA

Interviewee: Dick Borland, Accounts Receivable Manager
Date: MMDDYY
Time: HH:MM (A.M. or P.M.)
Place: Room number 111
Purpose: Determine current credit checking policy

2 min.	Begin the interview: Introduction Thank Mr. Borland for the use of his valuable time. Restate purpose of the interview.
5 min.	In general terms, how would you describe the current credit checking policy?
4 min.	What conditions determine whether a customer's order is approved for credit?
3 min.	What decisions are made once these conditions are evaluated?
2 min.	How are customers notified when credit is not approved?
2 min.	Who is responsible for making credit checks?
2 min.	May I have permission to question those responsible as to how they perform the credit checking process?
3 min.	End the interview: Thank Mr. Borland for his cooperation and tell him that he will receive a summary of all interviews pertaining to current credit checking policy.
23 minutes	for primary questions.
7 minutes	for secondary questions.
30 minutes	allotted for interview [the interview must be completed by HH:MM (A.M. or P.M.)]

in agreement or look puzzled. Each of these behaviors will initiate a corresponding behavior in the interviewee, and silence will motivate him or her to keep talking.

Planning the Order of the Questions

The sequence in which primary questions are asked should be planned. The most common sequencing formats are: funnel and inverted funnel.[2] These interview sequencing formats are depicted in Figure 5.10, and explained below.

The Funnel Format With the funnel format, the interviewer starts with open-ended questions, and then, using closed-ended questions, gradually works down to the specific information wanted. An example funnel format is depicted in Figure 5.11. Notice that the interview agenda is preplanned even to the time allotted to each primary question and to secondary (i.e., follow-up) questions.

[2] Gary Mitchell, *How to Interview Effectively* (New York: American Management Association, 1980).

The Inverted Funnel Format With the inverted funnel format, the interviewer starts with specific closed-ended questions and gradually draws the interviewee out to the point where he or she will respond to and expand on answers to open-ended questions. This sequencing format is shown in Figure 5.12. Notice that we have changed the identification section of this example from that of the previous example simply to show you an alternative.

The Psychology of Interviewing

The psychology of interviewing pertains to relationships between people. To some extent interviewing is an art and, accordingly, it does not always proceed as planned. Following is a list of counterproductive behaviors that interviewees may manifest and ways to correct them.

Behavior of Interviewee	Corrective Action
Appears to guess at answers rather than admit ignorance.	Ask probing questions. After the interview, validate answers that are suspect.
Attempts to tell the interviewer presumably what the interviewer wants to hear instead of the correct facts.	Avoid putting questions in a form that implies the answer. Validate answers that are suspect.
Gives the interviewer a great deal of irrelevant information or tells stories.	In a friendly but persistent fashion, bring the discussion back into the desired focus.
Stops talking if the interviewer begins to take notes.	Put the notebook away and confine questions to those that are most important. If necessary, come back later for details.
Attempts to rush through the interview.	Suggest coming back later.
Expresses satisfaction with the way things are done now and wants no change.	Encourage the interviewee to elaborate on the present situation and its virtues. Take careful notes and ask questions about details.
Shows obvious resentment toward the interviewer, answers questions guardedly, or appears to be withholding facts.	Try to get the interviewee talking about some self-interest or his or her previous experience with analysts.
Gripes about his or her job, associates, supervisors, and unfair treatment.	Listen sympathetically and note anything that might be a real clue. Do not interrupt until the list of gripes is complete. Then, make friendly but noncomittal statements, such as: "You sure have plenty of troubles. Perhaps the new system can help with some of them." This approach should bridge the gap to asking about the desired facts.
Acts as a crusader and is enthusiastic about computers, new ideas, and techniques.	Listen for desired facts and valuable leads. Do not become involved emotionally or enlist in the interviewee's campaign or personal agenda.

Figure 5.12 Interview using the inverted funnel format.

INTERVIEW FORM

Person interviewed: Karyn Jones Title: Inventory Supervisor

Purpose: To gather study facts about the proposed inventory control systems project.

Interviewed by: John Sellers Title: Analyst

Date: MMDDYY Time: HH:MM Place: Office 201

Follow-up interview summary sent on: MMDDYY

Response received from follow-up interview summary: MMDDYY

4 min. Begin the interview:
Explain who you are, what the purpose of the interview is, what the systems project is about, and what contribution the interviewee will make in the development of a new system.
Make sure that you have a correct understanding of the interviewee's job responsibilities and duties. A typical question is: "It is my understanding that your job is . . . (a brief job description). Is this correct?"

1 min. How many individual inventory items do you stock? ____________

1 min. Do you sell nonstocked items?
☐ yes ☐ no

1 min. Do you have multiple warehouses?
☐ yes ☐ no

1 min. Do you have an inventory numbering system?
☐ yes ☐ no

1 min. If yes, what is the maximum length of your inventory code number?
____

1 min. How many vendors does the company buy inventory items from?
____

1 min. Do prices change frequently?
☐ yes ☐ no

4 min. What costing method do you use for inventory? ____________

8 min. What kind of inventory control reports would you like to receive in the future?
____

5 min. End of interview:
At the end of the interview, summarize the main points of the session, thank the interviewee, and indicate that you will return if you have any further questions.

28 minutes for primary questions.
7 minutes for secondary questions.
35 minutes allotted time for the interview

Customizing Questions by Developing Preinterview Profiles of Interview Subjects By developing preinterview profiles on interviewees, the interviewer can plan the interview effectively. Such profiles provide the interviewer with a basis for creating questions about the interviewee's position within the organization, job responsibilities, and activities.

There are two basic sources of information to draw on for developing preinterview profiles. One source is what people say about the prospective interviewee. A second source is company records or documents, such as résumés and organization charts.

Figure 5.13 provides an example of a traditional organization chart and brief profiles on the president, vice president of marketing, and manager of sales. The brief annotated profiles of each person make a rather dry organization chart come to life and reveal the function of each manager. They give an insight into the roles that these people play in the organization. Such insights place the interviewer on a better footing than going into the interview cold.

Recording and Evaluating the Interview

After the interview is over, the interviewer should record and evaluate its results. Finally, the interviewer should summarize the results and report them to management and the interviewees, at least for verification and clarification. Reporting the results of the interview to interviewees gives them a sense that what they said was important enough to be recorded.

Recording the Interview The responses to closed-ended questions are easily recorded on the interview form. Responses to open-ended questions, however, generally require a great deal of active listening and note-taking. Active listening means paying close attention to what the interviewee is saying and not making judgments or jumping to conclusions before the interviewee has had his or her say. Active listening also means being able to paraphrase and summarize everything effectively. Using tape recorders presents two disadvantages. If the interviewer relies on the tape recorder, he or she will have to play it back, thus doubling time spent interviewing. Further, if the interviewee is nervous about being taped, this will present a barrier to communications. In most instances, therefore, traditional note-taking is the accepted way to record various points, observations, and answers to questions. As when taking notes during a lecture in school, the interviewer must guard against too much detailed note-taking and thereby losing the broad ideas and meaning of the responses being given.

Evaluating the Interview The purpose of evaluating an interview is to ensure improvement. By examining their effectiveness at interviewing, interviewers can learn their strengths and weaknesses.

Consider three areas in evaluating an interview:

- The interview itself
- The effect of the interview on the interviewee
- Self-evaluation of techniques and results achieved

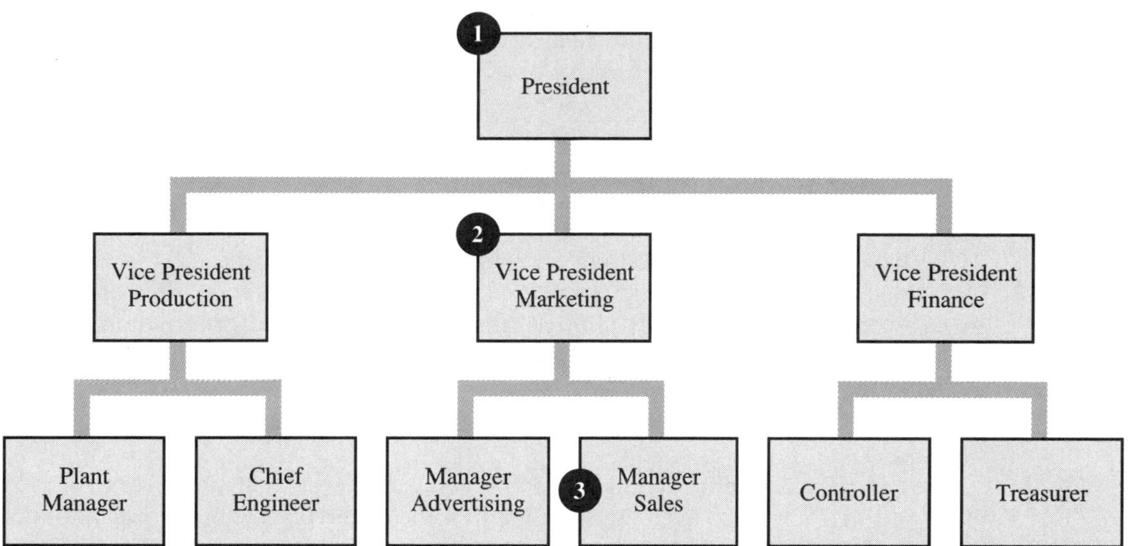

1

Cary Jones, Office 120. Is in charge of the organization, but practices decentralized management. Started in sales 20 years ago, is sales oriented, and is in the field 60 percent of the time calling on major customers and opening up new markets. Makes most decisions in committee meetings. Permits vice presidents "to run their own ships."

2

Henry Nunez, Office 215. Establishes sales quotas, sets priorities, and defines sales territories. Has spent the majority of his time the past six months working on a strategic sales model. Works closely with advertising, and devotes a great deal of time in Washington, D.C., on development of large government purchases. Holds a meeting of all staff the first Monday of each month.

3

Sylvia Amato, Office 360. Hires, fires, and establishes training programs for salespeople. Sets quotas among salespeople. Devotes about 50 percent of her time to selecting and training new salespeople. Has often pleaded for help from information systems to provide better sales performance information.

Figure 5.13
Traditional organization chart with annotated profiles.

First, the interview itself. How well were the questions phrased and sequenced? Were needed study facts attained? Were planned purposes achieved? Did any communication barriers occur, and if they did, how well were they handled?

Second, the effect of the interview on the interviewee. How accurate were the preinterview profiles? Did the questions fit these profiles? Was the interviewee secretive or defensive? Open?

Third, self-evaluation. Were conditions, including the attitude of the interviewer, conducive to good interviewing? Was control of the interview maintained by using tact to end ramblings and irrelevant comments? Was the interview completed within its allotted time? Was there good progression of the interview? Did the interviewer have the necessary skills and knowledge to deal with the interviewee?

Sampling

SAMPLING is the application of certain procedures to less than 100 percent of the items in a survey to evaluate or estimate some characteristic of the population. Sampling is particularly useful where determining characteristics or values of all items would be too time-consuming.

Nonstatistical and Statistical Sampling

In gathering study facts, analysts may use either nonstatistical sampling or statistical sampling, or both. Both types of sampling can provide analysts with many study facts.

The difference between the two types of sampling is that statistical probabilities are used to control sampling risk in statistical sampling. In nonstatistical sampling, analysts determine sample size and evaluate sample results entirely on the basis of judgment and their own experience. Thus analysts may unknowingly use too large a sample in one area and too small a sample in another. A properly designed nonstatistical sample may, however, be just as effective as a statistical sample for most sampling tasks during systems analysis. Statistical sampling is beyond the scope of this book. It requires a great deal of study for one to become proficient in its application.

Steps in a Sampling Plan

Steps in a nonstatistical sampling plan are basically the same as they are in a statistical sampling plan, but without the use of mathematical formulas and statistical tables.

Step 1. Determine Sampling Objectives. Analysts want to ascertain the value or characteristic of a population. For example, an analyst may want to know how long it will take to process 40,000 sales transactions or how many purchase orders are for more than $20,000.

Step 2. Define the Population and the Sampling Unit. The population must be appropriate for sampling objectives. For example, if the analyst wants to know the number of purchase orders for $20,000 or more, the population will be the file containing purchase orders.

The sampling unit is an individual element in the population. A sampling unit, for example, may be a record or a field within a record. Each sampling unit must fit the purpose of the survey. For the purchase order example above, the sampling unit would be the DOLLAR AMOUNT field in the purchase order records.

Step 3. Specify the Characteristic of Interest. A characteristic for the purchase order example may be defined as "dollar amount of $20,000 or more." The analyst may also wish to define another characteristic, such as "dollar amount of $20,000 or more containing approval."

Step 4. Determine the Sample Size. In statistical sampling, formulas and tables are used to determine the sample size. In nonstatistical sampling, it is not necessary for analysts to use formulas or tables to determine sample size. Analysts may simply state, according to their judgment, that a sample size should be, for example, 40 sampling units.

Step 5. Determine the Sample Selection Method and Execute It. Once sample size has been determined, a method of selecting sampling units from the population must be chosen. The key objective is to select sampling units in a way that results in a representative sample of the population. Thus all items in the population should have a chance of being selected. Three sampling methods considered here are:

- Random number sampling
- Systematic sampling
- Block sampling

To use random number sampling (also called simple random sampling), analysts must have a basis for relating a unique number to each sampling unit in the population, such as purchase order number. Then, either by reference to a table of random numbers or to a computer random number generator, a selection of numbers can be made to select the individual sampling units that make up the sample.

Systematic sampling consists of selecting every *n*th sampling unit in the population from one or more random starts. The interval between sampling units is called the skip interval. When a single random start is used, the skip interval can be computed by dividing the population size by the sample size. For example, when a sample of 50 is to be obtained from a population of 4000, the skip interval is 80 (4000/50). The starting point in this method of selection should be a random number that falls within the interval from 1 to 80. Then select this sample unit, skip 79 units and select the next one, skip 79 units and select the next one, and so on, until 50 sampling units are selected.

Block sampling consists of selecting sampling units within a specified time period or space dimension. For example, the sample may consist of all purchase orders processed during a one-week period. Or a block sample may be space-dimensional, such as inches or feet. For example, assume that Joe is in a warehouse containing 60 feet of cardboard file boxes (30 boxes 2 feet in length for each box) containing thousands of purchase orders processed last year, and Joe has 20 minutes to estimate the number of purchase orders for $20,000 or more processed last year. Joe randomly pulls a block of purchase orders about 6 inches thick from one of the file boxes and fingers through this 6-inch stack of purchase orders. From this stack he counts 10 purchase orders that are in the amount of $20,000 or more. He concludes that about 20 "big ones" are in 1 foot of purchase orders, and since there are 60 feet of purchase orders, then there must have been about 1200 purchase orders processed last year in the amount of $20,000 or more (20 purchase orders of $20,000 or more per foot of purchase orders × 60 feet of purchase orders). The accuracy of this inference, however, depends on how representative Joe's sample is of the population of purchase orders processed last year, together with such factors as the dates spanned by the sample orders.

Step 6. Evaluate the Sample Results and Make an Inference. If the sample results are satisfactory, analysts can estimate values or evaluate characteristics of the population represented by the sample. Considering a sample, an analyst may make an inference about the population, such as: "I'm 95 percent confident that the value of a characteristic of the population is *X*, plus or minus some relatively small amount." For example: "I'm 95 percent confident that there were 1200 purchase orders processed last year in the amount of $20,000 or more, plus or minus 100 purchase orders." The validity of this statement made by Joe will be open to question by nearly every sane observer, including trained statisticians. Remember, however, that Joe was given only 20 minutes to come up with an answer using nonstatistical sampling.

Observing

Another technique used to gather study facts is **OBSERVING** people performing various aspects of their jobs. Observing has many purposes. It permits the analyst to determine what is being done, how it is being done, who does it, when it is done, how long it takes, where it is done, and why it is done. As Yogi Berra said: "You can observe a lot by watching." Moreover, observing also allows the analyst to participate in procedures being performed by employees. If, for example, the analyst is examining the procedures for handling incoming shipments on the loading dock, that analyst might complete the receiving reports while a truck is being unloaded. With this kind of hands-on experience, the analyst may find out that the forms are improperly designed or that not enough time is available to do all the procedures. By attaining this hands-on experience, the analyst can often determine better and quicker ways of doing something.

To maximize the observing technique, the guidelines below should be followed.

Getting Ready

Before observing begins, the analyst should:

1 Identify and define what is going to be observed.

2 Estimate the length of time this observation will require.

3 Secure proper management approval to conduct the observation.

4 Explain to the parties being observed what will be done and why.

Conducting the Observation

Observing can be conducted most effectively by the analyst following a few simple rules:

1 The analyst should become familiar with the physical surroundings and components in the immediate area of the observation, as well as the nature and purpose of the work done there.

2 While observing, the analyst should periodically note the time.

3 The analyst should note what is observed as specifically as possible. Generalities and vague descriptions should be avoided.

4 If the analyst is interacting with the persons being observed, the analyst should refrain from making qualitative or value-judgment comments.

5 The analyst should show proper courtesy and heed safety regulations when conducting observing.

Analysts may not have much time to observe operations. In these situations, sampling is a technique that can be used effectively to reduce the time spent observing and still gather reliable facts. For example, if an analyst wants to find out how long it takes to process 50,000 customer orders in the shipping room, he or she might measure the time required to process a sample of 25 customer orders and extrapolate the expected time to process 50,000 orders.

$$25 \text{ orders require } T \text{ time}$$

$$\frac{T}{25} = \text{time per order}$$

$$50{,}000 \times \frac{T}{25} = \text{time for } 50{,}000 \text{ orders}$$

Documenting and Organizing Observation Notes

Following the observing period, the analyst's notes and impressions should be formally documented and organized. The analyst's findings and conclusions should be reviewed with the person observed, his or her immediate supervisor, and perhaps another systems analyst.

CONCLUDING SYSTEMS ANALYSIS AND COMMUNICATING THE FINDINGS

Throughout the systems analysis phase, the analyst should maintain extensive communications with users, the project manager, and other project personnel. On a continuing basis this communication effort includes:

- Feedback to persons interviewed, or observed, as to what the analyst understands
- Verification with user personnel as to the findings in related functions or activities that the analyst identifies

- Periodic status meetings to inform the project manager and other project personnel about progress, status, and adherence to schedule

Preparing the Systems Analysis Report

The documented deliverable produced from the systems analysis phase is the Systems Analysis Report. It (as well as other documented deliverables) should be prepared in a professional manner using bond paper and desktop publishing or CASE technologies. Following are some rules that help in constructing professional documented deliverables:

- Divide the report into meaningful chunks of information. Use appropriate report headings and section headings. Use a cover page to identify the report. Use a table of contents if the report contains over three pages.
- Use good-quality paper heavy enough to make letters and numbers stand out clearly. Normally, bond paper of 20-pound weight is sufficient. White, off-white, or buff colors are acceptable.
- If the sequence of a list is random, or arbitrary, use a symbol, such as a dash, bullet, arrow, or asterisk at the beginning of each entry in the list. If the order of entries in the list is important, use numbers. Also order data according to their importance.
- Maintain at least 1-inch margins all around.
- Paginate the report.
- Use plenty of spacing between lines, and use block letters and underlining for highlights. Different colors may be used to draw attention, but do not overdo the use of these differentiating devices.
- Use legible, nondistracting fonts. Avoid fonts with fancy type and flourishes.
- Unless models, such as data flow diagrams or entity relationship diagrams, are an integral part of the report body, place these models and other supporting details in an appendix and develop an easy-to-use pointer system to direct the reader to this detail as needed.

A well-prepared Systems Analysis Report defines and summarizes the findings produced from gathering and analyzing study facts. The major headings and content of this report include the following:

1 ***Reasons and Scope of Systems Analysis*** Relate reasons and scope back to the Systems Project Proposal Form (see Chapter 4). If, through analysis, reasons and systems scope have changed, these changes should be clearly stated and documented.

2 ***List of Major Problems Identified*** In some instances, systems analysis may not progress as planned, or constraints may arise that were not anticipated. For example, study facts may not be forthcoming because of lack of cooperation or the unavailability of some users. Or it could be that certain resources that were originally thought to be available are nonexistent.

3 ***Complete Statement and Definition of User Requirements*** Systems proposals or other reasons for initiating systems analysis are very general. They give only a broad definition of user requirements. The primary purpose for conducting systems analysis is to define user requirements precisely. The Systems Analysis Report should therefore include specific user requirements defined in terms that everyone can understand.

4 ***List of Critical Assumptions*** In many cases, certain aspects about the new system are not known at this stage of systems development. For example, at this particular time it may not be known for certain that a telecommunication network backbone will be in place by the time the systems project, which relies on the network backbone, will be ready for implementation. The analyst is therefore advised to state that his or her critical assumption is: "For systems project *X* to continue and the proposed systems project to be implemented by date *Y*, the planned telecommunication network backbone proposed during systems planning must be in place and working by date *Z*."

5 ***Recommendation*** Depending on the experience and progress that his or her team members, especially the systems analyst, has enjoyed with the new systems project to date, the project manager makes a statement as to the feasibility of the project and its worthiness. If, for example, any of the TELOS feasibility factors are considerably lower than estimated during systems planning, the project manager may recommend that the project be abandoned. If, on the other hand, the project manager deems the systems project, as presently constituted, to be feasible and worthy of continuation, he or she includes a recommendation to indicate this conclusion.

The Systems Analysis Report is directed to users, general management, systems management, and the steering committee (if one exists). The report permits these people to determine how well user requirements have been defined and to see if systems work is progressing as well as was originally thought. Also, if systems planning has been conducted, these people will want to ascertain if the systems project seems to still be supporting PDM strategic factors.

Presenting the Findings Orally

Simply handing out documented deliverables is not enough. The recipients of these written reports want eye contact. They want to hear directly from the project manager, systems analyst, and any other team members who played a major role in preparing the documented deliverable. Bringing together all interested parties in a meeting at one time and walking through the report enables all users, managers, and systems professionals involved in the new system to make necessary adjustments before moving into the general systems design phase. Also, this meeting gives the project team a chance to solve major problems encountered during systems analysis and to determine if the critical assumptions are at least in the ballpark.

An oral presentation of each documented deliverable is required for clear and comprehensive communication. Three methods that can be used to make this oral presentation are:

- Memorization
- Reading
- Extemporaneous delivery

Each has its place, but usually the extemporaneous method is the most effective.

- ***Memorization*** A memorized presentation is somewhat effective. It gives the presenter a feeling of security, but it sacrifices freedom and freshness. It stifles spontaneity. One will feel as if one is in a straightjacket during question-and-answer periods. Moreover, the method is extremely time-consuming to implement.
- ***Reading*** Reading from reports can be described in two words—ho and hum; it is a sleeping pill. People can read the reports themselves. The oral presentation is meant to encourage two-way communication. Oral presentations provide an opportunity to explain aspects of the system that are not easily explained by written reports and modeling tools. Reading the report does not meet these objectives. Unless followed by an extended exchange of ideas and questions and answers, reading a report wastes everyone's time.
- ***Extemporaneous*** The extemporaneous method is the best way to present your reports. This is the most versatile and expressive method of delivery. The presenter is likely to be spontaneous and energetic, able to adapt to unplanned topics and situations, and to respond to the audience. This method gives the best opportunity to maintain eye contact and establish rapport.

Using Audio and Visual Aids in Oral Presentations

The use of modeling tools should be an integral part of the presentation and help support it. Appealing to recipients' visual and audio senses simultaneously increases their attention and understanding. Other visual aids that may be effective are pictures, blackboards and flip charts, films, audio aids, videotapes, prototypes, full-scale models, and samples.

The Four Possible Outcomes of Systems Analysis

The following are four alternative outcomes of the systems analysis phase:

Abandon the Project

This outcome means that no further work is to be performed on the systems project and that resources will be directed toward other systems projects. This outcome may result because a particular systems project does not live up to its earlier estimated TELOS feasibility factors or PDM strategic factors, or both.

Abandonment of the systems project may also emanate from major problems that cannot be resolved. Another reason could be reshuffling of systems priorities by management or the steering committee, which results in the present systems projects being scrapped.

Postpone the Project
This outcome may result from a broad array of circumstances. For the time being, management may direct resources to other systems projects with higher priorities. Perhaps installation of the telecommunication backbone has been delayed, causing delay of the present systems project. Certain key users may be on vacation or absent for several weeks, thus causing a temporary postponement of SDLC.

Change the Project
This outcome means that significant aspects of the original systems proposal have to be modified significantly. Such changes may involve a major expansion or contraction of the systems scope. Or it may mean that user requirements are significantly different from those anticipated earlier, causing the need for more or less systems resources.

Continue the Project
This outcome means that the systems project will proceed as proposed in the Systems Analysis Report.

REVIEW OF CHAPTER LEARNING OBJECTIVES

The major goals of this chapter were to enable each student to achieve five learning objectives. We now summarize the responses to these learning objectives.

Learning objective 1:
Relate the systems analysis phase with other phases of SDLC and discuss why the application of modern methodologies, modeling tools, and techniques helps to reduce consumption of systems resources and increases the likelihood of successful systems development.

Systems analysis follows systems planning and precedes general systems design. The systems analysis phase is used to define and describe user requirements in detail, given specified time and budget constraints. In general systems design, the next phase, conceptual systems design alternatives that best satisfy user requirements are developed and presented. In this way, systems analysis answers the what question and general systems design answers the how question. Indeed, until user requirements specifications are complete, general systems design cannot be performed effectively. Successful systems analysis depends on:

- Comprehensive study facts
- Techniques to gather these study facts

- Complete user involvement
- Modeling tools to analyze study facts
- Iteration

Learning objective 2:
State the reason for defining the systems scope and list three sources of study facts.

Defining the scope of the system under development enables systems professionals to determine its boundaries and work on those things that are relevant to it. Study facts about the system under development come from three primary sources: the current system, internal sources, and external sources.

1 ***The Current System*** The purpose of gathering study facts about the current system is to develop a thorough understanding of its strengths and weaknesses and the way it works. Such an understanding can provide a springboard for development of the new system.

Several disadvantages exist in gathering study facts about the current system. First, the current system may be irrelevant now, and it will be totally scrapped when the new system is implemented. So it is waste of time to study it. Second, study facts about the current system may create mental barriers that cloud the vision of systems professionals or misguide them.

2 ***Internal Sources*** Users are typically the most important source of study facts. They are the ones who provide information that is used to specify user requirements, the basis for all systems work that follows systems analysis. Other internal sources of study facts are documents that describe how the company is structured, what the company is or has been doing, and what the company plans to do.

3 ***External Sources*** Sources of these study facts are the library, seminars, learning centers, and other systems. Often, these sources generate ideas and solutions that may not be forthcoming from the other two sources.

Learning objective 3:
Describe the techniques used to gather study facts, especially those used to fill in gaps after JAD and prototyping sessions have been conducted.

Three major techniques used by systems professionals to gather study facts are interviewing, sampling, and observing.

1 ***Interviewing*** This technique is the planned verbal exchange of information between an interviewer and interviewee in a question-and-answer format for a stated purpose.

2 ***Sampling*** This study-fact gathering technique is used to reach a conclusion about a population. Analysts use sampling when:

- The nature of the item under investigation does not demand a complete count.

- A conclusion or decision must be made about the value or characteristics of a population.
- The time and cost to make a complete count of the population are too great.

3 ***Observing*** Observing is a useful way of confirming or correcting study facts collected by other study-fact gathering techniques. The processes of a current system, for example, may differ from documentation describing the system or from what people or policies say that the system is supposed to do.

Learning objective 4:
Discuss the preparation and presentation of the Systems Analysis Report.

For the most part, the main purpose of documented deliverables and their presentation is to present findings and recommendations professionally. The Systems Analysis Report is such a documented deliverable. Its importance cannot be overstated, because it represents the foundation for building the system under development. The Systems Analysis Report includes five major components or sections.

1 The report should state the reasons and scope of systems analysis. This section indicates that the project is on the right track and headed in the right direction.

2 If major problems have been encountered, they should be explained. With this information, such problems may be solved.

3 The heart of the Systems Analysis Report is the definition and specification of user requirements. In many situations, modeling tools are especially well suited to specifying user requirements graphically.

4 Critical assumptions are documented to allow all parties to know that the system is being developed in an environment where not everything that will affect the system is absolutely clear or known. For instance, funds have not been made available yet for the acquisition of certain resources that the new system will use. Critical assumptions are that necessary funds will be forthcoming and systems work will continue based on that assumption.

5 It is the professional responsibility of the project manager or the systems analyst to make an objective recommendation based on the perceived values of the TELOS feasibility factors and the PDM strategic factors associated with the new system.

Written, oral, and graphic communication skills are all necessary for communicating documented deliverables effectively to all participants in the development of new systems. Oral presentation is especially effective for walking all parties through the report. The extemporaneous method is recommended for presenting all documented deliverables orally.

Learning objective 5:
List and discuss alternative outcomes of the systems analysis phase.

Completion of the systems analysis phase represents a major milestone in SDLC. Often, one finds that what was proposed during systems planning or what was requested by the user differs significantly from what was originally thought. Perhaps the systems scope has been expanded or contracted. Possibly, problems have been encountered that cannot be solved. Or it could be that systems analysis indicates that the new system is even more important than was originally estimated, and management wants to proceed with systems work at full speed.

The Systems Analysis Report leads to one of four outcomes:

1 The systems project may be abandoned because it did not meet expectations.

2 The systems project may be postponed temporarily because of a potpourri of reasons. As soon as conditions change, the systems project will continue.

3 Because the gathering and analyzing of study facts led to discoveries of new opportunities, the scope and goals of the original systems project is changed to accommodate the new discoveries.

4 The Systems Analysis Report is in line with what was originally thought, and the systems project continues on the same course.

SYSTEMS ANALYSIS CHECKLIST

The principal objective of systems analysis is to define user requirements. The following checklist states, in summary form, how to conduct this important SDLC phase.

1 Use modern methodologies, modeling tools, and techniques.

2 Use JAD and prototyping sessions to gain user involvement in the system under development and to elicit user requirements.

3 Use modeling tools such as DFDs, ERDs, STDs, decision tables, decision trees, and other tools to model user requirements and define systems scope.

4 Use CASE technology to help facilitate the modeling of user requirements. CASE tools cover a broad range of capabilities, from tracking and managing SDLC, to aiding systems professionals in enterprisewide modeling in conjunction with systems planning, to analyzing and designing models that underlie the enterprisewide model, to prototyping and screen painting, to generating software code in accordance with design models. CASE takes on the maintenance or redesign of old systems applications that were designed manually. The central repository (also called dictionary or knowledge base) contains all the documented deliverables, models,

prototypes, and other systems documentation. The repository stores and keeps current all information about the system. Some CASE systems will help tackle particular phases of SDLC. Others will go from planning to implementation—and then on into maintenance of the new system and redesign of old applications. At last count, there were more than 50 vendors offering products that could be classified as CASE tools. Some are more comprehensive and feature-rich than others. All, however, aid systems professionals in the systems analysis phase, such as creating DFDs, ERDs, STDs, and prototypes.

5 Use the current system (in some instances), internal sources, and external sources to gather study facts.

6 Use interviewing, sampling, and observing as study-fact gathering techniques.

7 Upon completion of the systems analysis phase, prepare a Systems Analysis Report in good, professional form using desktop publishing.

8 Present the Systems Analysis Report orally, using the extemporaneous method.

KEY TERMS

Iceberg effect

Interviewing

Observing

Sampling

Specialists with advanced tools (SWAT) teams

Study facts

Systems Analysis Report

Systems scope

REVIEW QUESTIONS

5.1 What is the SDLC phase that precedes the systems analysis phase? What is the SDLC phase that follows it? Discuss the relationship of the systems analysis phase to both of these phases.

5.2 Why is the systems analysis phase sometimes called the foundation phase?

5.3 What are study facts?

5.4 What is meant by the expression "iceberg effect"?

5.5 Explain how modern methodologies, modeling tools, and techniques help reduce the iceberg effect.

5.6 Explain how JAD can be used both to get users involved in systems development and to define user requirements.

5.7 Explain how prototyping elicits hard-to-define user requirements.

5.8 Explain how a DFD can be used to define systems scope.

5.9 Discuss the pros and cons of analyzing the current system. List the internal sources of study facts. List the external sources of study facts.

5.10 Discuss how the following are used to analyze study facts and model user requirements:

- DFD
- ERD

5.11 Define interviewing and its purpose.

5.12 Define and give examples of the following:

- An open-ended question
- A closed-ended question
- A primary question
- A secondary question

5.13 How does the interviewer encourage the interviewee and keep the interview moving?

5.14 Describe two interviewing or sequencing formats. Give a brief example of each.

5.15 List the behavioral pitfalls in an interview and ways to correct such behavior. Give an example of each.

5.16 What is the purpose of a preinterview profile? Describe how to obtain information for preparing preinterview profiles.

5.17 Explain why interviews are recorded and evaluated.

5.18 What is sampling? Explain its use as a technique for gathering study facts.

5.19 What is the primary difference between statistical and nonstatistical sampling?

5.20 Give the steps in a nonstatistical sampling plan. Give an example of each step.

5.21 Name and describe three sampling techniques. Give an example of each.

5.22 Explain how observing is used to gather study facts. How does observing supplement the other study-fact gathering techniques?

5.23 Explain the purpose of the Systems Analysis Report. Name and discuss ways to enhance the form of documented deliverables such as the Systems Analysis Report. What are the major items that make up the substance of this report?

5.24 If the Systems Analysis Report is professionally prepared, why should an oral presentation about it be made?

5.25 Define the three methods used to make an oral presentation. Give an example of each. What method is recommended for making oral presentations of reports during systems development? Why is this method recommended?

5.26 Describe oral presentation aids and indicate why they are used.

5.27 What are the possible outcomes of conducting systems analysis? Give an example of each.

CHAPTER-SPECIFIC PROBLEMS

These problems require exact responses based directly on concepts and techniques presented in the text.

5.28 You have acquired knowledge about an order-entry and inventory control system as follows: When a customer initiates an order, the order is edited for an order file. Edited orders are checked for credit against a customer master file. Orders with verified credit are processed against an inventory master file to allocate inventory and to determine if any inventory items have reached or exceeded their reorder points. Those that have are ordered from vendors. Allocated orders are extended and calculated. Inventory reports are generated for the inventory manager, accounts receivable reports go to the credit department, invoices and packing slips go to the shipping department, and the customer receives an acknowledgment. For those items that were ordered from vendors, when they are delivered, receipts are edited, and the edited receipts are used to update the inventory master file.

Required: Create a DFD to organize and analyze these study facts.

5.29 Following are four decision constructs:

1 If fewer than 400 units are sold, the salesperson's commission is 1 percent of total sales.

2 If between 400 and 499 units are sold, the salesperson's commission is 2 percent of total sales.

3 If 500 or more units are sold and the salesperson has been with the company more than one year, the salesperson's commission is 4 percent of total sales.

4 If 500 or more units are sold and the salesperson has been employed by the firm for one year or less, his or her commission is 3 percent of total sales.

Required: Prepare a decision tree that organizes these decision constructs properly.

5.30 According to your study facts for a job cost system, you have determined the entities that make up this system are: customer, job order, direct materials, direct labor, and manufacturing overhead. A customer may have multiple job orders. For each job order, there are charges for direct materials, direct labor, and manufacturing overhead.

Required: Prepare an ERD that organizes and clarifies these study facts.

5.31 The following paragraph represents study facts gathered from an interview between Stanley Goodman, systems analyst, and Mary Amato, credit manager at Tricor. These study facts represent Mary's thoughts on how the credit procedures would operate.

All orders received at Tricor will have a credit check performed. All orders received from new customers must be forwarded to the credit manager. All orders exceeding $1000 must be forwarded to the credit manager. If the dollar amount of an order plus the present accounts receivable balance for that customer exceeds the credit limit assigned to the customer, the order must be sent to the credit manager. All orders from customers with past-due accounts receivable balances must be forwarded to the credit manager. Orders from customers coded to class of trade 100 are not sent to the credit manager unless the account is on credit referral, the order exceeds $10,000, or the present accounts receivable balance is greater than $50,000. Orders that pass the credit check are sent directly to the shipping department.

Required: Prepare a decision table to organize and clarify these study facts.

5.32 You have been assigned to develop an airline reservation system. During analysis, you compiled the following notes: If the request is for economy-class but no seats are available and the customer will not accept an alternative, refer the customer to an alternate flight. If the request is first-class and first-class is available, issue a first-class ticket and reduce first-class available. But if first-class is unavailable and economy-class is available and acceptable, issue an economy-class ticket and reduce the economy-class available. If the patron requests economy-class but none is available and first-class is acceptable but none is available, refer the patron to an alternate flight. If, on the other hand, first-class is available, issue a first-class ticket and reduce the first-class available. If a flier requests first-class and none is available and the flier will not accept economy-class even if economy-class is available, refer the flier to another flight. Also, under the same conditions, except that economy-class is unavailable, again refer the flier to another flight. Finally, if the request is for economy-class and economy-class is available, issue an economy-class ticket and reduce the economy-class.

Required: Use a decision table to organize and clarify these notes.

5.33 You are analyzing a payroll system and you have collected the following study facts: Hourly employees are paid weekly. This requires input of time sheets and other data to process payroll checks and a payroll register and to update the payroll file. No data are entered to process salaried employees. Their payroll checks and payroll register are processed at the end of the month. The payroll file is also updated at this time.

Required: Prepare a Warnier–Orr diagram to organize and communicate these study facts. (For a review of Warnier–Orr diagrams, return to Chapter 2.)

5.34 "Read sales order" is an iteration module, and it has the following sequence modules subordinate to it: customer edit, product data, check ordered quantity, and ship order. "Valid order" and "special request" are the selection modules of product data. Similarly, "process filled order," "process partly filled order," and "process unfilled order" are the selection modules of "check ordered quantity." Finally, "prepaid order" and "unpaid order" are the selection modules of "ship order."

Required: Use a Jackson diagram to organize, clarify, and analyze these study facts. (For a review of the Jackson diagram, return to Chapter 2.)

5.35 Following are manifestations of undesirable behavior of an interviewee:

1 Appears to guess at answers.

2 Attempts to tell the systems analyst what he or she wants to hear rather than the facts.

3 Gives the analyst a good deal of irrelevant information and tells stories.

4 Stops talking when the systems analyst begins to take notes.

5 Attempts to rush through the interview.

6 Interviewee doesn't want to change anything.

7 Shows hostility and seems to withhold information.

8 Gripes about his or her job, co-workers, and unfair treatment.

9 Exudes fetish for gadgets and new technology.

Required: Present ways to change these undesirable behaviors.

5.36 Following are five questions that one may ask during an interview:

1 How many people work in the shipping department?

2 Please describe your past experience with local area networks.

3 How many flatbed trailers do we have?

4 What do you think of the new financial reporting system?

5 What kind of information do you need to allocate equipment costs?

Required: Identify each of the preceding questions as either open-ended (O) or closed-ended (C).

5.37 Following are five questions that one may ask during an interview:

1 You know that file is too large, don't you?

2 Who fills out the receiving report during the first shift?

3 Where did you buy that useless keyboard?

4 You couldn't prepare that form via a terminal, could you?

5 When was the last time your tape jammed?

Required: Identify each of the preceding closed-ended questions as neutral (N), loaded (LO), or leading (LE).

5.38 Study the following questions:

A Is this monthly report necessary for variance analysis?

B How do you feel about the new payroll system and retirement program?

C What is the cost per mile to run our large trucks?

D Which report is most responsive to your needs?

E From which vendors do we purchase our steel products?

Required: Arrange the preceding questions in a funnel sequence. Then arrange them in an inverted funnel sequence.

Funnel	**Inverted Funnel**
1. ________________	**1.** ________________
2. ________________	**2.** ________________
3. ________________	**3.** ________________
4. ________________	**4.** ________________
5. ________________	**5.** ________________

THINK-TANK PROBLEMS

These problems call for a feasible approach rather than a precise solution. Although the problems are based on chapter material, extra reading and creativity may be required to develop workable solutions.

5.39 You want to know how many individual inventory items are stocked and if the company sells nonstocked items. You also want to get an estimate of file growth. For setting up a data dictionary, you also want to know inventory description, inventory numbering system if one exists, largest quantity of inventory units on hand, and the highest-priced item. Also, you want to know if the company has multiple warehouses and if more than one vendor is involved in purchasing any particular inventory items. You want to know if prices change frequently and what costing method is used, such as FIFO, LIFO, or weighted average. You also want to know what reports are required from the new system.

Required: Develop an interview plan for this system under development. Make any assumptions you deem necessary to develop the interview plan.

5.40 Following is an example of a poor interview. SA stands for systems analyst and U stands for user.

SA: Hey, Joe, have you got a few minutes? I need to ask you some questions about the new sales reporting system.
U: I guess so, Harry. Maybe a few questions. You really caught me at a bad time.
SA: We've had a flu bug running around our office. I was out a couple of days. Glad to hear it hasn't hit you.
U: Well, I haven't got it yet, knock on wood.
SA: Do you have any particular needs concerning sales information?
U: Well, I really need. . . .
The SA is eyeing Joe's bowling trophies on a shelf behind Joe's desk.
SA: I didn't know you bowled. How many trophies have you won?
U: I've won about twelve, but I . . .
SA: I used to be quite a bowler myself. I've won fifteen trophies.
U: That's great. I started . . .
SA: Now back to that sales reporting system. Do you want sales by units or dollars?
U: I would like both in addition to sales by customer and salesperson.
SA: Do you want the report at the end of the month or week?
U: For it to be effective, I need it daily.
SA: Well, I'll see what I can do. I'll check with you later.
U: Will I get some results from you as to the status of the sales reporting system, or some kind of prototype? How will I know . . . ?
SA: I'm bogged down in paperwork, but I'm sure I know what you want. Maybe we can bowl together sometime.

Required: Comment on this interview and make recommendations on how Harry could improve his interviewing skills.

5.41 Following are some typical reasons to initiate systems analysis:

1 Customer complaints concerning poor-quality merchandise.

2 Inability of the shipping department to meet shipping schedules.

3 Inaccurate invoices being sent to customers.

4 High level of obsolescence in raw materials.

5 Excessive amount of returned goods from customers.

Required: Describe the study-fact gathering techniques and analyzing tools you would use for each project. Give examples. Make any assumptions you deem necessary in formulating your answers.

5.42 Following are situations systems professionals might encounter in performing systems analysis:

1 Defining the procedure followed by the accounts payable clerk in performing bank reconciliations.

2 Determining the extent of coding errors in classifying oil field royalties in an oil and gas holding company.

3 Ascertaining the opinions of professors at a local college about the quality of their benefits package.

4 Determining the productivity of a section of data-entry workers.

5 Determining the information needs of a portfolio manager of a $100 million mutual fund.

6 Determining the average number of purchase orders issued each month by a major automobile parts supplier.

7 Estimating the number of work orders issued last year for $100,000 or more. There are a total of 50,000 work orders filed in cardboard boxes stored in the local warehouse.

8 Attaining a consensus among executives and branch managers about the proposal to develop a centralized or distributed information system.

9 Gaining knowledge about how orders are filled in branch warehouses. The main source of this knowledge is the warehouse manager.

10 Trying to attain study facts about a new employee productivity report from a manager who is not really sure what he wants.

Required: For each situation, select an appropriate study-fact gathering technique. Give a brief justification for each situation.

SUGGESTED READING

Boddie, John. *Crunch Mode*. Englewood Cliffs, N.J.: Prentice-Hall, 1987.

Burch, John, and Fritz Grupe. "Methods and Tools for Upgrading Systems Development." *Information Executive,* Summer 1991.

Burch, John, and Gary Grudnitski. *Information Systems: Theory and Practice,* 5th ed. New York: John Wiley, 1989.

Carey, J. M., and J. D. Currey. "The Prototyping Conundrum." *Datamation,* June 1, 1989.

DeMasi, Ronald J. *An Introduction to Business Systems Analysis.* Reading, Mass.: Addison-Wesley, 1969.

Eisner, Howard. *Computer-Aided Systems Engineering.* Englewood Cliffs, N.J.: Prentice-Hall, 1989.

Glasson, Bernard C. *EDP System Development Guidelines.* Wellesley, Mass.: QED Information Sciences, Inc., 1984.

Gray, Paul, William R. King, Ephraim R. McLean, and Hugh J. Watson. *The Management of Information Systems.* Chicago: Dryden Press, 1989.

Keuffel, Warren. "House of Structure." *Unix Review,* Vol. 9, No. 2, February 1991.

Leslie, Robert E. *Systems Analysis and Design.* Englewood Cliffs, N.J.: Prentice-Hall, 1986.

Martin, Charles F. *User-Centered Requirements Analysis.* Englewood Cliffs, N.J.: Prentice-Hall, 1988.

Maxwell, Paul D. "Developing a Successful 'Deskside' Manner for Systems Analysts." *Journal of Systems Management,* April 1988.

Mitchell, Garry. *How to Interview Effectively.* New York: American Management Association, 1980.

Peter, Lawrence J. *Advanced Structured Analysis and Design.* Englewood Cliffs, N.J.: Prentice-Hall, 1988.

Stamps, David. "Taking an Objective Look." *Datamation,* May 15, 1989.

Viskovich, Fred. "From Anarchy to Architecture." *Computerworld,* April 25, 1988.

JOCS Case: Performing Systems Analysis

After receiving the information systems steering committee approval for a system to perform job-order costing, Mary Stockland hired Jake Jacoby, a systems analyst, to serve as project manager for the system. Jacoby was an analyst with a B.A. degree in computer information systems and six years' experience designing and programming manufacturing information systems. Jake Jacoby has spent six years working on pieces of systems for a very large manufacturing organization. He wants to manage all the systems development phases for a single project and is looking forward to working on the JOCS system. The reason he took the job with Peerless, Inc. was to enable him to understand the entire systems development cycle rather than just individual parts of the cycle. Many analysts enjoy great job satisfaction from participating in a project that can be handled by a small team, while others are thrilled by the prospect of conquering an extremely large and time-consuming project. Jake likes the idea of a project that he can get his arms around.

Figure 5.14 is a Gantt chart for the overall completion of the JOCS project. Jake and Mary have developed this chart primarily to make sure that the project continues to move ahead. At this point, the JOCS project has only

Figure 5.14
Gantt chart depicting JOS project schedule estimation.

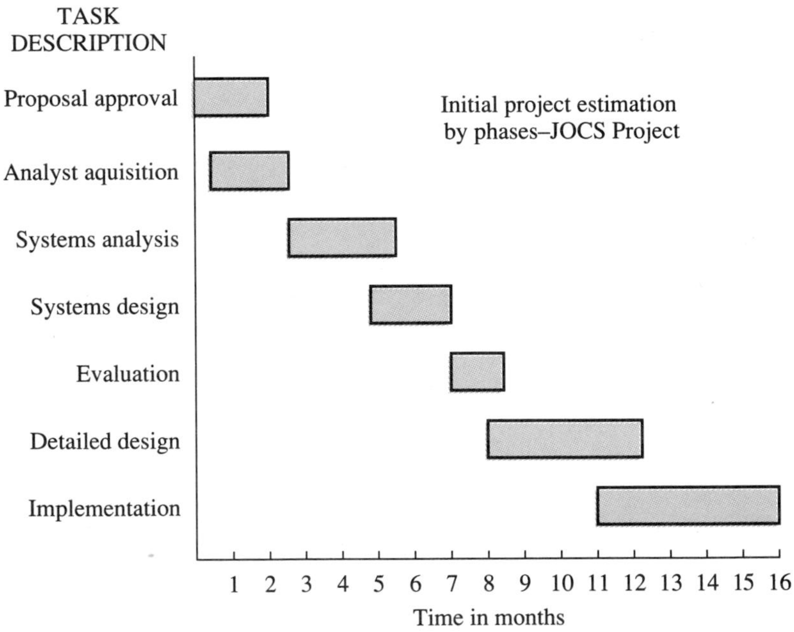

been approved for systems analysis, and the chart's projection of further steps will depend on project approval. The steering committee will have to approve further phases after Jake and Mary submit the Systems Analysis Report. The goal of systems analysis is to research a proposed system in depth, develop a comprehensive understanding of how a system currently works, and then prepare a report evaluating the potential of the system for further study.

Frequently, an analyst becomes overwhelmed when working on a given project because he or she is bombarded with a great deal of new information and must also listen to the wants and needs of the departments that will or are using the system. Jake wants to avoid analysis paralysis; he feels that the best way to do that is to have a deadline in front of himself constantly.

An analyst has to understand fully the system that is being analyzed. By the time Jake has finished this phase of the systems development life cycle, he will probably understand all aspects of job costing better than the people who are currently doing this task. To do this, he has to interview the system end users, survey all potential project participants, and model both the current and the proposed systems. Many aspects of his systems analysis can be completed with the knowledge that is available at Peerless. But the only way that Jake can provide insight into improving the current system at Peerless is by bringing additional knowledge to his task.

Luckily, Jake is already familiar with the manufacturing process. This is fortunate because he does not have to spend time understanding how marketing, engineering, purchasing, and manufacturing interrelate. He also understands that an order for a lifter is the culmination of many discussions between the customer and personnel within marketing and engineering. By the time an order is placed, the engineering department already has a very good idea of what a given lifter should look like. The engineering department then generates a bill of materials that is used as the recipe of all the components necessary to produce that lifter. The bill of materials is used by purchasing and manufacturing to purchase raw materials and divide the components for manufacture among workers within Peerless and subcontractors hired to produce a specific part of a lifter.

In addition to his past experience, Jake frequently reads magazines devoted to manufacturing systems. He reads all manufacturing case studies very closely so that he can learn how other users solve their systems problems. Sometimes he reads the case studies just to find out how to avoid certain systems problems. He also keeps current on applied computer technology by reading computer journals. One of the easiest ways he has found to stay current is by scanning computer equipment advertisements.

Jake will apply his experience and his current manufacturing system knowledge to the JOCS project at Peerless. His first step in systems analysis is to interview end users so that he can model the system. Figures 5.15a to 5.15d show the essential results of interviews conducted by Jake, systems analyst

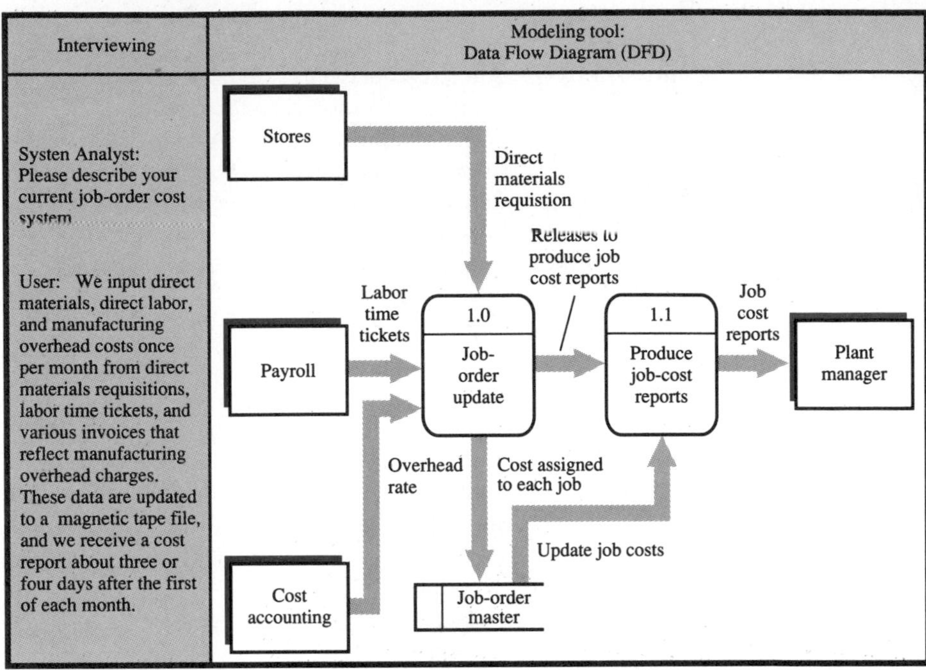
Interviewing
Modeling tool:
Data Flow Diagram (DFD)
Systen Analyst:
Please describe your current job-order cost system.
User: We input direct materials, direct labor, and manufacturing overhead costs once per month from direct materials requisitions, labor time tickets, and various invoices that reflect manufacturing overhead charges. These data are updated to a magnetic tape file, and we receive a cost report about three or four days after the first of each month.
Stores
Direct materials requistion
Releases to produce job cost reports
Labor time tickets
1.0
Job-order update
1.1
Produce job-cost reports
Job cost reports
Payroll
Plant manager
Overhead rate
Cost assigned to each job
Update job costs
Cost accounting
Job-order master

Figure 5.15
(a) High-level DFD based on key facts gathered during an interview.

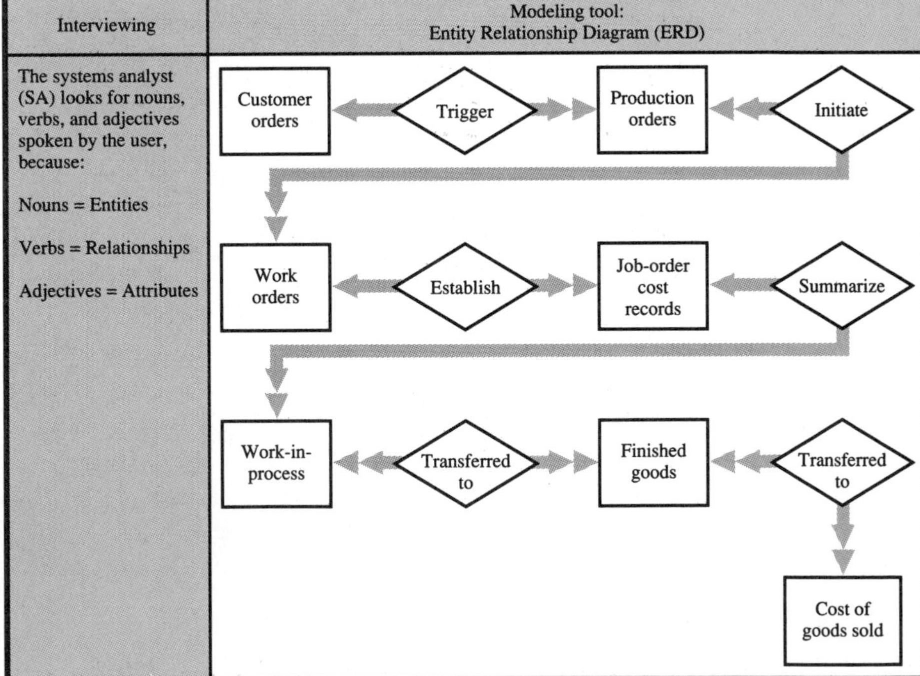
Interviewing
Modeling tool:
Entity Relationship Diagram (ERD)
The systems analyst (SA) looks for nouns, verbs, and adjectives spoken by the user, because:
Nouns = Entities
Verbs = Relationships
Adjectives = Attributes
Customer orders
Trigger
Production orders
Initiate
Work orders
Establish
Job-order cost records
Summarize
Work-in-process
Transferred to
Finished goods
Transferred to
Cost of goods sold

Figure 5.15
(b) High-level ERD of customer orders and production.

Figure 5.15
(c) Example of one attribute in a data dictionary. This attribute helps describe the work in process entity. Many other attributes will be defined and redefined before the system is finally implemented.

Interviewing	Modeling tool: Data Dictionary				
	Attribute Name	*Size*	*Type*	*Lower Limit*	*Upper Limit*
The systems analyst (SA) catalogs the system's attributes based on the data identified during the interview. SA: What is the range of values that the overhead rate might take? User: The overhead rate typically falls between 12 and 18 percent.	Overhead_Rate	Numeric	3	12%	18%

Figure 5.15
(d) Prototype developed during an interview.

Interviewing	Modeling tool: Prototype
SA: What kind of information do you need from the new system? User: For control purposes we need standard costs compared to acutal costs. SA: Here is a prototype of a job-order cost varience report. Is this what you need? User: Yes. That report will help us review variances and take corrective action.	(see report below)

Job-Order Cost
Variance Report

Date: MMDDYY
Job Order Number: XXXX
Job Description Model: XYZ

			Total variance	
Direct materials:				
Price variance:	$ 4,000	(F)		
Quantity variance:	12,000	(U)	$8,000	(U)
Direct labor:				
Rate variance:	$ 2,000	(F)		
Efficiency variance:	7,000	(F)	$9,000	(F)
Manufacturing overhead:				
Volume variance:	$12,000	(U)		
Controllable variance:	14,000	(F)	$2,000	(F)
		Total variance:	$3,000	(F)

U = Unfavorable variance
F = Favorable variance

Figure 5.16
Systems Analysis Report prepared by Jake Jacoby.

SYSTEMS ANALYSIS REPORT

To: Mary Stockland, CIO
From: Jake Jacoby, Project Manager
Subject: Findings from systems analysis of the job-order cost system (JOCS) project

Reasons for Analysis and Systems Scope
Systems analysis was performed to gain clearer knowledge of the current job-order cost system based on actual costs and to determine if standards can be implemented. The new system will require new computer hardware and software development.

General Performance Requirements
1. To improve timeliness of information.
2. To improve relevancy and accuracy of information.

Major Problems Identified
1. The present job-order cost system is inadequate. First, it is untimely. Second, no benchmarks (or standards) exist with which actual costs can be compared.
2. Present computer hardware and software are not sufficient to prepare job-order cost variance reports as outlined in the prototype.

Users' Requirements
1. Derive cost variances based on model for computing cost variances.
2. Provide daily job-order cost variance reports.
3. Integrate the new JOCS with other subsystems.

Critical Assumptions
1. Hardware will be available for online data entry and reporting.
2. Engineering will establish cost standards between August 12 and 24.

General Recommendation
At this stage, my recommendation is to continue systems development. Analysis to date indicates that it is feasible to develop JOCS systems applications that will meet users' requirements outlined in this report. I will prepare a General Systems Design Report by August 30 that will describe in broadbrush terms several design strategies that will meet the requirements presented in this report.

(SA), with users (U). This analysis provides Jake with enough knowledge to prepare a Systems Analysis Report and proceed to the general systems design phase.

Throughout the systems analysis phase, Jake maintains extensive communications with Mary Stockland, end users, and management. On a continuing basis, this communication effort includes feedback to persons being interviewed, or observed, as to what Jake understands about their functions. Jake conducts periodic status meeting to inform others of the status of the project and to gain consensus about what he is learning. The Systems Analysis Report he produces at the end of this phase is not a surprise to any of the project participants. Everyone has been informed during his analysis about what the report will contain. Figure 5.16 is an overview of the Systems Analysis Report submitted by Jake Jacoby to Mary Stockland for the JOCS project.

Chapter 6

General Systems Design

WHAT WILL YOU LEARN IN THIS CHAPTER?

After studying this chapter, you should be able to:

1 Define three systems design categories.
2 Discuss the use of rapid application development (RAD) in general systems design and describe RAD's four key elements.
3 Describe the structure-oriented design approach.
4 Describe the object-oriented design approach.
5 Compare the structure-oriented design approach with the object-oriented design approach.

INTRODUCTION

For a perspective of where we are in SDLC, review Figure 6.1. The subject of this chapter is general systems design, the third phase of SDLC. The general systems design phase involves a process performed to create alternative conceptual systems designs that fulfill user requirements. Why do we normally produce several general systems design alternatives? Constraints imposed on the project and an array of user requirements may interact in complex ways such that a number of designs are needed rather than a single design. The results of the general systems design phase are several plausible conceptual systems design deliverables compiled in the General System Design Report. These general systems design alternatives are ready for evaluation and selection of the optimum one for the detailed systems design phase (i.e., functional design).

Systems analysis and general systems design depend on each other. The study facts gathered, analyzed, and modeled during systems analysis provide the basis on which general systems design alternatives are created. The systems analysis phase was investigative and discovery-oriented. Now, in the general systems design phase, systems professionals must often create new or different features of basic models developed during systems analysis.

The key to general systems design is to get things down on paper first without trying to refine the systems designs too early. The game plan: Interact with users, check with team members, check with technicians; redesign, check, check, and recheck, but don't try to develop low-level detail or minispecs during this phase. That comes later after one general systems design alternative has been chosen for implementation.

Figure 6.1
SDLC phases and their related chapters in this book.

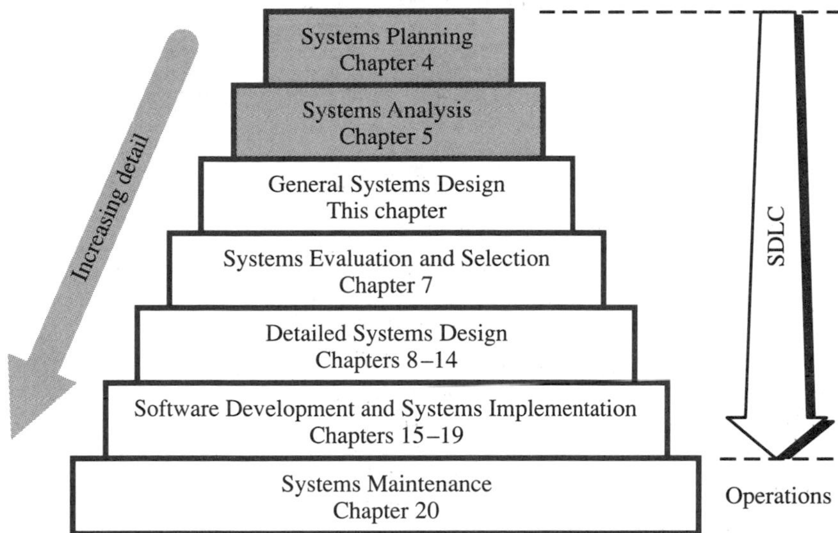

There are two broad approaches to systems design: **STRUCTURE-ORIENTED DESIGN** and **OBJECT-ORIENTED DESIGN**. The major portion of this chapter is devoted to these two design approaches. Before presenting them, however, let's first review systems design categories and rapid application development (RAD), a technique that applies to all phases of SDLC, especially general systems design.

WHAT ARE THE THREE CATEGORIES OF SYSTEMS DESIGN?

Systems designs can be divided into the three categories shown in Figure 6.2:

- Global-based systems
- Group-based systems
- Local-based systems

Global-Based Systems

Just as new factories must be built or completely overhauled, so must old, ineffective information systems be replaced with new, effective ones that enhance productivity, differentiation, and management (PDM) strategic factors. Such replacements are often caused by a new business strategy. They may, therefore, be large, complex enterprisewide information systems that process high volumes of transactions.

To design a **GLOBAL-BASED SYSTEM** requires a complete overhaul or replacement of all the systems design components. The following types of changes are common:

- Old output is changed from monthly tabular reports to online screen displays of two- and three-dimensional color graphics.

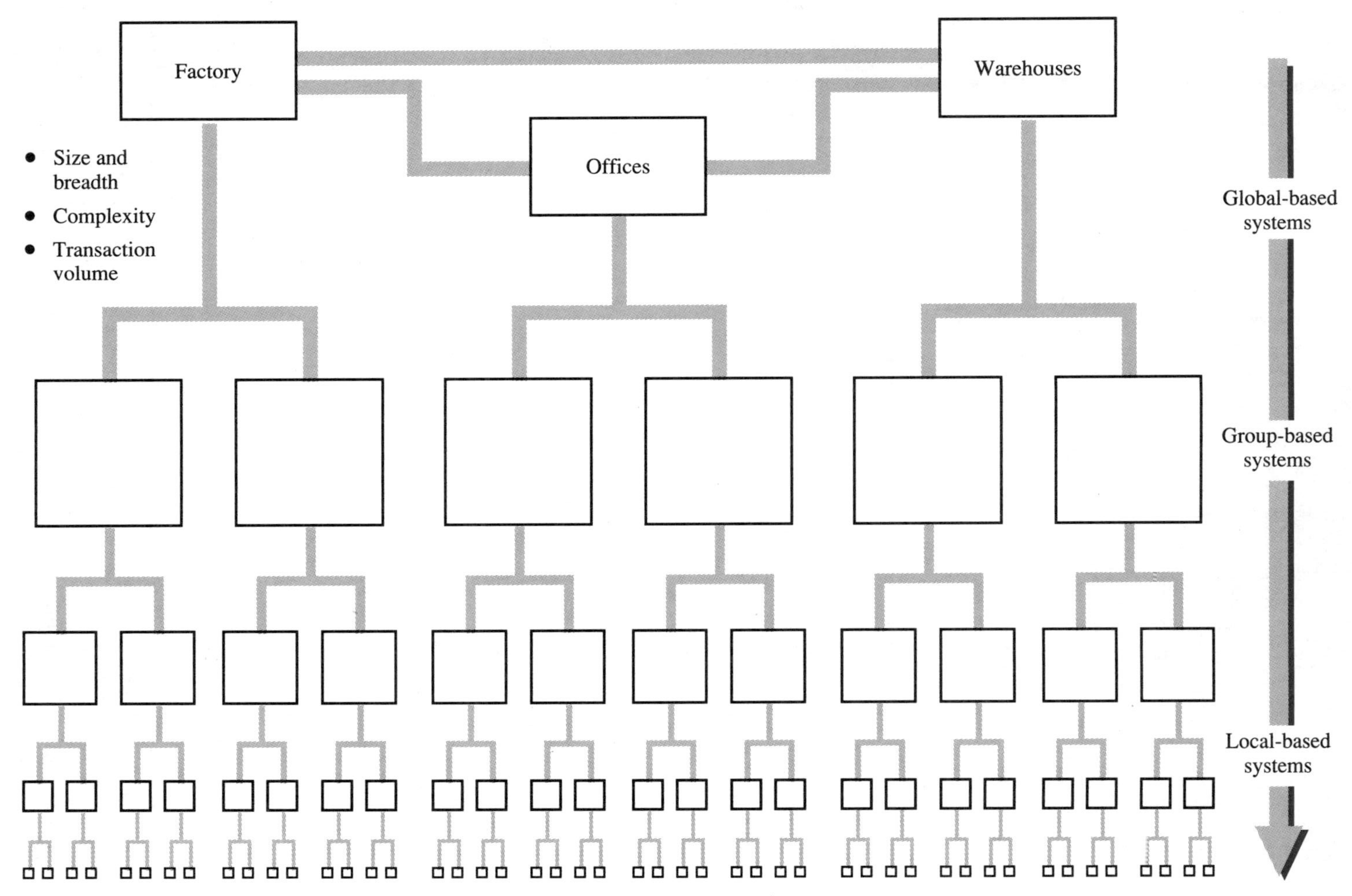

Figure 6.2 Categories of systems based on size and breadth, degree of complexity, and volume of transactions.

- New processes are developed.
- Input is captured via scanning devices rather than by pencil and paper.
- The old hierarchical database is converted to a new relational database with a standard query language.
- Various controls are installed, including uninterruptible power systems, disaster recovery plans, encryption devices, and biometric access control devices.
- A new technology platform, composed of an enterprisewide network topology to which a variety of computers and peripherals are connected, supports all of these systems design components.

To develop a global-based system may require several project teams under the direction of a chief information officer (CIO) or project director. A general systems design is created and coordinated among these teams and often presented to users in the form of a General Systems Design Worksheet illustrated in Figure 6.3.

The General Systems Design Worksheet brings together all the systems design components and gives a general narrative description of each. Several alternatives are presented to users for a walkthrough and review. Upon review of these alternatives, aspects of several alternatives may be combined to create a hybrid. One or two alternatives may be rejected, and one or more may be accepted as is. In any event, several general systems design alternatives, once thoroughly reviewed by users and all involved systems professionals, are ready for the acid test phase of the SDLC, the evaluation and selection phase covered in Chapter 7.

Group-Based Systems

The **GROUP-BASED SYSTEM**, although usually connected to a global-based system, services a branch office, department, or a special group of users in the enterprise. These groups have special needs to carry out their work and make proper decisions. Systems designers who are working on a group-based system should have a workable global-based system already in place. Therefore, designers may not be concerned directly with designing certain systems design components, such as the database and portions of the technology platform. The design focus is generally on output, input, process, control, and in some instances, a technology platform that pertains to the local group, such as a local area network.

Local-Based Systems

A **LOCAL-BASED SYSTEM** is typically designed for a few people, often only one or two, for a specific ad hoc application. A user may already have a microcomputer; he or she is planning to acquire a desktop publishing system. A systems professional is generally called in to work with the user in analyzing and design-

Figure 6.3
General Systems Design Worksheet showing one general systems design alternative.

GENERAL SYSTEMS DESIGN WORKSHEET

Company: Magna Manufacturing Project Director: Pam Smith

System: Corporate advanced system (CAS) Date: MM/DD/YY

Output:	**Input:**
Financial statements in tabular and graphical form Online displays of production variances Online invoicing to customers Daily inventory control reports: Usage patterns Reorders	Online transaction processing via scanning Online capture of production costs Online purchase orders from customers Online requisitioning for inventory items
Processes:	**Database:**
Cost-based and current value-based financial statements Variances computed by comparing materials, labor, and overhead standard costs with actual costs Electronic data interchange algorithm base on ANSI formats Statistical trend analysis for inventory and setting reorder points	Relational database with standard query language and online transaction processing capabilities Physical database change from magnetic tape files to magnetic disk files
Controls:	**Technology Platform:**
Disaster recovery plan Uninterruptible power system Input controls Processing controls Output controls Database controls Telecommunication controls Fire suppression system Access control: Password Biometric	Fiber optic enterprisewide backbone One mainframe Three superminicomputers Seventy workstations Two multiplexers Twelve large magnetic disk systems Four magnetic tape backup systems Three in-house developed software packages Four purchased software packages

ing requirements, evaluating different systems, acquiring one, and implementing it, along with its network and support.

Another good example of a local-based system is an **EXECUTIVE INFORMATION SYSTEM (EIS)** developed for two or three strategic-level decision makers, such as a chief executive officer (CEO), chief operating officer (COO), and the chief marketing officer (CMO).

USING THE FOUR KEY ELEMENTS OF RAPID APPLICATION DEVELOPMENT TO DESIGN SYSTEMS

The term **RAPID APPLICATION DEVELOPMENT (RAD)** was popularized by James Martin, a noted systems authority. Synergism is the idea behind RAD; that is, RAD combines elements that work together so that the total effect is greater than the sum of individual effects. We will use the Dillinson's Department Stores case to introduce RAD. Then we will present the four elements of RAD and examine it further.

General Systems Design Using Rapid Application Development (RAD): Dillinson's Department Stores Revisited

We introduced Chapter 5 with the systems analysis efforts of Dillinson's Department Stores, a nationwide chain with head offices in Chicago. In this chapter we rejoin Jim Montecucco, CIO at Dillinson's, and his systems managers, Brenda Clemens and Clarence Jankowsky, as they attempt to generate alternative conceptual systems designs. The information they gathered at the joint application development (JAD) session of Chapter 5 will be invaluable to them as they move into this next phase of the systems development life cycle.

Returning from the employee cafeteria, Brenda called to Clarence in the hallway, "Clarence! Are you headed to the meeting at a quarter after?" Clarence replied, "Well, yes, but I may be a few minutes late. I wanted to get enough copies of this article here for everyone at the meeting. I just came across it in this week's *InfoSys* magazine, and thought it might be applicable."

"What's it about? I don't suppose it has anything to do with CASE, does it?" Brenda asked as she winked at Clarence. She knew very well what a fanatic for CASE he could be. "Actually, it's about RAD, and I think it might just be the ticket to help us bring this project in on schedule. And yes, Brenda, it does involve CASE! Anyway, I'd better get these copies made before I miss the meeting altogether." Clarence headed for the copy room, while Brenda continued on her way to the conference room adjoining Jim's office, where the meeting had just begun.

As Brenda entered the room, Stephen Wells, Dillinson's CEO, was reviewing the Systems Analysis Report the group had completed last week. "It seems like you people have everything under control here," Stephen proclaimed. "I guess that JAD thing, or whatever you called it, was a big success!" Brenda nodded her head and smiled, although she was, as usual, underwhelmed by the insightfulness of Mr. Wells' comments. "Clarence should be here shortly," she remarked.

A few minutes later Clarence entered the room and distributed copies of the magazine article he'd been reading. "I mentioned this article to Brenda just a few minutes ago, but for the rest of you it's about a synergistic design methodology called

RAD, which stands for 'rapid application development.' Since we are running a little bit behind schedule with this project, my feeling is that RAD may streamline the general systems design phase and get us back on target." Jim queried, "Clarence, I understand how a synergistic approach can help improve the efficiency of a task by making the whole greater than the sum of the parts, so to speak. But what are the parts of RAD that could do this for systems design?"

Brenda interrupted, "Ah, I see here in the article it says that a JAD session is a key element of RAD. Maybe we've already got a head start on this!" "That's right," said Clarence. "Note that CASE tools are also a RAD element, as is prototyping. We've got some experience with those aspects of RAD, too."

Mr. Wells spoke up, "You folks explained JAD and CASE to me last time we met, but here's another acronym in this article—SWAT. Wasn't that a TV cop show?" "It sure was," said Clarence, "but as a RAD element it stands for specialists with advanced tools. A SWAT team would be a small, effective group of systems people like ourselves, equipped with all the latest CASE tools."

Jim mentioned that from his observations, there had been quite a bit of synergy already within the group. Darrell Laananen, the systems technician who was so capable with the CASE package at the recent JAD session, combined with Brenda and Clarence, had definitely been more effective than the three of them could have been working on their own. "Perhaps the RAD concepts are things we've already been using. We probably just need to recognize these elements formally within our group and make certain that we're using them effectively," he stated.

"Sounds good to me!" said Stephen, as he fumbled through the jar of peanuts on the table. "Just keep me posted on how things are going." Brenda replied, "We sure will, Mr. Wells. In fact, we'll be doing a lot of prototyping of the new system during this phase of SDLC, and we'll be calling in a number of users who will try out these prototypes and give us their reactions. When you get our General Systems Design Report, you'll see just how these prototyping sessions went." The meeting was adjourned, with the understanding that Jim, Brenda, and Clarence would begin right away with general systems design.

The team had read quite a bit in the systems journals about an object-oriented design approach, proclaimed to be a replacement for the structure-oriented approach they'd all been used to. Brenda was unsure if the team should embrace the object-oriented approach, although she knew the approach offered some advantages in terms of software code reusability. Jim made the decision here and opted for the structure-oriented approach. His decision was based on the fact that with rapid development as the goal, use of the structure-oriented approach they were all familiar with would require no extra time to learn a new approach. In addition, he felt that few standards were in place for the object-oriented approach, which might prove troublesome for the group later.

The structure-oriented approach would give the group a means of decomposing the design tasks in a top-down fashion. Two alternative subapproaches, the process-oriented approach for well-defined systems and the data-oriented approach for poorly defined systems, are possible as part of the overall structure-oriented design approach. Clarence suggested that the JAD sessions they'd conducted so far had pretty well defined the environment and requirements of the new system, and that the process-oriented approach would probably work well for the group. Jim agreed, and said "You know, we really have a good idea of all the inputs, processes, and outputs that will be involved with the new system. A lot of what we're doing here in the department store business is transaction-based. We're dealing with inventory, accounts payable, accounts receivable—all very stable, routine types of operations. The process-oriented approach is known to work well under these circumstances."

Over the next few weeks, Brenda, Clarence, and Darrell worked together to develop several alternative systems designs. Capitalizing on the elements of RAD helped to bring the team within three days of schedule by the time the general systems design phase was completed, which pleased Jim and Mr. Wells. Two modeling tools they found especially useful in this phase were the first-level data flow diagram (DFD) and the entity relationship diagram (ERD). They found the DFD to be especially useful in showing the processes of the system, while not getting bogged down in many of the finer details of systems operation. The team deliberately tried to keep their work at a general level, for they knew that detailed design work would come later once one of the feasible designs was selected.

The four key elements of RAD, as discussed in the sample case, are:

- Joint application development (JAD)
- Specialists with advanced tools (SWAT) teams
- Computer-aided systems and software engineering (CASE) tools
- Prototyping

Joint Application Development

JAD sessions are especially effective when global-based systems are under development because of the many different parties involved. On a less formal basis, JAD also works well for local- and group-based systems development.

The key word in JAD is *joint;* that is, users and systems professionals work together to analyze and design the system. Figure 6.4 shows three different design models:

- Designer's mental design model
- User's mental design model
- The conceptual systems design model described by a modeling tool, such as a data flow diagram (DFD), entity relationship diagram (ERD), decision table, screen prototype of a report, decision tree, and so on

The systems designer creates a conceptual systems design model from his or her mental design model. The systems designer's mental design model is formulated from past experience, expertise, study facts, and input from interacting jointly with the user. Ideally, the user's mental design model and the conceptual systems design model are equivalent. Should there be a difference, joint interaction and the design process are iterated (i.e., repeated) until the conceptual systems design model is equivalent to the user's mental design model.

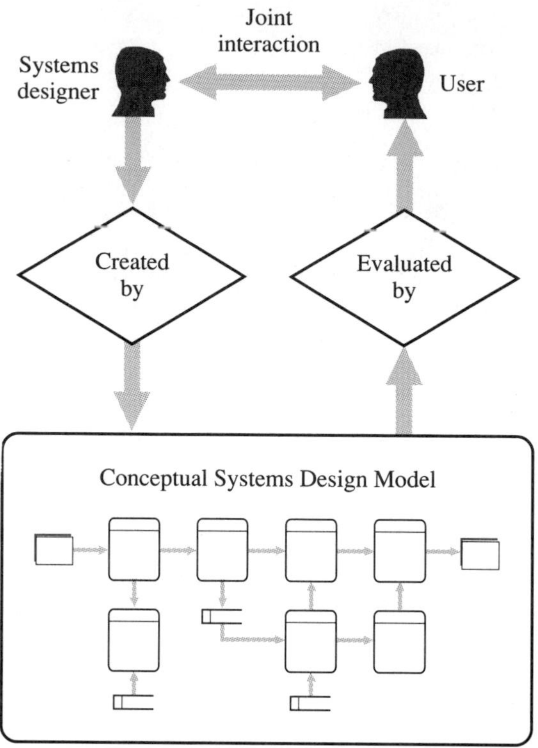

Figure 6.4
Systems designer and user interacting jointly to create conceptual systems design model.

Specialists with Advanced Tools (SWAT) Teams

SPECIALISTS WITH ADVANCED TOOLS (SWAT) TEAMS are generally composed of three or four highly skilled and motivated systems professionals armed with CASE tools. Formal research and practical experience indicate that small systems project teams are typically more productive than large systems project teams. Common sense tells us that bright, well-trained, highly motivated people are likely to be highly productive.[1]

CASE Tools

CASE tools are used by SWAT teams to increase systems development productivity and quality of work to:

- Add discipline to systems development
- Reduce design errors and omissions
- Reduce systems rework

Prototyping

Prototyping works hand in glove with JAD, in which users are shown what they're going to get, allowing them to react to it. CASE facilitates prototyping

[1] Merlyn Vaughan, "Is RAD 'RAD'?" *Software Magazine,* February 1991, p. 9.

by enabling SWAT teams to create screen designs, various models, and dialogues quickly, and to modify them while interacting with users.

With RAD, the evolving prototype is not thrown away; instead, it becomes part of the final systems design. Often, this approach exploits the 80:20 rule, in which 80 percent of user requirements can be met with 20 percent of the systems design. RAD identifies and quickly delivers the critical 20 percent. While users benefit from using the initial application, the SWAT team works on the rest of the system. Users' experiences with the initial part of the system help the SWAT team define needed changes that might have gone undiscovered otherwise. In this way the system is developed as it is used: that is, organically.

A variation of the 80:20 rule applied to systems development is DuPont's time box technique for systems projects that must be done in less than 90 days. The time box approach is as much a project management technique as a systems development method. DuPont reasoned that any effort longer than 90 days would be likely to miss the business opportunity, and it would also exceed time and budget estimates. An unexpected benefit, according to one of the time box's originators, Scott Shultz, was the creative explosion that occurred when users and developers were forced to separate a systems design's essential elements from the merely desirable ones.[2]

USING THE STRUCTURE-ORIENTED DESIGN APPROACH TO DESIGN SYSTEMS

We introduced the structured approach in Chapter 2, and the major focus of this book is on this approach. The structure-oriented design approach is based on the methodologies, modeling tools, and techniques of the structured approach. However, structure-oriented design is sometimes divided into two camps:

- The process-oriented approach
- The data-oriented approach

The goal of both the process-oriented and data-oriented approaches is to identify all data attributes that are needed by the system under development. The process-oriented approach accomplishes this goal by examining all existing input, output, and processing for a system. The data-oriented approach accomplishes this goal by examining the decisions that are made in a system, and then working backward to identify the data required to support those decisions.

Examining Current Input, Output, and Processing to Determine Data Needs

Systems professionals use the **PROCESS-ORIENTED APPROACH** by examining input, output, and processing of a given application to determine the data needs of the system. Applying the process-oriented approach, systems professionals look at all reports, display screens, calculations, and decisions needed for a process. The DFD is frequently used for the process-oriented approach.

[2] *Ibid.*, p. 8.

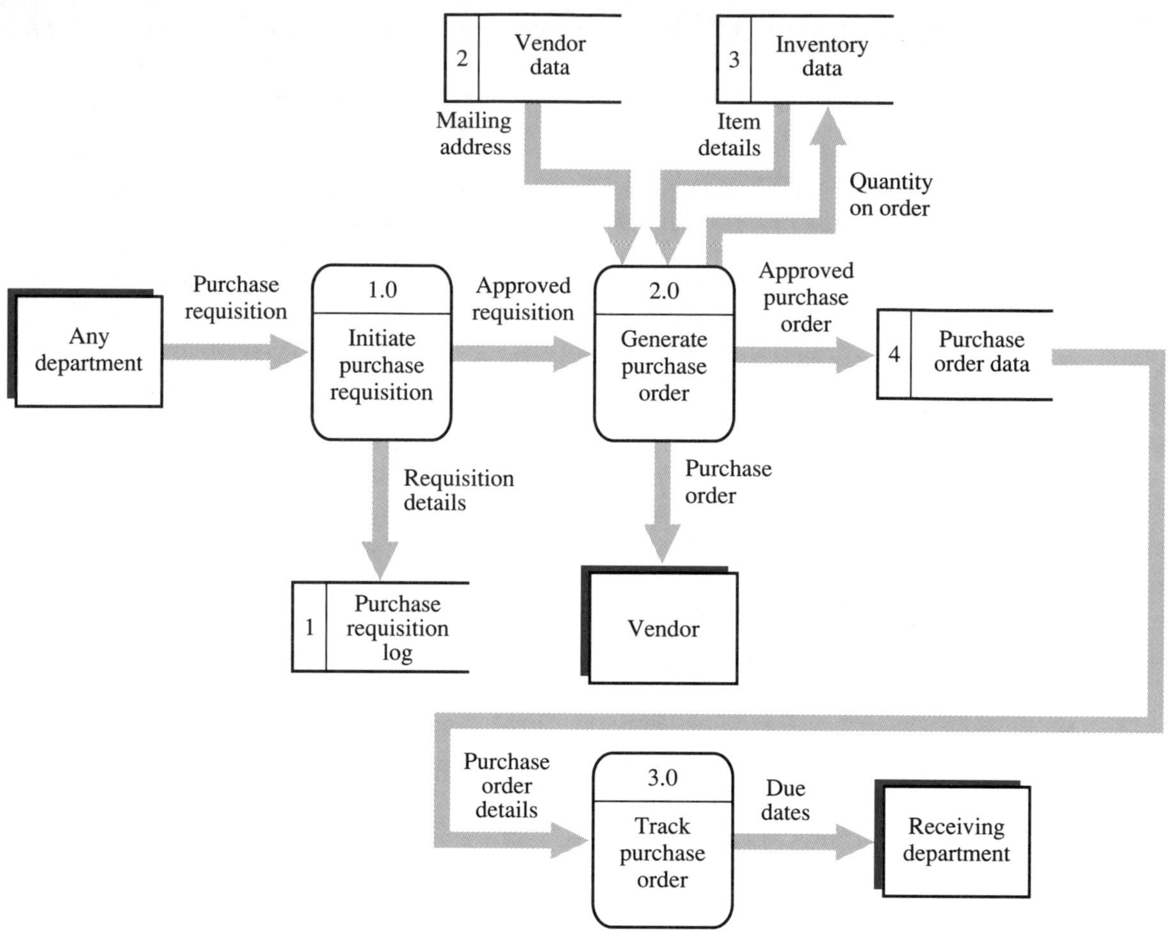

Figure 6.5
First-level DFD for purchasing system.

We will use the process-oriented approach to examine the data requirements necessary to support an organization's purchasing function. Figure 6.5 is a DFD for a purchasing system. From this diagram it is apparent that the overall input to the system is a purchase requisition from any department, and the output is a purchase order to be sent to a vendor. The system also uses vendor and inventory data for input. It also produces tracking data for the receiving department as output.

At this point, systems professionals gather all the necessary input and output documents. Figure 6.6 is a sample of a purchase requisition, and Figure 6.7 is a sample purchase order. From this information, systems professionals begin to devise the data needed to create the application.

For example, let's examine the purchase requisition form in Figure 6.6. Some of the attributes shown on this purchase requisition form are quantity required, minimum quantity required, part number, description, work order number, date required, and requisition date. Looking at Figure 6.7, the purchase order, we can see that some of the attributes for a purchase order include

BRUCE IND., INC. **MATERIAL REQUISION FORM** 64994

REQUISITION

QTY REQUIRED	MIN QTY REQUIRED	PN-DESCRIPTION-MANUFACTURER-OP/PROCESS & SPEC	FOR WO's	USED ON	DATE REQUIRED

Note: ______

❑ MRO NOT FOR PRODUCTION ❑ REQUISITION FOR KIT ONLY ❑ ALSO USED ONE ______

REQUISITIONED BY: ______ DATE: ______ APPROVED BY: ______ DATE: ______

MATERIAL CONTROL

	QTY	NOTE
❑ ON ORDER		
❑ FITTED THIS JOB		
❑ STOCK		
❑ REQUISITION FOR KIT ONLY		
❑ ENG. ALTERNATIVE		
❑ ONE-TIME SUBSTITUTE		

ALTSUS APPOVAL ENGINEERING BY: ______ DATE: ______ ALTSUS APPOVAL O.C. BY: ______ DATE: ______

CHECKED BY: ______ DATE: ______ FILLED BY: ______ DATE: ______

PURCHASING

DATE	VENDOR	PHS	CONTACT	QUANTITY	PRICE	EXTENSION	DELIVERY

Note: ______

ORDER PLACED WITH: ______ P.O. NO.: ______ QTY ORDERED: ______ DELIVERY PROMISE: ______

PLACED BY: ______ DATE: ______ APPROVED BY: ______ DATE: ______

FOLLOW UP

FOLLOW-UP	DATE	CONTACT	STATUS

Figure 6.6
Sample purchase requisition.

ORIGINAL

BRUCE *"WE LIGHT THE SKIES"* *INDUSTRIES, INC.*

P.O. Box 1700 Dayton, Nevada 89403 Ph: 702-246-0201
FAX: 702-246-0451 TWX: 910-395-0020
FED. SUPPLY CODE 17023
D-U-N-S 00-052-4612

PURCHASE ORDER NO.

VENDOR CODE

PURCHASE ORDER

TO:

SHIP TO:

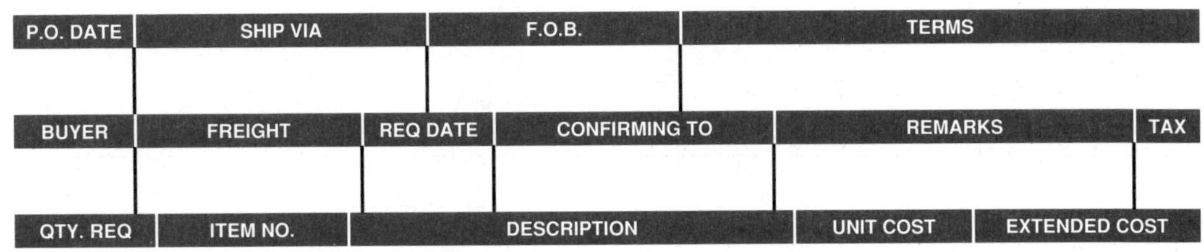

P.O. DATE	SHIP VIA	F.O.B.	TERMS

BUYER	FREIGHT	REQ DATE	CONFIRMING TO	REMARKS	TAX

QTY. REQ	ITEM NO.	DESCRIPTION	UNIT COST	EXTENDED COST

NEVADA RESALE PERMIT 1-106-443

BY: ______________________________

1) THE FOLLOWING QUALITY CONTROL CONDITIONS REFERENCED ON THE REVERSE SIDE APPLY: PARAGRAPH (3). 2) BY ACCEPTING THIS ORDER YOU ASSENT TO ALL THE TERMS AND PROVISIONS CONTAINED HEREIN. INCLUDING THOSE ON THE REVERSE SIDE. 3) MATERIAL SUBJECT TO INSPECTION AT OUR PLANT.

Figure 6.7
Sample purchase order.

Figure 6.8
List of some of the purchasing system attributes.

Purchase Requisition	Purchase Order
Requisition Number	Purchase Order Number
Quantity Required	Purchase Order Date
Minimum Quantity Required	Vendor Number
P/N Description	Vendor Name (To:)
For Work Order Number	Vendor Code
Used On	Shipping Instructions (Ship Via)
Date Required	Terms
Note	Requested Date (Req. Date)
Requisition Type Flag	Items Ordered
Requisitioned By	Quantity Ordered (Qty. Req.)
Requisition Date	Item Number
Approved By	Item Description
Approval Date	Unit Cost
Follow-up Date	Extended Cost

purchase order number, vendor code, purchase order date, shipping instructions, terms, and item ordered. (As you identify necessary attributes, you should write them down in a simple list format such as you see in Figure 6.8.) The attributes have been grouped under the entities "Purchase Requisition" and "Purchase Order," because it is obvious from the input/output documents that the attributes belong to these entities.

The process-oriented approach works very well if systems professionals know in advance the input, process, and output from a system. It also works very well if the data from each set of applications are essentially separate. This approach is effective for transaction-based applications such as accounts payable, accounts receivable, payroll, and inventory control. Each of these applications usually has a very stable set of input, process, and output.

But what if you are trying to design a system, but no one knows exactly what the system should produce? Or the output changes every few months to meet new demands or new situations? Or what if you are trying to identify the data needed to complete a process that depends on a variable set of processes? You will have to use a data-oriented approach.

Consulting with Future Users to Determine Data Requirements

The **DATA-ORIENTED APPROACH** is used by systems professionals working very closely with future users of the system. This approach is used when the processes of a system, as well as the input and output, are relatively undefined. In such cases, systems professionals must exert constant effort to determine, with users, how the system will be employed. The focus of the data-oriented approach is to determine data requirements for the decisions that will be based on the data.

To understand the data-oriented approach, let's expand our purchasing application to include another function. Instead of simply keeping track of all purchase requisitions and purchase orders, we also want to create a system that will help the buyers evaluate vendors—to measure input quality. We need to determine the data that have to be stored to assist a buyer making these vendor evaluations.

The first step in the data-oriented approach is to discuss potential decisions that will be made from the system with the users of the system. In this case, each buyer wants to be able to decide which is the best vendor from whom to purchase a given item. This decision is based on a set of prior decisions, including:

- Which vendor can provide the best price?
- Which vendor can deliver the best quality item?
- Which vendor has the best financing options?
- Which vendor can deliver the item when needed?
- Which vendor has a track record of on-time deliveries?

In this step, we have named specific decisions made by the buyer. We can now define the data necessary to make these decisions.

The second step in the data-oriented approach is to model the purchasing decision support system through the use of a flexible modeling tool, such as an entity relationship diagram (ERD). Figure 6.9 is an ERD defining the entities and the relationships between them in this systems design. (Figure 6.9 does not depict the people involved in this system, but they are added in the following

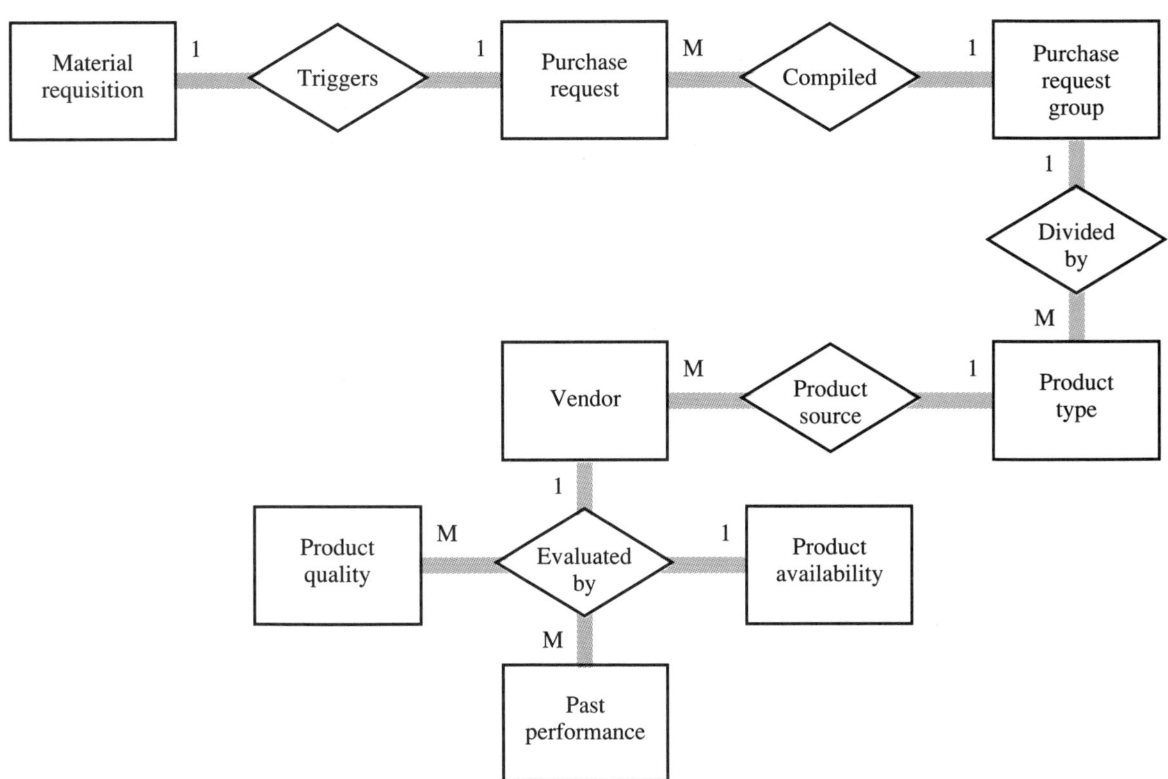

Figure 6.9 Entity relationship diagram for purchasing decision support system.

description.) The entity "Material Requisition," submitted by a department, causes the purchasing clerk to create a Purchase Request. The Purchase Request is in turn compiled by the purchasing agent into a series of requests called a Purchase Request Group. The purchasing agent then distributes the Purchase Request Group to the correct buyer. The buyer takes the compiled data in the Purchase Request Group and divides it into stacks by Product Type. The buyer is then able to identify which vendors would be applicable for a given Product Type. To evaluate the best vendor for a given Product Type, the buyer uses three distinct criteria:

- Product quality
- Past performance
- Product availability

The third step in the data-oriented approach is to divide each criterion into its attributes. The systems designer must ask the buyer what makes up each criterion in this system. Figure 6.10 shows the attributes that compose the criteria Past Performance and Product Availability.

The fourth and final step in the data-oriented approach is to break down the attributes identified above into their most fundamental data components. For example, the systems designer and buyer identified an attribute called Price Discount Calculation. To perform this calculation, subattributes such as Low Quantity Ordered, High Quantity Ordered, and Price are needed. This means that these subattributes must be stored in the system.

Creating the Data Dictionary for a System

Once the data are identified for a system using either the process-oriented or data-oriented approach, systems professionals use this information to create a DATA DICTIONARY. The data dictionary details each of the attributes and subattributes for a system and defines their relevant characteristics. The data dictionary includes the following characteristics about an attribute:

- Size
- Type
- Description
- Limits and exceptions

Figure 6.10
Attribute definition.

Past Performance	Product Availability
Item Description	Item Description
Quantity Ordered	Order Quantity
Price	Price
Order Date	Elapsed Delivery Time
Delivery Date	Delivery Method
Discount Offered	Price Discount Calculation
	Delivery Terms

- Ranges
- Security level
- Access privileges

An example of part of the data dictionary for the purchasing system under discussion is included in Figure 6.11.

Three aspects of data dictionaries require special note:

1 ***Changes*** The data dictionary is not a static document that is never changed. Preparing the data dictionary at this point is the task of formalizing all you have learned about the data for the system. Later, during database design, more work will be needed before attributes are finally specified.

2 ***Description*** Attribute names must describe the data being defined fully. For example, many abbreviations were used on the purchase requisition and purchase order shown earlier. Those abbreviations cannot be used in the data dictionary. Instead of Req_No, the field must be called Requisition_Number.

3 ***Order*** Attributes should be arranged in some form of standard order. Data dictionaries are sometimes arranged by alphabetic order of the attribute name, by alphabetic order of entity, or by order of where an attribute is used.

At the general systems design level, one cannot assume that the design process is complete. Normally, more refinement of the design is required during detailed systems design. More design refinement will often indicate that the data dictionary is not complete. Also, during database design, more modeling and other tasks will be needed before the data dictionary is fully specified.

Summary of Structure-Oriented Design

The goal of process-oriented and data-oriented approaches is to define all needed data attributes. Most systems have at least one component that is

Figure 6.11
Portion of a purchasing system data dictionary.

Attribute Name	Type	Size	Lower Limit	Upper Limit	Default Value	Characters	Where Used	Description
Purchase_Order_Number	Numeric	6	0	999999		9(6)	Purchase_Order	A number that makes each purchase order unique
Purchase_Order_Date	Date	8	1980	2005	Current Date	MM/DD/YY	Purchase_Order	The date a purchase order is used
Requisition_Number	Numeric	5	0	99999		9(5)	Purchase_Requisition	A number that makes each requisition unique

already well defined. Tracking purchase orders by date is an example. For such components, a designer can examine the input and output documents, the processes of the system, and the output requirements to determine needed data attributes. However, most new systems usually have at least one component that is not fully defined before systems development. The purchasing system, for example, included a new decision-support component to assist buyers in making decisions about vendors. The input and output for this system was not fully defined, because at present most of the input and output, and all of the processes, occur within the buyer's own mind. We used the data-oriented approach to identify each entity in the system and each entity's main attributes. A good systems designer will use both approaches in designing most systems applications.

USING THE OBJECT-ORIENTED DESIGN APPROACH TO DESIGN SYSTEMS

Object orientation has emerged as a driving design and programming paradigm. The aim of object orientation is to bring the software industry in line with the hardware industry, which employs reusable components.[3] Thus the main goal of object-oriented design is to design and build systems through the assembly of reusable software objects rather than writing software modules from scratch.

Designing and Building Systems Using Reusable Software Objects

For years, people have asked: "Why must every new systems project start from scratch? There should be a catalog of software modules from which designers could choose to create the new system." Such an approach would mean that less software would have to be written. Prewritten, pretested, and cataloged modules could be used in other systems. This approach would result in increased systems development productivity and higher systems quality.

The key to reusability is to catalog (in a library or database) objects that may be found by keywords, so that a designer can determine the likelihood that some existing object or objects will satisfy user requirements. Indeed, searching for such objects in a library or database and the immediate electronic downloading and prototyping of selected objects suggests a powerful, efficient way to design systems.

What Are the Key Elements of Object-Oriented Design?

Following are the chief elements of object orientation and the most popular terms used to identify them.

Objects

An OBJECT is anything that we deal with in the environment, such as a car, a computer, or a hamburger. An object exhibits certain behaviors. Object orientation provides the same concept in systems. In information systems, attributes

[3] Alex Lane, "Demystifying Objects—A Framework for Understanding," *Object Magazine*, May/June 1991, p. 62.

(also called data or data structures) and operations are **ENCAPSULATED** (pulled together) to create objects that behave in certain ways. For example, customer Charles Jones is an object encapsulating Mr. Jones' data and the billing operations performed on his data.

The term "operations" is used to describe how attributes are processed at the design level. The term "methods" (also called procedures, functions, and code) is used to describe the actual program written using an **OBJECT-ORIENTED PROGRAMMING (OOP)** language. Systems design focuses on identifying objects rather than specifying attributes and program code to manipulate them. Such specifications are made during software design and coding (see Chapters 16 and 17).

To users, objects behave in certain ways in response to **MESSAGES**. All access to objects is through messages that only specify what should be done, not how, leaving it to the objects to choose the operations that should be performed to respond to the messages appropriately. In this way, the combining (i.e., encapsulating) of attributes and operations enables objects to appear to possess intelligence. An object, then, is able to act on itself or to send messages to other objects.

Classes

The concepts of **CLASSES** and objects are tightly interwoven, for we cannot discuss an object without regard for its class. A class is a set of objects that share common structure and behavior. A class is a type; a single object is an instance of a class.

The Customer class in a company may be comprised of two instances: Wholesale Customer object and Retail Customer object. All customers in a company share common attributes, such as Customer_Number, Customer_Name, Address, Credit_Rating, and the like. The specific value of each attribute uniquely describes a specific customer. Common messages for making Customer objects behave in certain ways are: "Change Credit Rating of Customer 164 to B" or "Display Amount Owed by Customer 197."

From objects to classes we are building a logical set of relationships, or a hierarchy. In a tools-inventory class hierarchy, as illustrated in Figure 6.12, the **SUPERCLASS** Tools Inventory is made up of two **SUBCLASSES**: Type A and Type B. Specific tools-inventory items, such as Pick, Shovel, Hammer, and Saw, are subclasses of Type A and Type B.

Objects can inherit attributes and operations from other objects and can add more attributes and operations of their own. **INHERITANCE** enables the sharing of properties (i.e., attributes and operations) between classes while preserving their differences. Using inheritance between classes enables designers and programmers to reduce redundancy in attributes and operations.

For example, the Inventory object may contain attributes such as Item_Number, Total_Cost, Total_Price, and operations for manipulating these attributes. Each type of inventory inherits these properties and adds Inventory_Type, Location, and Vendor attributes combined with appropriate operations. Each Inventory object can inherit all the properties from its appropriate type and add such attributes as Description, Cost, Price, Quantity_Received,

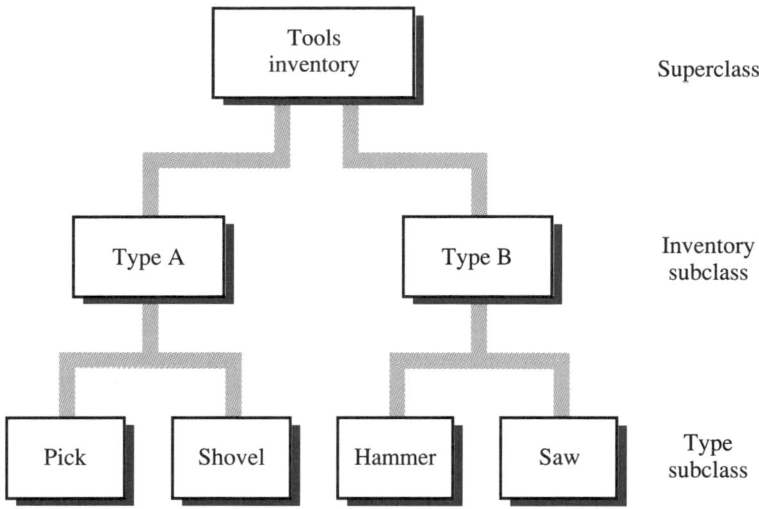

Figure 6.12
Hierarchy of tools-inventory classes.

Quantity_Issued, Quantity_On_Hand, and Rcorder_Point, plus appropriate operations for manipulating these attributes.

The ability of any objects to respond to the same message and of each object to implement it appropriately is called **POLYMORPHISM**. This ability greatly improves the flexibility of the object-oriented design approach. To explain polymorphism, let's look at the Inventory objects. One of their behaviors is to check Reorder_Point for reordering. The message: "Check Reorder Point and Display" defines "Check Reorder Point" as it applies to each inventory object. If you think about it, polymorphism has been around for a long time. For example, suppose that you are a FORTRAN programmer and you have two integer variables *IA* and *IB*. To calculate their sum you could write

```
ISUM = IA + IB
```

Similarly, if you have two real variables called *A* and *B*, you could calculate their sum by writing

```
SUM = A + B
```

Note that the "+" operator indicates integer addition and real addition. Any way you look at it, real addition is a very different operation from integer addition, yet using the symbol "+" for both integer and real addition actually makes the intent of both expressions clear. Such is the power of polymorphism.[4]

During the general systems design phase, designers normally do not concern themselves with the detail of attributes, operations and programming code, inheritance, and polymorphism. The specifics of these object-oriented elements are determined during software design that is abstracted from systems

[4] Tim Chase, "Everything You Wanted to Know About Object Oriented Programming But Were Afraid to Ask," *Interact,* February 1991, p. 26.

design. Then the software design is implemented by coding the objects using an object-oriented programming (OOP) language. For now, we are concerned about object-oriented design at the general systems design level, which involves high-level conceptual design. The next section presents steps to achieve such a design.

What Steps Are Required to Perform Object-Oriented Design at the General Systems Design Level?

The aim of object-oriented systems designers at the general systems design level is to understand the application domain (the scope of the system under development) so that a logical design can be created. At this level, essential object classes and their relationships are captured without introducing detailed design and implementation features that may prematurely restrict the high-level logical design. With study facts and user requirements collected from the systems analysis phase, the general systems design phase is the starting point for object-oriented design as it is with structure-oriented design. Object class properties, attributes, attribute values, operations, and methods are described in detail later during implementation (see Chapters 16 and 17). The conceptual model developed in this phase will be refined several times before it is finally implemented.

At this level, we perform four steps:

1 Identify object classes.

2 Identify relationships between object classes.

3 Identify major attributes, not detailed specifics.

4 Determine inheritance relationships and build a class hierarchy from them.

1: Identify Object Classes

The designer interviews users to identify candidate object classes. Nouns and noun phrases spoken by users while making requirements statements indicate object classes. For example, a librarian (user) may say, "We need a way to catalog and check out videos." "Videos" becomes an object-class candidate for the new cataloging and checkout system for the library.

Candidate object classes for any system under development are discovered in the same manner. We will use a tools-inventory ordering system as our object-oriented design example. Assume that interviews have been conducted, and a list of candidate object classes has been compiled. This is shown in Figure 6.13.

Our next task is to review this list and select only those object classes that are relevant to the systems application domain. We therefore eliminate all redundant or vague candidate object classes and candidate object classes that are actually attributes. The results of selecting relevant object classes are displayed in Figure 6.14.

If two or more object classes express the same information, the most descriptive should be selected. In the example, Customer is more descriptive than User and Client. Object classes should also be specific. For example,

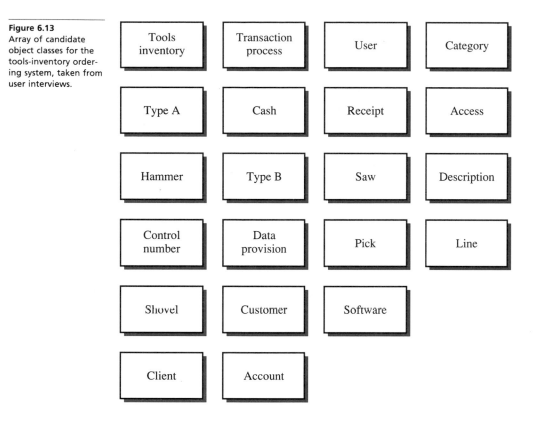

Figure 6.13
Array of candidate object classes for the tools-inventory ordering system, taken from user interviews.

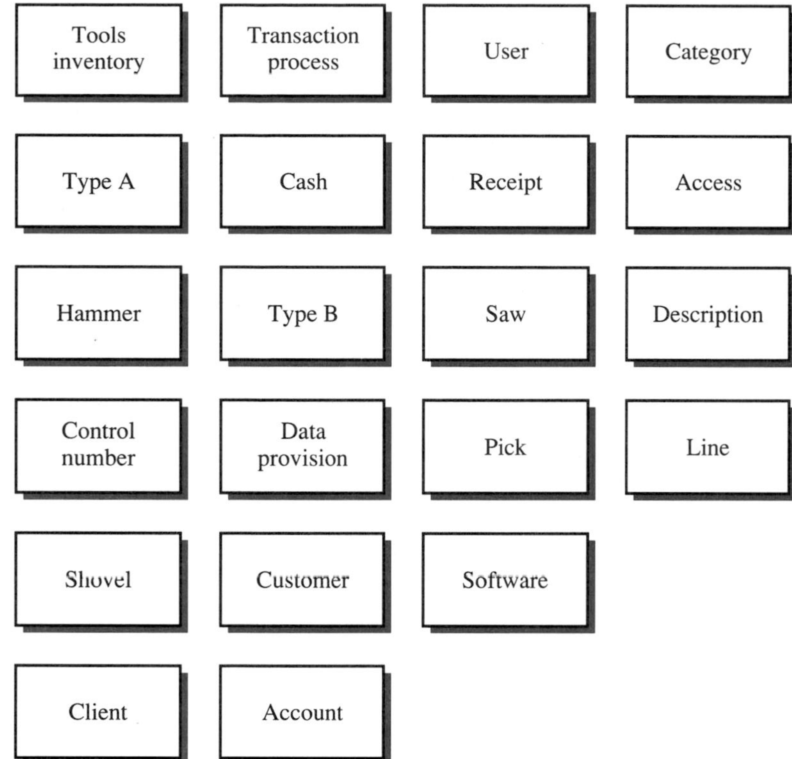

Category, Data Provision, Line, Software, and Access are vague in the systems application domain. Sometimes, things that may first appear as object classes can turn out to be attributes. For example, Cash, Receipt, Description, and Control Number will probably become attributes, as will a large number of other words to be defined during detailed systems design and implementation.

2: Identify Relationships

Any dependency between two or more classes is a relationship. A reference from one object class to another is a relationship: for example, "Customer places an order." A relationship corresponds to a verb in a requirements statement. Figure 6.15 portrays the relationships among the identified candidate object classes.

3: Identify Attributes

After relationships among object classes have been defined, it is necessary to identify major object class attributes. Attributes are properties of individual objects, such as Customer_Name, Description, Quantity_On_Hand, Quantity_Ordered, Price, and so on. Attributes are normally identified by adjectives of a requirements statement, such as "Tools-inventory items are identified by item number and description." Unlike object classes and relationships, attributes and attribute values are less likely to be fully described at the general

Figure 6.14
Array of object classes resulting from grouping object classes.

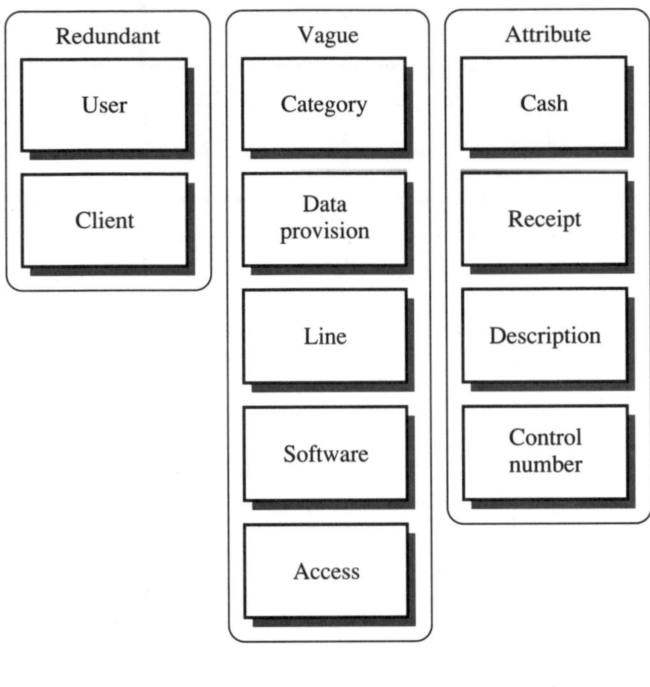

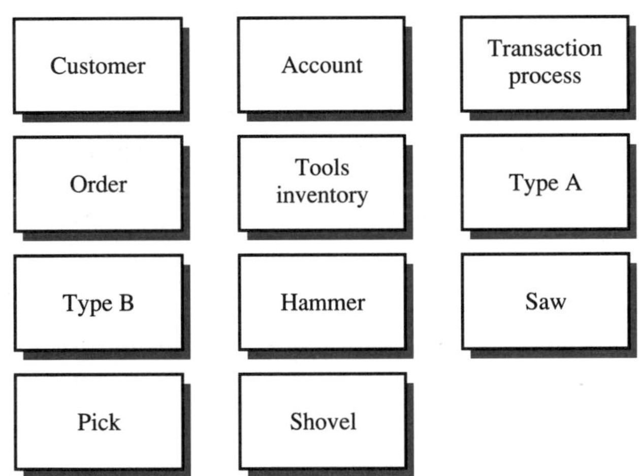

systems design level. As the general systems design model becomes more and more refined, so too are attributes. At this level, designers do not carry the discovery of attributes to excess. Describe the most important attributes first; fine details can be added later.[5]

[5] James Rumbaugh, Michael Blaha, William Premerlani, Frederick Eddy, and William Lorensen, *Object-Oriented Modeling and Design* (Englewood Cliffs, NJ: Prentice-Hall, 1991), p. 162

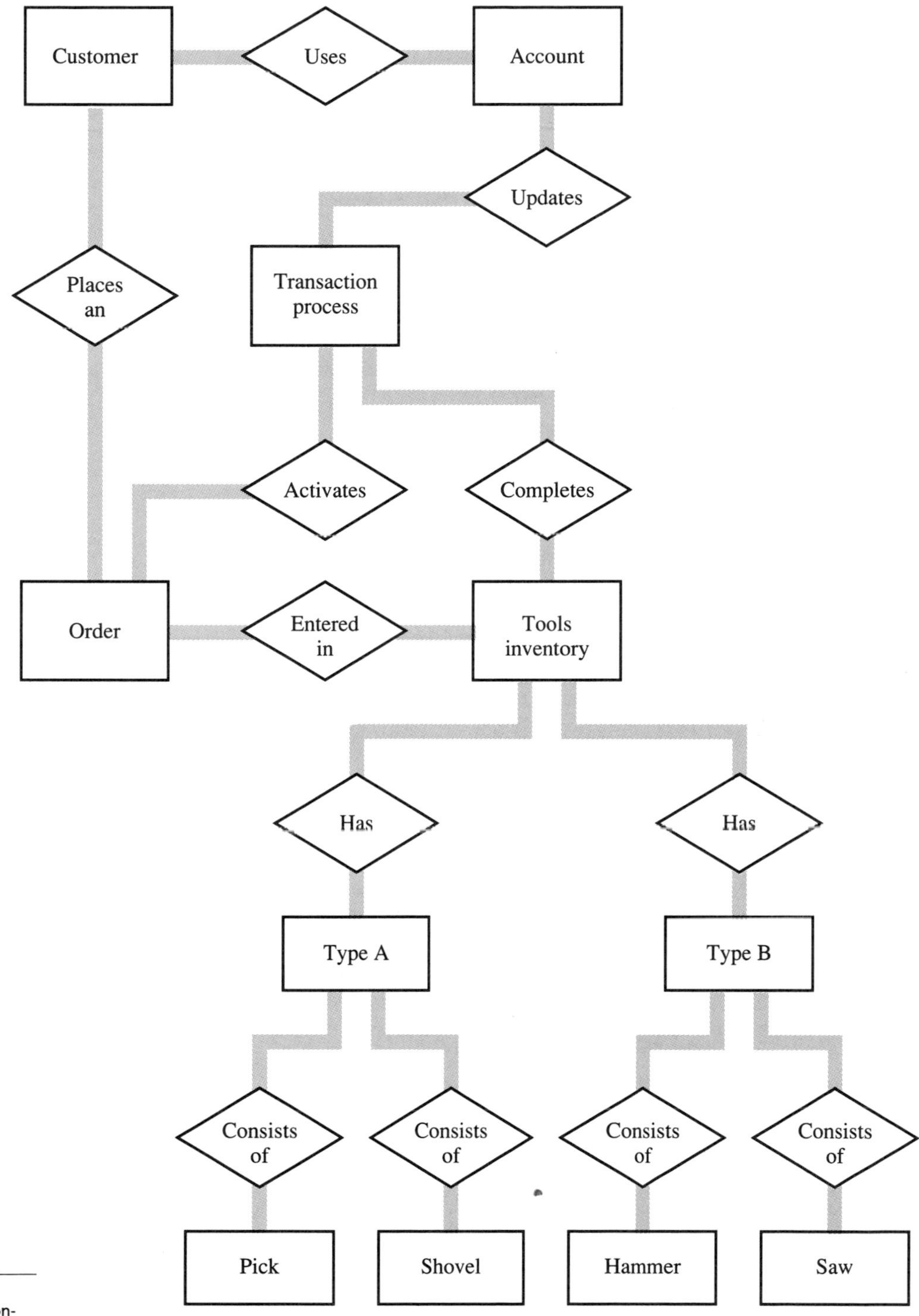

Figure 6.15
Object class relationships. Verbs in diamonds correspond to verbs in user-requirements statements.

4: Identify Inheritance and Build a Class Hierarchy

This step organizes classes by using inheritance to generalize common aspects of existing classes into a superclass (bottom up) or by refining existing classes into specialized subclasses (top down).

An excellent design approach is to create an object-oriented model by building a hierarchy of classes in which subclasses (i.e., more specialized object classes) inherit properties defined by superclasses (i.e, more generalized object classes). A subclass can inherit from one or more superclasses.

Generally, inheritance is discovered from the bottom up by searching for object classes with similar relationships, attributes, and operations. But sometimes classes have nothing in common but the relationship. Figure 6.16 presents the tools-inventory ordering system design model after adding inheritance.

The class hierarchy, shown in Figure 6.16, is composed of three superclasses:

- Order
- Transaction process
- Tools inventory

Customer is a subclass of both Order and Transaction Process. Thus Customer can inherit properties from both superclasses. This is referred to as multiple inheritance. Account is a subclass of Customer. This is a form of single inheritance. Type A and Type B are subclasses of Tools Inventory. Pick and Shovel are subclasses of Type A, and Hammer and Saw are subclasses of Type B.

What we have achieved to this point is a high-level conceptual object-oriented design. Before this design is implemented, several iterations will be

Figure 6.16
Class hierarchy.

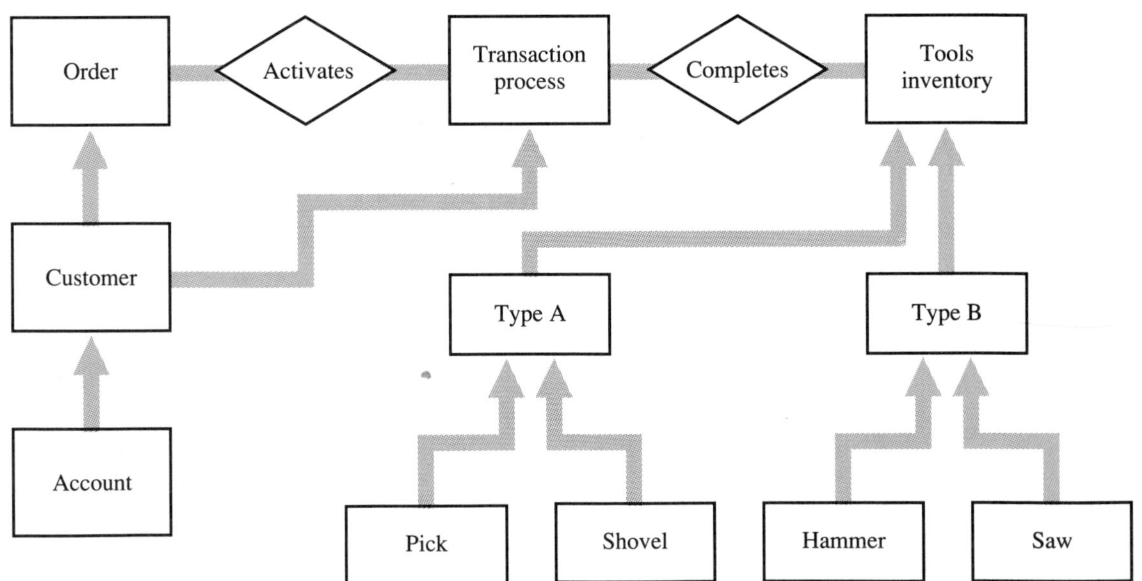

necessary to ensure that the systems design is workable. Then, with the systems design in hand, software design tasks specifying attributes and operations in detail will be performed. Then the software design will be coded to convert operations (procedure statements that both users and programmers understand) into methods (i.e., program code). Should the current information system have an object class library available, any of the object classes in the library that can be used in the systems application domain will be accessed and assembled along with those object classes that are created from scratch.

WHAT IS THE DIFFERENCE BETWEEN STRUCTURE-ORIENTED AND OBJECT-ORIENTED DESIGN?

Some people maintain that there is little difference or conflict between the structure-oriented and object-oriented design approaches. Still others say that structure-oriented design falls short of object-oriented design. In any event, we provide you with an analysis of both approaches in the following subsections.

Modularity: A Design Principle of Both Approaches

Both structure-oriented and object-oriented proponents extol the virtues of **MODULARITY**. It is a basic principle of good design, because it facilitates maintenance, reusability, reliability, and extendability of the system. We present a more detailed analysis of modularity in Chapter 16.

In structure-oriented design, a module is a unit of software code that performs a well-defined function. In object-oriented design, a module is an object that encapsulates attributes and program code to behave in a certain manner.

Modularity is the degree to which modules are standardized and independent, and show variety in use. If any systems design is composed of n modules, the number of intermodule connections (also called binding and coupling) should remain much closer to the minimum, $n - 1$, shown as (a) in Figure 6.17, rather than to the maximum, $n(n - 1)/2$, shown as (d). The reason: If there are too many connections between modules, the effect of a change or of an error may extend to a large number of modules. Ideally, the design structure of modules should strive to develop modules that stand on their own.

The modular structure in (a) shows an extremely centralized or authoritarian design, in which a central executive module communicates to every other module. Modular structure (c) is also authoritarian. It shows a top-down structure with a great deal of communication between the topmost module and lower-level modules. This design of modules is encouraged by the structure-oriented approach. Structure (d) has the greatest number of interconnections, and therefore the greatest potential for failures.

A less authoritarian design structure than (a) or (c) is (b). In this design structure, every module communicates only with its two immediate neighbors, but there is no central or top-level authority. Such a design scheme as (b) does not conform to structured, top-down design, as shown in (c). But it is a design structure that the object-oriented design approach will provide if it fits the real needs of users.

Figure 6.17
Types of module interconnections.

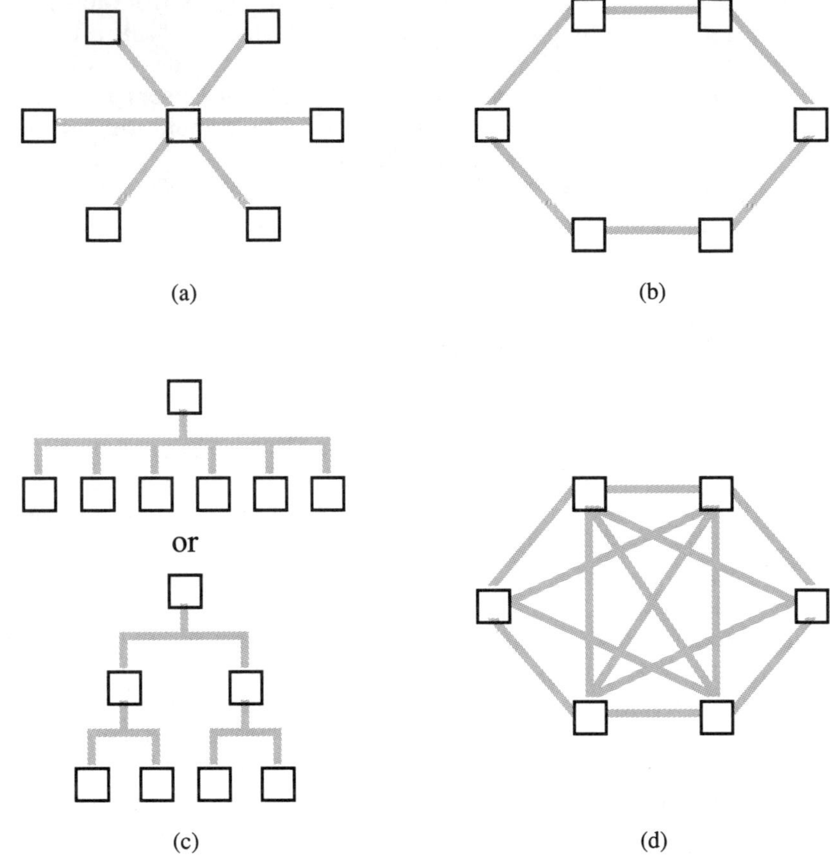

Controlling the amount and form of communication between modules is the essence of modularity, and the key objective of modularity is autonomous, stand-alone modules. Proponents of structure-oriented design say the same thing. But object-oriented proponents counter by arguing that the structure-oriented approach cannot build designs in which modules are highly independent, that is, very loosely tied to other modules.

Top-Down Versus Bottom-Up Design

A general view of how the structure-oriented approach uses top-down decomposition to design systems is presented in Figure 6.18. **TOP-DOWN DESIGN** directs analysts and designers to start with an abstract description of the new system and then refine this view in successive steps. Process 3 in the topmost DFD is decomposed into a detailed DFD at level 2. Assuming that this level 2 DFD gives sufficient detail for software design, process 3.3 is decomposed into a structure chart (a Warnier–Orr or Jackson diagram could also be used). The structure chart is a design of the software modules necessary to implement process 3.3. Software designs for the other processes are developed in a similar manner.

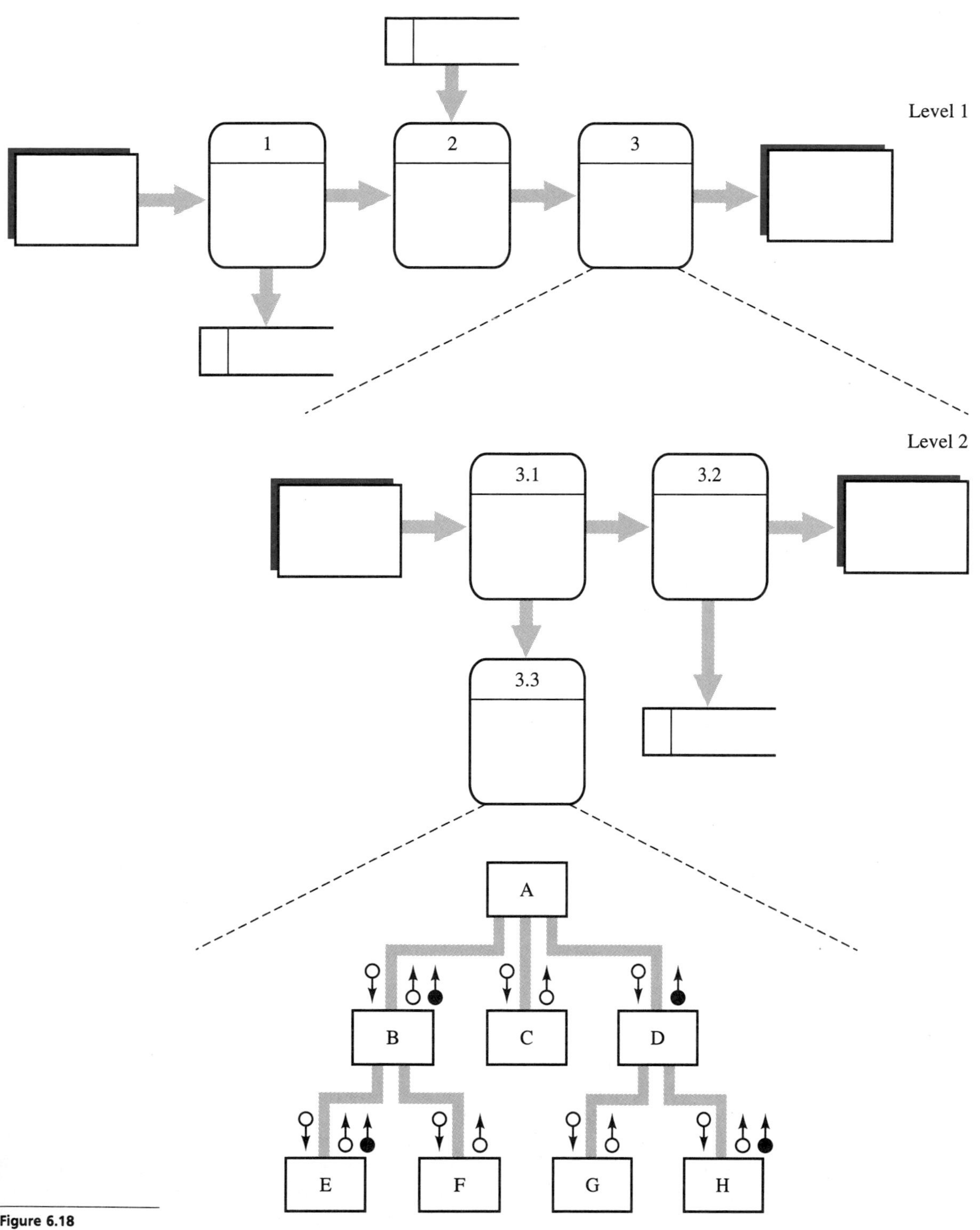

Figure 6.18
General view of top-down design.

Look at Figure 6.18 again. Notice that the structure chart shows control and input/output data being passed between modules, which is normal with structured, top-down design. These modules require:

- Opening files
- Reading data
- Indicating end of files
- Assigning values to certain variables
- Manipulating data
- Printing results

In our example, module A starts the process. It has access to all the data flowing throughout the design. The other modules send and receive data that belong to or derive from the data used by module A. This means there is relatively tight coupling and interplay among all the modules.

Object-oriented design adopts a neutral attitude toward leveling, such as top-down decomposition. The designer lists the various requirements that are applicable to the systems application domain, but defers for as long as possible the specification of the order in which these requirements may be applied. This approach is what some object-oriented authorities call the "shopping list" approach. That is, a list of requirements is determined during analysis; then object classes are assembled to see if this particular bundle of object classes meets user requirements. If not, several object classes are put back "on the shelf" and others are assembled. This prototyping-like process continues until a workable design is effected.

The design process of tying together a bunch of objects with a view to executing a result is called "assembly" by some authorities. The result of such an assembly of object classes is a designed system.

This **BOTTOM-UP DESIGN** process is analogous to building a new car. Objects, such as motors, instrumentation, power trains, wheels, chassis, and fuel tanks are available from a number of manufacturers. To put into production a new automobile design, you may start out by buying motors from manufacturer A, instrumentation from manufacturer B, wheels from manufacturer C, and so on. The only thing new that you may develop from scratch would be the body. Microcomputers are designed and developed much the same way. Many components under the skin of a microcomputer from vendor A are the same as those for a microcomputer from vendor B. Often, the only difference is in the cabinet and nameplate. The same modules (objects) are reused for each car or microcomputer. Extendability for cars is achieved by installing larger (or smaller) components such as motors, wheels, fuel tanks, storage components, and so on. Physical products, designed in this manner, are easy to maintain; standard components are reusable and reliable because of standardization and testing; and the products can be extended because of their modular structure. The same applies to object-oriented systems design.

Using Existing Object Class Libraries

The preceding discussion assumed that the designer had available an **OBJECT CLASS LIBRARY** and the object classes did not have to be developed from scratch and programmed. Otherwise, the SDLC goes through all phases from determining user requirements to general systems design to detailed systems design to software design and coding no matter which approach is used. (Indeed, if a library of structure-designed modules were available, the shopping list and assembling process used with the object-oriented approach could also be employed by the structure-oriented approach.)

The big promise of the object-oriented design approach is reusability. One aim of object-oriented designers is to create libraries so the systems professionals can look through object class catalogs, pick out appropriate object classes, retrieve them from object class libraries or databases, and assemble them to produce systems much as design engineers now select, retrieve, and assemble components to produce an electronic product.[6]

In the software world, such an approach has led to fruition of the object-oriented approach in window management objects, such as menus, dialogue boxes, and icons. We can, for example, open a document in a word processor by dragging an icon representing the document to the icon representing the word processor. The document is a text file object that displays itself in the window. Even the window itself is an object containing other objects. We can also access a graph object (e.g., pie chart, line graph), drag it to a marked area in the document, and incorporate it into the text document.

REVIEW OF CHAPTER LEARNING OBJECTIVES

The major goals of this chapter were to enable each student to achieve five important learning objectives.

Learning objective 1:
Define three systems design categories.

Systems design deals with three categories of systems: (1) global-based systems, (2) group-based systems, and (3) local-based systems.

A global-based system encompasses the entire organization. It is, therefore, large and complex with a high volume of transactions, possibly processed both by batch and real-time systems. A global-based system is the foundation information system for the entire company. Without it, the company could not survive.

Group-based systems are developed for units, segments, branches, or departments in an organization. Examples are:

- The group of cost engineers for a large construction company
- The billing department for a large hospital

---

[6] Scott Moody, "Library Classes," *Object Magazine*, May/June 1991, p. 15.

- The sports department of a university
- The financial planning group for a company

Normally, much of the technology platform—for example, a telecommunication network backbone and a number of workstations—may already be available.

Local-based systems are concerned with a specific end or purpose and are developed to meet the parochial needs of one or a few users. Such systems may be stand-alone for one user or connected to some part of a group-based or global-based system (e.g., the corporate database) for one or a few users. Examples are desktop publishing for a cost accountant or an executive information system (EIS) for a few strategic-level executives.

Learning objective 2:
Discuss the use of RAD in general systems design and describe RAD's four key elements.

The purpose of RAD is to create workable systems designs that fit user needs precisely and implement high-quality critical applications quickly. The four key elements of RAD are:

- Joint application development (JAD)
- Specialists with advanced tools (SWAT) teams
- Computer-aided systems and software engineering (CASE) tools
- Prototyping

Learning objective 3:
Describe the structure-oriented design approach.

The structure-oriented design approach applies modern methodologies such as SDLC and prototyping, modeling tools, CASE technologies, RAD, and various other techniques to employ an engineered, disciplined approach to systems development. The essence of the structure-oriented approach is top-down decomposition.

The structure-oriented design approach involves two subapproaches. The process-oriented approach is applied for well-defined systems in which the output, input, and process requirements are fairly well established. The data-oriented approach is used for ill-defined systems that support a variety of decision-making tasks.

Learning objective 4:
Describe the object-oriented design approach.

The object-oriented design approach uses all the methodologies, modeling tools, and techniques employed by the structure-oriented approach. The key objective of the object-oriented approach, at the general systems design level, is to identify:

- Object classes
- Relationships
- Attributes
- Inheritance

Learning objective 5:
Compare the structure-oriented design approach with the object-oriented design approach.

Both the structure-oriented design approach and the object-oriented design approach are based on the modularity principle. Software modules for the structure-oriented approach contain program code that access data from a file or database. Software modules, called objects for the object-oriented approach, encapsulate data attributes with program code.

Structure-oriented design applies top-down decomposition. If object classes are developed from scratch, the same approach is used in which object classes are designed at a systems level, refined into a software design, and finally coded with the use of OOP language. However, the ultimate goal of object-oriented proponents is to have, in an object class library or database, a complete catalog of object classes from which object classes are retrieved and assembled in a bottom-up manner to form a completed systems application. In this way, object classes are programmed once. They can then be reused many times.

GENERAL SYSTEMS DESIGN CHECKLIST

The purpose of the general systems design phase is to create several conceptual systems design alternatives that meet user requirements. Following is a checklist on how to conduct this phase.

1 Marshal sufficient systems design resources for the systems design category.

2 Employ rapid application development (RAD) using joint application development (JAD), specialists with advanced tools (SWAT) teams, computer-aided systems and software engineering (CASE) tools, and prototyping. CASE systems are used to verify the accuracy, consistency, and correctness of design solutions. For example, a data flow diagram (DFD) may contain an element (e.g., data store or process) that has not been defined previously. This situation would be flagged and corrected. This CASE verification feature can be likened to debugging features of programming language compilers.

3 If you apply the structure-oriented design approach, use the process-oriented approach for well-defined applications and the data-oriented approach for decision support systems that are more difficult to define.

4 If you use the object-oriented design approach in which the object classes are developed from scratch (i.e., not retrieved and assembled from an object class library), employ methodologies, modeling tools, and techniques that are used in the structure-oriented approach. In fact, the front-end work for both approaches is similar.

5 Use four steps for object-oriented design at the general systems design level:

Step 1. Identify object classes.
Step 2. Identify relationships.
Step 3. Identify attributes.
Step 4. Identify inheritance and build a class hierarchy.

KEY TERMS

Bottom-up design

Classes

Data dictionary

Data-oriented approach

Encapsulated

Executive information system (EIS)

Global-based system

Group-based system

Inheritance

Local-based system

Messages

Modularity

Object

Object class library

Object-oriented design

Object-oriented programming (OOP)

Polymorphism

Process-oriented approach

Rapid application development (RAD)

Specialists with advanced tools (SWAT) teams

Structure-oriented design

Subclass

Superclass

Top-down design

REVIEW QUESTIONS

6.1 What are the three systems design categories? Give an example of each.

6.2 What is a General Systems Design Worksheet? What is it used for?

6.3 What is the purpose of rapid application development (RAD)? List and briefly describe its four key elements.

6.4 Define a SWAT team and describe its purpose.

6.5 What is the 80:20 rule, and how does it apply to systems development?

6.6 Define *time box.*

6.7 Define the process-oriented approach and give its purpose.

6.8 Define the data-oriented approach and give its purpose.

6.9 Define the data dictionary and give its purpose.

6.10 Describe *reusability*. Explain how it can increase systems development productivity and quality of design.

6.11 Define:

- Object
- Encapsulated
- Messages
- Classes
- Inheritance
- Polymorphism

6.12 List and briefly describe the steps in performing the object-oriented design approach at the general systems design level.

6.13 Explain why modularity is the basic design principle for both the structure-oriented and object-oriented design approaches.

6.14 What is a class library, and what is its purpose?

CHAPTER-SPECIFIC PROBLEMS

These problems require exact responses based directly on concepts and techniques presented in the text.

6.15 Following are systems projects:

________________	Two tax accountants want you to help them review several tax-planning packages with the objective of installing one of them on their microcomputers.
________________	You are assigned to help develop a cost-estimating system for three engineers who work for a construction company.
________________	The billing department at Amax is planning on integrating various computing resources to be shared by all employees in this department.
________________	According to a comprehensive information systems strategic plan, Nucor plans to change from a centralized, batch system to a real-time system with computing resources distributed throughout the enterprise.
________________	A group of financial planners has hired you to develop a decision support system.

Required: Identify each of the preceding systems projects as either global-based (GL), group-based (GR), or local-based (LO).

6.16 After completing general systems analysis for Trident Company, you have developed in narrative form a conceptual description of one of the design alternatives. Following is this description: Orders from customers are scanned with block codes. Customer orders are processed through reorder and economic order quantity models. If any inventory items are to be reordered, the system automatically generates purchase orders indicating items to be ordered, quantity to be ordered, and the vendor. Before purchase orders are sent to vendors, a limit check is run against a table of reasonable order quantities for each inventory item (i.e., the order limit for check valves is 24). For this particular design alternative, you are recommending a relational database and a local area network.

Required: Prepare a General Systems Design Worksheet to organize your preceding narrative design.

THINK-TANK PROBLEMS

These problems call for a feasible approach rather than a precise solution. Although the problems are based on chapter material, extra reading and creativity may be required to develop workable solutions.

6.17 A specialist metal-plating company identified the need for a more rigorous method of pricing and estimating. Their established pricing and estimating techniques, and hence the quotations produced, relied heavily on the experience of particular specialists. The specialists subjectively interpreted a pricing policy formulated by the company. Inconsistencies in pricing occurred. They were caused by different interpretations of the policy and complexity of the processes involved. Complexity arose because of the diversity of components to be finished and the large number of subprocesses carried out (several hundred possible combinations). Pricing and estimating were further complicated by constant changes in trading conditions, particularly metal price fluctuations. Moreover, users had difficulty in articulating their requirements.

Required: Basing your answer on what you have learned to this point in the book, list the methodologies, modeling tools, techniques, and design approaches you would use to prepare a general systems design for the application above. Also, justify your choices.

6.18 Suppose that your task is to design a system to keep track of employees in a company. Assume that the company can hire or fire employees (they never quit), and the employees sometimes get raises. An interview that you had with a major user follows:

Systems designer: What do you need?
User: I need to keep track of the employees in this company.
Systems designer: What do you need to keep track of?
User: Their salaries and working hours.
Systems designer: No problem.

Required: Perform object-oriented design at the general systems design level. Make any assumptions that you deem necessary to create your design.

6.19 A person works for a company which has departments that manufacture products. A person can be a worker or a manager. A worker works on a project. A manager is responsible for projects; he or she also manages a department.

Required: Using an ERD, model the objects above and show their relationships. Also, identify a few of the major attributes for each object.

6.20 An object-oriented design includes Shape as a superclass, and Circle, Triangle, and Rectangle as subclasses. Solid Rectangle is a subclass of Rectangle.

Required: Draw the class hierarchy for a shape-drawing system. Also, discuss how inheritance and polymorphism may apply to this object-oriented design. What's the key message that involves each object?

SUGGESTED READING

Booch, Grady. *Object Oriented Design with Applications.* Redwood City, Calif.: Benjamin/Cummings, 1991.

Chase, Tim. "Everything You Wanted to Know About Object Oriented Programming But Were Afraid to Ask." *Interact,* February 1991.

Coyne, R. D., et al. *Knowledge-Based Design Systems.* Reading, Mass.: Addison-Wesley, 1990.

Evans, James R. *Creative Thinking in the Decision and Management Sciences.* Cincinnati, Ohio: South-Western, 1991.

Lane, Alex. "Demystifying Objects—Framework for Understanding." *Object Magazine,* May/June 1991.

McIntyre, Scott C., and Lexis F. Higgins. "Object-Oriented Systems Analysis and Design: Methodology and Application." *Journal of Management Information Systems,* Summer 1988.

Martin, James. "DSS Tools Help Build, Analyze Models to Make Decisions." *PC Week,* May 8, 1989.

Martin, James. "SWAT Teams Will Play Pivotal Role in '90s Development." *PC Week,* March 5, 1990.

Merlyn, Vaughan. "Is RAD 'RAD'?" *Software Magazine*, February 1991.

Meyer, Bertrand. *Object-Oriented Software Construction*. Englewood Cliffs, N.J.: Prentice-Hall, 1988.

Moody, Scott, "Library Classes." *Object Magazine*, May/June 1991.

Parsaye, Kamran, Mark Chignell, Setrag Khoshafian, and Harry Wong. *Intelligent Databases: Object-Oriented, Deductive Hypermedia Technologies*. New York: John Wiley, 1989.

Rumbaugh, James, Michael Blaha, William Premerlani, Frederick Eddy, and William Lorenson. *Object-Oriented Modeling and Design*. Englewood Cliffs, N.J.: Prentice-Hall, 1991.

Smith, Tom. "Object Lesson." *Workstation News*, March 1991.

JOCS Case: General Systems Design

After the steering committee examined the Systems Analysis Report for JOCS, they approved completion of the next phase in the systems development life cycle: that of general systems design. It was clear to both Mary Stockland, CIO, and Jake Jacoby, JOCS project manager, that the steering committee was committed to making this project work. Mary Stockland requested and received approval for funding for a team of three new systems analysts to complete this project and the other projects that would be undertaken in the future. Both Mary and Jake feel that a JOCS SWAT team would be the best approach to the design of this system. A SWAT team would work intensely and quickly to complete the evaluation and detailed design for the project. While Mary began the hiring procedure for the new team, Jake began the general design of the new system.

During systems analysis, Jake learned that Peerless had successfully installed a computer-integrated manufacturing (CIM) system. Jake determined that the CIM system is structured in a configuration illustrated in the top half of Figure 6.19. A small computer in the factory floor control room directs the processes necessary to produce a hydronautical lifter. This computer accepts the bill of materials, created by the engineering department, and controls the flow of materials that are required to manufacture a lifter. Needed components are loaded on an unmanned tote vehicle that is guided by the computer to the appropriate manufacturing worker. Manufacturing workers at Peerless include both people and robots. People are employed in tasks that require judgment and decision making, while robots are used to complete routine tasks that require precision or strength.

In this automated factory environment, robots scan bar-coded components. They are instructed by the computer to complete an assembly or perform a certain task. Human workers, at the start of their shifts and when they move to a different manufacturing workstation, pass a bar code wand over their ID cards. This procedure tells the computer that the employee is at work

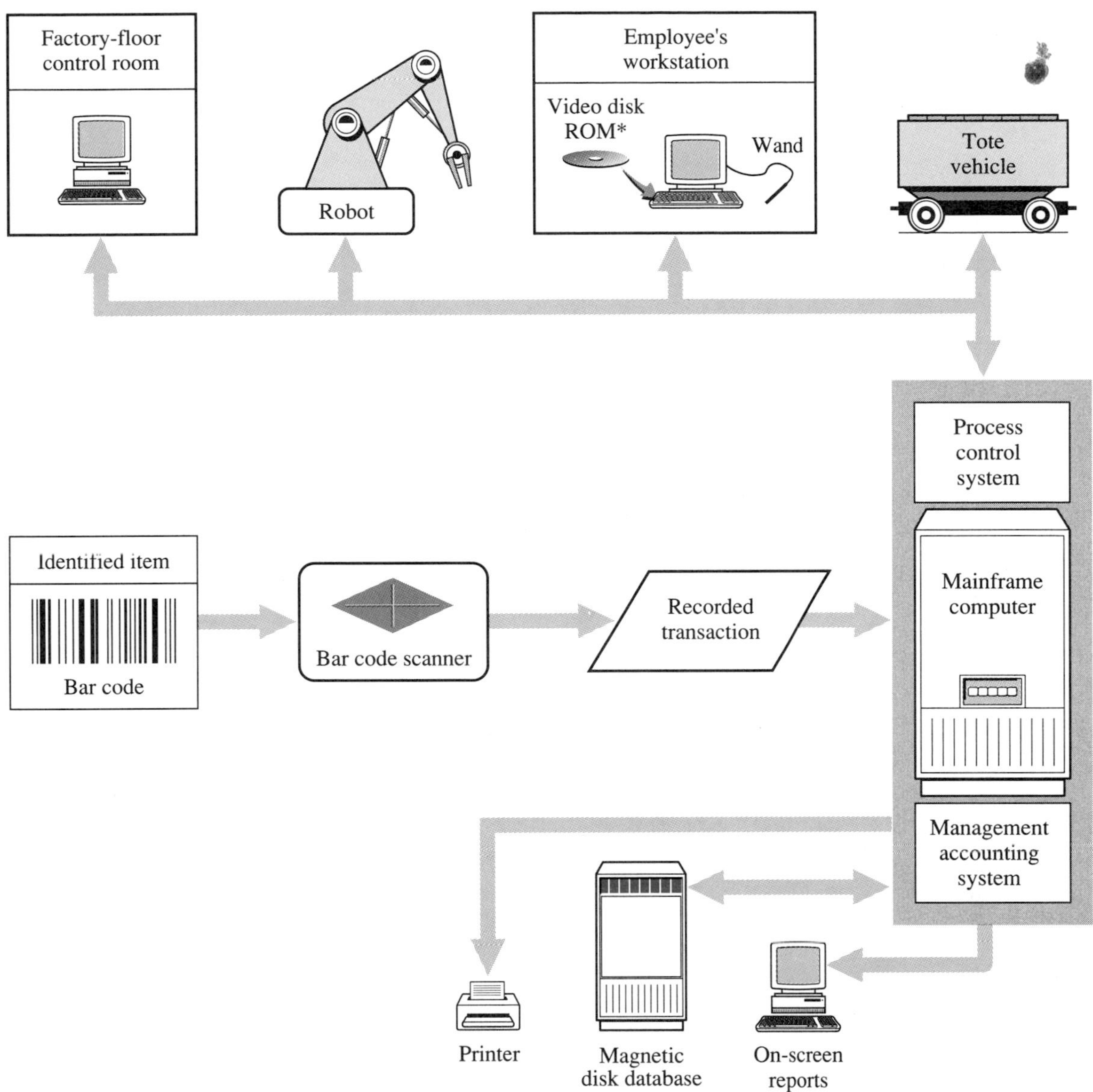

* Optical format Read-Only Memory, used to record instructions in video form

Figure 6.19
Computer-integrated manufacturing (CIM) system.

at a particular time, workstation, and job. If a human worker needs additional information about the job that he or she is about to start or is already engaged in, the worker can access help screens from the computer. The help screens provide video images that serve as tutorials and demonstrations of how a particular task is performed.

Jake must design the JOCS project to interface with the existing CIM system. He has drawn the bottom half of Figure 6.19 to show how JOCS will integrate with the CIM system by using data from that system. JOCS will accept the direct labor and material transactions captured on the CIM computer bar code scanner and will use these data to create managerial reports and screen-based displays.

To accomplish this integrating objective, Jake holds two brainstorming sessions with the future end users of JOCS. At first glance, JOCS appears to users to be an accounting system aimed at providing financial data for the company cost accountant. However, JOCS is really a global-based system, because it spans many areas of the company. JOCS has the potential to provide information for better control in purchasing raw materials, controlling manufacturing processes, and providing feedback to engineering, as well as simply controlling costs. As a result, Jake includes managers from the engineering, finance, manufacturing, and purchasing departments in his brainstorming sessions.

Management in the finance department wants JOCS to provide reports detailing direct materials variance, direct labor variance, volume variance, and controllable variance. They want to receive this information by specific job, by type of lifter, by specific customer, and by type of customer.

Management in the purchasing department would like to find out which vendors deliver high-quality materials and which vendors deliver on time. They would also like to find out failure rates, spoilage rates, and ease-of-work rates for all input materials.

Management in the engineering department would like to know if manufacturing personnel have to change their bills of material during production. They would also like to develop standard bills of material to apply past lifter design to figure design more easily. The engineering design manager would like

> a system that would ask us questions about the design of the lifter based on the specifications of the customer. It would be great if the computer asked us about dry-dock proportions, ship size, and usage. Then the computer could come up with the optimal configuration based on the probabilities we assigned from past lifter design.

Management in the manufacturing department has been very pleased with the CIM system. But in addition to this system they would also like to improve their control of job scheduling. Right now, there is no way to put the amount of time a job should take at a specific workstation into CIM and receive information about when that job will be completed. The production supervisor asked:

Would it be possible to get an updated estimate of completion for each workstation as a job progresses? Marketing always wants to know when a job will be ready to deliver. Most times I just have to estimate delivery based on how I think a job is progressing. If we could say, "This job has finished raw material cutting, taking 15 percent longer than it was supposed to, has finished milling, taking 10 percent less time than it was supposed to, and now will be completed in 35 more days," I wouldn't have to spend days making guesstimates.

Jake Jacoby writes down the design suggestions from each department and summarizes the potential suggestions in General System Design Worksheets. The four systems are depicted in Figures 6.20a to 6.20d.

Figure 6.20
(a) Basic JOCS general systems design.

GENERAL SYSTEMS DESIGN WORKSHEET

Company: Peerless, Inc. Project Manager: Jake Jacoby

System: JOCS—basic design Date: 07/15/91

Output:	**Input:**
Cost variance report by job	Direct material via bar-code scan
Cost variance report by type	Direct labor via bar-code scan
Cost variance report by date	Overhead: fixed and variable
Cost variance report by process	Standards: labor and materials
Display screens of above	
Processes:	**Database:**
Price variance	Possible relational
Quantity variance	Standard query language capabilities
Rate variance	Online transaction processing
Efficiency variance	Recovery and backup
Volume variance	Fourth-generation language
Controllable variance	
Applied manufacturing overhead	
Controls:	**Technology Platform:**
CIM data range checking	Expanded secondary storage
Validity data checking	Expanded communications network
Disaster recovery plan	5 additional workstations
Access control	2 additional printers
Database control	1 additional backup device
	JOCS software package
	Database management system
	CIM JOCS software interface
	CIM JOCS hardware interface

Figure 6.20
(b) Basic and purchasing JOCS general systems design.

GENERAL SYSTEMS DESIGN WORKSHEET

Company: Peerless, Inc.

Project Manager: Jake Jacoby

System: JOCS—basic design with additions for purchasing control

Date: 07/15/91

Output:

Cost variance report by job
Cost variance report by type
Cost variance report by date
Cost variance report by process
Display screens of above
Raw material usage by job
Raw material usage by vendor
Raw material spoilage by job
Raw material spoilage by vendor

Input:

Direct material via bar-code scan
Direct labor via bar-code scan
Overhead: fixed and variable
Standards: labor and materials
Direct materials purchases
Purchase orders
Receiving
Spoilage

Processes:

Price variance
Quantity variance
Rate variance
Efficiency variance
Volume variance
Controllable variance
Applied manufacturing overhead
Raw material usage
Raw material spoilage

Database:

Possible relational
Standard query language capabilities
Online transaction processing
Recovery and backup
Fourth-generation language
Prototyping tool

Controls:

CIM data range checking
Validity data checking
Disaster recovery plan
Access control
Database control

Technology Platform:

Expanded secondary storage
Expanded communications network
8 additional workstations
3 additional printers
1 additional backup device
JOCS software package
Database management system
CIM JOCS software interface
CIM JOCS hardware interface

Figure 6.20
(c) Basic and engineering JOCS general systems design.

GENERAL SYSTEMS DESIGN WORKSHEET

Company: Peerless, Inc. | Project Manager: Jake Jacoby

System: JOCS—basic design with additions for engineering feedback and projections | Date: 07/15/91

Output:	**Input:**
Cost variance report by job	Direct material via bar-code scan
Cost variance report by type	Direct labor via bar-code scan
Cost variance report by date	Overhead: fixed and variable
Cost variance report by process	Standards: labor and materials
Display screens of above	Anticipated bills of material
Anticipated bills of material vs. actual bills of material	Actual bills of material
Projected bills of material via system questioning: display and printer-based	Past bills of material

Processes:	**Database:**
Price variance	Possible relational
Quantity variance	Standard query language capabilities
Rate variance	Online transaction processing
Efficiency variance	Recovery and backup
Volume variance	Fourth-generation language
Controllable variance	Expert system tie-in
Applied manufacturing overhead	
Bills of material comparison	

Controls:	**Technology Platform:**
CIM data range checking	Expanded secondary storage
Validity data checking	Expanded communications network
Disaster recovery plan	5 additional workstations
Access control	2 additional printers
Database control	1 additional backup device
Expert system access	JOCS software package
	Database management system
	CIM JOCS software interface
	CIM JOCS hardware interface
	Expert system

Figure 6.20
(d) Basic and manufacturing JOCS general systems design.

GENERAL SYSTEMS DESIGN WORKSHEET

Company: Peerless, Inc. Project Manager: Jake Jacoby

System: JOCS—basic design with additions for manufacturing control Date: 07/15/91

Output:	**Input:**
Cost variance report by job	Direct material via bar-code scan
Cost variance report by type	Direct labor via bar-code scan
Cost variance report by date	Overhead: fixed and variable
Cost variance report by process	Standards: labor and materials
Display screens of above	Estimates: by job and workstation
Projected job estimate report	
Projected job to completion report	
Displays for job estimation	

Processes:	**Database:**
Price variance	Possible relational
Quantity variance	Standard query language capabilities
Rate variance	Online transaction processing
Efficiency variance	Recovery and backup
Volume variance	Fourth-generation language
Controllable variance	Prototyping tool
Applied manufacturing overhead	
Job estimation	
Workstation job estimation	

Controls:	**Technology Platform:**
CIM data range checking	Expanded secondary storage
Validity data checking	Expanded communications network
Disaster recovery plan	15 additional workstations
Access control	4 additional printers
Database control	2 additional backup device
	JOCS software package
	Database management system
	CIM JOCS software interface
	CIM JOCS hardware interface

Figure 6.21
General Systems
Design Report.

GENERAL SYSTEMS DESIGN REPORT

To: Mary Stockland, CIO
From: Jake Jacoby, Project Manager
Subject: General systems design strategy for the job-order cost system (JOCS)

GENERAL SYSTEMS OVERVIEW

Setting of standard costs, comparing actual costs with these standards, and reporting variances in a more timely manner help managers control costs. The job-order cost system (JOCS) based on this kind of variance reporting that establishes a management control loop is presented as follows:

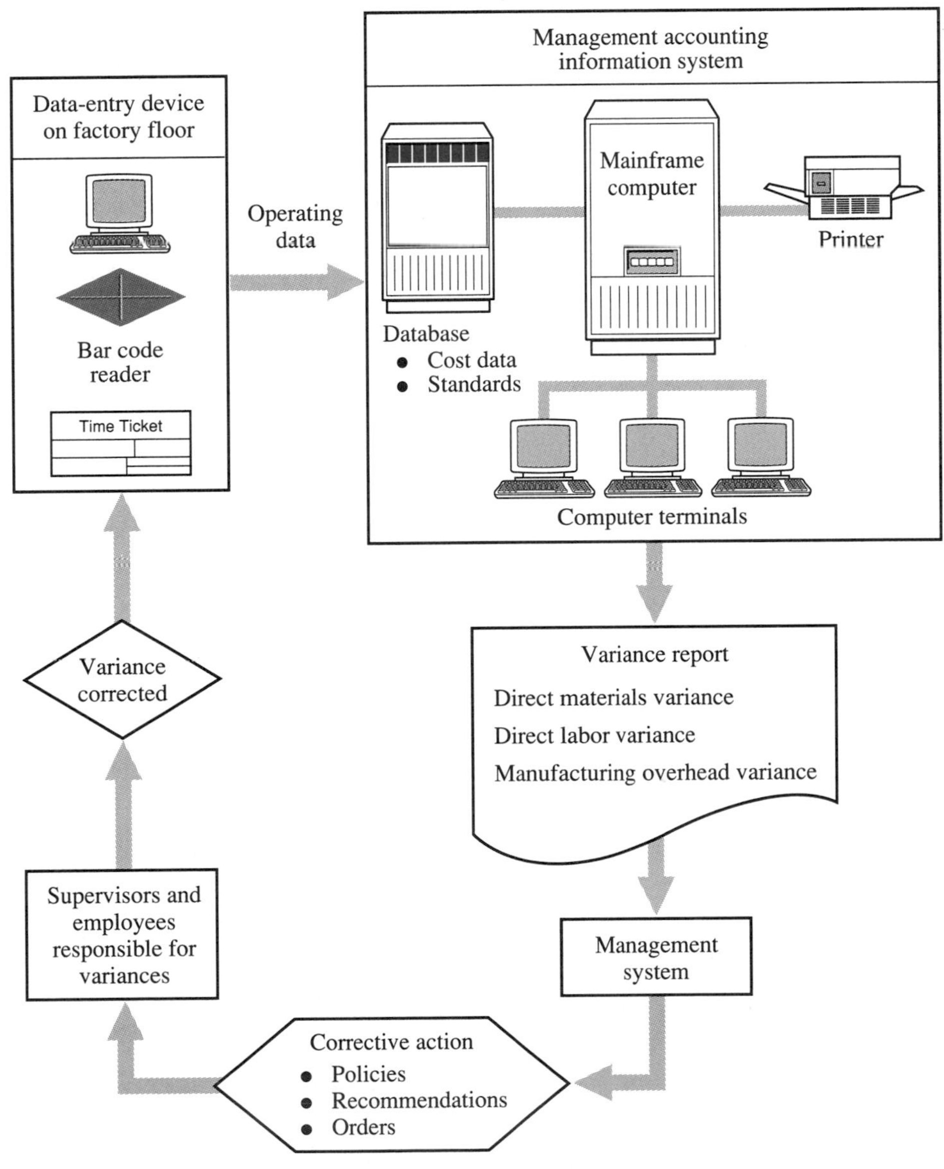

After Jake completes the general system design worksheets, he composes a memo to Mary Stockland detailing the results of the brainstorming sessions. The memo consists of a General Systems Design Report, shown in Figure 6.21, and the General Systems Design Worksheets. The memo from Jake to Mary is shown in Figure 6.22.

Figure 6.22
Overview of JOCS general systems design.

The control loop contains five steps. These steps and their proper sequences are:

1 Collecting production data from the plant floor. These data reflect time and quantity of the three cost elements of direct materials, direct labor, and applied manufacturing overhead.

2 Processing of production data and comparing these data to standards.

3 Reporting variances to managers who are responsible for the variances and who have authority to take corrective action.

4 Taking corrective action by managers by setting new policies, making recommendations, and giving specific orders to supervisors and employees directly involved in operations.

5 Changing tasks and activities by supervisors and operating employees to correct variances.

Chapter 7
Systems Evaluation and Selection

WHAT WILL YOU LEARN IN THIS CHAPTER?

After studying this chapter, you should be able to:

1 Discuss the role that technical, economic, legal, operational, and schedule (TELOS) feasibility factors play in determining systems value.
2 Discuss the role that productivity, differentiation, and management (PDM) strategic factors play in determining systems value.
3 Describe the maintainability, usability, reusability, reliability, and extendability (MURRE) design factors and show how they help to determine systems value.
4 Define cost/benefit analysis and describe its use in financially evaluating general systems design alternatives.
5 Prepare the Systems Evaluation and Selection Report.

INTRODUCTION

For a brief perspective of where we are in SDLC, review Figure 7.1. The subject of this chapter is systems evaluation and selection, the fourth phase of SDLC. The evaluation and selection phase entails a process by which systems value and costs and the benefits of alternative general system designs are compared, and one is chosen for detailed design. The evaluation and selection phase is an optimizing process that seeks not only a workable design, but also the best possible design for meeting user requirements.

Decisions to replace current systems with one of the general systems design alternatives are difficult. The purpose of this chapter is to provide a means by which management can make informed decisions during this critical phase in SDLC.

Informed systems evaluation and selection decisions come from two sources:

- **SYSTEMS VALUE**, which is measured by TELOS feasibility factors, PDM strategic factors, and MURRE design factors
- **COST/BENEFIT ANALYSIS**, which measures costs, tangible benefits, and intangible benefits of the system under consideration

RATING THE VALUE OF A WELL-DESIGNED SYSTEM

Today, Wall Street places more value on American Airlines' information system, named Sabre, than it does on American's core business. Sabre accounts

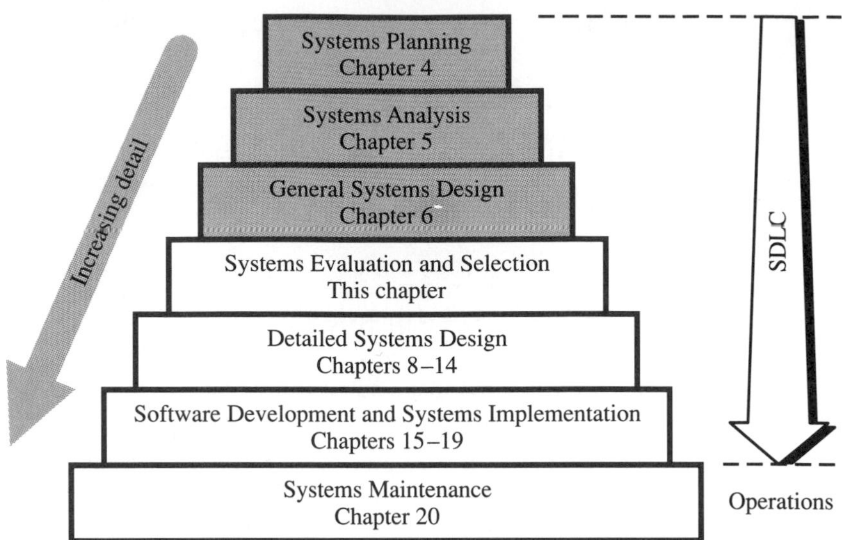

Figure 7.1
SDLC phases and their related chapters in this book. In Chapter 7 we learn how to evaluate and select systems based on systems value.

for over 50 percent of American's operating income. Yet when American was first considering the development of Sabre, the decision was neither easy nor obvious.

Whereas American's decision was the correct one, Chemical Bank has another story. Chemical Bank's aggressive implementation of Pronto, its electronic home-banking offering, resulted in almost no value or return to show for its multimillion-dollar investment.

Other companies are spending millions of dollars to replace entire information systems. Although cost/benefit analysis is useful in helping management make investment decisions, it does not tell the whole story. As David Freedman says, "Insisting on a dollars-and-cents analysis of every project can seriously damage an organization's competitiveness.[1]

For example, neither TELOS feasibility factors nor PDM strategic factors can always be tied to financial and accounting methods. Nor does cost/benefit analysis measure systems design value.

We can, therefore, estimate the value of a general systems design alternative by assigning ratings to three qualitative categories of factors:

- TELOS feasibility factors
- PDM strategic factors
- MURRE design factors

A three-dimensional view of the factors that define the value of systems is displayed in Figure 7.2. Descriptions of the factors of each category and procedures for rating each factor are presented next.

[1] David Freedman, "The ROI Polloi," *Chief Information Officer Journal,* April 1990, pp. 30–34.

Figure 7.2
Dimensions of systems value.

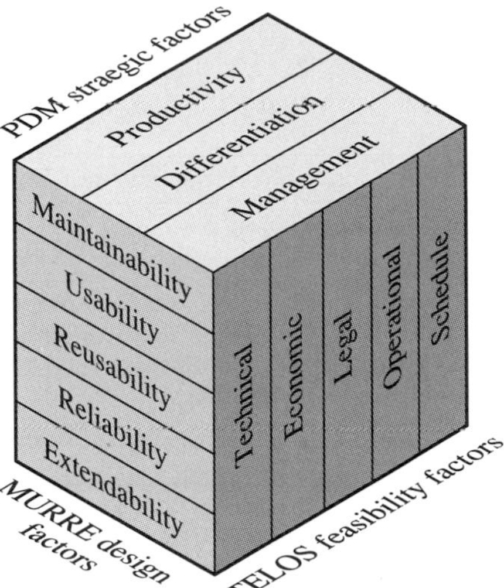

TELOS Feasibility Factors

During systems planning, systems professionals rated the technical, economic, legal, operational, and schedule feasibility factors of each systems project proposal to determine the likelihood of being able to develop these systems projects. The higher the TELOS feasibility factors score, the greater the chances were of completing a successful systems project.

During systems planning, rating of feasibility factors was very intuitive. It was based on broad descriptions of systems project proposals. But now that a particular systems project has progressed through analysis and several general design alternatives have been created (albeit at a conceptual level), feasibility evaluators have gained much more knowledge about the system under review. They can make more-informed ratings than was possible during systems planning, systems analysis, and general systems design.

Measuring the Risk in Designing and Implementing General Systems

The TELOS feasibility factors represent forms of systems development risks. The listing, presented in Figure 7.3, shows each TELOS feasibility factor, the highest-risk scenarios attached to each, and appropriate ratings.

The lower the ratings are for the TELOS feasibility factors, the higher the systems development risk. These ratings are used to measure and determine the inherent degree of risk in designing in detail and implementing general systems design alternatives. Such an evaluation helps the evaluators select the best design alternative for the systems project from the viewpoint of its potential of being fully developed and converted to operations. Whereas during systems planning, the degree of systems development risk was determined on each systems project proposal, here we are determining the riskiness of each general systems design alternative with the objective of selecting the one with the highest TELOS feasibility factors score.

Figure 7.3
TELOS feasibility factors and their relationship to highest-risk scenarios in systems development.

TELOS Feasibility Factor	Highest-Risk Scenarios	Appropriate Rating
Technical	The necessary supporting technology is unavailable.	0.0
Economic	The enterprise cannot afford the new system.	0.0
Legal	The new system will subject the enterprise to multiple lawsuits.	0.0
Operational	The new system will not meet user requirements, or the environment will change so much during the time of development that the system will no longer be operationally appropriate, or the enterprise's personnel do not possess the expertise to operate and use it.	0.0
Schedule	The scope of the systems undertaking, or its complexity, or its fit with the expertise of the systems project team, precludes successful completion of the project within a reasonable time frame.	0.0

Rating TELOS Feasibility Factors

Evaluators who assign feasibility ratings will normally include the project manager, several systems professionals who are not assigned to the systems project under review but who are knowledgeable about what it entails, and at least one user representative. Such a group of evaluators provides both the systems expertise and the objectivity necessary to assign informed and fair ratings. One method for conducting and documenting the ratings of TELOS feasibility factors is demonstrated in Figure 7.4. This TELOS Feasibility Factors Rating Worksheet includes each feasibility factor, the ratings assigned to each, and the final score for one general systems design alternative. If four general systems design alternatives were, for example, included in the General Systems Design Report, a worksheet would have to be prepared for each alternative. The ratings are determined as described below.

Rating Technical Feasibility In the TELOS Feasibility Factors Rating Worksheet, we have included a sampling of questions that each evaluator should ask and to which correct answers should be given. For example, under technical feasibility, if the new system is going to use established, well-known technology, the rating may be a 9.5 or 10.0. On the other hand, perhaps the technology is new to the company and the users, or nonstandard (i.e., to both the company and the industry); or it contains a first release by the vendor; or several vendors are involved; or it uses a highly complex networked system. Then any one or a combination of yes answers would tend to decrease the rating significantly below 10.0 (say, in the range 6.0 to 8.0). In our example, we determined that the particular general systems design alternative that is being evaluated will require new technology that is standard in the industry and has proven its ability to perform. So a rating of 9.0 is reasonable.

Another consideration in the technical feasibility area is the specific kind of technology required. For example, will the new computer technology be a

Figure 7.4
Worksheet for rating TELOS feasibility factors and computing a final score.

TELOS FEASIBILITY FACTORS RATING WORKSHEET

	Rating
Technical Feasibility	9.0
Uses established, well-known technology platform?	
New technology platform required?	
New technology is first release by vendor?	
More than one vendor involved?	
Complex networked system?	
Economic Feasibility	7.0
Total commitment from top management?	
Total development funds allocated and committed?	
Legal Feasibility	9.5
Corporate counsel satisfied with system as it relates to regulations and various privacy laws?	
Controls emphasized?	
Operational Feasibility	6.0
Narrow scope with few interfaces?	
New global-based system with many interfaces and users?	
Automates a known process?	
Automates a unique, unknown process?	
Involves knowledgeable and committed users?	
Schedule Feasibility	9.5
Total development time measured in: Hours? Days? Weeks? Months? Years?	
Systems project team uses advanced techniques?	
Systems project team has a high level of expertise and motivation?	
Total	41.0
Final score (41.0/5)	8.2

Rating scale:

0	5	10
Not feasible	Moderately feasible	Totally feasible

mainframe or a peripheral? Generally, a mainframe will cause greater impact on the system than will a new magnetic tape unit. At this particular phase in SDLC, specific kinds of technology may not be known, however, because often the general systems design alternatives are too abstract to describe the technology platform in detail. If such questions can be answered correctly, however, then by all means, they should be asked.

Rating Economic Feasibility Questions that should be asked about economic feasibility entail such things as top management's commitment to supporting complete development of the systems project with adequate resources. Without top management's support, it is difficult, if not impossible, to complete the system even though it may be feasible in terms of other factors. If top management indicates that it is still supporting the system, but funds have not been committed for its completion, the economic feasibility rating may range between 5.0 and 8.0, depending on the circumstances and the track record of top management in supporting past systems projects. If necessary funds have been committed, the rating may be 9.0 to 10.0.

In our example, funds have not been committed but top management assures the team that availability of funds will not be a problem. We, therefore, assign a 7.0 rating to economic feasibility.

Rating Legal Feasibility In some instances, the legality of a systems project is neither an issue nor a problem. The legal feasibility rating should receive a 10.0 rating. In other cases, if sensitive personal data (e.g., records in a hospital) are not safeguarded, the organization is vulnerable to lawsuits from violation of confidentiality and invasion of privacy laws. Or if designers do not design and install sufficient controls to guard against fraud, disasters, and other abuses, and such problems and abuses occur, stockholders and others may bring suit against the company and even the systems professionals who designed the system.

In our example, we determined that the general systems design alternative did not include any sensitive data that might be compromised. Moreover, the systems professionals working on the systems project are very controls-conscious. They therefore plan to design and install an array of specific controls to safeguard the system against errors, malfunctions, and other abuses. Consequently, we have assigned a rating of 9.5 to legal feasibility.

Rating Operational Feasibility A local- or group-based system is normally easier to operate than an enterprisewide system, because such systems are smaller and simpler and there are fewer people to train. An enterprisewide system that is a well-known standard system may, however, rate higher than a group- or local-based system that involves unique or experimental techniques.

The key to a high rating for operational feasibility is the availability of well-trained and committed users. Such users can help dissipate some of the negative impact that a unique or unproven system may otherwise cause.

The systems design alternative that we are evaluating in our worksheet example is a group-based system that is unfamiliar to several of the users. Moreover, several of the users are new employees and are not well trained as far as their jobs are concerned. Therefore, we rate operational feasibility at 6.0.

Rating Schedule Feasibility Can users rely on the schedule and completion dates revealed in a Gantt or PERT chart? Because such schedules and dates are estimates, they will be subject to error. The magnitude of the estimation error is a key consideration. If the system is completed well after its estimated completion date, it may be unacceptable to users.

If the general systems design alternative represents a simple, standard local-based system whose total development time is measured in hours or days, the estimated time required to design and implement the system should contain a small estimation error, which is the actual time minus the estimated time. Alternatively, if the system is enterprisewide and the total schedule is measured in years, the estimation error has a high probability of being larger. The risk of not meeting the estimated schedule for large, complex systems is much higher than it is for small, simple systems.

Using SWAT teams, JAD, and CASE tools can help reduce the time it takes to develop any category of systems. In our example, we are using a SWAT team and CASE technology; therefore, we strongly believe that the schedule will be met within plus or minus one week, so we give schedule feasibility a rating of 9.5.

Determining the Final TELOS Feasibility Factors Score

In our example, the sum of the ratings equals 41.0. This value is divided by 5, the number of feasibility factors, to give a final score of 8.2, which means that the general systems design alternative being evaluated is quite feasible, with a fairly low systems development risk. Or looking at it another way, if we were assigning a letter grade to this alternative, it would receive a B—good but not excellent.

Even when evaluators are confident that they have made an informed and proper evaluation of TELOS feasibility factors and the final score is acceptable (e.g., a score greater than 6.5 out of a possible 10.0), it is still necessary to manage systems development risk for the remainder of SDLC. In the course of making a formal evaluation of the TELOS feasibility factors, managers and systems professionals have identified areas of risk. This is the essential first step of managing systems development risk. To gain some insight into systems development risk and management of this risk, review the Bell Canada case.

Managing Systems Development Risk at Bell Canada

Bell Canada's current information system evolved without a coherent systems plan or integrated network and computer architecture design. Many of the programs are a dozen years old or more, and the overall system is inadequate to meet the challenges brought about by the changing marketplace, technological evolution, and changes in the telecommunications industry.

Many programs are incompatible, requiring the rekeying of data to integrate them. The company has the technical capability to develop future telecommunications products, but it may not be possible to market them intelligently or bill for them efficiently. Examples include 800-service and integrated services digital network (ISDN) billing. (ISDN is a family of standards that makes possible a wide variety of new and sophisticated business telecommunications services.)

Bell Canada initiated the Corporate Review Information System (CRISP). The system is scheduled for completion in 19XX, X years after project initiation. A key component will be a relational database management system (RDBMS). All business units were intimately involved in development of the CRISP systems plan and the presentation and evaluation of specific systems project proposals. Such involvement helped build enterprisewide support, thus reducing political risk to systems development and substantially increasing economic feasibility because of financial support given to CRISP.

Because CRISP will be a large, complex system, estimated at approximately $40 million to develop, management took prudent steps to identify and manage systems development risk. They divided CRISP into several large systems modules (or subsystem projects), none of which will cost over $2 million, thus reducing exposure, level of complexity, and magnitude of organizational change. This approach helped enhance TELOS feasibility factors.

The most visible user requirements (i.e., those that produce tangible outputs of immediate value) are to be completed first. Success on these subsystems will significantly enhance operational feasibility because the subsystems will establish an experience base. Also, the systems development group will establish a visible track record and experience that will facilitate technical feasibility, keep the systems project on schedule, and continue to foster political and economic support.[2]

PDM Strategic Factors

The steering committee rated systems project proposals during systems planning on their potential for increasing productivity, augmenting product and service differentiation, and improving management decision making. A good PDM strategic factors score, along with a good TELOS feasibility factors score, was sufficient to place a systems project proposal in the Systems Project Proposals Portfolio Statement for development.

The PDM strategic factors ratings during the systems planning phase were based on intuition to a great extent because few facts were available. But now, with more facts and greater knowledge of the systems project and several general systems design alternatives prepared for it, better-informed ratings are possible.

Determining the Potential for Systems to Increase and Enhance Their Value

Information systems always have the potential to become some of the most valuable resources of organizations. Designers of information systems can improve productivity by:

- Breaking down communication barriers between offices and operations through sharing resources and networking
- Designing easy-to-use and meaningful user interfaces
- Automating transaction processing and low-level tasks

[2] Eric K. Clemons, "Evaluation of Strategic Investments in Information Technology," *Communications of the ACM,* Vol. 34, No. 1, January 1991, pp. 31–32.

Designers of information systems can enhance product and service differentiation by:

- Automating and improving product design and production processes, thus giving customers modern and less expensive products and services
- Providing improved access to data regarding products and services
- Improving tracking and data management needed for improving quality
- Increasing the speed of response to customer needs and inquiries
- Generating improved product and service tracking and status information
- Increasing the breadth of information that can be used in decision making

For example, product and service differentiation can be enhanced by embedding the organization's system in both the vendor's and customer's system. Finally, information systems help all employees, but especially managers:

- Perform their tasks intelligently and efficiently
- Combat competitors effectively
- Innovate
- Reduce conflicts
- Adapt quickly to head-spinning changes in the marketplace

For a perspective on the promise of well-designed information systems providing strategic advantage, read this case:

Titan Company's Strategic Advantage

Titan Company makes a number of products for the energy industry, but it specializes in Christmas trees (gas or oil well control devices placed at the tops of wells). Titan has outperformed its industry by growing two times faster and earning four times more than the averages.

Titan's strategic advantage is quick delivery. Most competitors need four to six weeks to deliver custom or out-of-stock Christmas trees and other wellhead products. In fact, it takes some competitors over one week just to prepare a quote. Titan has organized and integrated its order entry, engineering, scheduling, and shipping systems for much faster response, usually within a few days to no more than two weeks.

In Titan's information system, order-entry workstations are linked directly into engineering, pricing, scheduling, and shipping systems. Thus Titan's salespeople can price and confirm delivery times for over 95 percent of orders while a customer is still on the phone!

Rating PDM Strategic Factors

Recall that members of the steering committee assigned strategic ratings during systems planning. Now they have the opportunity to see how well the systems project has progressed from its inception through general systems design. It is possible that the systems project, now in the form of general systems design alternatives, has acquired, through the best alternative, even more strategic potential during the systems planning phase. Or the verdict may be that it has less strategic potential. In any event, now is the time to find out. A PDM Strategic Factors Rating Worksheet, with a sampling of pertinent questions, is used to derive a strategic factors score for each design alternative. This worksheet is illustrated in Figure 7.5. The PDM strategic factors ratings are determined as described below.

Figure 7.5
Worksheet for rating PDM strategic factors and computing a final score.

PDM STRATEGIC FACTORS RATING WORKSHEET

	Rating
Productivity	9.5
Coordinated tasks?	
Easy and quick access to database?	
Automated transaction processing?	
Automated low-level tasks?	
Differentiation	5.0
Style, price, quality, or uniqueness of product or service affected?	
Inventory control improved?	
Order processing and tracing customer orders improved?	
Customers better-informed?	
Company's image enhanced?	
Management	9.5
Provides high-quality information in both form and substance?	
Supports budgeting and planning functions?	
Gives timely feedback about operations for corrective action?	
Gives status information?	
Gives early warning signals?	
Discloses conflicts?	
Assists decision-making tasks?	
Total	24.0
Final score (24.0/3)	8.0

Rating scale:

0	5	10
Poor	Fair	Excellent

Rating Productivity Productivity measures results against the cost of producing those results. To attain higher productivity, both the effectiveness and efficiency of a task or process must be improved. Effectiveness is expressed in terms of "doing the right thing." Efficiency is expressed as "doing the thing right." Killing a housefly with a sledgehammer may be effective because it produces the desired effect, but it is not very efficient. Installing an insect-repelling light may be efficient, but it is not very effective, because such a device does not generally have the strength or resources to produce the desired effect.

In our example, the general systems design alternative being evaluated includes a network connected to a database for quick access for performing effective tasks efficiently. Several low-level tasks will be supported by an expert system. In addition, an electronic messaging system will reduce the paper flow and keep all employees coordinated. Therefore, we rate productivity high by assigning a rating of 9.5.

Rating Differentiation If an enterprise can use an information system to differentiate its products and services, revenue may increase. Banks use a variety of information technology to service customers well, thus keeping them as loyal customers and at the same time winning new customers. Flexibility in manufacturing and quality of products is augmented by computer-controlled machines. Design and manufacturing data are stored in databases and communicated from one machine to another to control a complete machining process.

In our example, the design will provide better inventory control and help track customer orders. These features are important, but they will not differentiate our service significantly from that of our competitors, so we give this strategic factor a rating of 5.0.

Rating Management If information produced by the system is highly rated in both form and substance, the management function should be enhanced significantly. One of the major problems of managing anything is uncertainty. Managing is mostly predicting. To predict, managers need quality information in both form and substance. Because substantive information presented in an attractive form reduces uncertainty, managers receiving such information are able to make sound business decisions.

In our example, most of the old, voluminous computer printouts will be converted to online graphic displays. Relational databases with user-friendly query languages will enable managers to make ad hoc inquiries and receive immediate responses. Also, a decision support system (DSS) will be developed for a special group of financial planners. Because of these features, we give the management strategic factor a rating of 9.5.

Determining the Final PDM Strategic Factors Score

In our example, the sum of strategic factors ratings equals 24.0. This sum is divided by 3, the number of strategic factors, to give a final score of 8.0, which means that the general systems design alternative being evaluated adds significantly to the qualitative value of the systems project.

MURRE Design Factors

The **MURRE DESIGN FACTORS** include the following:

- **M**aintainability
- **U**sability
- **R**eusability
- **R**eliability
- **E**xtendability

These design factors relate directly to systems design quality.

Measuring the Quality of Systems Design

"Systems design quality" is a general term that represents the characteristics that distinguish the systems design and determines its merit or degree of excellence. Systems design quality is difficult to quantify, but it depends directly on the MURRE design factors. The higher these factors are rated, the higher the systems design quality will be.

To ensure that systems design quality is achieved, evaluators (who may be referred to as a quality assurance group) rate the MURRE design factors for each general systems design alternative.

Rating MURRE Design Factors

The evaluators use a MURRE Design Factors Rating Worksheet, displayed in Figure 7.6. This worksheet contains each design factor, its rating, and the final score for one general systems design alternative. Typical questions that evaluators may ask to derive reasonable ratings are included under each design factor as follows:

Rating Maintainability The question is not whether or not the system will have to be maintained. It will. The goal of the designer, therefore, is to design highly maintainable systems by:

- Creating standard data dictionaries containing standard data names
- Using standard programming languages
- Installing standard computer architectures
- Applying modular design containing highly independent modules
- Preparing comprehensive, clear, and current documentation

We devote Chapter 20 to maintenance, this very important attribute that is dependent on systems design quality.

The systems professionals working on the systems project under review in our worksheet example are very much committed to developing systems that are maintainable. They use standard hardware and software; they build independent modules; and they prepare excellent documentation. Therefore, a rating of 9.5 is assigned to the maintainability design factor.

Figure 7.6
Worksheet for rating MURRE design factors and computing a final score.

MURRE DESIGN FACTORS RATING WORKSHEET

	Rating
Maintainability	9.5
Standard data dictionary?	
Standard programming language?	
Standard computer configurations?	
High degree of modularity?	
Comprehensive, clear, and current documentation?	
Usability	9.0
High level of information and substance?	
Form of output fits users' cognitive styles?	
Reusability	5.5
Well-designed, independent modules?	
Modules applicable for future applications?	
High proportion of software modules reusable?	
Reliability	8.5
MTBF increased and MTTR decreased?	
Thorough reviews and walkthroughs?	
Effective controls?	
Rigorous testing procedures to be used?	
Fault-tolerant hardware?	
Disaster recovery plan to be established?	
Extendability	9.5
Is systems design flexible?	
Does systems design indicate an ability to grow and adapt?	
Total	42.0
Final score (42.0/5)	8.4

Rating scale:

0	5	10
Poor	Fair	Excellent

Rating Usability The major aspect of the usability design factor is directed toward human factors. This design factor generally has more to do with the ultimate success or failure of the new system than any other design factor. It is imperative that each end user perceive that the system is usable; otherwise, the fanciest, most sophisticated system using the latest technology will fail.

The product of the system that users want is information. But information has two dimensions: substance and form. Substance is provided by designing output that is relevant, accurate, and timely. Form relates to cognitive styles of users. The output must be in a form that is attractive and understandable. For example, a voluminous tabular report may contain excellent substance, but if busy users have to wade through the numbers to get to this substance, they may perceive the output as useless. On the other hand, if this same substance is converted to a two- or three-dimensional color-coded graph, the output may meet the test of both substance and form. Merely converting tabular reports to graphics will not, however, ensure usability, because some users' cognitive styles may prefer long lists of numbers. Knowing cognitive styles of users and designing outputs to match these styles is always a challenge.

The systems design being evaluated in the worksheet example is the result of extensive prototyping of screens and reports. The comments from users about the substance and form of the output are very favorable. Because of this, a rating of 9.0 is assigned to usability.

Rating Reusability This is the ability to use the same software, or other components of the system, for other applications while still meeting high levels of usability. This is a laudable goal in the design of any system, because it can significantly reduce the cost of future systems development. For example, if 50 percent of the software modules of the system under development can be used for new applications in the future, then at least 50 percent of software development costs and time spent for the new applications will be eliminated.

In our example, a larger-than-expected number of software modules will have to be written from scratch. At this time, it appears that few of them will be reusable for other applications because of the need to meet some unique user requirements. Therefore, a rating of 5.5 is assigned to reusability.

Rating Reliability Reliability refers to how dependably a system will perform its functions. **MEAN TIME BETWEEN FAILURES (MTBF)** is the quantitative measure of reliability and is expressed in months or years. MTBF is the average time the system is expected to operate before it fails. **MEAN TIME TO REPAIR (MTTR)** is a quantitative measure of maintainability and is expressed in seconds or minutes. The objective of the designer is to increase MTBF and decrease MTTR.

Two features that enhance a system's reliability are fault avoidance and fault tolerance. Fault avoidance reduces the probability that a system will fail. Fault avoidance is achieved by using modern methodologies, modeling tools, techniques, and system controls.

The term "fault tolerance" means that the system (e.g., the technology platform) has the ability to recover and perform processing tasks in the presence of various element failures. Fault tolerance procedures use redundant software and hardware elements and fault-detection devices to discover and to bypass the effects of failure so that the system can process tasks despite the failure of one or more elements in the system. A broader systems example of the fault-tolerance concept is a disaster recovery plan presented in Chapter 12.

The ideal goal is fault avoidance, but in the real world there is really no such thing as a 100 percent reliable system or a totally fail-safe system. Therefore, if we know we will experience failure, we will need three classes of recovery in a fault-tolerant system: full recovery, degraded recovery, and safe shutdown. Full recovery is an online system, for example, that provides continuous and complete operations even in a faulty environment. Degraded recovery enables selected applications to operate fully, or the total system to operate at below-standard performance levels until the system is repaired. Safe shutdown allows a graceful cessation of the system in order to terminate operations and tasks with minimal loss of data and no damage to hardware elements (such as head crashes on disks). The system can deliver appropriate messages and diagnostics to users and maintenance personnel, so that they can employ resources and techniques to restore the system quickly and efficiently.

MTBF in our systems design example receives a fairly favorable rating because of good fault-avoidance and fault-tolerant features. Because of excellent maintainability features, MTTR receives a very high rating. Also, the development and installation of a disaster recovery plan will have a positive impact on reliability. Therefore, we give a 8.5 rating to this design factor.

Rating Extendability Extendability gives the system a high degree of flexibility, so it can change or adapt easily to satisfy changing user requirements. A system that is implemented may work very well for a short time, but if it is a dead-end, inflexible system without the ability to grow and adapt, it may go belly up when user requirements change or increase.

Extendability is related to maintainability. In fact, it is difficult, if not impossible, to have extendability without maintainability. The extendability design factor, however, has more to do with improving the growth potential of the system and enhancing the adaptability to new requirements, whereas maintainability pertains more to keeping the system operating as it was originally designed to operate.

In our worksheet example, both planned software and hardware will possess the ability to grow and adapt to changing user requirements without the need for major conversions. In addition, because of the systems design's high marks for maintainability, the extendability design factor will also be enhanced. Thus we assign a rating of 9.5 to the extendability design factor.

Determining the Final MURRE Design Factors Score

In our MURRE Design Factors Rating Worksheet example, the sum of the five ratings equals 42.0. This sum is divided by 5, the number of design factors, to give a final score of 8.4, which means that the general systems design alternative being evaluated has good systems design quality and will therefore increase the systems project qualitative systems value.

EVALUATING AND WEIGHING THE COSTS AND BENEFITS OF SYSTEMS DESIGN

Now we turn our attention to a financial evaluation. Each general systems design alternative has a cost side and a benefit side. The least-cost general systems design alternative is not always the best design alternative. In the long run, the least-cost alternative may be the most expensive. As John A. Young, president and CEO of Hewlett-Packard, says: "There really isn't any right amount to spend on information systems. Many management teams spend too much time thinking about how to beat down the information system's cost instead of thinking about how to get more value out of the information they could have available and how to link that to strategic goals of the company." Clearly, system A, which costs $400,000 and produces $1,000,000 in benefits, is better than system B, which costs $300,000 and produces $500,000 in benefits.

A major objective of systems designers is to generate several general systems design alternatives, trying to produce the optimum design. Designers must juggle a number of features of a system to best meet user requirements at the lowest possible cost. This juggling act requires trade-offs. For example, placing workstations on everyone's desk may improve the users' ability to access information quickly, but the cost for doing so may far exceed the benefits.

> The heat is on to cut costs, boost cash flow, and generate fast, measurable returns on capital investment. That holds doubly true for information system projects, whether they involve extending local area networks to a new remote site or designing a hot new client–server application.
>
> It's not enough for an IS manager to do a cursory job of adding up numbers. The business managers want sophisticated cost-benefit analysis.
>
> For example, have the IS managers simply calculated the short-term payback on the project? Or have they calculated its net present value over time, which takes into account such factors as the cost of the capital needed to finance the project?[3]

What Are the Costs?

Costs fall into initial capital expenditures, to develop and implement the system, and recurring costs, to operate and maintain the system once it is implemented. Costs that represent the initial investment in systems are computing resources and systems development and implementation costs. Costs that recur over the life of the system are operations and maintenance costs.

Computing Resources Costs

Computing resources costs can range from the installation of a small PC-based software package costing a few dollars to a giant (50,000 or more square feet) data center, containing multiple mainframes and hundreds of peripherals and secondary storage devices, costing millions.

Systems Development and Implementation Costs

Systems development and implementation incur the following costs:

---

[3] Robert L. Scheier. "Cost Justification Is Crucial for IS Projects," *PC Week,* February 11, 1991, p. 67.

Systems Development Costs These costs involve systems planning, analysis, and design functions performed by systems professionals. In a large, complex system, other development-type personnel may include systems integrators, telecommunication experts, and a number of special technicians.

Equipment Installation Costs In some cases, the physical installation of equipment may present problems that require the use of special equipment such as cranes. Installing a system on the tenth floor of a building is not an easy task, but the cost is often overlooked until actual installation. Another charge not considered in some cost estimates is freight.

Programming Costs These costs are generally for customized application programs written by in-house or independent programmers. The cost estimate is based on hours required to write the program times the programmer's rate per hour plus overhead. If the programs are canned packages, the purchase or lease price plus installation fees represent the costs for this category.

Training Costs The training costs are incurred to prepare all users for the new system. Training procedures may range from brief overviews to intensive on-the-job training sessions.

Testing Costs A variety of tests are run on all the system components before converting to the new information system. This task requires a great deal of planning and preparation of effective test data. The costs can involve the salaries of in-house persons and fees charged by consultants.

Conversion Costs The costs of conversion depend on the degree of conversion, which, in turn, depends on how many applications in the current system are to be changed and how many are to be handled as in the past. Several factors need to be included in the estimate of costs of conversion:

1 Preparing and editing records for completeness and accuracy, as for example, when the records are converted from manual media to magnetic disk.

2 Setting up file library procedures.

3 Preparing and running parallel operations.

Operations and Maintenance Costs

Operations and maintenance costs include all elements necessary to keep the information system working after it has been implemented. These cost elements are:

Staff Costs These costs include the payroll for all employees in the information system and for occasional consulting fees. This staff may consist of:

- Chief information officer (CIO)
- Systems analysts
- Systems designers

- Accountants
- Programmers
- Systems engineers
- Computer operators
- Data preparers
- Database administrators
- Security officers
- Technicians
- Managers
- Clerical personnel

Supplies Costs As the system operates, it consumes supplies. These supplies include:

- Printer paper
- Ribbons
- Magnetic tape
- Magnetic disks
- Other general office supplies

These items are drawn from inventory. Their provision and costs will be subject to management control procedures.

Hardware Maintenance Costs Maintenance of a system may be performed by the company's own engineers and technicians, by vendor personnel, or by a combination of the two. In any event, hardware maintenance is a recurring expense.

Software Maintenance Costs These costs are incurred in debugging the system, adapting the system to meet new requirements, improving the system on behalf of users, and enhancing the system for operations.

Power and Light Costs After the initial electrical equipment for servicing the system is installed, there is a recurring charge, based on use, for power and light.

Insurance Costs For rented or purchased equipment, costs are incurred for insurance for fire, extended coverage, and vandalism. To safeguard against disgruntled employees who might be inclined to do injury to the system, costs will be incurred for a DDD (disappearance, dishonesty, destruction) bond.

Telecommunication Costs

- Use charges for telecom links, including leased lines, multiplexers, digital communications interconnect equipment, switches, and the like
- Move and change charges; service installation charge

Building and Building Service Costs

- Rent
- Maintenance
- Furniture

Security Costs In addition to routine security costs, a tornado, fire, earthquake, or flood can bring the information system to its knees or destroy it completely. Therefore, management may require a hot-site backup contract that will ensure a way to recover data and continue operations in case disaster strikes. Costs for this service must be included.

What Are the Benefits?

Benefits increase sales or decrease costs. Since sales minus costs equals profit, we could say that benefits include anything that increases the enterprise's profitability. For example, an online order-entry system can automate the ordering cycle, thus eliminating clerical tasks, which in turn reduces costs. Salespeople equipped with laptop computers can customize an insurance program for a client while interviewing the client, thus increasing sales. The laptop delivers benefits.

Normally, proposed general systems design alternatives possess both tangible and intangible benefits. In most cases, these benefits, especially intangible ones, are difficult to measure. But as some analysts say, "It is better to be roughly right than precisely wrong." So we should measure intangible benefits even though precision may be unattainable.

Tangible Benefits

TANGIBLE BENEFITS, sometimes referred to as direct benefits, are traced directly to the system. For example, in the current system, it may cost $2.00 to process a transaction, whereas the proposed system will process the same transaction for $1.50. Or an order-entry clerk may be able to enter twice as many sales orders with the new system. Or, because of better control and processing of accounts receivable, the organization's average cash balance may increase by $10,000. Or sales may increase 15 percent from connecting the information system to customers' ordering systems. Or errors may be reduced by 30 percent, resulting in a monthly cost savings of $20,000. Because these examples of tangible benefits can be directly attributed to a particular systems design, they are relatively easy to measure.

Intangible Benefits

INTANGIBLE BENEFITS cannot easily be traced to the system. An attempt should be made, however, to express in dollars and cents those that can be identified. For example, an analysis of customer sales might show that the organization is losing 9 percent of its gross sales annually because of inventory stockouts. The present system has an 85 percent customer service level, which means that on the average, 85 percent of all customer requests for stock items can be met by inventory on hand. The proposed system can achieve a 95 percent customer service level because of new inventory control models and a better database design. It is estimated that this expected increase in customer service will increase annual sales by 12 percent.

What about intangible benefits related to an improvement in human factors? For example, the proposed system may be ergonomically designed, with more pleasant working conditions. What is the payoff? Presumably, employee absenteeism and turnover will decrease and productivity will increase, which will result in a lower training cost and fewer job disruptions.

With the implementation of point-of-sale devices in a retail organization, average checkout time is reduced by 3 minutes, thus differentiating the retailer from competitors. What is it worth to the retailer to reduce average checkout time by 3 minutes? The benefits may be both tangible and intangible. Additional clerks may not be needed; this can be a cost-savings benefit. Sales may also increase because customers like the quicker service and are thereby attracted to this store. High quality and usable information from the new system will increase the efficiency and effectiveness of management. Better management produces increases in revenue, cost savings, or both.

One technique used for measuring both tangible and intangible benefits is the expected-value technique. It depends on estimates by systems professionals, users, and managers. These subjective estimates may be attained through interviewing.

The final survey produced by interviewing becomes the essential input for the expected-value technique. Analysts segment benefits into various ranges of cost decreases or revenue increases. For example, let's assume some measures assigned to the potential of point-of-sale devices increasing revenue. The revenue range is estimated to be $0.00 to $100,000.00, as shown in Figure 7.7. Each

Figure 7.7
Expected-value technique used to measure intangible benefits.

MEASURE OF INTANGIBLE BENEFIT OF INSTALLING POINT-OF-SALE DEVICES

Increase in Revenue	×	Probability of Occurrence	=	Expected Value
$ 0.00		0.05		$ 0.00
20,000.00		0.20		4,000.00
40,000.00		0.20		8,000.00
60,000.00		0.30		18,000.00
80,000.00		0.15		12,000.00
100,000.00		0.10		10,000.00
		Total expected value		$52,000.00

value in the range is assigned a probability which represents the likelihood that this amount of revenue will be produced. The result of multiplying each probability times its estimated revenue results is an expected value for that specific estimate. The sum of these expected values is the total expected value of installing point-of-sale devices, that is, an increase in revenue of $52,000.00.

Measuring Costs and Benefits

Now that we have defined costs and benefits and have explored ways to measure benefits, especially intangible benefits, we are ready to compile and organize these measurements in a model that will result in a valid cost/benefit analysis. Present value methods provide excellent models for such an analysis.

Net Present Value (NPV) Method

An investment in a systems project is viewed as the acquisition of a series of future net cash inflows (i.e., cash inflows minus cash outflows). The **NET PRESENT VALUE (NPV) METHOD** discounts all expected future net cash inflows to the present, using some predetermined cost of capital discount rate (also called a desired rate of return, cutoff rate, or hurdle rate).

The period of time over which these net cash inflows will be received is an important factor in evaluating the investment. A sum of cash to be received in the future is not as valuable as the same sum on hand now, because cash on hand today can be invested to earn income. If, for example, $1000 can be invested at 12 percent, the present value of $1120 a year from now is $1000 ($1120/1.12).

A "present value of $1" table, usually included in finance and accounting books, provides present value factors that can be multiplied by unequal future net cash inflows to discount them to their present value using a discount rate, which represents the expected return that management demands for a given level of investment risk. If the future net cash inflows are equal, a "present value of an annuity of $1" table can be used to obtain the appropriate discount factor. For example, if the initial investment of a systems project is $20,000 and the project is expected to yield equal after-tax net cash inflows of $5000 for seven years, the present value of this annuity of net cash inflows, discounted at 12 percent, is $22,820 ($5000 × 4.564), and the NPV is $2820 ($22,820 − $20,000). The 4.564 factor is obtained from a present value of an annuity of $1 table.

Present Value Index (PVI) Method

If general systems design alternatives being compared require different investment outlays, the systems design alternative with a greater NPV may not necessarily be the better one if it also requires a larger initial investment. To illustrate, NPV of $10,000 on systems design alternative A for an investment outlay of $100,000 is not as economically attractive as systems design alternative B's NPV of $8000 on an investment outlay of $40,000. In this situation, a **PRESENT VALUE INDEX (PVI) METHOD** should be used to rank the alternatives. The PVI places all competing alternatives on a comparable basis for the purpose of ranking them. The PVI of systems design alternative A is 1.10 ($110,000/

$100,000). The PVI of systems design alternative B is 1.20 ($48,000/$40,000). Thus systems design alternative B is ranked higher than systems design alternative A even though systems design alternative A's NPV exceeded project B's NPV by $2000.

Example of NPV and PVI Methods

Assume a general systems design alternative that we will refer to as System A. This system requires an initial investment of $100,000 and a discount rate of 10 percent. The expected useful life of the system is five years. The estimated net cash inflows (cash inflows from revenue-raising and cost-savings benefits less cash outflows from operations and maintenance costs) are: $35,000, $37,000, $25,000, $20,000, and $20,000, respectively. The NPV calculation is as follows:

Year	Present Value of $1 at 10%	Net Cash Inflow	Present Value of Net Cash Inflow
1	0.909	$ 35,000	$ 31,815
2	0.826	37,000	30,562
3	0.751	25,000	18,775
4	0.683	20,000	13,660
5	0.621	20,000	12,420
Total		$137,000	$107,232
Initial investment			100,000
Net present value (NPV)			$ 7,232

As we discussed earlier, if several general systems design alternatives requiring the same initial investment amount are being considered, the one with the largest excess of present value over the initial amount to be invested is the most desirable. If, on the other hand, the systems design alternatives involve different amounts of initial investment, it is useful to prepare a relative ranking of the design alternatives by using a PVI.

To illustrate the ranking of Systems A, B, and C by use of the PVI, assume that the NPVs are as follows:

	System A	System B	System C
Total present value of net cash inflow	$107,232	$86,400	$93,600
Initial investment	100,000	80,000	90,000
NPV	$ 7,232	$ 6,400	$ 3,600

The PVI for each general systems design alternative is:

	PVI
System A	1.07 ($107,232/$100,000)
System B	1.08 ($ 86,400/$ 80,000)
System C	1.04 ($ 93,600/$ 90,000)

The PVIs indicate that although System A has the largest NPV, it is not as attractive as System B in terms of the amount of NPV per dollar invested. System B requires an investment of only $80,000, while System A requires an investment of $100,000. (The possible use of the $20,000 available investment dollars if B is selected should also be considered before a final decision is made.)

COMPILING VALUE AND COST-BENEFIT DATA IN THE SYSTEMS EVALUATION AND SELECTION REPORT

The **SYSTEMS EVALUATION AND SELECTION REPORT** is the documented deliverable for the fourth phase of SDLC. This report compiles both systems value and cost/benefit evaluations as they are described in this chapter. A presentation of systems value grades based on TELOS feasibility factors, PDM strategic factors, and MURRE design factors for each general systems design alternative provides management and users with insight as to the inherent value of each general systems design alternative. Following the presentation of systems value grades is a cost/benefit analysis that calculates the NPV and PVI for each alternative.

The following three evaluation techniques represent a solid, informed foundation on which the optimum general systems design alternative can be selected:

- Systems value score
- Net present value (NPV) method
- Present value index (PVI) method

Normally, the three evaluation techniques should run in parallel; that is, a particular design alternative that scores high on systems value should also have a high NPV and PVI. If, for some reason, one design alternative has a high systems value score and a low NPV and PVI, and another alternative has a lower systems value score and a higher NPV and PVI, and the results based on this analysis are too close to call, selection will have to be made on other criteria that are beyond the scope of this book.

What follows is a Systems Evaluation and Selection Report involving two general systems design alternatives (System 1 and System 2). This report contains a systems value scoring summary; a cost/benefit analysis; and a current update of the schedule for the remainder of SDLC. A Gantt chart is used to schedule the detailed systems design phase. A PERT chart gives a schedule of the systems implementation phase.

SYSTEMS EVALUATION AND SELECTION REPORT

TO: The Steering Committee
FROM: Project Team A
SUBJECT: Evaluation of general systems design alternatives, System 1 and System 2

SYSTEMS VALUE SCORING SUMMARY

Scoring Factors	System 1[a]	System 2
TELOS feasibility factors	8.2	7.8[b]
PDM strategic factors	8.0	8.1
MURRE design factors	8.4	8.5
Total score	24.6	24.4

[a] System 1 was the alternative used to demonstrate systems value scoring in the chapter.
[b] The TELOS feasibility factors score for System 2 is lower because this system will involve a new, nonstandard procedure for handling manufacturing data processing.

COST/BENEFIT ANALYSIS

The net present value (NPV) and present value index (PVI) methods are used to evaluate Systems 1 and 2 on the basis of cost/benefit analysis.

The benefits, costs, and net cash inflow of both systems are described in the following matrix. Application of the NPV and PVI methods follow this matrix. Management discount rate (also called internal rate of return or cutoff rate) is 20 percent. The life expectancy of both systems is projected to be three years.

Benefits and Costs of System 1	Period			Benefits and Costs of System 2	Period		
	Year 1	Year 2	Year 3		Year 1	Year 2	Year 3
Tangible benefits:				Tangible benefits:			
Productivity through labor savings	$ 8,000	$10,000	$11,000	Productivity through labor savings	$13,000	$16,000	$18,000
Intangible benefits:				Intangible benefits:			
Differentiation through better scheduling	4,000	5,000	5,000	Differentiation through better scheduling	6,000	8,000	9,000
Management improvement through better control and decision making	5,000	7,000	8,000	Management improvement through better control and decision making	8,000	12,000	15,000
Total benefits	$17,000	$22,000	$24,000	Total benefits	$27,000	$36,000	$42,000

COST/BENEFIT ANALYSIS

Costs:				Costs:			
Computer resources	$10,000	—	—	Computer resources	$18,000	—	—
Systems development and implementation	7,000	—	—	Systems development and implementation	12,000	—	—
Operations and maintenance	8,000	$10,000	$12,000	Operations and maintenance	10,000	$15,000	$18,000
Total costs	$25,000	$10,000	$12,000	Total costs	$40,000	$15,000	$18,000
Net cash inflow	($8,000)	$12,000	$12,000	Net cash inflow	($13,000)	$21,000	$24,000

Period	Net Cash Inflow	System 1 Present Value Factor 20%	Present Value of Net Cash Inflow	Net Cash Inflow	System 2 Present Value Factor 20%	Present Value of Net Cash Inflow
1	($8,000)	.833	($6,664)	($13,000)	.833	($10,829)
2	12,000	.694	8,328	21,000	.694	14,574
3	12,000	.579	6,948	24,000	.579	13,896
	$16,000		$8,612	$32,000		$17,641

The net present value of System 2 exceeds the net present value of System 1 by $9029 ($17,641–$8,612). The present value index (PVI) of System 1 is $25,612 ($8612 plus the initial investment of $17,000) divided by $17,000, or $25,612/$17,000 equals 1.50. The PVI for system 2 is $47,641 ($17,641 plus initial investment of $30,000) divided by $30,000, or $47,641/$30,000 equals 1.59.

RECOMMENDATION

After studying the systems value grades and performing cost-benefit analysis, we hereby recommend that System 2 be designed in detail and implemented. Although its operational feasibility factor was rated slightly less than fair against a rating of fair for System 1, we believe that the difference will be offset once personnel become accustomed to the way that System 2 operates. Moreover, System 1's PVI is 1.50 against System 2's PVI of 1.59. The following Gantt chart represents the schedule for detailed design for System 2. The PERT chart that follows the Gantt chart is the schedule for implementation for System 2. (*Note:* The schedule for System 1 is approximately the same.)

Gantt Chart
Schedule for System 2 Detailed Design

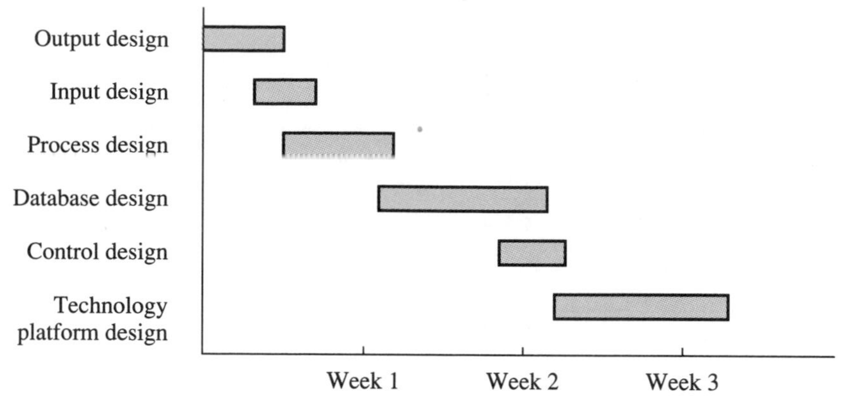

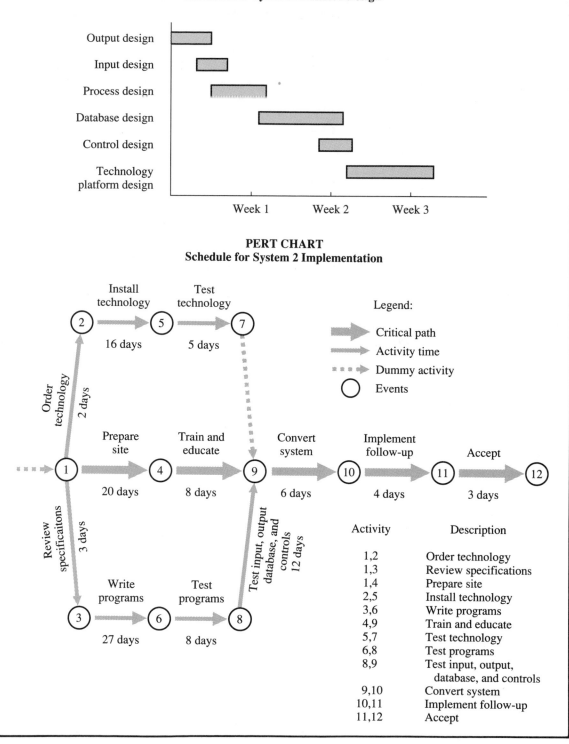

PERT CHART
Schedule for System 2 Implementation

Install technology
16 days
Test technology
5 days
Legend:
Critical path
Activity time
Dummy activity
Events
Order technology
2 days
Prepare site
20 days
Train and educate
8 days
Convert system
6 days
Implement follow-up
4 days
Accept
3 days
Review specificaitons
3 days
Test input, output database, and controls
12 days
Write programs
27 days
Test programs
8 days

Activity	Description
1,2	Order technology
1,3	Review specifications
1,4	Prepare site
2,5	Install technology
3,6	Write programs
4,9	Train and educate
5,7	Test technology
6,8	Test programs
8,9	Test input, output, database, and controls
9,10	Convert system
10,11	Implement follow-up
11,12	Accept

REVIEW OF CHAPTER LEARNING OBJECTIVES

The major goals of this chapter were to enable each student to achieve five important learning objectives. We will now summarize the responses to these learning objectives.

Learning objective 1:
Discuss the role that TELOS feasibility factors play in determining systems value.

All systems designs possess some degree of systems development risk stemming from technical, economic, legal, operational, and schedule feasibility factors. Before embarking on the arduous task of designing a system in detail, it behooves systems people to make sure that they are not only working on the best systems design, but also that they are designing in detail a system that is feasible with a high likelihood of success. High feasibility ratings increase the probability that a system will be developed and operated successfully.

Learning objective 2:
Discuss the role that PDM strategic factors play in determining systems value.

From top management's viewpoint, the reason for developing a system in the first place is to support and enhance productivity, product and service differentiation, and management functions. Clearly, any general systems design alternative that scores high on PDM strategic factors is of value to the organization.

Learning objective 3:
Describe the MURRE design factors and show how they help to determine systems value.

MURRE design factors measure maintainability, usability, reusability, reliability, and extendability. These factors provide systems design quality directly. A key design objective is to build systems that will be easily maintained after they are implemented. Systems must also be usable from the viewpoint of both form and substance of output. Where possible, systems should be designed with modules that are reusable. Systems must also be available for use at all times. Users cannot tolerate a system failing for even an hour or so each month. Therefore, MTBF must be measured in months or years, and MTTR must be measured in seconds or minutes. Most enterprises that systems serve are dynamic. They grow rapidly. Thus the systems design must also be flexible and extendable enough to enable the system to grow and keep pace with the growth of the enterprise.

Learning objective 4:
Define cost/benefit analysis and describe its use in evaluating general systems design alternatives.

Implementing and operating information systems require an investment from which future benefits are derived. Each general systems design alternative has definable costs and benefits. The objective of cost/benefit analysis is to determine whether or not the expected benefits to be derived from a particular systems design alternative are sufficient to justify its costs.

An investment in a system may be viewed as the acquisition of a series of future net cash inflows composed of cash inflow from revenue or cost reduction based on benefits less costs of operating and maintaining the system. The net present value (NPV) method computes the present value of these net cash inflows according to the company's discounted rate of return. If the present value of the net cash inflow expected from a proposed general systems design alternative, at a selected discount rate, equals or exceeds the amount of the initial investment, the design alternative is desirable.

When several general systems design alternatives all require the same initial investment, the one with the largest NPV is the most desirable. If alternative general systems designs involve different amounts of initial investment, it is useful to prepare a relative ranking of the alternatives by using a present value index (PVI). The PVI is computed by dividing the NPV by the amount to be invested.

Learning objective 5:
Prepare the Systems Evaluation and Selection Report.

The Systems Evaluation and Selection Report contains the following elements:

- Systems value scoring summary
- Cost/benefit analysis
- Recommendation
- A schedule for completing the recommended design alternative

The purpose of this report is to provide an informed basis on which the optimum general systems design alternative can be selected for detailed design and implementation.

SYSTEMS EVALUATION AND SELECTION CHECKLIST

The purpose of the systems evaluation and selection phase is to choose the optimum general systems design alternative. Following is a checklist on how to perform this phase.

1 With general systems design alternatives in hand, prepare to evaluate each and select the one that provides optimum systems value and financial benefit to the enterprise.

2 Evaluate each general systems design alternative in accordance with its TELOS feasibility factors, and compute a score.

3 Evaluate each general systems design alternative in accordance with its PDM strategic factors, and compute a score.

4 Evaluate each general systems design alternative in accordance with its MURRE design factors, and compute a score.

5 Evaluate each systems design alternative on its financial merits, using cost/benefit analysis.

6 After making the evaluations above, recommend the general systems design alternative with the highest systems value scores and the best financial return.

7 Prepare an updated schedule using project management techniques, such as Gantt and PERT.

KEY TERMS

Cost/benefit analysis

Intangible benefits

Mean time between failures (MTBF)

Mean time to repair (MTTR)

MURRE design factors

Net present value (NPV) method

Present value index (PVI) method

Systems Evaluation and Selection Report

Systems value

Tangible benefits

REVIEW QUESTIONS

7.1 What are the three sets of factors that determine the qualitative value of a general systems design alternative?

7.2 Give an example that should earn an excellent rating for each TELOS feasibility factor. Give an example that should receive a poor rating for each TELOS feasibility factor.

7.3 Give an example that deserves an excellent rating for each PDM strategic factor. Give an example that deserves a poor rating for each PDM strategic factor.

7.4 Give an example that should garner a high rating for each MURRE design factor. Give an example that deserves a low rating for each MURRE design factor.

7.5 Describe cost/benefit analysis. How is this method used to evaluate general systems design alternatives?

7.6 Explain why it's possible that System A that costs twice as much as System B may be the one that should be selected for detailed design and implementation. What is meant by the optimum general systems design alternative?

7.7 What are tangible benefits? Give three examples.

7.8 What are intangible benefits? Give three examples.

7.9 Demonstrate how the expected-value technique is used to measure benefits, especially intangible ones.

7.10 How are present value methods used in cost/benefit analysis? What is NPV? What is PVI?

7.11 What is the purpose of the Systems Evaluation and Selection Report? List and give a brief example of the elements that make up this report.

CHAPTER-SPECIFIC PROBLEMS

These problems require exact responses based directly on concepts and techniques presented in the text.

7.12 Review the following situations:

_______________ Top management wants to know the strategic value to the company of a particular general systems design.

_______________ Managers and users are disappointed with the design quality of previous systems. They want you to evaluate the design quality of two general systems design alternatives presently under development.

_______________ Management is concerned about systems development risk of a general systems design alternative that will be based on unproven technology.

Required: Insert in each blank the appropriate systems value evaluation method.

7.13 Review the following expenditures:

1 Purchase of a mainframe for $4,000,000.

2 Rental charges of $6000 per month for a network of workstations.

3 Installation of one-way emergency doors for $75,000.

4 Freight charges of $10,000 for a mainframe.

5 Wages of $1200 for a programmer to change an application program in order to comply with changes in a tax law.

6 Wages of $2500 for systems analysts to evaluate the technical feasibility of a new system.

7 Wages of $4800 for a special instructor to train a class of users.

8 Charges of $1000 to repair four workstations.

Required: Classify each of the preceding expenditures as computing resources, systems development and implementation, or operations and maintenance costs.

7.14 Estimated benefits produced by System A are expected to generate a stream of net cash inflows of $59,700, $65,400, $65,400, and $30,000 over the next four years, the useful life of the system. An initial investment of $180,000 is required. The discount rate for the company is 12 percent, and the present value of $1 at 12 percent is 0.893 (year 1), 0.797 (year 2), 0.712 (year 3), and 0.636 (year 4).

Required: Determine the net present value (NPV) of System A. Give your recommendation to accept or reject the proposal, and explain your recommendation.

THINK-TANK PROBLEMS

These problems call for a feasible approach rather than a precise solution. Although the problems are based on chapter material, extra reading and creativity may be required to develop workable solutions.

7.15 Windsong is a mail-order company that distributes a large array of novelty items. Analysts at Windsong are trying to develop a system that will combine cellular telephones and touch-screen equipment and link this technology via satellites to their order-entry system. Presently, their order-entry system is supported by a local area network of PCs. When asked what systems development risks the new system may incur, John Viola, chief systems analyst, said that the new system is state-of-the-art. Viola stated that costs for such a system are presently unknown. Viola recommended that a more comprehensive customer database could be sold to other mail-order companies to help offset costs. As Viola said, "A high-quality mailing list is worth its weight in gold."

Most of the employees at Windsong, especially those who process customer orders, have historically resisted any change in how they process orders. John Viola and his staff are trying to circumvent this resistance and implement the system before the order-processing personnel have a chance to stop it. How long it will take to develop and implement the new system ranges from one to four months.

Required: Evaluate the systems project above by using a TELOS Feasibility Factors Rating Worksheet. Justify your ratings and score.

7.16 A new systems design that is being evaluated at Nitec will enable accounts receivable personnel to decrease the billing cycle 30 percent, thus increasing cash flow sufficient to prevent Nitec from having to make a number of short-term loans. Customers of Nitec will have to pay their invoices quicker, without receiving additional services, such as better order tracking or faster delivery of orders. Managers will receive a monthly report that will help them manage cash better and make better production scheduling decisions.

Required: Prepare a PDM Strategic Factors Rating Worksheet based on the preceding information and justify your ratings and final score.

7.17 A SWAT team at Newmont has developed a systems design alternative using a structured, modular approach supported by a standard data dictionary, standard programming language, and sound documentation. Over 60 percent of the modules can be used for future applications. Users are very much right-brain people who find spatial output in the form of images to be very attractive. Screen prototypes of the systems design being evaluated include both two- and three-dimensional color-coded graphs, including timely, accurate, and relevant information. The MTBF is measured in years and the MTTR is measured in minutes. The system will be able to grow faster than the growth rate of Newmont.

Required: Prepare a MURRE Design Factors Rating Worksheet and justify your ratings and final score.

7.18 Jon Nehru, systems analyst for Maximum Company, has just completed his Systems Evaluation and Selection Report. Included in this report is some information concerning the acquisition of a computer configuration. This information is listed as follows:

1 The purchase price of the computer is $1,200,000. Maintenance expenses are expected to run $60,000 per year. If the computer is rented, the yearly rental price will be $370,000, according to an unlimited-use rental contract with free maintenance.

2 Mr. Nehru believes it will be necessary to replace the configuration at the end of five years. It is estimated that the computer will have sale value of $120,000 at the end of the five years.

3 The estimated gross annual savings derived from this particular alternative computer configuration are $450,000 for the first year, $500,000 the second, and $550,000 each of the third, fourth, and fifth years. The estimated annual expense for operations is $190,000, in addition to the expenses mentioned earlier. Additional nonrecurring costs are $70,000.

4 If Maximum decides to rent the computer instead of buying, the $1,200,000 could be invested at a 25 percent rate of return. The present value of $1.00 at the end of each of five years, discounted at 25 percent, is:

End of Year	Present Value
1	$0.800
2	0.640
3	0.512
4	0.410
5	0.328

Required: Considering the foregoing financial considerations alone, which method of acquisition do you recommend?

7.19 One of the general systems design alternatives for Odell Drug Company will rely heavily on the object-oriented design approach. Users have viewed the proposed graphic output, and they are very pleased with both content of the output and its format. Both fault-avoidance and fault-tolerance features have been employed in the design. The system has a growth potential of 150 percent, significantly more than the growth potential of Odell Drug Company.

The company's current computing resources are more than sufficient to handle the new system. Three new workstations, however, must be acquired, which according to the vendor's sales representative, are compatible with the present mainframe and operating system. Compatibility has not been tested. Management has assured the systems project team that money will be available for detailed design and implementation, but no funds have yet been budgeted. A new accounting system was abandoned last year because of insufficient funds required for its completion. The distribution of prescription drugs is under tight regulation by the Federal Drug Agency (FDA) and various state agencies. Complete audit trails, documentation, and record retention are imperatives for Odell to be in compliance with the various regulations. No one on the systems project team has sufficient expertise in this highly sensitive area, and legal counsel has not been consulted. A large number of inventory control personnel have used a Kardex system (a card-based manual system that keeps track of inventory items) for years. These people are computer-illiterate. Although a few employees have indicated some reluctance to change to a new inventory control system, they are impressed by the fact that the new system will be easy to use and will eliminate a lot of "pencil pushing." The systems project team has an excellent record of meeting its schedules.

Management prefers the payback method rather than the NPV method in its cost/benefit analysis. The payback method is computed by dividing the net cash inflow per year into the initial investment. Management's policy is that all projects must have a payback period of 2.5 years or less to be selected for implementation. The inventory control system will require an initial investment of $150,000, and the annual net cash inflow from benefits of this system is estimated to be $75,000.

The new system requires detailed design of output, process, and input that will require two months. One month after this detailed design task begins, detailed design of the database and controls will commence; it will require three months. At the same time, the workstations will be installed and tested. This activity will require one-half of a month. After all the foregoing activities are completed, training will begin. It will require one month. Testing will also begin at the same time and will require two months. After testing, conversion will take place. It will require one-half month. The systems project team uses the Gantt chart for scheduling.

Required: Prepare a Systems Evaluation and Selection Report based on the preceding information. Assume that the system just described is the only one being considered; that is, other design alternatives have not been developed. Also, include a recommendation in your report if the system should be accepted for detailed design and implementation, or rejected.

SUGGESTED READING

Amadio, William. *Systems Development, A Practical Approach.* Santa Cruz, Calif.: Mitchell Publishing, 1989.

Awad, Elias M. *Systems Analysis and Design,* 2nd ed. Homewood, Ill.: Richard D. Irwin, Inc. 1985.

Burch, John, and Gary Grudnitski. *Information Systems: Theory and Practice,* 5th ed. New York: John Wiley, 1989.

Clemons, Eric K. "Evaluation of Strategic Investments in Information Technology." *Communications of the ACM,* Vol. 34, No. 1, January 1991.

Connolly, James. "It Costs How Much?" *Computerworld,* September 19, 1988.

Emery, James C. *Management Information Systems: The Critical Strategic Resource.* New York: Oxford University Press, 1987.

Hartley, Ronald V. *Cost and Managerial Accounting,* 2nd ed. Boston: Allyn & Bacon, 1986.

Horngren, Charles T., and George Foster. *Cost Accounting, A Managerial Emphasis,* 6th ed. Englewood Cliffs, N.J.: Prentice-Hall, 1987.

Matz, Adolph, Milton F. Usry, and Lawrence H. Hammer. *Cost Accounting Planning and Control,* 8th ed. Cincinnati: South-Western, 1984.

Rivard, Edward, and Kate Kaiser. "The Benefit of Quality IS." *Datamation,* January 15, 1989.

Scheier, Robert L. "Cost Justification Is Crucial for IS Projects." *PC Week,* February 11, 1991.

Scott, George M. *Principles of Management Information Systems.* New York: McGraw-Hill, 1986.

Sen, Tarun, and James A. Yardley. "Are Chargeback Systems Effective? An Information Processing Study." *Journal of Information Systems,* Spring 1989.

Warren, Carl S., and Philip E. Fess. *Principles of Financial and Managerial Accounting.* Cincinnati: South-Western, 1986.

JOCS Case: Evaluating the Systems Design Alternatives

Jake Jacoby worked with the end users of the future JOCS system to develop four design alternatives for the new system. These alternatives include:

1 A basic system providing job costing financial data.

2 An expansion to the basic system, including purchasing data to control vendor quality.

3 An expansion to the basic system, including a complex expert system to be used by the engineering department for new lifter design analysis.

4 An expansion to the basic system, including job estimates and workstation estimates used to project manufacturing status for the marketing department to discuss with customers and the manufacturing department to schedule resources.

Figure 7.8
JOCS System Evaluation and Selection Report.

SYSTEMS FINANCIAL EVALUATION SUMMARY

	System 1	System 2	System 3	System 4
Net Present Value (discounted at 15%)	**$411.00**	**$878.00**	**($28,925.00)**	**$11,922.00**
Present Value Index (discounted at 15%)	**1.00**	**1.00**	**0.97**	**1.04**

(*Note:* Detailed computations to derive the values above are not part of this report.)

Recommendation
After studying the systems value grades and performing cost/benefit analysis, I recommend that System 4 be designed in detail and implemented for the JOCS project.

System 1 had the highest TELOS and MURRE ratings, but had a sharply lower PDM score. This is because System 1 does not provide the level of product differentiation that can be offered through the other alternatives. Financially, Systems 1 and 2 provide similar opportunities. System 2 has a net present value more than twice that of System 1, but the present value index for the two alternatives is the same. This is a result of the higher initial investment of System 2. System 2 would require more funds in both computer resources and systems development than System 1 without generating comparatively greater benefits.

System 3 has relatively low TELOS and MURRE ratings because of the complexity and individuality of the project. Most of the systems development for this alternative could not be used for other projects slated for the future. In addition, System 3 generates a negative net present value after four years of operation. As a result, System 3 would be a poor investment at this time.

System 4, on the other hand, has relatively average TELOS and MURRE scores. This means that this alternative is completely feasible in comparison with the other project alternatives. What makes System 4 distinctive are its PDM rating and financial outlook. System 4, with its project estimation component, will help differentiate our product and provide better managerial control. As a result, the benefits obtained from this alternative are financially brighter than those of the other alternatives.

SYSTEMS EVALUATION AND SELECTION REPORT

To: Mary Stockland, CIO
From: Jake Jacoby, Project Manager
Subject: Evaluation of general systems design strategies for the job-order costing system (JOCS) and selection of optimal strategy.

SYSTEMS VALUE SCORING SUMMARY

	System 1	System 2	System 3	System 4
TELOS feasibility factors				
Technical	**10.0**	**9.0**	**6.5**	**9.0**
Economic	**10.0**	**8.5**	**7.0**	**9.5**
Legal	**9.0**	**7.5**	**6.0**	**9.0**
Operational	**8.5**	**8.0**	**6.0**	**8.0**
Schedule	**9.5**	**9.5**	**8.5**	**9.5**
Total TELOS Score	**47.0**	**42.5**	**34.0**	**45.0**
TELOS Rating	**9.4**	**8.5**	**6.8**	**9.0**
PDM strategic factors				
Productivity	**9.0**	**9.5**	**9.0**	**9.5**
Differentiation	**3.0**	**7.0**	**9.5**	**9.5**
Management	**5.0**	**8.0**	**7.0**	**9.0**
Total PDM Score	**17.0**	**24.5**	**25.5**	**28.0**
PDM Rating	**5.6**	**8.2**	**8.5**	**9.3**
MURRE design factors				
Maintainability	**9.5**	**9.5**	**7.0**	**9.5**
Usability	**10.0**	**9.0**	**7.0**	**8.5**
Reusability	**10.0**	**9.0**	**7.0**	**9.0**
Reliability	**9.5**	**9.5**	**7.5**	**8.5**
Extendability	**10.0**	**9.0**	**7.0**	**9.0**
Total MURRE Score	**49.0**	**46.0**	**35.5**	**44.5**
MURRE Rating	**9.8**	**9.2**	**7.1**	**8.9**

Figure 7.8 (continued)

Jake must perform extensive analysis to determine the optimal design alternative for the JOCS project. Since each department was involved in the design process, their individual managers already feel a strong desire to have their particular system designed and implemented. However, it is vital that Jake select the alternative that has the greatest chance of success as well as the best return on resource investment for Peerless. Jake is especially interested in choosing an alternative that will evolve easily with the changing needs of the company. To accomplish this, he scrutinizes the reusability and extendability of the four alternatives.

Jake's final report to Mary Stockland is shown in Figure 7.8. In his report, he includes an evaluation of TELOS, PDM, and MURRE scores. In addition, he performs financial analysis with NPV figures and PVIs. His final recommendation is discussed in the report. At this point, Jake is ready to enter the detailed systems design phase with his SWAT team.

Part III

Detailed Systems Design

As the following exhibit shows, the systems design components are interrelated to form an integrated information system. Each component must interact with other components to produce desired results. If any component fails, the entire system is adversely affected. Each depends on the other to work efficiently. It is what all the systems design components do together that makes the system work.

To gain an understanding of how to design these systems components, we devote a chapter to each component. Each chapter allows you to focus on specific items related to a particular component. Chapter 8 treats output, the information system's finished product. Chapter 9 illustrates how input forms are designed and data-entry devices are used to capture and get data into the

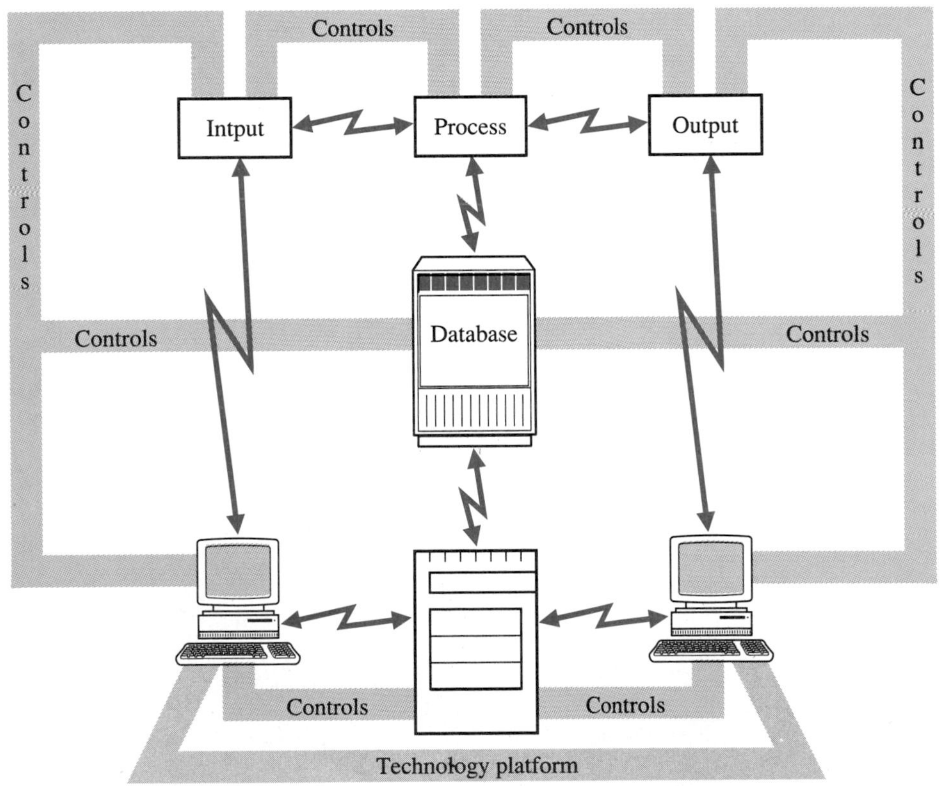

system. Chapter 10 shows how process designs are created to convert input to output. Chapter 11 presents techniques for designing databases. Chapter 12 covers an array of controls necessary to safeguard the information system from a host of threats. Chapter 13 deals with ways to design various networks. Chapter 14 demonstrates how to design a variety of computer configurations. The JOCS case continues to apply, in a real-world environment, selected topics from each chapter.

Chapter 8
Designing the Systems Output

WHAT WILL YOU LEARN IN THIS CHAPTER?

After studying this chapter, you should be able to:

1 Define high-quality and usable information, and state why output is the controlling systems design component.
2 Explain the use of hierarchical reports, and design at least one.
3 Discuss and design comparative reports.
4 Describe and design monitoring reports.
5 Discuss the fundamentals of designing display screens.
6 Illustrate the use of graphs, and state why they often add to output's usability.
7 Describe how tables and matrices can be used to convey information effectively.

INTRODUCTION

The purpose of this chapter is to learn how to design OUTPUT in detail and to enhance its usability (see Figure 8.1). Why start with output? Because output is the controlling systems design component. All the other components are developed and designed to produce useful output. It makes little sense to implement a system that produces reams of computer printouts that no one uses. Output must be designed to satisfy end users; otherwise, the systems project will be unsuccessful.

Output design produced by the general systems design phase, discussed in Chapter 6, is typically conceptual and not precisely formatted. Moreover, several general output design alternatives are presented for evaluation and selection. At this point, we know which general output design has been selected, so we are ready to design it in detail, including formatting and pulling together all the details that must be dealt with to convert output to its final form to be used by end users in an operating environment. How to deal with those details and design high-quality, usable output is the subject of this chapter.

PRODUCING HIGH-QUALITY INFORMATION FOR MAKING DECISIONS

In many companies, responsibility and accountability are diffused both vertically and horizontally. A single decision may involve five to ten vertical man-

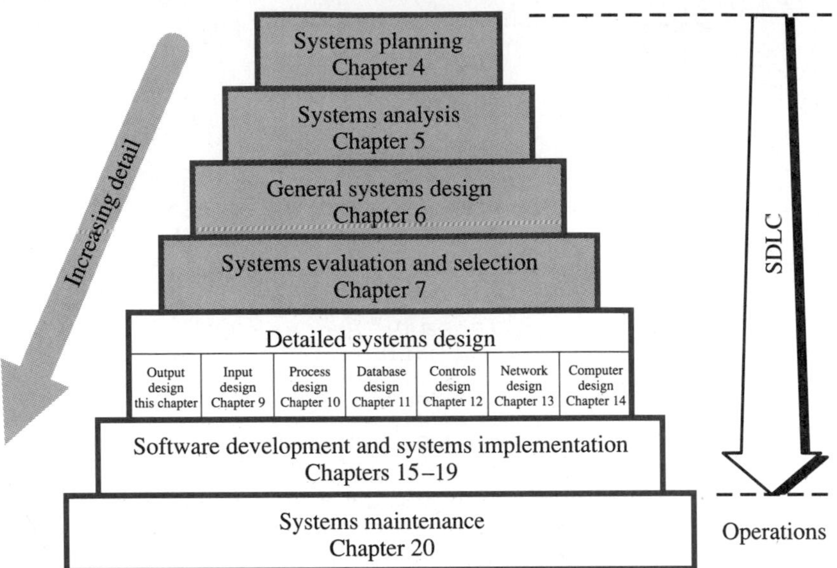

Figure 8.1
SDLC phases and their related chapters in this book. In Chapter 8 we focus on all varieties of output and the vehicles that display it.

agement levels and ten to fifteen horizontal staff and line managers. In these kinds of companies, senior managers may receive from 500 to 1000 hard-copy pages of computer-generated reports daily. Managers complain, and rightfully so, that they are being inundated with reports. A number of studies indicate that executives look at only 3 percent (or less) of the computer printouts delivered to them, which means that 97 percent or more of the printouts and all the resources required to produce them are wasted.

The human mind is easily boggled by even relatively small amounts of data. The purpose of output design is to turn mountains of data into quality and usable information. The goal is not to produce data, not even information, really. The ultimate goal is proper decision making. High-quality, usable information supports proper and streamlined decision making. What constitutes high-quality usable information?

- ***Accessibility*** This characteristic enables users to get at information via easy-to-use interfaces supported by menus and natural languages.
- ***Timeliness*** The information must be delivered in time so that users can take action. As an executive once said, "We need enough advance warning to steer around the iceberg, not a damage report."
- ***Relevance*** Relevant information makes a point and is void of trivial and superfluous details. Any report that contains anything beyond what is needed by the user is irrelevant.
- ***Accuracy*** This characteristic is not the same thing as precision. A report may show sales to be precisely $173,427.19, but be inaccurate. Accuracy means that the report is free from error. (Precision is the degree of refinement of the figures in the report. It can be customized easily to fit users' preferences.)

- ***Usability*** Usability of information means that the form of output fits the mental model or cognitive style of the user. Some users like tabular reports with lots of numbers and detail; others like graphs, tables, and matrices with limited detail. Some are numbers people; others are big-picture people. Moreover, usability equates to quality of the information less the level of effort and frustration the user must endure in assimilating the information.

Report designers provide a product—the report format—to the end-user clients who will use the report. Keeping in mind the differences in cognitive style, designers must ask their clients what format will appeal to them: what will be most helpful and most likely to be used.

To collect format data, designers can employ a form (see Figure 8.2). They can mail the form, or they can complete it during interviews with clients. Note that the questions are focused on the users, not on the system, the data, or the designer.[1]

CREATING REPORTS THAT CONDENSE DATA FOR DIFFERENT MANAGEMENT LEVELS

HIERARCHICAL REPORTS condense, aggregate, and level data for a management hierarchy to permit managers at all levels to receive the information that meets their specific requirements without having to sort through irrelevant details. Data need to be pared down to meaningful numbers and images. The executive is looking for trends, tendencies, and patterns, all of which may be obscured by too much detail. Executives, for example, normally don't want to know how much a department is spending on paper clips or floppy disks. They want to know whether the department is meeting its goals, and if not, why.

Several types of reports are used to provide the right information to the right user. Two popular report types are:

- Filter report
- Responsibility report

Filter Report

A **FILTER REPORT** is designed to permit the separating of selected data elements from the database so that each decision maker can obtain the level of detail appropriate to his or her individual needs. Traditionally, data are filtered at each superior/subordinate level in an organization.

In a construction company, for example, an awareness of actual costs incurred is an important aspect of each manager's job regardless of the manager's position in the organization, as shown in Figure 8.3. The president of the

[1] James B. King II, Richard J. Palmer, and Marvin W. Tucker, "Improving Information System Economy," *Internal Auditor,* August 1991, p. 32.

Figure 8.2
Report-design form. Questions 1 to 10 apply to existing reports. Questions 11 to 24 apply to proposed reports.

REPORT-DESIGN SURVEY

Date: ______________ Person completing survey: ______________________

Unit or person who will be using report: ______________________

Title of report, including ID number: ______________________

Purpose of report: ______________________

Questions 1–10 apply to existing reports; 11–24 to proposed reports.

1. Do you need the information provided in this report? Y _____ N _____

   If no, give details of what you do need in questions 10–24.

2. Is this report supplied to people who need it now? Y _____ N _____

   If no, who should get the report? ______________________

   Who should not get the report? ______________________

3. Do you get the report when you need it? Y _____ N _____

   If no, how often do you need the information updated? ______________

   Please respond to each of the following prompts using a scale of

   5 = I am completely satisfied.
   1 = I am completely dissatisfied.

4. The information in this report fulfills my needs for the purpose for which the report was designed. _______
5. The data reported are accurate. _______
6. The format is easy for me to interpret and to use. _______
7. I know where the data in the report come from. _______
8. I know where and how to get backup information. _______
9. The people who produce this report are available, courteous, technically knowledgeable, and helpful. _______
10. Please state what added content, features, or associated services you would like to make this report most useful to you:

____

____

Figure 8.2 (continued)

The following questions apply to reports you do not receive at present but would like to receive.

11. What is the subject of the proposed report? ____________________

12. What is the primary source of data? ____________________

13. How will you use the data? ____________________

14. How often do you need the data? ____________________

Please check the format in which you prefer to receive data.

15. Tables with all data and supporting notes. ________

16. Bar graphs with full supporting notes. ________

17. Pie charts with full supporting notes. ________

18. Matrices with full supporting notes. ________

19. Tables, bar graphs, pie charts, or matrices as appropriate, but without supporting notes. ________

20. Another format or combination of formats. (Please attach an example or a sketch, with details.) ________

21. Not sure; arrange a consultation and a complete design-planning session. ________

Consider how you will use the report itself, and answer the following questions to help us format the data in the most effective manner.

22. Do you plan to project the report onto a large screen for group viewing?
Y ______ N ______

23. Would you like subsets of the report created so that you can tailor presentations or distributions? Y ______ N ______

24. Do you have any specialized needs for using the report, such as translating data from one format to another or interpreting data in terms not immediately apparent from the display (e.g., do you need upper and lower control limit displays for all trend lines)? If you have specialized needs, please describe them below.

____

____

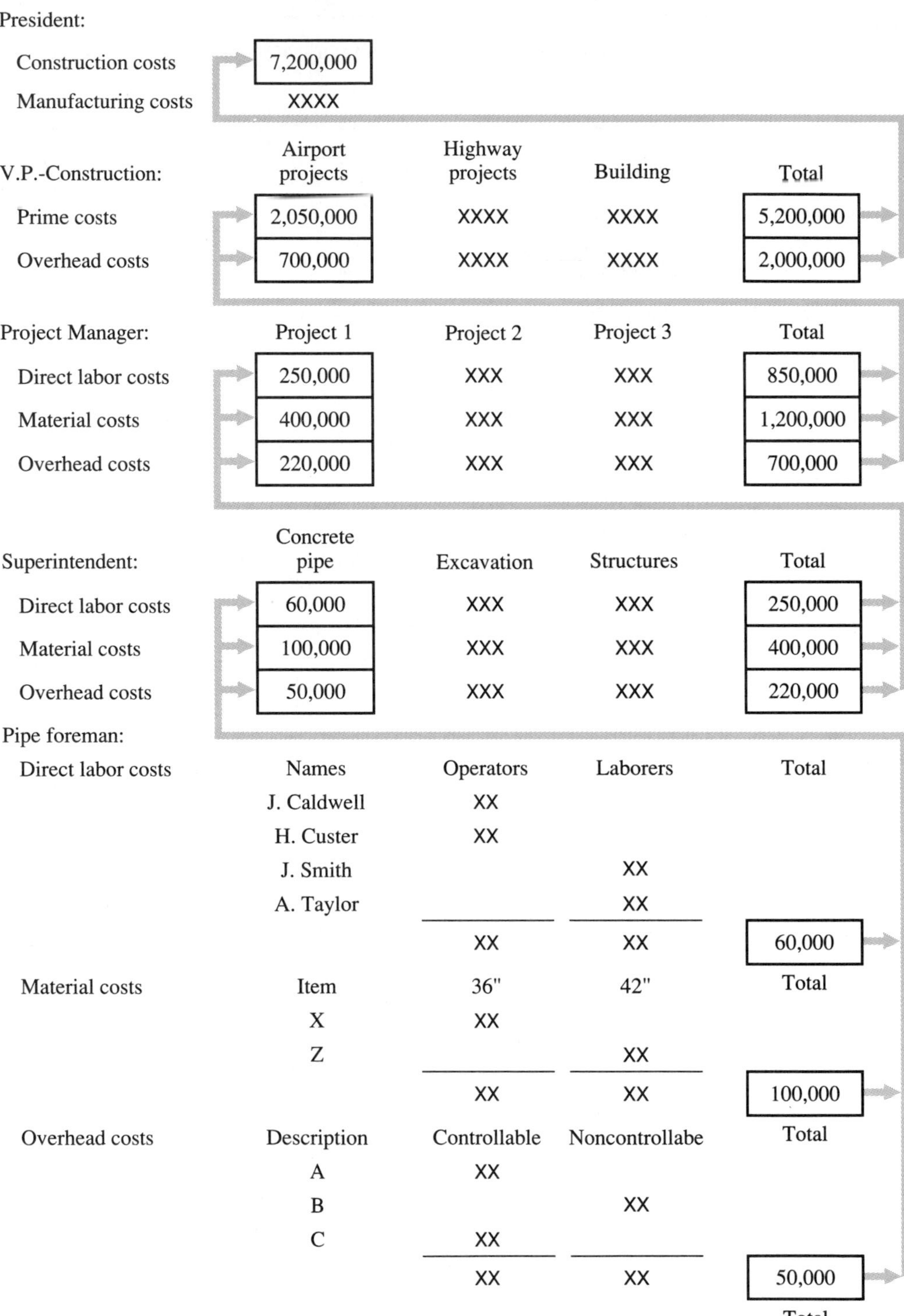

President:
Construction costs
7,200,000
Manufacturing costs
XXXX
V.P.-Construction:
Airport projects
Highway projects
Building
Total
Prime costs
2,050,000
XXXX
XXXX
5,200,000
Overhead costs
700,000
XXXX
XXXX
2,000,000
Project Manager:
Project 1
Project 2
Project 3
Total
Direct labor costs
250,000
XXX
XXX
850,000
Material costs
400,000
XXX
XXX
1,200,000
Overhead costs
220,000
XXX
XXX
700,000
Superintendent:
Concrete pipe
Excavation
Structures
Total
Direct labor costs
60,000
XXX
XXX
250,000
Material costs
100,000
XXX
XXX
400,000
Overhead costs
50,000
XXX
XXX
220,000
Pipe foreman:
Direct labor costs
Names
Operators
Laborers
Total
J. Caldwell
XX
H. Custer
XX
J. Smith
XX
A. Taylor
XX
XX
XX
60,000
Material costs
Item
36"
42"
Total
X
XX
Z
XX
XX
XX
100,000
Overhead costs
Description
Controllable
Noncontrollabe
Total
A
XX
B
XX
C
XX
XX
XX
50,000
Total

Figure 8.3 Using the system as a filter to report construction costs.

firm is likely to be concerned with the total costs incurred in a given period. The vice president responsible for construction might require a further breakdown of total costs into prime costs and overhead costs. Each lower level of management would require a correspondingly higher level of detail concerning costs related to its activities alone.

Responsibility Report

The first step in designing a **RESPONSIBILITY REPORT** is to decide who is responsible for what, and to tailor the report to match the responsibility. In Figure 8.4, we show responsibility levels of the CEO, marketing manager, and media specialist. At the highest level, the CEO has responsibility for performance of the total enterprise as disclosed in the financial statements. For example, the CEO (along with the board of directors, stockholder groups, and financial analysts) is vitally interested in the income statement. At a lower level, the marketing manager is interested in advertising expenses, because he or she is responsible for these expenses. At a still lower level, the media specialist is interested in specific media expenses, because he or she is responsible for these expenses.

CREATING REPORTS THAT COMPARE DATA

COMPARATIVE REPORTS enable managers and other users to examine two or more items to establish similarities and dissimilarities. With this comparison, the user is in a better position to make a rational decision. Types of comparative reports include:

- Horizontal report
- Vertical report
- Counterbalance report

Horizontal Report

Balance sheets and income statements represent periodic financial reports that summarize thousands of transactions and data elements into output for a variety of users. These reports follow conventional classification and layout, but their usability is somewhat limited when they are considered individually and in only one dimension. Users can often gain a clearer picture by seeing comparisons. One way to do this is to design a **HORIZONTAL REPORT**. The amount of each item on the most recent reports is compared with the corresponding item on one or more earlier reports. The increase or decrease in the amount of the item is then listed, together with the percentage of increase or decrease. The analysis can be applied to almost any kind of report, but balance sheets for 1991 and 1992 are used for an illustration in Figure 8.5.

Vertical Report

A percentage analysis can also be used to compare a component part with the total within a report. To enable the user to do this, vertical analysis calls for the

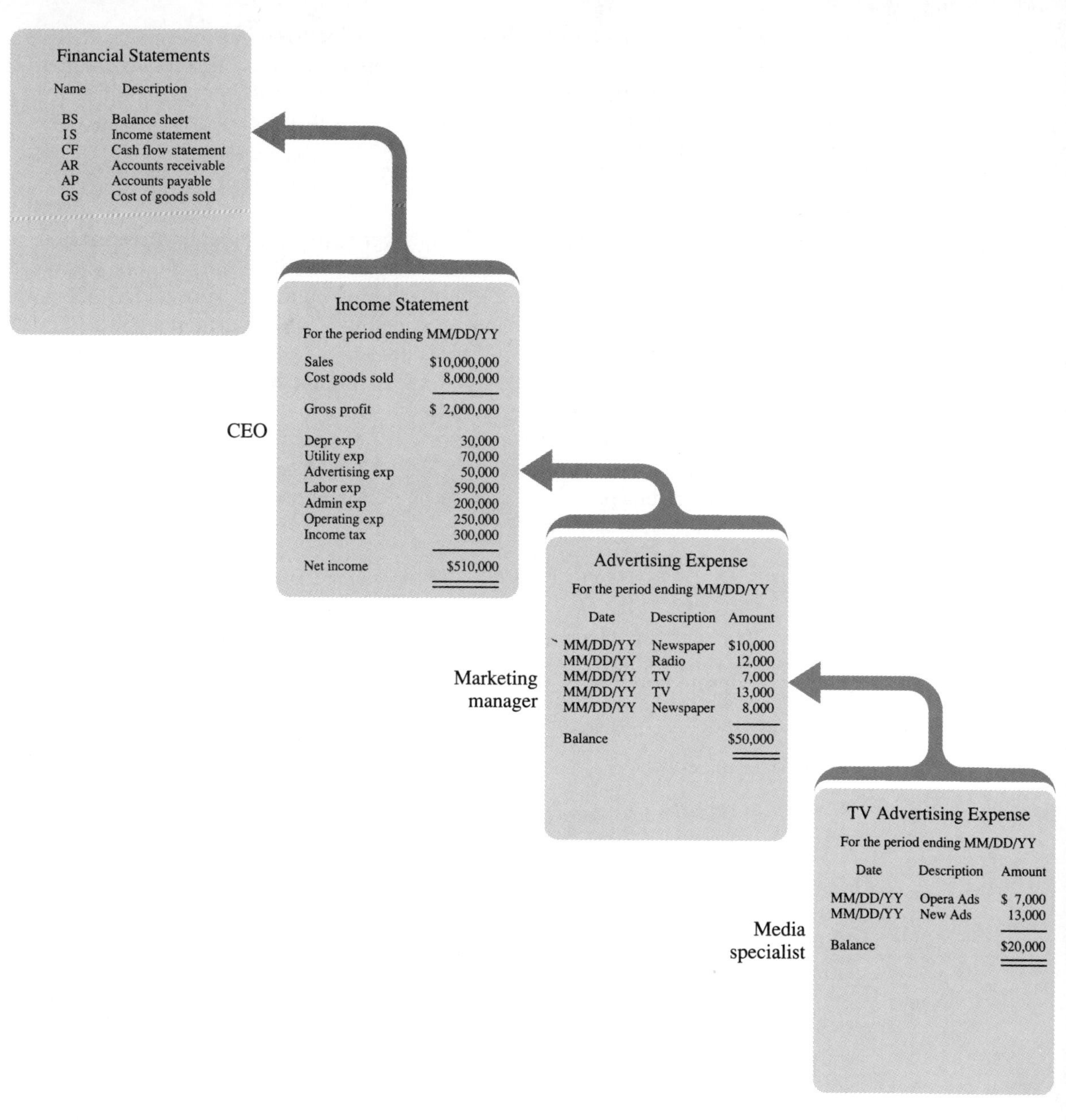

Figure 8.4
Responsibility levels in an organization hierarchy with accompanying responsibility reports.

design of a **VERTICAL REPORT**, as shown in Figure 8.6. In using an income statement, for example, each item is stated as a percentage of net sales.

In a **COUNTERBALANCE REPORT**, all sides of a situation under consideration are compared. For example, worst-, moderate-, and best-case scenarios, which help planners judge a project's riskiness, are valuable information for executives in decision making. The counterbalance report, depicted in Figure 8.7, is pre-

Figure 8.5
Comparative horizontal balance sheet report.

SILICON INDUSTRIES
COMPARATIVE BALANCE SHEET—HORIZONTAL ANALYSIS
FOR FISCAL YEARS 1991 AND 1992

	1991	1992	Amount	Percent
Assets				
Current assets	$ 40,000	$ 60,000	$ 20,000	50.0
Plant and equipment	200,000	250,000	50,000	25.0
Total assets	$240,000	$310,000	$ 70,000	29.2
Liabilities				
Current liabilities	$ 20,000	$ 30,000	$ 10,000	50.0
Long-term liabilities	50,000	20,000	(30,000)	(60.0)
Capital	$ 70,000	$ 50,000	$(20,000)	(28.6)
Common stock	$150,000	$200,000	$ 50,000	33.3
Retained earnings	20,000	60,000	40,000	200.0
Total liabilities and capital	$240,000	$310,000	$ 70,000	29.2

pared for executives who are planning to buy Mars Company, a small manufacturer of communication satellites. The report shows that under the worst conditions, the buyout target would suffer a loss of $15,000,000 for fiscal year 1993. Under moderate conditions, the projected profit is $22,000,000, and under best conditions, the projected profit is $88,000,000. Obviously, management will need additional information before making a final decision, but this counterbalance report is a key piece of information that enables management to see all sides of the proposed investment.

Figure 8.6
Comparative vertical income statement report.

PORTLAND COMPANY
COMPARATIVE INCOME STATEMENT—VERTICAL ANALYSIS
FOR FISCAL YEARS 1991 AND 1992

	1991		1992	
	Amount	Percent	Amount	Percent
Sales	$120,000	103.0	$160,000	101.2
Sales returns	4,000	3.0	2,000	1.2
Net sales	$116,000	100.0	$158,000	100.0
Cost of goods	86,000	74.1	100,000	63.3
Gross income	$ 30,000	25.9	$ 58,000	36.7
Selling expenses	$ 14,000	12.1	$ 15,800	10.0
Administrative expenses	8,000	6.9	11,060	7.0
Income before taxes	$ 8,000	6.9	$ 31,140	19.7
Income taxes	4,000	3.4	14,000	8.9
Net income	$ 4,000	3.4	$ 17,140	10.8

Figure 8.7
Counterbalance report.

MARS COMPANY
INVESTMENT ANALYSIS
($000 omitted)

	For Fiscal Year 1993		
	Worst Case	Moderate Case	Best Case
Sales	$90,000	$120,000	$180,000
Operating expenses	85,000	80,000	75,000
Gross margin	5,000	40,000	105,000
Selling expenses	10,000	10,000	12,000
Administrative expenses	10,000	8,000	5,000
Projected profit (loss)	($15,000)	$ 22,000	$ 88,000

CREATING REPORTS THAT MONITOR VARIANCES IN DATA

Using the information system as a monitor provides managers with action information. The system does the work of looking at activities and megabytes of data, ferreting out significant variables, continuously comparing actual events with expected events, and keeping track of hundreds of variates.

Generally, **MONITORING REPORTS** show a variance or divergence from a standard, budget, quota, plan, or benchmark. These kinds of reports can be classified as:

- Variance reports
- Exception reports

Variance Report

A labor performance report, as demonstrated in Figure 8.8, is a good example of a **VARIANCE REPORT**. Labor performance reports are designed to compare stan-

Figure 8.8
Labor variance report.

DEPARTMENTAL DIRECT LABOR COST VARIANCE REPORT

Department: Cooler Assembly **Supervisor:** J. Jones

Production: 40, Model z units **Date:** 01/12/94

Operation	Actual Cost	Standard Cost	Variance*	Reasons
Motor	$16,925.00	$16,500.00	$425 over 2.6%	Reboring hangers
Fan	3,000.00	3,060.00	60 under 2.0%	Skilled group
Freon	5,675.00	5,220.00	455 over 8.7%	Overtime
Total	$25,600.00	$24,780.00	$820 over	

* Expressed as a percentage of standard cost, such as: $425/$16,500 = 2.6%.

dards with actual results attained. In our example, the actual direct labor costs in the assembly department for January 12, 1994 are compared to standard labor costs (i.e., what labor should have cost). Variances are computed, and the reasons for the variances are given. Such a report to the cooler-assembly supervisor aids in controlling labor costs.

Usually, variance reports are generated due to an elapse of time or the completion of a particular process. For example, a variance report is prepared at noon and 5:00 each workday or at the end of a process.

Exception Report

An **EXCEPTION REPORT** is similar to a variance report. With exception reporting, some quota or limit is set for a process or activity. For example, in a sales department, each salesperson is assigned a sales quota, and the sales manager reviews only those who are well above or below their quota. The sales manager assumes that the salespeople are operating satisfactorily when sales are within, say, plus or minus 10 to 20 percent of their sales quota. The difference between a variance report and an exception report is that a variance report is based on the passage of time or the completion of a process, regardless of the size of the variance. An exception report is generated only when some process or activity falls outside a predetermined limit or quota.

For an example of how exception reporting works, review Figure 8.9. This figure represents the upper limit and lower limit of the sales quota for J. Johnson. During the 12-month period, Johnson exceeded the upper limit of the sales quota for Month 2 by $10,000. This exceptional performance was sent to Johnson's boss to flag positive exceptional performance.

In Month 8, however, Johnson's sales performance fell below the sales quota's lower limit by $5000. Because Johnson's sales for Month 8 fell below the lower limit, the preparation of an exception report is triggered, and the report is promptly sent to Johnson's boss, just as it was when Johnson exceeded the upper limit in month 2, except that this time sales performance is negative.

FUNDAMENTALS OF DESIGNING DISPLAY SCREENS

We now turn our attention to designing the medium that will convey some of this output: the computer screen. Techniques for designing screen reports can also be adapted to designing paper reports.

Screen Organization

Top-to-bottom, left-to-right scanning minimizes eye movements through the screen and allows users' cognitive powers to be utilized to their fullest. Two key elements of any screen are captions and data fields. A caption is a name that identifies the contents of a data field.

Captions should be spelled out fully using an uppercase font (i.e., a particular set of type all of one size and style) and displayed in normal density. A

Figure 8.9
The exception reporting process.

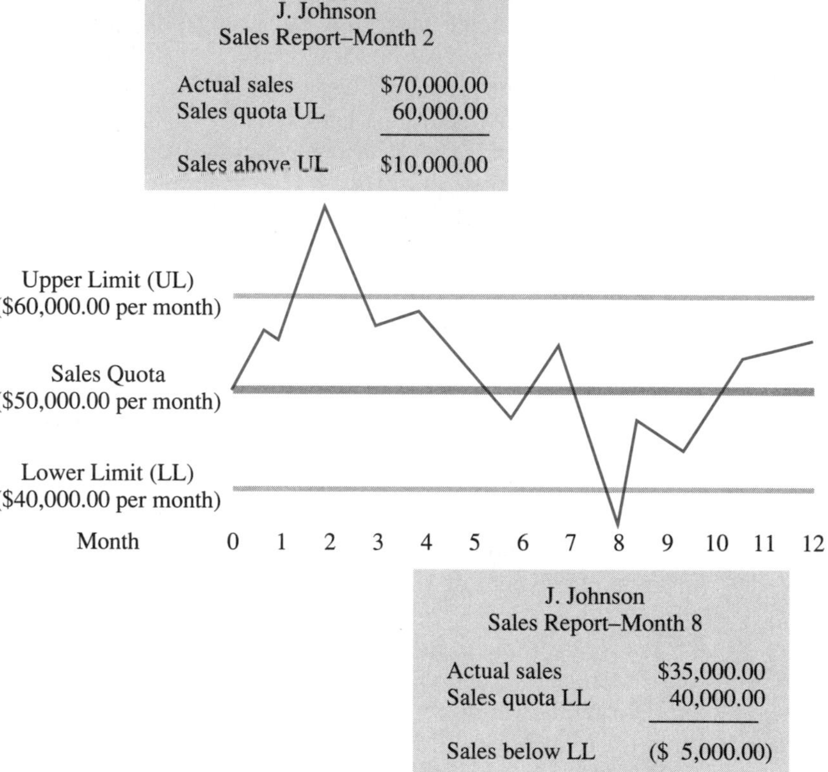

lowercase font may be used for long, very descriptive captions. For single data fields, locate captions to the left of the data field and separate by a unique symbol (a colon is the recommended symbol) and one space, such as

```
DEPARTMENT:xCUTTING
```

For multiple-occurring data fields, locate the caption one line above the column of data fields, such as

```
       PRODUCTION COSTS
DIRECT MATERIALS
DIRECT LABOR
MANUFACTURING OVERHEAD
```

For data fields, left-justify text and alphanumeric characters:

```
NAME:       MARY JONES
DEPARTMENT: SYSTEMS
```

Right-justify lists of numeric data:

```
SUBTOTAL: $1973.40
HANDLING:   200.00
TAX:         47.20
TOTAL:    $2220.60
```

Try to display long contiguous strings of numbers or alphanumeric characters in groups of three, four, or five characters, such as

```
5416 7811 0895 1877
```

For long lists, leave a space line between groups of related data, or about every five rows. Never exceed seven rows without inserting a space line. An example of how to break long lists into groups follows:

```
ANVIL
BRACKET
CLASP
DIE
ELEVATOR

FIRE DAMPER
GAUGE
HYDRAULIC PUMP
INTEGRATOR
JACK
```

Caption and Data Field Justification

Caption and data field justification can be achieved in either of two ways. Approach 1 results in both captions and data fields left-justified into columns. For example:

```
DRIVER NAME:    JOE DOAKES
LICENSE NUMBER: 145G HNN 499
EXPIRES:        12/31/94
```

Approach 2 right-justifies the captions up against the left-justified data field columns, such as

```
   DRIVER NAME: JOE DOAKES
LICENSE NUMBER: 145G HNN 499
       EXPIRES: 12/21/94
```

Headings

For section headings, locate the section heading on a line above associated data fields, and indent captions a minimum of five spaces from the start of the heading. Spell out fully in an uppercase font and display:

```
ASSETS
xxxxxCURRENT:       $899,420.00
     FIXED:       $1,566,748.12
     INTANGIBLE:     $75,000.00
```

Locate row headings to the left of the topmost row of related data fields, separate from the adjacent caption by a greater-than symbol, and separate the symbol from the heading by one space and from the caption by three spaces:

```
CATTLEx>>xxxHEREFORD: 5420xxxxx SIMMENTAL:      470
            ANGUS:    6210      BRAHMAN:       1800
HORSE >>    QUARTER:  2214      THOROUGHBRED:   419
```

or

```
RANCH ANIMALS
xxxxxCATTLE >>      HEREFORD: 5420     SIMMENTAL:      470
                    ANGUS:    6210     BRAHMAN:       1800
     HORSE >>       QUARTER:  2214     THOROUGHBRED:   419
```

For field group headings, center the field group heading above the captions to which it applies. Relate to these captions by a dashed line ended by pointed brackets. Spell out fully in an uppercase font and display:

```
<----------------------------COMPUTER USERS---------------------------->
    USERS                          DEPARTMENT                    ID NUMBER
JIM HARBAUGH                       ACCOUNTING                 ACCT 717 491
BETTY TERRENCE                     MARKETING                  MARK 941 877
MARSHA UMBRERO                     ENGINEERING                ENGR 643 225
```

Spacing

To separate columns visually, leave a minimum of five spaces between the longest data field in one column and the leftmost caption in the adjacent column:

```
MAKE:  EVEREX STEPxxxxxCORE SEEK: 10.0  MILLISECONDS
MODEL: 486/20          MIPS:       5.9
```

With section headings, leave at least five spaces between the longer data field in one column and the section heading in the adjacent column:

```
PERSONAL COMPUTER        BENCHMARK RESULTS
    MAKE:  EVEREX STEPxxx    CORE SEEK: 10.0 MILLISECONDS
    MODEL: 486/20            MIPS:       5.9
```

If space constraints exist, vertical lines may be substituted for spaces:

```
                  |
MAKE:  EVEREX STEPx|xCORE SEEK: 10.0 MILLISECONDS
MODEL: 486/20      | MIPS:       5.9
                  |
```

Without group headings, leave at least three spaces between the columns of data fields:

```
COMPAQ        DESKPRO 486/20Exxx$4500.00
DELL          SYSTEM  310       $3800.00
EVEREX STEPxxx486/20            $3900.00
```

With group headings, leave at least three spaces between columns of related data fields and leave at least five spaces between groupings:

```
<------COMPUTER USERS------>xxxxx<-----COMPUTER ASSIGNED----->
    USERS         DEPARTMENT        MAKE            MODEL
JIM HARBAUGH      ACCOUNTING     DELL          SYSTEM 310
BETTY TERRENCE    MARKETING      COMPAQ        DESKPRO 486/20E
MARSHA UMBREORxxxENGINEERING     EVEREX STEPxxx486/20
```

Title and Screen Identifier

A screen title should appear on all screens. It should describe the screen's contents and be spelled out fully in an uppercase font. Locate it in a centered position toward the top of the screen. For multiscreens, also include a screen identifier, such as *** JBCST 3 *** for JOB COST REPORT, Screen 3. Locate the identifier in the upper right corner one line above the title.

Color

Including color can add a new dimension to screen usability. If used properly, color can organize data, focus attention, accentuate differences, and make displays more interesting.

For best discrimination, select no more than four to six colors widely spaced on the color spectrum, as follows:

Color	Approximate Wavelengths (in Millimicrons)
Red	700
Orange	600
Yellow	570
Yellow-green	530
Green	500
Blue-green	490
Blue	470
Violet	400

The visual spectrum of wavelengths to which the eye is sensitive ranges from 400 to 700 millimicrons.

Refer to the following guidelines when using color:

1 For best discrimination between items, use red, yellow, green, blue, and white.

2 Use bright colors to emphasize, and nonbright colors to deemphasize. The brightness of color from most to least is white, yellow, green, blue, and red.

3 To emphasize separation, use contrasting colors, such as red and green, blue and yellow.

4 To convey similarity, use similar colors, such as orange and yellow, blue and violet.

5 Use no more than two or three colors plus white on a screen at one time. More than that makes the screen too busy for user comfort.

6 Use warm colors, such as red, orange, and yellow to force attention, to indicate action, or to evoke a response.

7 Use cool colors, such as green, blue-green, blue, and violet to provide status or background information.

8 Use desaturated or spectrum center colors, such as yellow and green, for text material.

9 Use darker, spectrally extreme colors for backgrounds. Red, blue, or black make good display backgrounds. Blue is especially good because of the eye's increased sensitivity to it in the periphery.

Conform to user expectations. A common expectancy, for example, is: red = loss or stop or danger, yellow = caution, and green = gain or go or normal. Consistency should be maintained in color usage. Changing color meanings will lead to difficulties in interpretations and errors. For example, if red means a problem or loss on one screen and something positive or a profit on another screen, or on the same screen, confusion among users will occur.

Contrast is generally the single most important design factor when making color choices. Dark colors on dark backgrounds and light colors on light backgrounds have low contrast levels and are less legible. For example, red and dark blue do not show up well on black. However, red on yellow or black on yellow or dark blue on white or white on dark blue are very readable color combinations.

CREATING GRAPHS TO ILLUSTRATE NUMERICAL DATA QUICKLY

GRAPHS provide a way to illustrate numerical information that can be comprehended quickly. Graphs make relationships between numbers visible by turning quantities into shapes. Presenting output in image form reinforces the cliches, "A picture is worth a thousand computer printouts," and "Seeing the problem is solving the problem." Four of the best tools to prototype graphs for users are spreadsheets, CASE tools, DBMS, and fourth-generation languages. More about DBMS and 4GLs appear later in the book.

Graphs can be divided into categories based on the kind of information they convey.[2] Some categories well suited to design output for business information systems are:

- Scatter graphs
- Line graphs
- Bar graphs
- Sectographs
- Picturegraphs

Scatter Graphs

SCATTER GRAPHS (also called **SCATTERPLOTS** or **SCATTER DIAGRAMS**) clearly reveal trends in the underlying data. Review the four *x, y* columns of data at the top of Figure 8.10. It is quite difficult to gain insight or detect a trend in these data. The

[2] Hilary Goodall and Susan Smith Reilly, *Writing for the Computer Screen* (New York: Praeger Publishers, 1988), p. 103.

I		II		III		IV	
X	*Y*	*X*	*Y*	*X*	*Y*	*X*	*Y*
10.0	8.04	10.0	9.14	10.0	7.46	8.0	6.58
8.0	6.95	8.0	8.14	8.0	6.77	8.0	5.76
13.0	7.58	13.0	8.74	13.0	12.74	8.0	7.71
19.0	8.81	9.0	8.77	9.0	7.11	8.0	8.84
11.0	8.33	11.0	9.26	11.0	7.81	8.0	8.47
14.0	9.96	14.0	8.10	14.0	8.84	8.0	7.04
6.0	7.24	6.0	6.13	6.0	6.08	8.0	5.25
4.0	4.26	4.0	3.10	4.0	5.39	19.0	12.50
12.0	10.84	12.0	9.13	12.0	8.15	8.0	5.56
7.0	4.82	7.0	7.26	7.0	6.42	8.0	7.91
5.0	5.68	5.0	4.74	5.0	5.73	8.0	6.89

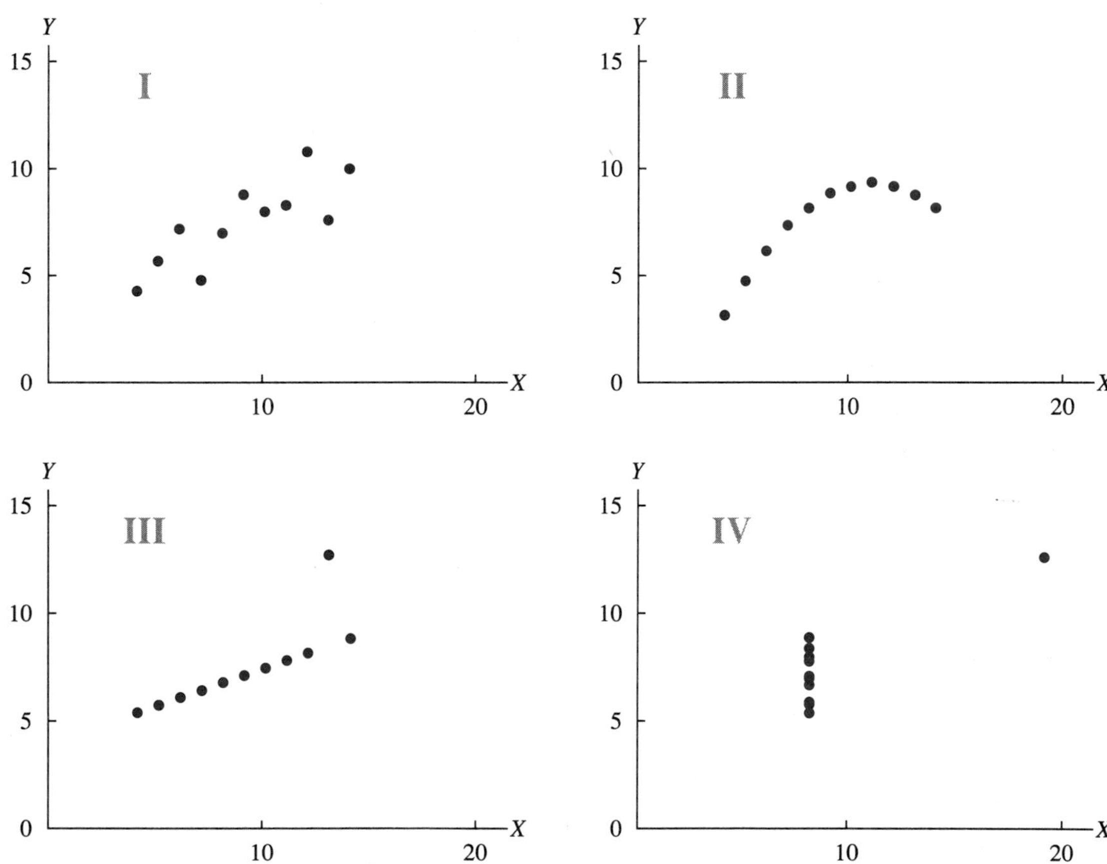

Figure 8.10 Data trends and behavior revealed in scatter graphs.

scatter graphs at the bottom of the figure, however, indicate plainly the behavior and trend of the plotted data.[3]

Line Graphs

LINE GRAPHS show fluctuations over time. These fluctuations are visualized by users by means of a rising and falling line that shows highs, lows, rapid movement, or stability. Relatively thin lines should be used so that points of reference are not obscured. Contrasting color can be used for emphasis more effectively than can bold lines. Care must be taken to adjust the size of the grid in a line graph after the points of reference are plotted to ensure maximum visibility of the information. Too many grid lines, however, obstruct a clear view. Too few lines make information difficult to read.[4] Examples are shown in Figure 8.11.

Figure 8.11
Misuse of grid lines.

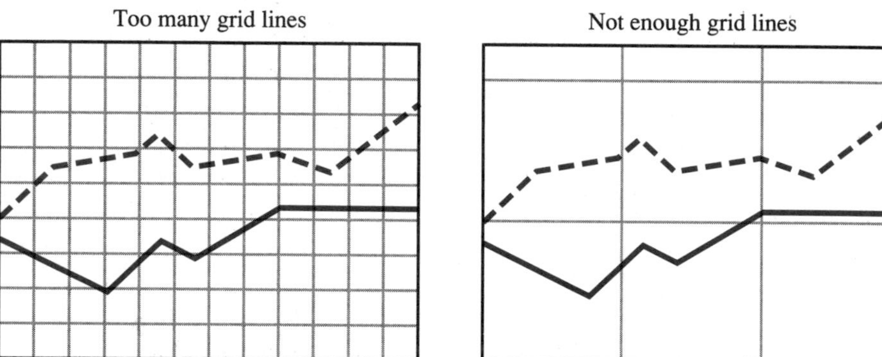

Figure 8.12 shows sales of products A and B in units at Starling Company for the fiscal year 1993. Notice how the grid helps facilitate reading of the lines at various points on the graph.

Bar Graphs

BAR GRAPHS show proportions or quantities related to each other. The two types of bar graphs are:

- Horizontal bar graph
- Vertical bar graph

In each graph, amounts are presented by bars placed on a grid. The length or height of the bars represent amounts determined by the grid's scale.

Unlike line graphs, bar graphs emphasize totals at specific points rather than fluctuations of numbers over time. Bar graphs should be used only when the amount of numbers to be compared is relatively small. If too many comparisons are made, the bars will be too thin to have much visual impact.[5] In some

[3] F. J. Anscombe, "Graphs in Statistical Analysis," *American Statistician,* February 1973, pp. 17–21.

[4] Goodall and Reilly, *op. cit.,* p. 104.

[5] Ibid., p. 105.

Figure 8.12
Line graph.

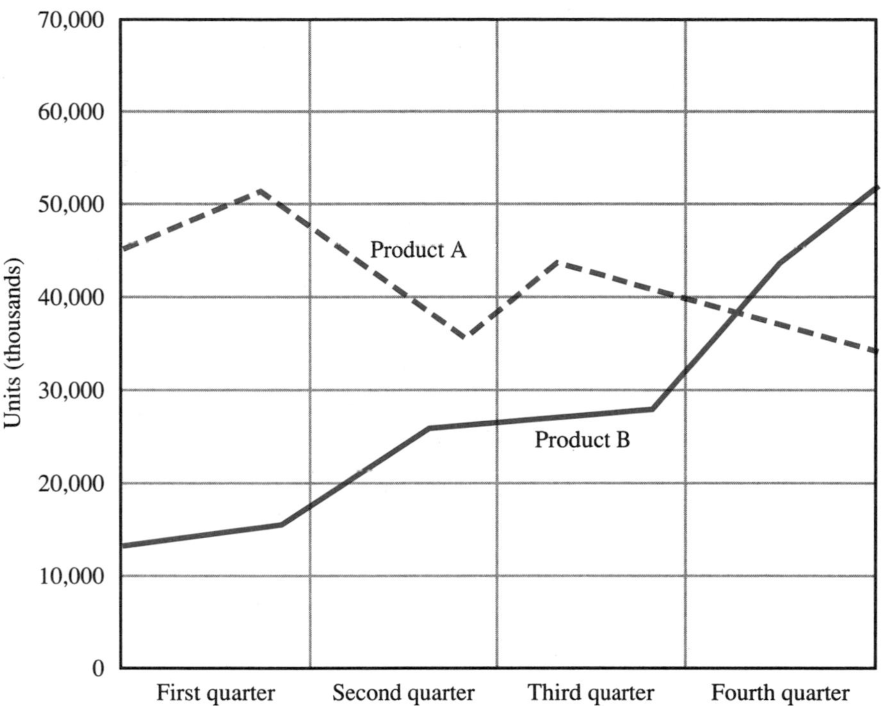

designs, bars can be differentiated by filling them with contrasting textures and colors.

Horizontal Bar Graphs

A **HORIZONTAL BAR GRAPH** compares different items during the same time frame. For example, the horizontal bar graph in Figure 8.13 discloses the number of students by professional schools who graduated from Omni University in May 1993.

Vertical Bar Graphs

A **VERTICAL BAR GRAPH** measures the same item compared at different periods of time. For example, the number of pizzas sold at Smalltown, home of State University, during the fall semester, is shown in Figure 8.14.

Sectographs

SECTOGRAPHS show how total amounts are divided up. Two popular sectographs are the

- Pie chart
- Layer chart

Figure 8.13
Horizontal bar graph.

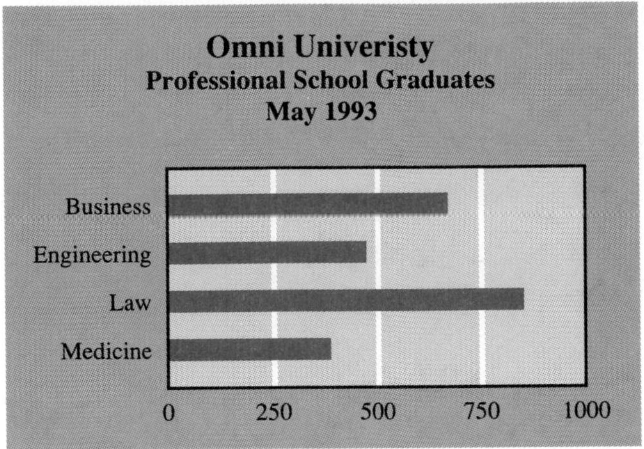

Figure 8.14
Vertical bar graph.

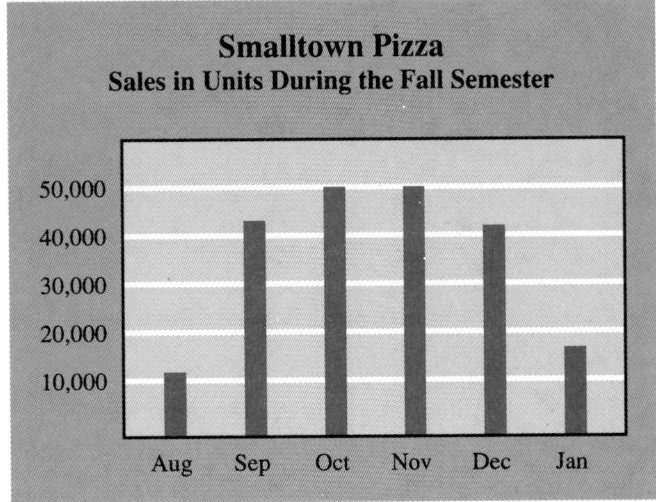

Pie Charts

In its simplest form, a **PIE CHART** is a circle that has been segmented into two or more pieces each of which represents a certain percentage of the whole, proportional to its contribution to the whole. The proportional amount is immediately apparent to any viewer. The visual impact is strong and memorable. The pie chart in Figure 8.15 shows the income statement items for Acme Company.

Pie charts can be combined with a bar graph to display additional information. For example, in Figure 8.16, a pie chart shows percentages of production costs, and a bar graph discloses the breakdown of manufacturing overhead costs.

When pie charts are used to design output, adhere to the following guidelines:

- Use fewer than ten wedges to provide adequate differentiation and to avoid confusion.

- Include labels and numbers to identify each wedge.
- Use contrasting colors or textures to intensify comparisons.
- Explode wedges slightly from the remainder of the pie for emphasis and clarity.

Figure 8.15
Pie chart.

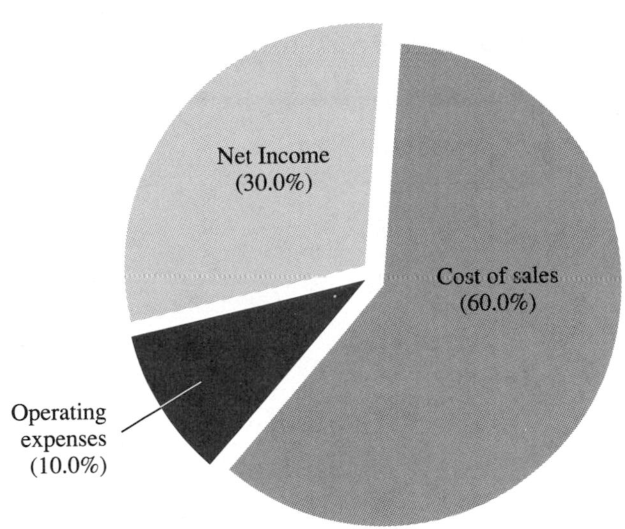

Layer Graphs

A **LAYER GRAPH** is created like a line graph, but areas between the lines represent quantities and add up to a total amount. The layers are differentiated by filling each with a different color or texture.[6] Figure 8.17 is a layer graph that shows sales by salesperson over four quarters at Palace Department Store.

Picturegraphs

In a **PICTUREGRAPH**, columns of little signs or icons are used in place of bars as one finds in a bar chart. Each picture represents a certain quantity of the item illustrated; see the picturegraph in Figure 8.18, which shows the number of pickup trucks sold over a four-year period in Suffolk County.

Picturegraphs do catch the user's eye. They are best used to convey a general impression of quantities rather than precise information. In preparing picturegraphs, use the following design guidelines:

- Use familiar pictures or symbols. Unfamiliar pictures or symbols must be learned.

---

[6] Ibid., p. 108.

- Design pictures and symbols that convey their intended meaning precisely.
- Design for efficiency. In some instances, a picturegraph may consume more display space than text or numbers.
- Symbols chosen must be easily distinguishable from other symbols. The users' ability to discriminate alphabetic or alphanumeric information is much more powerful than their ability to discriminate among a large number of symbols.

CREATING TABLES AND MATRICES THAT HIGHLIGHT, COMPARE, AND INSTRUCT

Tables and matrices combine features of conventional tabular reports and graphs. Both are applicable for highlighting relationships, showing comparisons, and giving instructions.

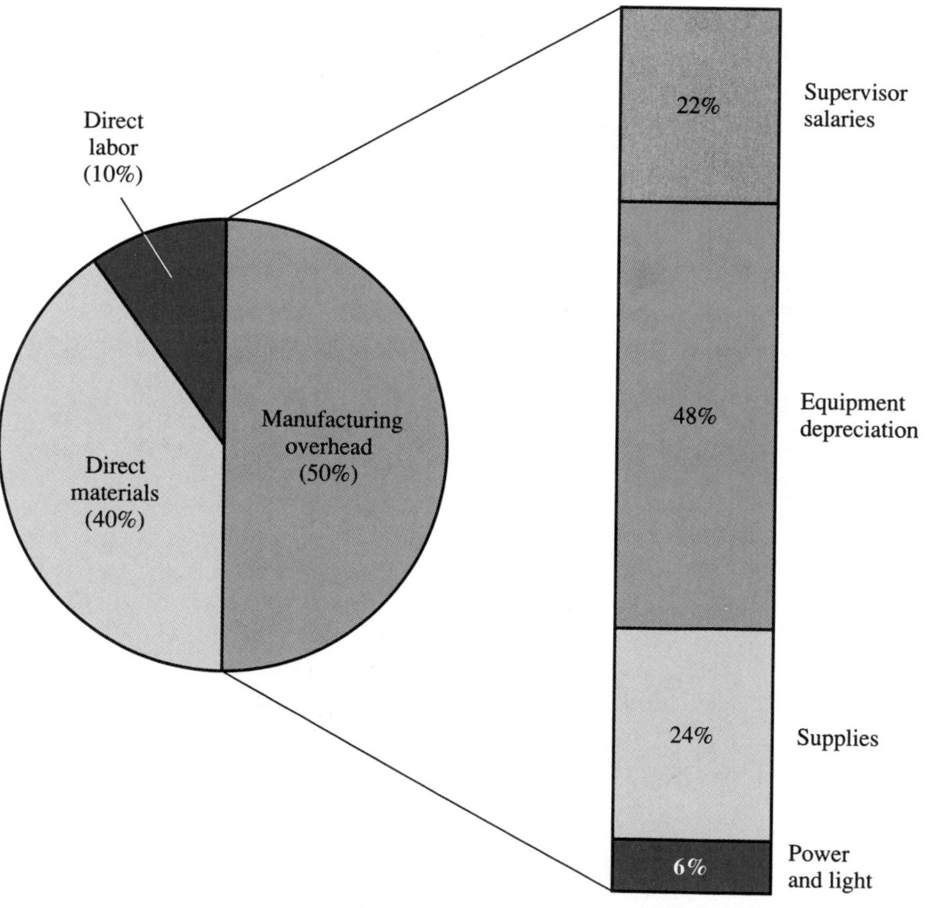

Figure 8.16
Pie chart combined with bar graph.

Figure 8.17
Layer graph.

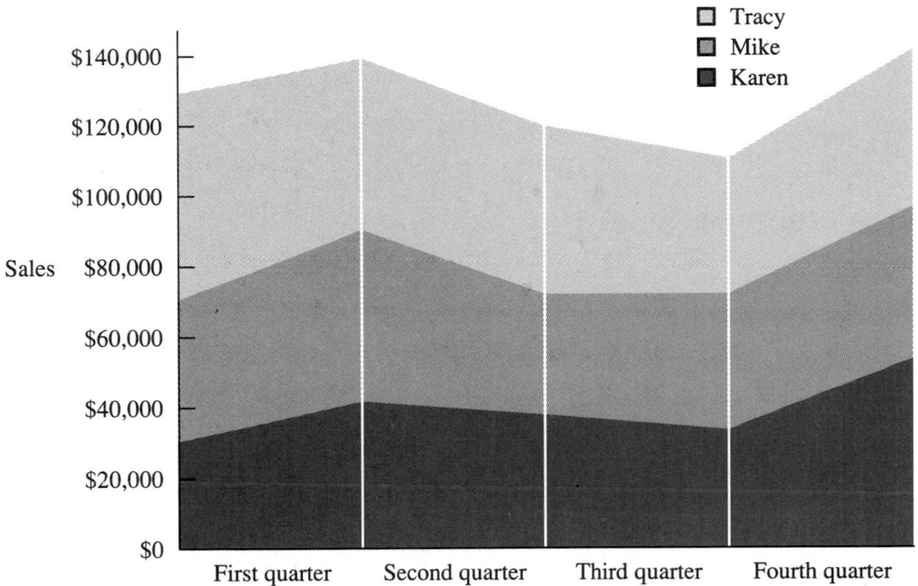

Tables

A **TABLE** is made up of columns of numbers with subject titles arranged on a grid, such as the one demonstrated in Figure 8.19. This table is used to compare features of three laptops under consideration. Such a table is an excellent means of providing executive summary information.

Matrices

A **MATRIX** is a rectangular arrangement of elements into rows and columns. It is an excellent device to show relationships among elements. In Chapter 6, we discussed objects and attributes belonging to certain objects. In Figure 8.20, we

Figure 8.18
Picturegraph.

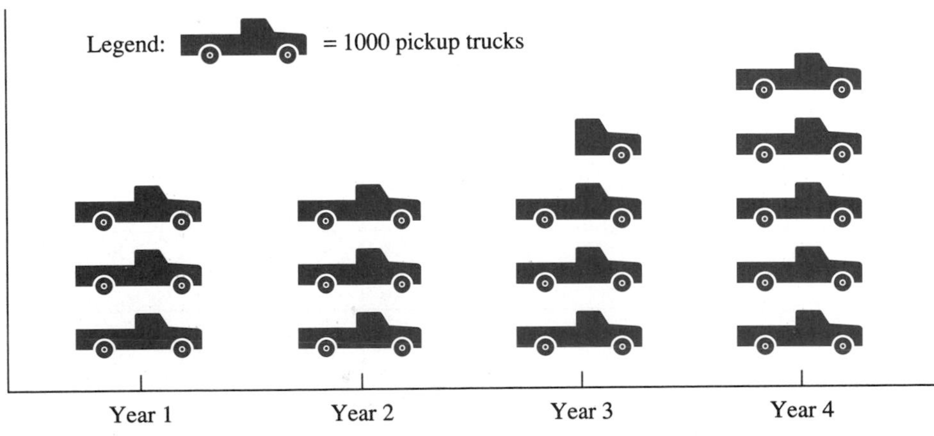

Figure 8.19
Table showing evaluation of laptops offered by three vendors.

GENERAL EVALUATION OF LAPTOP VENDORS

	Vendor A	Vendor B	Vendor C
Portability	Excellent	Excellent	Good
Battery Life	Fair	Good	Good
Screen	Fair	Excellent	Excellent
Keyboard	Excellent	Excellent	Good
Expandability	Fair	Excellent	Good
Performance	Good	Excellent	Good
Service and support	Excellent	Excellent	Good
Price	Excellent	Good	Excellent
Overall value	Good	Excellent	Good

Figure 8.20
Matrix showing attributes used in an order system.

	Object				
Attribute	Billing	Customer	Inventory	Order	Shipping
Billing_Address	X	X			
City	X	X			
Credit_Balance		X			
Credit_Limit		X			
Customer_Number		X		X	
Customer_Name	X	X			X
Date	X			X	
Date_Received				X	
Date_Shipped	X				X
Description			X	X	X
Invoice_Number	X				
Item_Number	X		X	X	X
Order_Packed				X	
Order_Shipped					X
Order_Status				X	
Purchase_Number	X		X	X	
Quantity			X	X	X
Shipping_Address		X			X
Size			X	X	
State	X	X			X
Street	X	X			X
Zip	X	X			X

use a matrix to document which attributes belong to which objects in an object-oriented order system.

REVIEW OF CHAPTER LEARNING OBJECTIVES

The major goals of this chapter were to enable each student to achieve seven important learning objectives. We will now summarize the responses to these learning objectives.

Learning objective 1:
Define high-quality and usable information, and state why output is the controlling systems design component.

High-quality information possesses the following characteristics:

- Accessibility
- Timeliness
- Relevance
- Accuracy

Usability of output is based on the following formula:

usability = (quality of information)
− (effort + annoyance + frustration in assimilating the information)

High-quality and usable information must be both substantive and in a form that appeals to users' cognitive styles.

Learning objective 2:
Explain the use of hierarchical reports, and design at least one.

Hierarchical reports are tailored to the appropriate user level. These reports include filter and responsibility reports. A filter report screens out extraneous data and data too detailed for particular users. A responsibility report includes information that pertains to tasks or decisions for which users are responsible.

Learning objective 3:
Discuss and design comparative reports.

Comparative reports allow users to examine results side by side or point by point in order to establish likenesses and differences. Types of comparative reports are: horizontal, vertical, and counterbalance reports. Horizontal reports permit analysis of increases or decreases in corresponding items in comparative statements. Vertical reports show the relationship of the component parts to the total within a single statement. A counterbalance report shows all sides of an item or situation under review.

Learning objective 4:
Describe and design monitoring reports.

The system is designed to keep track of some process or activity. If the process or activity fails to perform as expected, a report is produced automatically to inform managers of this condition. Such monitoring reports include variance and exception reports.

A variance report discloses deviations between actual and standard at the end of a time period or process. An exception report displays an activity or process only when such an activity or process has exceeded or fallen below certain limits.

Learning objective 5:
Discuss the fundamentals of designing display screens.

A well-designed screen format and paper report are essential in output design. The design objective is human ease in locating and understanding information. Scanning is made easier if eye movements are minimized, required eye movement direction is obvious, and a consistent pattern is followed. Colors can also be used to direct users' attention and convey information.

Learning objective 6:
Illustrate the use of graphs, and state why they often add to output's usability.

If detailed and precise text and numbers are required, conventional tabular reports, using both screen and paper, must be used. Graphs, however, can be used in a number of situations in which mountains of data can be converted to pictures with which users can quickly see the problem or condition.

Learning objective 7:
Describe how tables and matrices can be used to convey information effectively.

Tables are excellent devices to show a breakdown of data. Matrices organize output into a rectangular arrangement of elements into rows and columns. Whenever many items of data are to be presented or related, consider the use of tables and matrices as output design devices.

OUTPUT DESIGN CHECKLIST

Following is a checklist on how to perform output design. Its purpose is to remind you of major aspects of output design.

1 Emphasize quality of output by making it accessible, timely, relevant, and accurate.

2 Match the cognitive styles of users with appropriate forms of output. For users with a penchant for detail and lists of numbers, create tabular and textual reports. For users who like images and patterns, use graphs, tables, and matrices.

3 Design filter reports to show appropriate levels of detail. Use responsibility reports to tailor reports for specific areas of responsibility.

4 Design horizontal reports for comparisons over time. Design vertical reports to compare elements within a report. Design counterbalance reports to disclose good, bad, and likely results.

5 Design variance reports to compare actual results with standard results at the end of some process. Design exception reports to signal when a process or an activity falls outside predetermined limits.

6 For screens (and paper reports), format captions and data fields to increase readability.

7 Use colors to contrast, draw attention, emphasize, deemphasize, and complement output.

8 Design a scatter graph to indicate the behavior and trend of plotted data.

9 Design a line graph to show highs, lows, rapid movement, or stability of data over time.

10 Design bar graphs to emphasize totals at specific times. Horizontal bar graphs compare different items during some time frame. Vertical bar graphs measure the same item compared at different periods.

11 Design pie charts to show proportions of segments relative to the whole.

12 Design a layer graph to disclose relative quantities to the total over time.

13 Design picturegraphs to convey general quantities of the item illustrated over time with icons or symbols.

14 Design tables and matrices to show breakdowns of information or systematic arrangements of data in rows and columns for ready reference.

KEY TERMS

Bar graphs
Comparative reports
Counterbalance report
Exception report
Filter report
Graphs
Hierarchical reports
Horizontal bar graph
Horizontal report
Layer graph
Line graphs
Matrix
Monitoring reports
Output
Picturegraph
Pie chart
Responsibility report
Scatter graphs (scatterplots or scatter diagrams)
Sectographs
Table
Variance report
Vertical bar graph
Vertical report

REVIEW QUESTIONS

8.1 In detailed design, why start with output?

8.2 What determines output's quality and usability?

8.3 What's the purpose of hierarchical reports? What are the types of hierarchical reports? Give a brief example of each report.

8.4 What are the types of comparative reports? Give a brief example of each type of comparative report.

8.5 What are monitoring reports used for? What are the types of monitoring reports? Give a brief example of each type of monitoring report.

8.6 Explain why the proper design of display screens or paper reports, including organization, caption and data field justification, headings, and spacing, is important to the user.

8.7 Explain why colors are sometimes used in screen design.

8.8 Explain how to achieve strong discrimination between items on the screen, using colors.

8.9 Discuss how graphs can increase output usability.

8.10 Name the graph styles and state the purpose of each.

8.11 What are tables and matrices used for when you are designing output?

CHAPTER-SPECIFIC PROBLEMS

These problems require exact responses based directly on concepts and techniques presented in the text.

8.12 Study these situations:

____________ Users want to compare financial data over the past three years.

____________ Management wants a report that shows the worst-case and best-case scenarios of a proposed investment.

____________ Management wants a report that discloses when operating costs have exceeded predetermined limits.

____________ The plant manager wants a report that highlights deviations from standards at the end of each milling process.

____________ The construction superintendent wants a report showing total construction costs for each project. The pipe foreman wants construction costs that are traceable to the pipe-laying project.

____________ The controller wants a report that uses percentages to show the relationships of different parts to the total within an income statement.

________________ Management wants the new system to classify and compare reports of cost information according to specific areas of responsibility in the company.

Required: For each situation, insert the type of report you would design to meet the users' needs.

8.13 At Mecom, the drill press foreman is responsible for the following expenses: supervision, $4000; machine setup, $2000; and rework, $1500.The superintendent is responsible for the following production expenses: factory office, $10,000; machining, $30,000; assembly, $20,000; and drill press expenses. The vice president of production is responsible for the following total factory expenses: general office, $40,000; engineering, $60,000; and production expenses.

Required: Prepare responsibility reports for the preceding data, and indicate their relationship by drawing a line connecting the reports.

8.14 A partial comparative balance sheet for Dover Company is stated as follows:

Assets	1994	1993
Current assets	$390,000	$288,000
Equipment	500,000	467,000
Other	60,000	105,000
Total	$950,000	$860,000

Required: Convert this partial comparative balance sheet into a horizontal report showing increase or (decrease) in dollar and percentage terms.

8.15 A condensed income statement for Associated Company appears as follows:

Net sales	$1,000,000
Cost of goods sold	700,000
Gross profit	$ 300,000
Expenses	250,000
Net Income	$ 50,000

Required: Convert this condensed income statement into a vertical report expressing all items in the income statement as percentages of net sales.

8.16 Milestone Publishing is planning to buy New Horizons, a regional publishing company. The worst-case estimates are advertising revenue of $10,000,000, operating expenses of $8,850,000, selling expenses of $2,500,000, and administrative expenses of $3,000,000. The best-case scenario is advertising revenue of $16,000,000, operating expenses of

$9,000,000, selling expenses of $2,000,000, and administrative expenses of $1,500,000.

Required: Prepare a counterbalance report to display the preceding worst- and best-case scenarios.

8.17 Vulcan Manufacturing's purchase agent bought 4000 pounds of product A for $5.00 per pound. The standard price for product A is $4.60 per pound. Also, 2000 pounds of product B were purchased for $3.00 per pound. The standard price for product B is $3.20 per pound.

Required: Design a variance report to disclose the foregoing data.

8.18 A system produces a sales-by-salesperson report like the following:

Sales in Units by Salesperson Report
MM/DD/YY

	Sales Performance		
Salesperson	**Quota**	**Actual**	**Variance (U,F)**
Todd Bailey	**1500**	**1000**	**500 U**
Jane Craig	**1000**	**980**	**20 U**
Jim Elliott	**1400**	**1200**	**200 U**
Joe Ford	**1600**	**1640**	**40 F**
Sue Hampton	**1200**	**1290**	**90 F**
Amy Klein	**1100**	**1090**	**10 U**
Bill Zerga	**1300**	**1900**	**600 F**

Required: Redesign the preceding report to increase the level of decision support it provides by using an exception report. Upper and lower control limits are 100 units.

8.19 ______________ A physician wants to compare how the number of reported sunburn cases and the number of sunscreen units sold vary over the last year.

______________ The chief inspector wants you to design a simplified way of showing a weekly inspection schedule. Disclose each day of the week; inspection tasks performed on specific days; and the time the tasks are performed so that junior inspectors can easily understand the schedule.

______________ The chief financial officer wants to show operating expenses, administrative expenses, marketing expenses, and research and development expenses over the past year. The four categories of expenses will add up to the total expenses and will be differentiated by filling each expense category with a different texture.

_______________ The company stocks five different types of motorcycles. The sales manager wants to know how many units of each type were sold in May.

_______________ A sales manager wants to know how many units of a specific tennis racket were sold the first six months of this year.

_______________ A plant manager wants to know the percentage of each manufacturing cost to the total.

_______________ A fast-food restaurant offers six different sandwiches. Overall, the restaurant uses ten ingredients, such as chicken, ham, turkey, lettuce, and so on. Each sandwich requires a specific set of ingredients. For example, Ham Deluxe requires a sesame seed bun, ham, mayonnaise, tomato slices, and lettuce. Chicken Supreme requires a regular bun, chicken, mayonnaise, onion slices, and lettuce. And so on.

Required: Insert in the blank provided what type of device—graph, table, or matrix—you would design to meet the users' needs.

8.20 Following are some data pertaining to Bagdad's operations from January through June:

Month	Labor Hours	Utility Expenses
January	34,000	$700
February	30,000	640
March	36,000	660
April	39,000	720
May	42,000	600
June	33,000	670

Required: Based on the data above, prepare a scatter graph.

8.21 Marvel Company has two divisions that sell products A, B, C, and D. Division 1 sales for the fiscal year are broken down as follows: product A, 20 percent; product B, 15 percent; product C, 25 percent; and product D, 40 percent. Division 2 sales for the fiscal year are broken down as follows: product A, 35 percent; product B, 30 percent; product C, 20 percent; and product D, 15 percent.

Required: Draw a pie chart for each division showing its proportional sales of products A, B, C, and D.

8.22 The marketing manager at Stratus Company requires sales performance information on three salespersons: Brown, Jones, and Smith. Brown's quota was 3000 units; so far, she has sold 2400 units. Jones' quota was 2300 units; so far, he has sold 2600 units. Smith's quota was 2800 units; so far, she has sold 3400 units.

Required: Draw a bar graph showing the sales performance of Brown, Jones, and Smith. *Hint:* Use a bar to show sales quota and a bar to show sales to date for each salesperson.

8.23 Following are the sales of Saturn Company for 1993:

	Product A	Product B
Jan.	900	1700
Feb.	1000	1900
Mar.	1400	1000
Apr.	1200	2000
May	1800	900
June	1600	1000
July	1700	800
Aug.	2100	700
Sept.	2200	600
Oct.	1800	900
Nov.	1500	1000
Dec.	1200	2000

Required: Sketch a line graph to display these sales data.

THINK-TANK PROBLEMS

These problems call for a feasible approach rather than a precise solution. Although the problems are based on chapter material, extra reading and creativity may be required to develop workable solutions.

8.24 To be displayed on a screen are policy number; driver, including name, sex, birth date, marital status, occupation, driver's license number, state, and years licensed; vehicle, including make, year, model, and horsepower; and type of claim, including liability, collision, or comprehensive.

Required: Design a claim report screen to display clearly the preceding data. Include appropriate headings, captions, and data fields.

8.25 A screen for Autotech Insurance Company is titled Personal Automobile. The total screen design includes three pages. On the first page is a left section heading called Policy. Included in this section is one caption called Policy Number. The value in the field is 974211. A section heading below Policy is called Driver. Captions in this section are:

Name (Mary Jones), Sex (Female), Birth Date (10/15/54), Marital Status (Married), Occupation (Writer). A right section called Vehicle is horizontally aligned with the Policy section. It contains the following captions: Year (1991), New/Used (Used), Make (Volvo), Model-Number (740GL). Section headings are included and captions follow the right-justified approach. Visual separation of the left and right sides of the screen is aided by a vertical dashed line.

Required: Design a display screen for the foregoing.

SUGGESTED READING

Adams, Lee. *Supercharged Graphics, A Programmer's Source Code Toolbox*. Blue Ridge Summit, Pa.: TAB Books, 1988.

Ain, Mark S. "From Punch In to Payroll Automatically." *Information Strategy: The Executive's Journal,* Summer 1989.

Anscombe, F. J. "Graphs in Statistical Analysis." *American Statistician,* February 1973.

Friend, David. "EIS: Straight to the Point." *Information Strategy: The Executive's Journal,* Summer 1988.

Galitz, Wilbert O. *Handbook of Screen Format Design,* 3rd ed. Wellesley, Massachusetts: QED Information Sciences, Inc., 1989.

Goodall, Hilary, and Susan Smith Reilly. *Writing for the Computer Screen*. New York: Praeger Publishers, 1988.

Johnston, Stuart J. "Multimedia: Myth vs. Reality." *Infoworld,* February 19, 1990.

King, Karl G., and Raymond W. Elliott. "In Plain English, Please." *Journal of Accountancy,* March 1990.

Martin, James. "EIS Helps Managers Gain Insight into Factors for Success." *PC Week,* April 24, 1989.

Parsons, Andrew. "Better Data Worse Decisions?" *Information Strategy: The Executive's Journal,* Summer 1989.

Robey, D., and W. Taggunt. "Human Information Processing in Information and Decision Support Systems." *MIS Quarterly,* June 1982.

Schneiderman, Ben. *Designing the User Interface: Strategies for Effective Human-Computer Interaction*. Reading, Mass.: Addison-Wesley, 1987.

Smith, L. Murphy, and James A. Sena. "Spreadsheets That Know How to Use a Computer." *Financial and Accounting Systems,* Spring 1991.

Spencer, Cheryl. "Visual Decision Making." *Personal Computing,* January 1987.

Tufte, Edward R. *The Visual Display of Quantitative Information*. Chesire, Conn.: Graphics Press, 1983.

Whieldon, David. "Computer Graphics: Art Serves Business." *Computer Decisions,* May 1984.

Zilber, Jon. "Information, Please." *MacUser,* December 1988.

JOCS CASE: Output Design

At the end of Chapter 7, Jake Jacoby analyzed four different design alternatives for the JOCS system. He submitted his evaluation and recommendation to the steering committee so that they could decide which alternative should be pursued. The committee agreed to approve Jake's recommendation of System Alternative Number 4. This alternative included the basic job costing system discussed in previous chapters and also added a component to estimate the time and costs it would take to finish a hydraulic lifter after production was started. The committee felt that the information produced from this additional component would increase the overall value of JOCS.

While Jake was performing systems evaluation and selection for JOCS, Mary Stockland was conducting preliminary interviews for a systems SWAT team. Mary and Jake now work together to decide the personnel of the JOCS SWAT team. They select a relatively small development team of four people, because research from other application projects has shown that large numbers of people frequently slow the development process due to communication difficulties. They choose four people who have a combination of computer and application expertise. The people selected by Mary and Jake are:

> Christine Meyers, an analyst with ten years' computer programming experience and an associate degree from a community college. She has worked in a variety of organizations, with applications background in manufacturing, distribution, and retail businesses.
>
> Tom Pearson, an analyst with six years' experience in a combination of programming, analysis, and operations. He has worked for two small manufacturing companies during the past six years, performing a variety of different tasks. He has taken many different continuing education classes at a university level, but has not pursued a degree.
>
> Carla Mills, a cost accountant who has worked for Peerless for the past three years. She has worked in the accounting area of the finance department, reporting indirectly to Burt Flanders. She is very familiar with the systems currently in place at Peerless. Before being employed at Peerless, Carla completed her bachelor's degree in accounting from a university.
>
> Cory Bassett, a recent CIS graduate from a local university. She has just graduated with a bachelor's degree in business administration with an emphasis in computer information systems.

Jake, as project manager, meets with his team to discuss the approach to output design for JOCS. Jake states: "I want to create a system that really produces information for decision making. Instead of simply generating the standard reports that you see listed in a cost accounting textbook, I want each of us to find out what information management needs to make decisions."

Carla speaks up: "That should be easy. All you have to do is ask people around the company what they want to see. Everybody knows what they need to do their job."

Christine frowns and says, "I don't think that's true. People seem to have a vague idea of the information they need, but they don't even have a clue about what format is required. In fact, sometimes they don't even know what kind of information we can give them."

Carla disagrees. "Nobody really asks the users what they want to see. If we ask, we can design the output exactly to their specifications."

Cory tentatively enters the discussion, saying, "In college, they told us if you design a system exactly to one person's specifications, sometimes you can have problems in the future. Sometimes, it's a good idea to produce pretty standard output. I mean, what do we do when that person who approved the specs quits, and we have a new person who doesn't like or understand the system?"

Tom chuckles and says, "Yes, that's the problem. It's much easier just to come up with screens and reports that make sense to us and let the users get accustomed to them. We can spend some time on the back end of the project training users how to use our system. We have to get this project moving, and I don't like to waste time finding out every little piece of information people want to get from our system."

Jake steps in at this point and says, "Wait a second. This is not our system. The JOCS system is being created by us for the rest of the company. It's their system. Some good points have been raised here. We can't design output for any one person's whim; also, I don't want to waste our time and resources producing output that nobody is going to use. We can't depend on all of the users' understanding what can be produced as output from a computer. So what do we do?"

Carla, still convinced that users can design their own output, says, "Well, I can sure tell you the standard reports that are needed from this system. In fact, I could list some pretty important screens, too."

Christine smiles and says, "That's a great idea. How about if we pick Carla's brains and come up with the general output from the system? Then we can split up the team, go out to the users, and get some feedback. We could get together, do a little redesign, and then have a meeting with the key users."

Tom says, "Sounds good to me. Then we will have a starting place for the users. I hate to go out to people and ask, 'So what kind of information do you want to see?' It's much easier to say, 'What do you think about this report? What additional information would help you?' . . . I like this approach."

The SWAT team decides to spend some time over the next week working from the systems analysis performed by Jake and using Carla's knowledge to develop a preliminary draft of output for JOCS. They are especially concerned with matching up the style of output with the type of user. They want to provide detailed reports for users who understand and think in numbers, while providing graphics for people more suited to that style of output. They know that the output won't mean anything unless it is truly relevant and usable to the end user. For example, Figure 8.21 shows two different views of labor variance. The top half of the figure presents labor variance by employee within the fitters department. This report is very detailed, is destined to be used for day-to-day labor control, and is most suitable for a supervisor within

Figure 8.21
Labor variance report.

Labor Variance Report

Department: Fitters — Date: MM/DD/YY

	Actual	Budget	Variance
Bartff, Jim			
Regular:	30:00	40:00	10:00
Overtime:	3:00	0:00	(3:00)
Cherrington, Bob			
Regular:	32:00	25:00	(7:00)
Overtime:	4:00	0:00	(4:00)
Gelinas, Joe			
Regular:	40:00	40:00	0:00
Overtime:	5:00	10:00	5:00
White, Bob			
Regular:	44:00	49:00	(4:00)
Overtime:	10:00	8:00	(2:00)
Totals			
Regular:	146:00	145:00	(1:00)
Overtime:	22:00	18:00	(4:00)

Rate variance: \$145.00 (U) = (\$15.00 – \$14.00) × 145 hours
Efficiency variance: \$14.00 (U) = (146 – 145) × \$14.00

Wages as percentage of the current week's production costs

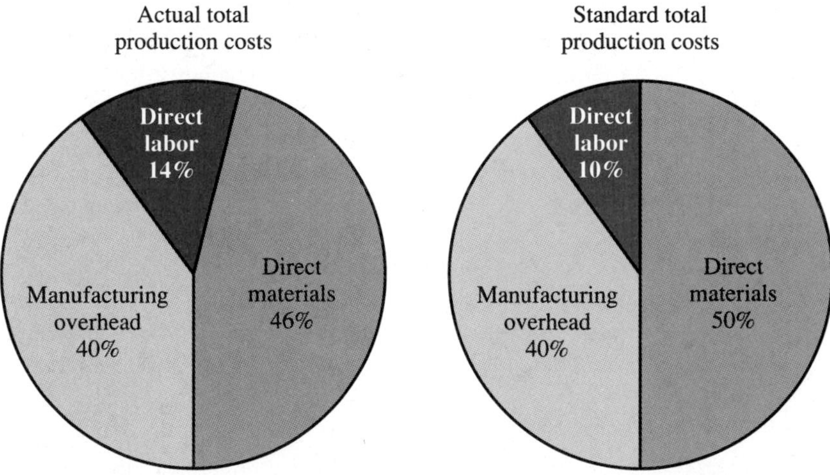

that department. The bottom half of the figure shows a graphic depiction of labor variance. It shows total wages as a percentage of the week's production costs, which is an excellent way to understand the week's costs. It is best suited for managers within the manufacturing and finance departments.

The SWAT team, with prodding from Carla, is working hard to design output that can help managers make decisions *before* problems occur rather

Figure 8.22
List of employees who have worked more than a specified number of hours approaching overtime.

APPROACHING OVERTIME REPORT

Miller, Steve		Welder
Hours worked:	34:00	
Epstein, Susan		Assembler
Hours worked:	36:00	
Lopez, Todd		Assembler
Hours worked:	33:00	
White, Bob		Fitter
Hours worked:	38:00	

than simply telling managers what has happened after the fact. Figure 8.22 is an example of an early warning report. This report will tell supervisors in the manufacturing and finance departments during midweek about the hours that employees are working. Decisions can be made then, before overtime has been incurred, about the cost-effectiveness of having employees work more hours. This report, used in conjunction with Figure 8.23, will enable supervisors to adjust work schedules to minimize overtime. Figure 8.23 allows a department supervisor to get a quick glance of working hour coverage without having to sort through the scheduled hours of each employee.

Figure 8.24 is a standard job-order cost report that will be generated from JOCS. This report provides a summary of the costs associated with a particular job. It is necessary for accounting for jobs. The data used for this report will also be used as a base for future job estimates. The SWAT team was planning to use a standard columnar report, detailing information for many

Figure 8.23
Staffing coverage report.

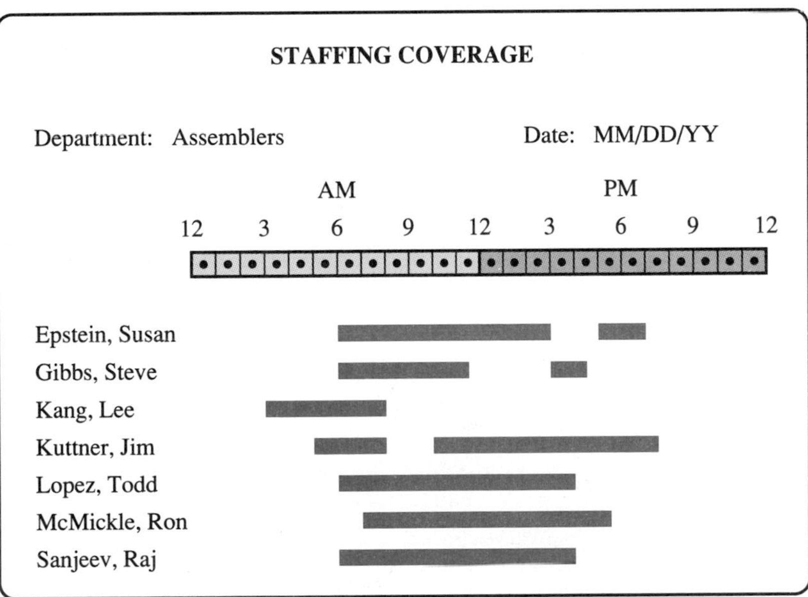

jobs on one report page. Instead of simply designing a columnar report, Cory suggested using blocks of columns for better report readability. Her suggestion led to the grouped data shown in Figure 8.24. Direct material cost, direct labor cost, and applied manufacturing overhead are well-separated by horizontal and vertical lines on this report. The cost summary is shown at the bottom of the report, making it a very distinctive section for the reader. The

Figure 8.24
Job-order cost report.

JOB-ORDER COST REPORT

Run Date: MMDDYY

Job Order No. 128-4

Job Description: Model 47-F Compressors

Units Ordered: 2
Date Ordered: ____
Date Promised: ____
Date Started: ____
Date Completed: ____
Units Completed: ____

CUSTOMER: Saline Dredging Company
Linkbelt Road
Houston, Texas

Direct Material Cost

Date	Requisition Numbers	Amount
MMDDYY	476	$ 4,000.00
MMDDYY	498	12,000.00
	Total	$16,000.00

Direct Labor Cost

Date	Hours	Rate	Amount
MMDDYY	28	$15.00	$ 420.00
MMDDYY	10	18.00	180.00
MMDDYY	70	20.00	1,400.00
		Total	$2,000.00

Applied Manufacturing Overhead

Direct Labor Hours	Predetermined Overhead Rate	Amount
400	$10.00 per hour	$4,000.00

Cost Summary

Direct Material	$16,000.00
Direct Labor	2,000.00
Applied Manufacturing Overhead	4,000.00
Total	$22,000.00
Unit Cost	$11,000.00

headings are detailed at the top left and right of the page. She feels that the report sections are more readable when separated by lined blocks. In addition, parts of this report could be displayed on a screen very easily without having to change the overall design.

Each department in the manufacturing area can access a screen such as the one shown in Figure 8.25. This is a variance display that focuses exclusively on a given area, supporting the principle of highlighting information that is under the control of that given area. This display shows the responsibility of a given area using an easily understood graphical format.

The SWAT team designs the sample output from the system and then discusses it with the future users of the system. Some of the output will be modified, while additional reports and screens will be created. A prototyping tool such as a screen designer will be used to show future users what the displays will actually look like on a computer screen. Since display screens can be difficult to visualize for many people, a screen designer helps bring life to a flat creation.

After some of the output has been designed and approved by the users, the SWAT team holds a meeting with the users to present the output in both printed and prototype form to gain more feedback about the system. A system such as JOCS is going to be evolving even after its implementation, so the SWAT team hopes to simply identify and design most of the output during this phase of the SDLC.

Figure 8.25
Direct materials and manufacturing overhead variance screen.

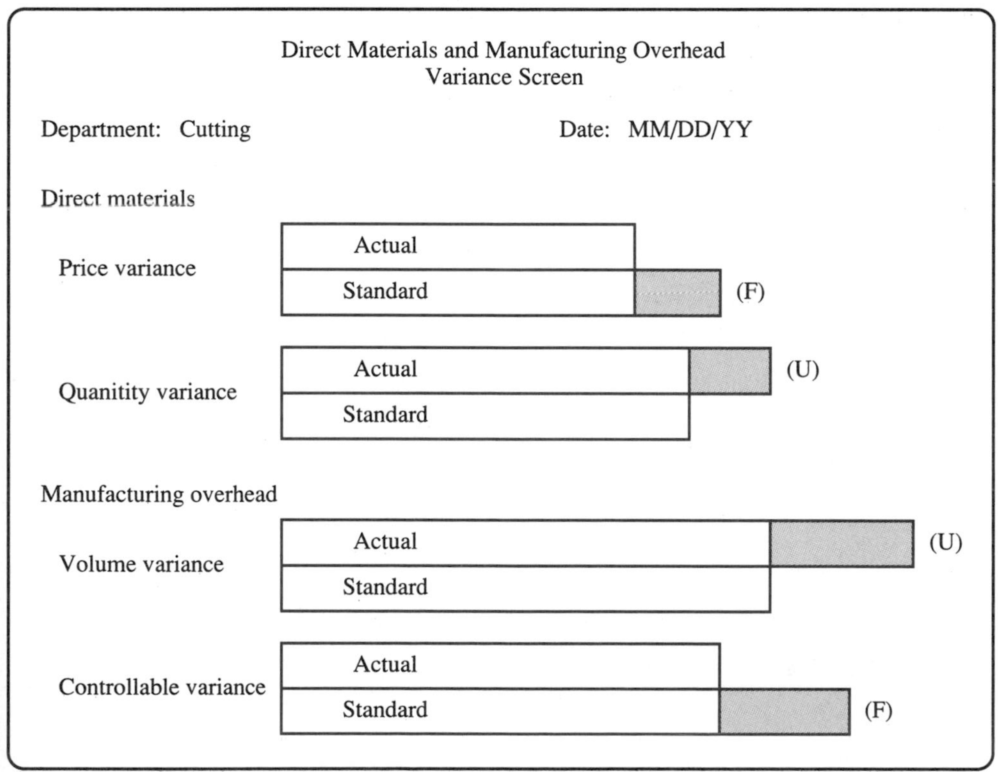

Chapter 9
Designing the Systems Input

WHAT WILL YOU LEARN IN THIS CHAPTER?

After studying this chapter, you should be able to:

1 Explain how to design paper forms and source-document-based data-entry screens.
2 Describe electronic forms, and state their advantages over paper forms.
3 List direct-entry devices.
4 Explain the use of codes for input design.
5 Discuss the menu interface, and explain how menus are designed to aid users to interact with the system.
6 Describe natural language interfaces.

INTRODUCTION

The purpose of this chapter is to treat **INPUT DESIGN** and discuss the use of input devices (see Figure 9.1).

Input starts the flow of data through a system. As with other systems design components, input design must be carefully planned and carried out in order to convert raw data (input) into usable information (output).

The following activities apply to raw data or input data:

- Insert into, delete from, or update the database, which is later used to produce output.
- Combine with other data from the database to produce output.
- Enter and process directly to output without combining it with other data.
- Initiate an action or perform a task.
- Carry on a dialogue with the system.

A variety of media and methods are used to capture and input data so they can be used properly, including the following:

- Paper forms combined with data-entry screens
- Electronic forms
- Direct-entry devices

- Codes
- Menus
- Natural languages

Each medium or method is discussed in the following sections.

DESIGNING PAPER FORMS

Like Mark Twain's prematurely announced demise, the death of paper forms, predicted for years, has been highly exaggerated. Paper mills continue to be built, and billions of trees are harvested each year worldwide to produce an ever-increasing quantity of paper. A significant portion of this paper is used for various business forms to capture and input data into systems for processing. The next time you are in any kind of organization, even a highly automated one, observe the flow of paper forms. The flow may be greater than you think. It would be quite surprising to find any kind of system, especially a global- or group-based one, that has not created the need to design a paper form.

PAPER FORMS are physical carriers of data. Events take place, transactions occur, and actions are taken. These activities generate data that must be captured and entered into systems for processing. A great many of these data-entry activities could be done by keying or scanning, but some companies prefer to fill in paper forms to make a hard-copy audit trail—or to accomplish some other aim, often misunderstood or hidden.

In some firms, however, the form *is* the business; for example, insurance policies, stock certificates, mortgage documents, or loan agreements. Forms

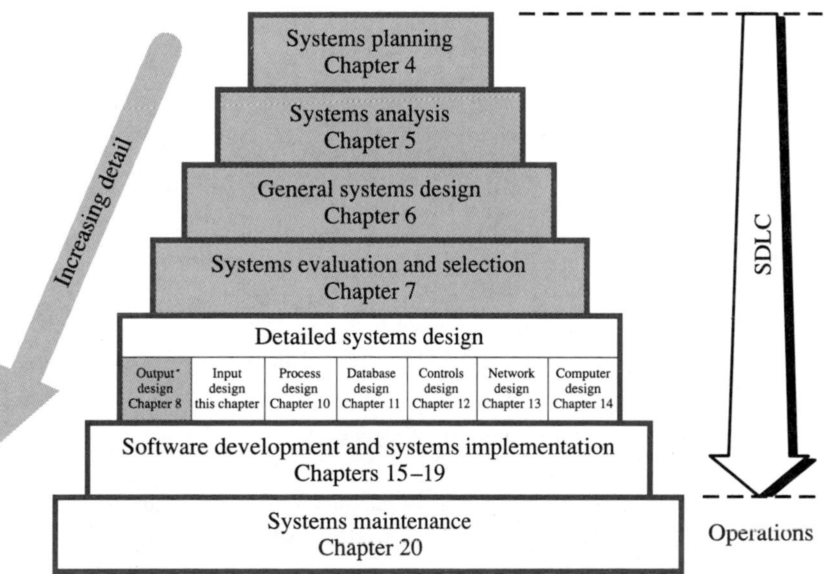

Figure 9.1
SDLC phases and their related chapters in this book. In Chapter 9 we examine the design of input (raw) data in detail.

that support a business include work orders, requisitions for materials, application forms, purchase orders, sales orders, invoices, time tickets, and so on. Support forms are the main concern of this chapter.

The discussion above means that as a systems professional, you are likely to become involved in designing paper forms, either as the designer or as the one who specifies and approves their design.

Selecting the Paper

Forms may be printed on paper of different colors, grades, and weights. Colored paper or colored printing on white paper is used to distinguish between copies and to enable sorting and routing of copies. Conventional color preferences are:

Order of Copy	Color
First	White
Second	Yellow
Third	Pink
Fourth	Blue
Fifth	Buff

A number of factors must be considered when selecting the paper to support your form:

- The length of time the form will be kept
- The appearance of the form
- The number of times it will be handled (e.g., average of six times per day for one month)
- How it will be handled (e.g., gently, roughly, abusively; folded and carried by the user)
- The ease of use and convenience afforded to users (e.g., coated papers may be hard to write on)
- Its durability if long-term filing may be required
- Its environmental exposure (e.g., grease, dirt, heat, cold, moisture, acids)
- The method used for filling in the blanks (e.g., handwritten or machine prepared)
- Its security against erasures

The longer a form will be kept, the better it has to look. The more abuse it is subjected to, the better the grade of paper required.

Determining the Size

The most common size form that will fit standard equipment, file forms, filing cabinets, typewriters, and so on, is 8½ by 11 inches. Try to make your paper forms this size. If you want a smaller paper form, specify one-half the standard length, or 8½ by 5½ inches. For card forms, the standards start with 8 by 10 inches. A half-card form is 8 by 5 inches. Nonstandard sizes can create a number of problems in preparing, handling, and filing forms. Moreover, because suppliers stock standard sizes, using an off-standard size can increase the cost of paper and cards.

Creating Forms that Make Multiple Copies

In forms design, **MANIFOLDING** means that multiple copies are made with one writing. Most manifold forms are action forms, such as sales orders, invoices, statements, bills of lading, purchase orders, shipping orders, and receiving reports. Usually, a number of people and functions are involved in the action that is initiated, such as shipping goods to a customer. Therefore, this action must be started in several places at once and must be coordinated throughout the process.

Also, with multiple forms, zoning (discussed in the next subsection) is required so that the user can fill in the first page of the form and have the same data appear on all or select copies of the set. Also, on a receiving report, for example, you want quantities to appear on the copies for accounting and inventory control, but you do not want them to appear on the copy for the receiving department, because you want to force the receiving clerk to count the items received. (If you show the quantities, the receiving clerk may be tempted simply to check off the items without actually counting them.) As another example, a set of manifold forms to ship an order to a customer is made up of invoice copies, shipping orders, bill of lading copies, acknowledgment, packing slip, and a mailing label. Although some of the data will appear in all copies, other data will appear in only a few copies. For example, only the customer name and address would appear on the mailing label, not prices and quantities. By zoning and using carbons, you can make data appear or not appear on selected parts in the manifold set.

If one of the copies must be on heavy paper stock, it should be the bottom (last) copy because each carbon has a cushioning effect, and the typing or handwriting becomes lighter and harder to read with each copy. Use a lightweight paper for copies below the original, except for the last one.

Dividing Forms into Zones of Data

Zones divide the form into logical blocks that contain related data, as presented in Figure 9.2. The **ZONING** concept can also be used for designing electronic forms, discussed later in this chapter.

The form should contain a name that tells what it is and what it does. Make the title as long as necessary to describe how the form is to be used. If the form is used by others outside the organization, also include their organization names and addresses.

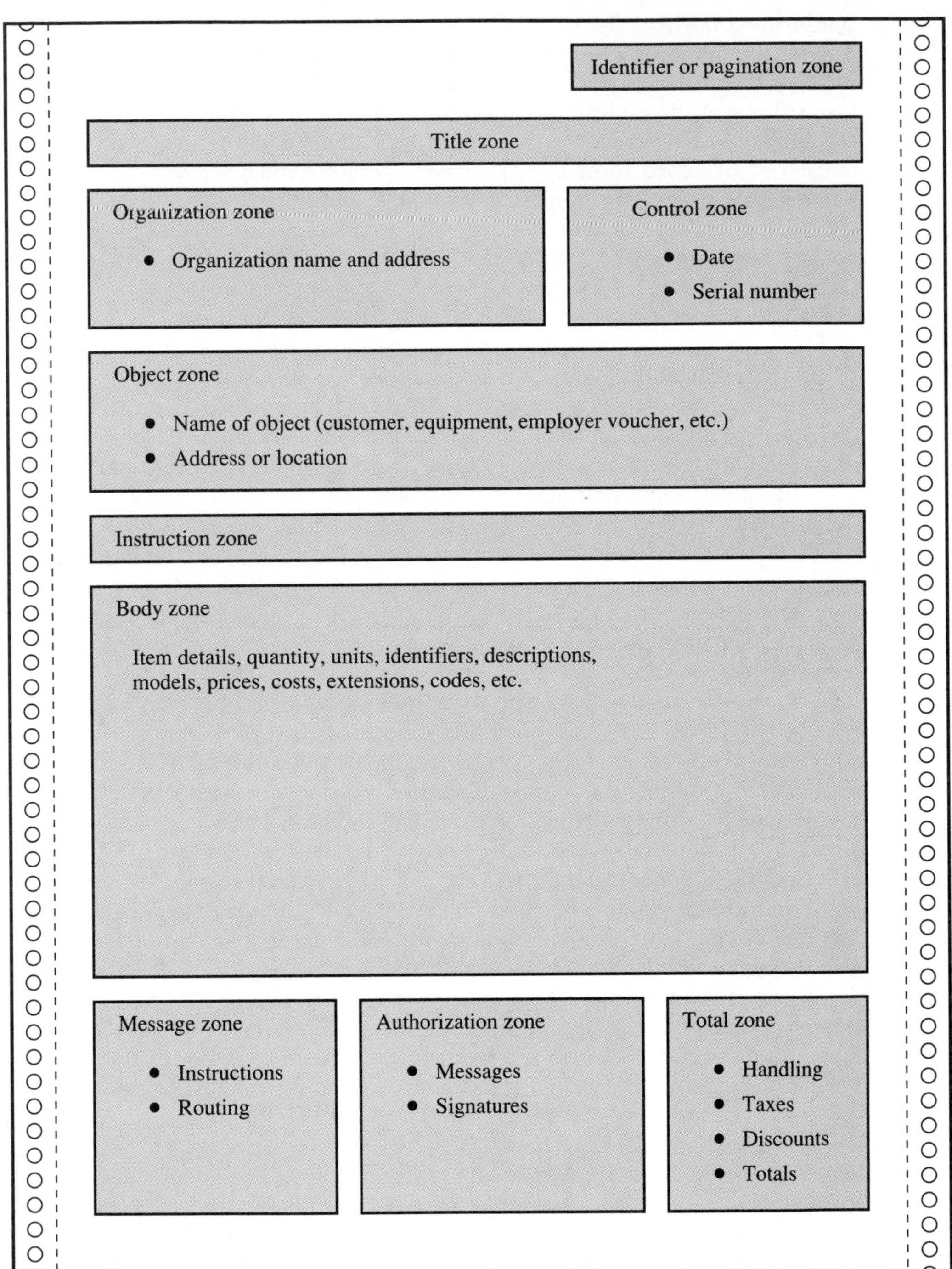

Figure 9.2
Form zones.

Creating Meaningful Titles

Good titles are often no longer than two or three words. Consider the following form titles:

Form Title	Comment
SALES	Too general
SALES ORDER	Better
CLAIM FORM FOR AN AUTOMOBILE ACCIDENT	Too long
AUTOMOBILE CLAIMS FORM	Better

Embedding Forms Instructions in the Design

Make your forms self-instructing; do not provide detailed instructions on how to fill out the form on a separate document or in a procedures manual. Only in a very few instances should such detailed instructions be necessary. Definitive captions are the key to making forms self-instructing. If you need further instructions, however, place such instructions at the top and within the zone to which they relate. If the instructions apply to the entire form, place them at the top of the form so that the user will see the instructions *before* filling out the form. Routing instructions (e.g., "Send buff copy to accounting department"), however, should be placed at the bottom of the form.

Using Lines, Boxes, and Captions

Each zone is made up of one or more lines or boxes and **CAPTIONS**. Lines, also termed rules, are used to guide the user's eyes to read and write data into each **DATA FIELD**. Boxes are used for the same purpose. Lines and boxes serve as boundaries for the data field into which data are entered. Captions identify these data fields. Common line, box, and caption combinations are demonstrated in Figure 9.3.

If the form is not used as a **SOURCE DOCUMENT** (i.e., the original paper form on which data are stored) to be keyed for computer entry, design paper forms using boxes with captions inside the upper left corner. The box design gets the captions up and out of the way. It can reduce overall form size by upward of 30 to 50 percent. It also makes data entry uninterrupted from left to right. Also, using lines can cause problems of positioning for typewriter-inscribed data.

The recommended approach for source-document forms is the caption above the box. The box maximizes visibility and reduces eye movement. It also provides the best cognitive match between the source document and data entry screen. (The **DATA-ENTRY SCREEN** is the video screen on the workstation that displays elements of the source document.)

Data entry and transcription are error-prone processes. The larger the number of different documents and screens that must be processed, the slower will be the overall data-entry rate and the greater the number of input errors. Therefore, proper caption design for source documents is imperative.

Figure 9.3
Common line, box, and caption designs.

- Captions before a line

NAME: ______________________________

ADDRESS: ______________________________

TELEPHONE: ______________________________

- Captions after a line

______________________________ NAME

______________________________ ADDRESS

__________________ TELEPHONE

- Captions above a line

NAME:

____

ADDRESS:

____

TELEPHONE:

____

- Captions below a line

____

NAME

____

ADDRESS

____

TELEPHONE

- Captions inside a box

NAME:

ADDRESS:

TELEPHONE:

- Captions above a box

NAME

ADDRESS

TELEPHONE

or

NAME

ADDRESS

TELEPHONE

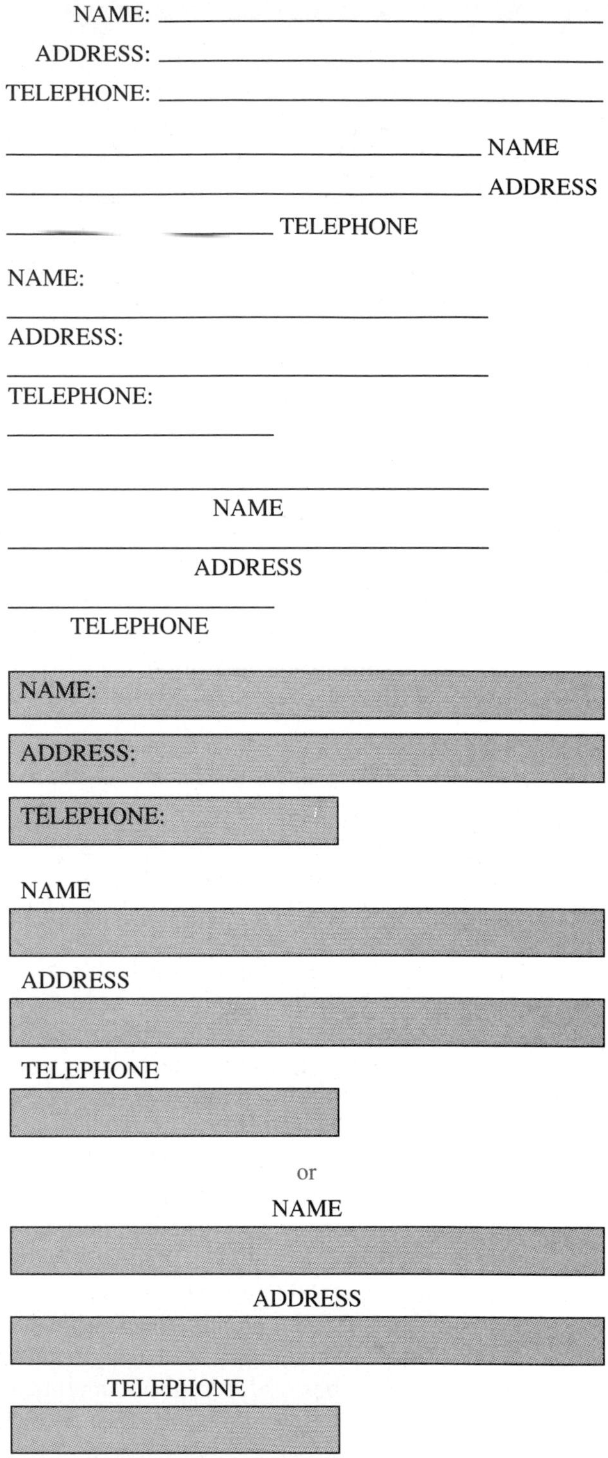

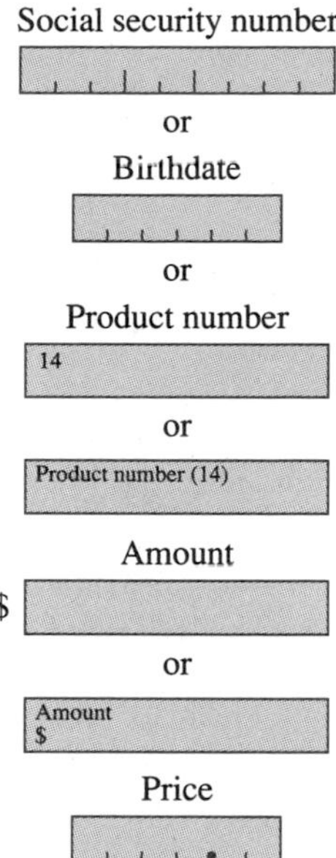

Figure 9.4
Methods of indicating field size, dollar values, and decimal point placement.

Data Field Indicators

Figure 9.4 illustrates methods of indicating field size, dollar values, and decimal point placement.

Alternative Selections

The most common ways to indicate alternative selections are ballot box checking and circling. Ballot box checking is normally used when responses are separated into categories (see Figure 9.5). Each alternative is located to the right of its ballot box. An x or check is used to select the proper alternative.

Circling is used when the form is completed by hand (see Figure 9.6).

Ballot box checking and circling rely on the user's power of recognition rather than recall; recognition is usually stronger than recall. Also, legibility is improved because these methods greatly reduce the amount of writing required. The less a form has to rely on handwritten characters, the better.

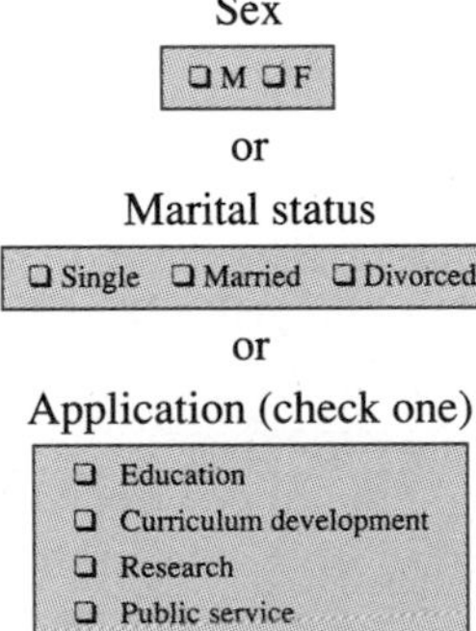

Figure 9.5
Examples of ballot box checking.

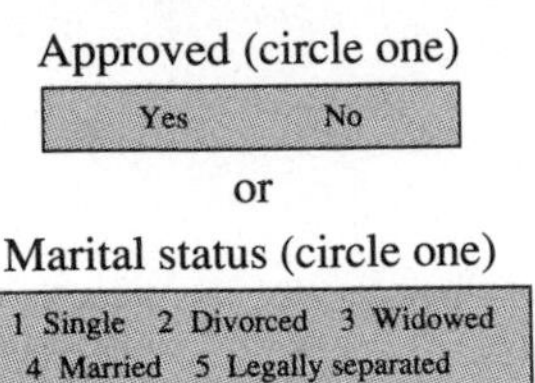

Approved (circle one)
Yes No
or
Marital status (circle one)
1 Single 2 Divorced 3 Widowed
4 Married 5 Legally separated

Figure 9.6 Examples of ballot box circling.

Spacing

A commonly used all-purpose spacing standard is termed 3/5 spacing. The 3 means the number of lines per vertical inch; the 5 means the least number of handwritten characters that fit in one horizontal inch. These spacing dimensions are shown in Figure 9.7.

For example, assume that a caption box requires 20 characters, and the designer does not know how it will be filled in. It may be typewritten with elite or pica characters, or it may be filled in by a clerk, or in some cases, it may be filled in by a worker in the field. What is the proper spacing for this caption box? It will be ⅓ inch high and 4 inches long. These dimensions will permit enough space for the box to be filled in by any method.

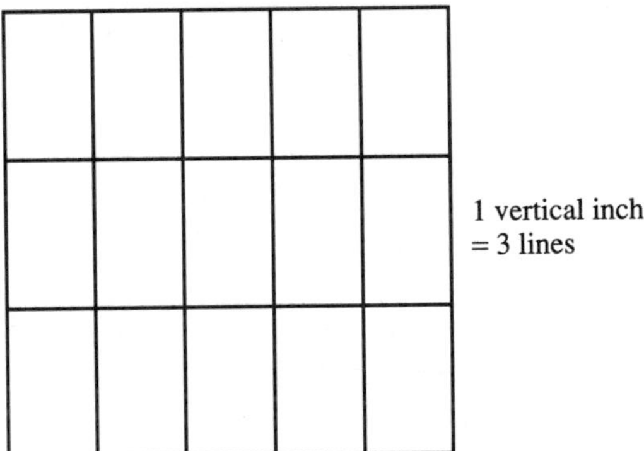

1 horizontal inch = 12 elite characters
10 pica characters
8 characters handwritten by clerks
5 characters handwritten by nonclerical workers

Figure 9.7 Conventional all-purpose 3/5 spacing.

Sequencing

Layout of captions and data fields should follow a logical sequence (top to bottom, left to right) that the user can follow easily (see Figure 9.8).

Designing Data-Entry Screens For Source Documents

When keying is performed from a source document dedicated to the DATA-ENTRY SCREEN, the entry person will focus attention on the source document, and the design of the data-entry screen must relate to the design of the source docu-

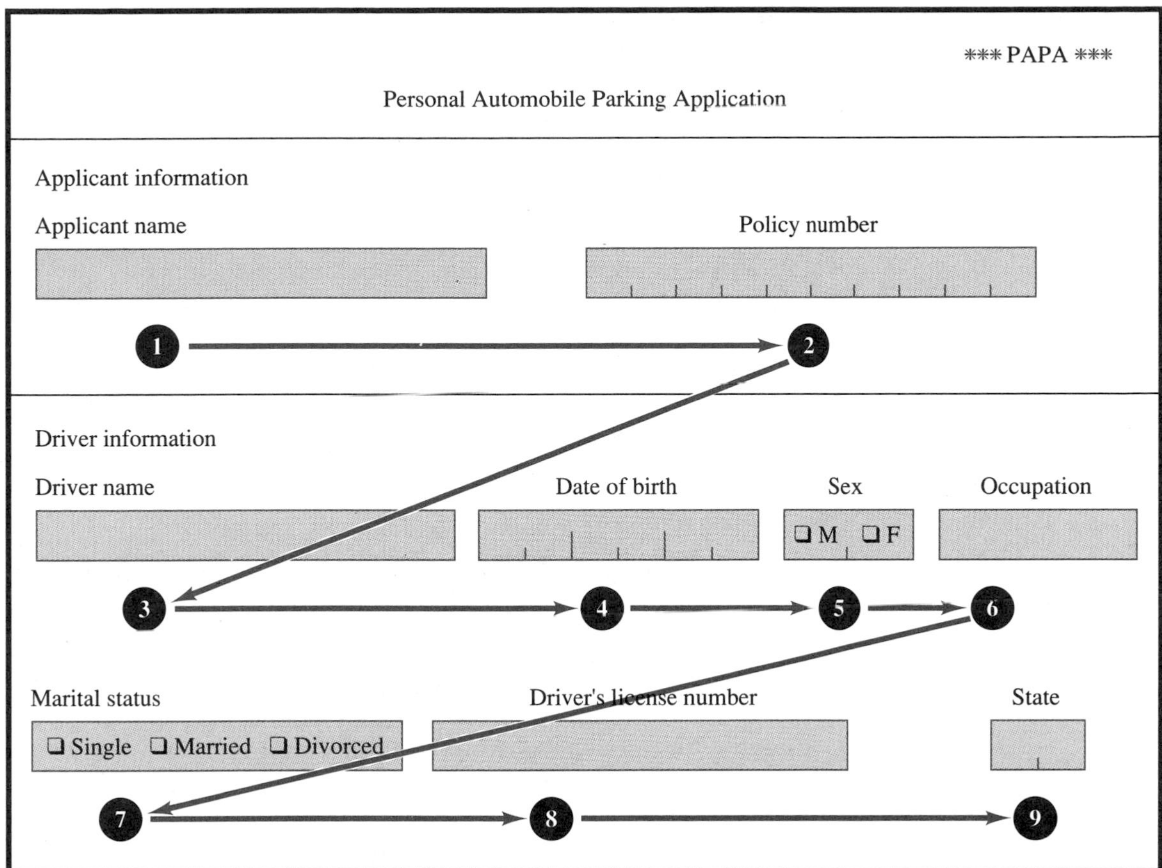

Figure 9.8 Properly sequenced source document for data entry.

ment. If the two designs are coordinated, users can complete data entry in 20 to 40 percent less time, and with 20 to 50 percent fewer errors, than would be the case if the designs were not coordinated. Moreover, reduced eye movement between source documents and data-entry screens reduces fatigue.

Creating Data-Entry Screens that Reflect the Source Document

CASE systems include a screen-design tool used to define the screen interface. A flexible screen painter permits prototyping of the data-entry screen design. Through prototyping, users can preview the presentation and the flow between screens and source documents before processing logic and database designs are developed. The key aspect of designing source-document-based data-entry screens is to create screen images of their associated source documents. Captions and data fields on the screen should be located on the same line and in the same order as captions and fields on the source document. In this way, the data-entry user follows the sequence of the source document. Ideally, keying should never require eye movement from the source document to the screen, except to edit and correct errors.

Using Abbreviations in Captions on Data-Entry Screens
Abbreviations (see Figure 9.9) are shortened forms of a written word or phrase used in place of the whole. For output screens, electronic forms (discussed later), and paper forms, abbreviations are not recommended, but for data-entry screens that are based on a source document, abbreviations for captions, data, and commands can lead to efficiencies.

Captions on the data-entry screen play a supportive rather than a main role in the data-entry process. Therefore, abbreviate captions and display at normal intensity. Abbreviations should not exceed five characters. Separate two or more abbreviations by hyphens. The total size of a caption should not exceed eight characters.

The most common abbreviation methods are:

- Make subjective judgments to decide on a logical representation of the full word.
- Retain the first and last letters of the word and delete all or more of the letters in between.
- Delete all vowels in the word.
- Retain the first few letters of the word and truncate the rest.

The last method (truncation) generally gives the most consistent results, but no method has been found to be superior to any other. Training users to recognize the abbreviations will give best results. To reduce problems in creating and remembering abbreviations, a standard dictionary of abbreviations is a valuable aid for designers in any screen design activity.

Figure 9.9
Examples of captions and their abbreviations.

Caption	Abbreviation	Caption	Abbreviation
Account	Acct	Hours	Hrs
Amount	Amt	Male	M
Average	Avg	Manager	Mgr
Balance	Bal	Merchandise	Mchse
Care of	c/o	No	N
Check	Ck	Paid	Pd
Credit	Cr	Quantity	Qty
Debit	Dr	Received	Rec'd
Department	Dept	Signature	Sig
Discount	Disc	State	St
Each	Ea	Weight	Wt
Female	F	Yes	Y
Freight	Frt	Zip Code	Zip

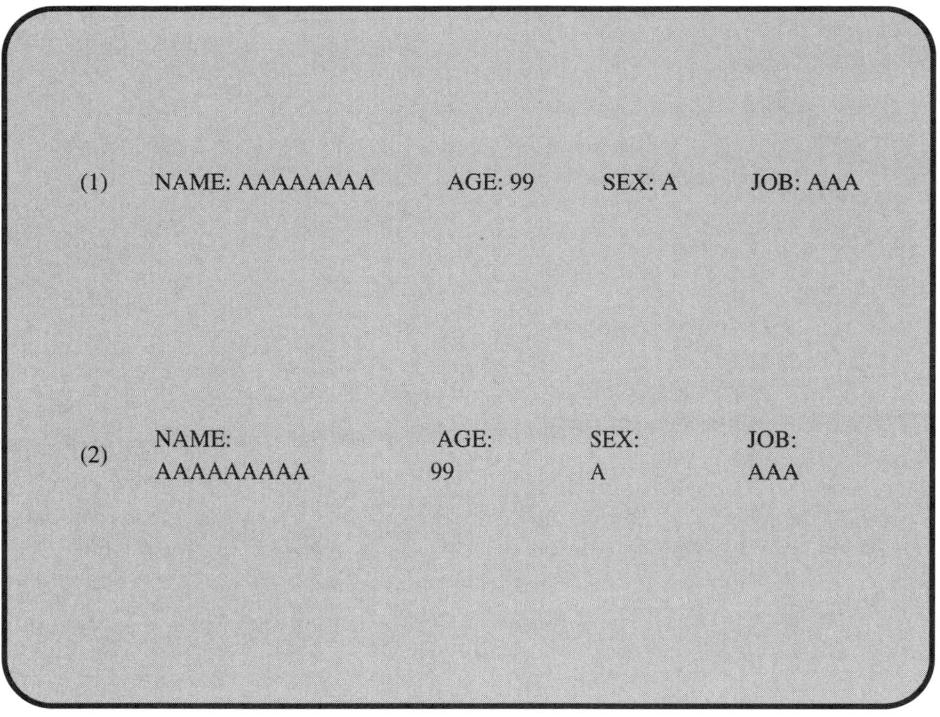

Figure 9.10
Caption and date field formats.

Caption and Data Field Formats

Two practical ways exist to relate a caption to its data field. Consider Figure 9.10. In Format 1, the caption comes just before the data field. In Format 2, the caption is above the data field. Format 1 is recommended for single data fields for the following reasons:

- It conforms to the normal left-to-right reading sequence.
- It permits easy field alignment, location, and scanning.
- It provides the best compromise between caption clarity and use of screen space. In the figure, for example, Format 1 uses one line of the screen, but Format 2 uses two lines of the screen.
- It provides a good match with source documents, which are usually in single-line format.
- On a crowded screen, the horizontal format provides better visual differentiation between captions and data than the vertical format. The caption-above format can cause captions and data to merge visually.

For multiple-occurrence fields, locate the caption to the left of data fields and separate by a colon (or other unique symbol) and one space. Separate each data field by a colon (or other unique symbol) with one space on each side of it, such as

```
STU-CLS:x_ _ _x:x_ _ _x:x_ _ _x:x
```

Break up long fields through the use of slashes, dashes, spaces, or other common delimiters, such as

```
DT.      _ _/_ _/_ _
TEL-NO: (_ _) _ _ _ - _ _ _ _
```

For horizontal spacing, leave a minimum of three spaces between one data field and the caption of the following field. One space is acceptable if space constraints exist. For an example, see the following:

```
REQ-NO: _ _ _ _xxxITEM-NO: _ _ _ _ _
```

For vertical spacing between rows of fields on a screen, spacing conventions follow the source document. If source document rows are single spaced, the screen rows are also single spaced. If gaps exist on the source document, use at least one space line to indicate each gap and relative spacing on the screen.

For section headings, locate the heading directly above its associated data fields and spell out fully. Indent related field captions or row headings a minimum of three spaces from the beginning of the section heading, such as

```
    STUDENT INFORMATION
     NAME:_ _ _ _ _ _ _ _ _ _ _ _ _
xxxBTH-DT:_ _/_ _/_ _
      MJR:_ _ _ _
```

For row headings, locate to the left of the first row of associated data fields and abbreviate or spell out fully. A meaningful convention used to designate row headings is the use of two greater-than symbols. These symbols direct attention to the right and indicate that everything that follows refers to this category.

```
FIXED ASSETSx>>xxxLAND:x_ _ _ _xxxBLDG:x_ _ _ _
```

For field group headings, center the heading above the captions associated with it and spell out fully. Relate to the captions by a broken dashed line ended by less-than and greater-than symbols:

```
    < ---------LEASED TRUCKS--------->
    CONV: _ _ _ _xxxCAB-OVR: _ _ _ _
```

or

```
              LEASED TRUCKS
   TYPE            EFF-DT           EXP-DT

_ _ _ _       _ _ _ _ _ _ _     _ _ _ _ _ _ _
_ _ _ _       _ _ _ _ _ _ _     _ _ _ _ _ _ _
```

For multiscreen transactions, locate a page number or screen identifier in the upper right-hand corner. This number may simply be page *n* of *x*, such as *** PAGE 1 of 3 ***, or it may incorporate a mnemonic code that is a contraction of the screen title. For example, a screen title, TRUCK LEASING FORM, could incorporate *** TRKLS02 ***, which identifies the type of screen and the second screen in the series.

To help you assimilate the foregoing material, we provide you with a paper form for a medical claim and its complementary data-entry screen (Figure 9.11). The data-entry screen is used by a person who keys the data from the source into the computer system for processing.

WHAT ARE ELECTRONIC FORMS?

When there is no dedicated source-document form from which keying is performed, the primary visual focus of the user will be on the screen itself. In such a situation, screen design takes on an even more important role.

ELECTRONIC FORMS are data-entry screen designs that are used without a dedicated source document. Electronic forms can replace all paper forms in the company. Electronic forms are designed on a digitizer or the video display terminal (VDT) screen of a CASE system, using the same components that paper forms designers use:

- Zoning
- Instructions
- Lines, boxes, and captions
- Data field indicators
- Sequencing design guidelines

MANAGING ELECTRONIC FORMS

After a form is designed and approved, it is stored on magnetic or optical media and transmitted to workstations. Typically, electronic forms are displayed on a workstation screen and filled in by users' operating keyboards. If a hard copy of an electronic form is needed, it can be accessed from the database and produced by a printer.

An electronic forms system is shown in Figure 9.12. The forms designer creates an electronic form, following good forms-design principles. A developing form can be displayed simultaneously throughout the network for suggestions and final approval. Users review the form on a what-you-see-is-what-you-get (WYSIWYG) basis. When the form is approved and completed, it is transmitted to the mainframe for storage, control, transmission, and processing.

The mainframe and appropriate software control provide security and management for all forms and form processing elements. This electronic forms

Figure 9.11
Source document and its associated data-entry screen.

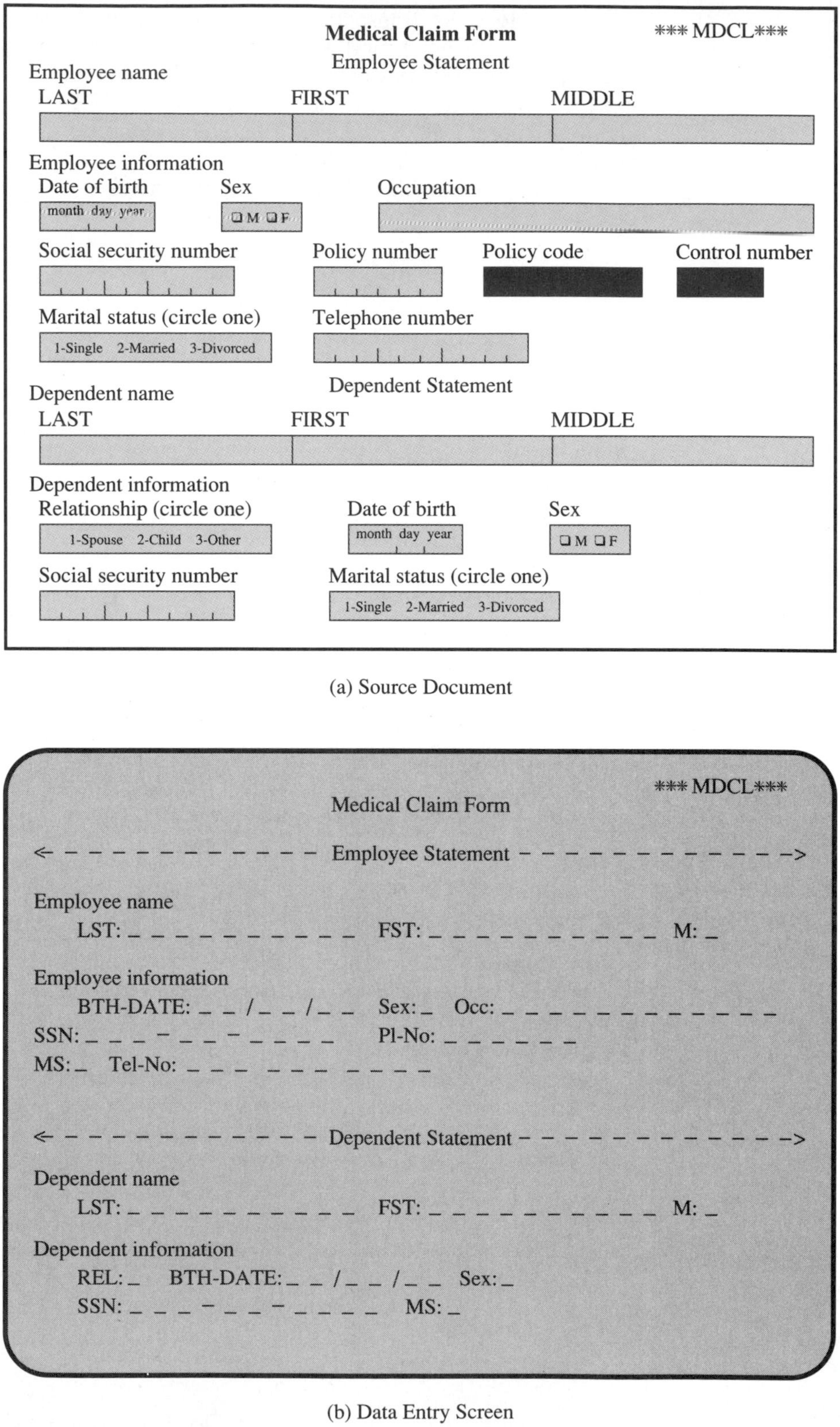

Medical Claim Form *** MDCL***

Employee Statement

Employee name

LAST FIRST MIDDLE

Employee information

Date of birth (month day year) Sex (☐M ☐F) Occupation

Social security number Policy number Policy code Control number

Marital status (circle one): 1-Single 2-Married 3-Divorced

Telephone number

Dependent Statement

Dependent name

LAST FIRST MIDDLE

Dependent information

Relationship (circle one): 1-Spouse 2-Child 3-Other

Date of birth (month day year) Sex (☐M ☐F)

Social security number

Marital status (circle one): 1-Single 2-Married 3-Divorced

(a) Source Document

*** MDCL***

Medical Claim Form

<- - - - - - - - - - - - - Employee Statement - - - - - - - - - - - - ->

Employee name

LST: _ _ _ _ _ _ _ _ _ _ _ FST: _ _ _ _ _ _ _ _ _ _ _ M: _

Employee information

BTH-DATE: _ _ /_ _ /_ _ Sex: _ Occ: _ _ _ _ _ _ _ _ _ _ _ _ _ _

SSN: _ _ _ - _ _ - _ _ _ _ Pl-No: _ _ _ _ _ _ _

MS: _ Tel-No: _ _ _ _ _ _ _ _ _ _

<- - - - - - - - - - - - - Dependent Statement - - - - - - - - - - - - ->

Dependent name

LST: _ _ _ _ _ _ _ _ _ _ _ FST: _ _ _ _ _ _ _ _ _ _ _ M: _

Dependent information

REL: _ BTH-DATE: _ _ /_ _ /_ _ Sex: _

SSN: _ _ _ - _ _ - _ _ _ _ MS: _

(b) Data Entry Screen

Figure 9.12
Electronic forms system.

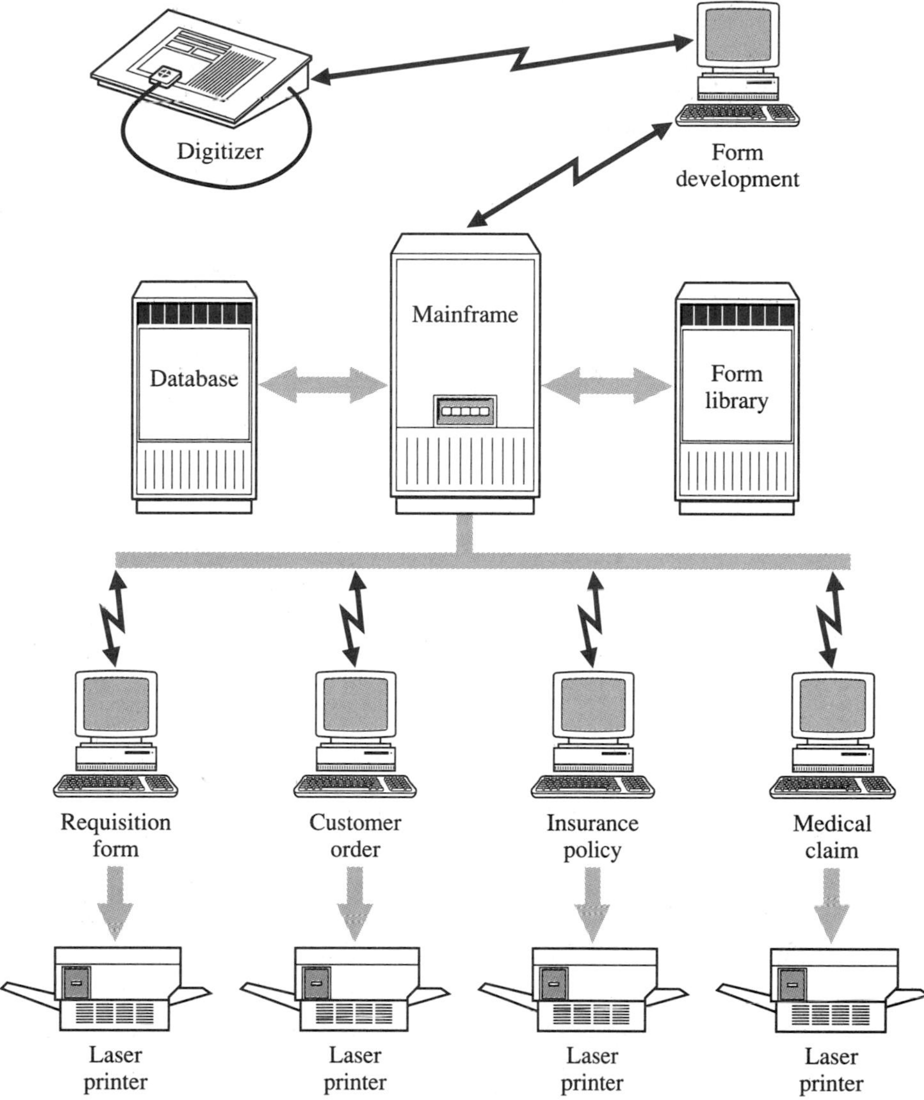

management system tracks and keeps statistics on form usage and protects the form library through a hierarchy of passwords. This password protection assures that employees will use only the forms intended. It also protects special form elements, such as authorized check signatures, from improper use.

Form-processing elements are distributed over a telecommunication network to remote offices. Changes in forms are centrally controlled and updated throughout the system automatically. With this kind of system, the company does not have to destroy warehouses full of preprinted paper forms when such changes occur.

Smart Electronic Forms

SMART ELECTRONIC FORMS show users how to fill in the form, provide online messages and instructions, perform calculations, and transmit the data for additional processing. Some examples follow:

Guide Users The smart form guides the user from box to box on the form, assuring that no data are left out, while at the same time filling in certain data fields automatically. It also defines the mode and length of each data field and rejects invalid data. Error messages appear on the screen telling the user what is wrong and how to correct errors. If a user gets stuck, a help key brings up a message window with detailed instructions on what is required in that data field.

Perform Calculations The electronic form can be instructed to add total costs—on requisition orders, for example. In a sales order it can add individual sales for total gross sales, compute taxes, retrieve the proper discount, calculate net sales due, and place each item in its proper space. In some applications, all the user has to do is put in a customer code and the amount and type of products ordered. The form inputs customer name and address, product descriptions, prices, extensions, handling charges, taxes, and totals.

Check Limits The forms designer can establish limits on a field. If net weight for a van cannot exceed 48,000 pounds, for example, any higher number will be rejected, and a window will automatically display the limit message.

Coordinate Processing Tasks Electronic forms can import to and export data from other applications and databases. It can trigger other forms work. For example, the filling-in of a customer's order triggers the preparation of other forms, such as shipping and acknowledgment forms. The system can also pull together other form elements. For example, to assemble an insurance policy, certain data about the insured are keyed. This input automatically accesses all form elements needed to complete the insurance policy, calling them up from the database.

Comparing the Costs of Electronic and Paper Forms

Visible and hidden costs relate to electronic forms as compared to paper forms. The visible costs of paper forms are the out-of-pocket costs for paper, typesetting, artwork, and printing. Costs are also incurred when paper forms require thousands of square feet of storage space in a warehouse. Moreover, money is tied up in a large inventory of paper forms. (Only one form's map or template of the electronic form needs to be stored.) Hidden costs also eventually affect the company's bottom line. Some of the hidden costs that can be reduced by using electronic forms are:

Costs of Running Out When a company runs out of preprinted paper forms, operations may be halted. Electronic forms never run out. Supply always equals demand.

Costs of Forms Obsolescence Needs change, laws change, and suddenly many preprinted paper forms aren't worth the paper they are printed on. Electronic forms greatly reduce this waste.

Costs of Inefficient Forms Pressure always exists to keep a paper form the way it is, even when it's not as useful as it could be. To create a new form costs money, so it's better to leave well enough alone. But it's easy to revise an electronic form to meet changing conditions and give users exactly the form they need.

Costs of Using the Wrong Form Murphy's law of forms is: If the wrong form can be used, it will be. With electronic forms, controls determine what people use a form, which form they use, and what they use it for. When a form is revised, employees have to use the new one, because the old one is no longer available. Use of forms can be restricted to certain people or departments by password control.

Costs of Forms Management and Enforcement When a company uses thousands of different forms, managing them is a large and expensive task. With electronic forms, the creating, managing, and processing of every form can be brought into a single, integrated system.

Costs Caused by the Speed Limit of Paper Every process in a company is subject to the speed limit of paper forms. Filling out electronic forms takes less time.

Costs Associated with Handling Data Twice With source-document paper forms, someone fills in the form, and then someone reads data off the form and inputs the data into the computer. Electronic forms end this transfer step, because data capture for the electronic form and data capture for computer processing are the same step.

Costs Caused by Data Float Data float is a function of time. It is the result of paper float, the time it takes a paper form to get from point A to point B. Electronic forms virtually eliminate data float, because data are input and transmitted electronically.

DESIGNING ELECTRONIC FORMS

To design a form that will take data from a source document, you must follow the format of the source document. With electronic forms, there is no need for a source document. However, some of the design guidelines for paper forms and their complementary data-entry screens can also be applied to designing electronic forms, except that you do not abbreviate captions for electronic forms. In this section we present design guidelines that are especially applicable to electronic forms, although subsections on data field design and message design for electronic forms can also be used for source-document-based data-entry screens.

Designing the Data Fields

The field is the basic component of screen design. A screen may be composed of only one field or upward of several hundred. Captions identify data contained in data fields. Certain special fields, such as titles, prompts, or instructions, do not require captions; they are sometimes termed labels or literals. Common alternative conditions of data fields are:

- Protected field will not allow keying of data into field. The field contents are for reading only.
- Unprotected field permits keying of data into the field.

- Numeric field allows keying of numeric data only (0 to 9, decimal point, and minus sign) in an unprotected field. If used with protected condition, "auto skip" is activated to move cursor over field.
- Alphanumeric field permits keying of alphanumeric data into an unprotected field. If used with protected condition, it requires manual tabbing to move cursor over field.

- Normal intensity displays data at normal intensity.
- High intensity displays data at a brighter-than-normal intensity.
- Nondisplay means that data such as user numbers and passwords in the field are not visible on the screen.

Common data field conditions are as follows:

- Numeric data are unprotected, numeric, and high intensity.
- Alphanumeric data are unprotected, alphanumeric, and high intensity.
- Descriptive or nonchangeable data are protected, alphanumeric, and normal intensity.

Justifying Captions and Data Fields

Two approaches are recommended for caption and data field justification, as presented in Figure 9.13. With Approach 1, left-justify both captions and data fields, and leave one space between the longest caption and the data field column. With Approach 2, left-justify data fields and right-justify captions to data fields, and leave one space between each pair of data field and caption.

The disadvantage of Approach 1 is that the caption beginning point is usually farther from the data field than it is with Approach 2. A multiple mix in caption length can cause some captions to be far removed from their associated data fields, greatly increasing eye movement and possibly making it difficult to match captions to data fields accurately. An advantage to Approach 1 is that the left-justified captions are generally pleasing to the eye.

The disadvantage of Approach 2 is that captions are not quite as readable as in Approach 1 because of the ragged left edge of the captions. Advantages are that the captions are always located close to their associated data fields,

Figure 9.13
Caption and data field justification.

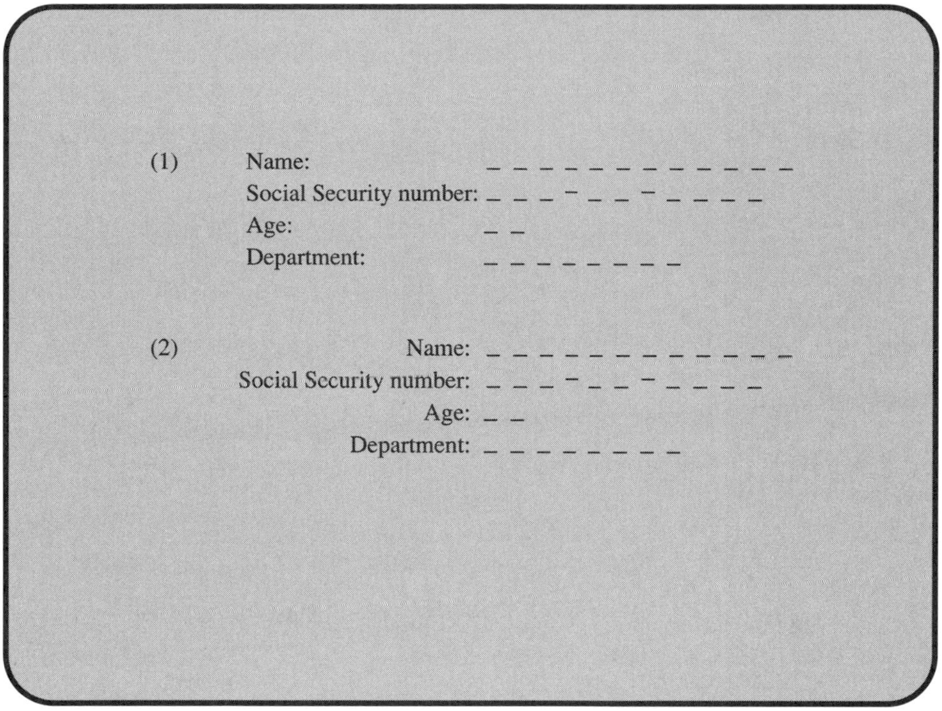

```
(1)     Name:                    _ _ _ _ _ _ _ _ _ _ _ _ _
        Social Security number:  _ _ _-_ _-_ _ _ _
        Age:                     _ _
        Department:              _ _ _ _ _ _ _ _

(2)                       Name:  _ _ _ _ _ _ _ _ _ _ _ _ _
        Social Security number:  _ _ _-_ _-_ _ _ _
                           Age:  _ _
                    Department:  _ _ _ _ _ _ _ _
```

thus minimizing eye movement between the two. Moreover, a screen with many captions normally has a balanced look if Approach 2 is used.

Because electronic forms are used to capture data and input it directly into the system, completion aids and prompts are often necessary. For example, see the following completion aids:

```
 RATE (99.99): _ _ _ _ _
ENTRY (DR,CR): _ _
DATE (MMDDYY): _ _/_ _/_ _
```

Prompts are instructions to the user on how to work with the screen being presented. They are similar to instructions on a paper form. Prompting messages should be in lowercase and located just preceding that part of the screen to which they apply. For example, see the following prompt:

```
(Fill in the following if the lessee is an owner-operator)
TRUCK NUMBER: _ _ _ _
       MODEL: _ _
  LEASE DATE: _ _/_ _/_ _
```

For source-document-based data-entry screen headings, three spaces are used for indentation. For electronic form data-entry screen headings, locate section headings on a line above related screen fields, and indent captions a minimum of five spaces from the start of each heading. Also, spell out fully in uppercase font and display in normal intensity. For example, see the following:

```
CLASSIFICATION
xxxxxFRESHMAN:    _ _ _ _
     SOPHOMORE:   _ _ _ _
     JUNIOR:      _ _ _ _
     SENIOR:      _ _ _ _
```

or

```
CLASSIFICATION
      FRESHMAN:  _ _ _ _
xxxxxSOPHOMORE:  _ _ _ _
        JUNIOR:  _ _ _ _
        SENIOR:  _ _ _ _
```

For subsection and row headings, see the following examples:

```
TRUCKSx>>xxxCABOVER: _ _xxxCONVENTIONAL: _ _
```

or

```
CARSx>>xxxFOUR-DOOR:   _ _xxxTWO-DOOR: _ _
          CONVERTIBLE: _ _   SPORTS:   _ _
```

or

```
COMPUTER PERIPHERALS
              PRINTERSx>>xxxLASER:       _ _ DAISYWHEEL: _ _
                            DOT-MATRIX:  _ _ LINE:       _ _
xxxxxSECONDARY STORAGEx>>xxxMAG TAPE:    _ _ CD-ROM:     _ _
                            MAG DISK:    _ _ DAT:        _ _
```

Two examples of the preceding design guidelines are provided in Figure 9.14. The top example is part of a job application form. The bottom example is an order form. In both examples, both captions and data fields are left-justified. The alternative approach is to right-justify the captions up against the left-justified data fields.

Designing On-Screen Messages

A system communicates with users via many kinds of messages, which consist of:

- Prompts
- Diagnostic messages
- Information messages
- Status messages

Figure 9.14
Examples of electronic forms.

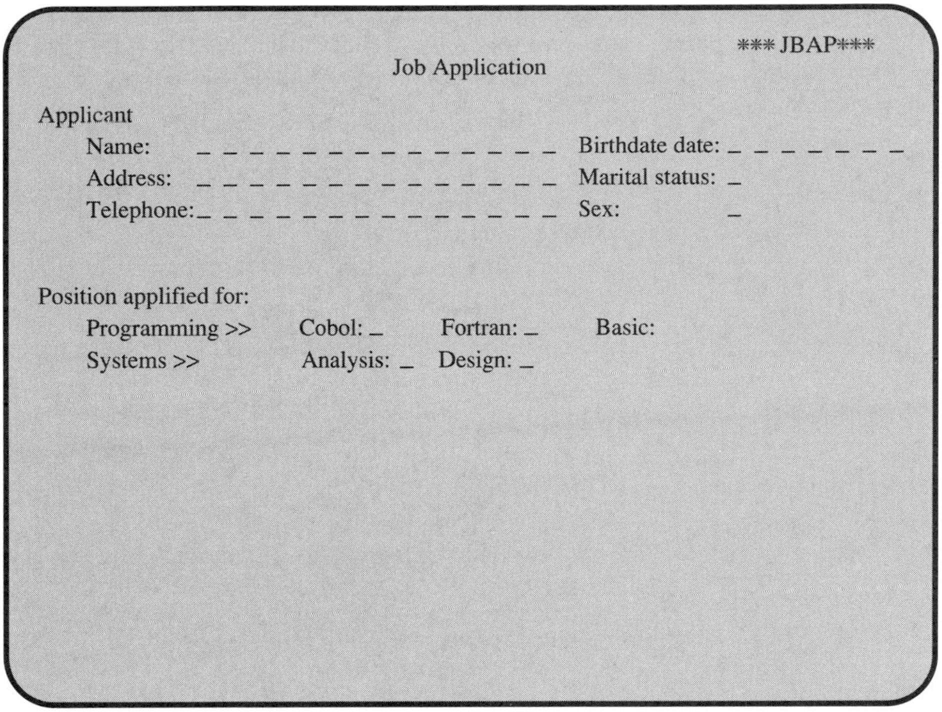

```
                                                       *** JBAP***
                         Job Application
Applicant
   Name:      _ _ _ _ _ _ _ _ _ _ _ _ _ _   Birthdate date: _ _ _ _ _ _ _
   Address:   _ _ _ _ _ _ _ _ _ _ _ _ _ _   Marital status: _
   Telephone:_ _ _ _ _ _ _ _ _ _ _ _ _ _ _  Sex:            _

Position applified for:
   Programming >>   Cobol: _     Fortran: _     Basic:
   Systems >>       Analysis: _  Design: _
```

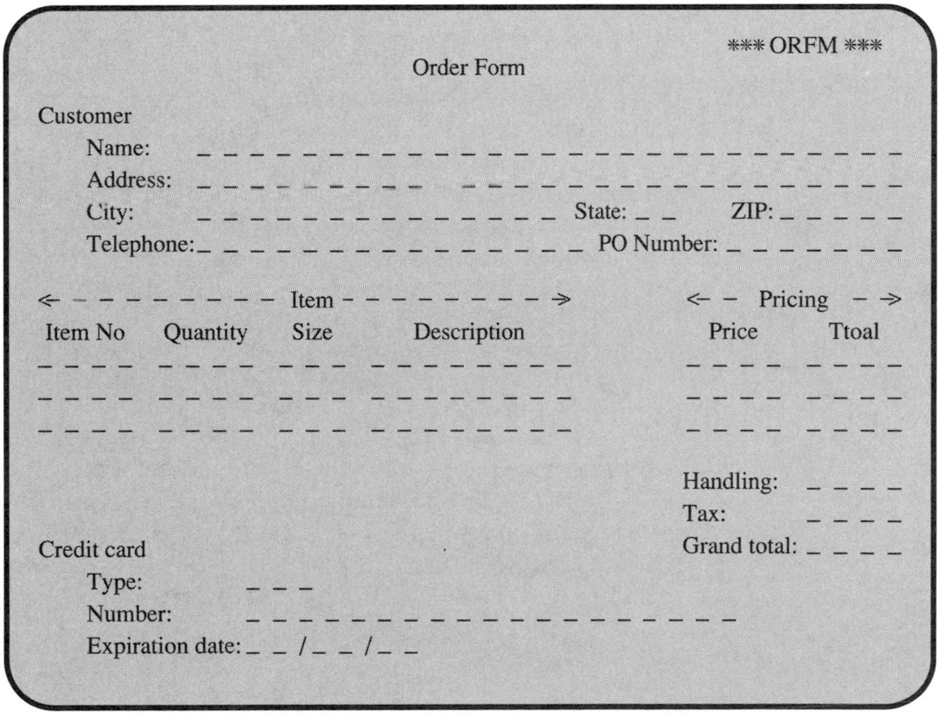

```
                                                       *** ORFM ***
                          Order Form
Customer
   Name:      _ _ _ _ _ _ _ _ _ _ _ _ _ _ _ _ _ _ _ _ _ _ _ _ _ _ _
   Address:   _ _ _ _ _ _ _ _ _   _ _ _ _ _ _ _ _ _ _ _ _ _ _ _ _ _
   City:      _ _ _ _ _ _ _ _ _ _ _ _ _ _   State: _ _     ZIP: _ _ _ _ _
   Telephone:_ _ _ _ _ _ _ _ _ _ _ _ _ _ _ _  PO Number: _ _ _ _ _ _ _

<- - - - - - - - - - Item - - - - - - - - - ->          <- - Pricing - ->
 Item No    Quantity   Size       Description            Price      Ttoal
_ _ _ _   _ _ _ _   _ _ _   _ _ _ _ _ _ _ _             _ _ _ _   _ _ _ _
_ _ _ _   _ _ _ _   _ _ _   _ _ _ _ _ _ _ _             _ _ _ _   _ _ _ _
_ _ _ _   _ _ _ _   _ _ _   _ _ _ _ _ _ _ _             _ _ _ _   _ _ _ _

                                                        Handling:    _ _ _ _
                                                        Tax:         _ _ _ _
Credit card                                             Grand total: _ _ _ _
   Type:            _ _ _
   Number:          _ _ _ _ _ _ _ _ _ _ _ _ _ _ _ _ _ _ _
   Expiration date: _ _ /_ _ /_ _
```

All messages should be simple and clear and immediately usable. They should be active, nonintimidating, nonpatronizing, and nonpunishing. The following guidelines will lead to appropriate message design:

1 Use simple, understandable words and brief statements that clearly communicate. Users thumbing through reference manuals to translate a message is unacceptable.

2 Use positive statements. For example, "Enter debit or credit code before updating accounts receivable" is more meaningful than "Do not try to update accounts receivable before entering debit or credit code."

3 Use active voice, because it's easier to understand than passive voice. For example, "Draw a process block by selecting PROCESS" is easier to understand than "The process block is drawn by selecting PROCESS."

4 Use attention-getting techniques such as the following:

   - Adding emphasis at several levels
   - Marking by underlining, enclosing in a box, or using bullets and asterisks
   - Including several character sizes and fonts
   - Blinking (with cursor or marker)
   - Applying color codes
   - Installing tones or voices for certain feedback

   Using too many of these attention-getting techniques, however, can diminish their effectiveness and create confusing screens. Novices need simple instructions, logically organized, that guide their actions. With expert users, subtle highlights and logical positioning are generally sufficient. Another way to get attention is to use error message displays. First, the error message should get the user's attention. Second, it should give instructions for error correction. Don't just report the problem with a negative TRY AGAIN, SYNTAX ERROR, or ILLEGAL. Rather, design informative and positive messages, such as YOUR RANGE OF CHOICES IS 1 THROUGH 5, USE MM/DD/YY FORMAT, or UNMATCHED LEFT PARENTHESIS FOUND.

5 Messages from the system to users should indicate that users are in control. The messages should be courteous, positive in tone, and nonintimidating. Don't use warnings or imperatives, such as ERROR, INVALID, ENTER DATA, ILLEGAL, and ABORT. Instead, use helpful messages such as READY FOR YOUR COMMANDS.

6 Suggestions that computers can think, know, feel, or understand should be reduced, if not abandoned entirely. Everyone knows it's a deception, and after several encounters it begins to wear thin.

7 Design less chatty messages and avoid superficial value judgments, such as "You're doing great. Keep up the good work."

8 Order words chronologically or in accordance with a proper sequence of events. A prompt should say, "Reference bill of materials and enter item number," rather than "Enter item number after referencing bill of materials."

9 Avoid words that smack of jargon or computerese. While not always bad, some words evoke user frustration and should therefore be avoided. Some suggestions are:

Avoid	Use
Abend	Cancel, End, Stop
Abort	Cancel, End, Stop
Boot	Start, Run
Execute	Complete
Hit	Depress, Press
Key	Enter, Type
Return Key	Enter, Transmit
Terminate	End, Exit, Quit

ENTERING DATA DIRECTLY

The hardware component normally used to enter data into forms is the keyboard. The keyboard is pervasive in information systems. However, because keyboard entry depends on human skill and effort, it is not always the most efficient and accurate way to enter data.

DIRECT ENTRY, sometimes referred to as source data automation, is a way of inputting data that does not require data to be keyed by someone reading from a source document or filling in an electronic form. Direct entry creates computer-processible data right on paper or computer media or inputs it directly into the central processing unit (CPU), thus increasing input efficiency and decreasing the possibility of errors being introduced, as might happen in the keying process.

Direct-Entry Devices

Some of the more popular direct-entry devices are presented next. The chief characteristic of these devices is that they input data directly to the computer.

Magnetic Ink Character Recognition (MICR) This device reads characters made of ink containing magnetized particles. These characters are printed at the bottom of a check or other document. MICR facilitates the processing of billions of checks in the banking industry.

Optical Character Recognition (OCR) Some OCR devices can read both printed and handwritten characters. They use photoelectric cells to identify characters. Once characters are imprinted on a document, such as retail price tags, cash register receipts, utility bills, phone bills, and various text documents, they need not be keyed again for entry into the computer.

Current OCR software can recognize a variety of typefaces and font sizes, handle typeset text, and flag unrecognizable characters. OCR software, however, is not yet 100 percent accurate, and there may be conversion errors and characters that OCR can't recognize. Despite these limitations, OCR devices are often much more cost-effective than manually rekeying data, especially large volumes of text, for computer processing.

Optical Mark Recognition (OMR) Sometimes termed "mark sensing," OMR devices sense the presence or absence of a mark. A wide variety of marks can be read, including marks made by computer printers, hard or soft pencils, and special typewriters. A typical use of OMR devices is the grading of answer sheets for multiple-choice questions.

Digitizer This device converts an image such as a photograph, map, or chart into digital data. The digital data can then be stored in a computer, processed, displayed on a screen, or printed on paper. Some large timber companies take aerial photographs of their timber holdings and digitize these data into computer storage media that help support inventory control applications.

Image Scanners These devices can identify typewritten characters or images automatically and convert them to electronic signals that can be stored and processed by the computer. The process of image scanning, sometimes called bit mapping, can identify different fonts by scanning each character with light and breaking it into light and dark dots, which are then converted to digital data. Many firms, such as accounting, law, and insurance companies, use image scanners to store thousands of documents.

Point-of-Sale (POS) Devices Sometimes called POS terminals, these devices capture data when a sale is transacted. POS terminals in some stores use wand readers and holographic scanners for reading and recording product codes. These terminals are usually connected to a central computer that performs processing for an entire chain of stores. The computer automatically tells the POS terminal what the price is; the terminal prints the price and the product name on the customer's receipt. The two principal codes (sometimes called bar codes) read by POS terminals are the Universal Product Code and Code 39.

Automatic Teller Machines (ATMs) These are specialized forms of POS devices; "point of action" (POA) may be a more descriptive term. ATMs permit users, at their convenience, to make deposits, loan payments, borrow money, and withdraw funds against a bank credit card. Some banks use smart cards, which look like an ordinary credit card but contain a small microprocessor. When the card is placed in the ATM's card reader, users enter their personal identification number (PIN). If an incorrect PIN is entered three times in a row, the card is automatically destroyed. The ATM wipes out the embedded program and data. This security feature makes smart cards safer to carry than conventional credit cards.

Mouse This is a hand-movable device that controls the position of the cursor on a video screen. The mouse can be described as a box with buttons on the top and a ball on the bottom. It is placed on a flat surface with the user's hand over it. As the user moves the box, the ball also moves, causing the cursor to move in a corresponding manner on the screen. When the cursor is moved to the desired option or instruction, one of the control buttons is clicked, causing the system to enter a value or perform a specific task. Such an entry is sometimes referred to as point-and-click.

Voice Recognition These devices create a natural interface between users and the system, and thus represent the ultimate in data entry for applications in which users' hands are occupied. Most voice recognition devices are speaker dependent, which means that each user must train the system to recognize his or her voice patterns. Words, phrases, and numbers are repeated three or more times, digitized, and stored as data on a disk. Speaker-dependent devices are typically used by commodity traders, quality control inspectors in factories, and baggage handlers in airports.

Voice recognition devices that are speaker independent—that is, devices that recognize the same words spoken by many different people—are rare. It is difficult to develop voice recognition devices to recognize continuous speech, because most people run their words together.

Document Image Processing

American business alone produces close to 1 trillion pages of paper per year, enough to blanket the earth, with pages to spare. It costs about $25,000 to fill a four-drawer file cabinet and $2160 to maintain that cabinet for a year. Also, about 3 percent of all documents are incorrectly filed or lost, and the average cost to recover a document is close to $120. Finally, the average user spends a total of four weeks per year waiting for documents to be located. The technology used to manage documents more efficiently and achieve near paperless offices is referred to as **DOCUMENT IMAGE PROCESSING (DIP)**.[1]

The components of a DIP system are:

- Scanners
- Storage on optical media
- Server
- Output via a video display terminal, printer, or fax

A DIP system may be for a single user or part of a network. On a network, each user may have a different purpose for accessing a document. Moreover, unlike a paper-based system, a DIP system permits a document to be accessed by

[1] David A. Harvey, "Catch the Wave of Dip," *Byte,* April 1991, p. 173.

more than one person at a time. To gain an understanding of how document image processing (DIP) works and what its benefits can be, review this case:

DIP at Worldwide Airlines

Worldwide was an airline faced with a paper crisis. Each of its 80,000 yearly purchase and repair orders required ten supporting documents. Additionally, all of these documents had to be stored for two years before being transferred to microfilm in the company's general archives. Millions of paper documents were stored in large automated filing cabinets. On an average day, 3000 documents had to be located and retrieved, resulting in the loss of many hours of employee time.

Management was growing frustrated with its inability to integrate electronically stored information with the paperwork and with the fact that paper documents were accessible by only one person at a time. After a great deal of analysis, a systems team installed a DIP system, such as the one shown in Figure 9.15.

Each document is scanned in by an operator, who then enters the appropriate index for document type, purchase order number, item number, and file number. The DIP software also date- and time-stamps each document. After these input tasks are completed, document images are transferred to optical disks that reside in an optical jukebox.

Users throughout the network have real-time access to documents. Multiple documents can be displayed on the screen, scaled, rotated (necessary if the document was scanned upside-down), zoomed to particular sections, and printed if desired.[2]

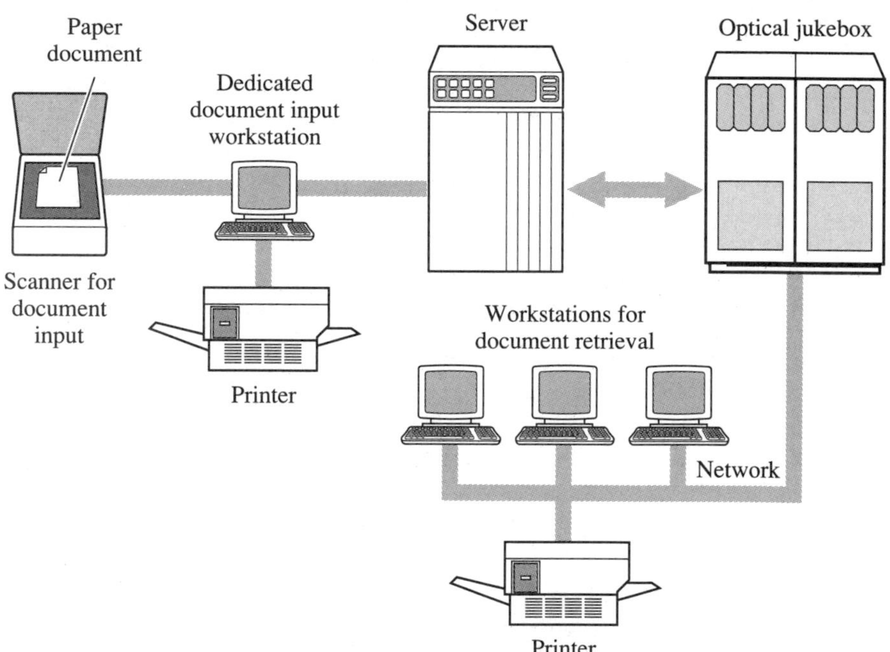

Figure 9.15 Document image processing (DIP) system.

[2] Based on David A. Harvey and Bob Ryan, "Practically Paperless," *Byte,* April 1991, pp. 185–190.

DESIGNING CODES THAT REPRESENT DATA

CODES, also called account numbers, identifiers, and keys, usually represent an item that is entered in a form to identify a particular item or to signify certain kinds of processing. For example, a number such as 45761 uniquely identifies a specific customer. A discount code A means that the customer's order is subject to a 10 percent discount. Or an employee code 32 means the person is a welder in the assembling department (i.e., 3 = welder and 2 = assembling department).

Codes represent a very important element used to classify and identify people, resources, documents, accounts, forms, events, and transactions into specific groups and for distinguishing these items within groups. Codes are comprised of the following:

- Numbers
- Letters
- Special characters
- Symbols
- Colors
- Sound

General Guidelines

A great deal of thought must go into the code design if it is to satisfy a variety of users. The following guidelines should be kept in mind when designing codes.

1 The code design must be flexible to accommodate changing requirements. It is costly and confusing to change the coding structure every few months. The coding structure should not be so extensive, however, that part of it will not be used for a number of years.

2 Standardization procedures should be established to reduce confusion and misinterpretation for persons working with the code. Some of the procedures that can be easily standardized in most systems are as follows:

   - Elimination of characters that are similar in appearance. For example, the letters O, Z, I, S, and V may be confused with the digits 0, 2, 1, 5, and the letter U, respectively.
   - Days and weeks should be numbered. For example, days are numbered 1 to 7 and weeks are numbered consecutively beginning with the start of the fiscal period.
   - The use of a 24-hour clock alleviates the AM/PM confusion.
   - Dates should be designed by digits using the "Year Month Day" format YYMMDD (in which September 18, 1992, becomes 920918). The MMDDYY format is also favored by many. The YY part of the format, however, will have to be expanded to YYYY, as explained next.

For the first 100 years, YYMMDD, DDMMYY, or MMDDYY formats worked well. When we get to the year 2000, data fields will sort to a position before 1901, not after 1999, where they belong. The reason for this is that we are only sorting the last two bytes of the date. The year 2000 is 00 and 1901 is 01. Hence new systems should use the full four bytes for dates: YYYYMMDD, DDMMYYYY, or MMDDYYYY. Otherwise, systems using the YY format will require a major reprogramming effort to prepare for the year 2000.

3 Where possible, letters that sound the same should be avoided (e.g., B, C, D, G, P, and T, or the letters M and N). In alphabetic codes or portions of codes having three or more consecutive alphabetic characters, avoid the use of vowels (A, E, I, O, and U) to prevent the inadvertent formation of recognizable English words.

4 The layout of the code itself should have parts that are of equal length. For example, a chart of accounts code should read 001–199 (for assets), not 1–199.

5 Codes longer than four alphabetic or five numeric characters should be divided into smaller segments (this is sometimes called "chunking") for purposes of reliable human recording and recall. For humans, 702-496-1358 is more easily remembered and more accurately recorded than 7024961358.

Code Structures

Codes are normally classified by the symbol arrangement used to identify the groups to which coded items belong. These structures are:

- Sequence
- Block
- Group

Sequence Codes

Items are numbered consecutively, such as checks, purchase orders, and invoices. These codes help to control the issue and use of such documents. Sequential codes can also be used to identify unique items in the database. An inventory item number or employee identification code, for example, must identify one and only one inventory item or employee.

The **SEQUENCE CODE** can identify an unlimited number of items with the fewest digits. New items are simply assigned the next number in the sequence. The main disadvantage of sequence codes is that they tell only the order in which the item appears in the sequence.

Block Codes

Blocks of numbers are assigned to each block of items, such as groups of accounts shown in the following **BLOCK CODE**:

```
1000-1999 Assets
                  1000-1049 Cash
                  1050-1079 Accounts receivable
                  1080-1089 Inventory
                  1090-1099 Prepayments
                  1100-1149 Fixed assets
                  1150-1199 Investments
     2000-2299 Liabilities
     2300-2399 Capital stock and retained earnings
     3400-3415 Sales
     4420-4999 Manufacturing expenses
     5410-5999 Marketing expenses
     7490-9999 Administrative expenses
```

Within each block of numbers, accounts are arranged in the order in which they are used in preparing financial statements. Unassigned numbers are reserved for the possible addition of new accounts in the future.

Group Codes

A **GROUP CODE** is arranged so that the interpretation of each succeeding symbol depends on the value of the preceding symbols, as shown in Figure 9.16(a). Block codes, as described in the preceding subsection, can be extended to form group codes. For example, conventional general ledger account numbers can be combined with other numbers to form a group code, as shown in Figure 9.16(b).

Figure 9.16
(a) Group code;
(b) block code extended for use as group code.

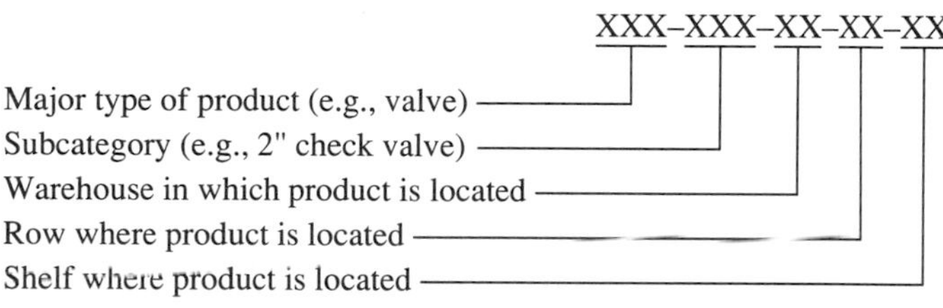

(a) Group code

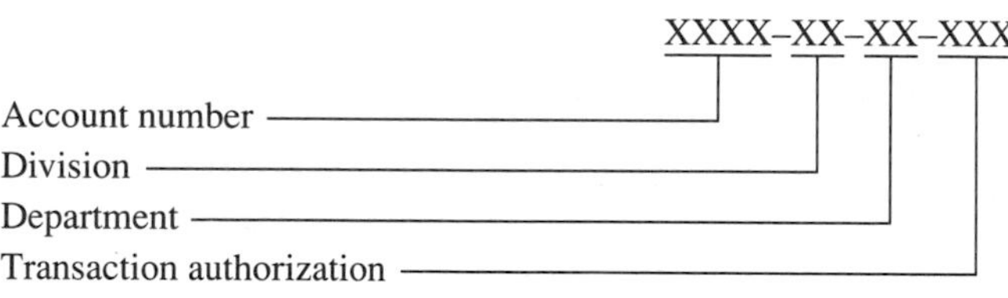

(b) Block code

The advantages of group codes are:

- They can capture more information than other types of codes.
- Users can memorize them easily because of consistent positioning.

- Computer sorting and retrieval are easily performed.
- Categories can be easily expanded.

The disadvantages are:

- Group codes can become quite lengthy.
- Maintenance is difficult when major changes are made.

Special Codes

Special kinds of codes used to classify items and increase efficiency in processing and identification are:

- Bar codes
- Color codes

Bar Codes

The ability to capture a transaction or event while it is occurring is one of the most powerful attributes of an information system. For example, **BAR CODES** used in the retail industry enable merchandisers to perform the following tasks:

- Install point-of-sale (POS) devices, which reduce checkout time.
- Increase inventory control.
- Eliminate the need for price marking of individual items.
- Improve resource and shelf allocation.
- Reduce the probability of human error, pilferage, and fraud through cash register manipulations.
- Produce a broad range of timely information for a variety of users.

Many effective applications of bar codes exist in other areas. For example, materials-control personnel use bar-coded labels and scanners in an integrated, online scheduling production control system. Each representative bar code is attached to a specific component and subassembly. These are monitored as they pass through production. The bar-code labels contain mnemonic codes and color codes (e.g., red means chassis, blue means motor block) for human reading and identification. Such a system provides an accurate count and control of materials. It also enables timely performance, scheduling, and tracking information.

Color Codes

For manual processing, **COLOR CODES** are used to help identify paper records fast. Applications include using color to file by year, by department, by project, by accounting use, and so on. Color by year (e.g., 10 color stripes, one for the last digit in each year) speeds the transfer of data to inactive files and helps avoid misfiling one year's data with another. Color by department helps avoid misdirection of information. Color by accounting use helps to separate information by function.

Colors are also used for filing sequences. In such a system, a different color is assigned to each digit, 0 through 9. Color coding for alphabetic filing is just as easy. Groups of letters are assigned specific colors. For example, in a doctor's office, alphabetical filing may be practical, eliminating the need for a cross-reference index. A variety of simple color-code systems are now available for alphabetic filing.

DESIGNING MENUS FOR ENTERING DATA

MENU-based interfaces give users a set of choices from which the desired selection is made. Users often forget, so a display of choices augments the users' memory and allows them to enter commands or data quickly by selection.

Menu interfaces minimize the amount of data that must be keyed in. They are especially appropriate for novice and occasional users. Menus, however, can be extremely boring for expert users, and can slow down the entire operation because of the wordiness and slow pace of somc selection methods.

If menus will be used repetitively by expert users, a fast-track option must be provided. This option permits expert users to bypass most or all of the menus and prompts and simply enter the required data. A system that combines full tutorial prompting and fast-track options can provide the best of both approaches. Examples of prompts:

- For normal tutorial mode, please press transmit.
- For fast-track mode, please enter user-ID and report number.

General Principles

Like captions on other screens, menu captions should be concise and meaningful. For menus as complete screens, choices should be left-justified, aligned into columns, and located in the center of the screen. When menus are included on other screens, locate the menu in the same location on each screen, and differentiate the menu from the remainder of the screen by enclosing it in a box or set it up on a horizontal line. See Figure 9.17.

Order lists of alternatives by their natural order if it is applicable. For lists with a small number of options (seven or fewer), order by frequency of occurrence or importance. Alphabetize long lists or short lists that have no obvious pattern or frequency.

Item captions may be identified by an ordinal or mnemonic code or be unidentified. See Figure 9.18. Nonmnemonic lettered codes are not recommended, because numbered ordinal codes can be searched much faster. But mnemonic codes can be searched a great deal faster than can ordinal numeric codes. Therefore, mnemonic codes are recommended for use whenever possible.

Menu as complete screen

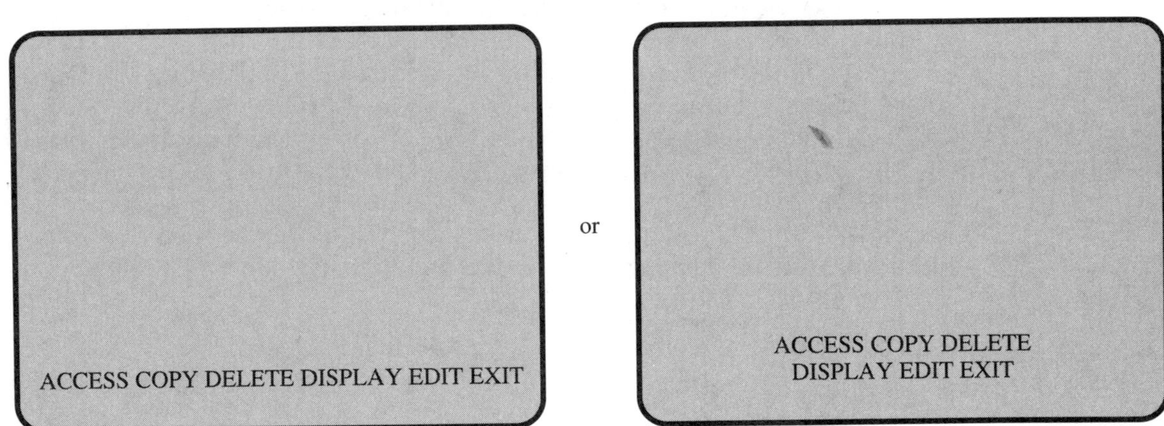

Menu on other screen

Figure 9.17
Caption alignment and location.

Common selection methods are:

- Keying
- Pointing
- Touching
- Voice input

For single selection fields, locate the selection field directly underneath the column of alternative codes (or first column in a multicolumn menu screen). Identify it with a unique and descriptive caption such as SELECTION or CHOICE, and separate it from the choice field by a colon and space.

```
          9xxBALANCE SHEET
         10xxINCOME STATEMENT
CHOICE:x_ _
```

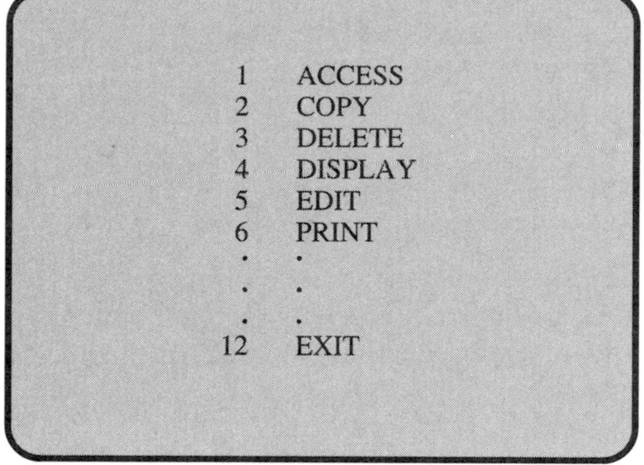

Ordinal identification

AC ACCESS
CO COPY
DE DELETE
DI DISPLAY
ED EDIT
PR PRINT
. .
. .
. .
EX EXIT

Mnemonic identification

1=ACCESS 2=COPY 3=DELETE 4=DISPLAY

or

1=ACCESS
2=COPY
3=DELETE
4=DISPLAY

Ordinal identification for space conservation

ACCESS
COPY
DELETE
DISPLAY
EDIT

No identification

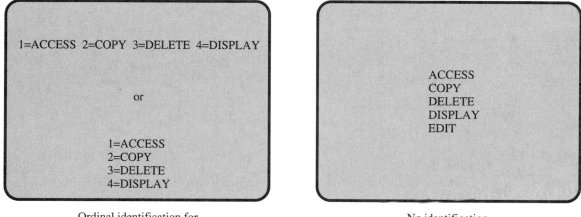

Figure 9.18
Item caption identification methods.

With multiple selection fields, the screen user moves the cursor to a selection field related to an item and keys a value to indicate his or her choice.

```
_xx1xxDIRECT MATERIALS REPORT
_   2  DIRECT LABOR REPORT
_   3  MANUFACTURING OVERHEAD REPORT
```

For small lists (seven or fewer), this method is quite efficient. As the number of items on a menu screen increases, this method, however, becomes less efficient as the cursor must be moved greater distances.

Embedded menus enable users to select an item in context. Item captions are selected by moving a selector box or cursor to the highlighted caption and clicking. A touch-screen version permits selection by touching the highlighted caption. An example follows:

```
     This menu demonstrates a DATA FLOW DIAGRAM for systems
analysis and a STRUCTURE CHART and PSEUDOCODE for detailed
design. Before the . . .
                                  .
                                  .
                                  .
To access bold or highlighted items, the user moves
the cursor to the first word, clicks or simply points.
```

Pointing and voice selection methods eliminate the need for keying. The user simply points to the selection or gives a voice command, such as DELETE, PRINT, or EXIT. The advantage of voice selection is that the users' hands and eyes are free to perform other tasks.

Creating Menus Using Visual, Touch, and Sound Cues

A vital consideration of the user/system interface is how well users can interact with the system through visual, touch, and sound cues. Some of the more prominent design techniques in this area are:

- Pull-down menus
- Nested menus
- Shingled and tiled menus
- Icon menus
- Touch menus
- Sound cues

Pull-Down Menus

PULL-DOWN MENUS are written onto the screen in a location that makes it appear as if they had been pulled down from the top of the screen. A main menu bar across the top of the screen provides the cursor choices that activate a particular pull-down menu. See Figure 9.19.

By moving a panning cursor across the main menu bar, the user can choose which pull-down menu is to be displayed. The pull-down menu is typically written onto the screen just beneath its namesake on the menu bar. A cursor is used to enable the user to make choices from the pull-down menu. Or a mouse can be used to point to the choice and click.

Nested Menus

The program logic that manages **NESTED MENUS** makes it necessary to pass through one menu before reaching a secondary menu, as illustrated in Figure 9.20. The secondary menu is, in effect, one of the choices available for selection from the earlier menu.

Our example is one you may be familiar with, because it can be found on many word-processing systems. A menu bar across the top of the screen offers four choices of broad menu categories. The user wishes to set the text style for the document he or she is creating in the word processor. Using a mouse, or

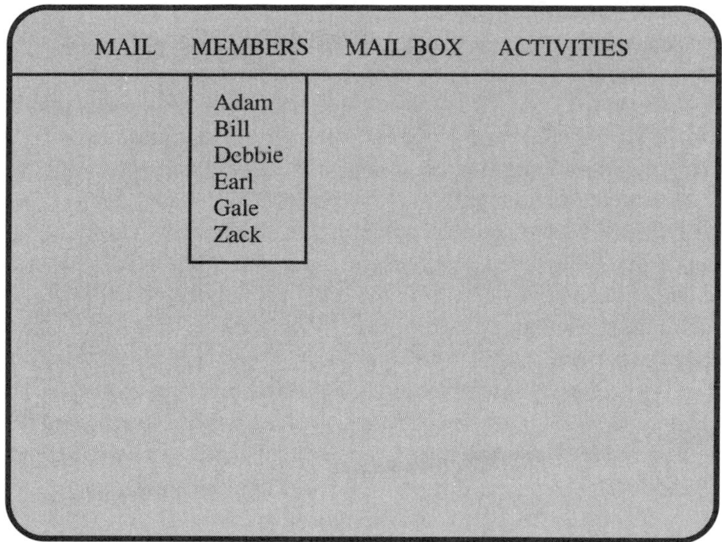

Figure 9.19
Pull-down menu.

perhaps a keyboard-controlled cursor, the user selects the TYPE option from the screen's menu bar, which causes the TYPE menu to be displayed. The user selects TEXT STYLE from this menu, and another menu appears. It offers four type fonts. The user selects one of these fonts (in this case SERIF); the computer displays a final menu offering five special effects. Here, our user has selected ITALIC as his or her choice from this menu.

The important attribute of nested menus is their hierarchical nature. In our example, for instance, the user could not have selected ITALIC text without first making appropriate choices from previous menus. Nested menus are an excellent method of presenting the user only those menu options that are relevant to the operation at hand.

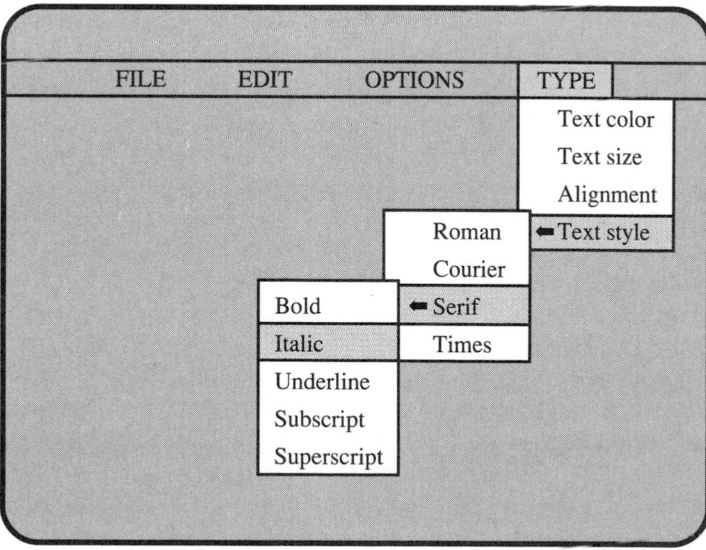

Figure 9.20
Nested menus.

Shingled and Tiled Menus

SHINGLED MENUS (sometimes called overlapping menus) are illustrated in Figure 9.21. These particular menus were derived from a popular software package, Microsoft Windows 3.0. Such design can display data, text, instructions, icons, or forms, so the user can perform multiple tasks or refer to several parts of text or images at the same time. Typically, the menu to the front is the most recent or current menu.

Menus help reduce short-term memory loads because they act as external memory that is an extension of the users' internal memory. Shingled menus also provide access to more data than would normally be available on a single screen of the same size by overwriting more important data on top of data that are less important at that time. Text from several documents can be reviewed, and portions can be selected, combined, and copied onto one menu. Moreover, the same thing can be viewed in several ways. For example, alternative drafts of a manuscript, different versions of a screen, or different graphics can be looked at for selection.

General guidelines for designing menus recommend that no more than six or seven menus be displayed at one time. **TILED MENUS**, as illustrated in Figure

Figure 9.21
Shingled menus and selection icons.

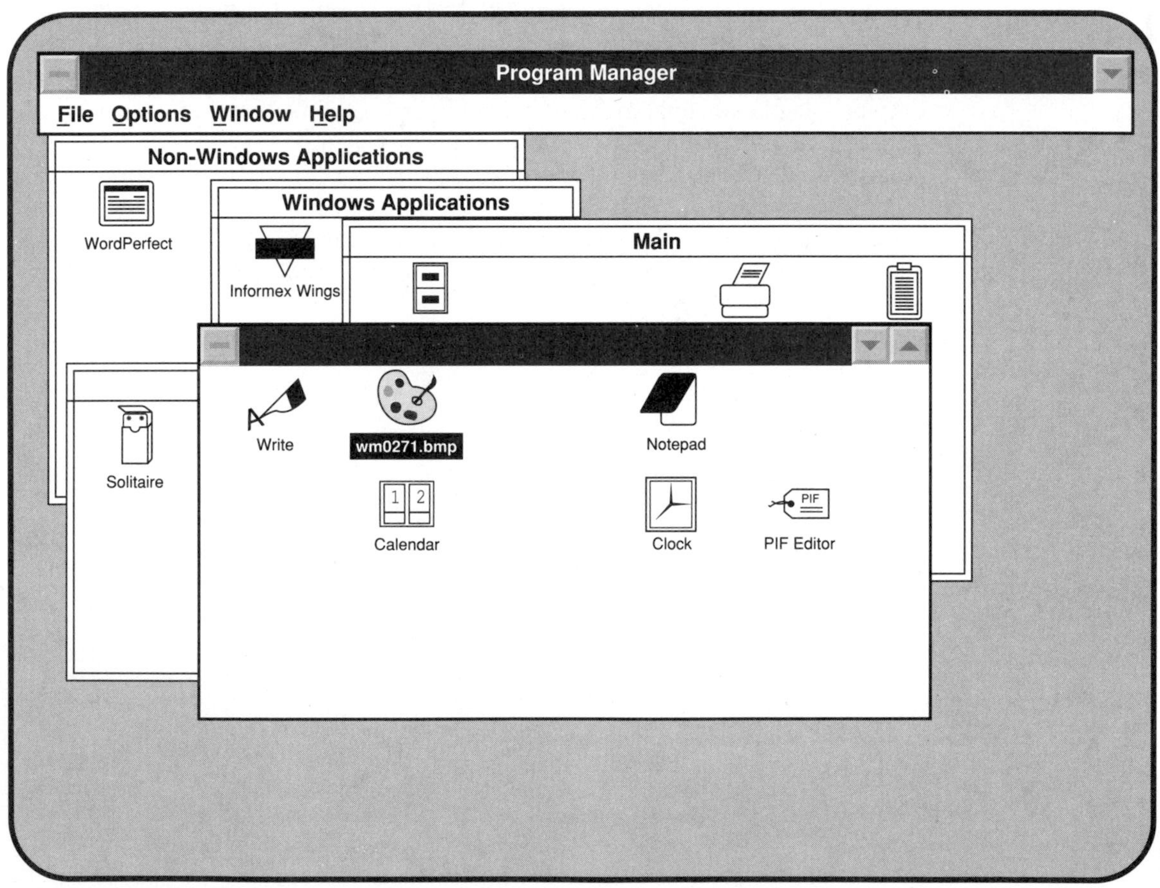

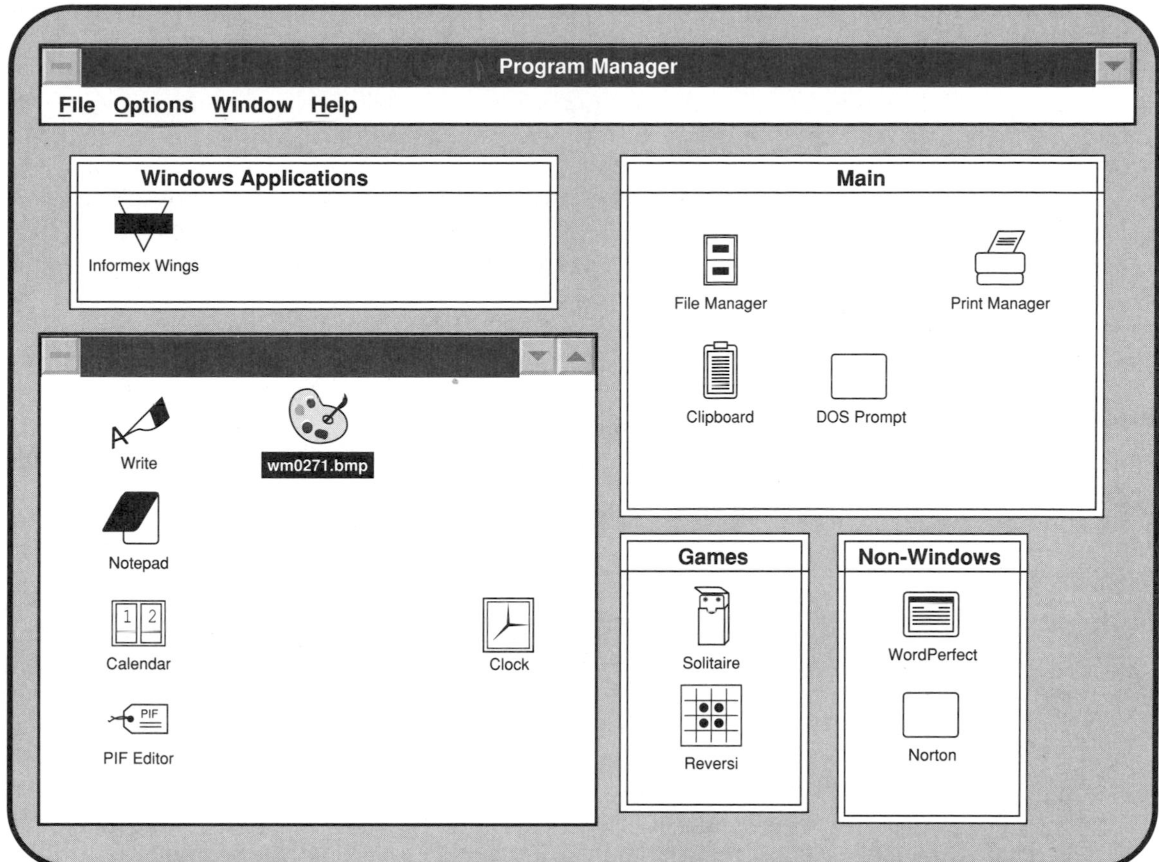

Figure 9.22
Tiled menus and selection icons.

9.22, should be designed for novice users in single-task activities. Shingled menus should be designed for expert users who are involved in multiple tasks.

Icon Menus

One way to design menus that are helpful for users who prefer to deal with images is to use icon-based design. Two vendors that provide extensive **ICON MENUS** are Apple and Microsoft (e.g., the Windows environment). Icons are associated with applications, as shown in Figures 9.21 and 9.22. Icons that are pictorial representations or symbols of the required application are selected by a key-controlled cursor or mouse, and the application is made ready for execution. Icons provide a level of user friendliness many people prefer. They can be applied to just about any of the types of menus we've discussed.

Touch Menus

Access to menu choices via a touch-screen interface is an excellent way to design the user/system interface for busy people. When users want the results of some function, they simply touch the icon or name of the function.

One example of TOUCH MENUS is illustrated in Figure 9.23. The main menu gives access to sales reports for any or all divisions; one simply touches their location on the map icon. Users can choose the way they want the reports displayed, that is, tabular, graphics, or both. If they want additional services, they simply touch any or all services chosen. They can go to the sales-by-region menu or exit the system. If they go to the sales-by-region menu, they can select any or all of the regional sales reports and have them displayed by tabular reports or graphics, or both. Also, additional services that are regional in scope are available. Users can either exit the system or go back to the main menu.

Sound Cues

A beep can be used to cue the user that some illegal function has been done or to draw attention to some message on the screen. A short chirp can be used to indicate the successful completion of an entry or task. With the advent of digital processor chips, voice message instructions or other SOUND CUES can be used to guide the novice user through a series of tasks.

DESIGNING SYSTEMS THAT RESPOND TO HUMAN LANGUAGE

A NATURAL LANGUAGE INTERFACE enables the computer system to understand human language. That is, the computer can accept ordinary human language input via voice or the keyboard and then perform the required task. This way, the system understands the user, so the user doesn't have to understand the system.

Conversational Dialogues

Permitting the user to carry on a conversation with the system represents a real breakthrough in designing friendly and easy-to-use user/system interfaces. An endless number of applications are waiting to be designed in this manner.

One example that helps an insurance company differentiate its service from its competitors is a system that delivers information to customers. The information delivered ranges from rebuilding costs and premium quotes to detailed coverage explanations. The system works by leading customers through a conversation directed toward gathering sufficient information for a premium quote. Along the way, it describes basic coverages, endorsements, and personal property schedules. Customers can interrupt the conversation at any time by asking questions in their own words, such as:

"In the event of loss, do I have additional living expense coverage?"
"Will this policy provide coverage for contents, and what amount?"
"Does this policy cover water damage?"
"What about my personal property away from premises?"

The system answers these questions specifically and gives additional details.

Database Interfaces

Many practical applications exist for interfacing with databases through English query or command languages. Such applications are especially useful in situations where queries or commands are ad hoc or unanticipated.

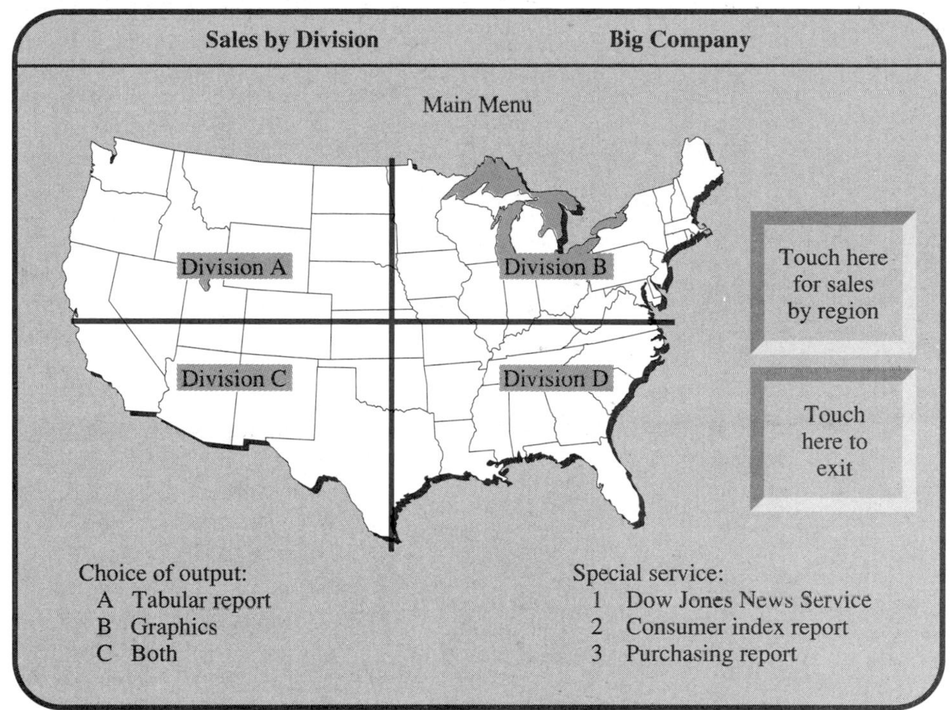
Sales by Division
Big Company
Main Menu
Division A
Division B
Division C
Division D
Touch here for sales by region
Touch here to exit
Choice of output:
A Tabular report
B Graphics
C Both
Special service:
1 Dow Jones News Service
2 Consumer index report
3 Purchasing report

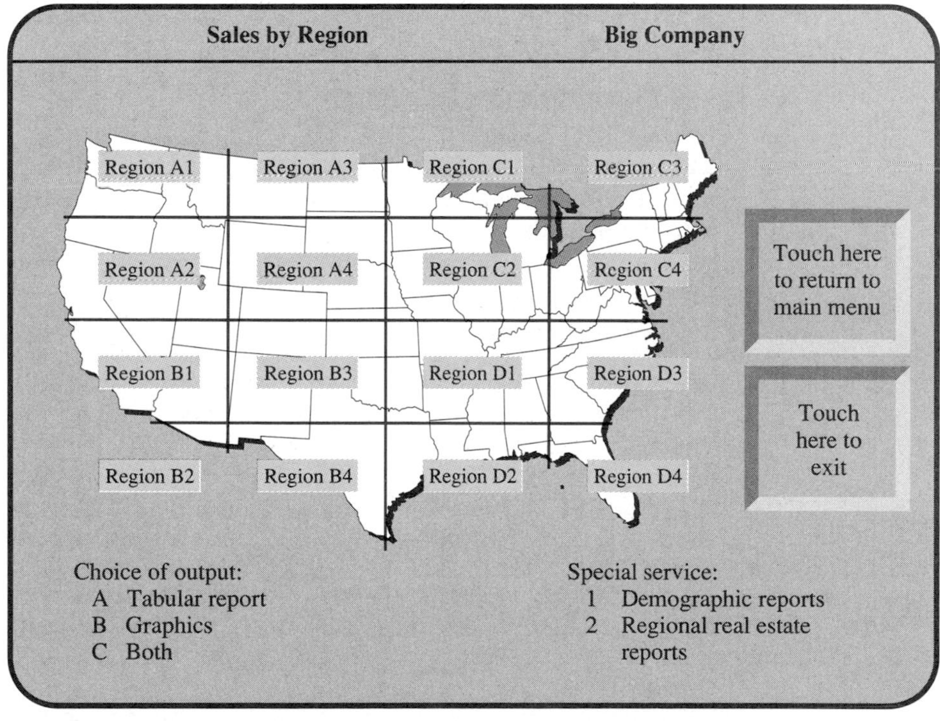
Sales by Region
Big Company
Region A1
Region A3
Region C1
Region C3
Region A2
Region A4
Region C2
Region C4
Region B1
Region B3
Region D1
Region D3
Region B2
Region B4
Region D2
Region D4
Touch here to return to main menu
Touch here to exit
Choice of output:
A Tabular report
B Graphics
C Both
Special service:
1 Demographic reports
2 Regional real estate reports

Some database management systems (DBMSs) have built-in linguistic knowledge to enable the computer to understand user input. The more robust systems come with large vocabularies, including:

Query or Command	Function Performed
"List analysts with salaries greater than $75,000."	Selection retrieval
"Display the accountants who work within division C in alphabetical order."	Sorting
"What's the average salary for engineers in division B?"	Arithmetic
"Increase the salaries of analysts in division D by 20 percent."	Selective updating

In some instances, the system may ask for verification. Following is such an example:

"How many boxes of product x do we have in warehouse 2?"
"Do you mean warehouse 2 in division A or division B?"

If the query or command contains an unknown word, the system permits the user to define or edit it. The system may do this using menus, such as the two depicted in Figure 9.24. The top menu tells the user that the word CODER is unknown. The bottom menu permits the addition of the new term as a synonym to PROGRAMMER. So, in the future, if the term CODER is used, the system will know what it means.

REVIEW OF CHAPTER LEARNING OBJECTIVES

The major goals of this chapter were to enable each student to achieve six important learning objectives. We will now summarize the responses to these learning objectives.

Learning objective 1:
Explain how to design paper forms and source-document-based data-entry screens.

In most systems projects, paper forms must be designed. Normally, these paper forms are source documents that require filling in by handwritten or typing methods. In turn, the filled-in source document is transcribed by an operator using a keyboard to convert the transaction data from human form (i.e., the source document) to computer form.

Designing paper forms involves:

- Selecting appropriate paper
- Sizing
- Manifolding
- Zoning

- Preparing instructions
- Captioning and describing data fields
- Spacing
- Sequencing

After a source document is designed, a data-entry screen is also designed to reflect the source document and provide a reference for the data-entry operator. Generally, abbreviations are used for captions and data-entry fields on the data-entry screen. The horizontal format is recommended for single data fields.

Figure 9.24
Menus for defining unknown terms.

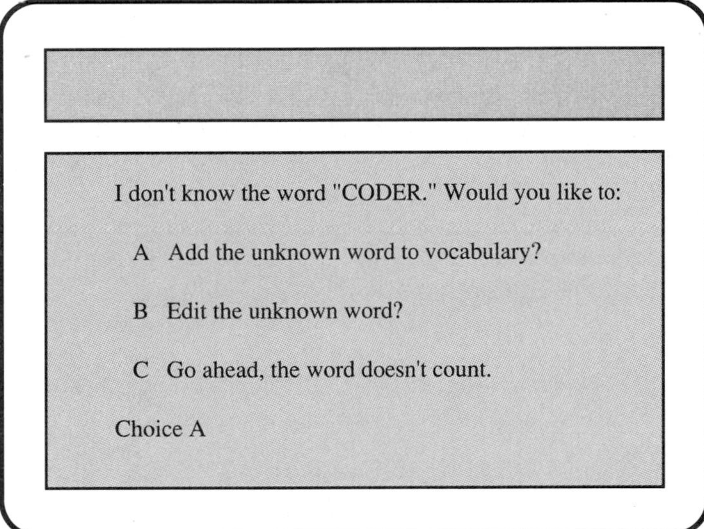

I don't know the word "CODER." Would you like to:
A Add the unknown word to vocabulary?
B Edit the unknown word?
C Go ahead, the word doesn't count.
Choice A

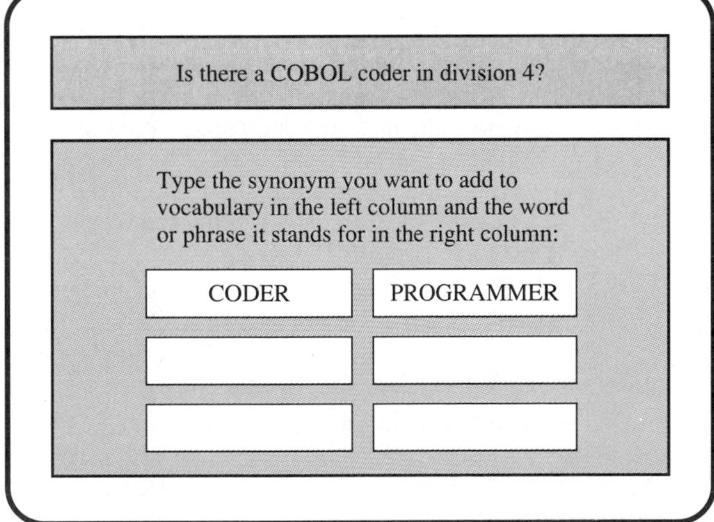

Is there a COBOL coder in division 4?
Type the synonym you want to add to vocabulary in the left column and the word or phrase it stands for in the right column:
CODER
PROGRAMMER

The use of headings, colons, greater-than symbols, and spaces help to format fields and give clarity to the data-entry screen design.

Learning objective 2:
Describe electronic forms, and state their advantages over paper forms.

With electronic forms, a dedicated source document is not required. Their screen design is the form that is filled in with one keying operation; that is, electronic forms are those onto which data are keyed. If data are captured from notes or manuals, the design must reflect the organization of these data sources. If the data are being provided by a person, such as a customer, it must include all the captions, data-entry fields, and instructions that are required to complete the sales transaction, and be organized in a clear, instructive, and logical manner.

Electronic forms have the potential of reducing data-entry costs when compared to paper forms. Electronic forms are easier to control in a forms library. Forms can quickly be transmitted to an authorized user's workstation for processing. A number of messages and instructions can be used to support the filling in of forms. Some of these ''helpers'' can be of the pop-up kind and therefore be unobtrusive; that is, they are displayed on the form at appropriate points to help users if they get stuck or make an error.

Electronic forms can also:

- Perform calculations
- Access certain data, such as customer names, product descriptions, and the current date (which reduces the amount of data users have to fill in)
- Perform editing automatically
- Make limit and reasonableness checks
- Coordinate processing tasks

Costs of electronic forms are less than those of paper forms in the long run, because paper forms can run out and cause business disruption, whereas electronic forms are almost always available. Electronic forms can be revised easily; controls can be built into an electronic forms system to provide strong access security and to ensure that the correct form is used; and the completion of a transaction is much quicker with electronic forms than with paper forms.

Learning objective 3:
List direct-entry devices.

Some of the more popular devices used to enter data directly are:

- Magnetic ink character recognition (MICR)
- Optical character recognition (OCR)
- Digitizer
- Image scanner

- Point-of-sale (POS) devices
- Automatic teller machines (ATMs)
- Mouse
- Voice recognition

Learning objective 4:
Explain the use of codes for input design.

As more data are entered from terminals and more queries are made into online databases, the use of codes (also called keys, identifiers, and account numbers) becomes increasingly important. Codes are used to:

- Condense data input
- Classify and identify data items
- Retrieve or select specific data items
- Allow one or more courses of action to occur according to the value stored in the code field

Code structures include:

- Sequence
- Block
- Group
- Special codes

A sequence code has no relation to the characteristics of the data except its ability to identify uniquely specific sets of data, such as a customer order or record. It is possible to code an unlimited number of items with the least number of digits. Little information, however, can be obtained from the sequential or serial code. In a block code, specific blocks of numbers are reserved for different classifications. Traditional accounting charts of accounts are excellent examples of block codes. Group codes are used to classify an item on the basis of major, intermediate, and minor classifications. A zip code for the U.S. Postal Service is a good example of a group code.

Special codes include:

- Bar codes
- Color codes

Bar codes can be used to identify and classify virtually any item. Bar codes are read by a bar-code reader or hand-held wand, and the data represented by the bar code are entered directly to the computer. Color codes help users identify and classify data items through the use of color-coding schemes.

Learning objective 5:
Discuss the menu interface, and explain how menus are designed to aid users to interact with the system.

A menu displays all the possible choices for a particular activity or process. Users simply select a number, mnemonic code, or icon that represents a particular choice. Menus are excellent devices for novice and occasional users, but they may be cumbersome for experienced users unless the menu allows them to leapfrog to specific tasks.

Menus should be properly titled and identified. They should contain concise, meaningful captions properly spaced and aligned. Common selection methods are:

- Keying
- Pointing
- Touching
- Voice input

Menu designers can use the following types of menus for specific applications:

- Pull-down
- Nested
- Shingled
- Tiled
- Icon
- Touch

Beeps, chirps, and voice messages help to augment menu designs.

Learning objective 6:
Describe natural language interfaces.

Natural language interfaces enable novice and occasional users to formulate a query for the system by expressing it in a conventional language, as if for interpretation by another person. The computer system then processes the query and generates the specified outputs. If the system detects an ambiguity, or cannot infer what the user means, it can enter into a dialogue with the user to resolve the uncertainty.

In a comprehensive natural language interface, the system recognizes natural languages in both text and spoken form. Output from the system may be voice, images, or text.

INPUT DESIGN CHECKLIST

Following is a checklist on how to design input for computer information systems. Its purpose is to remind you of key input design techniques.

1. For paper forms, select the correct size, weight, and kind of paper that will meet the form's demands, such as length of time it will be kept, appearance, how often it will be handled, how it will be handled, how it will be filled out, and its security.
2. Apply forms design techniques, such as manifolding and zoning.
3. Use proper titles, instructions, lines, boxes, captions, formats, spacing, and sequencing.
4. For data entry using a source document and a screen, make sure that the screen mirrors the source document design.
5. Use electronic forms for applications that do not require a source document.
6. For data-entry efficiency, use direct-entry devices, such as magnetic ink character recognition (MICR), optical character recognition (OCR), optical mark recognition (OMR), digitizer, image scanner, point-of-sale (POS) devices, mouse, and voice.
7. In applications that are document-intensive, consider the use of a document image processing (DIP) system.
8. Use sequential codes to identify items uniquely.
9. Use block codes and group codes to classify and identify specific groups of items.
10. Use bar codes to capture and input transaction data as soon as it occurs.
11. Use color codes to classify and differentiate items in a manual-based system.
12. Design menus for permitting users to input their selections easily.
13. Use a natural language interface for conversational dialogues and database queries.

KEY TERMS

Automatic teller machines (ATMs)
Bar codes
Block code
Captions
Codes
Color codes
Data-entry screen
Data field
Digitizer
Direct entry
Document image processing (DIP)
Electronic forms
Group code
Icon menus
Image scanners
Input design
Magnetic ink character recognition (MICR)
Manifolding
Menu
Mouse
Natural language interface
Nested menus
Optical character recognition (OCR)
Optical mark recognition (OMR)

Paper forms
Point-of-sale (POS) devices
Pull-down menus
Sequence code
Shingled menus
Smart electronic forms
Sound cues
Source document
Tiled menus
Touch menus
Voice recognition
Zoning

REVIEW QUESTIONS

9.1 What is a paper form? Give three examples of how paper forms are used.

9.2 List and briefly describe nine aspects of designing paper forms.

9.3 What is a source document? Explain how it is used.

9.4 Explain why a data-entry screen is used with a source document.

9.5 What are the key differences between an order-entry screen dedicated to a source document and a data-entry electronic form? Since they are both used to enter data, are there really any differences? Please explain.

9.6 Explain how to develop abbreviations. Why are abbreviations acceptable for data-entry screens dedicated to source documents, but unacceptable, except for obvious abbreviations, for paper forms such as source documents and electronic forms?

9.7 In general terms, explain how the data-entry screen is related to its source document.

9.8 Briefly describe an electronic forms system. Explain the role that each element plays in this system.

9.9 What are smart electronic forms? Could data-entry screens dedicated to source documents become smart data-entry screens?

9.10 Compare costs of electronic forms with costs of paper forms.

9.11 List and give brief examples of the guidelines for designing messages that are related to designing electronic forms.

9.12 List eight popular direct-entry devices.

9.13 Define document image processing (DIP).

9.14 What is a sequence code, and what is it used for? Give an example.

9.15 What is a block code, and what is it used for? Give an example.

9.16 What is a group code, and what is it used for? Give an example.

9.17 What is a bar code? Give three ways a bar code can be used. Why do we call a bar-code entry a point-of-sale (POS) device?

9.18 What is a color code, and what is it used for? Give an example.

9.19 Give a general definition of a menu interface for systems.

9.20 Explain pull-down and pop-up menus. What are they used for?

9.21 Define nested menus. What are they used for?

9.22 Define tiled and shingled menus. What are they used for?

9.23 Define icon menus. What are they used for?

9.24 Define touch menus. What are they used for?

9.25 Explain the use of beeps, chirps, and voice messages in designing menus.

9.26 Explain the value of a natural language interface.

9.27 In a natural language interface, explain how ambiguous or unknown terms are reconciled.

CHAPTER-SPECIFIC PROBLEMS

These problems require exact responses based directly on concepts and techniques presented in the text.

9.28 Here are some captions and data fields found on source documents being used in practice.

(a) List the business subjects you have studied

____

____

____

____

____

(b) Social Security Number ____________________

(c) Date: ____________________

Required: Redesign the three entries to show good design practices.

9.29 The Kokomo Company is developing an inventory control system. Suzy Wong is working on the Central Stores Supply Requisition paper form. The stores requisition number will serve as the identifier. In the Deliver To section, she wants to include department, building, room number, date, and department requisition number. The main body of the form will include Quantity Ordered, Description Unit, Unit Price, and Extension. These entries will result in Subtotal, Handling Charge, and Amount Due. Distribution instructions are: white to Accounting, yellow to Department, goldenrod to Central Stores, and pink returned with delivery.

Required: Design a paper form for Suzy that includes the preceding elements.

9.30 A portion of an automobile insurance application includes a major section on driver information composed of driver name, date of birth, sex, occupation, marital status, driver license number, state, years licensed, driver training, impaired driver, and code (code is not entered on the screen). Another major section is vehicle information, including make, new/used, vehicle identification number, horsepower, cylinders, use, and miles.

Required: Design an automobile insurance application source document and a data-entry screen for the foregoing captions and data fields.

9.31 Marigold Company is a startup mail-order business. A key input for processing is an invoice that includes four subsection headings: customer, item, charges, and credit card. Tied to customer subsection are name, address, and telephone number. Associated with item are catalog number, quantity, size, description, and color. Related to charges are price, handling, and total. Associated with credit card are kind, number, and expiration date.

Required: Design an electronic form data-entry screen with title, identifier, subsection headings, captions, and data fields.

9.32 Steve Goldberg, quality assurance manager for a car manufacturer, wants to develop a code that will help him to locate and isolate defective cars. After interviewing Mr. Goldberg, you have concluded that the code should include plant where car assembled, two digits; plant where subassemblers, such as body, chassis, rear end, transmission, and motor are manufactured, two digits for plant and one digit to identify subassemblies; subassembly inspector number, two digits; foreman number, two digits; line number, two digits; shift number, two digits, and date (MMDDYY).

Required: Design a group code that includes the classifications above. Describe each classification and show position.

9.33 A menu is to contain the following choices: bill, cancel, credit, debit, endorse, fill, grant, print, quote, and release.

Required: Using the captions above:

1 Design a menu with double columns assuming space constraints and single-selection fields.

2 Design a menu with double columns assuming no space constraints and multiple selection.

3 Design a menu assuming one column using mnemonic identification and single selection.

4 Design a menu with horizontal spacing and ordinal code.

THINK-TANK PROBLEMS

These problems call for a feasible approach rather than a precise solution. Although the problems are based on chapter material, extra reading and creativity may be required to develop workable solutions.

9.34 For the following proposed applications, recommend the kind of direct-entry device you believe to be most appropriate. Justify your recommendations.

1 Allow workers in factories to enter data while operating machinery. An example might be to signal to the system the completion of a fabricating operation.

2 A bank wants to offer its customers a secure way to interact with the bank's ATMs scattered throughout the city.

3 A large construction company wants to take aerial photographs of their huge piles of sand and gravel and convert these photographs to data stored on magnetic disks for inventory control.

4 An educational institution wants to score multiple-choice exams automatically.

5 Insurance forms will be prepared with selected character styles. The manager of the insurance company wants the data on these forms to be entered automatically.

6 Various items displayed on a video screen must be pointed to and selected quickly.

9.35 You are designing menus for two kinds of applications. Application 1 incorporates four single tasks performed by novice users. Application 2 involves six activities that require constant switching between tasks performed by expert users.

Required: Recommend the kind of menus, either shingled or tiled, that should be used for each application. Explain your recommendation.

9.36 The order form shown on page 368, for a mail-order record and tape business, is a source document from which computer operators enter order data into an order-entry system. It is critical that these data be entered in the exact sequence of the source document. Comment on the design of this source document. What improvements would you suggest?

9.37 A dentist's office uses the form shown on page 369 to gather information on new patients. This information is subsequently entered into the office's computer database by a secretary, in the exact sequence it appears on the form. How well designed is this source document? What is wrong with it? Redesign the dentist's form in such a way that it conforms to the guidelines presented in this chapter.

The Music Trader
3170 Palm Frond Drive
White Sands, NM 80657

Add's

State

Ctiy

Tel

ZC

Use ONLY black ink to complete this form!

	Price	Quanity	Item	Total
1				
2				
3				
4				
5				

Note: Use catalog number for item description. Indicate catalog date.

Total

Name

Cash ❑ Charge ❑

Dr. Edward R. Goldstein, Jr.

New Patient Information Form

Patient:	F	L	M
Spouse:	F	L	M

Social security number

Telephone | Occu.

Yes ❑ No ❑ Marital status

Yes ❑ No ❑ Employed?

Medical insurance underwriter

How were you referred to our office? [ ] [ ] Is this your first visit to our office

Address

City, State | Zip

Signature | Date

THIS FORM MUST BE TYPED

SUGGESTED READING

Adams, Lee. *Supercharged Graphics; A Programmer's Source Code Toolbox*. Blue Ridge Summit, Pa.: TAB Books, 1988.

Awad, Elias M. *Systems Analysis and Design*. 2nd ed. Homewood, Ill.: Richard D. Irwin, 1985.

Boer, Germain. *Classifying and Coding for Accounting Operations*. Montvale, N.J.: National Association of Accountants, 1987.

Galitz, Wilbert O. *Handbook of Screen Format Design*. 3rd ed. Wellesley, Mass.: QED Information Science, Inc., 1989.

King, Karl G., and Raymond W. Elliott. "In Plain English, Please." *Journal of Accountancy,* March 1990.

Peacock, Eileen. "Why Errors Occur in Accounting Systems." *Management Accounting,* August 1988.

Rich, Elaine. "Natural-Language Interfaces." *Computer,* September 1984.

Rivera, Christine. "Formbase: Unique Approach to Form Design, Filling Tasks." *Infoworld,* February 26, 1990.

Rivera, Christine. "Perform Is Among Leaders in Forms Management Field." *Infoworld,* February 26, 1990.

Schneiderman, Ben. *Designing the User Interface: Strategies for Effective Human–Computer Interaction*. Reading, Mass.: Addison-Wesley, 1987.

"Seven Hidden Costs of Every Form Your Company Uses." Carrollton, Tex.: Electronic Form Systems, 1987.

Smith, L. Murphy, and James A. Sena. "How to Design User-Friendly Data Base Applications." *Journal of Accounting and EDP,* Summer 1989.

"The Wonderful World of Color Makes Records Management Easier." *Information and Records Management,* October 1976.

JOCS CASE: Designing Input

After the SWAT team completed a large percentage of the output design for JOCS, Jake Jacoby began the first steps of input design. Jake realizes that systems development is a circular, or iterative process rather than a stepped process. In other words, a systems analyst doesn't really finish one complete phase of the cycle before starting a new phase. A good systems analyst is always looking at and incorporating the other phases of the cycle while working on specific tasks in a given phase.

While the other members of the SWAT team are designing the identified reports and screen-based displays, Jake is beginning to design the overall input for JOCS. The first step he takes in this process is to list the input sources:

- Direct material used in the manufacturing process is captured by the CIM (computer-integrated manufacturing) computer system.
- Direct labor used in the manufacturing process is also captured by the CIM computer system.
- Bills of material for a given hydronautical lifter are developed by the engineering department. They are available from the CIM computer system.
- Job order numbers are assigned by one of the cost accountants. They are maintained in a coded card system.
- Customer data are gathered by the marketing department on their standard paper-based source documents.
- Overall job data, including a job description, quantity ordered, and date received are gathered by the marketing department and delivered to the accounting area of the finance department.
- Inventory data for raw materials are maintained manually by the receiving department on their card-based system.
- Standards for each job are developed by the accounting personnel within the finance department, in conjunction with personnel from the engineering and manufacturing departments. They are maintained on their paper-based source documents.
- Standards for each manufacturing workstation are developed by the manufacturing department. They are not currently maintained with any kind of historical reference system.

After listing the input sources, Jake has a good idea of where data are located, who is responsible for a given item of data, and the format of the input data. Since direct material, direct labor, and bills of material are available from another computer, Jake must assign a member of the SWAT team to research a potential interface between the JOCS computer and the CIM computer. He thinks about who would be best suited for this task and assigns it in his mind to a team member with strong programming skills, Tom Pearson, before continuing with the process of input design.

Jake needs a way to get the rest of the data into the computer. Job and customer data can be input directly from the existing source documents without having to change much of the current format. Accounting standards for a given job can also be entered from existing source documents. Workstation standards, on the other hand, do not currently have a source document, so Jake decides to use an electronic form. Since no one is depending on a paper-based source document for workstation standards, Jake has the opportunity to install an efficient electronic form for this input. Jake anticipates that, in time, people in the accounting and marketing departments will come to rely on JOCS and will abandon their paper-based source documents. Jake feels it's better for an analyst to help people learn new ways to do things rather than force them to abandon their current systems and start using new systems. However, until the people decide they can get along without their existing forms, Jake plans to design input in accordance with the current source documents. This will make data input simple and will result in a system with few data-entry errors.

The next step in input design is done with the full SWAT team. Jake calls a meeting to discuss input design. He begins by congratulating the team on its timely completion of output design. He then discusses the input sources he has identified. Jake explains to the team that the next step is to design the overall structure of the JOCS online menus.

Christine has designed quite a few different menu systems during her past employment experience. She says, "I think a standard menu system designed for any kind of data-entry terminal would work well for JOCS. If we just design a simple approach, we won't have to worry about the kind of equipment being used." She draws out a couple of sketches (Figure 9.25) and shows them to the rest of the team.

Carla likes the look of the menus, but she says: "Not everyone will automatically understand terms like 'data update' and 'file maintenance.' Quite honestly, I think some of the people in accounting would be intimidated by words like 'terminate system.' That sounds as though the entire system will automatically be destroyed if that selection is chosen. I think we could divide the functions by department or functional area, rather than task."

She modifies Christine's sketches to explain to the group what she has in mind (Figure 9.26). Carla says: "Since JOCS will be used by different functional areas within the company, I think the system would be more understandable if we designed the menus on the basis of these functions. This way, accounting personnel can go directly to their set of menus to put in data, produce reports, and view display screens."

The rest of the team likes Carla's approach, because it makes the system seem a little easier to use for people without much computer experience.

Figure 9.25
Initial menu sketches for JOCS.

```
Date:  MM/DD/YY                          ** JOCSMN00 **

              JOB-ORDER COSTING SYSTEM
                      Main Menu

              1. Perform file management
              2. Generate printer-based reports
              3. View online displays
              4. Terminate system

              _  ENTER CHOICE
```

```
Date:  MM/DD/YY                          ** JOCSMNF1 **

              JOB-ORDER COSTING SYSTEM
                 File Maintenance Menu

              1. Update job order data
              2. Update customer data
              3. Update accounting standard data
              4. Update workstation standard data
              5. Update raw material inventory data
              6. Return to main menu

              _  ENTER CHOICE
```

However, Jake feels that the format doesn't take advantage of the computer equipment that is currently in place and the equipment that will be purchased in the future.

Jake explains what he knows about the use of computer data-entry equipment: "We aren't going to purchase nonintelligent data-entry terminals any longer. Kyle Bartwell did a survey of computer resources at the beginning of this year, and he found that there were more dollars tied up in microcomputer equipment than any other kind of computer resources in this

Figure 9.26
Updated menu sketches for JOCS.

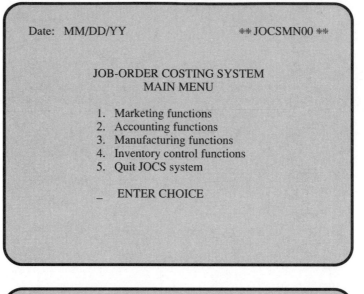

```
Date:  MM/DD/YY                                ** JOCSMN00 **

                JOB-ORDER COSTING SYSTEM
                       MAIN MENU

                1. Marketing functions
                2. Accounting functions
                3. Manufacturing functions
                4. Inventory control functions
                5. Quit JOCS system

                _  ENTER CHOICE
```

```
Date:  MM/DD/YY                                ** JOCSMNF1 **

                JOB-ORDER COSTING SYSTEM
               ACCOUNTING FUNCTIONS MENU

                1. Enter accounting standards
                2. View information on screen
                3. Print reports
                4. Return to main menu

                _  ENTER CHOICE
```

company. He gave Mary the directive to make better use of all those little computers sitting on everybody's desks. I think we should design a menu system that will take advantage of more sophisticated microcomputer technology."

Cory, excited by this approach, says: "Let's take Carla and Christine's design and turn it into a pull-down menu approach. In college, we learned how to design menus that can be used with a simple click of a mouse. We also learned how to provide pop-up help windows if the user needs more information about a given menu selection. That way we can take advantage of the

Figure 9.27
Pull-down menus for JOCS with function-key activated pop-up held window.

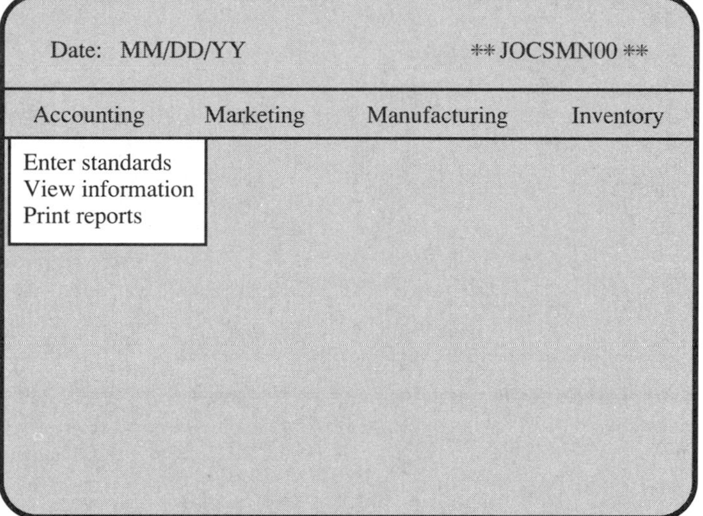

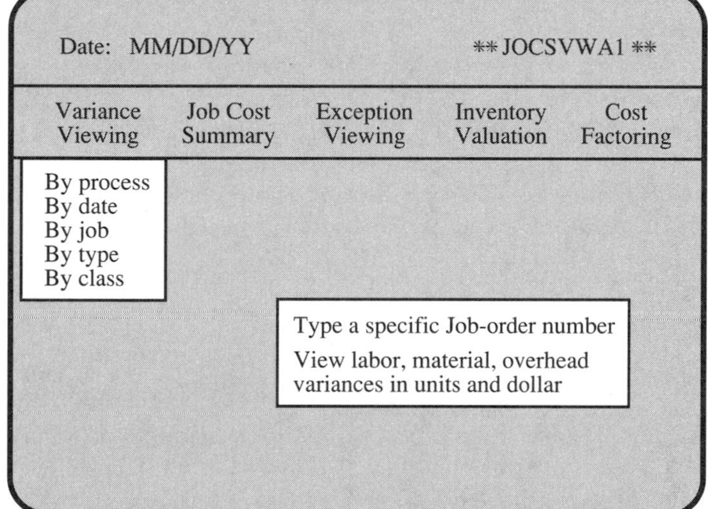

microcomputer screen and mouse to make JOCS really simple to use." She quickly changes the sketches into a pull-down menu approach (Figure 9.27).

Once the team agrees on a menu system, Jake begins to assign input design tasks to the team members. Tom is to research the CIM to JOCS interface possibilities, as discussed earlier. Carla and Cory are assigned responsibility for translating accounting and marketing source documents into computer input screens, while Christine is to design the electronic forms for manufacturing personnel to input job workstation standards.

Chapter 10

Designing the Systems Process

WHAT WILL YOU LEARN IN THIS CHAPTER?

After studying this chapter, you should be able to:

1 Discuss the three dimensions of process design.
2 Describe real-time process design, batch process design, and the use of process specification models.
3 Explain the use of equations in developing process specifications.

INTRODUCTION

The objective of this chapter is to present **PROCESS DESIGN**, which is the design component that specifies when and how things should be done to support user requirements (see Figure 10.1).

Because users have diverse information requirements, there must also be diversity in process design. For example, the following requirements are possible:

- The chief executive officer (CEO) of an automobile company is concerned with the trend of imports from Germany and Japan.
- The chief financial officer (CFO) of a service company wants a balance sheet and an income statement printed weekly.
- The chief operating officer (COO) of a manufacturing firm wants cost-volume-profit analysis reports and variances from budgets every Monday morning.
- Accountants want a system to update accounts receivable and process payroll periodically.
- Customers want to know the status of their orders immediately.
- Physicians want to know the test results of their patients online.
- The marketing manager wants to know the sales forecast and product sales trends for the next quarter.

Let's look at the need for diversity in three dimensions, as portrayed in Figure 10.2. First is the time dimension. Will the process design require an interactive interface and transactions processed as soon as they occur? Or will

Figure 10.1
SDLC phases and their related chapters in this book. In Chapter 10 we study the many facets of process design.

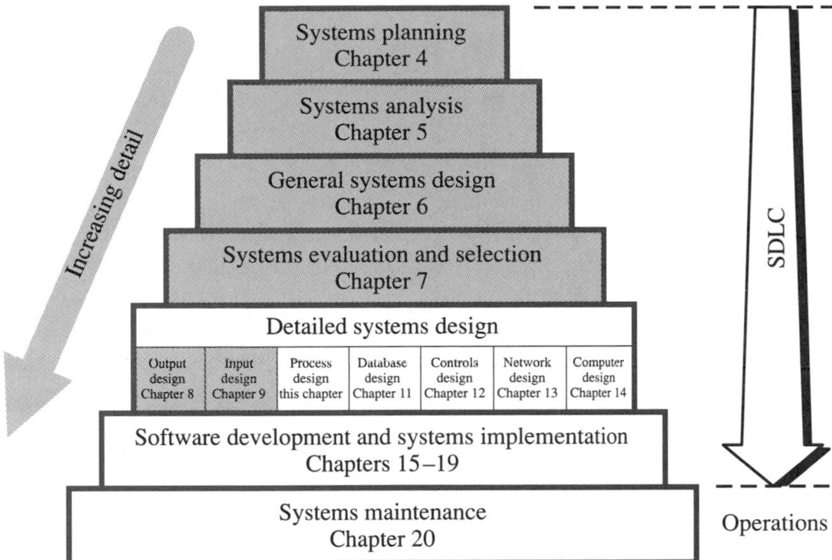

the process design not need an interactive interface, and allow transactions to accumulate over some time before they are processed?

The time dimension dictates the type of technology platform that will be designed and acquired. If the time dimension is real-time, the technology platform must be some kind of networked computer architecture with online access to a database. If, on the other hand, the time dimension is batch, the technology platform will typically be a stand-alone computer that processes sequential files periodically.

The third dimension deals with the modeling tools used to design real-time and batch process applications. These modeling tools generally include DFDs, STDs, and a host of process modeling tools.

This chapter treats the time dimension and modeling tools used to design real-time and batch processes. Chapters 13 and 14 present technology platform design.

Figure 10.2
Three dimensions of process design.

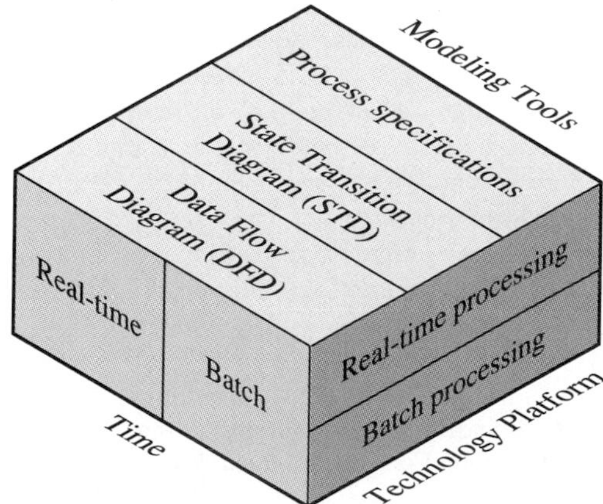

TIME DIMENSIONS

Some information systems are strictly real-time; others are batch. Most information systems, however, are composed of a hybrid of the two processes.

What Is Real-Time Processing?

In **REAL-TIME PROCESSING** the value of processing is a function of both the output (or results) of processing and the time at which output (or results) is delivered. Real-time processing includes an interactive interface that is dominated by interactions between the system and external agents, such as users, various mechanical or electronic devices, or other systems.

There are two kinds of real-time processes: hard and soft. Hard real-time processes are those in which the application fails completely if the process does not always meet the time constraint. Soft real-time processes are those in which systems performance is degraded if the time constraint is not always met, but the performance requirements can be fulfilled if the conditions are met with a response distribution. A typical hard real-time system is an aircraft flight control system, where loss of control occurs if the system does not meet time constraints. An example of a soft real-time system is an airline reservation system in which slow response results in degraded operations but not system failure.[1]

Applications at Innovative Products Company call for dynamic, time-dependent processing. Let's now review such applications at Innovative.

Innovative Products Company's Real-Time Processing System

Innovative Products Company offers hundreds of leading-edge products to a vast number of customers across the country. Some of these products, such as children's toys, have a life cycle of only one or two seasons.

Research and development personnel and the marketing staff must be very aggressive and creative. They must stay on top of product changes and be aware of sudden shifts in the marketplace. The manufacturing people must be able to change production procedures overnight to manufacture new commercial prototypes developed by research and development. Expert systems help to generate bills of materials for new product designs. The marketing staff must be apprised of such changes to formulate sales and advertising strategies. Logistics personnel must plan for materials flow and distribution of the new products to customers to take advantage of each selling season. Often new products are shipped via air freight or by overnight trucking firms. Such rapid shipment to diverse customers scattered throughout the country calls for online transaction processing (OLTP), split-minute decisions, and extremely tight coordination among disparate groups.

Top managers need decision support and information that will help them make tactical and strategic decisions. These managers must be able to interact with the system via dialogue and menu interfaces.

---

[1] John F. Muratore, Troy A. Heindel, Terri B. Murphy, Arthur N. Rasmussen, and Robert Z. McFarland, "Acquisition at Mission Control," *Communications of the ACM,* Vol. 33, No. 12, December 1990, p. 23.

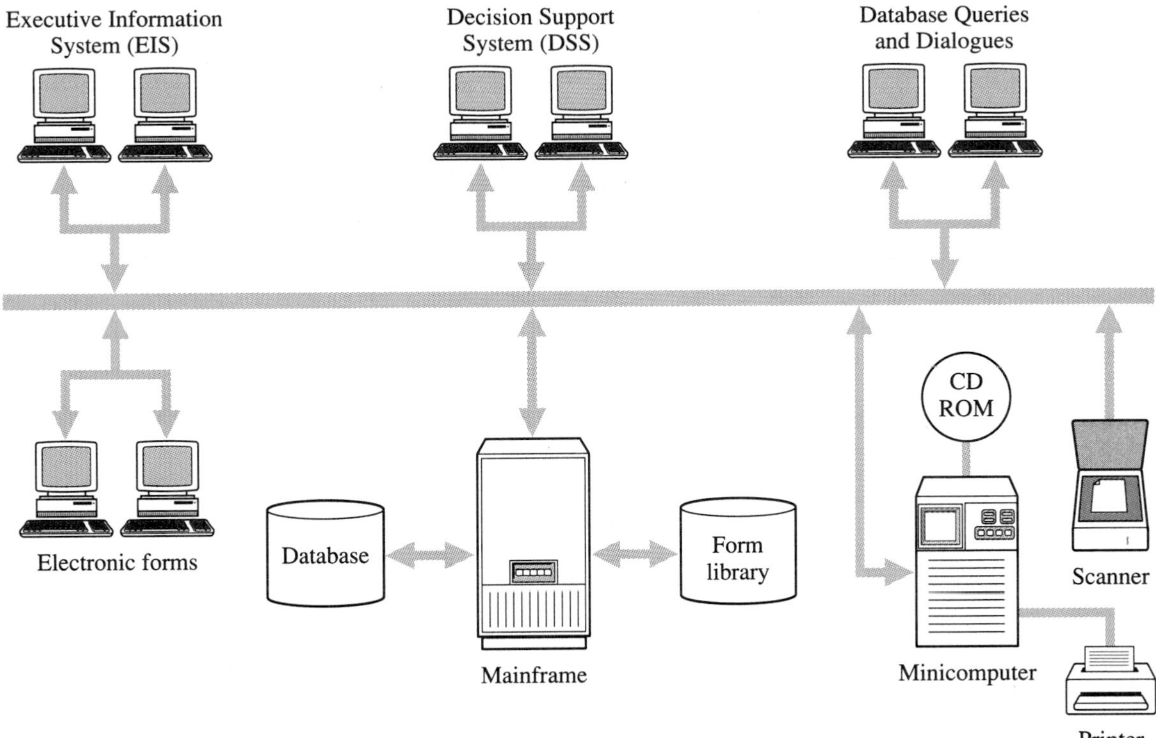

Figure 10.3
Real-time process design elements.

With a real-time process design, as demonstrated in Figure 10.3, transactions or queries are input as they occur, typically in a random manner, without presorting into batches. The media, such as magnetic disk and compact disk–read-only memory (CD-ROM) disk, on which data are stored, provide direct access for user dialogues, querying, and updating. Most of the users interact with the system through various workstations for various applications, such as executive information systems (EISs), decision support systems (DSSs), database queries and dialogues, and OLTP via electronic forms and scanning devices. Mainframes and minicomputers act as servers; that is, these computers coordinate and maintain the database or databases and control various devices, such as printers. The workstations, on the other hand, permit user interaction with the system and presentation of information.

Summary of Key Characteristics of Real-Time Processing

Real-time systems exhibit these key characteristics:

- ***Process Orientation*** Data are processed on a continuous basis, in contrast to the processing of data on a periodic basis. The information system acts as an integral part of the total operations of the organization.
- ***Online File Availability*** The database is online and available to users at all times for interacting, querying, and updating.

- ***Very Short Time Intervals*** Real-time processing eliminates the time interval between the point when users query the system or enter transactions and the time the system processes such queries or transactions. Thus the database always reflects current conditions.
- ***Constant Updating*** When a transaction occurs that requires a change in a master file in the database, a record is accessed from a file into the computer, updated, and written back to its original physical location. The original value of the record is lost or destroyed unless the updates are recorded on a transaction log in another file.
- ***Organization of Records for Rapid Access*** Records are stored and processed in a direct manner. To access a record, the read/write mechanism of the direct access storage device need only position itself at the physical location of the record being accessed.

What is Batch Processing?

BATCH PROCESSING is periodic in nature. It is based on a sequential input-to-output transformation. There is no ongoing interaction. An excellent example of a batch process is a payroll system that is processed, for example, once a week.

Is batch process design a relic from yesteryear? Is it a dinosaur waiting for the end? Is it bad design? Answers to these questions depend on what systems designers are trying to achieve. If users want decision support systems (DSSs), executive information systems (EISs), user/system dialogues, and databases that contain current information, then batch processing is not only a bad design for them. It is the wrong design, because it simply won't work.

But not all organizations want or need the kinds of applications alluded to above and in other chapters in this book. All they want is a process designed to handle traditional accounting applications, such as general ledger, payroll, accounts receivable, accounts payable, and fixed assets. These applications are periodic, so they lend themselves well to batch processing. For example, Mellow Foundry has such applications.

Mellow Foundry's Batch Processing System

Mellow Foundry manufactures aluminum wire primarily for one customer, Cascade Utilities, a large utility company serving the western United States. In addition, Mellow has about 200 customers who buy smaller quantities of aluminum wire. All of these customers have been trading with Mellow for years.

Mellow Foundry is an employee-owned company that enjoys relatively stable sales year after year. The design of the aluminum wire has never changed, and there is no indication that it ever will. Mellow uses one raw material item, which is ¼-inch aluminum bars that are extruded into wire. Wire is made to order rather than to stock; therefore, all wire is shipped to customers as soon as it is drawn.

Mellow does not want to grow and change its present operations. Management planning, controlling, and decision making, what there is of it, is fairly perfunctory. The critical data processing functions are the general ledger, payroll, accounts receivable, and accounts payable. The general ledger is updated monthly. Payroll is processed

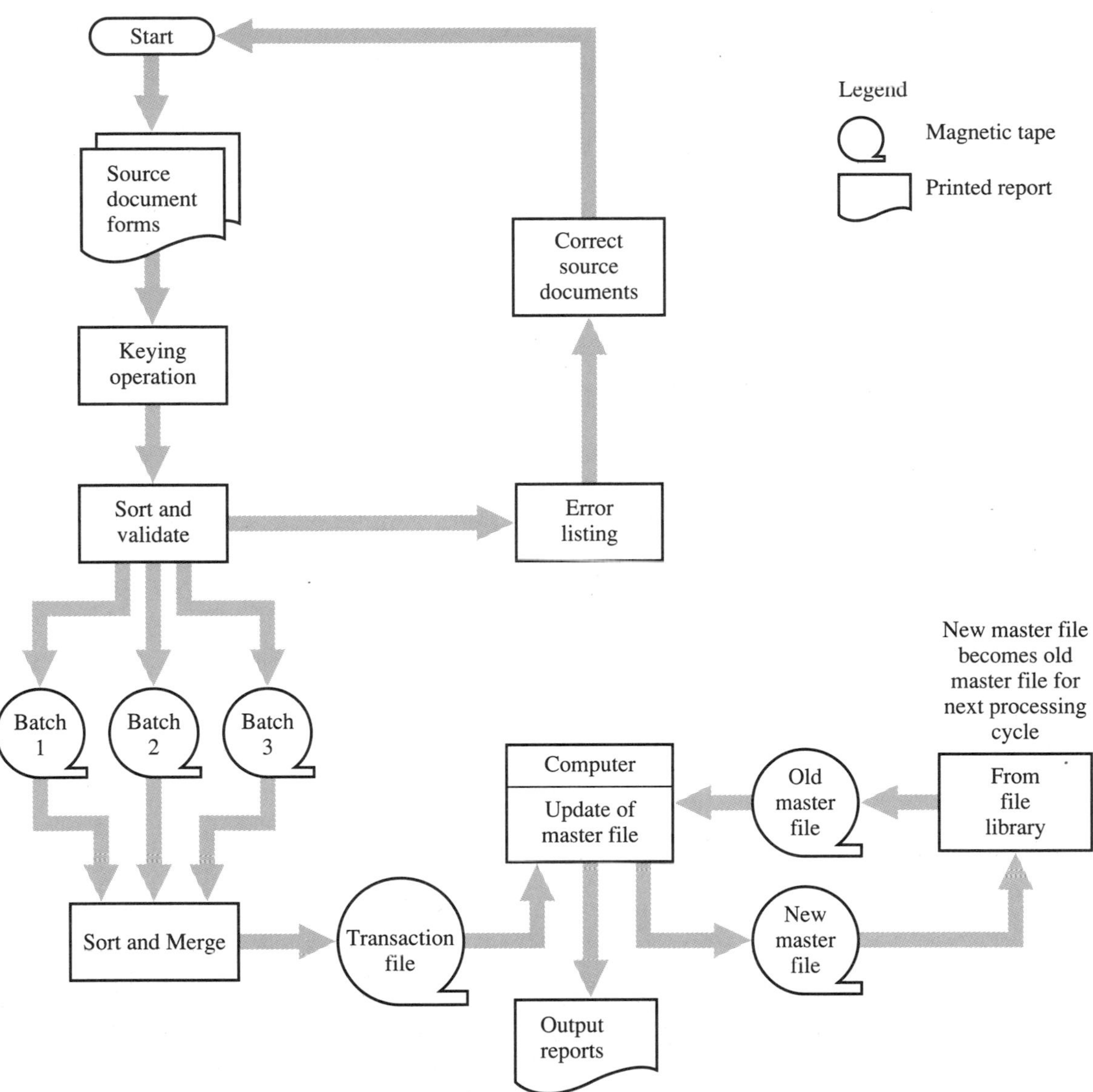

Figure 10.4
Batch process design.

weekly; accounts receivable are paid by customers in time to take advantage of Mellow's early-payment discounts; and accounts payable are processed during the last Friday of each month.

The work force is very stable. It has a big-family culture. All customers are well known by Mellow, since they've been doing business with each other for years. If by chance a payroll was ever processed late or a bill was not paid exactly on time, no one would be that upset over such a mishap.

What kind of process design does Mellow Foundry need: real-time or batch? A real-time process design for Mellow would be as out of place as a batch process design would be for a major airline. The kind of system that supports Mellow is illustrated in Figure 10.4.

Source documents are prepared that represent transactions or events such as sales orders. The data on these forms are keyed into machine-readable data. The data are sorted in accordance with a code, such as a transaction number or customer number, and checked to determine if there are any inaccuracies or omissions. Any forms that fail to pass the validation process are returned for correction. Validated batches of transactions are sorted, merged, and stored in a transaction file in the same order as the old master file of customer records stored in the file library, such as by customer number in ascending order. Both files are mounted and processed by matching a transaction code in the transaction file with a code in the old master file. Both files are read in sequence, with matches between the two files being found along the way. As each match is found, the computer has all the data necessary to process the matched transaction. When the computer gets to the end of the old master file, all transactions will have been processed and records updated to produce a new master file together with reports such as sales reports or documents such as customer invoices. The new master file is returned to the file library until the next batch of transactions is processed, when it becomes the old master file. The present old master file and old transaction file are stored off-premises in a bank vault for backup should the new master file be lost or destroyed.

DESIGNING REAL-TIME PROCESSES

Real-time processes are modeled both statically and dynamically. We will discuss them in turn.

Creating a Data Flow Diagram to Model Static Design

A system can best be understood by first examining its static design—that is, by a model of its processing at a single moment in time. A DFD is an excellent modeling tool for a process's static design. The dynamic nature of the process can be modeled with a STD. Let's use a DFD, as depicted in Figure 10.5, to specify what processes are involved in an automatic teller machine (ATM) cash dispensing application at a bank.

A user inputs a cash card at an ATM. The cash card contains the user's personal identification number (PIN). The user keys in a password (or code) and transaction details (e.g., cash amount requested). The system determines if the PIN and password are correct by matching those input by the user against those in a table of valid PINs and passwords. If there is no match, a rejection message is displayed on the ATM screen. If there is a match, transaction details are used to determine if the transaction should be authorized. For example, if the amount requested exceeds the user's credit limits, the transaction is rejected. If the transaction is authorized (i.e., accepted), it is processed, customer accounts are updated, and the user receives his or her cash along with a message and a receipt.

The DFD describes the flow of data from external inputs through processes and data stores to external outputs. The DFD specifies the results of processing without, however, specifying when or how they are processed. A STD specifies when; process specification modeling tools specify how.

Figure 10.5
Static design of an ATM cash dispensing program.

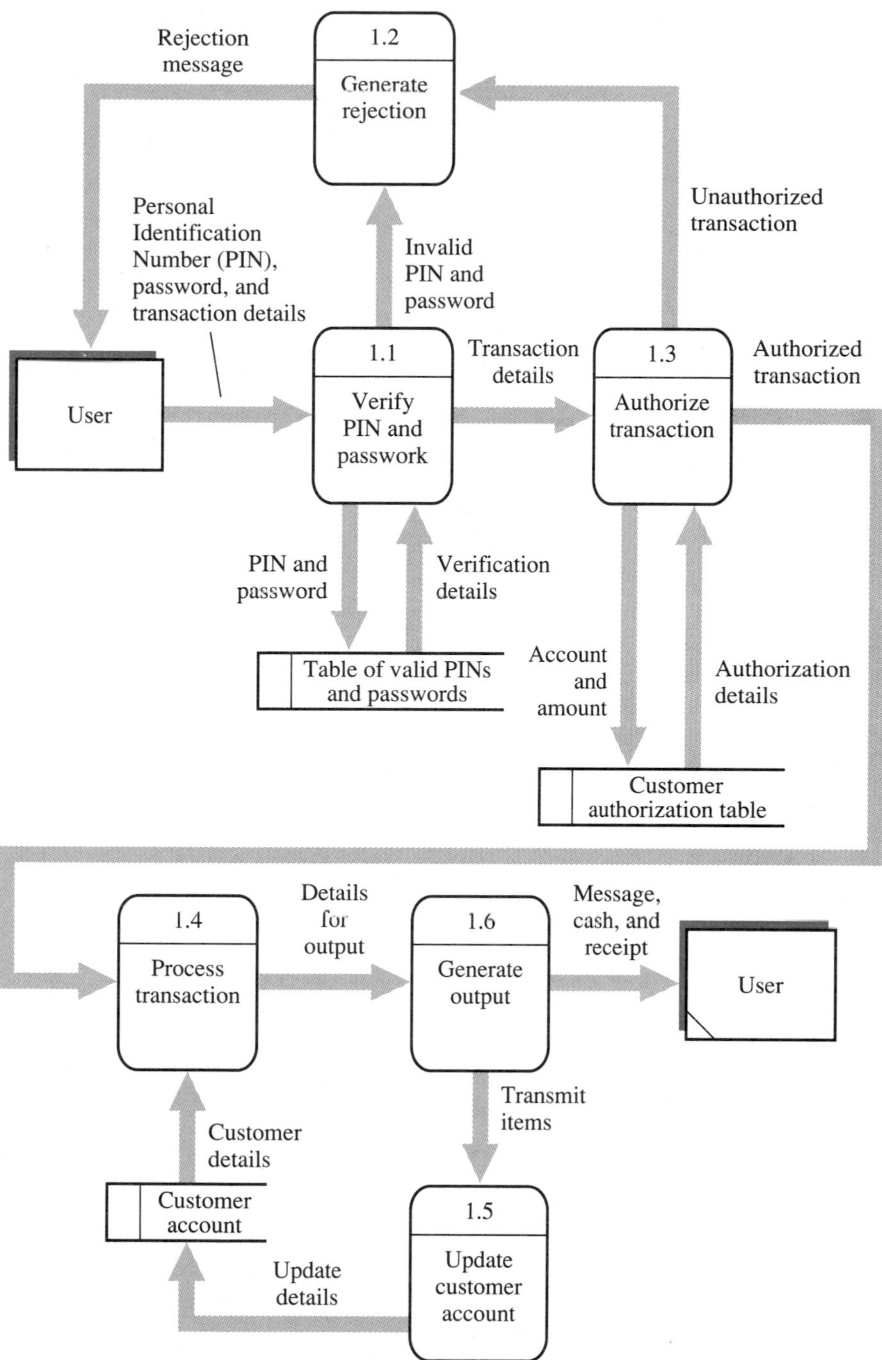

Creating a State Transition Diagram to Model Dynamic Design

Because the ATM cash dispensing processes specified in the preceding DFD must respond within a very brief time, perhaps only a few microseconds, the STD is used to model the dynamic behavior. As we learned in Chapter 2, rectangular boxes represent states that a system can be in. The arrows connecting the boxes show the change from one state to another, that is, state transition. Associated with each state transition is one or more condition (C), that is, the event that causes the state transition and actions (A). The action is a response, output, or activity that takes place as part of the change of state. Figure 10.6 displays a STD that provides a dynamic (i.e., a time-dependent) model of the ATM cash dispensing system.

Using Process Specification Models for Real-Time Processes

The DFD describes *what* the ATM system does. The STD describes *when* it is done. Process specification models describe **ELEMENTARY PROCESSES**, which are the most basic type of process. Elementary processes are detailed definitions of *how* something should be done to meet user requirements. As we know from Chapter 2, how processes are performed can be expressed by using one or a combination of process specification models, such as decision tables, decision trees, equations, and structured English. Also, process action diagrams and elementary process/entity matrices are excellent tools that can be used to define elementary processes.

For example, let's take the 1.3 Authorize Transaction process from Figure 10.5 and specify this process in structured English. Such process specifications read like this:

```
IF AMOUNT REQUESTED EXCEEDS CREDIT LIMIT
   THEN
      REJECT TRANSACTION
      DISPLAY "REJECTED" ON ATM SCREEN
      DO NOT INITIATE PROCESS TRANSACTION
   ELSE
IF AMOUNT REQUESTED DOES NOT EXCEED CREDIT LIMIT
   THEN
      ACCEPT TRANSACTION
      DISPLAY "ACCEPTED" ON ATM SCREEN
      INITIATE PROCESS TRANSACTION
```

Process Action Diagrams

Some CASE systems provide **PROCESS ACTION DIAGRAMS** that show elementary processes and detail steps within processes for a specific application. They provide the detailed process logic on which code generation will be based.

The process action diagram displayed in Figure 10.7 specifies the elementary processes and steps necessary to CREATE_STOCK_OF_PRODUCT, a process of an inventory control system. Examples of other processes are UPDATE_STOCK_OF_PRODUCT and DELETE_STOCK_OF_PRODUCT.

During development of the process action diagram, the CASE system asks pertinent questions based on the systems professional's knowledge of the in-

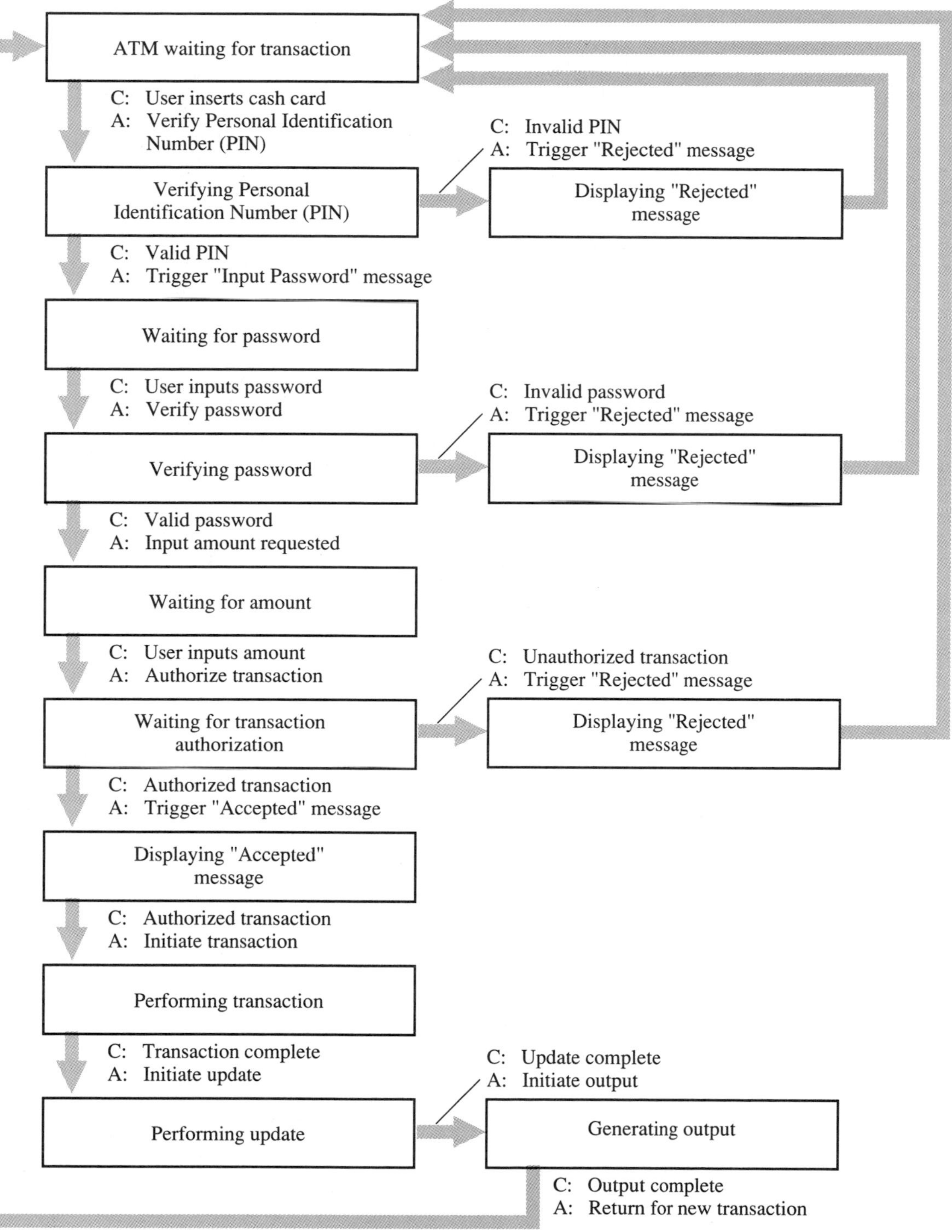

Figure 10.6 STD or dynamic design of an ATM cash dispensing program.

Figure 10.7
Process action diagram used to specify an inventory control system process.

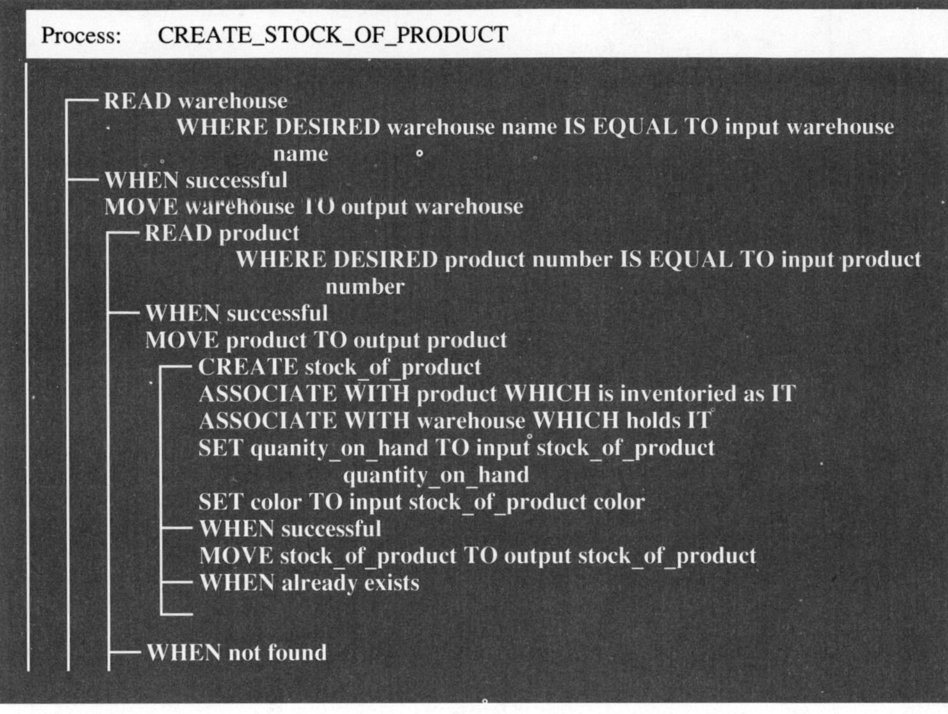

ventory control system. Basing its output on this dialogue, the CASE system includes the appropriate logic in the generated process action diagram.

Process: CREATE_STOCK_OF_PRODUCT
READ warehouse
WHERE DESIRED warehouse name IS EQUAL TO input warehouse name
WHEN successful
MOVE warehouse TO output warehouse
READ product
WHERE DESIRED product number IS EQUAL TO input product number
WHEN successful
MOVE product TO output product
CREATE stock_of_product
ASSOCIATE WITH product WHICH is inventoried as IT
ASSOCIATE WITH warehouse WHICH holds IT
SET quanity_on_hand TO input stock_of_product quantity_on_hand
SET color TO input stock_of_product color
WHEN successful
MOVE stock_of_product TO output stock_of_product
WHEN already exists
WHEN not found

Elementary Process/Entity Matrices

ELEMENTARY PROCESS/ENTITY MATRICES define the effects of elementary processes on entity types. They are excellent tools provided by some CASE systems that enable the systems professional to define elementary processes and to show how the processes relate to entities defined with an entity relationship diagram (ERD).

The elementary process/entity matrix, shown in Figure 10.8, describes some of the elementary processes of an order-entry system and the entities affected. This tool helps not only in designing processes, but also in designing databases (see Chapter 11).

Figure 10.8
Elementary process/entity matrix showing elementary processes of an order-entry system and how entities are affected.

	Entity				
Elementary Process	**Customer**	**Order**	**Order Line**	**Product**	**Stock of Product**
Receive order	U	C	C	U	
Check customer credit	U				
Check stock of product					R
Dispatch order		U			
Confirm order		U			
Update stock of product					U
Bill customer	U				

C = Create D = Delete U = Update R = Read only

DESIGNING BATCH PROCESSES

The STD is generally not used to model a batch process. On the other hand, the DFD is used extensively to model batch processes.

Creating a Data Flow Diagram to Model a Batch Process

The preceding ATM bank application design calls for DFD, STD, and process specification models. Processing monthly bank statements, on the other hand, can be designed using a DFD and process specification models only, because this kind of bank application is amenable to batch processing. Figure 10.9 illustrates such a design using a DFD.

Using Process Specification Models for Batch Processes

Process specification models are used for both real-time and batch process design. For example, 1.3 Calculate Service Charge requires a simple equation to specify how it is processed, such as

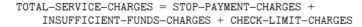

```
TOTAL-SERVICE-CHARGES = STOP-PAYMENT-CHARGES +
    INSUFFICIENT-FUNDS-CHARGES + CHECK-LIMIT-CHARGES
```

Figure 10.9
Batch process design for generating monthly bank statements.

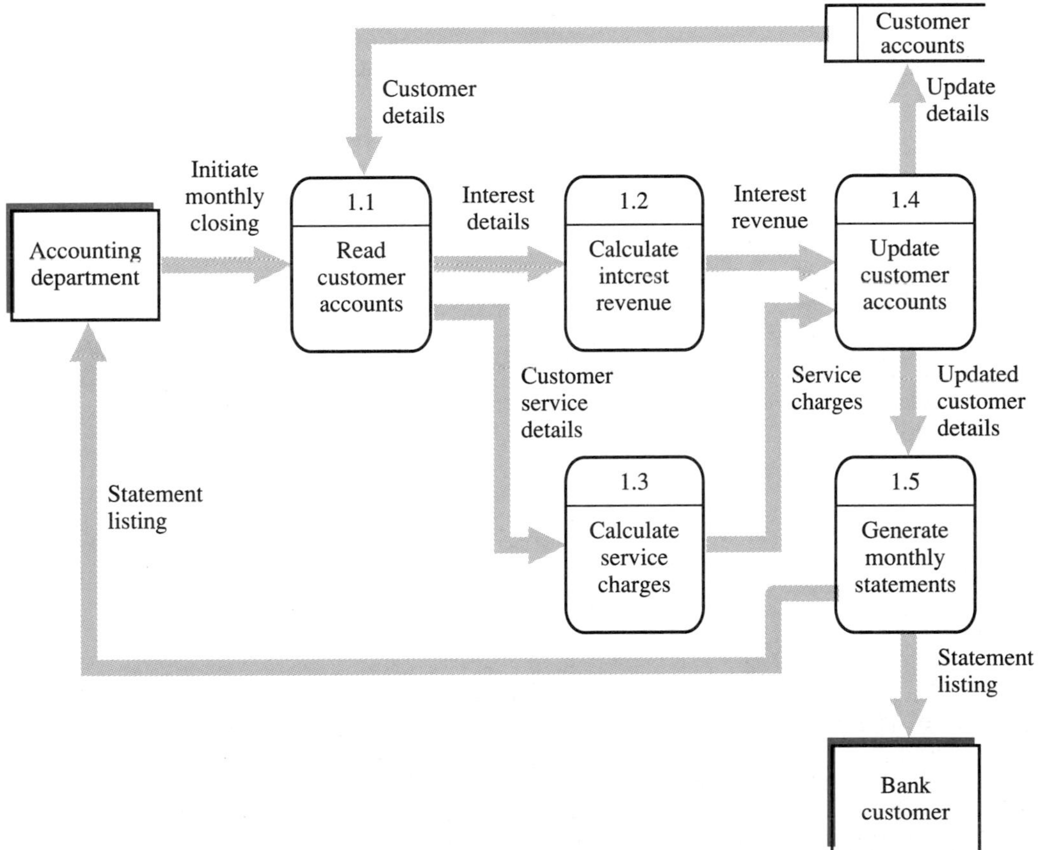

DEVELOPING PROCESS SPECIFICATIONS USING EQUATIONS

You were introduced to process specifications in Chapter 2. As you have seen in several chapters, process specifications have been developed using decision tables, decision trees, and structured English, as well as process action diagrams and elementary process/entity matrices. Equations also play a key role in developing process specifications for many user applications. In fact, most user requirements will include at least one equation of some kind. Following are some examples of the kinds of equations used by systems professionals to develop process specifications.

Transaction Equations

Some equations are nothing more than simple procedural models that describe how certain transactions are processed. In the following example, **TRANSACTION EQUATIONS** are integrated into structured English to describe how a sales order transaction is to be processed.

```
FOR EACH ITEM IN SALES-ORDER:
        COMPUTE ITEM-COST = ITEM-QUANTITY * ITEM-UNIT-PRICE
        ADD ITEM-COST TO ORDER-SUBTOTAL
  IF ITEM IS TAXABLE
     THEN
        COMPUTE SALES-TAX = ITEM-COST * SALES-TAX-RATE
        ADD SALES-TAX TO SALES-TAX-SUBTOTAL
        COMPUTE ORDER-TOTAL = ORDER-SUBTOTAL + SALES-TAX-SUBTOTAL
     ELSE
        ORDER-TOTAL = ORDER-SUBTOTAL
  ENDIF
ENDFOR
```

Basic Accounting Equations

BASIC ACCOUNTING EQUATIONS are used to prepare balance sheets, income statements, and measure performance. The fundamental characteristic of every balance sheet is that the total figure for assets always equals the total for liabilities and owners' equity. The equality of assets on the one hand and of the claims of the creditors and owners on the other hand is expressed in the equation

$$\text{assets} = \text{liabilities} + \text{owners' equity}$$

Other equations can be employed, such as

$$\text{debt ratio} = \frac{\text{total liabilities}}{\text{total assets}}$$

To determine net income, a business must measure for a given time period:

- The price of goods sold and services rendered to customers
- The cost of products and services used up

Accountants also state the following equation:

$$\text{income} = \text{sales} - \text{expenses}$$

Conversion of data to information may require more than one equation. For example, the average collection period, which is used to appraise accounts receivable, is computed in two steps:

1 Annual sales are divided by 360 days to get the average day's sales for the year.

2 Daily average sales are divided into the accounts receivable amount to find the number of days' sales "tied up" in accounts receivables.

The result is defined as the average collection period, because it represents the average length of time that the firm must wait after making a sale before receiving cash:

$$\text{daily average sales} = \frac{\text{annual sales}}{360}$$

$$\text{average collection period} = \frac{\text{accounts receivable}}{\text{daily average sales}}$$

Cost-Volume-Profit Equation

Costs can be divided into the two categories: variable and fixed. Variable costs react in direct proportion to changes in activity, but fixed costs remain the same within a specified range of activity.

By knowing cost behavior, we can simulate income that might be obtained with changes in the level of activity or volume by the following **COST-VOLUME-PROFIT EQUATION**:

$$I = (SP - VC)X - FC$$

After subtracting fixed costs (FC), income, I, is equal to the difference between the unit selling price, SP, and the unit variable cost, VC, times the number of units sold, X.

The cost-volume-profit equation provides a useful way to simulate the income factors of any organization. Three ways to increase income are:

- Increase selling price per unit.
- Decrease variable cost per unit.
- Increase sales volume.

A typical question from management may be: What would income be if we decreased selling price by 5 percent; variable cost remained constant; and we increased volume of sales by 20 percent? A variety of questions such as this, proposed by management, could generate information to enhance planning and decision making.

Budget and Performance Evaluation Equation

A budget is a plan of action, expressed in quantitative terms, that covers a specific time period. The key concept of a budget is structuring it in terms that equate to the responsibility of those charged with its execution. In this way, the budget is used not only as a planning device but also as a control device.

The **BUDGET AND PERFORMANCE EVALUATION EQUATION** used to determine budgeted performance over some time period is:

$$\text{budget variance} = \text{budgeted amount} - \text{actual amount}$$

Economic Order Quantity Equation

The **ECONOMIC ORDER QUANTITY EQUATION** is typically expressed as follows:

$$O = \frac{\sqrt{2QP}}{C}$$

where O is the order size in units, Q the annual quantity used in units, P the cost of placing one order, and C the annual cost of carrying one unit in stock.

In companies with a vast array of inventory items, the system can be designed to monitor and control this inventory. Reorder points can be set, and as inventory items are depleted, the system will automatically display when particular items have reached or exceeded their reorder points. In addition, the system can automatically display the optimum order size by processing an economic order quantity equation for each item in inventory.

Statistical Equations

To compute the average age of students attending Big Time University, add student ages and divide by the number of students. This process design calls for a simple **STATISTICAL EQUATION**, which is the arithmetic mean. The median age and the mode (i.e., the most frequently occurring age) can also be found using statistical equations. Many other statistical equations are available to systems professionals for process design. Two that help managers make special decisions are the **STRAIGHT-LINE EQUATION** and **EXPECTED-VALUE EQUATION**.

Straight-Line Equation

A straight line can be expressed in equation form as

$$Y = a = bX$$

with a as the fixed element and b as the degree of variability, or the slope of the line. From this basic equation and a given set of data, the estimated value of these data can be extrapolated. To fit the straight line to the data requires the solution of simultaneous equations, which are beyond the scope of this book.

Expected-Value Equation

Simply put, a probability means that there is some chance that a particular event will occur. Several events may be possible in a given situation. The sum of the probabilities of the events associated with a particular situation will always add to 1.0.

In some situations, decision makers must outline the range of events that in their judgment could happen, and assign to each event some probability of occurrence. The expected value is the sum of the probabilities of each event.

Assume that the Neptune Company, a manufacturer of ski boats, is trying to estimate the number of units of a new type of ski boat that will sell during the next fiscal period. The decision makers' approach should be to estimate the range of sales in units and the probabilities that such sales will actually materialize.

We will assume that the decision makers at Neptune have estimated, either through objective evidence gathered from various information sources, or from subjective evaluation, that a minimum of 1000 and a maximum of 3000 units will be sold. The probabilities that the decision makers have assigned to various levels of sales from a minimum of 1000 to a maximum of 3000 are given as follows:

Number of Units	Probability of Occurrence	Expected Value
1000	0.05	50
1500	0.15	225
2000	0.30	600
2500	0.35	875
3000	0.15	450
	1.00	
Expected number of units to be sold		2200

Therefore, management expects to sell about 2200 units of the new ski boat during the next fiscal period. Other budgetary, productive, and marketing decisions will be made based on the expected number of units to be sold.

The preceding equations represent just a few of those that systems professionals will use in developing process specifications. Many equations come from a broad range of disciplines, such as accounting, decision sciences, econometrics, and finance.

REVIEW OF CHAPTER LEARNING OBJECTIVES

The major goals of this chapter were to enable students to achieve three important learning objectives. We now summarize the responses to these learning objectives.

Learning objective 1:
Discuss the three dimensions of process design.

Companies operate differently and thus place different requirements on their information systems. Both internal and external users also have different requirements. This vast diversity of requirements causes systems professionals to consider at least three major dimensions of process design:

- Time
- Technology platform
- Modeling tools

If the process design calls for an interactive interface and transactions processed as they occur, the time dimension is real time. On the other hand, if the process design does not call for an interactive interface and transactions are accumulated over time, the time dimension is batch.

For real-time processing, the technology platform is a networked, online computer architecture. For batch processing, the technology platform is a stand-alone computer.

For real-time process design, both DFDs and STDs are used. For batch process design, a DFD is indicated. Both real-time and batch process designs require the application of process specification modeling tools to describe how the designs work.

Learning objective 2:
Describe real-time process design, batch process design, and the use of process specification models.

A real-time process should be designed statically, using a DFD, and dynamically, using a STD. A batch design, because it is essentially static, can be designed using a DFD. Again, both real-time and batch process designs must be described as to how they will work by using process specification modeling tools such as decision tables, decision trees, structured English, process action diagrams, and elementary process/entity matrices.

Learning objective 3:
Explain the use of equations in developing process specifications.

What is our return on total assets? What will be our income if we increase our selling price by 3 percent, decrease variable costs by 6 percent, and increase volume of sales by 9 percent? What's the economic order quantity for item C? These questions represent the users' information requirements that are met by integrating equations into process designs.

PROCESS DESIGN CHECKLIST

Following is a checklist relative to process design. Its purpose is to review the dimensions of process design and describe the use of modeling tools for process design.

1 Determine the time dimension of the process. Does the application call for real-time processing or batch processing?

2 If the process possesses real-time characteristics, use a DFD to describe it statically and a STD to describe it dynamically. Use process specification modeling tools to describe how it works.

3 If the process possesses batch characteristics, use a DFD to describe what it does and use process specification modeling tools to show how it works.

4 Use equations to specify certain processes.

KEY TERMS

Basic accounting equations

Batch processing

Budget and performance evaluation equation

Cost-volume-profit equation

Economic order quantity equation

Elementary process/entity matrices

Elementary processes

Expected-value equation

Process action diagrams

Process design

Real-time processing

Statistical equation

Straight-line equation

Transaction equations

REVIEW QUESTIONS

10.1 Name the three dimensions of process design and give a brief definition of each.

10.2 Define real-time process design, and give a brief example.

10.3 Explain why a real-time process is designed by using both DFDs and STDs.

10.4 Define batch process design, and give a brief example.

10.5 Explain why a STD is normally not used in batch process design.

10.6 What's the purpose of process specification modeling tools used in both real-time and batch process design?

10.7 What's the purpose of equations in process design?

CHAPTER-SPECIFIC PROBLEMS

These problems require exact responses based directly on concepts and techniques presented in the text.

10.8 Magna Manufacturing produces a large variety of oilfield and mining equipment. The total business is composed of six warehouses scattered throughout the southwest, with one located in Alaska. To serve its many customers, Magna must be able to provide accurate and current inventory information. Management at Magna must know immediately about changes in rig counts in the oilfield and the opening of new mining sites.

Required: Based on the time dimension, outline a process design, using Figure 10.3 as your guide, appropriate for Magna. Explain why you chose this particular design.

10.9 A state agency maintains two master files on two business categories: industrial master file and retail master file. During each month, inspectors gather a variety of data concerning the impact of these businesses on the environment. At the end of each month, data in the inspection documents are keyed into a transaction file. The state agency is allowed one week after the first of each month to convert the raw inspection data into environmental impact reports for the federal environmental protection agency (EPA).

Required: Based on the time dimension, outline a process design, using Figure 10.4 as your guide, appropriate for the system above. Defend your design.

10.10 Procedures for handling payroll at Zeus Construction Company are as follows. Salaried employees are paid a base salary whether they work less than, equal to, or greater than 40 hours. If they work less than 40 hours, however, an absentee report is generated. For hourly employees who work less than 40 hours, hourly wages are calculated and an absentee report is generated. If they work 40 hours, hourly wages are calculated. If they work greater than 40 hours, hourly wages and overtime wages are calculated.

Required: Prepare a decision table to specify the process described above. Then specify the process using structured English.

10.11 Pam Owen is working on an inventory control system for Santana Produce Company. She is trying to define procedures for changing order quantities in a clear and precise manner. The objective is to check recent sales activity, which can be fast or slow; the inventory level, which can be high or low; and whether the inventory item is perishable or not.

For perishable inventory with fast sales activity and high inventory level, no percentage change in order quantity is indicated. For perishable inventory items with low levels, order quantity is increased by 5 percent. For slow-selling perishable inventory items with high inventory levels, order quantity is decreased by 15 percent. For low-level perishable inventory, no change is indicated.

For nonperishable inventory items with fast sales activity and high inventory levels, increase the order quantity by 5 percent, but if the inventory levels are low, increase the order quantity by 15 percent. If, on the other hand, nonperishable inventory items have slow sales activity and the inventory levels are high, decrease the order quantity by 15 percent. But if the inventory levels are low, do not change the order quantity.

Required: Help Pam unscramble these written procedures by preparing a decision tree.

10.12 Item cost is calculated by multiplying the item quantity by the item unit price. The item cost is added to the order subtotal. A particular item may or may not be taxable, but if it is taxable, we have to compute the sales tax on that item. This calculation of taxes is achieved by multiplying the item cost by a sales tax rate. After the taxes have been computed, we add the sales tax on the item to the sales tax subtotal. Finally, we compute the total order by adding the order subtotal to the sales tax subtotal.

Required: Use structured English to make the foregoing procedures more concise.

10.13 This year's revenue (also called sales) for Miles Company is $10,000,000 with $8,000,000 in expenses. Miles has total assets of $40,000,000. The average inventory for this year is $5,000,000; the average receivables are $900,000.

Required: Use the proper accounting equations to compute the return on total assets for the board of directors of Miles Company. Also, for the managers, calculate inventory turnover, sales per day, and average collection period.

10.14 Big Crawler, a new tractor prototype that Centipede Company plans to manufacture, has an estimated selling price of $500,000 per unit with variable costs of $300,000 per unit and fixed costs of $100,000. You have encoded this cost-volume-profit accounting algorithm in FORTRAN so that management can ask what-if questions.

Required: Calculate the expected earnings for the following what-if questions: What if we sell 20 units? What if we sell 30 units and reduce variable costs to $200,000 per unit? What if we sell 10 units and variable costs are $300,000 and fixed costs are $600,000? (*Note:* Fixed costs are the same no matter how many tractors are sold. That is, variable costs relate to units, whereas fixed costs are the total amount for a period regardless of whether or not any units are sold.)

10.15 At Brakin Incorporated, the annual quantity of subassemblies used is 3000; it costs $10 to place one order of subassemblies; and it costs $0.80 to carry one subassembly in stock for one year.

Required: Compute the economic order quantity for subassemblies ordered by Brakin.

10.16 The cost of electricity to run Bama Company's machine is expressed in the following formula:

$$Y = \$95.00 + \$20.00X$$

where $20.00 is the variable rate for each 1000 machine hours of operating time and $95.00 is the fixed cost.

Required: It is estimated that Bama will expend 15,000 machine hours during March. Compute the estimated cost for electricity for March.

10.17 Managers at Barton Company have participated in estimating next year's sales of product X. In dollar sales, the range includes $50,000 at a 10 percent probability; $60,000 at a 15 percent probability; $70,000 at a 45 percent probability; and $80,000 at a 30 percent probability.

Required: Calculate the expected sales for product X.

THINK-TANK PROBLEMS

These problems call for a feasible approach rather than a precise solution. Although the problems are based on chapter material, extra reading and creativity may be required to develop workable solutions.

10.18 You have been assigned to help Telecommunications Network Trainers (TNT) to develop a seminar registration system. The registration part that you're working on can be in one of the following states: null, submitted, accepted, rejected, withdrawn, followed-up, and canceled. State transitions are: application received, seats available, seats not available, offer new place, transfer registration, customer cancels, sell new seminar, student fails to arrive/payment not received, follow up, and customer reapplies.

Required: Prepare a STD to show clearly and concisely the process described above.

10.19 The main path of control in dispensing cash at an automatic teller machine (ATM) is: reading a card, querying the user for transaction information, processing the transaction, dispensing cash, printing a receipt, and ejecting the card. Alternative flows of control occur if the customer wants to process more than one transaction or if the password is incorrect and the customer is asked to try again.

Required: Develop a STD and structured English that model this real-time process.

SUGGESTED READING

Arcidiacono, Tom. "Computerized Reasoning." *PC Tech Journal,* May 1988.

Bidgoli, Hossein. "DSS Products Evaluation: An Integrated Framework." *Journal of Systems Management,* November 1989.

Brody, Alan. "The Experts." *Infoworld,* June 19, 1989.

Burch, John, and Gary Grudnitski. *Information Systems: Theory and Practice,* 5th ed. New York: John Wiley, 1989.

Eckols, Steve. *How to Design and Develop Business Systems.* Fresno, Calif.: Mike Murach & Associates, 1983.

Krebs, Valdis. "Can Expert Systems Make Business Decisions?" *Information Strategy: The Executive's Journal,* Spring 1989.

Martin, James. "DSS Tools Help Build, Analyze Models to Make Decisions." *PC Week,* May 8, 1989.

Martin, James, and Carma McClure. *Diagramming Techniques for Analysts and Programmers*. Englewood Cliffs, N.J.: Prentice-Hall, 1985.

Muratore, John R., Troy A. Heindel, Terri B. Murphy, Arthur N. Rasmussen, and Robert Z. McFarland. "Acquisition at Mission Control." *Communications of the ACM,* Vol. 33, No. 12, December 1990.

Rumbaugh, James, Michael Blaha, William Premerlani, Frederick Eddy, and William Lorensen. *Object-Oriented Modeling and Design*. Englewood Cliffs, N.J.: Prentice-Hall, 1991.

Shepard, Susan. "Sophisticated Expert." *PC Tech Journal,* July 1988.

Teague, Lavette C., Jr., and Christopher W. Pidgeon. *Structured Analysis Methods for Computer Information Systems*. Chicago: Science Research Associates, 1985.

Umbaugh, Robert E., Editor. *The Handbook of MIS Management,* 2nd ed. Pennsauken, N.J.: Auerbach Publishers, 1988.

JOCS CASE: Designing the Processes

In Chapter 9, Carla Mills and Cory Bassett were assigned the task of designing input screens for the data that will be the responsibility of the accounting area within the finance department. Carla and Cory form a well-rounded team for this task. Carla's knowledge of accounting and her understanding of the procedures used by Peerless are a good balance to the instruction that Cory received in college about computer information systems. The combination of application expertise with computer knowledge is especially valuable when designing a new system.

While they are designing the input screens, they must also design some of the processes that will turn the input into the output designed earlier. Output, input, and processing are highly interrelated and frequently are designed together. These three components form the basis for any given system. They often dictate the rest of the systems design.

During output and input design, the SWAT team determined that the accounting area will be responsible for entering overall job-related data, such as the job order number, job description, date received, and date due into the computer. In addition, the accounting area maintains accounting standard data such as the budgeted labor hours, material dollars, and overhead application rate.

One of the first dimensions of process design that Carla and Cory must decide upon is that of update time for these data. They have the choice of updating accounting-related data when it is entered into the system (also termed real-time processing), or updating these data after a predetermined time period has elapsed (also termed batch processing).

Job-order data are used for all other aspects of JOCS. These data must be entered and kept accurate for any other data to be entered, or for information such as variance display screens to be created. Data in JOCS will be entered by a variety of personnel in departments at unknown intervals. For example, job-order data will be entered in the accounting area. These data must be in the system before direct material can be charged against that job. Direct material data, on the other hand, will be entered at unpredictable intervals from the CIM system. At any time, someone in the manufacturing department may access a display screen and may want to look at the status of a given job. The status of that job will be a reflection of data entered from those two different areas.

Since Peerless is relying on JOCS to help provide better customer service, better job-order information, and immediate access to data for early warning about the status of a given job, Carla and Cory think that data should be updated in real-time. They think that coordinating data entry between departments would be difficult and might end up defeating the objective of immediate access to JOCS information. If they tell the manufacturing department to access new jobs only on Monday of the week following a job starting in production, the manufacturing personnel may be unwilling to use the system in all situations. In such cases, people in the manufacturing department might not remember exactly when current data are available from the system. Since they wouldn't be sure when the data were accurate, they might not learn to depend on JOCS data.

Carla and Cory also think that the information generated from the system could help prevent problems if update was conducted in real-time. For example, if they tell the cost accountants that labor variances are accurate only on the Wednesday a week after the labor hours were incurred, the cost accountants may not be able to alert the supervisor in manufacturing in enough time to do something about potential cost overruns. To satisfy the need for timely, accurate data, Carla and Cory have decided that whenever a job order is entered in the accounting area, that job order should be available, within normal computer processing time, to all other users of JOCS.

Carla and Cory continue designing the accounting-related processes for JOCS by defining both the computer software and the people procedures. Figure 10.10 is an example of the job-order-entry procedure for an accounting clerk. This is a very general procedure outlining the overall process that will be followed by an accounting clerk to enter a new job order into the JOCS system.

Figure 10.11 is an example of the computer software procedure that will be used to enter job orders. Nowadays, detailed procedures to determine such things as job overhead application rate, or even a job-order number, are performed by computer software rather than by people. In the examples provided in Figures 10.10 and 10.11, an accounting clerk determines the job-order number, but the computer determines whether or not this number is already in the computer. Marketing and accounting management at Peerless are concerned about the structure of the job-order number changing over the next few years because the company plans to offer quite a few new machine styles. Since it would be easier to change a people procedure than a computer software procedure in JOCS, management decides to have a person determine the job-order number until the procedure becomes more constant. The

Figure 10.10
General job-order-entry procedure for clerical accounting personnel.

Title: Job-Order Entry Effective Date: 07-12-91

Instructions: ______________________________

Accounting Clerk: ______________________________

1. Receive green copy of the customer order from the marketing department.
2. Assign a job-order number in the following format:
   CC-TT-NNNN
   CC = Lifter Classification:
   Class A Lifter - 01
   Class B Lifter - 05
   Class C Lifter - 10
   TT = Lifter Type
   Hydraulic Light - 01
   Hydraulic Heavy - 02
   NNNN = numerical sequence starting with last two digits of the current year; for example, the third lifter produced in 1991 would be 9103.
3. Call production supervisor (has yellow copy of customer order) for job start date.
4. Contact cost accountant (has blue copy of customer order) for budget data:
   Direct Labor Regular Hours
   Direct Labor Overtime Hours
   Direct Material Dollars
   Direct Labor Total Dollars
5. Enter job-order data into JOCS: Job-Order Number, Job Description, Customer Number, Date Received, Date Due, Start Date, Quantity Ordered, Budgeted Regular and Overtime Hours, Budgeted Material, and Labor Dollars.

computer software procedure uses the job-order number to decide whether the transaction to be performed should add a new job order or change/delete an existing job order.

During process design, Carla and Cory also identify the accounting equations that will be used to generate the output designed in Chapter 8. The most important output from JOCS for the finance department concerns labor, materials, and overhead variances. In Chapter 8, both printer-based reports and screen-based displays were designed to highlight information about these variances. The accounting equations used to produce variance information must be isolated and written in a manner that will make it possible for translation into computer software when appropriate during SDLC. Each equation is devised and placed into the process design documentation compiled by the SWAT team.

Some of the variances determined by Carla and Cory are shown below.

Direct Materials Variances

$$\text{Price} = (\text{actual price} - \text{standard price}) \times \text{actual quantity}$$
$$\text{Quantity} = (\text{actual quantity} - \text{standard quantity}) \times \text{standard price}$$

Figure 10.11
Computer software procedure to perform job-order entry in structured English.

```
Enter a job-order number
Do while job-order number not = spaces
    Validate job-order number
    If valid job-order number
        Read database with job-order number
        If job order number in database
            Display job order
            Display change or delete request type
            If change
                Process change transaction
            Else
            If delete
                Process delete transaction
            Endif
            Endif
        Else
            Process add transaction
        Endif
        If key hit to terminate signifies a correct transaction
            Lock database
            Update record
            Unlock database
        Else
            Clear transaction area
        Endif
    Else
        Display message indicating incorrect job-order number
    Endif
    Enter a job-order number
Enddo
Transfer control to menu
```

Direct Labor Variances

$$\text{Rate} = (\text{actual rate} - \text{standard rate}) \times \text{actual hours}$$
$$\text{Efficiency} = (\text{actual hours} - \text{standard hours}) \times \text{standard rate}$$

Carla showed Cory how to devise the accounting equations shown above. Cory was unfamiliar with cost accounting variances. After Carla showed her how these variances were used to compare real costs incurred by a company to the expected costs, the equations above began to make sense to Cory. Cory had a much better understanding of how these equations fit into the entire picture of a job-costing information system after she helped to complete the process design for JOCS.

Chapter 11

Designing the Systems Database

WHAT WILL YOU LEARN IN THIS CHAPTER?

After studying this chapter, you should be able to:

1 Describe the properties of the relational database model.
2 List and explain the steps necessary to perform the relational database design process.
3 Explain the purpose of normalization and the first three normal forms.
4 Differentiate between the traditional file system approach and the database management system (DBMS) approach.

INTRODUCTION

In this chapter we discuss database design with emphasis on the relational database model. At this point in SDLC (see Figure 11.1), user requirements are well defined, or as much so as possible. Detailed design of output, input, and process are well defined. With these user requirements as input, the detailed database design process can be performed with a high degree of confidence that it will be complete and successful.

Learning the material in this chapter will not make you an expert on all topics of database design. Nor will you be able, even after closely studying this chapter, to design the comprehensive data requirements for a large system. It would be impossible to present enough material in a single chapter to accomplish these goals. (If you are interested in this subject, there are complete courses offered in the area of data structures, database design models, and database management systems.)

A CASE central repository (also called a project dictionary, knowledge base, dictionary, or central encyclopedia, depending on which CASE system one is referring to), as introduced in Chapter 3, provides many elements that help in database design. For example, the kinds of reports and screens described earlier will influence database design. Report and screen designs are integrated in the central repository along with their underlying data definitions. Also, entity relationship diagrams, data flow diagrams, and elementary process/entity matrices will aid database design. These models may already be stored in the central repository, checked for accuracy, completeness, and consistency. Thus systems professionals who have access to such CASE central repositories are well on their way to designing the database.

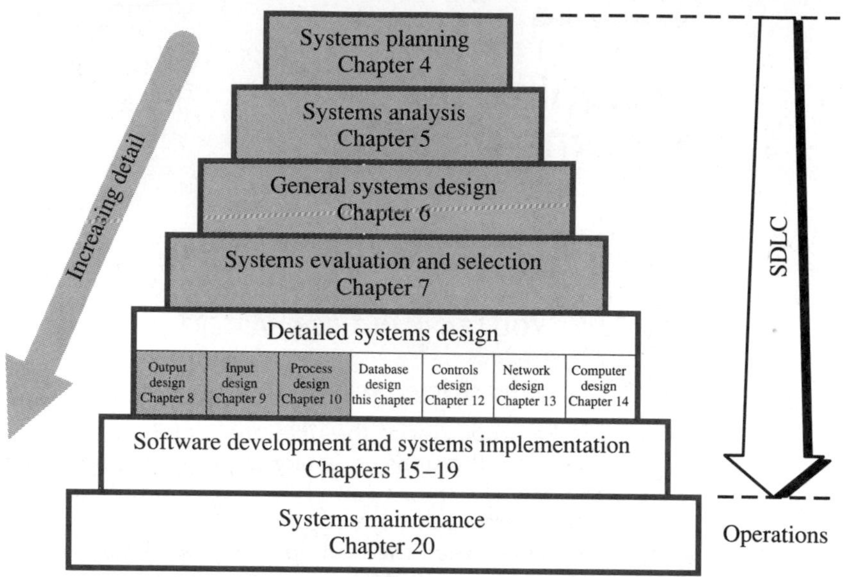

Figure 11.1
SDLC phases and their related chapters in this book. In Chapter 11 we look in detail at the design of a systems database.

OVERVIEW OF DATABASE MANAGEMENT SYSTEMS

In this chapter we study **DATABASE DESIGN**, that is, the process of determining the content and arrangement of data needed to support various systems designs. Unless user requirements are extraordinarily simple, database design occurs at two levels. At the first level, systems planning, analysis, and general design are performed to establish user requirements. This level of database design involves the front-end phases, independent of any specific database design or database management system (DBMS). At the second level (the one we're involved with in this chapter), general designs, such as high-level entity relationship diagrams, are transformed (or decomposed) into a detailed database design for a specific DBMS that will be used for implementation of the total system.

Three well-known database models are:

- Hierarchical model
- Network model
- Relational model

In the past, a large number of vendors offered **DATABASE MANAGEMENT SYSTEMS (DBMSs)** based on both **HIERARCHICAL MODELS** and **NETWORK MODELS**. These models were good for storing data, but they provided poor facilities for retrieving it. Today, the **RELATIONAL MODEL** is dominant. Most database software vendors, therefore, offer **RELATIONAL DATABASE MANAGEMENT SYSTEMS (RDBMSs)** software products. We use the relational database model in this chapter. We also present the traditional file system approach and compare it with the DBMS approach, especially the RDBMS.

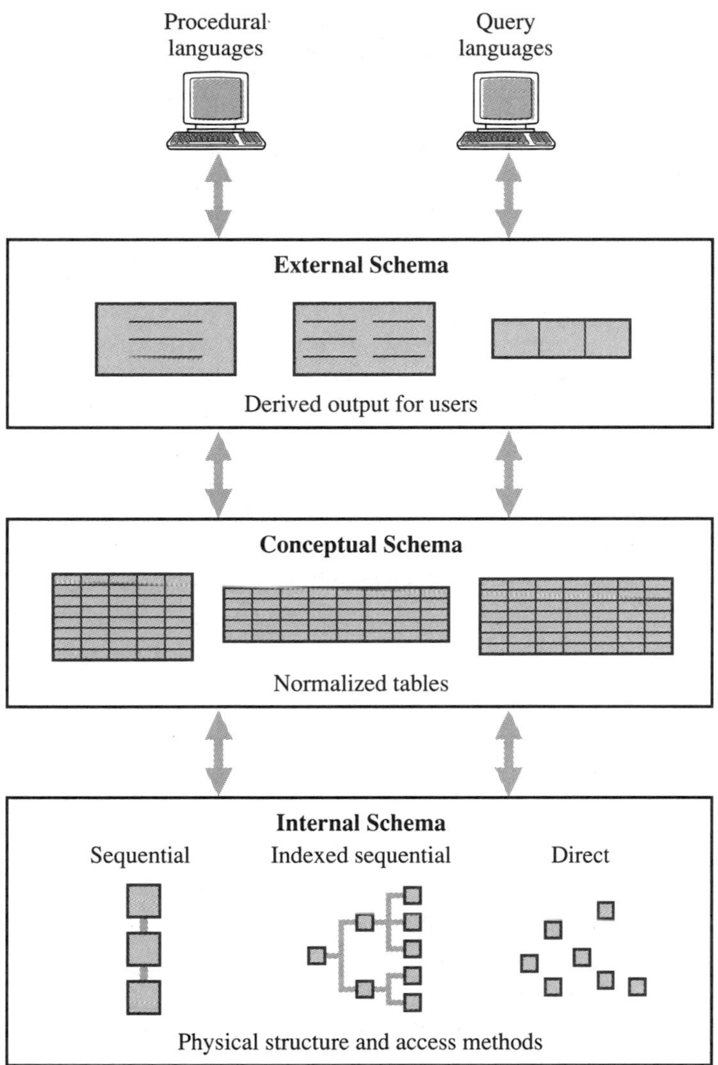

Figure 11.2
Three-schema structure of a RDBMS.

A RDBMS is built on the following three-schema structure, as illustrated in Figure 11.2. The schemata are:

- External
- Conceptual
- Internal

This layered structure defines the enterprise's data at three distinct levels. The external schema defines how users access and view the output from the RDBMS, independent of how the data are physically stored or accessed. Such access and manipulation is performed by users who employ procedural languages, such as COBOL, or query languages, such as **STRUCTURED QUERY LANGUAGE**

(SQL), the de facto standard language for a RDBMS. In some instances, a query language may be embedded in a procedural language to facilitate **ONLINE TRANSACTION PROCESSING (OLTP)** and similar applications. We will discuss query languages later in this chapter.

The conceptual schema defines the relational database model. It consists of a set of normalized tables. We discuss tables and normalization in the next sections. The conceptual schema is the design of the database, which is the main subject of this chapter.

The internal schema consists of the physical organization of the data (e.g., sequential, indexed sequential, and direct) in terms of physical data structures and access methods of the computer's operating system. The internal schema is usually covered in computer hardware courses.

WHAT IS A RELATIONAL DATABASE?

The relational model is based on the set theory of mathematics.[1] The structure is defined by **TABLES**. Tables are called **RELATIONS** in mathematical terms. Systems professionals often use the terms "tables" and "relations" interchangeably.

Properties of Tables

Each table in the relational model is composed of rows and columns. A column is called an **ATTRIBUTE**. A **DOMAIN** is a pool from which the values for an attribute must be chosen. Because multiple columns in the same table can be defined over the same domain, attribute names are defined for each column. Each attribute name in a relation must be unique.[2] The left-to-right order of the columns is immaterial. The order of the rows is immaterial. The intersection of any row and column contains a single value.

If there are 10,000 students to be represented in a relational database, the STUDENT table will contain 10,000 rows. If it requires 50 attributes, such as Student_Number, Student_Name, Department_Number, Classification, and so on, to describe students, the STUDENT table will contain 50 columns.

The formal definition of a table introduces properties of the table. First, duplicate rows are not permitted. To enforce this property, there must be at least one attribute or combination of attributes that uniquely identifies each row of a table. The attribute or combination of attributes that performs this task is called a **PRIMARY KEY (PK)**. Student_Number, for example, is a primary key that identifies each student uniquely.

Because PK values are the only way for users to address individual rows in a relational database, a PK with duplicate or missing values makes no sense.[3] How would you know if you've accessed the correct student if the Student_Number, say 434559764, accessed three rows, or three different stu-

---

[1] Cecelia Bellomo, "Relational Database Systems: Data Integrity," *Interact*, October 1990, p. 107.

[2] Ibid.

[3] Fabian Pascal, "Preventing Corruption," *DBMS*, September 1989, p. 44.

dents? And how could you find a specific row (or student) with no identifying value? Thus the second property of relational databases is that PK values should not have duplicates or be **NULL** (i.e., value unknown). Duplicate students make no sense, nor do unidentifiable ones.

The third property defines a **RELATIONSHIP** between two tables. If Table R2 has a **FOREIGN KEY (FK)** matching the primary key (PK) of Table R1, then for every value of FK there must be a matching value of PK, or the value of FK must be null. For example, assume an attribute Department_Number in the STUDENT table column and also in a DEPARTMENT table column. In the DEPARTMENT table, Department_Number is a primary key (PK) that uniquely identifies rows. In STUDENT, Department_Number values represent students assigned to departments (e.g., mechanical engineering, computer information systems, accounting). The Department_Number values in STUDENT refer to the departments to which a student is assigned; that is, each one is a value in the STUDENT table that references rows that uniquely identify departments in the DEPARTMENT table. Such columns, whose values point to primary key (PK) values, are foreign keys (FKs). Clearly, each FK value must match up with a value in the corresponding PK column. Otherwise, nonsensical operations could occur, such as a student being assigned to a department that doesn't exist. Thus the rule: FK values either reference existing PK values, or they are null.

The relationship between STUDENT and DEPARTMENT tables is demonstrated in Figure 11.3. Department_Number, a foreign key (FK) in STUDENT, references Department_Number, a primary key (PK) in DEPARTMENT. Smith and Jones are assigned to CIS; Brown and Burns are assigned to Accounting; Adams is assigned to English; and Atkins is assigned to Mechanical Engineering.

Relationships between tables can be one-to-one (1 : 1), one-to-many (1 : M), or many-to-many (M : N). The relationship between DEPARTMENT and STUDENT is one-to-many; that is, a DEPARTMENT can have many students assigned to it.

The final property of a relational database deals with data manipulation. Data manipulation defines a set of commands necessary to create tables and their data domains; select, project, and join tables; and insert, delete, and update data.

Using Structured Query Language

SQL is a standard database language used to query, manipulate, and update RDBMSs. As more and more organizations decide to consolidate their databases into enterprisewide systems, knowledge of SQL will become a requirement for database designers.[4]

Sample Tables

Four sample tables are presented in Figure 11.4 to illustrate the foregoing properties of tables and the application of SQL. The primary keys (PKs) for

---

[4] Rob Gerritson, "SQL Tutorial," *DBMS*, February 1991, p. 44.

Figure 11.3
FK to PK references.

STUDENT			
Student number	Student name	Department number	Classification
PK		FK	
437891240	Smith	145	Sophomore
434559764	Jones	145	Junior
392771428	Brown	174	Senior
392845210	Adams	136	Junior
422448914	Atkins	149	Junior
444984210	Burns	174	Sophomore

DEPARTMENT		
Department number	Department name	Department address
PK		
149	Mechanical Engineering	Thomas Building
174	Accounting	Business Building
145	CIS	Business Building
136	English	Hogan Building

CUSTOMER and ITEM tables are Customer_Number and Item_Number, respectively. The INVOICE table has one row for each invoice identified by Invoice_Number, the primary key (PK). The Customer_Number in INVOICE is a foreign key (FK), because it refers to the PK in CUSTOMER table. It assigns invoices to the customers who are obligated to pay them. The LINE_ITEM table contains two FKs: Invoice_Number identifies the invoice, and Item_Number identifies the item on the line item of the invoice. For example, Invoice 2 might look like the one illustrated in Figure 11.5.

SQL CREATE Command

The first SQL command to install tables and their attributes into a RDBMS is CREATE TABLE. In Figure 11.6, the four sample tables are created and

Figure 11.4
Sample tables.

CUSTOMER		
Customer number	Name	City
PK		
106	Nolan	Seattle
150	Smith	Los Angeles
264	Jones	Dallas
180	Adam	Atlanta
120	Brown	New York
190	Norman	Boston

ITEM			
Item number	Type	Color	Price
PK			
10	Hat	White	20.00
11	Dress	Blue	80.00
12	Shoes	Brown	50.00
13	Pants	Red	60.00

INVOICE		
Invoice number	Customer number	Invoice date
PK	FK	
1	106	4-15-92
2	106	5-19-92
3	264	5-25-92
4	190	4-15-92

LINE_ITEM		
Invoice number	Item number	Quantity
PK/FK	PK/FK	
2	10	10
1	11	5
2	13	6
4	13	4
3	10	12

Figure 11.5
Sample invoice composed of data accessed from the four sample tables.

INVOICE					
To: Nolan Seattle			Item number: Invoice date:		2 5-19-92
Item number	Type	Color	Quantity	Price	Extension
10	Hat	White	10	20.00	200.00
13	Pants	Red	6	60.00	360.00
				Total	560.00

defined under the control of a RDBMS. (Once the tables have been defined, some CASE systems will automatically generate the appropriate SQL CREATE commands.)

The tables are set up or declared by the CREATE TABLE command followed by a table name. Next come one or more instances of column name (or attribute name) and data type enclosed in parentheses. Each attribute after the first is preceded by a comma and optionally followed by NOT NULL. The last attribute (or column) in the table is followed by a closing parenthesis. The American National Standards Institute (ANSI) Database Committee definition of CREATE TABLE states that if a column does not contain either a NULL or NOT NULL specification, NULL is assumed. Some vendors of RDBMS, using their own versions of SQL, state that if a column does not contain either a

Figure 11.6
SQL CREATE commands used to set up tables in the RDBMS.

```
CREATE TABLE Customer
(Customer_Number        CHAR(3) NOT NULL,
 Name                   CHAR(10),
 City                   CHAR(15));

CREATE TABLE Item
(Item_Number            CHAR(2) NOT NULL,
 Type                   CHAR(10),
 Color                  CHAR(10),
 Price                  DECIMAL(6,2));

CREATE TABLE Invoice
(Invoice_Number         CHAR(2) NOT NULL,
 Customer_Number        CHAR(3) NOT NULL,
 Invoice_Date           DATE));

CREATE TABLE Line_Item
(Invoice_Number         CHAR(2) NOT NULL,
 Item_Number            CHAR(2) NOT NULL,
 Quantity               NUMERIC));
```

NULL or NOT NULL specification, NOT NULL is assumed. Thus specifying NOT NULL for primary keys (PKs), for example, is not necessary.

Domain specification is addressed only by the definition of valid data types. The following are the valid data types defined by ANSI:

```
CHARACTER (or CHAR)
INTEGER (or INT)
DECIMAL (or DEC)
SMALLINT
NUMERIC
FLOAT
REAL
DOUBLE PRECISION
```

Some vendors provide additional data types for their RDBMS. Examples are:

VARCHAR	Variable-length character field
TEXT	Variable-length columns of printable characters
MONEY	Stores dollars and cents
DATETIME	Holds dates and time of day
TIMESTAMP	Every time a row containing this column is inserted or updated, the column is updated automatically.

The domain DATETIME, for example, contains all the valid date and time combinations between midnight January 1, 1753 and December 31, 9999, inclusive. It does not allow invalid dates or times, such as February 30, 1992 at 14:70 P.M.

Query Commands

The core SQL query or search commands are:

```
SELECT (attribute 1, attribute 2,  ..., attribute n)
FROM (relation 1, relation 2,  ..., relation n)
WHERE (predicate)
```

Some typical queries and the SQL query commands required to answer them are demonstrated in Figure 11.7. CASE code generators will automatically generate SQL query commands.

A user can select values from one table based on values in another by using a subselect, which is always enclosed in parentheses.[5] For example, to list the invoices for Nolan, the user writes

```
SELECT Invoice_Number, Invoice_Date FROM Invoice
   WHERE Customer_Number IN (Select Customer_Number
             FROM Customer
          WHERE Name = 'Nolan');
Answer:   Invoice_Number
             1
             2
```

[5] Ibid., p. 49.

Figure 11.7
Examples of SQL query commands.

Query: Display the colors that begin with B.

```
SELECT Color FROM Item
     WHERE Color LIKE 'B%';
```

Note: The % symbol denotes wild card.

Query: List the number, name, and city of all customers.

```
SELECT Customer_Number, Name, City
     FROM Customer;
```

Query: Give a complete list of items.

```
SELECT * FROM Item;
```

Note: The asterisk accesses all attributes.

Query: List all customers, ordered by name.

```
SELECT Customer_Number, Name
     FROM Customer
     ORDER by Name;
```

Query: How many shoes do we have?

```
SELECT COUNT *
     FROM Item
     WHERE Type = 'Shoes';
```

Multitable queries are made by a join condition. The number of join conditions is always one less than the number of tables. For three tables, there should be two join conditions. For example:

```
SELECT * FROM Customer, Invoice
WHERE Customer.Customer_Number =
       Invoice.Customer_Number;
```

This query produces the following results:

Customer_Number	Name	City	Invoice_Number	Customer_Number	Invoice_Date
106	Nolan	Seattle	1	106	4-15-92
164	Jones	Dallas	3	164	5-19-92
106	Nolan	Seattle	2	106	5-25-92
190	Norman	Boston	4	190	4-15-92

The asterisk refers to all of the columns in the referenced tables, in this case, CUSTOMER and INVOICE tables. When the two columns have the same name, which is Customer_Number in this example, it is necessary to qualify the name by preceding it with the table name and a period. Generally, a join condition involves both a primary key (PK) and a foreign key (FK).

Update Commands

SQL uses three commands to change the rows in a relational database: INSERT, UPDATE, and DELETE. CASE code generators automatically generate SQL update commands. For example, to insert a new customer in the database, all the user would have to do is type the variable data (i.e., 144, Kelly, Pittsburgh), as follows:

```
INSERT INTO Customer
     VALUES (144, 'Kelly,' 'Pittsburgh');
```

This command adds a new customer to the CUSTOMER table. Notice that the order of data for the new customer follows the same order of Customer_Number, Name, and City attributes in the CUSTOMER table.

This command increases the price of Item_Number 11 Dress by 20 percent:

```
UPDATE     Item
SET        Price = Price * 1.20
WHERE      Item_Number = 11;
```

This command deletes Customer_Number 120 Brown from the CUSTOMER table:

```
DELETE FROM Customer
WHERE Customer_Number = 120;
```

DESIGNING A RELATIONAL DATABASE

The database design process described next can be mapped to the relational database model.[6] It uses the entity relationship diagram and the interviewing process, both of which have been presented in previous chapters. The essence of this database modeling process is to capture the workings of an application from the words that emerge from users during interviewing. The best way to determine entities of an application is to discuss that application with people who will operate and use it.

Model the Entities and Map to Tables

ENTITIES correspond to nouns, relationships are verbs, and attributes are adjectives, adverbs, or prepositional phrases. Words that are provided by users to describe their requirements are in turn used to prepare a logical data model with an entity relationship diagram (ERD). This is a natural way to go through the design process, because the resulting model reflects exactly the way the enterprise operates and is managed. The process used to get from a verbal description to a database is what we will be discussing.

The enterprisewide modeling done in Chapter 4 provides a high-level ERD for general guidance. Further refinements of this model and other models were made during analysis and general design. Some CASE systems will automatically create a "first-cut" relational database design based on high-level ERDs and refine the database design as the high-level ERD is refined. Moreover, CASE will ensure consistency between the general design and the detailed design.

Perhaps what you first define early on as an entity is really not an entity but an attribute. As we have already stated, an entity will be revealed as a noun and

[6] Al Foster, "A New ERA for Data Modeling," *DBMS*, November 1990.

an attribute will generally be revealed as an adjective. For example, a user may say: ''A purchase order number is received from a customer.'' You may in turn define ''purchase order number'' and ''customer'' as entities. It is likely that ''customer'' is an entity, but it is unlikely that ''purchase order number'' is one. More than likely, ''purchase order number'' is an attribute of the ''purchase order'' entity. It is also likely that ''purchase order number'' will become a primary key for the ''purchase order'' entity.

Let's assume, however, that you have already defined the entities for handling customer orders using an ERD. Entities from this ERD are mapped into tables as shown in Figure 11.8.

Now you have a set of tables. But notice that the VENDOR table has been eliminated by drawing an X through it. After further review, it has been determined that this entity is not within the scope of the system under development, which is concerned exclusively with handling customer orders. Although vendors supply inventory items that are sold to customers, the VENDOR entity is not needed for this particular application.

Figure 11.8
Model of entities mapped to tables.

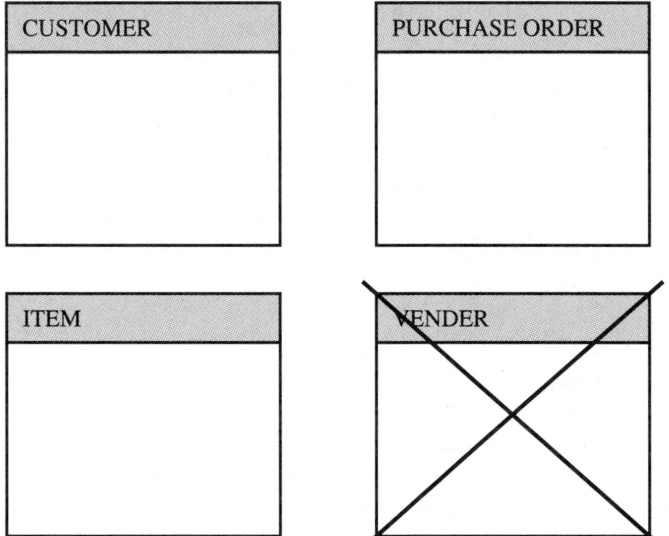

Designate Primary Keys

The next important step is to designate primary keys that uniquely identify data in each table. A primary key must serve as a unique identifier and must always be available. The tables defined in the first step now contain PKs as demonstrated in Figure 11.9.

Use of primary keys is nothing really new. Businesses were using keys (or codes) long before computers for uniquely identifying things such as customers, warehouses, invoices, trucks, and so on. The PKs in the example tables are represented by Customer_Number, Purchase_Order_Number, and Item_Number, along with sample data.

At this point, the database model is a collection of tables, each with one

Figure 11.9
Tables with primary keys (PKs).

CUSTOMER
Customer number
PK
0001 0002 0003 0004

PURCHASE ORDER
Purchase order number
PK
200 201 202 203

ITEM
Item number
PK
1000 1001 1002 1003

column, primary key, and sample data. One table has been eliminated because it was beyond the scope of application. But maybe the database model is still not complete. Maybe one or more entities have been overlooked. In our case, users, after further review and analysis, decide that a "carrier" entity should be included. (Carriers are the transportation companies that deliver purchased items to the customers.) So the database model is revised to include a CARRIER table, Carrier_Number as a PK, and sample data. The revised model is displayed in Figure 11.10.

Model Relationships Between Tables

When you feel somewhat assured that all entities for the system under development have been defined and mapped into tables with PKs, you are ready to go through every possible combination of tables and indicate any relationships between them. Remember that relationships are the verbs of the relational database model. For example: Customers (noun) submit (verb) purchase orders (noun).

After you determine that a direct relationship does exist between tables, determine what type of relationship it is. Is it one-to-one (1 : 1), one-to-many (1 : M), or many-to-many (M : N)? You can decide by asking two questions:

1 Can the entity described in Table A be linked to more than one of the entities described in Table B?

Figure 11.10
Revised model of tables.

CUSTOMER
Customer number
PK
0001 0002 0003 0004

PURCHASE ORDER
Purchase order number
PK
200 201 202 203

ITEM
Item number
PK
1000 1001 1002 1003

CARRIER
Carrier number
PK
100 101 102 103

2 Can the entity described in Table B be linked to more than one of the entities described in Table A?

If the answer to both questions is no, there is a one-to-one (1 : 1) relationship. If the answer to both questions is yes, there is a many-to-many (M : N) relationship. If the answer to one question is yes and to the other question is no, there is a one-to-many (1 : M) relationship.

In our example, there is a one-to-many (1 : M) relationship between CUSTOMER and PURCHASE ORDER, because one customer can have many purchase orders, but one purchase order cannot have many customers. The relationship is indicated by placing a foreign key (FK) in the PURCHASE ORDER table, as portrayed in Figure 11.11. The FK in the PURCHASE ORDER table matches the PK in the CUSTOMER table. In fact, a FK enforces the fact that a value for the FK must match a value for the PK of the other table. A one-to-many (1 : M) relationship also exists between PURCHASE ORDER and ITEM. There is a many-to-many (M : N) relationship between CUSTOMER and CARRIER. For this relationship, a new table, CUSTOMER/CARRIER, is defined. It contains a PK key that has been constructed from the two FKs. These two FKs match the PKs of the CUSTOMER and CARRIER tables that are involved in the many-to-many (M : N) relationship.

Figure 11.11
Model of entity relationships.

CUSTOMER
Customer number
PK
0001
0002
0003
0004

PURCHASE ORDER	
Purchase order number	Customer number
PK	FK
200	0001
201	0002
202	0003
203	0004

ITEM	
Item number	Purchase order number
PK	FK
1000	200
1001	201
1002	202
1003	203

CARRIER
Carrier number
PK
100
101
102
103

CUSTOMER/CARRIER	
Customer number	Carrier number
PK	PK
FK	FK
0001	100
0002	101
0003	102
0004	103

Model Attributes of Tables

An attribute is a named property of a table. These are the properties that users are interested in tracking, updating, and accessing. They're represented as additional columns in the tables. As already mentioned, attributes will typically be described by users during the interview process as adjectives, adverbs, or prepositional phrases. In Figure 11.12 we have added the attributes necessary to define each table fully. These attributes appear as named properties placed in columns of the tables.

Statements from users will serve as guides in discovering and describing attributes; for example:

CUSTOMER		
Customer number	Name	Address
PK		
0001	Cole	Boston
0002	Lear	Detroit
0003	Robbins	Atlanta
0004	Cook	Miami

PURCHASE ORDER			
Purchase order number	Customer number	Date	Dollar amount
PK	FK		
200	0001	4-12-92	700.00
201	0002	5-16-92	400.00
202	0003	5-17-92	595.00
203	0004	5-24-92	820.00

ITEM		
Item number	Purchase order number	Quantity in units
PK	FK	
1000	200	20
1001	201	10
1002	202	12
1003	203	16

CARRIER		
Carrier number	Name	Telephone number
PK		
100	Speedy	7024944852
101	Overnight	6014764444
102	Nationwide	2019876492
103	Overland	2128844025

CUSTOMER/CARRIER		
Customer number	Carrier number	Delivery date
PK	PK	
FK	FK	
0001	100	5-14-92
0002	101	5-17-92
0003	102	5-19-92
0004	103	6-02-92

Figure 11.12
Fully defined tables.

- "We need to have ready access to our carriers' telephone numbers."
- "We need the names of our customers."
- "We need to know when an order was delivered."

From these kinds of statements, attributes for all tables should be fully defined.

Normalize the Database Model

In the preceding design process, normalization was not consciously considered. The aim of the design process we have just gone through is to wind up with tables that are free of potential problems, sometimes called update, insert, and delete anomalies. Some CASE systems provide normalizing tools along with their database design tools.

Prepare a Data Dictionary

A **DATA DICTIONARY** is a central repository of the enterprise's data. Most RDBMS offer the data dictionary as an automated software tool. Most CASE systems also offer a data dictionary tool that automatically creates a complete data dictionary of the relational database.

A data dictionary has a number of uses, depending on the phase of the SDLC. Portions of the data dictionary may have been captured during enterprisewide modeling, systems analysis and general systems design, or during output, input, and process design. Here we assume that the data dictionary is an automated means for defining tables, attributes, and relationships of the RDBMS. The data dictionary includes such characteristics as an attribute's size, type, description, limits, ranges, security level, and access privileges. For example, the data dictionary output for an attribute named Purchase_Order_Number is shown in Figure 11.13. This output is only one of several ways to format data dictionary output.

Figure 11.13
Sample data dictionary output.

Attribute name	Length		Type	Where used
	Min.	Max.		
Purchase_Order_Number	3	3	Numeric	Purchase_Order
Description: A number that identifies each purchase order				

WHAT IS NORMALIZATION?

NORMALIZATION is a technique for optimizing relational database designs and freeing them from potential problems or anomalies. Normalization is also applicable to other database models. Simply put, it involves breaking data in tables down into smaller tables until each attribute in each table depends only on the key or keys within the table.

A poorly designed relational database design which is not normalized will cause problems for as long as the database is installed. At best, it will be inefficient to run and difficult to maintain. At worst, users will find out after the database is installed that it is not producing what is required or it is generating inaccurate results.

Using Normal Forms to Analyze the Database

The relational database model provides a number of powerful analytical techniques that offer valuable assistance in designing and optimizing relational databases. These techniques are **NORMAL FORMS**, and the process of applying them is normalization.[7] The normal forms in this chapter are:

- The first normal form
- The second normal form
- The third normal form

These forms provide the database designer guidance and point out problems or potential problems.

The First Normal Form

The **FIRST NORMAL FORM (1NF)** requires elimination of repeating attributes or groups of attributes from a relation. Typically, a violation of the first normal form is so obvious that anyone but a novice would detect it immediately. We use a purchasing application, shown in Figure 11.14, for our example. The PURCHASE_ORDER and ORDER_ITEM relations are in first normal form because they do not contain repeating attributes.

The Second Normal Form

A relational database design is in **SECOND NORMAL FORM (2NF)** when it is already in first normal form and all non-key attributes are dependent on the primary key.

Figure 11.14
First normal form.

PURCHASE_ORDER							
PO number	Order date	Vendor number	Vendor name	Vendor address	Shipping data	Billing data	Order total
PK							

ORDER_ITEM					
Item number	PO number	Item description	Unit of measure	Quality	Cost
PK	FK				

[7] Tom Kemm, "A Practical Guide to Normalization," *DBMS*, December 1989, pp. 46–52.

The goal of second normal form is to get rid of any partial functional dependencies. Partial functional dependencies can occur only with **CONCATENATED** (combination of keys) keys. Second normal form requires that all attributes in a table be dependent on the *entire* primary key expression rather than on any part thereof.

For example, Figure 11.14 shows the ORDER_ITEM with a concatenated key made up of Item_Number and PO_Number. Looking at the rest of the attributes in ORDER_ITEM, you will see that two of the attributes are not dependent on the whole key. Keeping in mind that the primary key for this table contains both the Item_Number and the PO_Number, you should note that the Item_Description and Unit_Of_Measure are probably identified completely by Item_Number alone. Think about Item_Description. The Item_Description, for this application, is associated with the Item_Number. The Item_Description will not be different if the PO_Number changes. Yet, the Item_Description will be different if the Item_Number changes. The Item_Description is functionally dependent only on the Item_Number. This means that the Item_Description is only partially functionally dependent on the concatenated key field. It must be removed from this table to achieve second normal form.

Sometimes an analyst has to do a bit of research to know which attributes are dependent on another. In this case we knew that the Item_Description is always the same regardless of which purchase order was used to buy that particular item. But we might have to ask the end users if this is also true of the attribute Unit_Of_Measure. For example, if we purchase sheet metal, what measurement do we use? Do we always buy sheet metal per foot? Or do we sometimes buy it per meter? If we sometimes buy the sheet metal per foot and other times per meter, do we need to know the particular purchase order to know the unit of measure? If the measurement unit is different, depending on the purchase order, the attribute Unit_Of_Measure must stay in ORDER_ITEM. If the measurement unit is the same regardless of the purchase order, the attribute Unit_Of_Measure will be moved to the ITEM table portrayed in Figure 11.15.

For this application, we are going to assume that the Unit_Of_Measure is always the same for a given item. Unit_Of_Measure is thus fully functionally dependent on the key field Item_Number. With this assumption in mind, Figure 11.15 removes the partial functional dependencies and depicts the relational database design in second normal form.

The Third Normal Form

Normalization to the **THIRD NORMAL FORM (3NF)** involves eliminating transitive dependencies, that is, the dependence of any non-key attribute on any other attributes except the primary key (PK).

For example, examine PURCHASE_ORDER shown in Figure 11.15. This table contains, among others, the attributes PO_Number, Vendor_Number, Vendor_Name, and Vendor_Address. The attributes Vendor_Name and Vendor_Address are not dependent on the primary key, PO_Number. Instead, these two attributes are uniquely and completely defined by the attribute Ven-

PURCHASE_ORDER							
PO number	Order date	Vendor number	Vendor name	Vendor address	Shipping data	Billing data	Order total
PK							

ORDER_ITEM			
Item number	PO number	Quantity	Cost
PK	FK		

ITEM		
Item number	Item description	Unit of measure
PK		

Figure 11.15
Second normal form.

dor_Number. To achieve third normal form, these two attributes must be separated from PURCHASE_ORDER and placed into another table. Figure 11.16 shows these attributes removed from PURCHASE_ORDER and placed into a VENDOR table. Figure 11.16 is now a relational database design for the purchasing function with the data in third normal form.

Some other normal forms that one could consider include:

- Fourth normal form
- Fifth normal form
- Boyce-Codd normal form
- Domain-key normal form

But normalization beyond the third normal form is rarely necessary.

Figure 11.16
Third normal form.

PURCHASE_ORDER				
PO number	Order date	Shipping data	Billing data	Order total
PK				

ORDER_ITEM			
PO number	Item number	Quantity	Cost
PK	PK/FK		

ITEM		
Item number	Item description	Unit of measure
PK		

VENDOR		
Vendor number	Vendor name	Vendor address
PK		

Design and Normalization

The accompanying sample case provides a summary of database design and normalization:

Database Design at Athena Consulting

Tom Snyder and Sylvia Webster graduated from college together, where they majored in computer information systems. They both worked as systems analysts for a large bank, but they became bored and decided to form a systems consulting partnership and strike out on their own. Their entrepreneurial venture resulted in a small systems consulting firm named Athena Consulting. They are presently organizing their firm and readying it for future clients.

"We need to develop a simple time-billing database for tracking the amount of time spent on projects," said Sylvia.

My favorite for a relational database management system is the entity relationship approach," Tom said.

"Mine too," agreed Sylvia.

"It seems to me that the goal of tracking time worked on projects simply means two entities, projects and hours, with a one-to-many relationship," said Tom.

Tom sketched an ERD:

"I don't believe that's sufficient," said Sylvia. "Projects are conducted for clients, so some client data need to be stored in the database. Here, let me add a clients entity to your ERD."

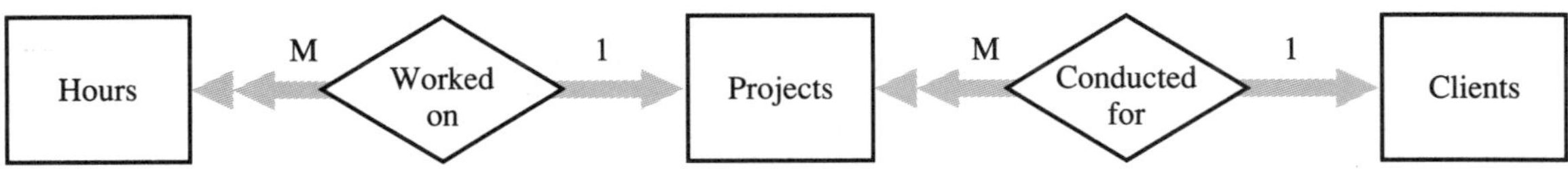

"I see what you mean," Tom responded. "If every project was for a different client, it would be reasonable to store client data in the project entity, but this will probably not be the case. Rather, we will conduct several projects for the same client. With my design, such a situation would result in data redundancy with the same client data related in different projects."

"Yes, and we don't want that," Sylvia responded. "Such redundancy would waste space and also create the possibility of inconsistent data if one instance of a client's data were updated and the others not."

"Yeah, normalization prevents those problems," Tom responded. "After further thought, I believe we need a Status table."

"What's the purpose of a Status table?" Sylvia asked.

"The Status table will be used as a lookup table to validate the Project-Status column in the Projects table and to store an expanded text description of the status of projects that will be used in our reports to our clients," Tom responded. "So Project Status is the PK in the Status table and an FK in the Projects table. OK with you, Sylvia.?"

"You bet! Now we're ready to map our normalized database design into our new relational database management system from Systems Database," Sylvia said excitedly.

"We've got everything defined—our tables, relationships, keys, attributes, all normalized ready to go," Tom responded.

The tables that Tom and Sylvia created are presented in Figure 11.17.

"Why aren't we using NOT NULL for our primary and secondary keys as well as other attributes?" asked Sylvia.

"Because our RDBMS assumes that all attributes are NOT NULL unless we say otherwise," Tom answered.

"Well, I believe we have a good database that will help us track our projects," said Sylvia.

"Now all we need are some clients," Tom said wistfully.[8]

```
CREATE TABLE Hours
(Project_Number          CHAR(5),
 Project_Description     CHAR(20),
 Start_Date              DATE,
 Hours_Worked            DECIMAL(4,2),
 Completion_Date         DATE);

CREATE TABLE Projects
(Project_Number          CHAR(5),
 Project_Description     CHAR(20),
 Project_Status          CHAR(2),
 Invoice_Number          NUMERIC(5),
 Invoice_Date            DATE,
 Invoice_Amount          MONEY,
 Check_Date              DATE,
 Check_Amount            MONEY,
 Client_Number           CHAR(5));

CREATE TABLE Clients
(Client_Number           CHAR(5),
 Client_Name             CHAR(20),
 Address                 CHAR(40),
 Contact_Person          CHAR(20),
 Phone_Number            CHAR(12));

CREATE TABLE Status
(Project_Status          CHAR(2),
 Status_Description      CHAR(500));
```

Figure 11.17 Tables created for Athena's relational database management system.

SYSTEMS FOR STORING AND MANIPULATING DATA

There are two general design approaches to storing and manipulating data in a computerized information system:

1 The traditional file system approach.

2 The database management system (DBMS) approach.

Traditional File Systems

The **TRADITIONAL FILE SYSTEM APPROACH** to data storage and access uses separate data files for each different application system. Designers of traditional file systems use terms such as "file," "record," and "field." A file is analogous to a table (or relation). A file is composed of records similar to rows in a table. The records are made up of fields, which are similar to attributes in a table. One of the chief differences between the traditional file system approach and the RDBMS approach is that, in traditional systems, data in File A is not logically

[8] This case based on Steve Roti, "Database Design," *DBMS*, February 1990, pp. 17–18.

related to data in File B, whereas the key feature of a RDBMS is relationship between tables and data within those tables.

In a traditional file system, data are created, updated, and accessed by individual programs written in a procedural programming language, such as COBOL, RPG, C, or Pascal. There is no special system of software to support the storage and access of data such as those in the database management system approach. All files are created and maintained separately. All programs written to access the files use the file access method of the operating system of a particular computer architecture to read and write data.

Data managed by a DBMS (or RDBMS) are typically stored on **DIRECT-ACCESS STORAGE DEVICES (DASDs)**, such as magnetic disk or optical disk devices. Data for a traditional file system are typically stored on magnetic tape for sequential batch processing or DASD for direct or indexed sequential access.

To understand how a traditional file system functions, let's examine the systems used at a videotape rental store (see Figure 11.18). Some of the data required for a videotape rental store include data about the vendors from which it purchases videos, the videos in inventory, customers who rent videos, and store employees. Each of these entities is depicted as a file in Figure 11.18. Using a traditional file system, programs must be written to create and maintain the files for each of these entities. Other programs must be written to access the data in these separate files. Both printer-based report programs and screen-based inquiry programs will probably be written.

This system works effectively and efficiently until there is a need to share data among files. Suppose that the manager of the store wants to know which vendor produces the best-quality videotapes and also releases the most frequently rented videotapes. This inquiry would require three different files to be

Figure 11.18 Traditional file system for videotape rental store.

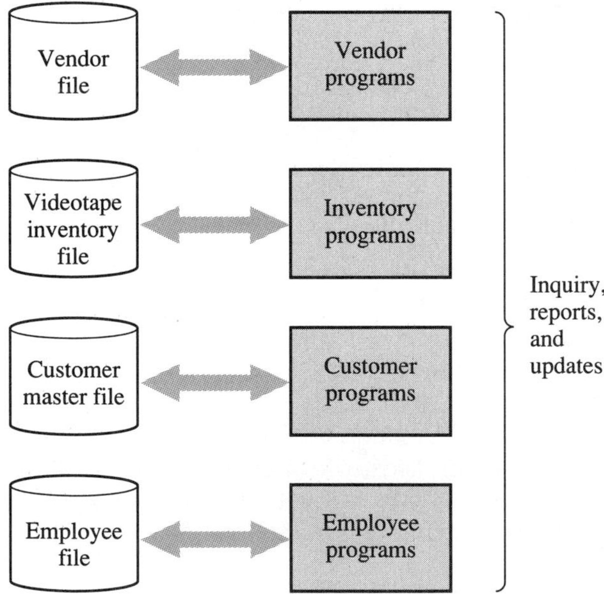

accessed. A program would have to be written to open the three files, search for the required information, and produce the output. Why? Because the data in the stand-alone files are not related as they are in a DBMS (or RDBMS).

Database Management System Approach

Figure 11.19 depicts the videotape rental store system using the **DATABASE MANAGEMENT SYSTEM (DBMS) APPROACH**. Each program now interfaces with the DBMS software rather than the individual data files. The data files are perceived by both users and programmers as an interrelated collection of data called a database.

What Makes Each Approach Different?

Either the traditional file system or the DBMS approach may be used successfully to store and access data for a business information system. No single approach should be used simply because the technology is available, or because the terminology sounds more "up-to-date." A systems professional must understand the advantages and disadvantages of the two approaches and should know when to use a given approach.

Figure 11.19
DBMS for videotape rental store.

Advantages of the Traditional File System Approach

The advantages of the traditional file system approach include:

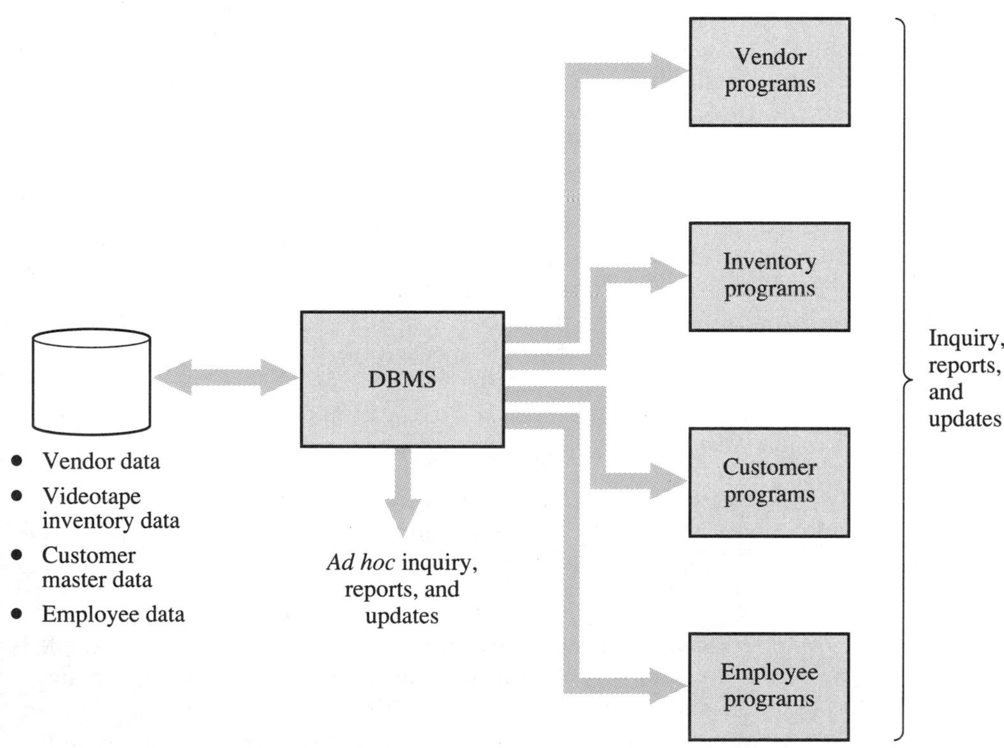

Simple Data Design It is very easy to create a normalized data model to support a single application or a small group of applications.

Fast Data Access The traditional file system relies on files organized on a sequential, indexed sequential, or direct basis. Programs have direct control of file access and can be written to optimize the time it takes to access and update records.

Inexpensive Supporting Technology Traditional file organization methods are supported by most existing operating systems and their accompanying file access methods. They do not require any additional software expenditures.

Disadvantages of the Traditional File System Approach
The disadvantages of the traditional file system approach include:

Lack of Data Relation Data are designed and implemented for a single application without concern for any effects on existing and future systems.

Redundant Data As additional applications are implemented, some of the same data will be stored in several files. For example, customer name and address may be stored in an accounts receivable file and also a marketing file. Redundant data have the potential for becoming inconsistent. For example, if customer name and address are located in more than one file, it is possible that a change to the address may be applied to only one of the files.

Lack of Standards Generally, there is no centralized set of software to support enterprisewide standards. More often than not, a traditional file system does not encourage a centralized data dictionary and does not force programmers and others to conform to any set of standards. Customer name may be Cust_Nm in file A and Customer_Name in file B.

New Application Productivity Lack of standards can result in poor application productivity. If a systems analyst or programmer must spend time searching old programs to see if the data required for a new application already exist in a file, or must check a file's organization to see if it is conducive to a new application, the programmer will probably spend more time in ferreting out these answers than in creating the new application.

Advantages of the Database Management System (DBMS) Approach
The advantages of the DBMS approach include:

Improved Data Integration Instead of designing the data requirements for a single application, the analyst examines groups of applications to develop the design. The DBMS then develops relationships among data sets for storage and access.

Increased Data Accessibility Data accessibility is a user's ability to get needed information from the database. A DBMS contains a high-level query language that allows one to obtain data from the database without having to write an application program. For a RDBMS, one such language, for example, is SQL.

Improved Data Integrity A database consists of an integrated set of files. Data redundancy can be minimized because the data resource is controlled by a single set of programs. The data dictionary helps to enforce a standard approach to data development. As a result, there is a greatly reduced likelihood of data inconsistencies.

Easier Application Systems Development and Maintenance Studies show that once a database has been designed and implemented, a programmer can code and debug a new application at least two to four times faster than with traditional files. With a DBMS, the programmer does not have to be aware of the actual structure, organization, and location of a file. Removing these needs from the application programmer reduces the cost of software development.

Improved Data Security Facilities Most organizations require some form of protection against unauthorized access to data. Data security prevents unauthorized access to data. Since the DBMS can control access to data entities, the security function is centralized and easily implemented with the use of a DBMS.

Logical and Physical Data Independence The DBMS is another software layer sitting between the application programmer and/or the end user and the data files. As a result, it is possible to separate how the data are perceived by these users from how the data are stored on a secondary storage device. An organization can change the data structure of the database without having to modify the application programs that access the data. This function reduces systems development and maintenance costs.

Disadvantages of the Database Management System (DBMS) Approach

The disadvantages of the database management system (DBMS) approach include:

Complex Data Design Designing a database model is complex and difficult for multiple, integrated systems. Systems professionals capable of designing such a database model are experienced, well-paid people. Thus labor expenses are typically much higher for the company that uses the DBMS approach than for one that uses a traditional file approach.

Slow Data Access for Some Applications If a user wants to access a specific record, say customer 14569, a traditional file system approach normally provides quicker access than DBMS. Also, if a company's systems applications are designed for batch processing, traditional file systems generally work more efficiently than DBMS.

Expensive Supporting Technology A DBMS is an expensive piece of software. A DBMS requires a sophisticated and complicated set of programs. An organization pays for these functions in the initial cost of the DBMS and the ongoing maintenance of the package. The complexity of a DBMS frequently requires an increase in staff to implement and maintain the package. A DBMS makes it easier to write applications, possibly allowing an organization to decrease or at least hold constant the application programming staff, but this cost saving can be offset by the increase in DBMS support staff. In addition, more computer hardware is almost always required to support the DBMS than file systems. Sometimes it is necessary to purchase another computer to execute the DBMS package efficiently. This DBMS disadvantage can be summed up as follows: A DBMS will cost an organization more than a traditional file system at the time of implementation, and it could cost more to maintain.

What Factors Affect the Selection of a System?

A systems analyst has to weigh both the advantages and disadvantages of the two approaches to select the optimal way to store and access data. Certain factors about an organization indicate the applicability of the DBMS approach. An organization should consider using a DBMS if:

- Application needs are constantly changing. An organization that is pursuing new markets, or is functioning in a highly competitive marketplace, or is pursuing mergers with other companies will have different data requirements from a relatively static organization.
- Ad hoc and intermittent inquiries are frequent and normal. A company with a large percentage of employees participating in end-user computing should see if an enterprisewide DBMS and query language could be installed to satisfy information needs.
- Many departments share the same data to satisfy their information needs.
- There is a need to reduce programming lead times and decrease program development costs.
- There is a need to improve the consistency of data.

Not all organizations should adopt the DBMS approach. An organization should use the traditional file system approach if:

- There is very little need to share data among users.
- The initial and ongoing expenditures for a DBMS are too costly to be afforded by the organization.
- The organization operates in a static internal and external environment. An organization experiencing little to no change usually requires only very standardized transaction processing. This processing is most efficiently completed with a traditional file processing system.
- Fast online access to specific records and efficient batch processing are dominant applications that may make the overhead of a DBMS intolerable. This factor assumes that it would be impossible to increase the hardware of the system to compensate for the additional DBMS requirements.
- Data planning cannot, for any reason, be coordinated across departmental boundaries. A traditional file system does not require that separate departments work together to plan their future data needs. If there is a lack of

communication between departments, or very strong animosity between individual areas, it will be almost impossible to use the DBMS approach successfully.

The two approaches are, however, not mutually exclusive. It is possible to have an organization that wants to take advantage of the strengths of both approaches. For example, a systems professional designing the data requirements for a mail-order clothing company might decide to use a traditional file system for inventory data to take advantage of the access speed of direct organization files. Batch processing applications, such as payroll, may also use the traditional file system approach with sequential files. However, for tracking customer sales and making ad hoc inquiries, such as: "What customers purchased more than 500 red widgets in June?" the DMBS approach is applicable. Thus the systems professional responsible for database design would use both approaches to optimize data storage and access for the enterprise.

REVIEW OF CHAPTER LEARNING OBJECTIVES

The major goals of this chapter were to enable each student to achieve four learning objectives. We will now summarize the responses to these learning objectives.

Learning objective 1:
Describe the properties of the relational database model.

A relational database model is perceived by the user and database designer as a collection of tables comprised of columns and rows. During preliminary design, such as when an entity relationship diagram (ERD) is used, these tables may be referred to as entities or relations.

Relationships are implemented through common columns in two or more tables, using attributes known as primary keys (PKs) and foreign keys (FKs). Attributes of a relational database model are derived from a domain of data types.

There are several ways of manipulating a relational database. One pervasive language is referred to as the structured query language (SQL).

Learning objective 2:
List and explain the steps necessary to perform the relational database design process.

Generally, when systems professionals begin the relational database design process, it is assumed that the planning, analysis, general design, and evaluation and selection phases of the SDLC have been completed. In addition, output reports, input forms and transactions, and processes have been designed in detail. If these requirements are in hand, the following steps can be performed with confidence:

Step 1. Model the entities and map to tables.
Step 2. Designate primary keys for each table.

Step 3. Model relationships between tables.
Step 4. Model attributes from the domain of data types for all tables.
Step 5. Normalize the relational database model.
Step 6. Prepare a data dictionary.

Learning objective 3:
Explain the purpose of normalization and the first three normal forms.

Normalization is a process for:

- Optimizing the database design
- Preventing insert, delete, and update anomalies
- Providing database design stability over time

Normalization analytical techniques are called normal forms; that is, relational database model experts have identified anomalies of data relationship in three normal forms:

- First normal form (1NF)
- Second normal form (2NF)
- Third normal form (3NF)

A relation is in 1NF if it does not contain repeating groups. That is, if it has no repeating attributes and no duplicate rows, each attribute is named, and the primary key (PK) is not null. A relation is in 2NF if it is in 1NF and no non-key attribute is functionally dependent on only a portion of the PK. A relation is in 3NF if it is in 2NF and contains no transitive dependencies.

We have identified three normal forms. Other normal forms exist, but 1NF, 2NF, and 3NF are the ones most likely to be encountered.

Learning objective 4:
Differentiate between the traditional file system approach and the database management system (DBMS) approach.

The systems professional weighs the enterprisewide objectives and a variety of other factors to determine whether to use the traditional file system approach or the DBMS approach. A traditional file system is an approach using the file access method of the operating system. It supports the standard file organizations of sequential, indexed sequential, and direct formats. A DBMS (or RDBMS) approach adds an additional layer of software to the file access method, providing centralized data standards and increased methods of data access. A systems professional weighs the advantages and disadvantages of the two approaches to make the decision of which to use. In addition, an analyst looks at the following conditions:

- The amount of flexibility required for the new system
- The access methods needed
- The amount of communication between departments

- The types of inquiries that will be made
- The number of ad hoc inquiries made by end users

In many organizations, both approaches are used to leverage each one's advantages.

DATABASE DESIGN CHECKLIST

The database design process involves converting user requirements to a database model that will support these requirements. Although there are several database models to choose from, the most popular model in use today is the relational database model.

There are two general approaches to database design: the traditional file system approach and the DBMS approach. Although the traditional file system approach is used effectively and efficiently for several kinds of applications, we have concentrated on the DBMS approach, especially the RDBMS. The following is a checklist on how to design a relational database for a specific application.

1 Understand the properties of the relational database model and SQL.

2 With most of the front-end SDLC work achieved, perform the following relational database design steps:

Step 1. Model the entities and map to tables.
Step 2. Designate primary keys.
Step 3. Model relationships between tables.
Step 4. Model attributes of tables.
Step 5. Normalize the database model.
Step 6. Prepare a data dictionary.

KEY TERMS

Attribute

Concatenated

Data dictionary

Database design

Database management system (DBMS) approach

Database management system (DBMS)

Direct-access storage devices (DASDs)

Domain

Entities

First normal form (1NF)

Foreign key (FK)

Hierarchical model

Network model

Normal forms

Normalization

Null

Online transaction processing (OLTP)

Primary key (PK)

Relational database management system (RDBMS)

Relational model

Relations

Relationship

Second normal form (2NF)

Structured query language (SQL)

Tables

Third normal form (3NF)

Traditional file system approach

REVIEW QUESTIONS

11.1 Define a RDBMS and describe its schemata.

11.2 List and briefly define properties of the relational database model.

11.3 Define PK and FK. State what each is used for.

11.4 Name two languages that manipulate data in a RDBMS.

11.5 Name and briefly define the steps necessary to perform the relational database design process.

11.6 What's a table? An attribute?

11.7 How are relationships between tables formalized?

11.8 What's the general purpose of normalization?

11.9 What's the purpose of a data dictionary?

11.10 Define 1NF, 2NF, and 3NF.

11.11 Give three advantages and disadvantages of a traditional file system approach. Do the same for a database management system (DBMS) approach.

11.12 Give three reasons why a traditional file system approach should be used. Give three reasons why a DBMS (or RDBMS) approach should be used.

CHAPTER-SPECIFIC PROBLEMS

These problems require exact responses based directly on concepts and techniques presented in the text.

11.13 Heavenly Chocolate Boxes is a chain of candy stores. Following are some notes you collected from Heavenly's owner:

> "I started out with one small store four years ago, and now I have 20 stores and 200 employees in a three-state area. I specialize in various chocolate items, and I need to know which items are sold in which states and which aren't."

Required: Model the tables, table names, and primary keys necessary to fulfill Heavenly's needs.

11.14 Using the relational database design in Figure 11.4, perform the following exercises:

(a) Write the SQL query for: "What are the item numbers on invoice number 2?"
(b) Write the SQL query to display the colors that begin with B.
(c) Write the SQL query to display the items that have been ordered by Nolan in Seattle.

(d) Insert data for a pink shirt, part number 16, with a price of $10.00.
(e) Change the price of white hats to $30.00.
(f) Delete all data about brown shoes.

11.15 SQL's great strength lies in its ability to create complex queries with relatively few commands. Single SQL statements can replace hundreds of lines of procedural program code. For example, analyze the following SQL command:

```
SELECT    Customer.Customer_Number,
          Name, Order_Number
    FROM  Customer, Orders
    WHERE Customer.Customer_Number =
          Orders.Customer_Number;
```

Required: State the purpose of the SQL command above.

11.16 Following are the steps, in no particular order, performed to develop a relational database model:

_______________ Model relationships

_______________ Normalize tables

_______________ Prepare data dictionary (In many instances, a large portion of the data dictionary has already been prepared, but here we're assuming that it has not.)

_______________ Model attributes

_______________ Designate primary keys

Required: Insert the correct step number in the blank provided. Define each step, and explain its purpose.

11.17 You've modeled two tables: ORDER and CUSTOMER. You want to establish a relationship to identify which orders belong to which customers.

Required: Draw the two tables and insert the proper keys to identify this relationship. What kind of a relationship is it? (A 1 : 1, 1 : M, or M : N?) Do not consider other attributes that might be modeled for these tables. Use only the keys.

11.18 A table is said to be in third normal form (3NF) if every attribute in the rows of that table is dependent solely on the primary key. In other words, data cannot be dependent on any other attribute except the primary key.

You have defined the following attributes: Item_Number, Color, Quantity_On_Hand, Warehouse_Number, Warehouse_Location.

Required: Model and normalize three tables: ITEM, WAREHOUSE, and an ITEM_LOCATION table that links the two tables.

11.19 Normalization to the second normal form (2NF) involves tables that have composite primary keys; that is, primary keys that are constructed by concatenation of two or more foreign keys. Second normal form (2NF) requires that all attributes in such a table be dependent on the entire primary key rather than any part thereof. For example, Part_Number and Warehouse_Number are concatenated to form the primary key of INVENTORY table. Inventory_Item, however, is dependent only on Part_Number rather than the concatenation of Part_Number + Warehouse_Number. Thus this situation creates a partial dependency.

Required: Give two examples of partial dependency and show how they are eliminated.

11.20 Normalization to the third normal form (3NF) involves eliminating transitive dependencies, that is, the dependence of any non-key attribute on any other attributes except the primary key.

Required: Give an example of a transitive dependency and how it is corrected.

THINK-TANK PROBLEMS

These problems call for a feasible approach rather than a precise solution. Although the problems are based on chapter material, extra reading and creativity may be required to develop workable solutions.

11.21 The following data attributes would probably be stored for a student in a college:

student number, course number, course name, student name, student address, student telephone, number of credits for course, grade received in course, major, instructor name, instructor's office, student occupation, semester, name of college, advisor.

Required: Design a relational database model in third normal form for the attributes above. Your relational database model should show all the tables, attributes, relationships, and primary and foreign keys. Make any assumptions that you deem necessary to complete the problem.

11.22 The following data attributes would probably be stored for an employee in a commercial personnel system:

employee number, employee name, address, city, state, zip code, telephone number, emergency number, department number, department name, supervisor name, pay rate, pay period, number of dependents, optional child care plan, optional pension plan, optional stock purchase plan, date of last review, rating on last review, spouse's name, last degree earned, date of degree, current rank, previous ranks.

Required: Design a relational database model in third normal form for the data attributes above. Answer any questions you may have about dependencies yourself. Make any assumptions you deem necessary to complete the problem.

11.23 The Starlight Video Rental Company is planning on installing a complete customer information system. Starlight rents videotapes to individuals. It has four stores spread across three different cities. Starlight currently uses a stand-alone microcomputer in each location with a computerized system to handle all accounting and financial applications, such as accounts payable, payroll, inventory control, general ledger, and financial statements. These applications are composed of prepackaged software using a traditional file system.

At this time, store personnel in each location manually keep customer rental records for that specific store. Each store has the same system: Customers walk in the store, pick a video from the empty video boxes sitting on a shelf, and present their rental cards. An employee gets the video, records the rental in a notebook rental log, and collects the money from the customer. The inventory number and rental date are the key field for the rental log. At the end of the day, an employee takes the rental log and transfers each rental to a customer's specific card. The customer's telephone number is the key field. Each customer has a card in a 3 × 5 index card file.

Customers also purchase videos. The purchase is recorded in a different notebook, the purchase log, but the purchase is transferred to the same customer card as that used for rentals at the end of the day. Purchases are noted on the 3 × 5 card with a (P). Management would like to use the customer cards to determine (1) which movies are more popular in a given location, (2) which customers rent the most movies, (3) whether or not to install a bonus-point system (35 rentals and you get a free rental), and (4) if certain very regular customers have specific movie preferences.

Required:

(a) Identify the data attributes necessary to create a customer information system for the Starlight Video Rental Company.

(b) Create a relational database design model in third normal form for the data attributes you identified in the step above.

(c) Would you recommend a traditional file system or a database management system approach for the Starlight Video Rental Company? Why?

SUGGESTED READING

Bellomo, Cecelia. "Relational Database Systems: Data Integrity." *Interact,* October 1990.

Foster, Al. "A New ERA for Data Modeling." *DBMS,* November 1990.

Gerritson, Rob. "SQL Tutorial." *DBMS,* February 1991.

Kemm, Tom. "A Practical Guide to Normalization." *DBMS,* December 1989.

Loomis, Mary E. S. *The Database Book.* New York: Macmillan, 1991.

McFadden, Fred R., and Jeffrey A. Hoffer. *Database Management.* Menlo Park, Calif.: Benjamin/Cummings, 1990.

Pascal, Fabian. "Preventing Corruption." *DBMS,* September 1989.

Peck, Charles. "Tuning Oracle SQL Statements." *DBMS,* September 1989.

Pratt, Philip J., and Joseph J. Adamski. *Database Systems Management and Design,* 2nd ed. Boston: Boyd & Fraser, 1991.

Roti, Steve. "Database Design." *DBMS,* February 1990.

Vasta, Joseph A. *Understanding Database Systems Management and Design,* 2nd ed. Boston, Mass.: Wadsworth, 1991.

JOCS CASE: Database Design

During systems analysis, Jake Jacoby and his SWAT team identified the processes to be computerized and documented those processes using both data flow and entity relationship diagrams. A tentative data dictionary was prepared as an analysis tool to better understand the JOCS application under review. After the analysis was complete, Jake chose the best overall design for the JOCS application and had his team design the output, processes, and input. The SWAT team worked with the future users and had many meetings while designing the output, processes, and input for the JOCS project.

As we discussed earlier, Jake, as the project leader, keeps looking ahead and planning the design tasks ahead of his team. While the SWAT team finalizes the input design, Jake begins the tasks associated with database design.

The JOCS system has well-defined input, processes, and output. The formats of the reports, display screens, and input screens are known and understood by both the users and Jake. Figure 11.20 is a copy of the worksheet Jake composed to list attributes. Jake lists the attributes as he finds them while examining the supporting analysis documentation. This list is large, but each of the attributes is necessary to support the JOCS system. This list also contains many of the same attributes. Jake believes that it is better to list all the attributes as he finds them, then eliminate duplications, rather than to check the list continually to see if he has repeated some attributes. To create some order in this list, Jake identified entities for the system and then regroups his list by those entities, eliminating duplicate attributes. Figure 11.21 is the same worksheet showing all entities, attributes, and relationships.

Next, Jake applies the process of normalization. Figure 11.21 provides an initial view of the attributes organized by entity with identified primary keys and foreign keys before normalization. Jake looks for repeating groups

Figure 11.20
JOCS data attribute worksheet.

Job-Order Number
Job Description
Customer Number
Date Received
Date Due
Start Date
Quantity Ordered
COGS Percent
Customer Number
Average Cost
Last Cost
Minimum Stock Level
Maximum Stock Level
Type Code
Customer Telephone Number
Customer Contact Name
Last Order Date
Last Completion Date
Actual Materials
Actual Regular Hours
Actual Overtime Hours
Overhead Application Rate
Employee Number
Operation/Department
Hourly Rate
Time Started
Time Stopped
Pieces Made
Date of Operation
Standard Hourly Pay Rate
Quantity Used
Date Used
Budgeted Material Dollars
Description
Cycle Count Number
Unit of Measure
Lead Time in Days
Economic Order Quantity
Quantity on Hand
Quantity on Order
Quantity on Reserve
Standard Cost

Original Quote
Engineering Changes Cost
Billed to Date
Budgeted Regular Hours
Budgeted Overtime Hours
Budgeted Labor Dollars
Budgeted Material Dollars
Overhead Application Rate
Customer Name
Customer Address
Customer City
Customer State
Customer Zip Code

Actual Labor Dollars
Total Budgeted Dollars
Total Actual Dollars
Labor Burden to Date
Material Burden to Date
Percent Complete
Total Cost
Variance Regular Hours
Variance Overtime Hours
Variance Labor Dollars
Variance Material Dollars
Date Last Update of Totals
Due Date
Standard Hourly Pay Rate
Budgeted Regular Hours
Budgeted Overtime Hours
Budgeted Regular Pay
Budgeted Overtime Pay
Budgeted Material Dollars
Overhead Application Rate

in his attributes to take the list into first normal form. He notices that in the entity OVERALL JOB, the following attributes will be repeated: Operation/Department, Hourly Rate, Time Started, Time Stopped, Pieces Made, Date of Operation, Item Number, Quantity Used, and Date Used for each job. In the top part of Figure 11.22, you can see the unnormalized logical group OVERALL JOB. Figure 11.22 demonstrates how Jake moves through the normalization process to create a logical data model for the entity OVERALL JOB.

Figure 11.21
JOCS data attribute worksheet grouped by entity.

OVERALL JOB

Job-Order Number**
Job Description
Customer Name
Date Received
Date Due
Start Date
Quantity Ordered
COGS Percent
Operation/Department
Hourly Rate
Time Started
Time Stopped
Pieces Made
Date of Operation
Item Number
Quantity Used
Date Used

INVENTORY

Item Number**
Description
Cycle Count Number
Unit of Measure
Lead Time in Days
Economic Order Quantity
Quantity on Hand
Quantity on Order
Quantity on Reserve
Standard Cost
Average Cost
Last Cost
Minimum Stock Level
Maximum Stock Level
Type Code

** Represents Primary Key

JOB BALANCE

Job-Order Number**
Original Quote
Engineering Changes Cost
Billed to Date
Budgeted Regular Hours
Budgeted Overtime Hours
Budgeted Labor Dollars
Budgeted Material Dollars
Budgeted Total Hours
Actual Materials
Actual Regular Hours
Actual Overtime Hours
Actual Labor Dollars
Total Budgeted Dollars
Total Actual Dollars
Labor Burden to Date
Material Burden to Date
Percent Complete
Total Cost
Variance Regular Hours
Variance Overtime Hours
Variance Labor Dollars
Variance Material Dollars
Date Last Update of Totals

OPERATION STANDARDS FOR A JOB

Job-Order Number**
Operation/Department Number**
Description
Due Date
Standard Hourly Pay Rate
Budgeted Regular Hours
Budgeted Overtime Hours
Budgeted Regular Pay
Budgeted Overtime Pay
Budgeted Material Dollars
Overhead Application Rate

CUSTOMER

Customer Number**
Customer Name
Customer Address
Customer City
Customer State
Customer Zip Code
Customer Telephone Number
Customer Contact Name
Last Order Date
Last Completion Date

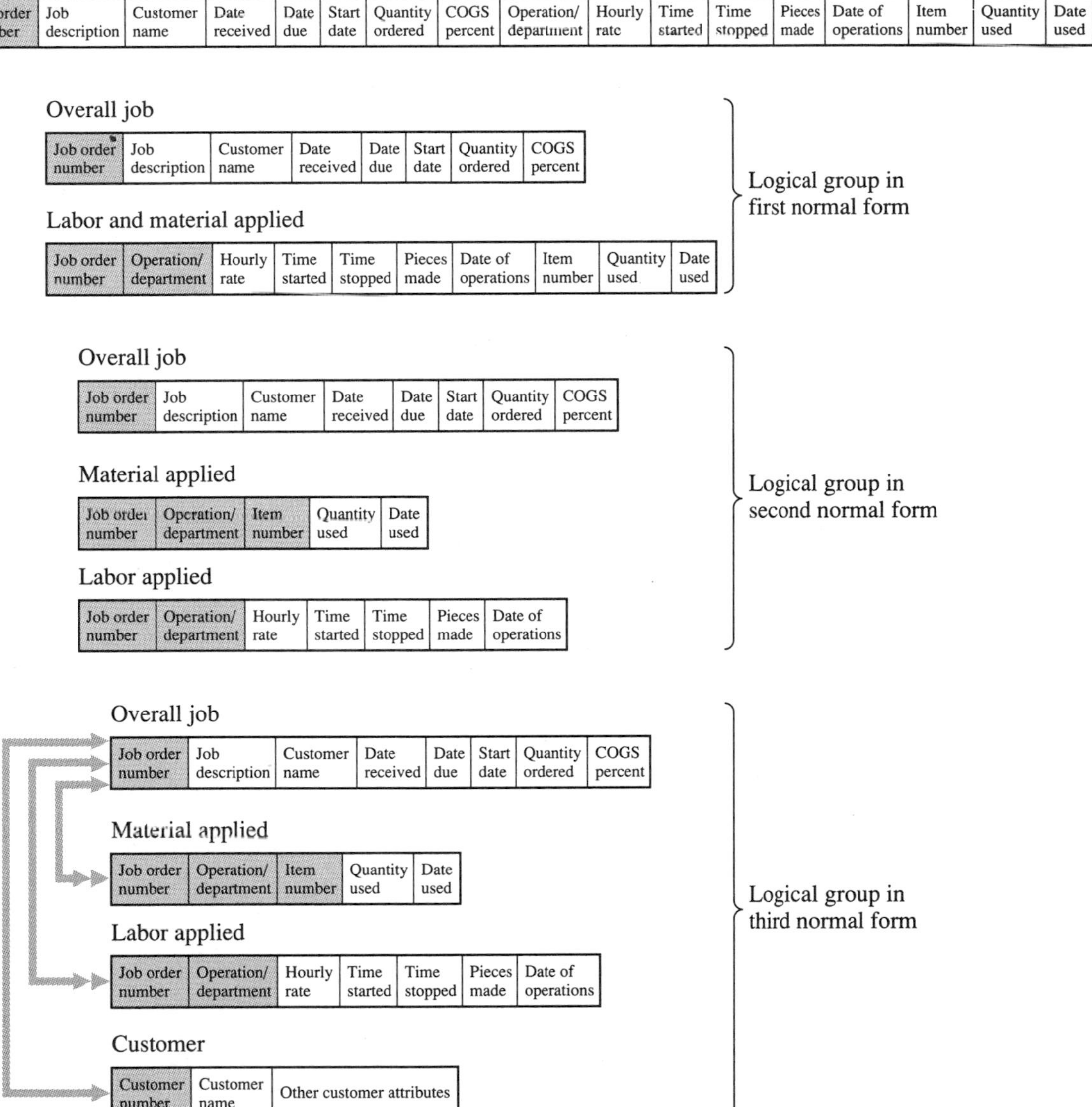

Figure 11.22 Creating a normalized logical data model for the OVERALL JOB entity.

Jake has to make some assumptions to complete normalization correctly. Some of his assumptions include:

1 Single quote is made for a job. It will not change during the production of that job.

2 Engineering change dollars will be added during the production of a job.

3 Only one operation, such as cutting, is performed in a given production department.

4 Materials are purchased to inventory rather than to a specific job.

5 All materials needed for a job are available in inventory.

6 All processes for a job are completed in-house.

7 Costs are collected and evaluated by both department and job.

8 An employee payroll system and a purchasing system are already in place.

Normalization generally requires that assumptions about an organization be made. During the analysis phase of SDLC, most systems analysts become fully aware of the functions of an organization. If, however, an analyst gets to the normalization steps of database design with some remaining questions concerning these functions, the analyst should always consult with the targeted users of the system before making any general assumptions. Jake checked his assumptions with the users to make sure that they are correct.

Jake must now choose the best approach for the JOCS system. Current technology will easily support either a traditional file system or a DBMS. The factors affecting his decision are as follows:

1 The strategic mission for information systems is to assist in the return of the company to profitability and to increase market share.

2 The JOCS system is to provide detailed, online information regarding job status and current costs. The system is to permit ad hoc inquiry as well as standardized reports. The emphasis is on immediate response to varied inquiries.

3 Peerless is functioning in a competitive, changing business environment.

4 JOCS will have to interface with existing accounting, purchasing, and payroll systems.

5 JOCS should be operational as quickly as possible to help achieve corporate objectives.

A traditional file system would provide faster online access to specific records than a DBMS, and would permit much easier interface to currently existing systems, since the currently existing systems are based on a traditional file system. A traditional file system would also require a smaller investment in hardware and software resources. However, the need for ad hoc reporting and access to cost and variance data in variable format would best be done with a DBMS. A DBMS could also be implemented more quickly than could a traditional file system. The competitive, changing business environment of Peerless also indicates a need for a DBMS. Weighing each of these factors, Jake chooses a DBMS. He realizes that special programs will have to be written to interface a DBMS with existing systems, and he also knows that the need for fast access speed will influence the implementation of the system.

Chapter 12

Designing Systems Controls

WHAT WILL YOU LEARN IN THIS CHAPTER?

After studying this chapter, you should be able to:

1 Give an overview of threats to information systems.
2 Explain input controls and demonstrate how they can be used to filter out erroneous input data. Also, design an audit trail of input data for a real-time system.
3 Discuss destructive and fraudulent software to which the process systems component is vulnerable and explain how to prevent this kind of software from entering the system.
4 Apply both real-time system output controls and batch system output controls.
5 Present database controls.
6 Present technology platform controls, both for mainframes and personal computers (PCs).
7 Discuss access controls and explain why biometric controls are the most effective controls against unauthorized access.

INTRODUCTION

The purpose of this chapter is to show how to design and maintain a system of controls that will protect the information system from various threats (see Figure 12.1). Some of the serious threats to information systems are:

- Human error stemming from errors of commission and omission and lack of training
- Destructive and fraudulent software
- Wiretapping and output interception by unauthorized people
- Corruption of, or loss of, databases
- Technology platform failure caused by poorly designed environment, fire, degraded power supply, and theft
- Unauthorized access

Such threats are ever present, as suggested by Figure 12.2. Effective controls to prevent loss from these threats can be classified into three broad categories:

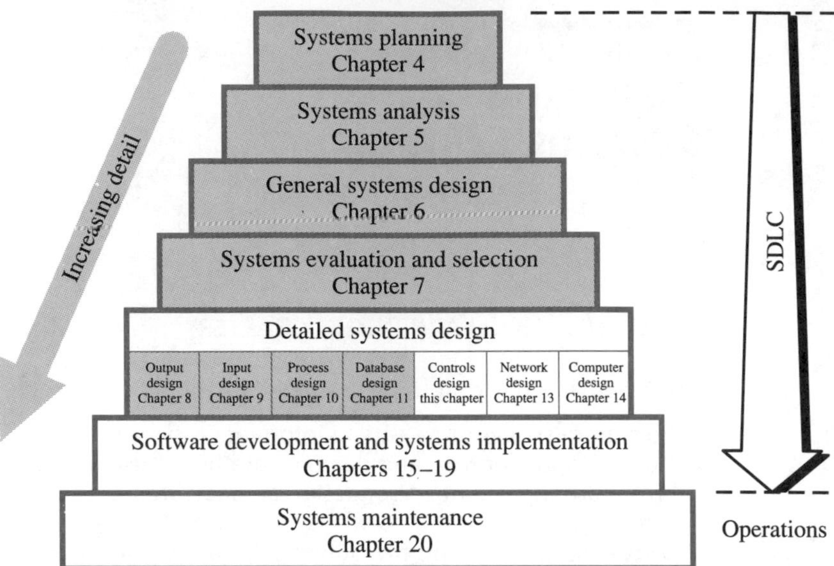

Figure 12.1
SDLC phases and their related chapters in this book. In Chapter 12 we discuss how to design protective systems controls.

- Preventive
- Detective
- Corrective

PREVENTIVE CONTROLS help keep threats from happening. **DETECTIVE CONTROLS** help spot threats and thereby reduce or eliminate exposure to them. **CORRECTIVE CONTROLS** help the system recover from the occurrence of threats. All of the input, process, output, database, technology platform, and access controls presented in this chapter contain one or a combination of these categories.

CONTROLLING THE INTEGRITY OF THE DATA ENTERED IN THE SYSTEM

As you learned in Chapter 9, there are two basic ways to input data: source document-based methods and direct-entry methods (e.g., electronic forms, touch menus, voice, and scanning). Direct entry reduces the number of transcriptions and can incorporate feedback instructions to help ensure immediate validation. Because there is less human intervention and more online validation with direct entry, the risk of errors is reduced.

The input component is responsible for bringing data into a system. Such input must be controlled to ensure that it's authentic, accurate, and complete. **INPUT CONTROLS** include:

- Code controls
- Input validation controls
- Input identification controls

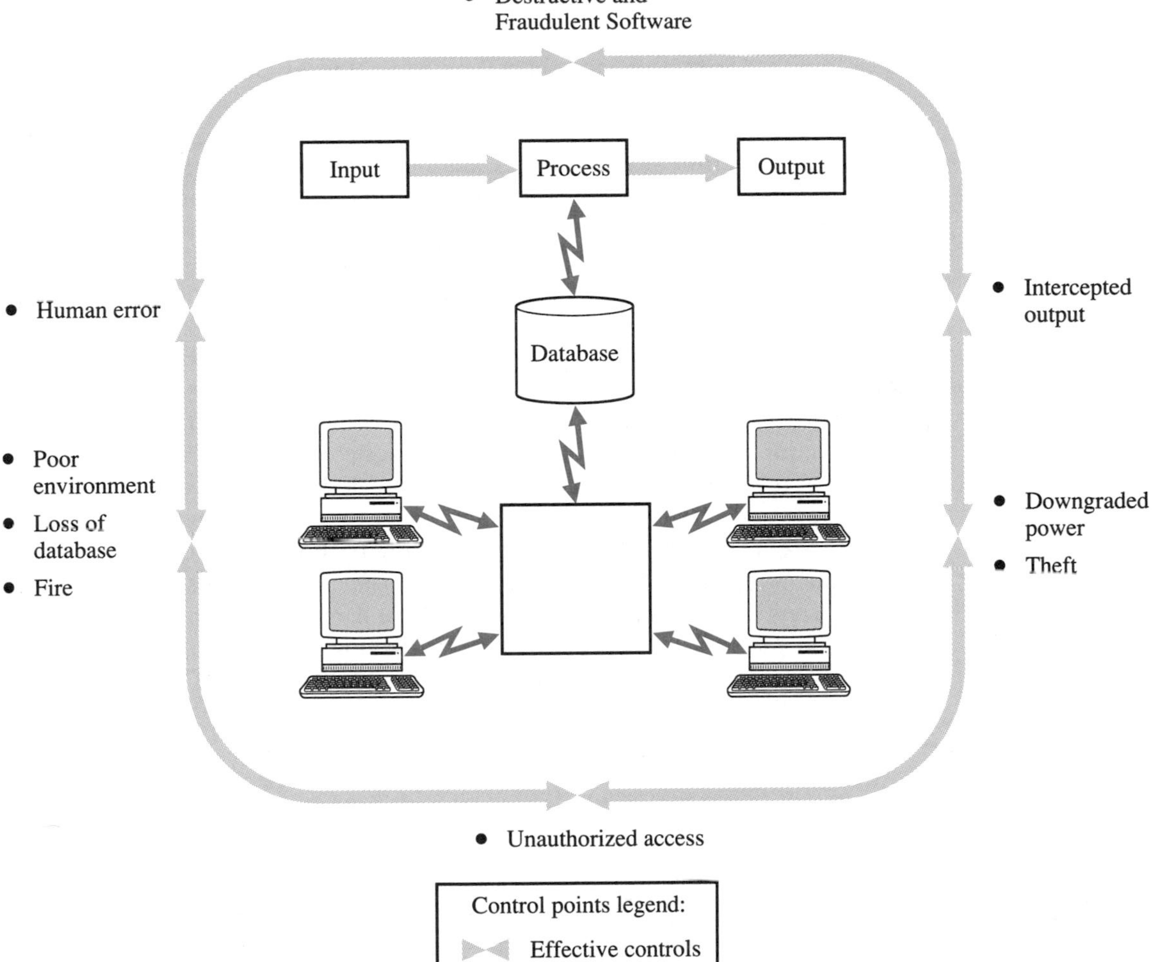

Figure 12.2
Threats to an information system and control points. The figure suggests effective controls, the subject of this chapter.

- Batch controls
- Audit trail controls

(We end this section with a presentation on how to handle input errors.)

Controlling Code Errors

CODE CONTROLS check for code errors. Code errors fall into five types:

- Addition, where an extra character is added, such as 14351 is coded as 143519
- Truncation, where a character is omitted, such as 14351 is coded as 1435
- Transcription, where a wrong character is recorded, such as 14351 is coded as 64351

- Single transposition, where adjacent characters are reversed, such as 14351 is coded as 41351
- Double transposition, where characters separated by more than one character are reversed, such as 14351 is coded as 15341

Any of these errors made in key codes can have serious consequences. For example, adding or truncating characters in a bank depositor's account number may result in depositing cash to someone else's account. Transcribing the wrong characters in an account number for a creditor may result in a payment being made to someone who has no legal claim to the payment. Transposing characters in a part number may result in a large quantity of incorrect parts being sent to a job. One control used to guard against these types of errors is a check digit.

A **CHECK DIGIT** is a redundant digit added to a code that permits the accuracy of other characters in the code to be checked. The check digit may be a prefix character or a suffix character, or it may be inserted somewhere in the middle of the code.

There are many ways to calculate check digits. For one example, let's use the account number 14351 as a typical code assigned to a customer of First National Bank. For all account numbers at First National Bank, we will use modulus 10 and 1–2 weighting (many organizations use modulus 11 and the weight factors, from right to left, 1, 2, 3, 4, 5, . . .); that is, the account number 14351, and all the other account numbers at First National, are computed as follows:

$$\begin{array}{ccccc} 1 & 4 & 3 & 5 & 1 \\ \times 1 & \times 2 & \times 1 & \times 2 & \times 1 \\ \hline 1 + & 8 + & 3 & +10 + & 1 = 23 \end{array}$$

After multiplying the 1–2 weights times each character and adding the products, we get 23. We divide this sum by the modulus 10 as follows:

$$\frac{23}{10} = 2 \text{ with remainder } 3$$

We then subtract the remainder from the modulus and the result constitutes the check digit:

$$10 - 3 = 7$$

The check digit 7 is added to the code as a suffix (or as a prefix or inserted in the middle of the code). The resulting self-checking code is 143517. Every time an account number is entered, it is recalculated to make sure that its check digit is equal to the preassigned one.

Check digits involve overhead in terms of one or more redundant characters carried through the system and the algorithm required to calculate and validate the check digit. Therefore, use of check digits should be limited to critical codes such as account numbers. For such codes, the check digit is precalculated and assigned as part of the code. Where possible, the computer should make the assignments of check digits.

Identifying Data Errors Before Processing

INPUT VALIDATION CONTROLS are used to identify errors in data before the data are processed. As a general rule, data should be validated as soon as possible after capture and as close as possible to the source of data. Controls to validate input data can be performed at three levels:

- Field checks
- Record checks
- File checks

Checking Field Data

Each field should be checked by the program to make sure that it contains appropriate data. The following FIELD CHECKS should be designed into the program:

Field Check	Description
Missing data or blanks	Parts shipped from different warehouses should contain both a part number and a warehouse code. Part number 1738 shipped from warehouse A should be coded as WA1738. The code field containing only the part number 1738 would be in error. Normally, a field should not contain data with intervening blanks, such as 99b9.99. (*Note:* b stands for blank)
Alphabetics and numerics	An alphabetic field such as a customer name should contain all alphabetic characters. For example, AC4E is in error if the name is ACME. Computational fields and various code fields should contain all numbers. A price field containing 64A.21 is in error.
Range	A number of fields can be specified to contain values that fall within an allowable range. For example, the pay rate range in a particular company is $10.00 to $20.00 per hour. Any value in the pay rate field outside this range is in error.
Check digit	Is the check digit valid for the code in the field? For example, the code 18743 was assigned 6 as its check digit. Each time the code 187436 (i.e., the code with its appended check digit) is processed and the check digit recalculated, it should equal the preassigned check digit, which is 6.
Size	If variable-length fields are used and a set of permissible sizes is defined, does the field delimiter show the field to be one of these sizes? If fixed-length fields are used and spaces are defined between fields, do the spaces contain blanks?

Checking Records

For records, the validation part of the program makes the following RECORD CHECKS:

Record Check	Description
Reasonableness	Even though a field may pass a range check, the contents of another field may determine what is a reasonable value for the field. For example, $14.50 may fall within the pay rate range, but is not reasonable for welders in division 2, who are supposed to be paid $18.50 per hour. Therefore, employees with code W2 (welders in division 2) are assigned a pay rate of $18.50 per hour.
Sign	The contents of one field may determine which sign is valid for a numeric field. If the record represents a credit to a general ledger account, it should have a negative sign for the amount field. If it's a debit, the amount field should have a positive sign.
Size	If variable-length records are used, the size of the record is a function of the sizes of the variable-length fields. The permissible size of variable and fixed-length records may also depend on a field indicating the record type.
Sequence	In a batch system, records are normally sorted in ascending order based on a code number such as customer number. Both master file records and transaction records must be in the same order for correct processing.

Checking Files

FILE CHECKS ensure that the correct file is processed. The following types of file checks can be applied:

File Check	Description
External label	Each physical file, such as an accounts receivable tape file, should contain a label that people can clearly identify to prevent accessing and mounting the wrong file. For example, a file label should contain: name of the file, creation date, retention date, and the number of physical files, such as 1 of 3.
Header internal label	This label is stored on the physical file for the computer to read. It contains the same information as the external label, so if an operator misreads the external label, the computer will detect it as the wrong file and abort the job and display that a wrong file has been mounted.
Trailer internal label	This label may contain the number of records in the file, various control totals, and an end-of-file (EOF) indicator if it is the last physical file in a series of physical files.

Controlling and Validating Sources of Data

There are thousands of stories about how companies have been defrauded by paying fictitious vendors or employees. Such frauds can be reduced, if not eliminated, by creating identification (or validation) tables and comparing each

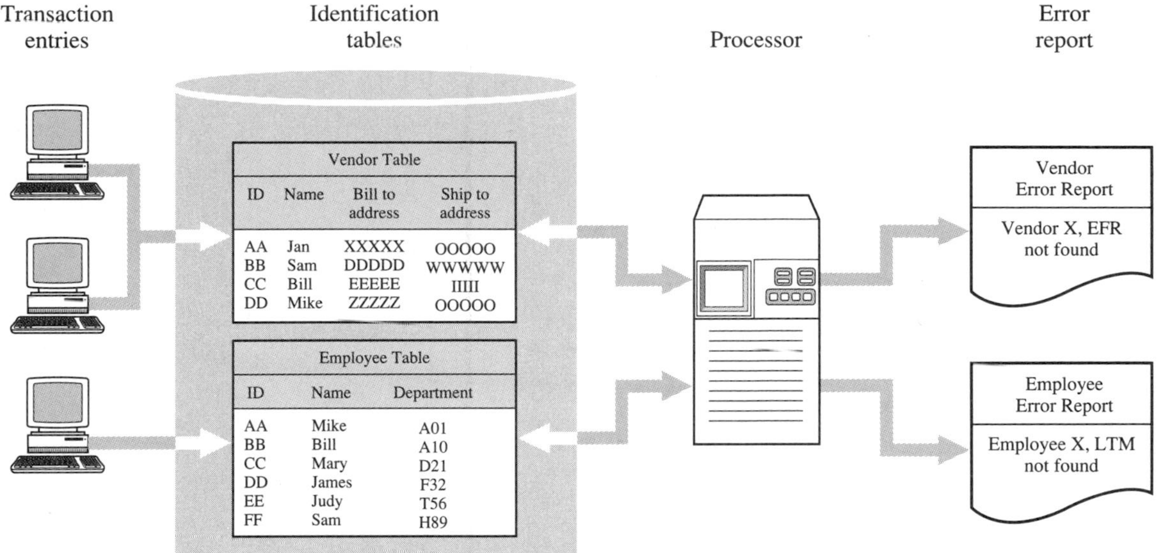

Figure 12.3
Use of identification tables to validate transactions.

transaction with authenticated entries in the identification table. Such **INPUT IDENTIFICATION CONTROLS** procedures are presented in Figure 12.3.

All authorized vendors, for example, are assigned specific identification numbers and descriptions. Before any vendor's record can be accessed for update or payment, the transaction must contain this identifying data, and it must match data in the identification table exactly. If the program is used to pay an invoice, the payee identification number or name should be checked against a file of approved suppliers. For shipments, ship-to addresses should be checked against a file of acceptable customer ship-to addresses for the items being shipped. Identification of employees is handled in a similar manner.

Using Batch Controls to Protect Data Integrity

Some of the simplest and most effective controls over data capture, preparation, and entry are **BATCH CONTROLS**. Various controls can be exercised over the batch to prevent or detect errors. In such cases, the design of batch controls is required to ensure the following:

- All source documents are processed
- No source documents are processed more than once
- An audit trail is created from input through processing, to final output

For an example of batch controls, refer to Figure 12.4. Sales orders are source documents that are accumulated and time and date stamped. At 4:00 P.M. a batch from each branch store is sent to the batch control clerk, who batches sales orders into groups of 25 or less and prepares a batch control ticket (or cover sheet) with a branch number and unique batch number for each batch,

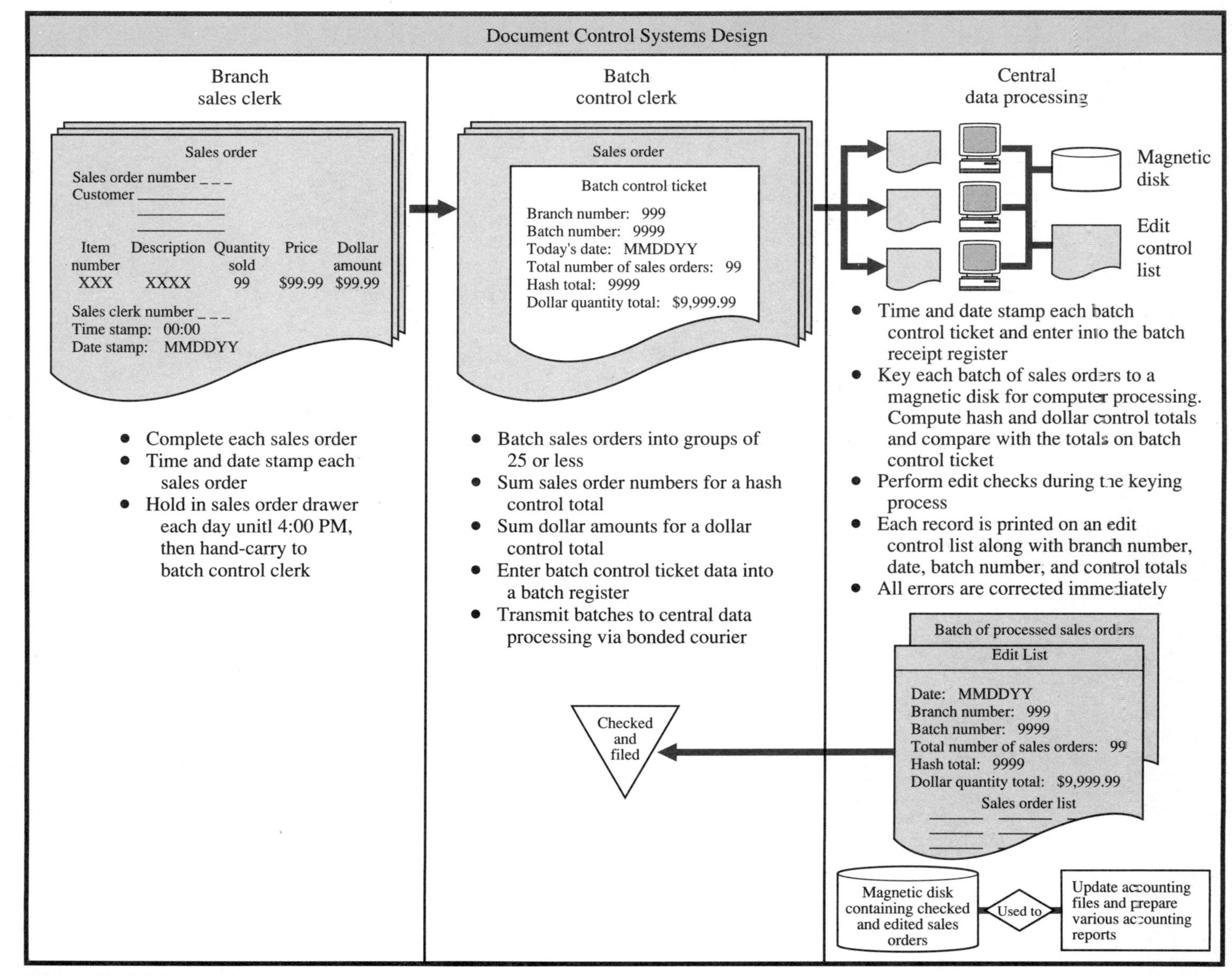
Document Control Systems Design
Branch sales clerk
Sales order
Sales order number _ _ _
Customer
Item number Description Quantity sold Price Dollar amount
XXX XXXX 99 $99.99 $99.99
Sales clerk number _ _ _
Time stamp: 00:00
Date stamp: MMDDYY
• Complete each sales order
• Time and date stamp each sales order
• Hold in sales order drawer each day unitl 4:00 PM, then hand-carry to batch control clerk
Batch control clerk
Sales order
Batch control ticket
Branch number: 999
Batch number: 9999
Today's date: MMDDYY
Total number of sales orders: 99
Hash total: 9999
Dollar quantity total: $9,999.99
• Batch sales orders into groups of 25 or less
• Sum sales order numbers for a hash control total
• Sum dollar amounts for a dollar control total
• Enter batch control ticket data into a batch register
• Transmit batches to central data processing via bonded courier
Checked and filed
Central data processing
Magnetic disk
Edit control list
• Time and date stamp each batch control ticket and enter into the batch receipt register
• Key each batch of sales orders to a magnetic disk for computer processing. Compute hash and dollar control totals and compare with the totals on batch control ticket
• Perform edit checks during the keying process
• Each record is printed on an edit control list along with branch number, date, batch number, and control totals
• All errors are corrected immediately
Batch of processed sales orders
Edit List
Date: MMDDYY
Branch number: 999
Batch number: 9999
Total number of sales orders: 99
Hash total: 9999
Dollar quantity total: $9,999.99
Sales order list
Magnetic disk containing checked and edited sales orders
Used to
Update accounting files and prepare various accounting reports

Figure 12.4 (OPPOSITE PAGE)
Document control systems.

date of the batch, number of sales orders within the batch, hash total (a grand total of any number such as sales order serial numbers), and financial total (a grand total or totals calculated for each field containing dollar amounts).

The batches are transported to a central data processing system where each batch control ticket is time and date stamped and entered into the batch receipt register. Batches are then distributed to data-entry personnel for keying batches onto magnetic disk (or magnetic tape) for computer processing. Intelligent terminals compute batch totals and perform various edit checks (in cases where check digits are used, each digit is calculated to see if it matches the preassigned check digit), and all errors are corrected immediately. An edit list containing date of completion, branch number, batch number, hash total, dollar total, and a records list is sent back to the batch control clerk to be checked and filed. In some systems, the completed batches are canceled by a stamp or by perforation to make sure that they are not inadvertently transported to the data processing center and processed again. The magnetic files presumably contain error-free data ready for computer processing.

Not shown in our example are spaces for signatures (or initials) required in some batch control systems. Typical signatures include the signature of each person who:

- Prepared the batch
- Checked the batch
- Transported the batch
- Keyed the batch
- Prepared the edit list
- Checked and filed the completed batches

Controlling Errors Through Audit Trails

If source document-based data input is used, as one finds in a batch system, the AUDIT TRAIL data should be present on the source documents that are prepared. For example, the preceding subsection on batch controls demonstrates an excellent source document-based audit trail.

In a real-time system where some type of direct entry is used, the audit trail should capture transactions and store them on magnetic media, such as magnetic disk or tape, sometimes referred to as a transaction log, as shown in Figure 12.5. For audit and control purposes, the transaction log should show where the transaction originated, at what terminal, when, and the user number. For example, in an insurance company the transaction log supports all entries to the general ledger accounts. An entry into the general ledger that debits accounts receivable and credits written premiums is simultaneously recorded in the transaction log. It contains the following detail support:

- User, terminal, and user identification number
- Time of day

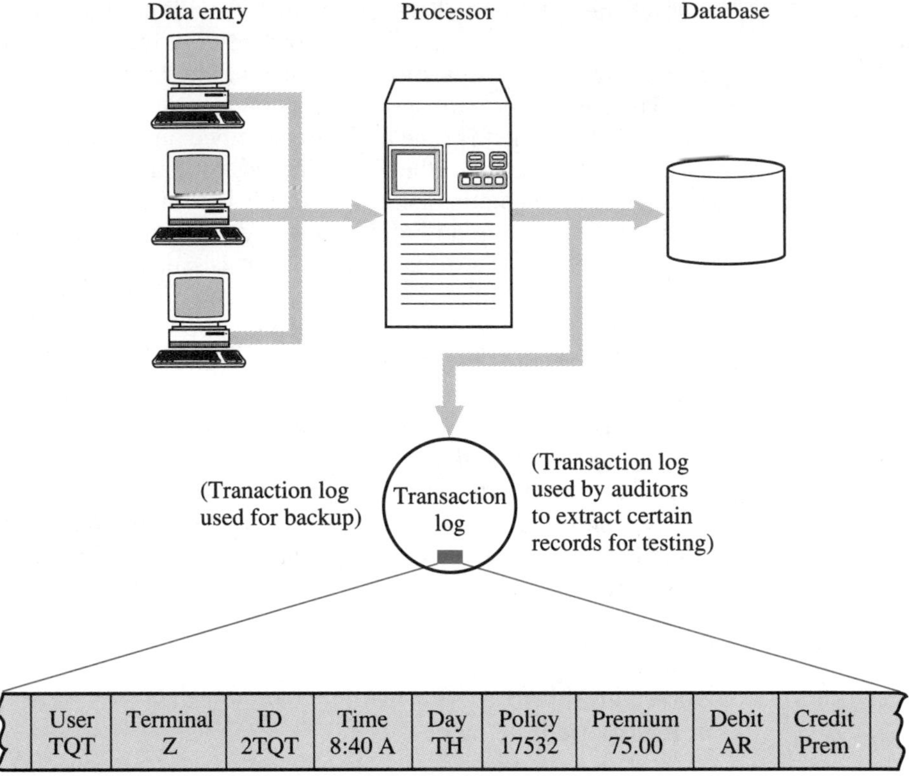

Figure 12.5
Transaction log for a real-time system.

- Day of the week
- Policy number
- Premium
- Other identifying data

Not only is such a transaction log used for audit purposes, but it can also be used for backup and recovery in the event of mishap.

Correcting Data Errors

Control over input data must be performed to ensure the following:

- All data are entered into the system.
- All errors are corrected.
- Errors are corrected only once.
- Patterns of errors are identified.

A general outline of how these tasks are accomplished is shown in Figure 12.6.

Error reports should clearly identify each error; that is, a field, record, or file error. In some systems, a space is designated for the signature of the person

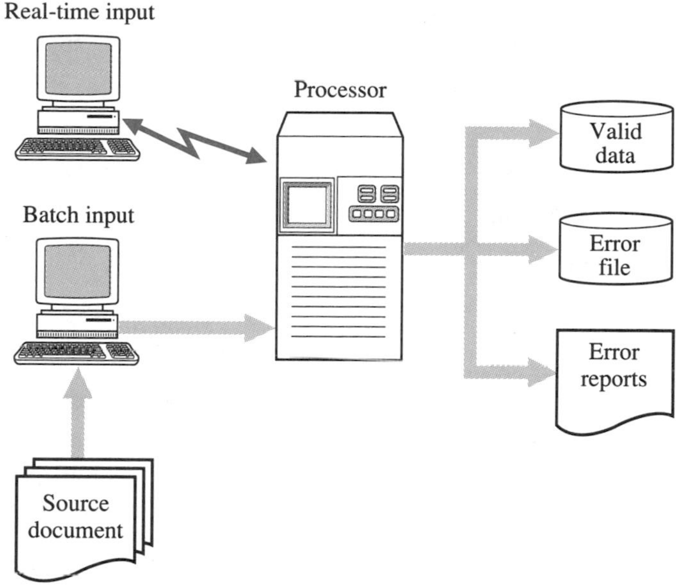

Figure 12.6
System for handling input errors.

correcting the error. This helps ensure that all errors are corrected, corrections are not duplicated, and an audit trail exists for corrected errors.

Once an error is corrected, it should be removed from the error file so that duplicate resubmissions are rejected. Ideally, error reports are used to make corrections. If they contain signatures, duplications should be avoided.

To reduce the cause of errors, a pattern of errors can be prepared that shows the frequency of different errors. This pattern may indicate that more training is required for those responsible for gathering, preparing, and inputting data. Or it may show that the input systems component needs to be redesigned.

The key to designing and implementing controls that will be executed by computer programs is to design the controls and then write the code pertaining to the controls while the programs are being developed, not after the programs have been written. Otherwise it is difficult, if not impossible, to retrofit controls in software that has been developed and that is presently operating. All the input controls just covered must be embedded in the program code. For example, check digits are calculated by the appropriate program module; validation and identification controls are checked by certain program modules; control totals are computed and displayed; and transaction logs are updated.

GUARDING AGAINST UNRELIABLE SOFTWARE

Our specific concern in this section is with the reliability of the application software itself. Although we devote a great deal of time later in the book to software designing, coding, and testing techniques, in this section we present descriptions of special types of unreliable software. Ways to detect unreliable software are discussed in Chapter 18. These types of software are often destructive and, in some cases, fraudulent. **DESTRUCTIVE AND FRAUDULENT SOFTWARE** contains:

- Salami techniques
- Trojan horses
- Logic bombs
- Worms
- Viruses

Salami Techniques

A programmer in a bank could calculate interest for customer accounts and round down each of these calculations. The balances of less than a penny could be placed in the programmer's personal account. This is a variation of the **SALAMI TECHNIQUE** applied to fraudulent programming procedures. If the programmer takes a small slice, no one realizes that anything is missing until the programmer has absconded with a sizable aggregate of money. Then it's usually too late to do anything about it.

Trojan Horses

A **TROJAN HORSE** is a set of program procedures that will sometimes perform unauthorized functions but will usually allow the program to perform its intended purpose. For example, a programmer may develop a program to process customer files correctly. Yet within this program is a set of procedures that will extract certain data from records according to predetermined criteria and will write the data on a tape file that the programmer has designated as a test file. This test file, holding valuable customer information, is removed from the premises and sold by the programmer to a mail-order company for a large sum of money.

Logic Bombs

A **LOGIC BOMB** does something destructive or fraudulent either unintentionally or intentionally. For example, a logic bomb may be a poorly written program that destroys data or files when it is executed. On the other hand, a logic bomb may be intentionally implanted by a programmer to take effect on a particular date, when an event has occurred, or when a certain condition exists.

There are numerous examples of logic bombs. One occurs where a programmer has inserted procedures into a payroll program that cause the application to destroy critical files if the programmer's employee number does not appear on the payroll input data. Thus, if the programmer were ever fired, files would be destroyed to avenge the termination.

Worms

A **WORM** is a program that replicates (i.e., rewrites itself repeatedly) and spreads, but it does not require a host, nor does it attach itself to a program—a worm is a stand-alone program. Worms usually spread within one computer or network of computers; they are not spread by the sharing of software, as viruses (discussed next) so often are. Worms are capable of transmitting themselves over a modem link. The best-known example of a worm is the 1988 Internet worm, which infected and disabled at least 6200 UNIX-based computers and disrupted, directly or indirectly, the work of more than 8 million government and university employees in a single day. Worms may or may not destroy data. The Internet infection affected no data; computers were disabled because the rogue program took control of all available memory, making it impossible for the computer to do anything.

Viruses

A **VIRUS** is a self-replicating (i.e., rewriting and reproducing itself) segment of program code. When the virus is introduced into the computer, whether by insertion of an infected disk or by a disgruntled programmer, the virus code takes control of the computer long enough to copy itself into memory and onto the hard disk. When other software applications are subsequently launched, the virus checks to see if the new software has been infected yet. If not, the virus copies itself onto that software, infecting it.

Although the risk of viral infection cannot be totally eliminated, the following antiviral procedures should be developed and communicated to users:

- Write-protect master disks as soon as they are taken out of the shrink wrap.
- Restrict downloading of files and programs from electronic bulletin boards.
- Use diskless workstations to decrease ways to input data from outside sources.
- Use software that has been certified virus-free, and purchase software from reputable vendors.
- Forbid the use of disks brought from employee's home or the use of pirated copies.
- Watch for unauthorized program changes, and test thoroughly authorized changes.
- If booting (i.e., loading) is done from a floppy disk, use only a locked disk that is known to be virus-free.
- Employ virus vaccines, also known as virus filters, that are designed to detect viruses before they can infect either programs or data.

CONTROLLING THE INTEGRITY OF THE SYSTEMS OUTPUT

In Chapter 8, our focus was on designing output that is appealing in form and substance. In this chapter, we stress controls that seek to ensure that the integrity of such output communicated to users is preserved.

Two major considerations affect the kinds of controls needed over output. First is the level of sensitivity of information. Information may be categorized as top secret, secret, restricted, and public. **TOP-SECRET OUTPUT** is very sensitive, such as medical records or plans to buy another company. Secret output is also sensitive, but if disclosed to the wrong people it would not affect the enterprise as severely as top-secret output. Examples are customer information, salaries, and stockholder registries. **RESTRICTED OUTPUT** is circulated to a number of users within the enterprise but is not to be disclosed to anyone outside the enterprise. Examples are various production and cost reports. **PUBLIC OUTPUT** is disseminated to public agencies, such as the Securities and Exchange Commission (SEC), or to stockholders.

The second factor that affects the choice of output controls is the process design. That is, are outputs produced by batch or real-time systems?

The level of the sensitivity factor (i.e., top secret, secret, restricted, or public) dictates the degree and stringency of controls. The process design factor determines the types of controls.

Using Real-Time System Output Controls to Protect Data Integrity

In a real-time system, output is generally printed or displayed at users' terminals or workstations. The user/system interface is online with the system, and users interact directly with the system to obtain the output required. From a **REAL-TIME SYSTEM OUTPUT CONTROLS** perspective, the major concerns are preventing unauthorized:

- People from intercepting the transmission of output from the system to the user
- Viewing of output displayed on the terminal
- Removal of output from terminals that have removable storage devices

Figure 12.7 presents these three critical areas, where control must be exercised over real-time output.

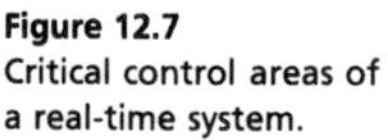

Figure 12.7
Critical control areas of a real-time system.

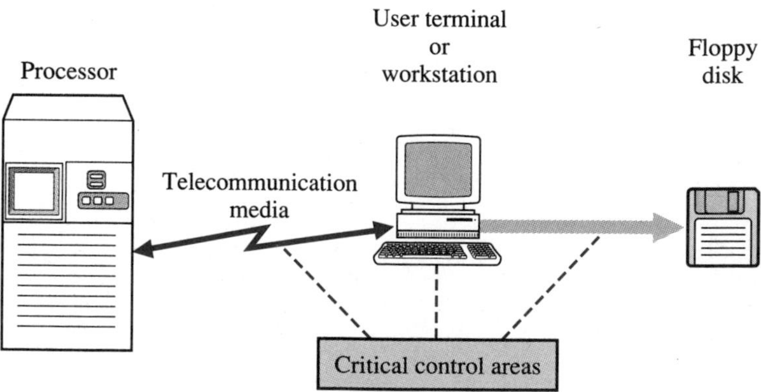

Telecommunication Controls

If systems designers want to reduce the chance of unauthorized persons intercepting top secret and secret information, **ENCRYPTION** techniques (also called cryptosystems) must be installed. Encryption techniques scramble data into secret form so that the original meaning is not understood by anyone but the intended user. Prominent cryptosystems fall into one of two categories: single-key or double-key.

The Single-Key Data Encryption Standard (DES) Cryptosystem IBM developed DES in 1977. Its function is illustrated in Figure 12.8. The secret key of this cryptosystem is known by both the sender and receiver. Sender A supplies a cleartext message and a secret key to an encipher (encrypt) algorithm, which produces the ciphertext. The ciphertext is transmitted and deciphered (decrypted) by receiver B using the same key that produced the ciphertext.

The Double-Key Public Key Cryptosystem The double-key public key system shown in Figure 12.9 is an alternative encryption method. Unlike the DES, the public key system uses two different but mathematically complementary keys, one for encryption and the other for decryption. Because the decryption key cannot be derived from the encryption key, a user can let the encryption key be public and keep the decryption key secret to ensure secure messages. The user can also keep the encryption key secret and make the decryption key public in order to sign documents digitally.[1]

Assume that A wishes to send a top-secret message to B using the public key system. A encrypts the message using B's published key. The only one who can decrypt the information is B, using his or her secret key.

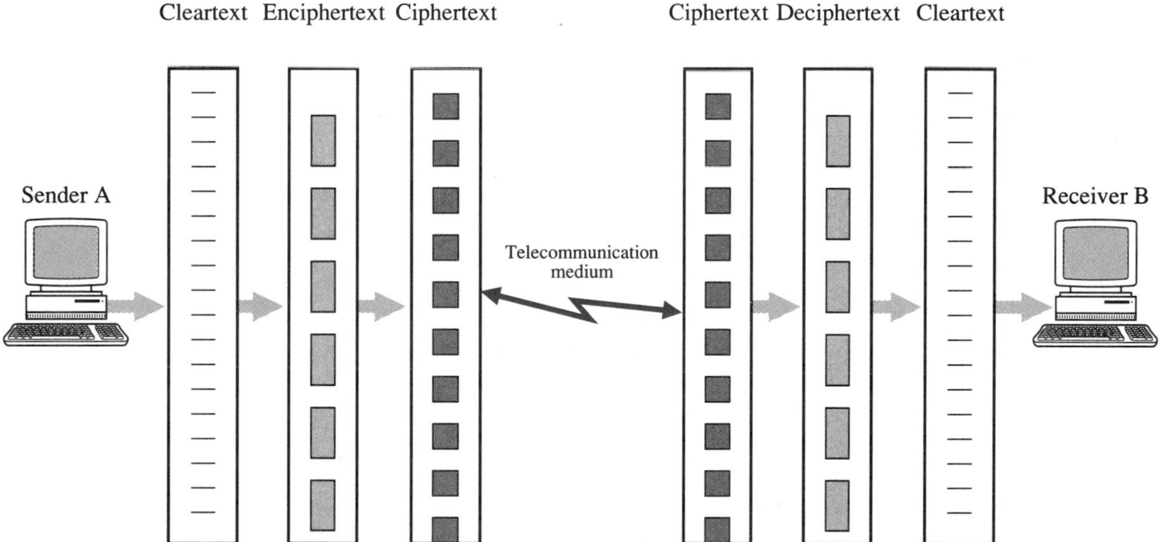

Figure 12.8 Data Encryption Standard (DES) encryption system.

[1] Steinberg, Steve, "A Student's View of Cryptography in Computer Science," *Communications of the ACM,* Vol. 34, No. 2, February 1991, p. 16.

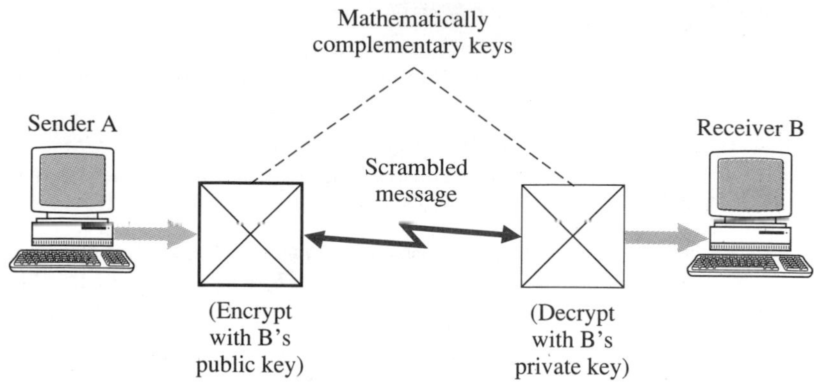

Figure 12.9
Public key encryption system.

SPOOFING is the main area in which the public key system is vulnerable. For example, C can send a message to B and call himself A. The message appears authentic to B because it is encrypted by B's public key and decrypted by B's secret key. The spoofer, however, can be shut out by using a combination of keys to ensure authentication. For example, A encrypts the message with A's secret key and B's public key. B then decrypts the message with B's secret key and A's public key. This way, the sender is uniquely identified, and B is assured that the message was originated by A.

Terminal Controls

Unauthorized scanning of information displayed on the terminal screen can be controlled in several ways:

- Place each terminal or workstation in a separate room with tight access controls.
- Use hoods on terminals.
- Display information at a low intensity.
- Blackout screen automatically after some period of nonuse.
- Position terminals so users sit with their backs to the wall.

These methods can help safeguard the screen from viewing by unauthorized people. Should a user leave the terminal, he or she should log off.

Least-privileged access provides access authority necessary to perform assigned tasks. For example, a terminal located in a warehouse easily accessible by various personnel may be given few access privileges and permitted to perform only low-level tasks on the network. For example, payroll data would be prohibited from displaying on the warehouse terminal.

If users have access to unsupervised or private printers, sensitive information can be printed and easily taken offsite. Therefore, the network should be configured for shared printers. Also, a printer log should be installed and one person given authority and responsibility of separating and routing output back to authorized users.

Terminal authentication ascertains that the proper terminal is connected and operating before transmission is accepted. Authentication software checks to determine transmission type and terminal ID and location. In some situations, private and dedicated transmission lines are used to connect certain terminals. Periodically, the computer disconnects a terminal and redials the terminal number to ensure that the user is reaching the system from the proper terminal. If it is a bogus terminal, the callback will occur in the legitimate location, not in the bogus terminal location.

The terminal or terminals regularly used by a person to access the system, the days of the week, and times of the day a person normally accesses the system are predictable. As a consequence, the security software should assure that all three conditions—valid user, valid terminal for this user, and valid day of week and time of day for this user—are met before allowing access to sensitive information.

Floppy Disk Controls

The microcomputer-to-mainframe connection is usually very good from a user design perspective. From a control design perspective, it can present some real risks. Data downloaded from the corporate database, changed, and then uploaded back to the corporate database can corrupt the corporate database. Data downloaded from the corporate database and copied on floppy disks also present some real risks of sensitive information falling into the wrong hands. Also, viruses often enter the system via floppy disks. To prevent these problems from happening, diskless workstations are recommended, because users are prevented from copying sensitive data from the mainframe database and taking these data offsite. There is no way to do so. Neither can users make unauthorized copies of programs or insert unauthorized data or programs.[2]

Using Batch System Output Controls to Protect Data Integrity

If we concentrate strictly on output, batch systems generally require more controls over output, because batch output typically involves lots of hardcopy reports. Thus output flows through more hands from the point at which output is produced to where the end user receives the output.

Figure 12.10 illustrates all the possible stages through which a batch-generated report may pass during the output process. Some reports may not pass through every stage. For example, some reports may be printed directly rather than spooled, some reports may not need decollating and bursting, and some reports may be distributed directly to the end user. Identifying all the stages through which a report may pass provides systems designers with key points where controls can be applied. A full set of controls, however, may be applied only for top-secret and secret output. In any case, all the **BATCH SYSTEM OUTPUT CONTROLS** that should exist at each stage are described below.

---

[2] Bob Francis, "A Secure Future for Diskless Workstations," *Datamation,* November 15, 1990, p. 54.

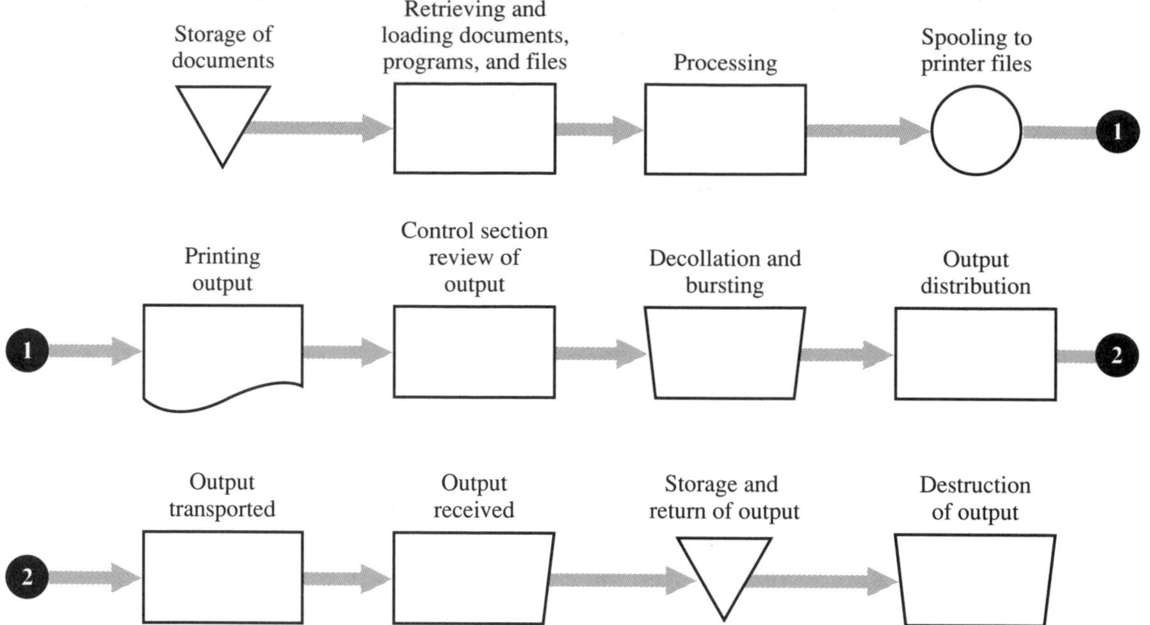

Figure 12.10 Stages in processing batch-generated reports.

Document Controls

Generally, all systems, batch or real-time, print output on a variety of documents, some of which carry top-secret information and output that can be misused. Some of these documents are checks for paying employees or creditors, stock certificates, and various negotiable instruments.

Such documents should be stored in a secure place and accounted for in the same way that expensive inventory items are accounted for and controlled. All vital documents should be prenumbered. Only authorized users should have access to documents. For top-secret documents, a dual-custody and dual-signature procedure requires that two authorized users must jointly retrieve and sign for such documents.

Loading Controls

Documents are issued to authorized personnel only. Programs are stored under the control of a program librarian package. Batch files are stored in a secure site under the control of a human librarian. Like documents, files are issued to authorized personnel only and accounted for. After processing, files are returned to the file library immediately.

Processing Controls

The main concerns here are that the correct version of the program performs the processing and that no unwarranted interventions occur. No business transaction should ever be entered by computer operators. Only computer operators should operate the computer. Access to computer facilities should be restricted to authorized personnel only. The console log should be in the form of a continuous paper printout. The console log gives a running account of all the messages generated by the computer and of all the instructions and entries made by the computer operator during a particular processing run.

Spooling and Printer File Controls

If a report cannot be written directly to a printer, the output is spooled, and a printer file is created. When the printer becomes available, spooling software accesses the printer file and prints the report. This spooling output to printer files provides an opportunity for unauthorized modification and copying of reports. Therefore, tightly controlled access to the spooling software and printer file is recommended.

Printing Output Controls

Controls over printing have two goals:

1 Prevent operators and others from scanning top-secret information printed on the reports.

2 Ensure that only the required number of copies of reports are printed.

The operator's procedures manual should state clearly the number of copies to be printed. Also, the printing of unauthorized copies can also be controlled by how many copies of sensitive documents are issued in the first place. Moreover, if control totals, such as the number of payroll checks to be printed, are designed in the process component, the likelihood of printing unauthorized copies is reduced.

In case of top-secret information, it may be necessary to prevent operators from viewing any report contents. This control objective can be achieved in several ways. The report can be printed at a remote location under tight access control. The report program can print several covering pages to permit operators to perform printer housekeeping functions before the sensitive contents of the report are printed. Special multipart documents can be purchased with the top copy blacked out so that the contents cannot be read. Authorized users tear off the unreadable top copy. The duplicate contains the readable output. Some authorities suggest completely filling a print line with characters so that the authorized user has to apply a template to a page to detect report characters from characters used to obfuscate the report contents.

Review Controls

The control section should perform general checks on printed reports. For example, the reports should be scanned for obvious errors, such as missing title, time, date, and an end-of-report indicator.

Decollation and Bursting Controls

Any chances of making copies or removing pages from reports should be eliminated. All reports should be transported directly to and from the decollation and bursting facilities. Upon return of reports to the control section, all reports should be checked to see if they are complete. Moreover, all carbon paper should be shredded so that the imprint of the reports cannot be read.

Distribution Controls

Reports may be delivered directly to end users, or reports may be placed in lockers and retrieved by users having keys to the appropriate lockers. Or an authorized courier can be used to deliver reports to remote locations. End users may have to pick up top-secret reports in person and sign for them.

Storage and Retention Controls

Usually, a retention date is specified for certain output. Each type of report should be considered with respect to future reference needs, such as supporting title to property; payments made to creditors; claims against outside parties; and requirements by government agencies. Until the retention date has expired, reports should be filed and stored in a secure location. Records of retention are themselves sensitive. They require attention during systems design.

Destruction Controls

When reports are no longer useful, they should be destroyed. With the use of classification procedures and retention dates, reports can be indexed in a **TICKLER FILE** (i.e., a file that brings matters to the attention of people in a timely manner). Thus reports can be destroyed on an automatic, routine basis. For security and confidentiality, reports should be shredded or cremated.

PROTECTING THE DATABASE

This section examines **DATABASE CONTROLS** particularly applicable to the database systems component:

- IBM's DB2 security model
- Concurrency controls
- Encryption techniques
- Backup and recovery controls

DB2 Security Model

IBM's DB2 security procedures provide an excellent user control model for databases. The DB2 security philosophy is: "If the object was defined within DB2, then DB2 will control access to it. Any activity involving a DB2 object, resource, or function is a privilege requiring authorization."

Implicit authorization is given through the CREATE command. Explicit authorization or loss of authorization is assigned through GRANT or REVOKE commands, respectively.

Following initial installation of the database, or anytime thereafter, the database administrator can grant authority for specific operations to specific users; for example, GRANT SELECT, UPDATE (CUST NAME) on CUST

TABLE to SAM. Sam is able to read and update all customer names in the customer file, but nothing else. Some of the privileges include create, display, reorganize, alter, delete, insert, and recover.

Privileges can be given and also taken away. The general form of the REVOKE command is: REVOKE privileges ON resources FROM userid.

Concurrency Controls

Concurrency means that two or more users have access to the same data at once. Although this is a desirable design feature in a shared real-time system, it increases the chance of conflict. For example, two salespeople may be trying to sell the last gizmo in inventory at the same time. Only one can win. Such a situation requires **CONCURRENCY CONTROLS**.

In some advanced relational databases (e.g., IBM's DB2), an automatic locking, conflict detection, and resolution feature is available. The user who first accessed the item of contention locks it, assuming ownership until he or she commits the transaction or cancels it. The system detects the "younger" user, who is trying to acquire update control of the same item, and causes that transaction to wait until the current transaction has been completed.

Encryption Techniques

Although many people view encryption techniques as used exclusively to protect data sent over transmission media, it also applies to protecting the privacy of databases. Portable storage media, such as removable disk packs and magnetic tapes, can be protected by implementing a secure encryption device in a disk drive or tape drive controller. The data are automatically encrypted each time they are written and decrypted each time they are read by an authorized user.

Database Backup and Recovery Plans

For batch systems, a three-file backup and recovery plan, called the **GRANDFATHER-FATHER-SON RECOVERY PLAN**, shown in Figure 12.11, is generally used. The son, or most current master file, is onsite. The father is in the vault and the grandfather is stored in a secure area offsite. If the son is lost, the father and previous transaction file are rerun to recreate the son. If the father is lost, the grandfather and previous transaction file are rerun to recreate the father.

For real-time systems, there is always a trade-off between frequency of dumping (copying) and logging (accounting for and copying each transaction that updates the file). Frequent dumping provides fast recovery, but is expensive and time-consuming. Infrequent dumping and depending on the log for recovery is slow but relatively inexpensive. A balance must be struck between these two extremes. A logging and dumping design is presented in Figure 12.12.

In any event, dumping can be complete or residual. Complete dumping simply means that everything in the database is copied. The residual dumping method copies all records that have not changed since the last residual dump, and the transaction log contains only before- and afterimages of the records that have changed. If the current file is lost, the last residual dump contains all

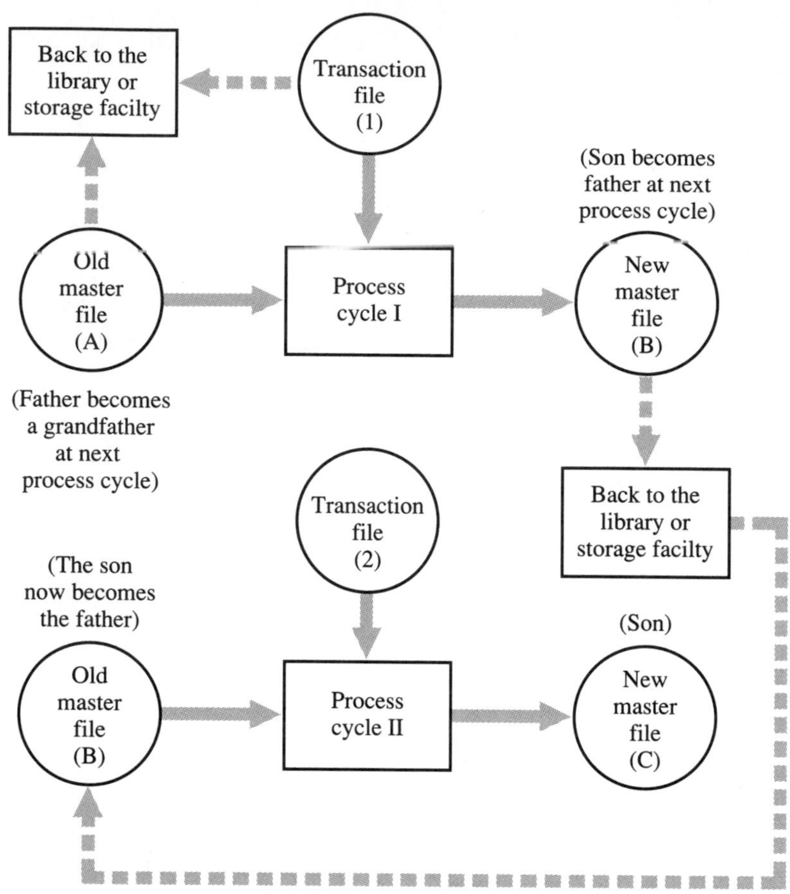

Figure 12.11
Grandfather-father-son file recovery plan for a batch system.

the records that have not changed since the last dump. These records plus afterimages of the records that have changed recreate the current file.

Offsite Backup

A company backs up files, programs, and documentation offsite in secure areas to protect itself against threats of human beings and nature. Two methods are:

- Remote secure storage facility
- Televaulting

Remote Secure Storage Facility

A number of entrepreneurs have converted bomb shelters, old government buildings, abandoned mines, and caves into **REMOTE SECURE STORAGE FACILITIES**. Many of these facilities are converted to James Bond–like environments.

Huge limestone caves 15 to 20 acres in size have been used by security firms to construct vaults, buildings, roads, and even artificial parks. Such facilities are actually underground cities devoted to secure storage. Copies of the

Figure 12.12
Real-time system that logs transactions and periodically dumps all or a portion of the database.

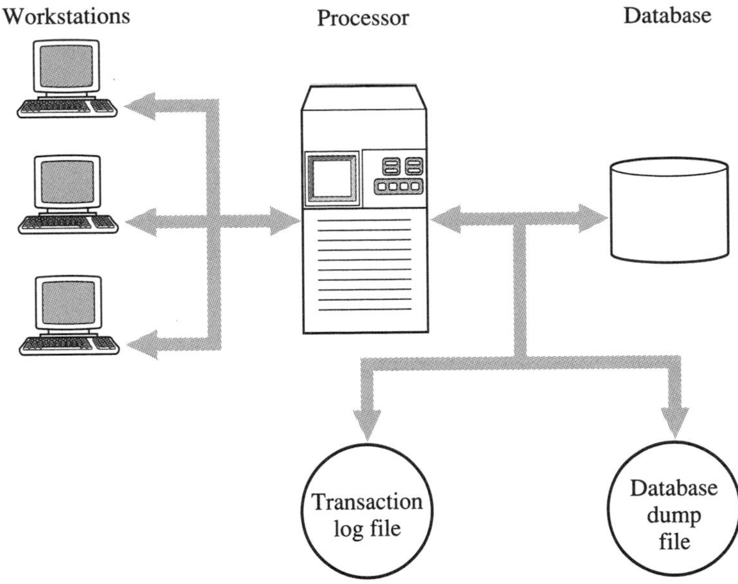

database (as well as other sensitive documents and programs) are carried to these remote sites for safekeeping.

Televaulting

No matter how often database backup is performed, the data are only as secure as the last backup completed, and even the most careful manager could lose a day's worth of data in a disk crash. For enterprises with mission-critical applications such as brokerage houses, airlines, and banks, loss of any data is unacceptable, because data loss and downtime could significantly threaten the very survival of such enterprises. These enterprises need a fault-tolerant backup system such as a **TELEVAULTING** system, which is the transporting of important data to backup sites electronically.

Typically in televaulting (also called electronic vaulting), the computer system is linked to a secure remote location by a high-speed transmission medium. Every workstation and mainframe is connected to the offsite electronic vault, and every transaction is transmitted online and in real-time. This method eliminates the risk of physically transporting disks, tape files, and documents to the offsite facility. With televaulting, backed-up data can be accessed any time, and the backed-up data are always current.

Onsite Backup

Another form of fault-tolerant backup system is called **DISK-MIRRORING**. It is similar to televaulting except that data are not transmitted to a remote site but transmitted to a local, secondary disk.

Disk-mirroring software writes data on two separate servers simultaneously, so if one disk crashes, the system automatically switches to the "mirror'' or secondary disk with no loss of data and no system downtime. More and more information systems must be up and running 24 hours a day, so disk-mirroring is critical.[3]

PROTECTING THE HARDWARE TECHNOLOGY

Our main objective in this section is to present **TECHNOLOGY PLATFORM CONTROLS** that are particularly applicable to mainframes and personal computers. We begin with mainframe controls.

Mainframe Data Center Controls

For any company, a mainframe data center will be the most expensive and difficult to build. Control features add to expense and building complexity. Some control features that help create a secure mainframe data center follow:

- ***Physical Location*** The data center must be located away from processing plants, radar towers, gas and water mains, congested highways, airports, high-crime areas, and geological faults. The location should be well above the flood level and should provide proper drainage.
- ***Construction*** Generally, the data center should be a one-story block building with limited entry and exit points. Power and communication lines should be underground. Air intakes and windows should be covered with wire mesh. All ducts should contain fire dampers.
- ***Filtration*** A surgically clean site is a must for sensitive computer technology. Special filtration systems can handle paper and ink dust and paper mites that accumulate in the ceilings in heavy printing environments.
- ***Air Conditioner*** Every computer manufacturer expects its equipment to be maintained in a suitable environment, typically at a temperature of 70 to 75 degrees Fahrenheit and a relative humidity of 50 percent. If a site permits the standards to slip, vendors will invalidate warranties.
- ***Backup Water*** The municipal water supply may not be sufficient in case of a fire. The site may therefore require an additional water source and storage tanks. Generally, a water tower is located near the data center to provide chilled water to the computer site in case of emergency or for extra cooling.
- ***Emanation Protection*** Computers and peripherals emanate sensitive information that can be intercepted by unauthorized people. Two methods are used to prevent emanation of data:

  1 Containment encloses the data center and equipment with emanation-proof materials.

---

[3] Sally Winship, ''Disk-Mirroring Products Offer True Fault Tolerance,'' *PC Week*, February 4, 1991, p. 112.

2 Source suppression makes circuit boards and wiring in the equipment itself emanation-proof.

Specifications for source suppression is found in the federal government's **TRANSIENT ELECTROMAGNETIC PULSE EMANATIONS STANDARD (TEMPEST)** security guidelines. TEMPEST products limit electronic emanation levels, making it difficult to use sensing equipment to tap data transmissions. Although these products are currently used almost exclusively by the government sector, some vendors are targeting their offerings to business, especially the financial sector.

Fire Suppression Systems Over half of all companies that suffer fires go out of business. Computers can be replaced, but unless it is properly protected, critical data can be lost without a trace. The loss of an accounts receivable master file, for example, is enough to cause most companies to go belly up. A fire suppression system to guard against such disasters emphasizes early detection and containment.

Uninterruptible Power Supplies It is difficult, if not impossible, to run current-day computers, telecommunications switching equipment, and electronic peripherals on commercial-grade power with its inherent problems. Power problems include flickers, sags, surges, brownouts, blackouts, and frequency variation. Equipment used to handle several different combinations of these problems include voltage regulators, surge suppressors, isolation transformers, power conditioners, and motor generators. **UNINTERRUPTIBLE POWER SUPPLIES (UPS)**, however, are the only systems that protect computers against all of these power problems. Currently, three types of UPS are available:

- Static
- Rotary
- Hybrid

Each has advantages and disadvantages, but rotary and hybrid systems are generally more rugged and expensive.

Static systems rely primarily on batteries. Rotary systems use a generator as backup. Such systems are usually driven by a diesel engine for reasons of economy and safety. Hybrid systems include batteries and a generator. When the batteries begin to run low, the generator kicks in to continue generating power as long as fuel is available. Rotary and hybrid UPS are used for situations in which extended downtime cannot be tolerated.

Personal Computer (PC) Controls

Some of the mainframe controls can be applied to personal computers (PCs), and vice versa. In this subsection, however, we want to emphasize controls especially applicable to PCs.

Environmental Controls PCs must be located in a contained and locked site away from public or easily accessible areas. Never leave PCs unattended when they are switched on. Establish after-hours access control for authorized personnel. Restrict access to file libraries and lock diskettes in secure storage. Prohibit smoking, drinking, and eating near PCs. Locate PCs away from magnets, telephones, and radios.

In sensitive areas, deploy video camera surveillance and alarms. Have night watchmen check the site randomly. Insist that defective parts, such as memory chips, be turned in after hardware maintenance.

Maintain a physical inventory of PCs. Keep a list offsite of descriptions, serial numbers, and most recent movements or additions. Use labeling devices that contain permanent identification tags. Use property transfer forms, sign-off sheets, and passes to control PC movement. Conduct a periodic equipment inventory audit.

Do not place a PC on a primitive multiplug power line shared by other electrical equipment, because this can subject PCs to voltage fluctuations that can damage chips and destroy data. In most environments, voltage regulators and surge suppressors are required. Uninterruptible power supplies may or may not be a requirement for PCs. If the exposure is high and the processing critical, a UPS for PCs may be justified.

Physical Controls Physical controls include motion detectors, keylocks, and physical locking devices. Motion detectors will set off an alarm if the PCs in which they are installed are moved slightly. Keylocks secure the PC chassis and keyboards, and they can also be used to make sure that the system doesn't boot. Keylocks are thus a good first-line defense. Even if a keylock secures the PC's chassis, the system may still be vulnerable to theft. If office security is weak, PCs and peripherals can be lifted and walked out the door. Therefore, the PC itself should be secured to the desk by a cable or bolts. For monitors or printers that lack secure bolts, a plate-mounted cable fixture that is permanently attached with cyancrylate glue can be used.

Database Controls Set up a diskette library and keep an up-to-date diskette log. Code the labeling of diskettes with sensitive content and require a standardized labeling procedure for other diskettes. Make backup copies regularly. Never leave idle diskettes or cartridges unattended. Store them in the library when not in use.

Creating a Disaster Recovery Plan

The following case should give you an appreciation for **DISASTER RECOVERY PLANNING**.

Disaster Recovery Planning: Thinking the Unthinkable

Several years ago, Marla Cavuto worked as senior systems analyst for a midsized hospital supply company located in San Francisco. Marla had been employed at the company only a few months when an earthquake hit, destroying most of the company's buildings and obliterating the information system. Some of the critical files, such as the accounts receivable master file, were backed up and stored in a remote warehouse, but it, along with all its contents, were destroyed by a fire triggered by the earthquake.

After the disaster occurred, the company tried to continue business, but it became impossible to reconstruct its data, documents, and programs. It filed for bankruptcy protection, but eventually went out of business. Since then, Marla has worked as an adjunct professor at a university in San Francisco and as a part-time systems consultant.

Today, Marla is being interviewed for the senior systems analyst position at Medico, a hospital supply company located in Houston. Carl Armstrong, chief financial officer (CFO), asked Marla what she would try to do if hired by Medico.

"As I've been told, Medico is planning to make a major systems replacement. I also understand that this undertaking has been given a name: that is, information systems for competitive advantage, or ISCA for short. I'm very excited about being part of this project. Along with strategic systems planning, I would strongly recommend disaster recovery planning."

"Why disaster planning?" asked Carl. "Aren't the preventive controls that we design into the system as it is being developed sufficient?"

"I agree that controls should be designed into the system as it is being developed, but this is not enough. If a disaster, such as an earthquake, hurricane, flood, or fire strikes, the company must have in place a plan to keep the business running. To keep the business running requires that all or a portion of the information system continues to operate. If it's destroyed and there's no disaster plan in place, the business will not be able to service customers, process payrolls, and the like. Also, all of its data, documentation, and programs may be lost, without any chance of recovery."

"You really seem to be convinced that disaster planning is necessary," said Carl.

"I believe it is imperative," said Marla. "The risk is too great without a disaster recovery plan. I can speak from personal experience."

"Would you mind telling us about your experiences?" asked Bill Lawson, internal auditor.

"As you learned from my earlier conversations with you, the company I worked for, very similar to Medico, went out of business because of the earthquake. Had we had a disaster recovery plan in place, I believe the company would still be in business today. I pleaded with top management to install a disaster recovery plan as soon as I was employed. Unfortunately, my pleas fell on deaf ears."

"What actually happened?" asked Bill.

"Well, the earthquake struck without warning. No preshocks, no alert from seismologists. It lasted 15 seconds, during which time the energy equivalent of a hundred 20-kiloton atomic bombs was released. The peak of the Transamerica Pyramid swayed three feet each way, as did two support towers on the Golden Gate Bridge. The Bay Bridge was knocked out. Fires broke out in the Marina District, where we had our files backed up. The building in which our processors, files, and various peripherals were located was destroyed."

"I can see your point about the need for disaster planning," said Carl. "It can happen to us."

"Maybe not an earthquake, but a hurricane, flood, fire, or some other disaster can occur. If it does, the disaster recovery plan will provide you with a means to continue operating," Marla responded. "Since the earthquake, I've thought about a disaster recovery plan in terms of a will. It forces you to think about the unthinkable. What should be done when the disaster is occurring? What should happen after the disaster has occurred? How will the company recover from the disaster?"

"Once a disaster recovery plan is developed, what else should be done?" asked Bill.

"I'm glad you asked this very important question," Marla responded. "The disaster recovery plan, once developed, should be tested. People should practice their assigned tasks. Drills should be conducted. You must make sure that your plans actually do what they say they will; otherwise, they are just documents sitting on shelves collecting dust, or as some people say, they are just shelfware."

"Marla, we appreciate your presentation of disaster recovery planning," said Carl. "You've really opened my eyes. We will, I'm sure, become very dependent on our new ISCA. Downtime will no longer be acceptable. You've certainly sold me on the idea of disaster planning."

When a disaster strikes, it will happen suddenly and unexpectedly. Very little warning time for research and planning will exist. Quick decisions must be made. The disaster recovery plan helps to ensure that the right decisions are made and the enterprise survives.

Components of a Disaster Recovery Plan

The **DISASTER RECOVERY PLAN** contains four subplans:

- Prevention plan
- Contention plan
- Contingency plan
- Recovery plan

Prevention Plan A complete system of controls is designed into the information system as it is being developed, or controls are installed in an old system. In either case, the types and levels of controls installed are a function of risk assessment. This plan details how to protect the system and to avoid the disaster in the first place.

Contention Plan The prevention plan cannot guarantee prevention. Therefore, a plan must be developed to prescribe how to react and carry on while the disaster is occurring. The first few hours of a disaster are the most critical and chaotic.

Contingency Plan This plan describes how the company will operate and conduct business while recovery efforts are taking place, no matter what the condition of its information system. The plan contains alternative and emergency procedures. The contingency plan enables the company to survive and operate. Emphasis is on any disruption that threatens business continuity.

Recovery Plan The disaster has occurred, and it has damaged the system. The recovery plan includes procedures for restoring the system to full operations. The long-range recovery plan may entail anything from acquiring some new peripherals to the construction of a new data center complete with new equipment.

Preparing a Contingency Plan

The primary focus of this subsection is on how to prepare the contingency plan. What follows are the steps to do this.

Step 1. Identify Critical Functions. The first step in developing a contingency plan is to identify functions necessary to keep the company operating. Typical critical functions include:

- Transaction and order processing
- Accounts receivable
- Accounts payable
- Inventory control
- Pricing and billing
- Payroll

Copies of all these applications, including programs, data, and documentation, should be stored in a secure place offsite.

Step 2. Select Contingency Options. One simple option is fall-back to manual methods for some applications. Or a critical function, such as payroll, may be performed by an outside service. More comprehensive contingency options include:

- Company-owned backup facility
- Reciprocal agreement
- Hot site
- Cold site
- Mobile data center

If a company has a centralized computer system (e.g., large data center), a **COMPANY-OWNED BACKUP FACILITY** is duplicated somewhere distant from the present system. If a company has a large network with multiple sites and a site suffers a disaster, another site can process the injured site's critical and necessary functions. Company-owned contingency options such as these require a great deal of redundant equipment in the technology platform, and they are more expensive than other contingency options. If the mix of computer and network equipment is large and complex, such an option may be the only one totally feasible. A **RECIPROCAL AGREEMENT** may be made with an organization that has a comparable system. Reciprocal agreements have built-in difficulties, including possible changes in the configurations after the agreement is signed, insufficient capacity, and scheduling problems.

HOT SITES are secure facilities that provide system configurations that mirror the organization's system. These hot sites are owned by companies (e.g., Comdisco Disaster Recovery, Inc.) that provide clients with backup processing facilities. Hot sites are scattered around the country but linked together by a high-speed network. In some cases, client companies install direct connections between their mainframes and those at a hot site. The information at the backup site is continuously updated. If disaster strikes, the latest information is waiting at a hot site. A networked hot site system works much better than stand-alone hot sites. With a stand-alone hot site system, if a particular backup facility is unavailable at the client's preferred hot site, the client has to travel to the remote hot site and set up backup operations there. Hot sites are one of the most popular contingency options because they provide one of the fastest recovery strategies, and they may be tested periodically without business disruption.

COLD SITES, sometimes called shells, are simply buildings located within the vicinity of clients. Cold sites contain the necessary outlets and connections ready to receive a computer configuration. Cold sites can be built by an organization for its own use or subscribed to from a commercial cold site owner. Cold sites are less expensive than hot sites, but they may take several days to configure.

A **MOBILE DATA CENTER** is a system on wheels, generally housed in a large van that is transported to disaster sites. The front section of the van is for meetings, living quarters, and office space. The middle section has built-in workstations and supplies. The back section contains the main processors, disk drives, and printers. Mobile data centers remain on alert around the clock in a central, secure location.

Testing the Plan

Effectively, the system doesn't have workable contention and contingency plans until they have been tested. Moreover, all plans should be maintained on a continuous basis to represent the organization's most current conditions. It is better to know what the problems are during testing rather than to learn about them during a disaster. Analysts and auditors recommend that organizations test their plans twice a year without advance warning.

Passive testing approaches include observations, inspections, checklist reviews, and walkthroughs. Active testing includes physical tests that can be simulated, such as drills and rehearsals, or live tests that consist of tiger teams and break-the-glass or pull-the-plug procedures. Moreover, all personnel responsible for contention plan initiation and contingency plan activation should be trained and updated on a regular basis to ensure that these people are fully aware of the latest revisions. Full-blown disaster simulations are the best way to discover if a plan will truly pull the company through a disaster.

CONTROLLING ACCESS TO DATA

The term ''access control'' has been used or implied throughout this chapter as it relates to input, process, output, database, and technology. Therefore, we end this chapter with a presentation of **ACCESS CONTROLS**. Access controls are based on what users know, what users have, and what users are.

Granting Access Based on What Users Know: Passwords

Users are granted access to information system resources on the basis of what they know. A password is a unique word assigned to authorized users to gain access to the system. It may be combined with a personal identification number (PIN). Such an access control technique is traditional and widely used.

Other forms of unique knowledge can be used for access control. For example, the system may ask, "What is your husband's favorite hobby?" "What is your wife's favorite color?" Or the system may ask a question, such as, "What is the capital of California?" The "correct" answer may be Cleveland.

Users should be required to modify their passwords on a random, short-time cycle basis. This procedure can be programmed into the system so that when a user whose password has not been changed for a specified period of time accesses the system, the security software will not allow the user to proceed, or log off, until the user changes passwords.

Granting Access Based on What Users Have: Smart Cards

Access control is effected by a credit card–sized device with an embedded computer chip, sometimes referred to as a smart card. To use it, cardholders insert their cards into a terminal and enter their personal identification number (PIN). When an authorized PIN is entered, the card checks its program to see what authorization the user has. It then unlocks only the authorized functions. If an unauthorized person tries to fake the PIN and enters the wrong number three times in a row, the card disables itself permanently and has to be replaced. Other possession-based access control devices include keys to locks, badges, and encryption keys.

Granting Access Based on Users' Physiological and Behavioral Characteristics

If a user knows something, there is a fairly high probability that an unauthorized person can gain that same knowledge. If a user possesses something, it can be stolen. But it is difficult to impossible to steal someone's characteristics. Therefore, designing access controls on the basis of what authorized users are provides the strongest and most reliable controls.

Technically, the term **BIOMETRIC CONTROLS** describes these controls based on what users are. They are divided into physiological and behavioral characteristics, as depicted in Figure 12.13.

The physiological characteristics include the following:

- Hand geometry
- Retina signatures
- Fingerprints
- Body weight

Hand geometry devices measure, record, and compare finger length, skin translucency, hand thickness, or palm shape. Generally, they are used for physical access systems such as locked doors. Retina scans use a low-intensity light source to scan the circular section of the retina. Retina signatures of authorized users are recorded in a microprocessor. Typically, these devices are used for physical access control and data access applications. Fingerprint con-

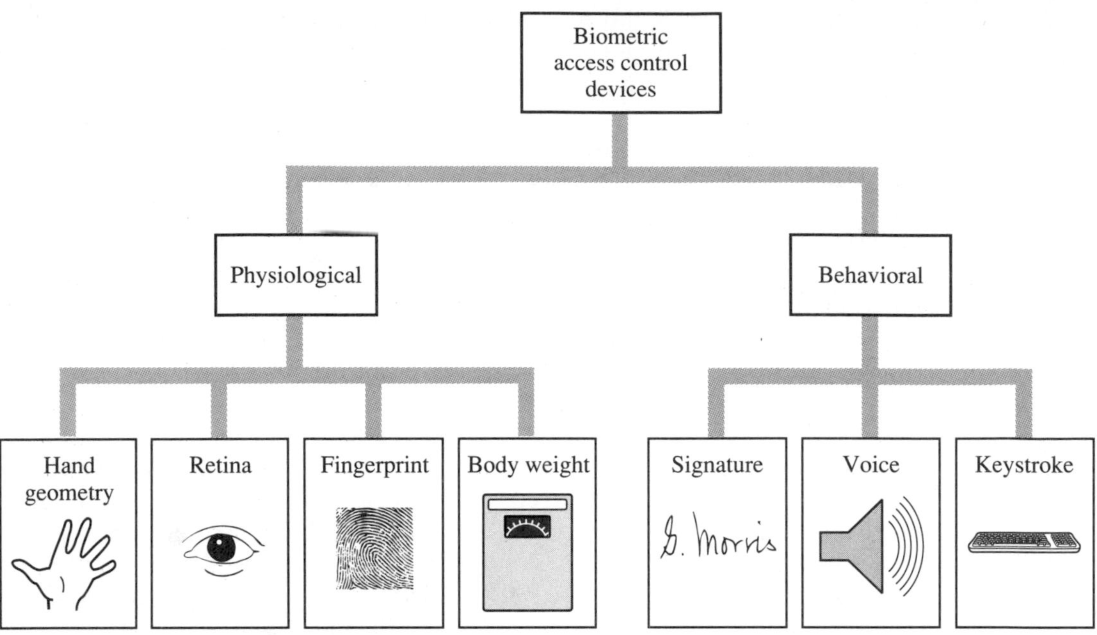

Figure 12.13
Types of biometric access control devices.

trol devices create templates of a fingerprint by measuring and digitizing the ridges, loops, and bifurcations unique to a user's print. They can be used for both physical and data access control. The security system can also include a device that weighs the person attempting to gain entry and then checks that weight against a security file. This device prevents tailgating, that is, another person trying to walk into the facility along with the authorized person.

The behavioral characteristics include the following:

- Dynamic signature verification
- Voice recognition
- Keystroke dynamics

Dynamic signature verification devices trace the pressure and speed of users' signatures, not the curvilinear features. They work well for both physical and data access control. Voice recognition devices make use of voice patterns and inflections to make unique voice templates. They are used primarily for physical access control. Keystroke dynamics identify individuals by their typing patterns and rhythms. These devices generally monitor character groups such as i-n-g, t-h-e, e-d, t-u-r-e, and t-i-o-n. Keystroke dynamics are typically used for workstation data access control.

Biometric devices may frustrate users. Most of the devices err on the side of conservatism, which means that they can at times prevent an authorized user from accessing the system.

When an authorized user is denied access, it is called an alpha or type 1 error. This rejection is obviously irritating to the user. But a more serious error from the viewpoint of tight control is a beta or type 2 error, in which an unauthorized person is permitted access. In systems that require high levels of security, a higher-than-average alpha error rate may be condoned to ensure that beta errors do not occur.

REVIEW OF CHAPTER LEARNING OBJECTIVES

The major goals of this chapter were to enable each student to achieve seven important learning objectives. We will now summarize the responses to these learning objectives.

Learning objective 1:
Give an overview of threats to information systems.

A system of controls seeks to safeguard the information system from a variety of threats, including:

- Human errors
- Destructive and fraudulent software
- Unauthorized disclosure of output to interception
- Power failure
- Theft
- Hostile environment
- Loss of databases
- Various destructive hazards, such as fires, floods, earthquakes, and hurricanes

Learning objective 2:
Explain input controls and demonstrate how they can be used to filter out erroneous input data. Also, design an audit trail for a real-time system.

The purpose of input controls is to screen out errors and prevent garbage in, garbage out (GIGO) from occurring. Input controls include:

- Coding
- Validation
- Identification
- Batch
- Audit trail

Use of check digits helps ensure that proper codes are being processed. Validation controls are exercised at three levels:

- Field
- Record
- File

Identification tables ensure that bona fide transactions are being processed. Batch controls are designed and implemented to ensure that all source documents are processed; no source documents are processed more than once; and there is a hardcopy audit trail from input, through processing, and to final output.

The audit trail for a batch system is produced by a trail of documents produced at each stage from input to final output. In a real-time system, however, such a document trail doesn't exist. Therefore, we must build one by copying all transactions onto a transaction log, usually a tape file. This transaction log serves two purposes:

1 It provides backup in case of a processing mishap.

2 It provides a complete record of transactions that can be processed easily by auditors.

If errors occur, they should be corrected immediately and a record of their correction should be made. An added, yet important, feature is a profile or pattern of types of errors that are occurring. Based on this error pattern, countermeasures can be effected to reduce such errors.

Learning objective 3:
Discuss destructive and fraudulent software to which the process systems component is vulnerable and explain how to prevent this kind of software from entering the system.

Destructive and fraudulent software has the potential of crippling or destroying an enterprise by performing malicious acts or stealing assets. Such software includes:

- Salami techniques
- Trojan horses
- Logic bombs
- Worms
- Viruses

All of these types of destructive and fraudulent software can be controlled by following the SDLC and structured programming procedures, designing a sound system of controls, and performing stringent software testing.

Learning objective 4:
Apply both real-time system output controls and batch system output controls.

The control of top-secret and secret output is imperative, whether it's produced by real-time or batch systems. The control of real-time output involves three critical areas:

- Transmission media
- Terminal
- Removable storage such as floppy disks

Encryption (i.e., cryptography) is a proven, practical way to protect telecommunication transmissions. Encryption techniques transform data in such a way that it becomes useless to a would-be interceptor. Two dominate cryptographic methods are the Data Encryption Standard (DES) and the public key system.

Sensitive information displayed on the screen can also be scanned by a snooper. To prevent such snooping, terminals should be placed in locations protected by strong access controls (e.g., biometric controls). Also, installing hoods and positioning terminals away from would-be snoopers adds a degree of protection. The terminal itself can be identified, authenticated, and authorized to perform certain tasks on a least-privileged access basis. A principal way to prevent downloading data from the database on removable storage media (e.g., diskette) is to install diskless workstations.

Batch-generated output requires many more intermediate controls than does real-time-generated output, because unlike real-time output, batch output flows through a large number of intermediaries before it reaches the end user. Controls must be established in the following areas:

- Where documents, programs, and files are loaded
- At the processing stage
- During spooling operations
- Where the documents are printed
- At the point where documents are made ready for distribution
- During transportation

After the output is received, it must be stored securely. Upon expiration of its retention date, the output must be destroyed by shredding or burning.

Learning objective 5:
Present database controls.

IBM's DB2 security model for databases is a good one to emulate. It has a well-designed method for establishing levels of authority and granting and revoking access and authorization privileges.

In modern databases that grant access to a large number of users, the problem of two or more users gaining access to the same data item at the same time is prevalent. Therefore, effective concurrency controls must be built into the DBMS to resolve such conflicts in an orderly manner.

Cryptographic methods are applicable for sensitive databases. Like telecommunication encryption techniques that scramble data to confound would-be interceptors and then unscramble the data for the end user, encryption techniques encrypt the database to make it useless for penetrators and, with the appropriate key, decrypt the data so that they can be read by end users.

Should an organization lose its database, it is at extreme risk of failure. Clearly, then, one of the most important aspects of database control is to design a proper and efficient backup system. For batch files, the grandfather-father-son file recovery plan is both effective and efficient. With real-time systems, there are countless procedures for backing up online databases. Normally, backup procedures include a combination of dumping certain records from the database and continuously logging transactions at the production site. Other companies use high-speed telecommunication media to transmit both the database and transactions to a remote, secure site. Others use disk-mirroring techniques.

Learning objective 6:
Present technology platform controls, both for mainframes and personal computers (PCs).

Mainframe data center controls include locating the center as close to a threat-free environment as possible and constructing it specifically for computer data processing. Ancillary controls include:

- Filtration devices
- Air conditioners
- Auxiliary water supply
- Emanation protectors
- Fire suppression system
- Uninterruptible power supplies (UPS)

PCs also require various environmental and ancillary controls similar to mainframes.

Disaster recovery plans are a key aspect of designing a complete system of controls for technology platforms. The contention plan and contingency plan are two critical subplans of the disaster recovery plan. The contention plan answers the question, "How do we deal with a disaster when it happens?" The contingency plan answers the question, "How do we make sure that we can continue processing when a disaster occurs?" Major contingency backup options are:

- Company-owned backup facility
- Reciprocal agreement
- Hot site
- Cold site
- Mobile data center

After a disaster recovery plan is designed and implemented, it must be updated to reflect changes in the organization and tested periodically to see if it works as intended.

Learning objective 7:
Discuss access controls and explain why biometric controls are the most effective controls against unauthorized access.

To a large extent, a sound system of controls is dependent on strong access controls to guard against intrusion by penetrators. Access control can be based on three general methods:

1 What a person knows.

2 What a person has.

3 What a person is.

Basing access control on biometric characteristics of a person gives by far the strongest control. These biometric characteristics include the following:

- Hand geometry
- Retina signatures
- Fingerprints
- Body weight
- Dynamic signature
- Voice recognition
- Keystroke dynamics

Access control devices for reading each of these characteristics are commercially available for quick and easy implementation.

CONTROLS DESIGN CHECKLIST

Following is a checklist on how to design controls to protect a system from a wide variety of threats.

1 Install a check digit system for important codes and account numbers.

2 Develop field, record, and file checks for validating input data.

3 Set up identification tables in computer storage to verify and validate sensitive or critical transactions.

4 Establish batch controls that check batched transactions from the time they occur until they are processed by the system.

5 Build an audit trail using computer storage media to record and log all transactions processed by the system.

6 Create a procedure to reconcile all processing errors.

7 Safeguard the system against destructive and fraudulent software, such as salami techniques, Trojan horses, logic bombs, worms, and viruses, by using proper software designing, coding, and testing procedures and anti-viral measures.

8 Employ encryption techniques, terminal and floppy disk controls, and batch controls to safeguard sensitive output.

9 Establish database authorization procedures, concurrency controls, encryption techniques, and backup to safeguard the database.

10 Install mainframe data center controls, fire suppression systems, uninterruptible power supplies, various PC controls, and a disaster recovery plan to protect the system's technology platform.

11 Implement access controls on the basis of what a person knows, what a person possesses, or what a person is physiologically or behaviorally. The strongest access control is based on what a person is.

KEY TERMS

Access controls
Audit trail
Batch controls
Batch system output controls
Biometric controls
Check digit
Code controls
Cold sites
Company-owned backup facility
Concurrency controls
Corrective controls
Database controls
Destructive and fraudulent software
Detective controls
Disaster recovery plan
Disk-mirroring
Double-key public key cryptosystem
Encryption
Field checks
File checks
Grandfather-father-son recovery plan
Hot sites
Input controls
Input identification controls
Input validation controls
Logic bomb
Mobile data center
Preventive controls
Public output
Real-time system output controls
Reciprocal agreement
Record checks
Remote secure storage facilities
Restricted output
Salami technique
Single-key Data Encryption Standard (DES) cryptosystem
Spoofing

Technology platform controls

Televaulting

Tickler file

Top-secret output

Transient Electromagnetic Pulse Emanations Standard (TEMPEST)

Trojan horse

Uninterruptible power supplies (UPS)

Virus

Worm

REVIEW QUESTIONS

12.1 List and give an example of prominent threats to information systems.

12.2 There are three categories of effective controls. Name and give an example of each category.

12.3 Briefly explain the nature of: field checks, record checks, and file checks. Give a brief example of each check.

12.4 Distinguish between a range check and a reasonableness check. Why is a reasonableness check not a field-level check?

12.5 Explain the purpose of input validation controls. Give an example.

12.6 Define and explain the use of control totals. Give a simple example of three types.

12.7 Define and explain the use of a transaction log. How does it provide an audit trail for a real-time system?

12.8 Why is it imperative to design controls into the information system as it is being developed rather than after the information system is implemented?

12.9 List and give an example of each type of destructive and fraudulent software.

12.10 List and briefly explain three antiviral measures.

12.11 Why does a batch output system usually require more controls than a real-time output system?

12.12 What are the three critical areas for controlling output in a real-time system? Give an example of how each area is controlled.

12.13 Name and briefly define the two prominent encryption techniques.

12.14 Explain why diskless workstations provide an added control feature, especially as it relates to control of output. Also, explain how diskless workstations safeguard the information system from viruses.

12.15 How can unauthorized viewing of output displayed at an online terminal be prevented?

12.16 Briefly describe four techniques that can be used to prevent an unauthorized viewer or operator from perusing the contents of a report during the printing process.

12.17 Why is it important for each page of a report to have a heading and a page number? What is the purpose of printing an end-of-report (or end-of-job) label immediately after the last entry on a report?

12.18 Why do printer ribbons and carbon paper present a control problem? What should be done with these items immediately after being used?

12.19 What should be done with reports once their retention dates expire?

12.20 Briefly explain the IBM DB2 security model.

12.21 Explain why encryption techniques may be used in some databases.

12.22 Explain how the grandfather-father-son file recovery plan works.

12.23 Define televaulting and give a brief example. Differentiate disk-mirroring from televaulting.

12.24 What is the purpose of TEMPEST products?

12.25 Specify and briefly explain the three types of UPS.

12.26 List and give examples of four PC controls.

12.27 List and briefly define the four subplans of a disaster recovery plan.

12.28 What distinguishes a hot site from a cold site?

12.29 What's the primary disadvantage of a company-owned backup facility?

12.30 What's the primary disadvantage of a reciprocal agreement contingency option?

12.31 Explain how a mobile data center contingency option works.

12.32 Explain why it's important to test contention and contingency plans. Distinguish between passive and active testing.

12.33 Describe three methods used to control people access. Which access control method provides the strongest control?

12.34 What's an alpha or type 1 error? What's a beta or type 2 error? What kind of errors are more likely to be condoned if the company seeks a tightly controlled system? Explain why.

CHAPTER-SPECIFIC PROBLEMS

These problems require exact responses based directly on concepts and techniques presented in the text.

12.35 Calculate check digits for assignment to the following codes:

Code	Modulus	Weight
85143	11	1-2-1-2-1
46624	10	6-5-4-3-2
67129	11	5-4-3-2-1

Transpose numbers in the codes and recalculate the check digits. Are these check digits the same as those assigned?

12.36 Identify the following types of destructive and fraudulent software.

1 Within an accounts receivable program, there is a module that collects names of customers according to selection criteria. The programmer who wrote this program sells this data.

2 A self-replicating segment of program code is inserted in a system via a technician's demonstration disk.

3 A module counts the number of payroll periods missed by employee number 78492. If the number of payrolls missed is greater than three, the module is programmed to erase the payroll master file.

12.37 Identify the component plan of the disaster recovery plan to which the following situations relate.

1 A way to keep the business running.

2 A way to deal with the disaster while it is occurring.

3 A way to increase the probabilities that the system will not suffer a disaster.

4 A way to regain the system's previous status, or an improved status, after a disaster occurs.

THINK-TANK PROBLEMS

These problems call for a feasible approach rather than a precise solution. Although the problems are based on chapter material, extra reading and creativity may be required to develop workable solutions.

12.38 Using structured English, and your own examples, write statements necessary for:

Reasonableness check
Range check
Self-checking digit
Field numeric check

For example, a structured English statement for a range check is

```
IF PAYROLL-AMOUNT IS GREATER THAN
      RANGE-AMOUNT, THEN PERFORM
      DISPLAY-ERROR-ROUTINE
```

12.39 The following data appear on a time sheet that you have designed for a new payroll system:

Field	Picture
Employee number	9(5)
Regular hours	9(2)
Overtime hours	9(2)
Commissions	9(5)V99
Sick time	9(2)
Vacation time	9(2)

Required: What field and record input validation checks do you propose to be designed into the payroll application program? Make up any parameter values you believe necessary. All fields are of fixed length.

12.40 A payroll application is being developed for Marvelous Department Store. Some of the fields in the payroll record are

Employee name
Employee number
Gross pay
Deductions
Net pay
Pay rate

Required: List and explain input validation checks that you think should be embedded in the payroll program. For example, would you recommend a hash total? If so, what field would you use to compute a hash total?

12.41 Following are common business transactions processed by systems:

A public utility company preparing utility bills based on meter readings.
An insurance company processing policyholders' application for maternity benefits.
A construction company preparing weekly payroll checks.
A university bookstore ordering textbooks for next semester.
The Internal Revenue Service (IRS) processing personal income tax returns.

Required: Describe the reasonableness checks that might be implemented in a computerized editing process in the preceding situations.

12.42 A new group-based system is being developed for the personnel department at Thor Enterprises. The firm employs 2000 people at various offices and construction sites. You are a controls expert on a SWAT team that is developing the new system. You have found from your analysis that various offices and construction sites send personnel change of status information to the personnel department in the head office. Change of status source documents are prepared after the supporting documentation has been checked. The batches of source documents are sent to the data center for updating payroll master tape files.

Required: Outline a batch control system for the new system.

12.43 Following is a list of scenarios in which errors or fraudulent activities took place:

1 A customer payment was enclosed along with the remittance advice showing $28.50, but it was entered into the computer system as $2850.00

2 The accounts receivable master file was incorrectly loaded on a drive that was supposed to hold the payroll master file. When the payroll program was run, it destroyed the accounts receivable file.

3 A payroll check issued to an hourly employee was based on 98 hours for a week's work instead of 40.

4 A chemical salesperson entered a customer's order from a portable terminal using a valid but incorrect product number, which resulted in a delivery of 16 drums of floor wax instead of insecticide.

5 A bank customer transaction was coded with an invalid customer account number and was not detected until the customer received the monthly statement, which did not show the transaction.

6 In the processing of payroll checks, the computer somehow omitted four of the total 3052 checks that should have been processed. The omission was not discovered until the checks were distributed by each department.

7 A fire destroyed the inventory master file, and a complete inventory had to be taken to set up a replacement file.

Required: For each scenario, recommend controls that would prevent or at least reduce the likelihood of the situation's recurring.

12.44 You have recently been hired to help Fourth National Bank design a system of controls for 900 terminals scattered throughout 12 branch banks in a large metropolitan area. Most of the terminals are for tellers, but others are for loan officers and other personnel throughout the bank's operations. All of these terminals, even including three in a supply room, are connected to a central processing hub located in the main bank building. Presently, the system is not using any access controls. Moreover, the central processor has no way of identifying and differentiating the terminals.

Required: Design a system of controls for the terminals at Fourth National Bank.

12.45 Mark Tick is an external auditor who has been assigned to audit the data center at Nept Company. Entering the computer room, Mark waved to the sole occupant, a machine operator who was hastily punching up cards and inserting them in a deck labeled "Payroll Source Code."

"Obviously, a valuable employee," mused Mark. "It's good to see someone putting forth some extra effort."

Mark poured a cup of coffee, and, as he started to count the petty cash, placed it on top of the 4-foot-high stack of dust-covered disk packs. Noticing that the hot cup was causing the plastic top of the disk to bend a bit, he pulled a few cards from the deck labeled "Daily Sales Update," which was lying on the console, to use for a coaster.

Mark noticed that the machine operator, having run the unnumbered payroll checks through the check signer, was separating the carbons from the checks. The machine hummed smoothly, giving the operator a chance to have a smoke and discuss with two mailboys who had just entered how much the various vice presidents were being paid. Mark was impressed with the operator's concern for neatness, displayed by his having run the console log sheets through the shredder as soon as he finished the payroll run. Mark drew an appreciative smile from the operator as he quipped, "Nobody could make anything out of the gobbledygook the typewriter just printed, so better to destroy it than get buried under it."

Mark glanced at the bulletin board and immediately got an indication as to how well organized the data center manager was and that he was nobody's fool. The two signs that impressed him the most read:

> This is a data processing operation, not a delivery service. All output for the current week will be placed on the big table in the cafeteria before 4 P.M. each Friday. Help yourself.
>
> To expedite processing and cut down on unnecessary paper shuffling, all documents rejected by the computer because of out-of-balance controls or invalid data will be immediately corrected and re-entered by the machine operators.

The data center manager came in and introduced himself to Mark. He apologized for being away so long. He explained that he had had a hard time finding a garden hose long enough to reach into the data center through a hole in the plywood partition separating it from the adjacent boiler room.

"Good idea," Mark said approvingly. "A lot cheaper than buying fire extinguishers for the data center."

"Well, how does the place look?" the manager asked, perspiring slightly in the 90-degree heat.

"Great!" said Mark. "There will be only one item in my report. There is the serious matter of the 23-cent unexplained shortage in your $5 petty cash fund."

Required: Write a complete critique of Mark's audit.

12.46 Your specialty is designing and testing disaster recovery plans for large commercial banks. You have been hired by the board of directors of First Interstate Bank to review their disaster recovery plan. Lawrence Kleinrock, the person who was in charge of installing the present disaster recovery plan, has answered "yes" to every question on your review questionnaire. For example, Lawrence indicates that a reciprocal agreement has been signed between First Interstate and Stateside Bank, and their system is available on short notice for First Interstate's use. Also indicated by his responses are that all individuals are aware of various contingency procedures; that emergency procedures are installed for recreating lost files and other critical items; that emergency procedures are established for fire, sabotage, power failures, and natural disasters; and that emergency drills are conducted periodically.

Required: Prepare a report for the board of directors at First Interstate as to how you would go about testing the disaster recovery plan developed by Lawrence Kleinrock.

SUGGESTED READING

Beamguard, Bud. "Disaster Planning." *Interact,* January 1991.

Brill, Alan E. "Computer Fraud: What You Can Do to Prevent It." *Computers in Accounting,* June 1988.

Brown, Bob. "Disaster Recovery Company Links Backup Sites Into Net." *Network World,* January 19, 1990.

Burch, John G. "Disaster Recovery Plan: A Moral and Professional Responsibility." *Internal Auditor,* June 1989.

Chalmers, Leslie. "Fighting the Common Virus." *Journal of Accounting and EDP,* Spring 1990.

Fiderio, Janet. "Voice, Finger and Retina Scans: Can Biometrics Secure Your Shop?" *Computerworld,* February 15, 1988.

Flach, Joseph P. "Increasing Programming While Preventing the 'F' Word." *DP&CS,* Fall 1987.

Francis, Bob. "A Secure Future for Diskless Workstations." *Datamation,* November 15, 1990.

Gondek, Chris. "Establishing Information Security." *Management Accounting,* April 1989.

Graggs, Tuseda A. "Foreign Virus Strains Emerge as Latest Threat to U.S. PCs." *Infoworld,* February 4, 1991.

Hirsch, Steven A. "Disaster! Could Your Company Recover?" *Management Accounting,* March 1990.

Hunter, John. "Savvy Users Can Fight Threat of Viral Infection." *Network World,* July 17, 1989.

Johnson, Paul D. "Mark Tick's Data Center Audit." *EDPACS,* June 1974.

Joseph, Gilbert W. "Computer Virus Recovery Planning—An Auditor's Concerns." *Journal of Accounting and EDP,* Spring 1990.

Korzeniowski, Paul. "How to Avoid Disaster with a Recovery Plan." *Software Magazine,* February 1990.

Maher, John J., and James O. Hicks. "Computer Viruses: Controller's Nightmare." *Management Accounting,* October 1989.

Moad, Jeff. "Disaster-Proof Your Data." *Datamation,* November 1, 1990.

Perdue, Lewis. "Blanket Security." *PC World,* February 1989.

Perry, William E. "Recognizing and Controlling Computer Viruses." *Journal of Accounting and EDP,* Summer 1989.

Steinberg, Steve. "A Student's View of Cryptography in Computer Science." *Communications of the ACM,* Vol. 34, No. 2, February 1991.

Weber, Ron. *EDP Auditing Conceptual Foundations and Practice,* 2nd ed. New York: McGraw-Hill, 1988.

Winship, Sally. "Disk-Mirroring Products Offer True Fault Tolerance." *PC Week,* February 4, 1991.

Zajac, Bernard P., Jr. "Disaster Recovery." *Interact,* February 1990.

JOCS CASE: Designing Controls

"But nobody will use the system if we make it this complicated!" states Christine Meyers, one of the SWAT team members for the job-order costing system (JOCS) under development for Peerless, Inc. She continues with her argument by saying, "I have ten years' experience in information systems, and I think the extent of controls being discussed here is absolutely ridiculous. What's the use of creating all these controls for a system nobody will use anyway?"

"Christine, this system contains sensitive manufacturing data. We don't want our competitors to learn about our production process," says Carla Mills, another member of the SWAT team. "I've worked in accounting for three years, and we have much stricter controls than these. I think you are overreacting."

Corey Bassett tries to find a compromise between the two positions. "I think Christine has a point. If we make it too difficult for the end users to access the system, nobody will even try to log-on. If we make the input controls too lax, we won't have confidence in the correctness of our data. Let's discuss each of the controls individually."

Tom Pearson, the instigator of this argument, declares his position by stating, "Jake asked me to do a preliminary design of systems controls. So I listed the important control areas and outlined what I thought was necessary to make a secure system. What's wrong with these controls?"

The SWAT team is having a design meeting and looking at the control worksheet devised by Tom Pearson. The worksheet shown in Figure 12.14 contains only Pages One and Two of a four-page document. This worksheet, shown in Figure 12.14, does not contain the technology platform controls and disaster recovery plan also designed by Tom.

"I think the access controls are unnecessary," says Christine. "We aren't the CIA protecting national security or anything. This is just a manufacturing system to help track costs."

"We have customers and competitors wandering all over our building," says Tom. "Management has already decided that our building will remain open. Anybody could walk up to a workstation and access the database. We have to prevent people from looking at our data."

"What's wrong with a password system?" asks Carla. "I used passwords on the systems in college."

"So did I," says Cory. "The problem was that everybody always wrote down their passwords and then people read them and broke into systems. Sometimes people would choose really easy passwords that other people could just guess. We had a lot of security problems."

"This isn't college. We could make our users understand the importance of a secure system. I'll go for a password system," states Christine. "We could make sure that our operating system keeps the passwords in an encrypted format. I'll even make our users change their passwords every week. I'll even

INPUT CONTROLS

CIM (Computer-Integrated Manufacturing) Data Validation:

1 Validate job numbers, part numbers, operation numbers, and employee numbers. Should consider using a check digit for each of these critical identifying numbers.
2 Perform valid range checking on material quantity assigned to each operation for a given job. Create an exception report for any quantity falling outside of the valid range for immediate feedback to production supervisor.
3 Perform validation for total materials assigned to a given job. Create an exception report and contact production manager about any discrepancies.
4 Perform range checking on hours spent for each operation for a given job.
5 Validate total number of hours worked for each employee at the end of each day.

Online Human Input Data Validation:

1 Perform alphabetic and numeric checking on data.
2 Check current dates.
3 Program input into relevant fields such as dates, logical fields, and batch control numbers.
4 Validate customer numbers after input.

Audit Trail Controls:

1 Log each human transaction.
2 Log each transaction applied from the CIM system—might consider batching transactions from CIM and applying at specific times during the day.
3 Dump all data at the end of each day.

PROCESS CONTROLS AND OUTPUT CONTROLS—QUESTIONS TO BE ANSWERED AT THIS TIME

1 Will we use a secure networked operating system?
2 Check the interface between the engineering department's minicomputer and the new JOCS system. Will the two systems communicate?
3 If JOCS will interface with engineering's minicomputer, does the engineering minicomputer connect to the outside world through the Internet?
4 Will we allow outside communication to our system?
5 Do we want users to communicate through our system to the outside world?
6 Are we going to interface JOCS with the accounting system?
7 Does the accounting system connect to the outside world?

DATABASE CONTROLS

1 Buy a database management system (DMBS) with either field or record level lockout for concurrency controls.
2 DBMS should include automatic logging of transactions with a way to easily dump data.
3 Store grandfather-father-son dump of data at offsite storage facility.

ACCESS CONTROLS

1 Obtain biometric access for all online human interaction with JOCS computer workstations.
2 Purchase fingerprint recognition devices for each computer workstation.

Figure 12.14
Preliminary design of controls for JOCS.

go so far as to suggest they use passwords that aren't real words so that somebody can't write a dictionary program to compare the encrypted passwords with all the words in a dictionary. But let's please keep reasonable about security and controls."

Jake steps in at this point and says, "Tom was simply following my instructions to make JOCS secure. But you're right, Christine. Our users will be pretty intimidated about fingerprint-controlled access. Let's use an encrypted password system for access control. Now, what do you think about the input controls that Tom has identified?"

"They look right on target to me," says Carla. "We have had problems in accounting with CIM data, and these controls will prevent future problems."

"I agree with all the input controls except the check digits," says Christine. "Check digits will produce a couple of problems for us. First, we will have to change all of our job, operation, and employee numbers. That means that all the current systems using these numbers will have to be modified. Second, check digits take a lot of computer processing time. This is probably going to slow down response time unless we buy some pretty powerful computers."

"Oh, come on, Christine," states Tom. "All a check digit requires is a couple of calculations. A normal computer could do that in a few microseconds."

"Christine is right about changing all the other systems, though," says Carla thoughtfully. "I forgot about our current accounting systems. If we change those numbers, all the other systems will have to be changed, or else we will be working with incompatible numbers. I really think we should check with the other departments before we go ahead with that idea."

"OK, so we postpone check digits and discuss it at another meeting," says Jake. "Looks like we should also have a meeting to discuss process controls. JOCS isn't being designed in a vacuum. JOCS will have to interface with CIM, the minicomputer in the engineering department, and possibly even the current accounting systems. We better have another meeting to talk about the controls necessary for all these computers. Tom, please draw up a couple of diagrams showing all the potential connections we could have with these other computers. We will meet tomorrow morning and decide the best computer interface for JOCS. Then I will arrange a meeting with the other groups to talk about process and output controls. Everybody else can continue with their current design tasks."

"Do you guys at least agree with the database controls?" sighs Tom.

"Yeah, looks good!" agrees the rest of the SWAT team.

"But I want to talk about your disaster recovery plan," says Christine. "You know, the chances of a tornado here are phenomenally low, and you have provided for just about every kind of natural disaster. . . ."

"Christine," interrupts Jake, "let's save the discussion until we decide about our potential system interfaces. Our next meeting will affect the disaster recovery plan, too. Why is it that nobody can agree about the extent of computer controls?"

Chapter 13

Designing Networks

WHAT WILL YOU LEARN IN THIS CHAPTER?

After studying this chapter, you should be able to:

1 Describe the steps in designing networks.
2 Define basic networking elements.
3 Discuss twisted pair cable, coaxial cable, and fiber optic cable.
4 Discuss designing of local area networks (LANs).
5 Describe services for designing wide area networks (WANs).
6 Explain how microwave and satellite transmission media are used for designing wide area networks (WANs).

INTRODUCTION

The subject of this chapter is network design (see Figure 13.1). A well-designed network can fundamentally change the way an enterprise conducts its business. A network enables an organization to bring its systems applications out to where the work is being performed. Networks permit distributed users access to centralized information and centralized users, such as senior management, access to distributed information. Such access augments management decision making. In many cases, the new strategic business plans of enterprises are empowered by the network design outlined in an enterprisewide model developed during the systems planning phase of the SDLC. The process of designing a network is perhaps one of the biggest and most important tasks performed by systems professionals. After all, a network represents a long-term commitment to the organization's productivity, differentiation, and management (PDM) strategic factors.

STEPS IN DESIGNING NETWORKS

The purpose of this section is to explain the general steps that systems professionals perform in designing networks. The following presentation is technology-independent. The technology required to support network designs is covered elsewhere in this chapter.

Step 1: Segment the Enterprise. The first step in constructing the basic building blocks of the network is to segment the enterprise by geography, by department, by building, or by floor—whatever is required by logic. The en-

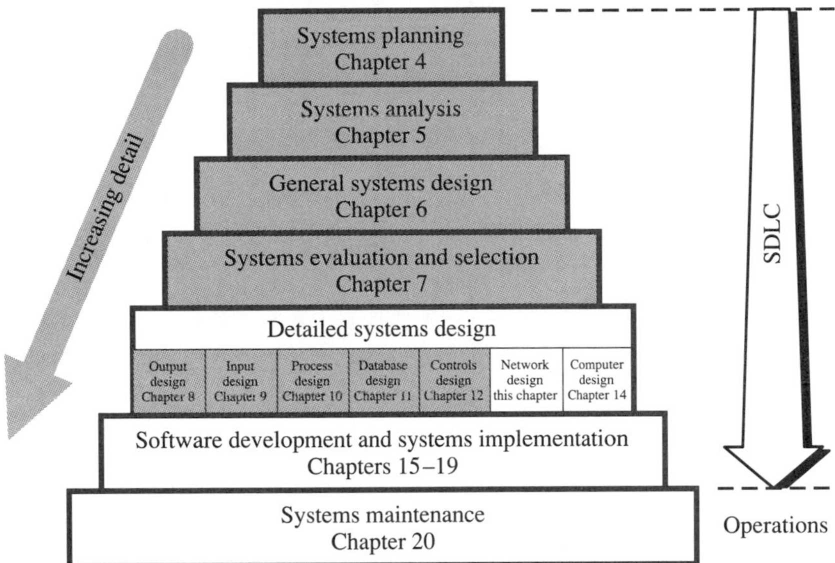

Figure 13.1
SDLC phases and their related chapters in this book. In Chapter 13 we look at types of networks and aspects of their design.

terprisewide model provides an excellent base from which to work. Figure 13.2 shows how Taffy Candy Company is segmented by building and by department into three segments:

- Warehouse
- Office
- Factory floor

Step 2: Create a Model Local Area Network. The next step is to create a **LOCAL AREA NETWORK (LAN)**, which is an interconnection of information technology that serves a segment of an organization. Because of the readily available twisted pair telephone wiring in Taffy's office, the office segment is chosen as the logical place to begin the network design. This layout is shown in Figure 13.3.

Once the details for the office LAN are worked out, we move on to the warehouse segment, not worrying too much yet about how the segments will be interconnected. This detail will be worked out after we have designed each segment's LAN.

Next, the factory floor LAN segment is designed. It was suggested that a fiber optic cable (not shown) be used as the transmission medium for the factory floor because of its resistance to electromagnetic interference prevalent there. We discuss fiber optic cable later in this chapter.

Step 3: Evaluate the LANs to Determine If They Are Appropriate for Each Segment Throughout the Enterprise. The network design for Taffy begins to resemble the one in Figure 13.4. All three segment networks have been designed on paper for further evaluation and approval. The warehouse segment's main job is receiving and shipping inventory. One key warehouse requirement

Figure 13.2
Segmenting by building and by department.

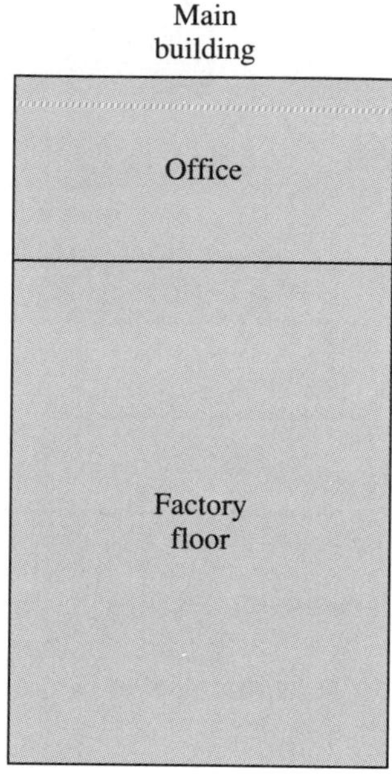

is that its LAN has the ability to read bar codes and track inventory. The office LAN will be required to service administrative and accounting tasks. The plant manager and his or her staff will need to prepare manufacturing production schedules. The controller will need to generate financial reports. The accounting area will need to record business transactions and perform product costing. The factory floor LAN must be able to read bar codes and track inventory as units move through four production processes:

Figure 13.3
Network design for the office segment.

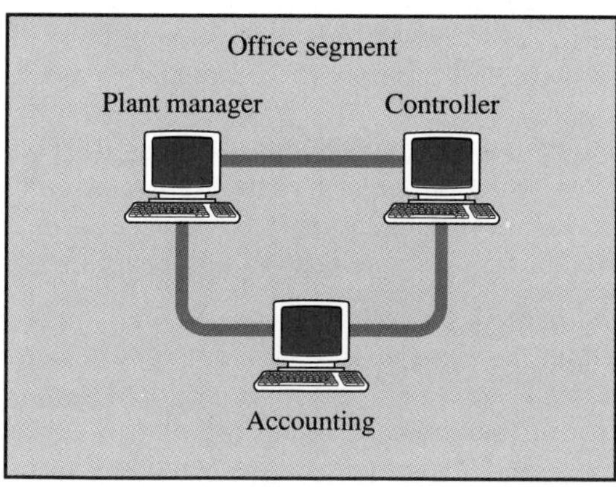

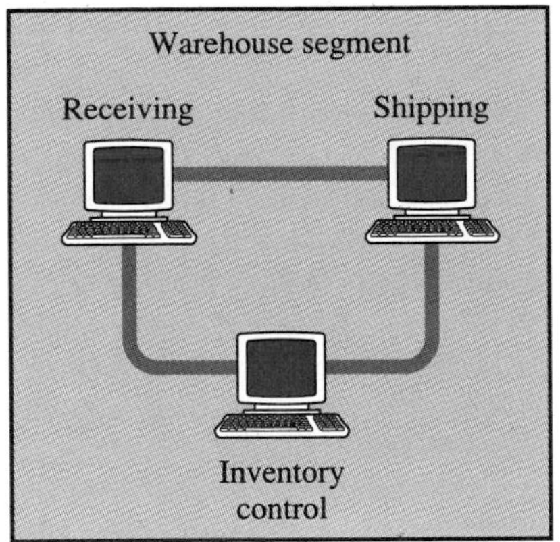

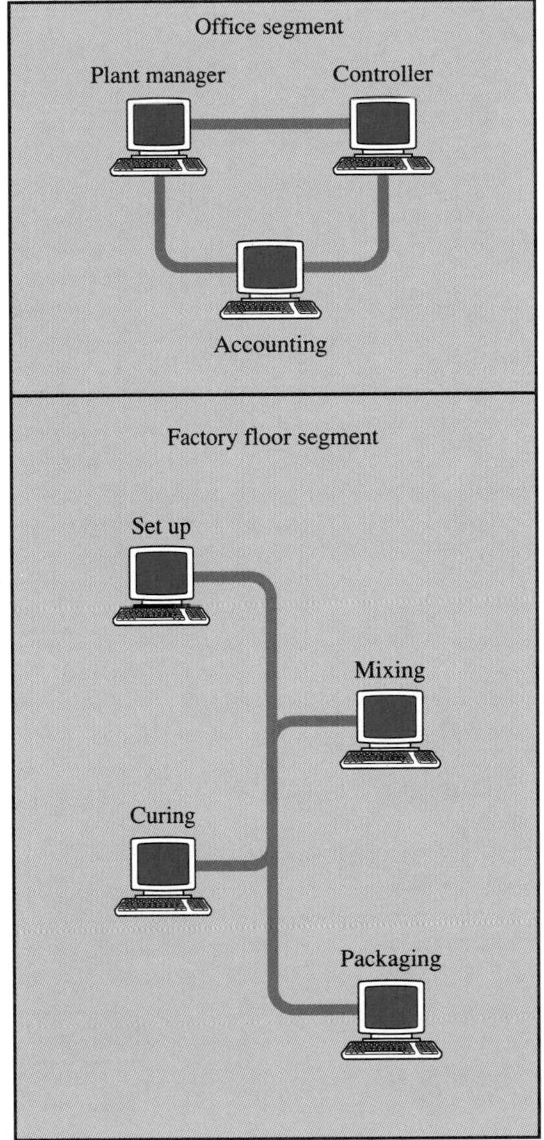

Figure 13.4
LAN design for each segment.

- Setup
- Mixing
- Curing
- Packaging

Step 4: Interconnect Network Segments. After each segment's LAN design has been approved, internetworking schemes are proposed. There are many ways to interconnect networks. **BRIDGES** and **GATEWAYS** are the most common systems used for interconnection. In Figure 13.5, bridges are used to interconnect the segments' LANs.

Bridges are used to connect different LANs. Gateways are used to connect LANs to **WIDE AREA NETWORKS (WANs)**. Both bridges and gateways perform various translation tasks according to the differences between networks. The final internetwork design serves as a guide for installation of cable and other telecommunication hardware and software to make up the network.

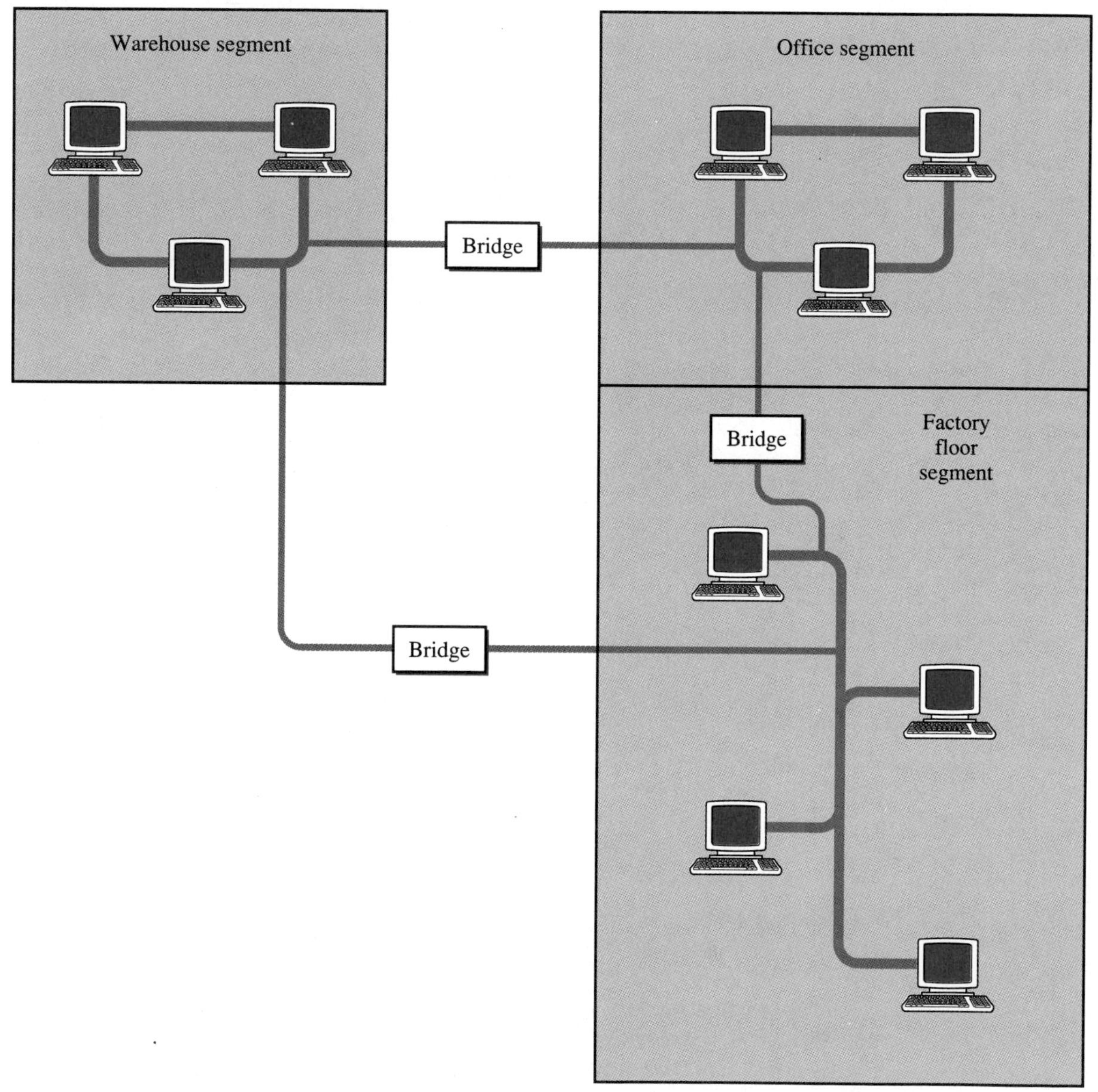

Figure 13.5
Internetwork design.

NETWORKING ELEMENTS

The enterprisewide model becomes the basis for designing the network. You build the network on paper or screen first. Once the design is approved, the next step is to install the network backbone. Then hardware and software networking elements are connected to the backbone to form the technology platform, which is composed of an integrated network plus a computer architecture. Let's now become familiar with some basic networking elements.

Types of Transmission Media

There can be one or a combination of **TRANSMISSION MEDIA** types used to build a network. These include the following:

- Twisted pair cable
- Coaxial cable
- Fiber optic cable
- Microwaves
- Satcllites

Transmission Signals

All messages sent from one node (i.e., a device connected to the network backbone) to another are transmitted as either an analog signal or a digital signal. Digital signals are discrete, binary signals consisting of two states, on or off.

Analog signals are in the form of a sine wave. The top of the wave is called the peak, and the depression is called the trough. The distance between the peak and trough is the amplitude of the signal. Each complete wavelike motion of the analog signal is called the signal's frequency. The number of frequencies (also called cycles) per second is represented in hertz (Hz).

Transmission Modes

There are three **TRANSMISSION MODES**:

- Simplex
- Half duplex
- Full duplex

Simplex is one-way transmission. Television is a good example of simplex communication. The main node, a transmitter, sends a signal, but it does not expect a reply. The receiving nodes cannot send a message back to the transmitter. In half-duplex mode, a sending and receiving node communicates over the same transmission medium, but only one node can send at a time. While one node is in send mode, the other is in receive mode. In full-duplex mode, both ends of the transmission medium can send and receive at the same time.

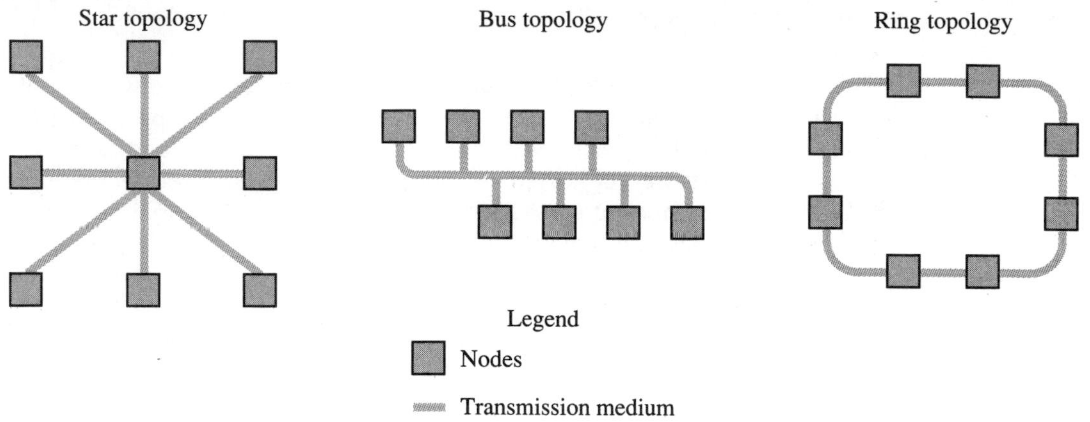

Figure 13.6
Three network topologies.

Network Topologies

Transmission media configured in a particular way is termed a network backbone, that is, the foundation of a network design. Various nodes connected to the network backbone in a specific configuration are referred to as the **NETWORK TOPOLOGY**, such as star, bus, or ring (or some variation), as illustrated in Figure 13.6. Think of a network topology as an architectural drawing of the basic network elements.

In a star topology, all nodes are connected to each other and controlled by a central node or hub (possibly a computer). A bus topology connects all nodes by a single transmission medium. A ring topology consists of nodes arranged in a ring pattern. Messages are transmitted from the sending node, node by node, until the receiving node receives the messages.

Network Interface Cards

The **NETWORK INTERFACE CARD (NIC)** represents the electronic circuitry that connects a workstation to a network. Usually in the form of a circuit board, it fits into one of the expansion slots inside a computer and makes the physical connection between the computer and the network backbone wiring. It works with the network operating system and the computer operating system to send and receive messages on the network. The NIC also determines the network's topology.

Network Servers

Peripheral management functions are usually centralized and executed by special-purpose computers, which are referred to as **NETWORK SERVERS**, as depicted in Figure 13.7. When several requests for resources arrive at the LAN server at the same time, the server places the requests in a queue and services them in sequence. While the server is processing requests, it must also handle administrative tasks, such as denying unauthorized requests and maintaining the hard disk directory. When many active workstations are connected to the server and all are sending requests, a throughput bottleneck can quickly occur at the

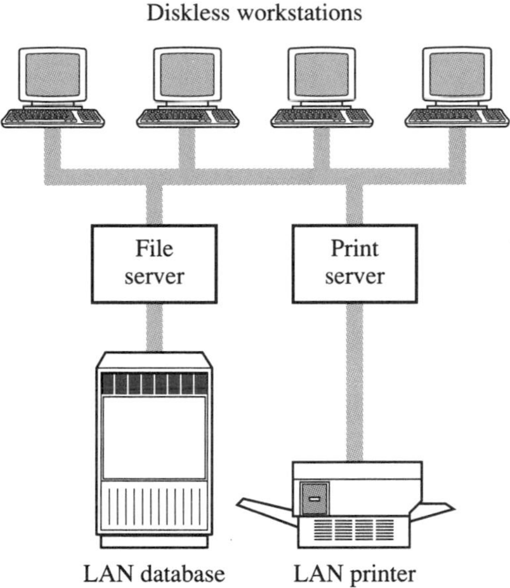

Figure 13.7
Local network with file and printer servers.

network server. To avoid such a bottleneck, the throughput of the server should be matched to the needs of the network design.

In some high-volume, group-based network designs, a number of servers may be required, such as those presented in Figure 13.8. All servers are connected to a high-speed backbone permitting various workgroups to share databases and printers and communicate with one another. All the workgroups have access to the two major group-based print and database servers. Specific workgroup workstations are connected by workgroup servers. New applications and workgroups can be added or deleted with little difficulty and disruption.

Modems

A **MODEM** (MOdulator/DEModulator) is an interface device that transmits data from computer devices at one site to another modem connected to computer devices at the other end of an analog telephone line, as depicted in Figure 13.9. The objective of modems is to prepare data signals produced by a computer so that they are compatible with an analog telephone channel, and the analog telephone signals are compatible to a computer at the other end. Therefore, a modem must be able to convert data back and forth between digital (ones and zeros) signals used by computers and analog (continuously varying amplitude) signals used by analog telephone channels.

Multiplexers

A **MULTIPLEXER (MUX)** is a device that permits line sharing in which data from several terminals are interleaved on a common higher-speed line, as demonstrated in Figure 13.10. The multiplexing process is similar to the way in which vehicles merge into a single high-speed lane from several lower-speed on-ramps.

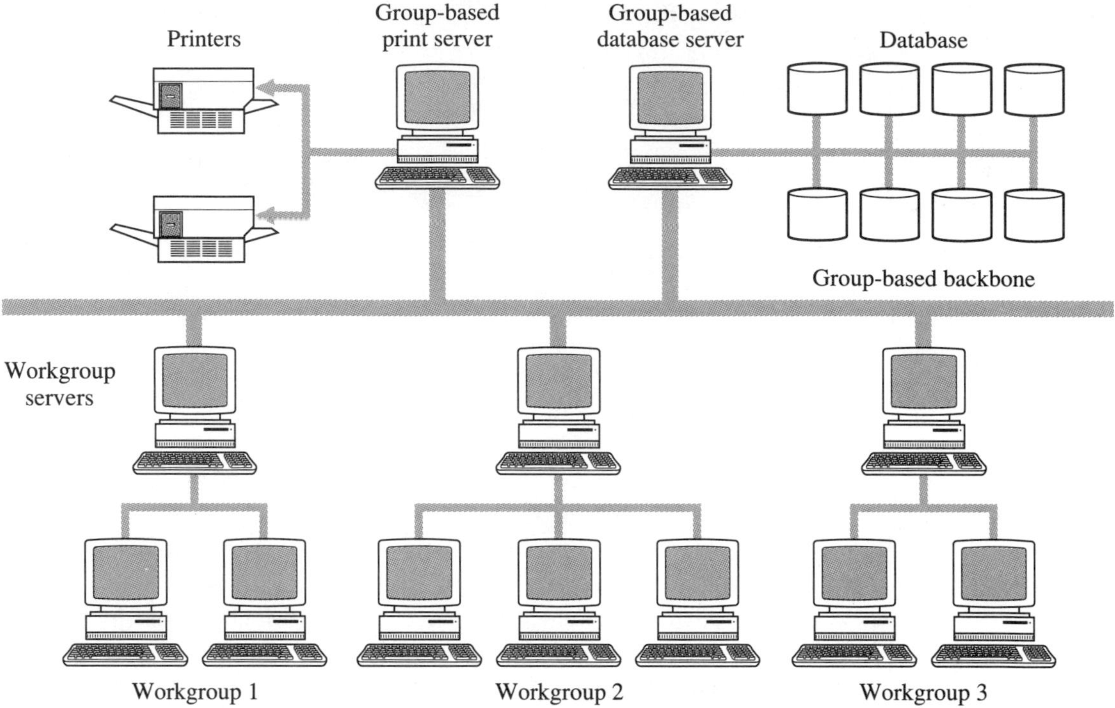

Figure 13.8
Group-based network using a number of servers.

Time-division multiplexers (TDMs) interleave the character data in a fixed, invariant sequence, enabling the receiving multiplexer to distribute the incoming data stream to the appropriate output. This approach is inefficient when a terminal user is not keying in data, because the time slot assigned to that terminal remains unused. Statistical time-division multiplexers (STDMs) take advantage of the predictable periods of inactivity in an interactive terminal system, creating a more efficient line use than is possible with TDMs.

TDM works well when a massive amount of bandwidth is allocated immediately to a particular transaction and then released immediately so that other users may access the same bandwidth. It does not meet the need for bursts of large data streams interspersed with low-level data traffic, such as one finds in most LANs. This is the forte of STDMs.

Network Operating Systems

The network design elements, such as the network backbone, network interface cards, and server equipment are all hardware and are relatively easy to understand because of their physical nature.

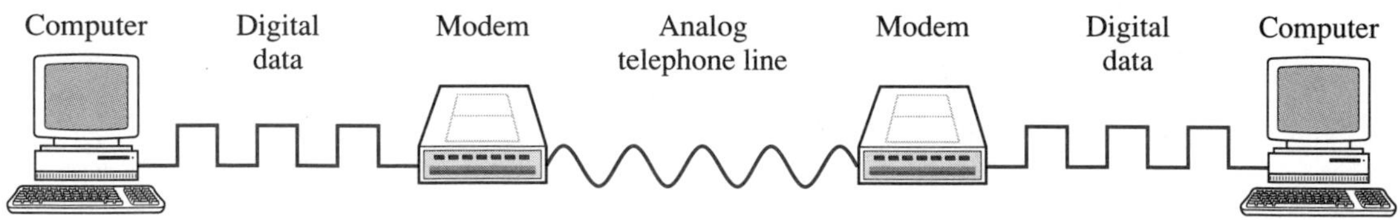

Figure 13.9
Modem operation.

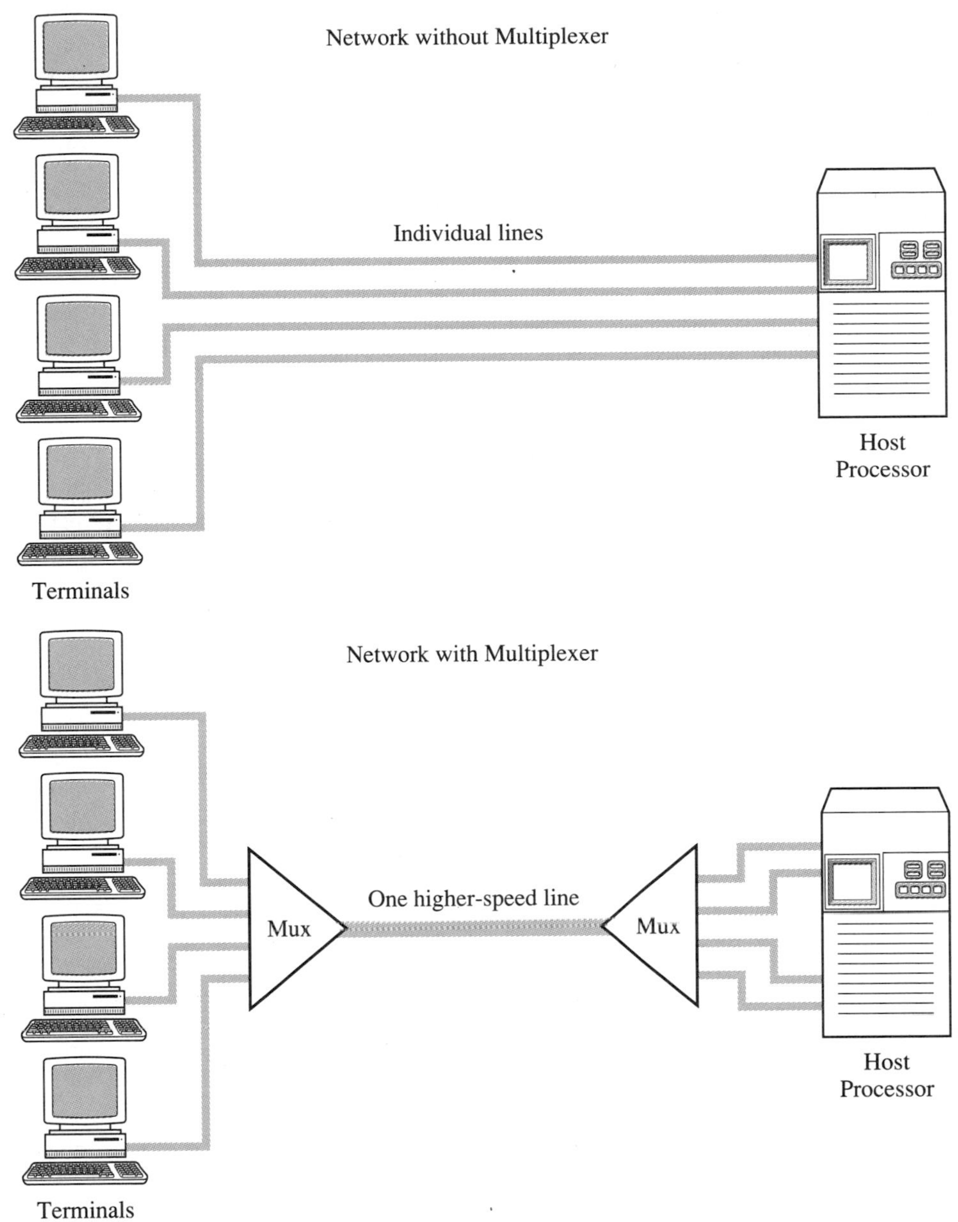

Figure 13.10
Network with and without multiplexer (mux).

The **NETWORK OPERATING SYSTEM** (or network management system), on the other hand, falls into the software category. The functionality of the network, however, is provided by the network operating system. File transfers, security and access control, print spooling, and E-mail are just a few examples of the functions the network operating system provides to the network and the applications running on it. Essentially, the better the network operating system chosen, the better the network design.

Protocols

PROTOCOLS are to smoothly running networks what language translators are to communication at the United Nations, or what traffic lights and road signs are to traffic systems. Telecommunication network protocols are developed to interconnect heterogeneous networks, architectures, and devices, and achieve interoperability. Two dominant open (public) network protocols are:

- Open systems interconnection (OSI) reference model
- Transmission control protocol/internet protocol (TCP/IP)

The dominant proprietary network protocol is IBM's systems network architecture (SNA).

Open Systems Interconnection Reference Model

The standard of standards is the International Standards Organization's seven-layer OSI reference model. No doubt a need exists for a worldwide telecommunications standard, and many authorities believe that OSI is the best candidate by far. Indeed, virtually every major systems vendor as well as the federal government supports it. In fact, all networks acquired by the federal government must conform to the government OSI profile (GOSIP). The private sector is also demanding OSI-compliant products.

The OSI reference model is composed of seven layers, as depicted in Figure 13.11, in which different kinds of network activity are taking place. Each layer builds on the layers beneath it, requiring the services of the lower levels to function. Each layer can communicate only with the layer directly above and the layer directly below.

Figure 13.11
OSI reference model.

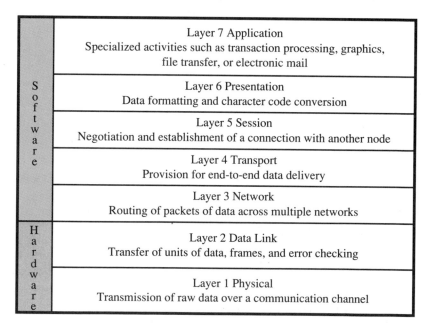

Software	Layer 7 Application Specialized activities such as transaction processing, graphics, file transfer, or electronic mail
Software	Layer 6 Presentation Data formatting and character code conversion
Software	Layer 5 Session Negotiation and establishment of a connection with another node
Software	Layer 4 Transport Provision for end-to-end data delivery
Software	Layer 3 Network Routing of packets of data across multiple networks
Hardware	Layer 2 Data Link Transfer of units of data, frames, and error checking
Hardware	Layer 1 Physical Transmission of raw data over a communication channel

Layer 1, the physical layer, provides transmission of raw bit streams between systems. It sends and receives the signals that are transmitted via transmission media, such as twisted pair, coaxial cable, fiber optic, microwave, and satellite. Layer 2, the data link layer, manages direct connections between nodes on a network. These two layers are hardware-based. The other five layers are implemented strictly in software.

Layer 3 is the network layer responsible for designating the route between nodes on the network. Layer 4 is the transport layer, performing error recovery by requesting retransmissions of garbled data and regulating the speed of data transmission. Layer 5, the session layer, establishes, manages, and terminates sessions between application programs, connecting the correct application on a workstation to the appropriate application on a server or mainframe.

Layer 6 is the presentation layer, responsible for converting code from a form readable by an application on a workstation into a form readable by an application running on a remote system. Layer 7 is the application layer, providing services directly to the end user.

Transmission Control Protocol/Internet Protocol

TCP/IP is a widely used and well-defined protocol. It is well suited for the internetworking of multivendor systems. TCP/IP works with many different architectures and standards.

Although TCP/IP is the current de facto standard for interoperability (i.e., the ability to connect and operate disparate systems as an integrated entity), it is viewed by many vendors as a stepping-stone to the OSI reference model. But TCP/IP will still be widely used for a long time to come. Indeed, some diehard TCP/IP proponents say that OSI will never replace TCP/IP. Other authorities say that they will coexist.

Systems Network Architecture

Systems network architecture (SNA) is IBM's primary protocol standard. SNA has become a de facto industry standard by virtue of IBM's control of over 80 percent of the mainframe market.

Like the OSI reference model, SNA is a seven-layer architecture. These layers include:

- Layer 1—the physical control that connects nodes physically and electronically.
- Layer 2—the data link control that transmits error-free data along the circuits.
- Layer 3—the path control that routes data in packets between source and destination.
- Layer 4—the transmission control that paces data exchange to match data processing rates at source and destination.
- Layer 5—the data flow control that synchronizes data flow between source and destination.

- Layer 6—the presentation service that formats data for different nodes.
- Layer 7—the transaction service that provides specific application services for the end user.

Electronic Messaging Protocols

One of the big advantages of companies designing networks is the ability to transmit electronic messages. X.400 is the international electronic messaging standard that enables users to exchange messages, such as E-mail, with anyone anywhere in the world. A great impetus for the use of X.400 is the rapid increase of global ventures and pan-national movements such as the European Community. X.400 has rich capabilities to move image, graphics, facsimile, and video.

The X.500 standard defines a method for building a distributed message directory, something like an electronic phone directory. This directory makes it easier for disparate E-mail systems to exchange messages across multiple networks.

X.12 is the American National Standards Institute standard for electronic data interchange (EDI). X.400 will operate with EDI to handle internetwork electronic messages, because X.12 proponents are interested in adding messaging to their list of features. X.12 will transport formatted text and leave the rest to X.400. All of these electronic messaging protocol standards work within OSI, TCP/IP, and SNA.

Front-End Processors

A **FRONT-END PROCESSOR (FEP)** is a fully functional computer system, usually with its own operating system and access method. It is used to offload transmission tasks and protocol overhead that would otherwise be handled by the computer or computers assigned to perform user applications.

For example, if TCP/IP is being used, a FEP supporting TCP/IP may be used to offload transmission processing overhead required to transmit and receive over the network and to manage protocol. Incorporating FEPs into network design will generally improve the price/performance ratio of the total system and provide increases in systems throughput.

TYPES OF TRANSMISSION CABLES

Cable transmission media carry signals via metallic or glass (or plastic) material. Twisted wire pairs, coaxial cable, and fiber optic cable are examples of popular cable transmission media.

Twisted Wire Pairs

TWISTED PAIR CABLE gets its name from the fact that its two wires have two twists per foot. They are stranded or solid; each wire has a solid-color or multicolored covering. The cable can contain over 300 multiple sets of wire pairs. It is covered with a protective jacket. Color code standards define the relationship between wire pairs in the cables and their connections. This color code standard simplifies service nationwide. For example, when a disaster occurs, repair crews from all over the country can work together to restore services quickly, splicing cables by using the common color codes.

Twisted pair is attractive to network designers because it is easy to install, has high mobility and flexibility, and is inexpensive compared with other media. Further, it is a mature technology with mature standards and is pervasive in all kinds of buildings. It is supported by a host of vendors and qualified cable installers.

The two types of twisted pair wiring are unshielded twisted pair (UTP) and shielded twisted pair (STP). UTP has its roots in the telephone industry. It is the familiar wire between the phone and the wall jack. STP emerged in the 1980s when IBM began to support it as an alternative to coaxial cable. In fact, STP is used primarily for IBM Token Ring LANs (discussed in a later section).

Coaxial Cable

COAXIAL CABLE continues to be used for some local area networks (LANs). Coaxial cable is especially applicable when electromagnetic interference is a problem and when economics prevents the use of fiber optic cable. However, if design and installation specifications are not followed to the letter, coaxial cable can cause the most network problems.

Selecting the Right Cable

The key objectives in selecting cable are:

- Choose the best cable for the job.
- Avoid impedance mismatches.

For each application one medium works best at the best cost. Do not use the highest quality and cost when less will do the job. One way to avoid impedance mismatches is to lay all cable from the same spool. When a large job requires more than one spool, get all the cable from the same manufacturer. Match cable properties exactly. Physical and electrical properties must be the same to avoid problems.

Laying Coaxial Cable at First State Bank

Joe Nept, systems analyst at First State Bank, was selected to lay the cable for a 200-workstation local area network (LAN) on three floors.

Marsha Magee, CIO, told Joe: "We don't have anyone on our staff who knows very much about network design. This is unfortunate, because networking is becoming such an important, integral part of systems design."

"What is there to network design? Big deal. Lay the cable, hook the hardware to it, and you're in business," Joe said smugly.

"Well, I don't know as much about telecommunications and network design as I should," said Marsha. "But in any event, we need someone to lay coaxial cable for our planned LAN. Do you think you can handle the job, Joe?"

"Nothing to it," said Joe. "It's as good as done."

Joe's new LAN is up and running. The network is currently running smoothly. But as the weeks go by and more users access the network, problems develop. Some users notice a slowdown when accessing and updating the database or preparing reports. Still others are getting error messages or garbled data. Before too long, Joe is spending most of his day troubleshooting the network or listening to angry users.

Marsha calls Joe and says, "Joe, I need to speak with you immediately about the network."

Joe arrives at Marsha's office and is introduced to Gary Hartley, a network consultant.

"Joe, I've hired Gary to assist you with our network problems. You and Gary are to troubleshoot the network and correct our problems and develop rules for successful network installation for future reference," Marsha said.

After extensive troubleshooting, Gary discovers three reasons for the network's problems. First, Joe had cheated on his cable lengths, with several segments exceeding specified lengths. Second, cable segments from the old LAN had different electrical properties than did the new cable. And third, the cable runs had several bends that exceeded the allowable bend radius.

During a meeting with Marsha, Gary said, "I believe Joe and I have uncovered and corrected the network's problems. Designing and installing a cable plant (or network backbone) requires strict adherence to standards and guidelines. I've prepared a list of guidelines for successful cable installation in the future."

Gary hands Marsha and Joe a copy of the following:

Guidelines for Successful Cable Installation

1. Create a blueprint for cable layout and cable-device attachment location points.
2. Label every cable connector and cable segment.
3. Select high-quality cable capable of doing the job.
4. Avoid impedance mismatches by laying all cable from the same spool. Should a job require more than one spool, get all cable from the same manufacturer and, if possible, from the same manufacturing batch.
5. If extensions are made to present cable, match cable specifications precisely.
6. Ground cable to earth, preferably through the building ground from a central point.
7. To avoid kinks or sharp bends in the cable, put the cable spool on a reel and pull from the bottom of the spool.
8. Pull cable in bunches to a central distribution point and then fan out from that point. Hang the bunches with cable hangers or trays.
9. Keep the cable at least six inches away from power wiring and lighting fixtures. Run cables perpendicular to power wiring with at least six inches between the two should the cable cross power wiring.
10. Don't lay cable in places where people can step on it or roll over it with chairs.
11. Properly terminate cable ends and connectorize cable.

Fiber Optic Cable

FIBER OPTIC CABLE, also called optical fiber, is any filament or fiber made of dielectric materials that are used to transmit laser- or light-emitting diode (LED)-generated light signals. Fiber optic cable is made of glass or plastic.

The basic structure of fiber optics is a glass core surrounded by a glass clad. Light signals travel through the core, while the clad acts as a kind of wall to prevent the escape of light from the core. Fiber optic cable has reinforcing fibers surrounding the glass to take the strain off the glass. Outside the reinforcement is a plastic cover. Light travels in only one direction in a fiber optic cable, so a pair of cables or a single cable with two strands of glass is required for each link.

Fiber optic's advantages are:

- It works in environments with high electrical interference, as in factories.
- It has high data-carrying capacity and data transfer rate.
- It takes up little physical space compared with twisted pair and coaxial cables.
- It is easier to maintain than other media.
- It can carry signals for much longer distances than twisted pair and coaxial without the need for repeaters because of its low signal loss.

Optical fiber is one of the most secure transmission media available. Tapping the signal carried in the cable is extremely difficult for would-be interceptors.

The negative aspects of fiber optic cable are:

- Poor flexibility
- Weaker pulling tensions
- Cost

Extra care must be taken to avoid kinking and excessive bending. Maximum allowable pulling tensions may also be lower than for twisted pair and coaxial cables. For some applications, twisted pair cable is much less expensive than fiber optic cable, as well as coaxial cable.

Making a Transmission Medium Choice

Martin Teague is a systems analyst who specializes in network design. He has been hired by Landfill Operators, which is planning to build a large landfill in Orlando, Florida. The information system planned for the new landfill facility will help automate the recycling and processing of millions of gallons of waste each year.

In a single month, four remote pumping stations process a combined 300,000 gallons of leachate, depending on the amount of rainfall. These pumping stations are to be connected to programmable logic controllers (PLCs), which tell the remote pumping stations when to turn on or off. Any disruption that might shut down the pumping devices running at maximum capacity could result in an environmental disaster.

Martin Teague is holding a meeting to determine the best transmission medium to be used to network the pumping stations to computer control devices.

Pam Womack, a project team member, said, "I believe twisted pair will be the least expensive and simplest cable to use."

Keith Clark, another project team member, said, "Why not a copper coaxial cable? This kind of cable is traditionally used for industrial-type applications."

"Fiber optic cable has several major advantages over coaxial or twisted pair cable that make it ideally suited for applications such as the landfill facility," Martin responded.

"Like what?" chimed in John Carroll, another project team member.

"First, fiber optic cables are exempt from electrical disturbances, such as lightning, which tend to paralyze copper cables. Fiber optic cables depend on lightwaves which are immune to electromagnetic interference generated by rotating machinery, lightning, and high-voltage power lines, pervasive in this area. This is an important factor since Florida has the highest lightning rate in the country. If a thunderstorm were to disrupt communication and shut down a pumping station, the maintenance staff would have to control the operation of the pumps manually. Second, fiber provides much lower attenuation than copper cables. Unlike copper networks, data can be communicated over fiber more accurately and over much longer distances without the need for power boosters. This will be the case at the landfill facility, where cable runs in some cases exceed 8000 feet," responded Martin.

"I've always heard that splicing fiber optic cable requires expensive equipment and highly trained specialists," John said.

"That was true at one time," Martin said. "Recent improvements in fiber cable splicing technology have helped to reduce installation costs. Splicing kits are easy to use and require virtually no training, decreasing the time, and consequently the high labor cost, of early fiber optic cable installations."

"What about connectors?" asked John.

"The beauty of connectors is that they provide easy reconfiguration capabilities with simple removal and insertion."

"It looks like fiber optic cable is our choice," said John.

Everyone on the project team agreed.

DESIGNING LOCAL AREA NETWORKS

Companies that want their employees to perform their tasks in coordinated workgroups, communicate with each other, and share a variety of resources are prime candidates for local area networks (LANs). A key characteristic of a LAN is that all nodes on the network are treated as equals in what is called peer-to-peer communications network.

Another LAN characteristic is that all nodes on the network are contained within one department or building, or possibly, within several buildings in close proximity to each other (e.g., a campus). In some companies, multiple LANs are interconnected at a single site. Or LANs may be linked to form wide area networks (WANs).

Selecting Commercial LAN Topologies

LANs are configured in network topologies. Most LANs contain primarily workstations, network servers, hard disks, erasable optical disks, and printers. The dominant LAN topologies are:

- Ring
- Bus

■ Star

Probably the two most popular commercial LANs are IBM's Token Ring and DEC's Ethernet.

Token Ring

IBM's Token Ring network is actually a combination star and ring topology that uses the token-passing access method. It runs at 4 or 16 Mbps. With fiber distributed data interface (FDDI), Token Ring will run at 100 Mbps. Although FDDI was originally designed for fiber optic cable, FDDI today will also accommodate twisted pair and coaxial cables.

A Token Ring configuration is illustrated in Figure 13.12. The Token Ring runs on twisted pair or fiber optic cabling. The topology consists of a series of multistation access units (MAUs), also referred to as media access units, that are linked together in a ring. Up to eight devices can be connected to each MAU. The MAU, then, becomes the hub of the star portion of the network.

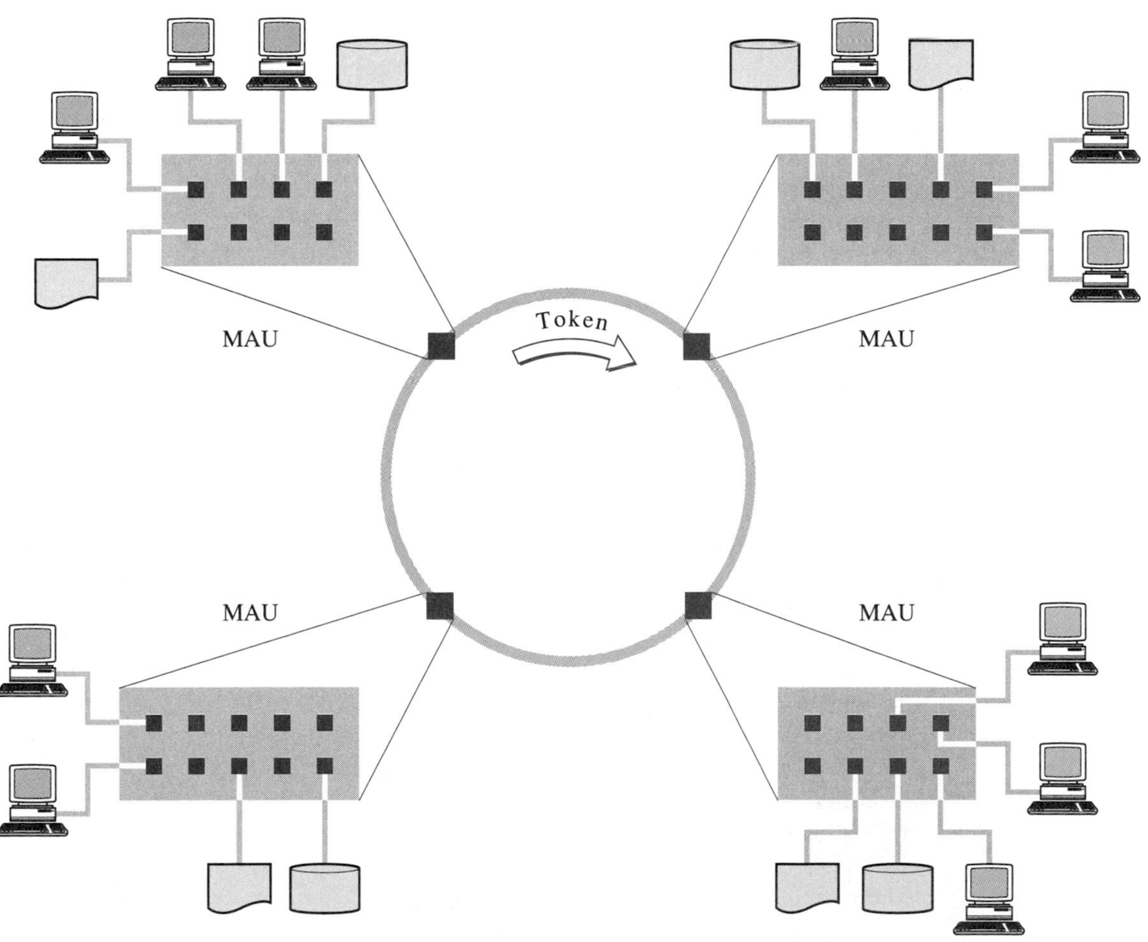

Figure 13.12
Token Ring LAN.

Because several MAUs can be linked together, many devices can be joined within a single network using twisted pair.

This combination of a ring and star topologies, also known as a star-wired ring topology, is used by a large number of vendors, including IBM. In this kind of topology, various devices can be easily and quickly connected to or disconnected from the MAU. This approach provides far greater control over security and network management than does the normal ring topology. Moreover, the star-wired ring topology is considered to be one of the most fault-tolerant topology designs.

The advantages are twofold. First, only one cable is needed from each node (or station) on the network in a single, centralized location. Telephone systems are wired the same way, and IBM recommends that the same conduits and wiring closets be used for both. This design requires more cable than it would take just to connect successive nodes, but it makes it much easier to add new nodes and remove old ones. The second advantage is that it's easy to bypass an inactive or malfunctioning node at the MAU by connecting the upstream node directly to its downstream neighbor.

The token in IBM's Token Ring is a special type of data packet that is originated by a workstation called the monitor. Any station (or node) on the network can be the monitor, but there can be only one monitor at a time. The token circulates freely around the ring until it is claimed by a node wishing to transmit. It says to this node, "It's your turn to send data if you want to." The token is changed to a "busy" status by the transmitting node, which adds data and address intended for another node on the net. The busy token delivers the data to the receiving node. After delivery, the token circulates freely until it is captured again.

Ethernet

Ethernet is a LAN offered primarily by DEC. Ethernet uses a bus topology, which is portrayed by a single line, as illustrated in Figure 13.13.

All devices are equal sharers of the bus. Because all network devices are connected to the same medium, some contention for the bus is inevitable. For instance, contention could occur when two or more workstations on the network try to broadcast onto the bus at the same time.

To prevent interference that occurs whenever signals from different sources collide on the bus, a listening strategy called carrier sense multiple access/collision detection (CSMA/CD) is used. It is a "listen before talking" communications method. When collisions are detected, each sender immediately stops broadcasting. Each sender backs off and waits a random period of time before trying to broadcast again. This method is like a conversation taking place at a party where everyone is trying to talk at the same time. A potential speaker, if he or she has good manners, usually waits for a pause in the conversation before speaking. If two or more people try to speak at once, both stop and try again in a moment.

Using a Premises Distribution System in Designing LANs

Today, when systems professionals plan for and design a cable network backbone for a building or campus in a local area, a **PREMISES DISTRIBUTION SYSTEM (PDS)** is

Figure 13.13
Ethernet LAN.

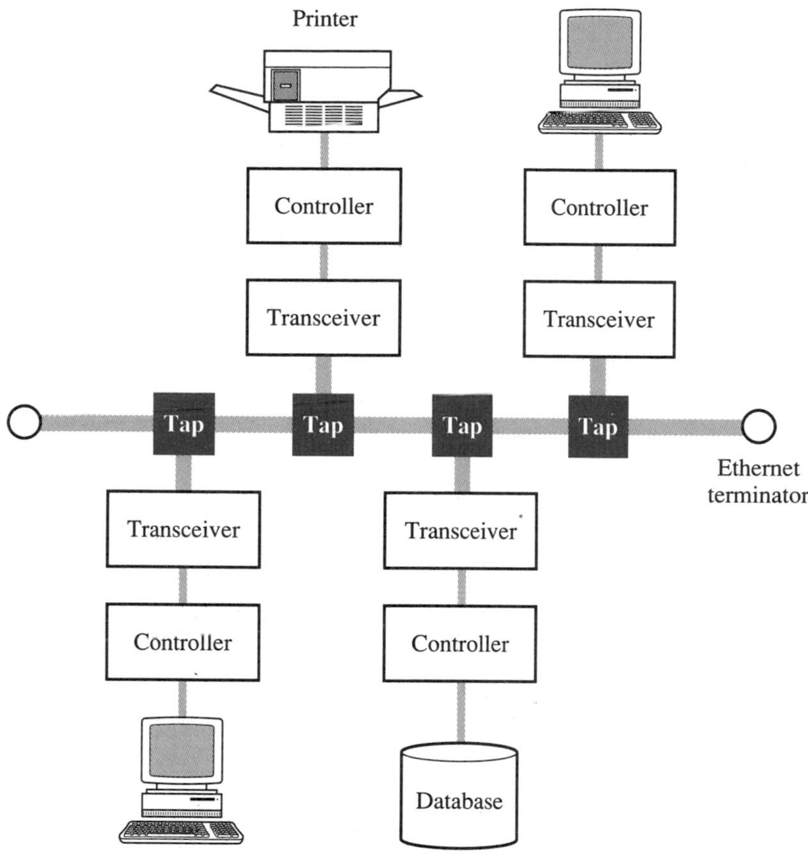

normally used. PDS is a consequence of experience gained by the telephone industry from years of designing and putting in place premises-wide telephone cable backbones.

Several good reasons exist for designing a well-planned premises-based cable backbone. First is its ability to adapt easily to future changes and support new system applications. Second, it provides a network that is easy to manage. Time and money are saved in troubleshooting and maintenance because the connectivity paths are known and can be monitored. That is, a description of the network topology, which lists all nodes and their connection points, is well documented and easy to understand. If there is a problem, mean time to repair (MTTR) is measured in minutes at most.

What Is the Premises Distribution System?

The premises distribution system (PDS) is a structured cabling design method formalized and extensively used by AT&T. It is used for connecting telephones, private branch exchanges (PBXs), host computers, local area networks, personal computers, workstations, and various office equipment (e.g., copiers and facsimile machines), and for linking these systems to outside net-

Figure 13.14
Premises distribution system (PDS).

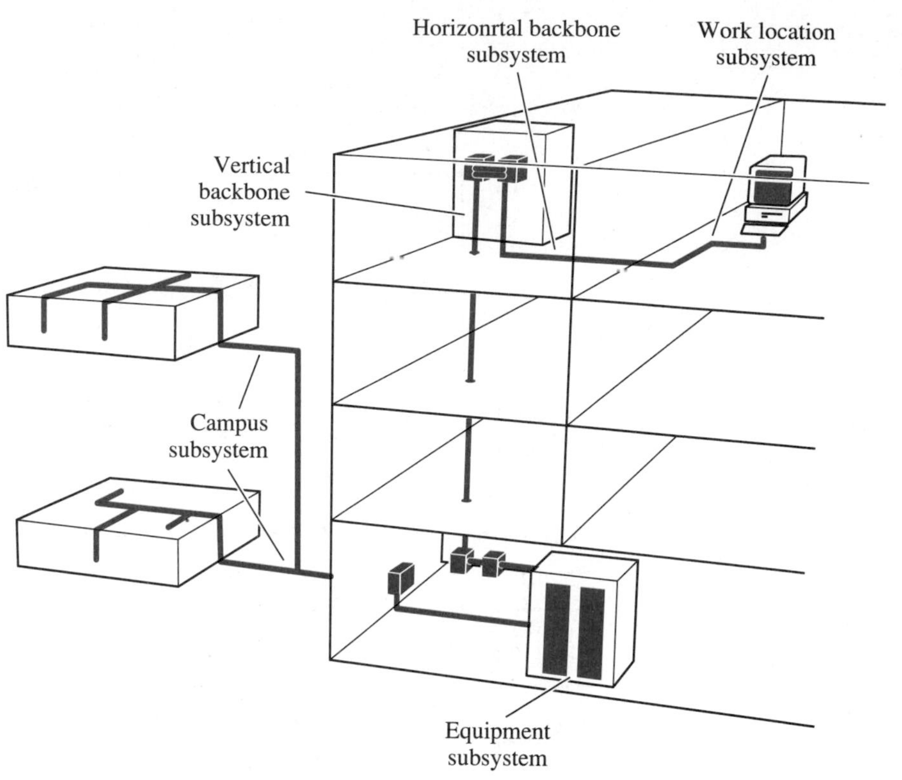

works, wide area networks, and public networks. PDS can support integrated voice, video, and data over a single transmission medium, such as twisted pair, which is independent from the sending and receiving devices it serves. Figure 13.14 illustrates PDS, which is based on a hierarchical star topology. PDS uses a set of cabling subsystems to design a full cabling infrastructure that can quickly and easily be reconfigured to adapt to changes in technology or the system environment.[1]

Five PDS Cabling Subsystems

The following material describes five PDS cabling subsystems that are relevant to this chapter.

The Vertical Backbone Subsystem This vertical backbone subsystem provides interconnection of the main wiring closet with floor closets. A prime cabling design factor is to keep cabling as centralized as possible and minimize the number of wiring closets to make future interconnections simpler. The vertical backbone, sometimes called the riser backbone subsystem because it is frequently located in risers in the building, is typically twisted pair or fiber optic cable.

[1] Dan Sarto and Greg Campbell, "An Inside Look at Premises Wiring," *LAN Technology,* February 1990.

The Horizontal Backbone Wiring Subsystems These horizontal backbones extend the vertical backbone subsystem to user workstations. They are always located on one floor and terminate at one end in an information outlet or wall jack. A horizontal backbone subsystem is required for each floor that has a work location.

Work Location Wiring Subsystem The work location wiring subsystem consists of cables, patch cords, or fiber used to connect an end-user device, such as a workstation, printer, or facsimile machine to the horizontal backbone. This area is where the end users' workstations are connected to the system.

Administrative Subsystem The administrative subsystem is in the wiring closet that consists of the cross-connects, which provide multiple connections, interconnections, and a method of linking other subsystems. It, in addition, can include equipment such as bridges that connect two LANs.

Campus Subsystem The campus subsystem is usually a fiber optic or copper cable, either twisted pair or coaxial, that provides the connection between multiple buildings similar to a campus environment. Within each building, the horizontal and vertical backbones are designed the same as in the main building.

The modular and star topology design aspects of PDS enable incremental growth of networks, allowing new systems equipment and LANs to be connected easily and quickly. Reconnection to an information outlet and an adjustment in the wiring closet is all that is required. It also helps eliminate duplication of expensive cable and the pulling of more cable, and it assists in planning for further systems growth. By far the most important cabling precedent to follow is to use a centralized wiring closet design. A central wiring closet makes tie-ins very simple. The PDS, with a star topology, regardless of the LAN's specific topology, minimizes the number of wiring closets.

Connecting LANs

An important element in connecting LANs is a bridge, as shown in Figure 13.15. In simple terms, a bridge is a software and hardware combination that connects different LANs (such as IBM's Token Ring to DEC's Ethernet) into a single network. Technically, bridges pass data from one LAN to another, translating protocols if necessary. The more translations the bridge has to make, the more sophisticated and costly it is.

In case of remote LANs linked by public or private telecommunication networks, gateways are used to connect LANs to this network to form a wide area network (WAN), as demonstrated in Figure 13.16. Gateways answer the need to integrate geographically dispersed LANs to transport E-mail and to share distributed databases and interactive applications. In this way, the services and benefits of LANs can be multiplied on an enterprisewide basis.

What Are Wireless LANs?

WIRELESS LANs are a special form of LANs in which nodes transmit data via radio signals. Speed of transmission of this radio technology is competitive with that

Figure 13.15
Bridge that links Token Ring and Ethernet LANs.

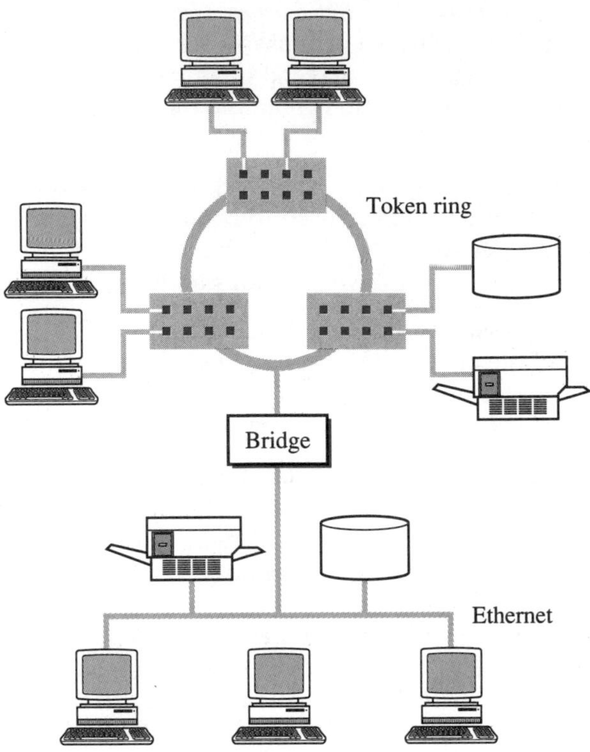

of Ethernet and Token Ring cable-based LANs. Also, infrared light-based Ethernet and Token Ring versions are now available. Wireless LANs are particularly well suited for small distances, such as in a building or a department within a building. Because anyone with the appropriate receiver device can capture the signal transmitted from one node to another, encryption may be necessary.

Wireless LANs provide an alternative to cabling. They may even be an incentive for enterprises reluctant to network computer hardware because of the problems associated with cable. Businesses that move their employees frequently or that have structural problems in their buildings are also turning to wireless LANs. Wireless technology is also applicable for connecting mobile robots to a LAN. Among the arguments for wireless LANs are the following:

- Allow for flexibility in locating and moving personal computers
- Avoid labor and other charges for installing cable
- Lower maintenance costs because there are no cables to maintain or reconfigure
- Permit older buildings that do not have ducts for cabling to install LANs
- Provide an alternative in factory and other environments where wiring may be impractical

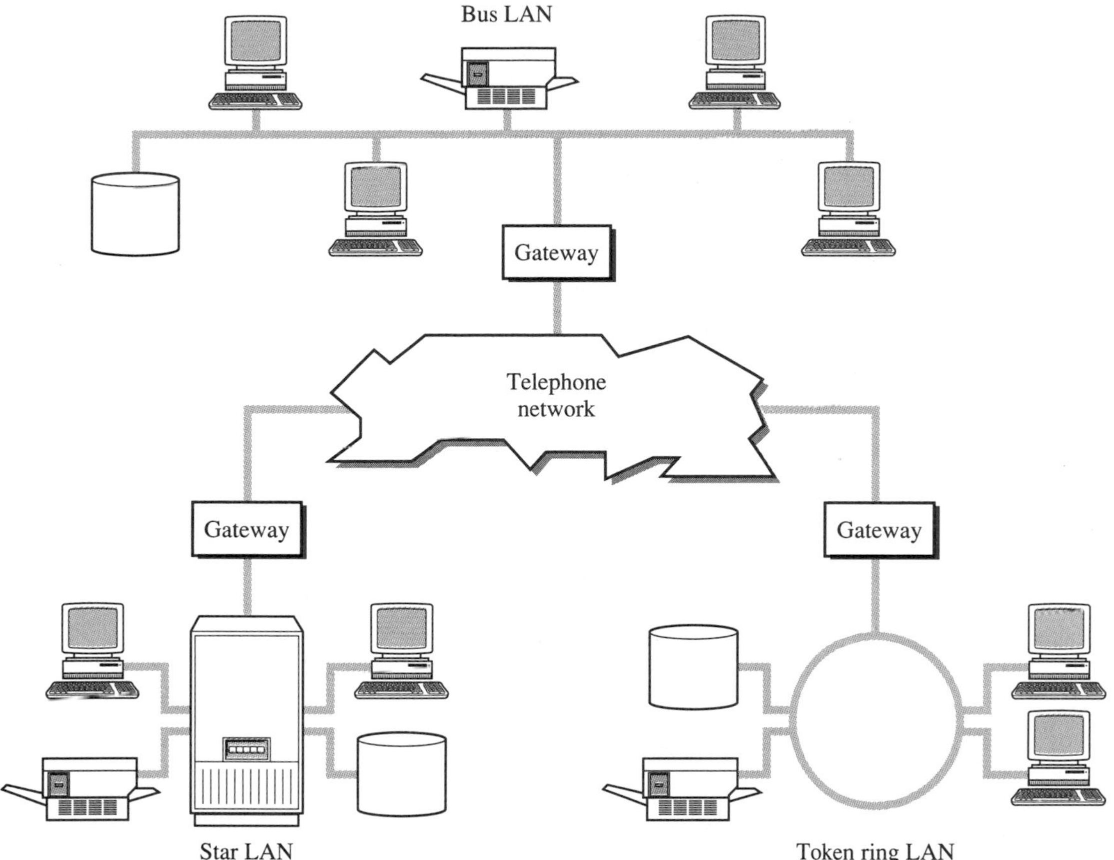

Figure 13.16
Use of gateways to connect LANs to form a wide area network (WAN).

On the down side, wireless LANs:

- Lower performance below that of cable-based (especially fiber optic cable) LANs because of limited bandwidth
- May be subject to interference from other electrical equipment

SERVICES FOR DESIGNING WIDE AREA NETWORKS

All the phone companies provide T-span, X.25 packet-switching, and integrated services digital network (ISDN) services designed for long-distance high-speed telecommunications. These services provide wide area network (WAN) design solutions and LAN-to-WAN internetworking.

T-Span

To obtain access to lines that support speeds beyond those available in standard dial-up or leased telephone lines, a firm may actually buy a bundled set of digital lines. These digital private-line services are tariffed by the phone companies and referred to as **T-SPAN SERVICES**.

T1

T1 is a term used to describe time-division-multiplexed data carrier T-span service, including a cable and its hardware, used to transmit digital signals at an aggregate rate of 1.544 Mbps. T1 can transmit a large volume of information in data, video, or voice form across great distances at high speed. It can do all this at a potentially lower cost than that provided by traditional analog service.

A T1 line consists of one four-wire circuit providing 24 separate 64-Kbps logical channels, as portrayed in Figure 13.17. Telephone carriers can combine individual channels from various customers until they have a "full" T1 line, that is, 24 circuits simultaneously carrying digital data. Digitization is changing the technical and economic considerations for private-line services because the data flow can now be provided without the need for dedicated lines. An integrated service unit (ISU) replaces the modem in the T1 arrangement. An ISU

Figure 13.17
Typical fractional T1 circuit access.

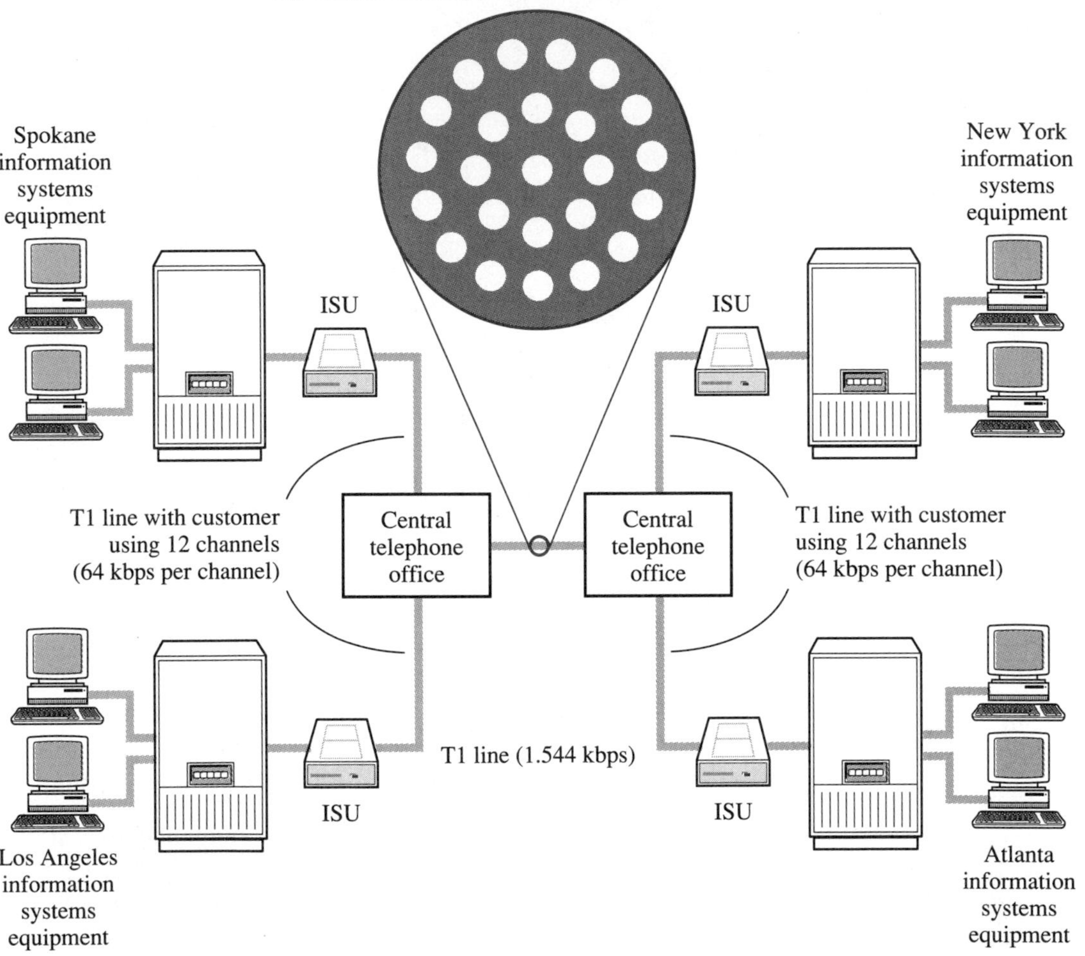

combines the channel service unit (CSU) and the data service unit (DSU); it is sometimes called a CSU/DSU.

Fractional T1

Fractional T1 service is aimed at customers who don't need all 24 channels of a "full" T1 line. Fractional T1 service offers the use of one or more of the individual channels within a T1 line; the customer pays only for the channels used.

An excellent application of fractional T1, in addition to its normal data transmission, is to split a T1 line into two half T1 circuits as a network backup element in designing a disaster recovery plan. By splitting the T1 line, the enterprise can protect itself against an interexchange carrier line failure.

X.25 Packet-Switching Network

X.25 is the international standard for wide area packet-switching networks. The providers of **X.25 PACKET-SWITCHING NETWORKS** (i.e., a network that transmits a packet or group of bits that are electronically routed or switched like box cars in a railroad system) offer a variety of options to the network designer. For example, a leased line can provide LANs high-speed access to a global X.25 packet-switched network. The packet network, illustrated in Figure 13.18, can then be used on an as-needed or a pay-as-used basis.

This network design approach can provide the maximum bandwidth at the least cost. It is of growing strategic importance for firms that seek high-speed transmission during limited periods of each day.

Besides E-mail, file transfers, and many interactive applications, another arena in which X.25 packet-switched network service will play a key role is in electronic data interchange (EDI). EDI will probably have a great deal to do with how business is conducted between organizations in the future, and, therefore, will dictate how networks are designed.

Integrated Services Digital Network

INTEGRATED SERVICES DIGITAL NETWORK (ISDN) allows a user to plug a device into the network anywhere in the world and achieve instant communications with any other device in the network. By using the telephone company's ISDN services, enterprises can access worldwide digital transmission devices and services at reasonable costs. For example, a credit-granting employee in a small business can make online credit card checks with a remote computer without tying up the telephone for sales orders. Compared to conventional transmission methods, ISDN technology offers simplicity of operations, increased data integrity, and higher transmission speeds. Moreover, it provides the base for an interesting portfolio of voice, video, and data services, such as teleconferencing, E-mail, and automatic dialing.

Designing a WAN for an Airline Company

An airline gives us an excellent model for designing a typical WAN, as shown in Figure 13.19. The airline company's four mainframes at corporate headquarters are connected via an X.25 backbone to IBM Token Ring networks in

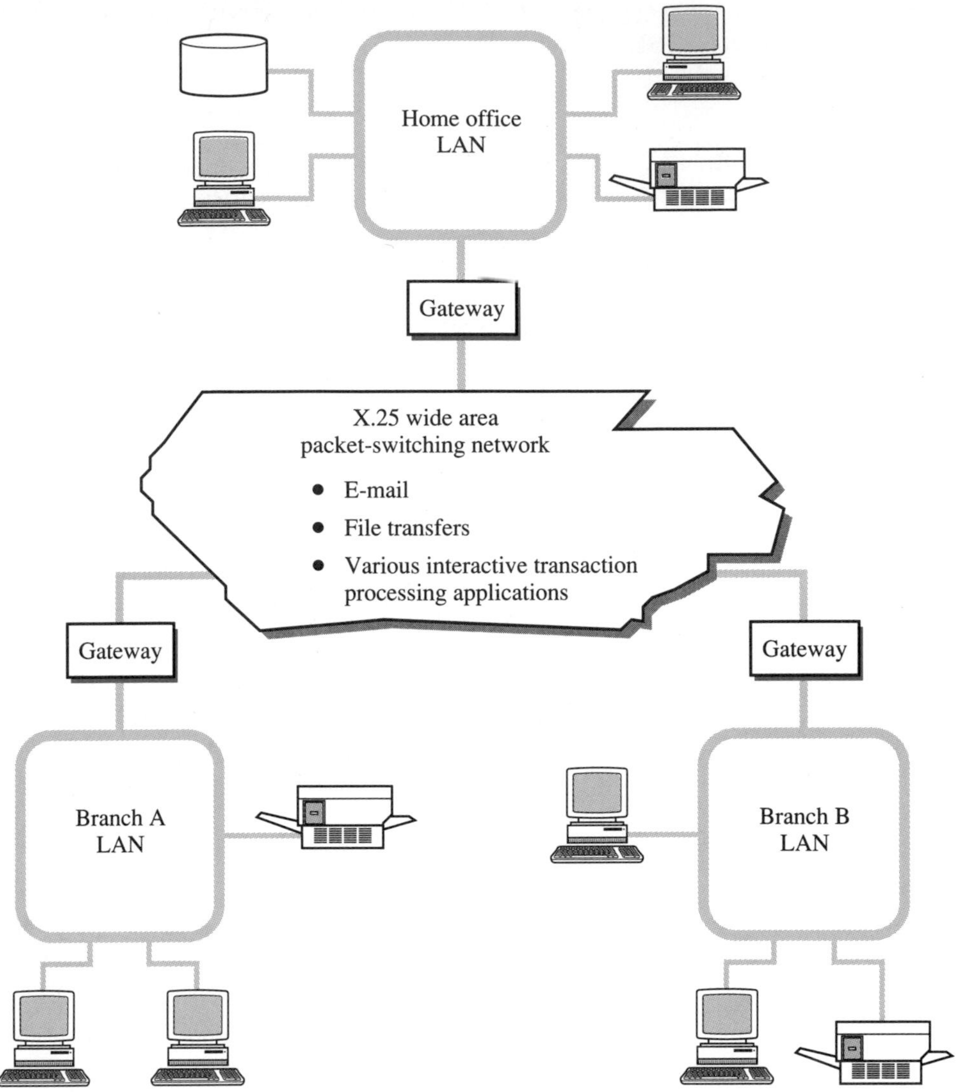

Figure 13.18
X.25 packet-switching network.

airports around the world. The objective of this WAN is to streamline baggage tracking and to increase business by improving customer service.

USING AIRWAVE TRANSMISSION MEDIA TO DESIGN WIDE AREA NETWORKS

Popular airwave (or space) transmission media include:

- Microwaves
- Satellites

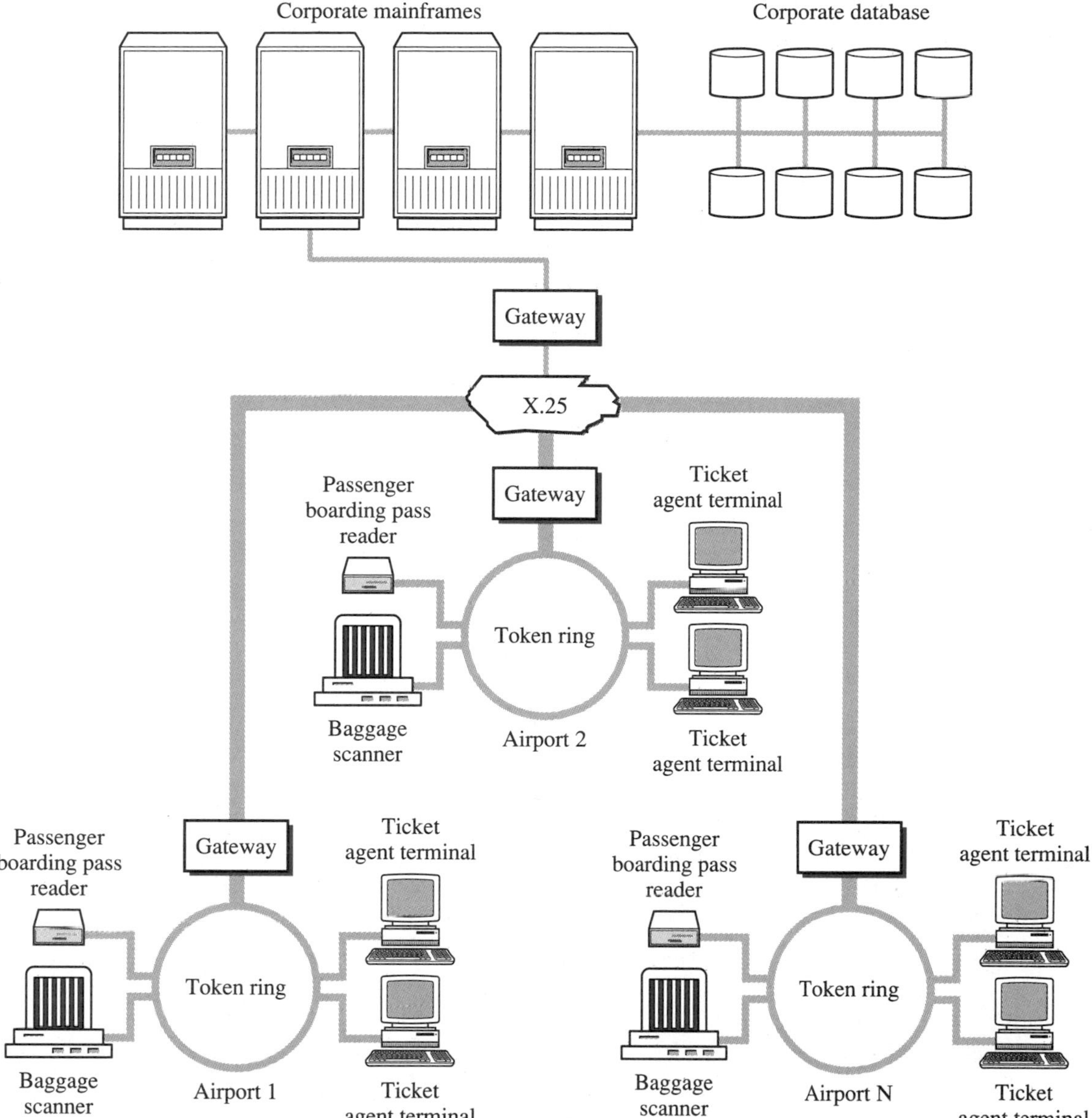

Figure 13.19
Wide area network (WAN) for an airline company.

With these media, the signal radiates from the sending device and continues forever onward through space. The signal becomes progressively weaker, or attenuates, as it gets farther from its source.

Microwaves

MICROWAVES are very high frequency radio waves that travel in a line-of-sight path. The end locations of a microwave system consist of a transmitter and receiver to send and receive telecommunications to the next site. Between the end locations are repeater sites which receive an incoming signal and retransmit it to the next repeater. Microwave systems carry large quantities of voice and data traffic.

Fiber optic, coaxial cable, or twisted pair are generally recommended for the transmission media backbone for a private network if the buildings are located on private property. If, however, a street or highway or property owned by another party must be crossed, microwave may be the answer. Rights-of-way take time to attain and are normally expensive. Also, cable laid on another's property runs the risk of being accidentally dug up and damaged. Microwave is, however, an airwave medium and does not require rights-of-way, as demonstrated in Figure 13.20.

Microwave communication is a highly reliable transport medium where point-to-point, line-of-sight links are needed. Microwave is especially well suited for large-volume, high-speed bulk data transfer between two discrete points, such as between buildings on campus or businesses in a metropolitan

Figure 13.20
Use of microwave to cross private property and other obstacles.

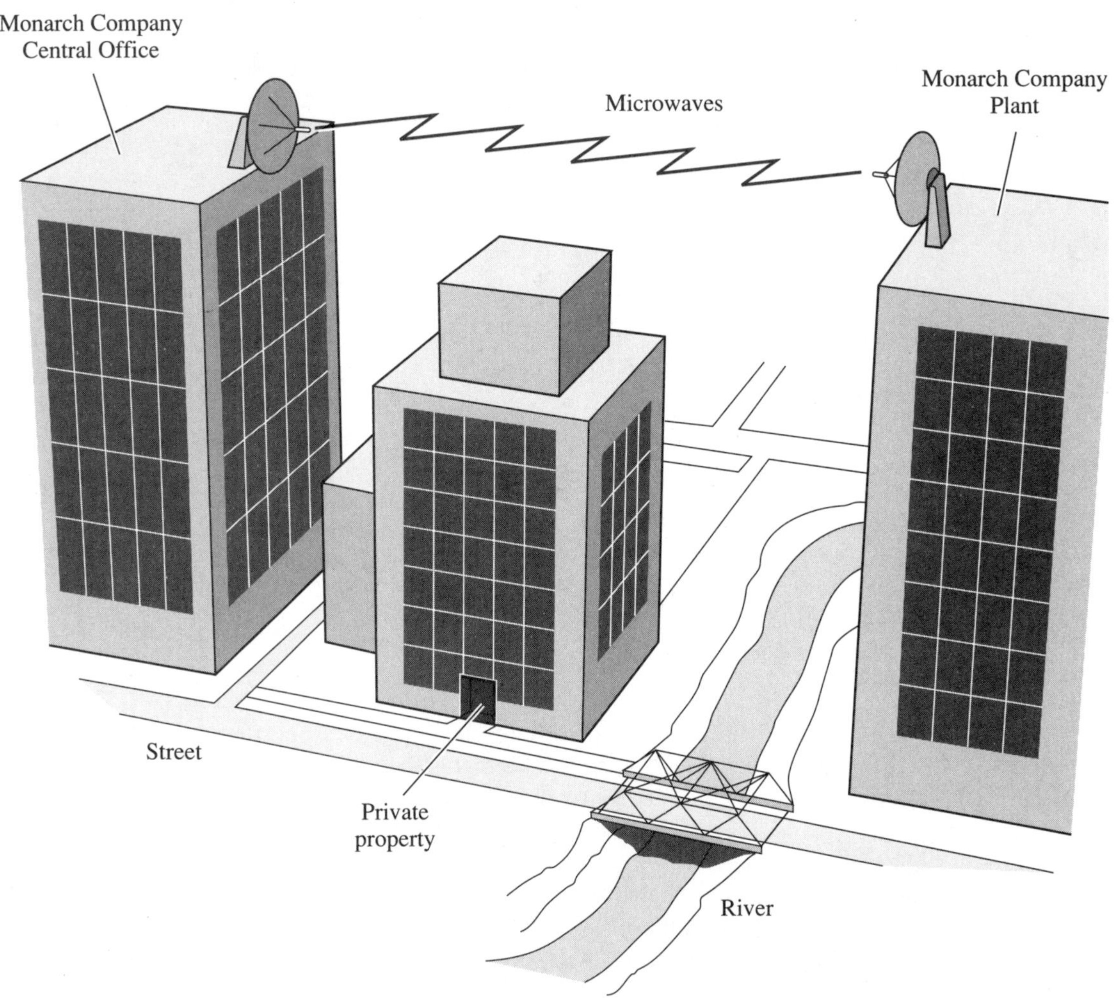

area. The disadvantages are that the repeater towers require line-of-sight placement, which is not always easy or inexpensive to achieve, and transmission is adversely affected by atmospheric conditions, such as lightning, rain, fog, snow, and birds.

Satellites

SATELLITES are telecommunication devices that orbit the earth and send and receive signals to and from sites on earth. A satellite system, as portrayed in Figure 13.21, contains several receiver/amplifier/transmitter sections, called transponders, each operating at a slightly different frequency. Individual earth stations send narrow beams of information called uplinks to the satellite. The satellite functions as a relay, receives the signal, amplifies it, then retransmits it to earth on a beam of a different frequency called a downlink.

The satellite orbits the earth at a constant speed and in the same direction that the earth rotates on its axis, so that the satellite's position is stationary with respect to a location on earth. This is called fixed or geostationary orbit.

Satellite circuits feature cost of service, which is independent of distance and thus most cost-effective for long-distance applications. In some instances, satellite transmission offers superior performance over terrestrial transmission media.

A negative characteristic of satellite transmission is propagation delay,

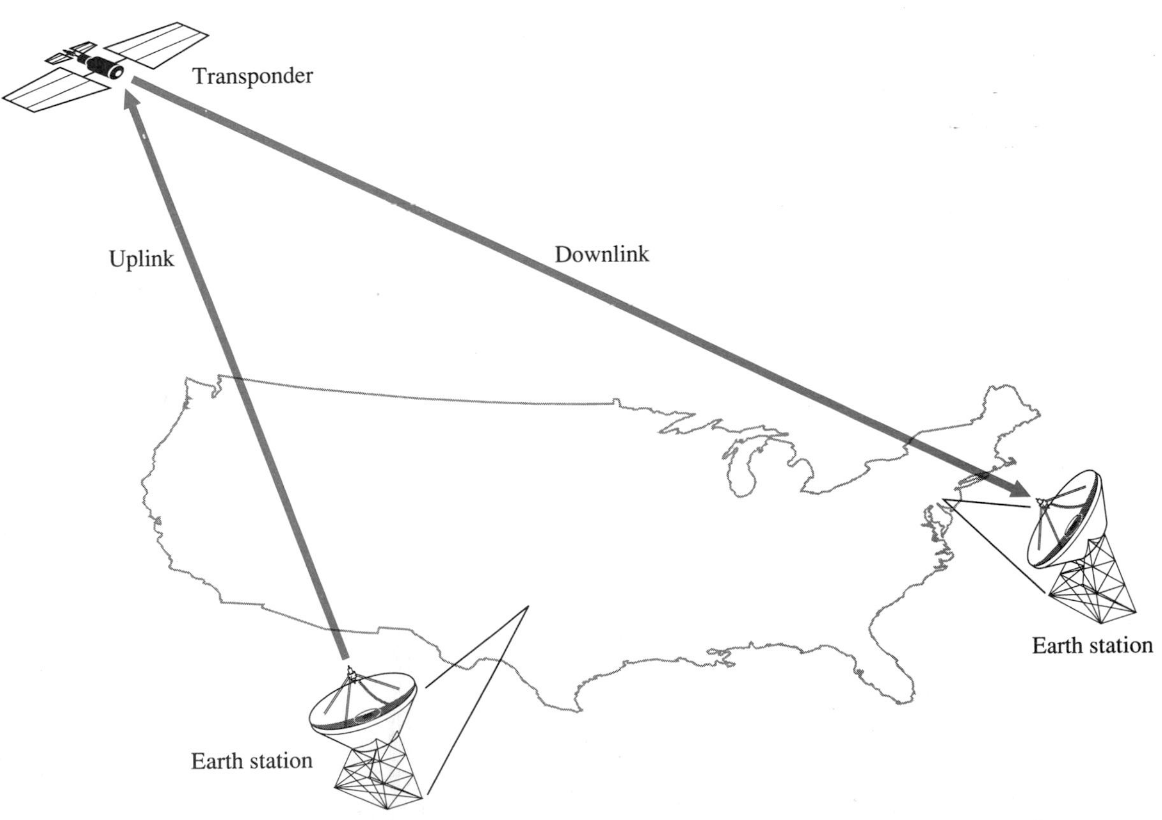

Figure 13.21
Satellite system.

which is the time the data signal takes to travel the length of the circuit, measured at approximately the speed of light. Because satellites orbit the earth at an altitude of about 22,000 miles, a signal must travel a minimum of 44,000 miles from the sender to the receiver. This takes much longer than the same signal sent by way of terrestrial transmission media. Selecting a satellite as a transmission medium must be planned by using protocols and defining applications to minimize this propagation time-delay characteristic. Failure to do this will result in poor throughput and response time.

VERY SMALL APERTURE TERMINAL (VSAT) network services, a special application of satellite transmission, have become widely available for companies that want to design private networks to replace phone company services or serve as backup networks to other networks. The VSAT services industry has established a large number of hubs that the service providers own and operate for multiple customers. Video and data are the primary traffic on VSAT. Although voice traffic is generally not considered the prime driving application for VSAT, voice transmission is adequate if needed because the quality of satellite voice is almost as good as terrestrial voice.

Reliability is one of the key design factors of VSAT. Suppliers of VSAT are so sure of their service reliability that most offer performance guarantees. These guarantees are typically set at 99.5 percent total network availability, or better. Also, VSAT suppliers will include 48-hour or better partial service restoration from major disasters at no extra charge.

Using VSAT at Big Value

Recently, Big Value has connected its chain of 300 retail stores scattered throughout the south and midwest onto a very small aperture terminal (VSAT) network. The purpose of installing the VSAT network is to improve customer service and reduce expenses. Before the network, Big Value had to transmit data manually or through mail services and obtain credit card authorizations through a time-consuming dial-up process.

"Networking wasn't something that was emphasized by previous management," said Paul Knight, director of information systems. "With our new VSAT network, credit card authorizations are cut from as long as 2 minutes to about 8 seconds. The VSAT network also provides broadcast video applications. For example, video is used to broadcast presentations from top management to employees at remote sites, make new product announcements, and for training programs. Moreover, all users are linked to enterprisewide inventory and pricing information."

REVIEW OF CHAPTER LEARNING OBJECTIVES

The major goals of this chapter were to enable each student to achieve six learning objectives. We will now summarize the responses to these learning objectives.

Learning objective 1:
Describe the steps in designing networks.

Ideally, network designers should start with an enterprisewide model as a general pattern or guide for designing an enterprise's network. Then the enterprise should be logically segmented. An appropriate local area network (LAN) should be designed for each segment. If the enterprisewide model calls for internetworking, LANs for each segment should be interconnected using bridges or gateways.

Learning objective 2:
Define basic networking elements.

Networking elements include:

- Transmission media
- Transmission signal
- Transmission mode
- Network topologies
- Network interface cards (NICs)
- Network servers
- Modems
- Multiplexers
- Network operating systems
- Protocols
- Front-end processors (FEPs)

Transmission media carry a digital or analog signal. Transmission mode of the signal is simplex, half duplex, or full duplex. Popular network topologies are star, bus, or ring. The network interface card (NIC) carries the electronic circuitry that connects a computer to a network. Network servers, usually file and print servers, coordinate and manage access to databases and printers. Modems permit digital devices to communicate with analog devices, and vice versa. Multiplexers permit line sharing in which data from several terminals are interleaved on a common higher-speed line. Like a computer operating system, a network operating system is the software that manages the network and provides its functionality. Protocols enable disparate networks and devices to communicate with each other. A front-end processor offloads networking tasks from main computers so that they can devote all of their processing power to user applications.

Learning objective 3:
Discuss twisted pair cable, coaxial cable, and fiber optic cable.

Twisted pair cable is traditional telephone copper cable. It is an excellent and relatively inexpensive transmission medium for many network applications. Coaxial cable supports large bandwidths and allows high-speed data rates. Fiber optic cable is usually made of a glass core with a glass or plastic cladding covered with a plastic jacket. As a transmission medium, fiber optic is attractive because of its huge bandwidth and ability to transmit data at very high speeds. Because fiber optic cables carry light rays, they are free from problems of electromagnetic interference.

Learning objective 4:
Discuss designing of local area networks (LANs).

A LAN interconnects a variety of nodes using nonpublic transmission media within an area of 5 to 10 miles, often considerably less. Usually, a LAN serves a department in a company, such as the sales or accounting department.

Two prominent commercial LANs are IBM's Token Ring and DEC's Ethernet. Token Ring uses a data signaling scheme in which a special data packet, called a token, is passed from one node to another along a transmission medium (e.g., fiber optic) ring. When a node wants to transmit, it takes possession of the token, transmits its data, then frees the token after the data have made a complete circuit of the ring. Ethernet is based on a bus topology. All network nodes on the bus "listen" to all transmissions, selecting certain ones in response to address identification. A standard for designing a premises-based transmission media backbone is premises distribution system (PDS), championed by AT&T. Its main wiring subsystems are the vertical backbone, the horizontal backbone, and the campus backbone.

Wireless LANs are a special form of local area networks (LANs). LANs using wireless technology are installed in cases where using cable would present a problem.

Learning objective 5:
Describe services for designing wide area networks (WANs).

Three important services offered by the phone companies are:

- T-span
- X.25 packet switching
- Integrated services digital network (ISDN)

T-spans (or T-carriers) are time-division-multiplexed digital transmission lines. Probably the most popular T-span services are T1 and fractional T1. A T1 line consists of one four-wire circuit providing 24 separate 64-kbps logical channels with an aggregate data rate of 1.544 Mbps. Telephone carriers can combine individual channels from various customers until they have a "full" T1 line, all 24 channels simultaneously carrying data. Fractional T1 offers the use of one or more of the individual channels within a T1 line.

Public access data networks are operated by common carriers or telecommunication administrations to provide many customers with leased-line, circuit-switched, and packet-switched networks. The standard protocol for packet-switched networks is X.25.

Integrated services digital network (ISDN) is a service offered by telephone companies to enable users to transmit and receive, in digital form, a wide variety of information traffic. It allows almost any new telecommunication service to be introduced and operated efficiently without the massive investment required if a separate network were to be set up.

Learning objective 6:
Explain how microwave and satellite transmission media are used for designing wide area networks (WANs).

Microwaves and satellites are especially applicable for designing WANs. Microwave radio carries large quantities of voice and data traffic, and it is especially applicable where the company does not have a right-of-way to lay terrestrial transmission media. Satellites provide excellent transmission devices to span long distances and interconnect LANs with WANs. Very small aperture terminals (VSATs) are small and portable dishes that provide access to remote areas not serviced by telephone companies. VSATs can serve as a backup network or as an alternative to telephone transmission services.

NETWORK DESIGN CHECKLIST

Following is a checklist on how to perform network design. Its purpose is to summarize the chapter and give you a perspective on key network issues.

1. Use an enterprisewide model as a basic blueprint for network design.
2. Permit the enterprisewide model to guide you in dividing the enterprise into logical segments.
3. Draw on paper or screen an outline of the network backbone, and access and work locations for each segment. Make sure that a proposed LAN meets the segment's requirements.
4. Network the local area networks for each segment into an enterprisewide network, using bridges and gateways.
5. Understand the networking elements and the role they play in installing your network design.
6. For premises cabling, select one or a combination of these media: twisted pair cable, coaxial cable, and fiber optic cable. Understand each cable's advantages and disadvantages.
7. Designing local area networks will generally follow a ring, bus, star, or some variation thereof. Two popular commercial LANs are IBM's Token Ring and DEC's Ethernet.
8. Use premises distribution system (PDS) for laying cable throughout a building or building complex.
9. Should your LAN design be in an environment where cabling is not feasible, consider installing a wireless LAN.
10. For wide area networks (WANs) consider the application of T-span, X.25 packet switching, and ISDN.

11 For network designs that call for airwave transmission, use microwaves or satellites. For an alternative to telephone networks or as a backup network, consider VSAT.

KEY TERMS

Bridges
Coaxial cable
Fiber optic cable
Front-end processor (FEP)
Gateways
Integrated services digital network (ISDN)
Local area network (LAN)
Microwaves
Modem
Multiplexer (mux)
Network interface card (NIC)
Network operating system
Network servers
Network topology
Premises distribution system (PDS)
Protocols
Satellites
T-span services
Transmission media
Transmission modes
Twisted pair cable
Very small aperture terminal (VSAT)
Wide area network (WAN)
Wireless LANs
X.25 packet-switching networks

REVIEW QUESTIONS

13.1 List and briefly describe the general steps in designing networks.

13.2 List and briefly define each networking element.

13.3 Name and define three popular protocols.

13.4 Why is OSI called the standard of standards? Name and briefly describe each layer in the OSI reference model.

13.5 State the purpose of X.12, X.400, and X.500.

13.6 Define twisted pair cable.

13.7 List and briefly discuss cable installation guidelines.

13.8 Differentiate IBM's Token Ring LAN from DEC's Ethernet LAN.

13.9 Explain why a Token Ring is not really a ring topology. What is a MAU?

13.10 Describe the premises distribution system (PDS) and define its major subsystems.

13.11 Name and briefly describe two situations in which wireless LANs are particularly useful.

13.12 Define bridges and gateways.

13.13 Define T-span services and explain how they are used.

13.14 Explain X.25 packet switching and what it is used for.

13.15 Explain how ISDN is used.

13.16 Explain why microwave may be the only feasible transmission medium for some network designs.

13.17 Explain the advantages and disadvantages of satellite transmission compared to terrestrial transmission media.

13.18 Explain the use of VSATs for designing WANs.

CHAPTER-SPECIFIC PROBLEMS

These problems require exact responses based directly on concepts and techniques presented in the text.

13.19 For each of the following network situations, insert the name of the networking element that is applicable.

_______________ Outline of a specific network configuration.

_______________ One-way transmission.

_______________ Coordinate and manage a printer connected to a network.

_______________ Circuitry board that enables a workstation to be connected to a network.

_______________ Line sharing in which data from several terminals are interleaved on a common higher-speed line.

_______________ Convert digital signals to analog signals and back to digital signals.

_______________ A device used to offload networking tasks from the main computer.

_______________ A software package that manages the network and gives it functionality.

_______________ A method that allows disparate networks and devices to communicate with each other.

13.20 Review the following situations.

_______________ Conventional telephone wiring is already in place and is sufficient for the planned network.

_______________ The copper cable must possess the ability to carry large volumes of data (i.e., bandwidth) and be somewhat resistant to electromagnetic interference.

________________ The system must be able to carry signals with low attenuation and be resistant to lightning.

Required: Insert in each blank the appropriate transmission medium cable.

13.21 For each application described below, select an appropriate cable transmission medium.

1 Spruce Company is moving its LAN to a building that is already wired with conventional telephone cable.

2 Hamilton Investment will occupy the third, fifth, and twelfth floors of a twenty-floor building. Hamilton's vertical backbone will be installed in the building's riser. Within the building are Hamilton's competitors, some of whom would like to know about Hamilton's top-secret merger and acquisition operations.

3 Pathfinder's operations require a great many top-level meetings each week attended by managers from offices scattered throughout a large campus-like complex. The CEO has proposed to conduct meetings via videoconferencing.

4 Bigbang is a manufacturer of explosives that are easily detonated by small electrical charges. Managers of Bigbang are planning to install a transmission medium backbone on the factory floor for robot control and for general data processing.

13.22 Modern Iron Works' headquarters are in a three-story building with work locations on each floor. In addition, a large one-story building is adjacent to Modern's headquarters. This building houses a number of computer-controlled machines and workstations.

Required: Using PDS, design a premises-based transmission media backbone for Modern Iron Works.

13.23 The accounting department at Northstar is composed of three work groups: accounts payable, accounts receivable, and general ledger. A large database stored on magnetic disk is controlled by a LAN server. Also, two laser printers are controlled by a LAN print server.

Required: Draw a general network design for the accounting department at Northstar.

13.24 Noland Company has four workstations located in a remote warehouse connected to the headquarters host computer via low-speed analog lines. Management wants to employ some kind of line-sharing technique in which the line shared would be a high-speed analog line. Applications from the workstations have predictable periods of inactivity.

Required: Design a new network to support management requirements.

13.25 A Token Ring for Mastergard, Inc., will contain three MAUs. On MAU 1 will be two workstations. Connected to MAU 2 and MAU 3 will be two workstations, one printer, and one large external hard disk.

Required: Design this Token Ring for Mastergard. Also, connect this Token Ring LAN with an Ethernet LAN that contains four workstations, two printers, and one external hard disk. Both LANs are in the same building.

THINK-TANK PROBLEMS

These problems call for a feasible approach rather than a precise solution. Although the problems are based on chapter material, extra reading and creativity may be required to develop workable solutions.

13.26 Liberty Company is planning to connect its computer equipment in Philadelphia to a computer system in the Boston division. After performing steps in the network design process, it is concluded that a wide-bandwidth, private leased line that transmits data in digital format is required. A transmission rate of 64 Kbps via three channels is more than adequate.

Required: Name the data transmission service you recommend for Liberty. Also, sketch the wide area network, including connectivity devices.

13.27 Powerglide Company is headquartered in Miami, and an assembly plant is located in Los Angeles. Management desires a high-speed connection to the LA plant to transmit data, video, and voice. Results from your analysis indicate that three logical channels carrying an aggregate data rate of 192 kbps will be adequate for the next three years, at which time another logical channel of 64 kbps will be needed.

Required: Name the type of communication service you recommend for Powerglide. Also, design the network you envision. How many logical channels will Powerglide need for the first three years? How many at the beginning of the fourth year?

13.28 You have been serving as systems consultant for Spasmodic Industries, which has several branch offices and warehouses scattered around the country. You are trying to develop a network design that fits Spasmodic telecommunication needs, which are widely distributed, lightly used applications. Joe Frugal, CEO, said, "I don't want to spend much money on data communications. I expect to spend money for linking to the network and use only, like the postal service, for which there is no charge for the availability of service, just for letters stamped and posted." In fact, Joe's postal service description is apt, because Spasmodic's transmission messages can be segmented into small packets and routed through a shared network in a way similar to the way the postal service forwards letter packets.

Required: Name the kind of network service and protocol you recommend for Spasmodic and make a general design of the network.

13.29 Tracy is a large department store chain that has four data centers that house one mainframe at each center. The data centers are located in the eastern, midwestern, southern, and western regions of the country, respectively. Each data center is stand-alone and isolated from the others. A new management team at Tracy practices a decentralized management approach. This team wants the data centers to be interconnected by high-speed digital circuits to achieve cooperative processing.

Required: Design the network topology that you deem appropriate for Tracy, and label the transmission media that will handle the data traffic.

13.30 Webster Medical Center is located in downtown Sacramento. The hospital, with a number of workstations, is in a building on the north side of Crescent View Drive, a four-lane highway. On the south side of this thoroughfare is Webster's administration building, where a mainframe and various peripherals are located. Administrators want the computing resources in both buildings connected.

Required: Select a feasible way to connect the computing devices and sketch your recommendation.

13.31 Buggy-Wash, a nationwide car wash company, intends to install a voice, data, and video network that will permit centralized supervision of remote sites. Buggy-Wash's central headquarters, where centralized supervision will be conducted, is located in Chicago. Car wash sites are scattered throughout the country.

Required: Design a nonterrestrial network that connects all the remote car wash sites to Buggy-Wash's central headquarters.

13.32 Ocean Carriers plans to develop some kind of automated identification system to track containers that it transports. Managers at Ocean Carriers hope that this new system will give them a competitive edge by improving customer service and by reducing operating expenses. Ocean Carriers distributes and transports containers in Asia and North America as well as between the two continents, using a combination of ocean liners, trains, and trucks.

Miriam Singleton, CEO of Ocean Carriers, wants some kind of microchip tag attached to containers, container carts, trucks, and rail cars which can be scanned by reader devices at Ocean Carriers' ports and terminal facilities. She believes this approach will help employees at sites expecting shipments to use the data to prepare carts and trucks to handle incoming containers more efficiently. Moreover, customers can access Ocean Carriers' mainframe via their computers to track shipments. To date, the company has had to input cargo data manually. Ms. Singleton expects the new system to replace this labor-intensive and error-prone method of tracking containers.

Required: Design a network for Ocean Carriers.

13.33 Lone Star Industries enterprisewide model, prepared during the systems planning phase of SDLC, shows the following: corporate headquarters with a mainframe that controls an accounting LAN and a financial and accounting database; a bus LAN for the main plant next to corporate headquarters; a remote plant containing a bus LAN, star LAN, and ring LAN; and a warehouse containing a ring LAN. Data packets are transmitted between corporate headquarters and the remote sites periodically.

Required: Design a wide area network for Lone Star.

13.34 Odyssean Industries is a multinational manufacturer that has recently extended operations to eastern Europe. Odyssean is a division-based enterprise headquartered in Chicago. In each division, there are branch and field offices and manufacturing facilities, all of which have highly integrated networks. Also, all divisions, including the overseas division, are connected by a combination of satellite, T1, and packet-switching transmission services.

Required: Design a wide area network for Odyssean.

13.35 Deepwell Energy Company is based in Houston. In addition to drilling platforms in the Gulf of Mexico, Deepwell plans to set up drilling platforms in the jungles of South America. Management wants all drilling platforms connected to corporate headquarters in Houston. The network must carry voice, facsimile, and data to keep drilling operations coordinated. Receiving and transmitting devices must be small enough to fit in the back of a pickup truck or on top of a drilling derrick, and they must have the ability to be set up or taken down in 10 to 15 minutes. Deepwell cannot depend on the phone company, because one does not exist where Deepwell's drilling platforms are located.

Required: Design a wide area network for Deepwell.

SUGGESTED READING

Briere, Daniel. "Digitization Advances Private-Line Services." *Network World,* January 1, 1990.

Brown, Bob. "Shipper's Tracking System to Yield Edge." *Network World,* November 14, 1988.

Casatelli, Christine. "Open Sesame." *Network World,* April 25, 1990.

Cashin, Jerry. "TCP/IP Is Now in Your Stores, While OSI Has Future Promise." *Software Magazine,* August 1989.

Coursey, David, and Jodi Mardesich. "Motorola to Unveil Fast Wireless LAN Technology." *Infoworld,* October 22, 1990.

Desmond, Paul. "Car Wash King Oversees Empire Via Integrated Net." *Network World,* November 14, 1988.

Eichten, Chuck. "Networking by the Numbers." *Interact,* January 1991.

Fisher, Sharon. "Five Ways Networks Pay Off." *PC World,* March 1991.

Freund, Mark. "Design Rules and the Single LAN." *LAN Technology,* February 1990.

Glass, Brett. "Understanding the OSI Reference Model Standard." *Infoworld,* April 16, 1990.

Hancock, Bill. *Network Concepts and Architectures.* Wellesley, Mass.: QED Information Sciences, 1989.

Hatfield, W. Bryan, and Michael H. Coden. "Designing Today's Fiber LAN to Support FDDI Tomorrow." *LAN Technology,* February 1990.

Khosrowpour, Mehdi. *Microcomputer Systems Management and Application.* Boston: Boyd & Fraser, 1990.

Knight, Robert. "To Sell to Sears, EDI Is a Must." *Software Magazine,* March 1991.

Kousky, Ken. "Bridging the Network Gap." *LAN Technology,* January 1990.

Leeds, Frank, and Jim Chorey. "Cutting Cable Confusion: The Facts About Coax." *LAN Technology,* March 1991.

Leeds, Frank, and Jim Chorey. "Twisted-Pair Wiring Made Simple." *LAN Technology,* April 1991.

Maxson, Mark A. "Building a Corporate Network from Heterogeneous LANs." *LAN Technology,* January 1990.

Morris, William T., and Tony E. Beam. "How to Install Fiber-Optic Cable." *LAN Technology,* December 1990.

Salamone, Salvatore. "Internal EDI Helps Sales." *Network World,* April 1, 1991.

Sarto, Dan, and Greg Campbell. "An Inside Look at Premises Wiring." *LAN Technology,* February 1990.

JOCS CASE: Designing the Network

At the end of Chapter 12, Jake Jacoby, JOCS project manager, assigned Tom Pearson, JOCS SWAT team member, the task of outlining the computer interfaces required for the JOCS system. Tom was to create this outline because the SWAT team members could not agree on the level of controls necessary for JOCS. During a meeting of the SWAT team to discuss controls, it became clear that the team members were unsure about which of the existing computers would interface with the JOCS system. This confusion helped to fuel the disagreement among team members about controls, and Jake hoped that better clarification of the JOCS system placement would help the team reach consensus.

Unfortunately, this didn't happen. Tom did outline the possible computer interfaces. The computers involved include:

1 Multiuser system located in the engineering department: This system consists, briefly, of 25 workstations, five plotters, two printers, a UNIX operating system, and a variety of software, including a computer-aided design (CAD) system linked together through an Ethernet-based communications system.

2 Public access data network accessible through the multiuser system located in the engineering department: The personnel in engineering access the Internet telecommunications network in order to share information with people doing related work in engineering. In addition, the Internet connection is used to access supercomputer computing resources when required for complex engineering problems, such as modeling and statistical analysis. The Internet connection allows engineering personnel to access a variety of computer systems.

3 Computer-integrated manufacturing (CIM) system located in the manufacturing department: A small computer used to gather direct labor and direct material data, as well as control the robots used to assemble hydronautical lifters.

4 Multiuser system located in a centralized data processing facility: This system is used by the finance department to track accounting transactions. The system is capable of supporting 45 workstations. There are currently 20 workstations attached to the system, two printers, and 700 megabytes of disk. The system is a stand-alone star configuration running a proprietary operating system and a centralized computer, with each of the workstations communicating directly with the centralized computer. This system is not linked to any other systems.

As Tom attempted to outline the required computer interfaces for JOCS, it became clear that he would need more information about the future uses of this system. Frequently, the pieces of systems development are highly interdependent. For example, to define the process and output controls required for JOCS, Tom must understand how the existing computer systems will interrelate with JOCS. To define how the systems will interrelate, Tom has to understand what systems will be used in the future. The result is that the SWAT team must begin network design before they can finish the design of their process and output controls.

The first step in network design is to define the applications that will run on the system. Most companies have existing systems, such as the ones described above at Peerless, and can't install a new system without taking into consideration what is already in place. In addition, most companies also plan to use their systems for a given period of time and must plan their network to support future applications that will be installed. Management at Peerless, Inc., has plans that call for a strong growth rate over the next ten years. The network designed for even a seemingly small system, such as JOCS, must be able to accommodate the projected growth of the company.

To define the applications that will run in the future along with JOCS, Jake calls for a meeting of the SWAT team with Mary Stockland (CIO for Peerless, Inc.). Jake wants Mary to explain the systems that she envisions will be installed at Peerless within the next five years.

Mary hands out a general diagram, Figure 13.22, and begins to explain her plans for the future.

> I have outlined six critical systems to be installed over the next five years. In addition to JOCS, we will install an electronic data interchange system to process transactions more efficiently with our vendors and our customers. This system will require that we maintain contact with current public access data networks. Either we will create a new ISDN connection, or we will use our current Internet connection to perform EDI. I also plan to install a companywide electronic mail system along with a complete office automation system. This system will require that all departments be linked into a compatible backbone. From these two systems will grow a marketing information system, an engineering support system, and a financial decision support system.

Mary went on to describe the types of communication needed for Peerless:

> We need a communications network that will do the following:
>
> 1 Maintain an easy connection with public networks.
>
> 2 Allow us to eventually interface all systems within the company.
>
> 3 Permit the company the flexibility to use different vendors without having to change our data communications backbone.
>
> A manufacturing company cannot afford to have a series of stand-alone systems. Each separate department is really an interrelated portion of the entire picture of our company. As a result, all communications will interface, if not immediately, in the future. One of my major goals for this company is to ensure that data can be shared and integrated between departments.

Figure 13.22
Premises-based transmission media backbone.

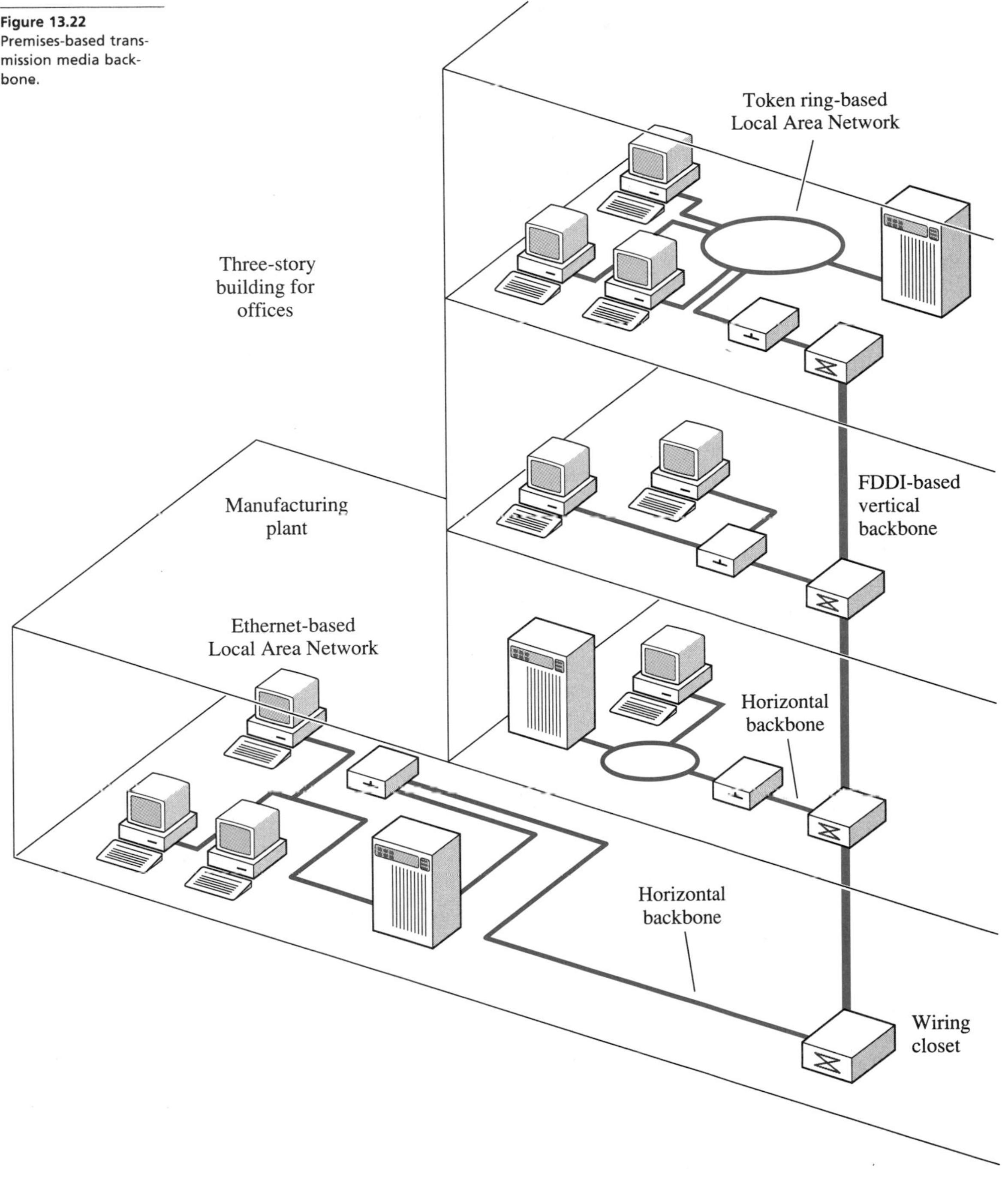

Jake looked at her diagram and commented, "This diagram primarily details a premises-based transmission media backbone for the company. You have detailed the overall direction for communications."

Mary agrees, "That's right. We needed a place to start. I started by evaluating the best communications backbone for our future system. As you can see from the diagram, we will install a fiber optic backbone according to the fiber distributed data interface (FDDI) standard. This will allow us the flexibility to integrate a variety of different types of computers, while also providing the bandwidth necessary to process our transactions quickly."

Tom, examining the diagram, says, "Right now, it looks like you have only defined the vertical backbone for the system. Are you planning on enforcing standards for each new application?"

Mary shrugs her shoulders and says, "Yes and no. The horizontal backbone will fit FDDI standards, too. However, we already have an Ethernet system, and a proprietary network in place. I am not planning on removing any of our current systems. Our current systems are doing their jobs; we just have to tie them together. Our backbone is not going to limit development. It will help us develop standard systems that are able to be integrated. I want each application system to have a set of options we can choose from when designing new systems. For example, when an application dictates the use of a local area network, our backbone will work best with an Ethernet connection or a Token-Ring network."

Jake says, "So what we have to do is recommend a network for JOCS that will fit within a FDDI communications backbone. It must be installed with our current technical expertise, and it will satisfy the needs of the JOCS users. It sounds like the rest of the systems will be integrated for us."

Mary laughs and says, "Well, don't expect it tomorrow. Networks are expensive and complicated. We are installing the vertical backbone immediately. We will be linking the CIM system first to the vertical backbone through a fiber optic horizontal backbone. The JOCS system will be next, since you need CIM data as input to JOCS. After both of those are working, we will link the system in engineering. It may be two years before we get the accounting system on our network. I'm having a little trouble convincing the vendor of our accounting system that it's a good idea to use a variety of different computing equipment."

The meeting ends and the SWAT team continues with the process of designing the network for JOCS. The second step of network design is to define the location of the workstations. At least five workstations will be located on the production floor in the manufacturing department, while another ten workstations are destined for accounting personnel in the finance department. After measuring the distances, the team discovers that they cannot use a local area network for the workstations located in the manufacturing department. However, they can have local area networks for the workstations located in the finance department and the one in the manufacturing department. They can then link the two networks together.

The third step of network design, analyzing data volume and transfer rate and estimating peak transmission activity is also completed. The team counts the number of characters that will be transmitted from the CIM system each hour and the number of input/output characters that will be transmitted from both the manufacturing and accounting departments each hour. They use these counts to estimate the required bandwidth. The team also estimates the peak times for data transmission. JOCS will require a bandwidth of 8 Mbps, growing to 16 Mbps over the next three years, to satisfy current needs.

After analyzing the data derived during network design, the team elects to install an Ethernet-based local area network for the following reasons:

1 Personnel in engineering are already familiar with the technology and are willing to assist in the installation.

2 It will be simple to interface with existing systems.

3 Ethernet can easily handle the bandwidth requirements.

4 The distances between workstations are suitable for an Ethernet network.

5 The technology exists to link the manufacturing and finance departments together.

6 Equipment will be interchangeable with that of the engineering department. This will allow potential price discounts on purchased equipment.

Chapter 14
Designing Computer Architectures

WHAT WILL YOU LEARN IN THIS CHAPTER?

After studying this chapter, you should be able to:

1 Discuss using computers in architecture designs.
2 Discuss how systems professionals evaluate and acquire computer architectures.

INTRODUCTION

Not all new systems will require a new computer architecture. The purpose of this chapter is to treat computer architecture design and explain how to evaluate and acquire it when a new system demands one (see Figure 14.1).

End users drive the design of systems output, input, process, control, and database. These system design components in turn drive network and computer architecture design. A computer architecture design, from a design viewpoint, is a set of hardware and software building blocks and their relationship to one another. The network backbone enables systems professionals to design various computer architectures (or configurations).

USING COMPUTERS IN ARCHITECTURE DESIGNS

Computers are generally divided into three groups:

- Mainframes
- Minicomputers
- Microcomputers

Let's listen in on Clara and Mark's debate about computer architecture design at Nexon.

Debating Computer Architecture Design at Nexon

Clara Wells, controller at Nexon, asked, "Aren't mainframes dinosaurs?"

"No, not really," said Mark Ricardo, chief systems designer at Nexon.

"I've read about companies saving millions of dollars by replacing old dumb terminal-to-host mainframes with microcomputer-based LANs," said Clara. "Can't we reap similar benefits with our planned replacement system?"

Figure 14.1
SDLC phases and their related chapters in this book. In Chapter 14 the focus is on computer architecture, based on the network architecture described in Chapter 13.

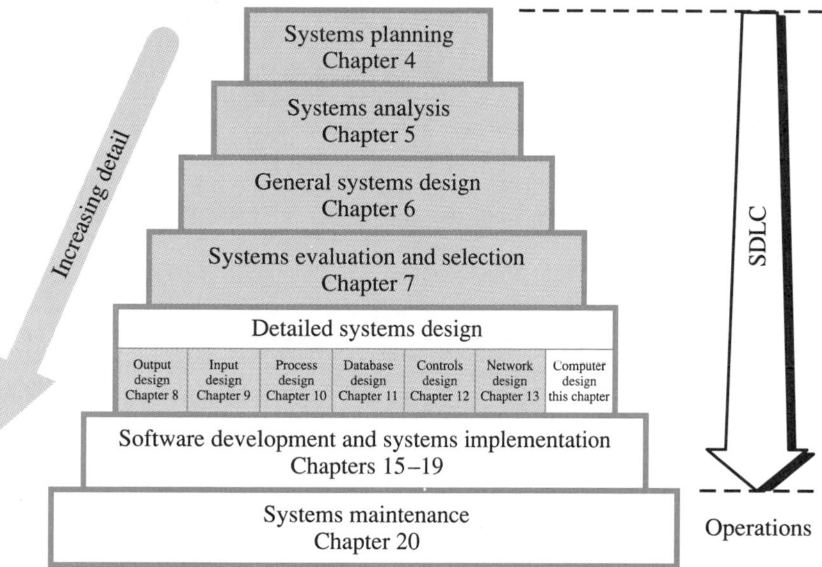

"In some cases, yes. In other cases, no," said Mark. "During the past few years, many companies have replaced mainframes with LANs, or have connected microcomputers to mainframes and distributed many applications to end users. But for many business applications, the mainframe is still the only workable platform."

"I don't know if I quite understand what you're saying. Either a mainframe is applicable or it's not," said Clara, somewhat frustrated.

"It's not that simple," Mark responded. "What is unclear is the dividing line that determines when an application should be run on a mainframe and when a smaller computer is appropriate. A transportation company has a similar problem in trying to determine the optimum-size truck for certain jobs. A pickup is applicable for light loads and an eighteen-wheeler is appropriate for big loads."

"Well, it may be cheaper to buy several pickup trucks and distribute the big load among them," said Clara.

"Maybe so, but what if your big load consisted of a 50,000-pound bulldozer? How would you distribute this load among several pickup trucks?" asked Mark.

"I see what you mean," said Clara. "Application size is the most obvious dividing line."

"Yes. Remember that some of our planned applications will work with more than three gigabytes of online storage and pull together data stored in 30 locations," explained Mark.

"Obviously, a LAN is not powerful enough to support applications that large," said Clara.

"The number of users also serves as a dividing line," said Mark. "When the number of users reaches into the thousands, a mainframe better be involved."

"Also, the kinds of applications implemented will clearly be important," said Clara.

"That's right," responded Mark. "A LAN with a microcomputer server simply does not work fast enough to support applications that require a lot of disk space. Also, in some cases, backing up data can take hours. A LAN server would take hours to sort one or two million records. Report generation can also stretch LANs. Applications calling for simple reports work well on LANs. If users, however, are generating complicated reports, then such applications are going to require a much more powerful server; that is, a mainframe, or, possibly, a minicomputer."

"So, you're saying that systems applications drive the kind of computer architecture design that is developed?" Clara asked.

"Precisely," said Mark. "Systems and software plans and designs should always dictate the kind of technology platform that a company acquires, not the other way around."

"What about some of our old mainframe applications?" asked Clara. "What do we do with them?"

"Those applications are well-designed programs that are still doing what they are supposed to do. Software may become obsolete, but it doesn't wear out. It would not be cost-effective to rewrite these programs for a LAN; that is, assuming the applications could run on a LAN. Moreover, some of these applications are very sensitive and top secret. LAN security and control is not as reliable as mainframe security. Because it's difficult to secure each microcomputer, there are situations when we're not sure who is working with corporate information. With a mainframe, security comes from limited access to the machine."

How are Mainframes, Minicomputers, and Microcomputers Used in Designing Computer Architectures?

Too much attention may be paid to computer sizes and titles. They are just one set of categories of building blocks used in designing computer architectures. The processor box drives software and data. The software models applications. Systems professionals develop applications, and applications are implemented to meet user requirements.

If management wants a highly centralized system, such as a host-based architecture (i.e., a centralized mainframe) provides, design of the computer configuration is fairly simple. Typically, one or several mainframes are installed at the company headquarters. All mass storage devices and printers are also installed there. Connectivity, if any, to the host mainframe is provided by terminals (usually dumb terminals) and a transmission medium.

If, on the other hand, management wishes to distribute systems resources out to end users to achieve a **COOPERATIVE-BASED ARCHITECTURE**, computer architecture design becomes challenging, requiring connectivity of various nodes via a well-designed network.

A key design tenet of **INTEROPERABLE** cooperative processing is to make optimum use of all resources, assigning the right application to the appropriate level of computer power. For example, mainframes are assigned the responsibility of processing transactions that have already been edited, validated, and formatted into an appropriate structure by micros. The mainframe processes the transactions and updates the database—something mainframes do very well.

Permitting local workstations to prepare transactions for mainframe processing and updating the database is cost-effective design. Also, the workstation provides a much friendlier user interface. For example, in an airline reservation system, the mainframe tells the local workstation the type of aircraft involved and the seats taken. The workstation then accesses a graph of the aircraft from its local files from a LAN server. The graph presents a seating chart to the agent. The agent points to the seat(s) desired, and a properly formatted transaction zips off to the mainframe to update the database.

Local Area Network-to-Mainframe Cooperative Design

Many companies have replaced host-based architectures with new LAN-to-mainframe architectures. For an example of this design approach, review the Mutual case.

Offloading Applications at Mutual Life Insurance Company

"The systems plan for our replacement system and new applications mandates offloading the new application designs to PC LAN platforms from the mainframe," said Debra Winthrop, senior systems analyst at Mutual Life Insurance Company.

"Why the change from a mainframe-based architecture?" asked Bob Sawyer, the new CFO, who previously worked for another insurance company.

"Offloading applications reduces costs and provides end users with a better user interface. Furthermore, it makes it easier for my people to develop new applications," Debra responded.

"So, you're saying that PCs on LANs provide better presentation mechanisms for users?" asked Bob.

"That's exactly it," said Debra. "The PC LANs are connected to the mainframe, and the mainframe is used as a middleman to provide a central point for distribution and control." Debra pointed to the architecture design shown in Figure 14.2.

"So updates and big batch jobs are handled by the mainframe and the end users perform their local tasks and interact with the mainframe or other end users as needed?" asked Bob.

"Yes, updates are sent immediately from the mainframe to target users on a need-to-know basis. The mainframe also runs big batch jobs at predetermined times, generally late at night," said Debra.

"I certainly can see the advantage of this design. The end users' local applications are developed quicker, clients are served better, and the offloading of applications effectively downsizes the total system," said Bob.

"That's right," said Debra, "Rather than installing two or three expensive mainframes, we have distributed the system to permit PCs to do what they do best and the mainframe to do what it does best. This way, we have a better system for users and at less cost."

"I see what you mean," said Bob. "The downsizing provides a scalable investment in both hardware and software, allowing us to implement small-scale, manageable, and efficient hardware over a period of time. This is different from large-scale centralized mainframe platforms that have to be replaced almost entirely when we upgrade."

Personal Computer-Based Cooperative Design

Some companies have eliminated the old mainframe-based architectures and moved all applications to a personal computer (PC)-based network. For example, the following case explains how this **DOWNSIZING** (designing in smaller size) was done at Blockbuster Video.

Figure 14.2
Computer architecture design at Mutual Life Insurance Company.

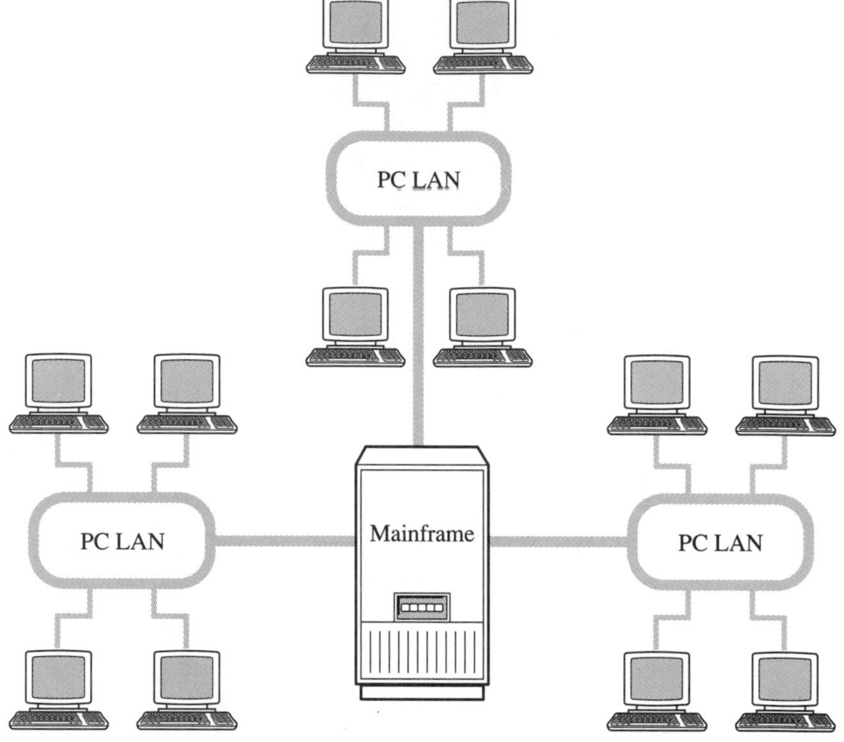

Downsizing at Blockbuster Video

Blockbuster is a video rental chain based in St. Louis. The organization's 150 video stores have a policy that any tape rented from any store in the chain can be returned to any store. This policy requires keeping track of which titles are in which stores.

For several years, the application ran on a large mainframe every night after store hours to keep track of Blockbuster's massive videotape inventory.

Sara Bond, new director of information systems at Blockbuster, wanted to get rid of the mainframe and downsize the system; that is, move mainframe applications onto workstation-based networks.

The result was a 150-node network of PCs with a minicomputer server, as shown in Figure 14.3. The new system has saved Blockbuster approximately $3.2 million annually in maintenance and support costs. Moreover, productivity has increased, and videotape tracking information is more accurate and timely.

Interoperable Enterprisewide Cooperative Design

Interoperable computer architectures include multicomputers in which applications are distributed to them in an optimal fashion. It is based on global, enterprisewide systems strategy with an emphasis on decentralized management. Interoperable cooperative-based design mirrors the way the enterprise operates, or should operate.

This design philosophy starts with the assumption that the basic design element is the user application and that the network and the computer building

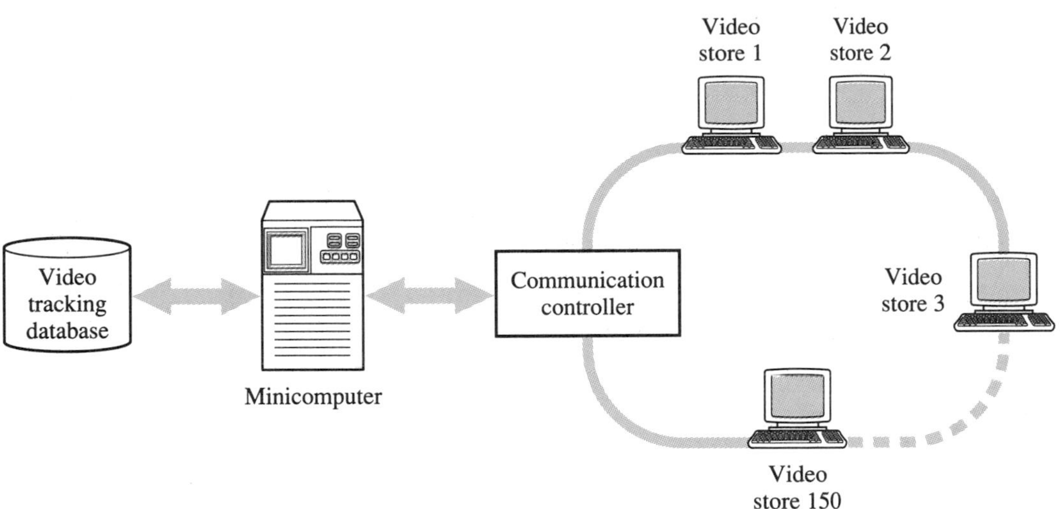

Figure 14.3
Computer architecture design at Blockbuster Video.

blocks are configured in such a way to execute user applications in the most efficient and effective manner possible. Whether an application is processed by a mainframe, minicomputer, or microcomputer is not the important thing. What is important is that the architecture work as a single integrated entity. The following case describes an application of interoperable cooperative-based design.

Developing an Interoperable Cooperative-Based Architecture at Simco

Simco, a large steel fabricator, has relied for years on a fragmented, dispersed information system. Each business area runs its own system, leading to redundancies in software, computers, and data, as well as instances of conflicting information. There was not a policy in place for addressing and formulating a strategic information systems plan.

James Kirby, Simco's CIO, was recently hired to develop a systems plan to coordinate Simco's disparate systems and replace obsolete applications.

James decided to leave the computer devices scattered throughout Simco in place and provide an integrated, interoperable system through networking. That is, all the computer devices were to be linked together by an enterprisewide network. The computer architecture design called for the centralization of the corporate database, with provision of ready access for end users throughout the company. The computer architecture design proposed by James is demonstrated in Figure 14.4.

After the new system had been operating for about six months, James had a meeting with Teri Brooks, Simco's CEO.

"Congratulations on the new system," said Teri. "Everyone seems to like it very much. It's giving end users access to different kinds of applications and information regardless of their location. Do I understand that you refer to this as interoperability?"

"Yes," said James. "It's a rather unwieldy term used to describe the main feature of an integrated, enterprisewide system. In it, users in different segments of the company work with each other in a cooperative manner. The basic driver for interoperability is ready access to applications and information by all end users throughout the company. It seems to me that in the previous system, our LANs focused on isolated

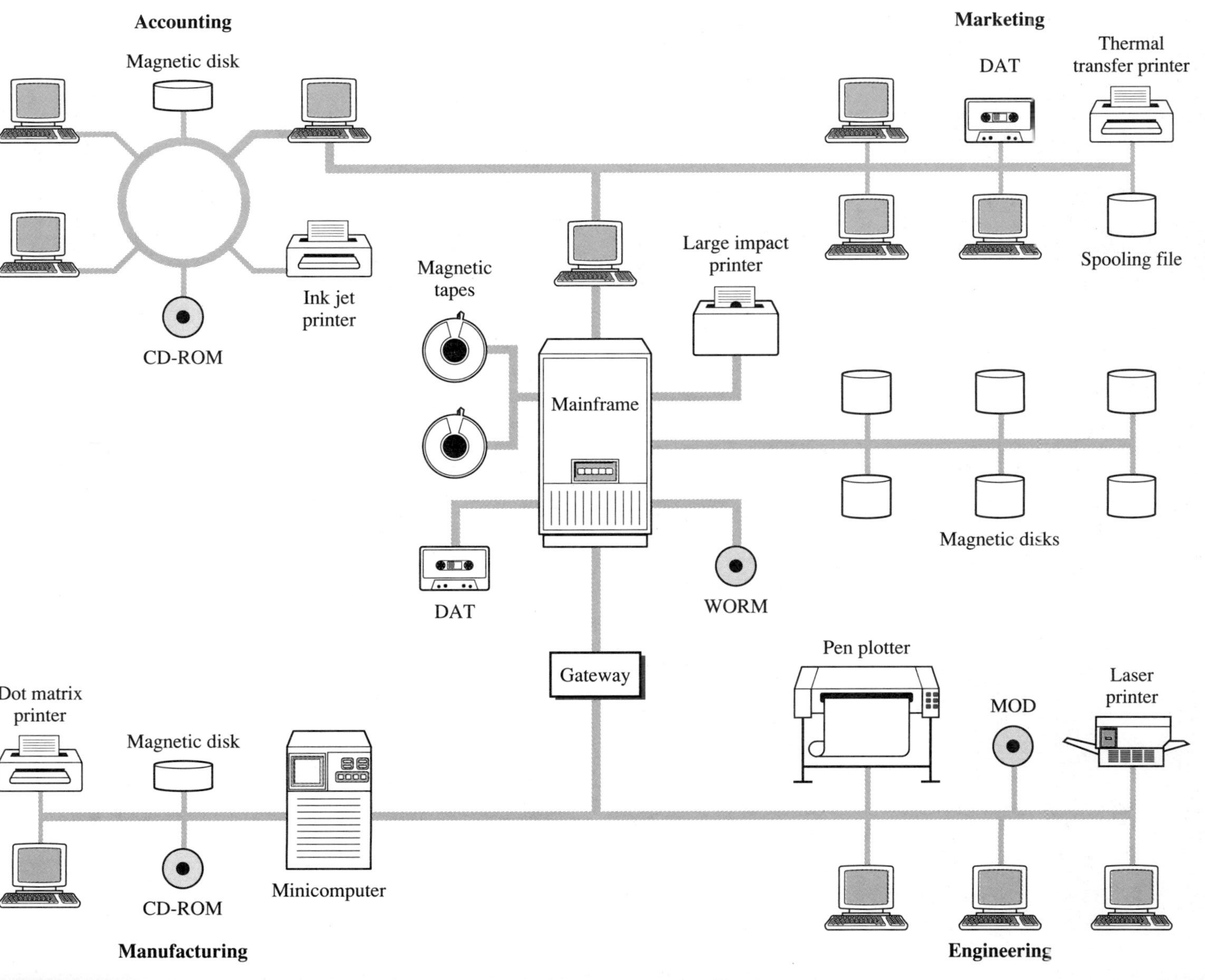

Accounting
Magnetic disk
Ink jet printer
CD-ROM
Marketing
DAT
Thermal transfer printer
Spooling file
Magnetic tapes
Large impact printer
Mainframe
Magnetic disks
DAT
WORM
Gateway
Dot matrix printer
Magnetic disk
Minicomputer
CD-ROM
Manufacturing
Pen plotter
MOD
Laser printer
Engineering

Figure 14.4 (Opposite Page)
Interoperable cooperative-based architecture design.

departmental needs rather than strategic interoperability with other segments of the company. As our systems plan showed, we needed an enterprisewide system with mainframes, minis, and micros networked together working in unison."

"Where does the term 'interoperability' actually come from?" asked Teri.

"The word was first used in the late 1960s by multinational military groups, such as NATO. The idea was that cooperating forces should be able to share key resources, rather than being able to maintain separate stocks of what ought to be common supplies. The same idea carries over to designing computer architectures. It induces communication and cooperation between groups in the company, and it spreads the cost of computing," James answered.

"That's a simple yet powerful concept. I can see how it can effectively relate to computer architecture design," said Teri.

"Yes, an interoperable, enterprisewide computer architecture is probably the most user-driven computer architecture design," said James. "To develop it, you've got to know and respond to what users want. Moreover, the architecture must be flexible, almost like a telephone or an electric supply system. Users want to be able to run an application on whatever machine they want, and wherever they happen to be. To achieve interoperability and user flexibility requires linkages among all nodes on the network and portability of applications to any node on the network."

EVALUATING AND ACQUIRING THE COMPUTER ARCHITECTURE

It is important to design the right computer architecture. It is equally important to evaluate and acquire that architecture. The first step of this process is to prepare a **REQUEST FOR PROPOSAL (RFP)**, copies of which are sent to vendors to guide them in preparing and submitting their product proposals. The second step is to review vendors' proposals. The third step is to rate each candidate vendor's proposed computer architecture on general performance and design criteria. Computer architectures that remain are benchmarked. Finally, negotiations are made with vendors that offer high-performance computer architectures. The vendor who can provide the best price and terms is the winner, and this vendor's computer architecture design is acquired by leasing or purchasing.

What Is a Request for Proposal?

An RFP is a document sent to vendors inviting them to recommend an architecture that they believe best supports the systems design requirements. By using an RFP, the analyst is trying to learn before selecting a vendor what that vendor and its computer architecture will be like after the system is implemented. The minimum elements of an RFP are:

- Systems design
- General business information
- Price and acquisition methods
- Overview of the benchmark test
- General performance criteria
- Benchmark test management and schedule

Systems Design

The RFP should contain the systems design and general description of the computer architecture design that supports the systems designers have in mind. The design models and documented deliverables contained in the CASE central repository will provide this information.

General Business Information

Each vendor is requested to submit a business history, a statement of financial health, service and support policies, customer references, and the size of the vendor's installed base. Financial condition is a very important element that tells whether vendors are able to provide customer support and meet long-term obligations. Vendors whose financial condition is weak are likely to skimp on services and less likely to provide hardware and software enhancements. Support from the vendor is one of the most critical criteria in the evaluation process. The general advice is to make vendors prove their support claims and to get it in writing because, in the long run, software and hardware are no better than their support from the vendor. This support includes such things as:

- Availability of training facilities
- Installation support
- Systems development, conversion, and testing assistance
- Experience level and competency of vendor's personnel
- Duration and quality of support after installation

References from vendors' customers provide excellent evaluation information. Typical questions put to each vendor's customers are:

- "What level of training and installation support did you get?"
- "Do users like to work with the system?"
- "Did you have any hidden or unexpected costs?"
- "Does your vendor meet its contractual obligations?"
- "Does your hardware and software fit and support your systems design?"
- "On a scale of 1 to 10, rate your vendor's support and the technology's performance."
- "If you were starting over, what would you do differently with respect to this vendor?"

Price and Acquisition Methods

Full prices of the proposed computer architecture should be requested, along with the vendors' proposed financing arrangements and alternative acquisition methods, such as lease or purchase. All specimen contracts, such as sales agreements, leases, service and maintenance contracts, licenses, and other legal documents, should be examined thoroughly by the company's attorney.

Pricing schemes, financial plans, acquisition methods, and tax ramifications should be scrutinized by an accountant.

Overview of the Benchmark Test

The systems designers may wish to outline how the proposed computer architecture designs will be tested. If the plan is to benchmark, vendors should understand what their responsibilities are. Types of benchmarks should also be specified.

General Performance Criteria

General performance criteria include:

- Documentation
- Ease of use
- Quality of user interface
- Compatibility
- Benchmark results
- Vendor's personnel

At a minimum, the vendor's documentation should tell how to set up and use the system. It should include accurate diagrams. It should also include an index, table of contents, troubleshooting help, and customization information. Also useful are quick-start guides, menus, icons, and online tutorials. To a great extent, the better the documentation, the easier the system is to use. Other things that improve ease of use include help screens, preformatted input screens, editing and error checking, and menu-driven interfaces.

Generally, users are attracted to interfaces that display meaningful images and forms. Many graphical user interfaces (GUIs) provide such an interface. Different hardware, software, and network elements have the ability to work effectively and efficiently with each other without modification or the aid of special devices. For example, a LAN can talk to a mainframe. Compatibility is a key characteristic of a cooperative-based architecture.

BENCHMARK TESTS are run to determine the amount of time required to complete applications and transactions. Timing is extremely important in many applications, such as online transaction processing (OLTP).

Also important are the skill and dependability of vendor's personnel who will provide liaison between the vendor and user company and who will be involved in installing and maintaining the technology. The vendor must confirm that its liaison people have the skills required to perform these tasks. Their skills should not, however, be limited to technical proficiencies, but should include management, business, and human-relation skills.

Benchmark Test Management and Schedule

The person or persons who will represent the company requesting the proposals should be named, and the test site should be identified. Also, dates and times of tests should be specified. This schedule permits vendors to plan their participation in the benchmarking process.

Preparing an RFP

Systems designers have determined that the accounting department at International Enterprises will need a LAN that will be connected to the corporate headquarter's mainframe. The accompanying RFP is a copy of the RFP that International sent to LAN vendors.

TO: Vendors
FROM: International Enterprises
DATE: January 4, 1993

Systems Design

International Enterprises is in need of a LAN that includes eight diskless workstations, two laser printers, a server, a large removable hard disk, and a network management system. For this reason, International Enterprises invites vendors to submit design solutions that fulfill this need.

General Business Evaluation

Before the test, International Enterprises would like to receive by mail a history of each vendor, audited financial statements for the current fiscal period and at least three previous fiscal periods, information on its service and support policy, the system's documentation, three customer references, and the size of the system's installed base.

Price and Acquisition Methods

International Enterprises will need to determine the exact price and acquisition methods for the tested system. Prices must be quoted on the basis of a purchase price and a leasing price. For this reason, each vendor must supply, at the start of the test, a complete written inventory of the components in the proposal and its most recent prices for purchase and for leasing.

The final price will include all hardware and software components that are needed to make the system work. International Enterprises, with the aid of a vendor representative, will confirm that this list accurately reflects the tested computer architecture and, for future reference, will take photographs of each product.

Overview of the Benchmark Test

The company recognizes that there are many different possible designs for any given problem. A network backbone is provided to which your hardware and software can be connected in whatever manner you recommend. The test itself will involve running a benchmark test similar to network applications that are to be encountered in the day-to-day operations of International Enterprises.

First, the benchmark will be introduced into the network. You, the vendor, will be familiarized with the test site to set up your system and then asked to leave. You will then be asked to return to the test site to observe the benchmark in operation. The goal of the benchmark test is to determine which vendor's system runs the benchmark most efficiently and effectively.

During the test, each participating vendor will be permitted to have at most two representatives present. At least one of these should be a highly skilled technical person capable of dealing with any problems that might arise. Vendors are free to take their systems with them when they complete the benchmark test.

General Performance Criteria

After each test, the proposed systems design will be judged by a panel of corporate end users on the following criteria:

1 Clear and concise online documentation

2 Ease of use for end users

3 Ease of use as a development environment in customizing LAN-to-mainframe linkage

4 Quality of the user interface

5 Transaction processing time

6 Skills, knowledge, and performance of vendor personnel

Benchmark Test Management and Schedule

Jerry Laventhal, chief technical systems analyst, will act as the network administrator. He will conduct the benchmark tests at International Enterprises' corporate headquarters in Cincinnati, Ohio. International Enterprises and each vendor will agree on a date and a time in advance. Two tests will occur each day from April 14 through April 19 of this year. Morning tests will run from 9 A.M. to 12 P.M. Afternoon tests will run from 1 to 6 P.M. Lunch and refreshments will be provided. Vendors should plan to arrive the day before the day of their test and be prepared to stay one extra day in case of technical difficulties.

Comparing Proposed Computer Architectures

Sorting through the marketing hype and obtaining objective performance information in today's computer market is a challenge for any systems professional.[1] Over the years, benchmarks have been developed for the purpose of comparing one computer architecture to another objectively, but no single benchmark provides the best measurement for all systems. Some kind of benchmarking is necessary, however, if systems designers are to make an informed choice. In fact, the only feasible way to tell how a computer architecture is going to work before it is acquired is to benchmark it. A standard benchmark also provides at least a measure of uniformity of performance between computer architectures enabling systems designers to compare apples to apples.

Processor-Bound Benchmarks

Some selected benchmarks that test the internal speed of processors include MIPS, Whetstone, Sieve, and Dhrystone, as displayed in Figure 14.5. Median

---

[1] Ed Paulson, "Standard Benchmarks Fail to Dispel Controversy," *HP Chronicle,* February 1991, p. 10.

Figure 14.5
Test results of processor-bound benchmarks.

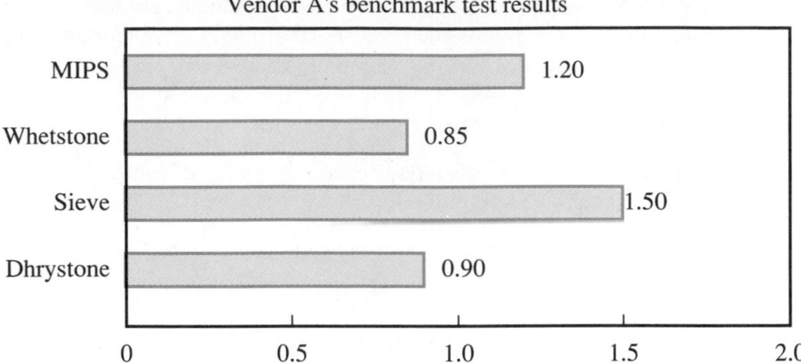

scores are equalized to 1. Results are relative numbers, where higher is better. A score of 2.0 is twice as good as the median. A score of 0.5 is half as good.

- ***Millions of Instructions per Second (MIPS)*** The MIPS benchmark calculates the execution of a command off a processor. MIPS are used to compare similar processors. MIPS aren't used quite as widely as they once were, because there's no agreement among different vendors on a standardized method for computing a MIPS score.
- ***Whetstone*** This benchmark measures floating-point operations and other purely numerical functions on small to midsized computers. Whetstones are usually written in FORTRAN. They are commonly used in the scientific and engineering industries.
- ***Sieve*** This benchmark is a prime-number-generating algorithm. It measures a processor's speed at performing double-precision integer arithmetic.
- ***Dhrystone*** This benchmark measures the types of computations encountered in typical business programming rather than number crunching. To some extent, it has characteristics of an I/O-bound benchmark, but it is still oriented more to internal processing operations.

Input/Output-Bound Benchmarks

Relative to engineering and scientific applications, most business applications are much more I/O-bound. Many I/O-bound applications necessitate the concentrated effort of various computing resources to perform operations. For example, OLTP benchmarks are far more demanding and far more difficult to execute than are processor-bound benchmarks. Unlike these processor-bound benchmarks, which test only a small portion of the total computer architecture's capabilities, OLTP benchmarks exercise virtually all significant hardware and software resources, including the following:

- Computer operating system
- Network operating system

- Database servers
- Database software
- Storage hierarchy
- Protocols
- Gateways
- Bridges
- Multiplexers
- Controls

In many respects, a strong OLTP benchmark is the ultimate test, because if a transaction rate is pushed to the limit, bottlenecks may be discovered in both hardware and software.

Generally, OLTP benchmark results are stated in transactions per second (TPS). This figure should also be combined with response time (RT). For example, the average response time for all transactions, or the percentage of all transactions meeting a specified response time, may be 90 percent under 1.5 seconds. The results of a benchmark may state that a particular computer architecture is capable of processing 40 TPS, with 99 percent at a RT of 1.2 seconds.

What is meant by "transaction," however, differs considerably among industries. A lottery transaction may mean the capture of wager data. A passenger ticket transaction for an airline may require a great deal more interaction and systems resource interfacing before the transaction is completed. In banking, each transaction may pass through several controls (e.g., check-digit algorithm) and involve several database accesses, each requiring record locking. It is not meaningful to compare TPS and RT ratings from benchmarks that use such significantly different transaction profiles.

The Transaction Processing Council (TPC) was formed to generate support across the industry for a systemwide transaction performance standard, TPC-A.[2] More than 30 major vendors are members of the TPC, and all accept TPC-A. TPC-A uses an exponential weighting function to randomize transaction input times. This function captures reality more accurately than others. TPC-A test results require a comprehensive independent audit report.

Essentially, a TPC-A OLTP benchmark testing rating indicates that a computer system configured in a certain way and tested under specific conditions will process a certain number of transactions per second for an estimated cost ($/TPS). The tests are conducted; they do not necessarily reflect how the system will perform in a particular company. The ratings are good for comparing one system to another. A number of systems professionals use the TPC-A ratings as an initial evaluation of computer systems to eliminate those that clearly do not meet performance standards. Then they will benchmark the remaining

---

[2] Ibid.

candidate systems under near-operating conditions to determine what level of performance can be expected for the money and time invested.

The Transaction Processing Council's TPC-B benchmark is intended to provide a level playing field that will help users select databases by relative performance. TPC-B is a benchmarking test for the database industry that tests only the database management system (DBMS) performance. TPC-B may become more widely used than TPC-A, because it is aimed at software performance and is less expensive to run.

Negotiating the Price and Contract Terms

At this point in the evaluation and acquisition process, only a few candidate vendors remain. It is now time to negotiate with these vendors to determine which one gives the best deal based on price and contract terms. The asking price and general provisions are already known. Negotiations set the selling price and provide most favorable contract terms for the buyer.

Contract Compared to RFP The final contract should contain key elements of the original vendor proposals that were submitted in accordance with the RFP. Especially important is to include the price and detailed specifications of software and hardware to disclose exactly what is to be delivered and a guarantee that the technology will perform according to these specifications and the benchmark results.

Delivery Date and Acceptance Test Two important points to include in every contract are a delivery date and an acceptance test. Delivery date should be stated for all components, including a "time is of the essence" covenant with the delivery dates.

Some 60 to 90 days prior to the delivery date, the systems designers should furnish the vendor with the acceptance test required by the contract. It is in the vendor's interest to see that the computer configuration and supporting software pass the test before it is shipped.

Warranties Typically, the remedy for a breach of contract is repair or replacement by the vendor within a specified time at the vendor's expense. An added remedy requires the vendor to compensate the company for downtime. For purchased technology, the warranty period may run from 30 days to a year after acceptance. For leased technology, the warranty should be in effect throughout the lease and on the company site.

Maintenance and Service During the warranty period, which may range from 30 days to one year, the vendor performs virtually all maintenance free of charge. When the warranty expires, the company can get service protection in the form of renewable one-year agreements from the vendor, vendor-authorized service centers, or third-party maintenance companies. Charges may run from 10 to 15 percent of the technology's purchase price annually for an on-site service contract.

Provisions for maintenance and service contracts should include MTTR. In this case, MTTR stands for mean time to respond. For example, the vendor will provide a one-hour response time between 8 A.M. and 6 P.M., Monday through

Friday, excluding specified holidays. Level 1 service is for routine repairs and module replacement. Level 2 service involves the solution of complicated problems provided by a regional service center. Response time for level 2 service is 24 hours or less. [The other MTTR (mean time to repair) would be an even stronger provision.]

Acquiring the Selected Computer Architecture

The two most popular methods of acquiring computer architectures are lease or purchase. The inescapable question is whether it is more profitable to lease or purchase. The following material covers the advantages and disadvantages of each method.

Leasing

A **LEASE** is a contract by which the lessor, or owner of the computer equipment, conveys the computer equipment to a lessee, or the user, for a specified term and a specified rent. The benefits of leasing include the fact that the risk of obsolescence and the risk of lost market value remain with the lessor. Leasing also provides a great deal of flexibility to the lessee, because the lessor can switch computer equipment and make upgrades much easier than with purchased equipment. Leasing is generally the only option when systems designers and users feel uncertain about a vendor's future products. Also, from a financial standpoint, no initial capital outlay is required and leases of up to five years can be structured as operating leases and thus need not be capitalized on the books of the corporation. In some companies, the decision to lease rather than purchase computer equipment can rest on one factor: the amount of cash or credit available for purchase. Some companies simply don't have the cash or credit to buy.

The negative aspect of leasing is that total cost of the computer architecture can be high, especially if the lease term is not well planned. Additionally, ownership benefits accrue to the lessor, not the lessee.

Today, a large number of lessors are in the business to buy, sell, and lease new or used mainframes, midrange computers, peripherals, personal computers, and software. Some of these lessors are turnkey (i.e., install and supply complete systems) vendors who can meet all the technology needs of most companies.

Lessors go through all the machinations of buying and disposing of computer equipment. Therefore, lessees don't have to hire or assign personnel to perform this rather specialized task. This intrinsic service generally results in better computer equipment management and frees up lessees to concentrate on things that they do best.

Purchasing

With the **PURCHASE** method, the company buys the computer architecture, takes title to it, and depreciates it over its useful life. Usually, a purchaser wants to acquire the computer equipment when it is first released, keep it over its entire life cycle, and have total control over it.

If the technology is early in its life cycle, and if the company is cash rich without a sufficient number of investment alternatives to use this cash, it may be to the company's advantage to buy instead of leasing the computer system. Also, if the company is experiencing low growth, it can keep the same computer architecture for a long time without the fear of running out of capacity. As a rule of thumb, it is generally more profitable to purchase if the same computer architecture will be kept for more than three years.

Other Alternatives

Traditionally defined as one type of facilities management, **OUTSOURCING** involves an enterprise turning over its information systems hardware, software, and personnel to an outside vendor which then supplies the information system as a service to the enterprise for a fee. The advantage of outsourcing is to use information resources on a service basis while the enterprise focuses on operating its business; that is, it concentrates its effort on what it does best. As Henry Pfendt, director of information technology at Kodak said, "We're in the photographic, pharmaceutical, and chemical businesses, not data communications."[3]

Other enterprises opt for modular or selective outsourcing for specific offerings, such as training or maintenance services. The idea is to optimize the system's efficiency by acquiring specific services or resources based on a buy-versus-build analysis; that is, if a service or resource is cheaper to buy rather than develop or do it themselves, they buy, or vice versa. For example, payroll processing is required by all businesses. But it may not make economic sense to spend resources on such an application because there is no strategic advantage. Moreover, an outsourcing vendor may be able to process the payroll at less cost because that's what they are set up to do, and they are consequently experts at it.

INSOURCING is just the opposite of outsourcing. Enterprises that practice insourcing develop, acquire, and manage all their information system resources and provide processing to their divisions and subsidiaries as a vendor in competition with outsourcing vendors for their services. Insourcing often sparks an entrepreneurial approach to systems development. Some enterprises have information systems that not only service all users within the enterprise, but also sell their services to other businesses. For example, J.C. Penney and American Airlines both have information systems that service all users within these enterprises, and also sell services (as outsourcers) to other businesses. In fact, these systems are viewed as a business within a business.

REVIEW OF CHAPTER LEARNING OBJECTIVES

The major goals of this chapter were to enable each student to achieve two important learning objectives.

[3] Sheila Osmundsen, "Kodak Gives Exposure to Outsourcing," *Digital News,* November 12, 1990, pp. 1, 27–33.

Learning objective 1:
Discuss using computers in architecture designs.

Kinds of computers available for developing computer architectures that support business information systems are:

- Mainframes
- Minicomputers
- Microcomputers

Processors may work as stand-alone central units in host-based architectures. However, a well-planned and well-designed network backbone that connects a variety of computers in an optimal fashion may be the most efficient and effective computer architecture design.

Learning objective 2:
Discuss how systems professionals evaluate and acquire computer architectures.

After the systems designers have developed a computer architecture design, or possibly several alternatives, they are ready to prepare and submit requests for proposals (RFPs) to a number of vendors. To this point, all systems designs, including the computer architecture design, have been conducted without regard to specific vendor products. In this way, systems designers did not permit vendor product tails to wag systems design. But now a decision must be made to select a vendor's specific computer architecture that meets the design performance objectives and provides the best platform for the new system. This decision process starts with the RFP.

Once proposals from vendors are received, the following three levels of evaluation should be conducted:

- The first level of evaluation is to review the proposals thoroughly, evaluate vendor personnel, interview vendor customers, review vendor documentation, determine testing procedures, assess vendor financial condition, and rate vendor legal and business procedures.
- The second level of evaluation estimates vendors' proposed computer architectures on the basis of general performance criteria.
- The third and final evaluation level tests proposed computer architectures or their facsimiles under close-to-real operating conditions. These tests are called benchmarks.

When a full evaluation has been made of all vendors, the likelihood is that only a few vendors that can provide desirable computer architectures remain. Assuming that the remaining computer architectures are fairly equal at this point, the winning vendor is the one that can provide the best price and contractual terms under either a lease or purchase arrangement.

Outsourcing is another way to acquire computing resources by transferring some or all computer and telecommunications resources, as well as day-to-day systems operation and personnel, to a vendor. Insourcing, on the other hand, provides information system services throughout the enterprise. Some information systems that practice insourcing also provide services for a fee to external businesses. In this way, these information systems operate as businesses within businesses, or as profit centers.

COMPUTER ARCHITECTURE DESIGN CHECKLIST

Following is a checklist on how to perform computer architecture design and evaluate and acquire a specific computer architecture from its vendor.

1 Gain an understanding of mainframes, minicomputers, and microcomputers and identify what each does best.

2 Configure computers, storage media, and other peripherals in a cooperative design.

3 After a particular computer architecture design is developed that will support the new system, prepare a RFP and send copies to candidate vendors.

4 Subject vendor proposals to a stringent evaluation process to identify the most cost-effective computer architecture.

5 Consider outsourcing and insourcing alternatives.

6 Acquire the selected computer architecture through lease or purchase.

KEY TERMS

Benchmark tests

Cooperative-based architecture

Downsizing

Insourcing

Interoperable

Lease

Outsourcing

Purchase

Request for proposal (RFP)

REVIEW QUESTIONS

14.1 What are the three main types of computers used in business information systems?

14.2 Discuss the special forte of mainframes, minicomputers, and microcomputers.

14.3 Explain the purpose of a RFP.

14.4 What's the purpose of TPC-A?

14.5 Discuss the advantages and disadvantages of lease versus purchase.

14.6 Define outsourcing and insourcing.

CHAPTER-SPECIFIC PROBLEMS

These problems require exact responses based directly on concepts and techniques presented in the text.

14.7 Following are typical systems applications:

________________	Spreadsheet analysis.
________________	Processing payroll for a large multinational enterprise.
________________	Processing accounts receivable for a large department or a small division of an organization.
________________	Providing a graphical user interface to end users.

Required: Insert the type of computer most appropriate for the applications above.

14.8 Following is a list of various situations:

________________	The process of moving systems from large computers to small computers.
________________	The ability to operate between disparate systems and equipment and to share common resources.
________________	A standard benchmark test for online transaction processing (OLTP).
________________	Acquiring a computer architecture by making periodic payments to the owner.
________________	A company turns over part or all of its hardware, software, and personnel to an outside vendor.

Required: From the following list of terms, select the one most appropriate to each preceding situation and insert it in the proper blank.

- Mainframe
- Software
- Whetstone
- Downsizing
- Outsourcing
- Interoperable
- Lease
- TPC-A
- RFP

THINK-TANK PROBLEMS

These problems call for a feasible approach rather than a precise solution. Although the problems are based on chapter material, extra reading and creativity may be required to develop workable solutions.

14.9 Sunnydale, a wholesale food distributor for fast-food restaurants, is planning to develop a system that will allow its customers to place orders and retrieve account data via terminals. Sunnydale has traditionally taken orders over the telephone, but it has decided to install an order-entry network to reduce the cost of supporting its growing customer base and to serve as a foundation for new customer services. To anchor the network, Sunnydale is thinking about installing a mainframe at the warehouse and one at the company's data center in its headquarters four blocks away.

Customers place orders from their terminals connected to the warehouse computer, which screens calls, formats transactions, generates shipping information, and acts as a front end to the data center computers. The data center processes the orders, updates the database, and transmits order-filling instructions to the warehouse.

Required: Outline a computer architecture design for Sunnydale. Include the kinds of transmission media, computers, and storage media you recommend. Also, prepare a RFP. Develop any assumptions you deem necessary.

14.10 Oracle Company wants you to design a computer architecture to support its new information system. Management wants the home office to house a mainframe and a special database management processor. All the major processing of formatted transactions will be handled by the mainframe, and the database management processor will control a high-volume database. Oracle has two divisions, each of which will house minicomputers that will be connected to the mainframe. Each division contains a warehouse and an administrative office. The warehouse will contain a bus LAN network topology of terminals and the administrative office will contain a LAN with terminals connected by a central hub. All LANs will be gatewayed to division minicomputers.

Required: Outline a design for this computer architecture platform. Also, state how you would evaluate vendors' proposed computer architectures. Make any assumptions you deem necessary.

14.11 Paulbilt is a nationwide manufacturer of long-haul trucks. Corporate headquarters are in Peoria, Illinois. Plant A is located in Washington. Plant B is located in California. Plant C is located in Massachusetts. Plant D is located in Georgia. A corporate database is stored on removable magnetic disk packs and controlled at the Peoria plant, but access is granted to all plants via a high-speed digital leased circuit. Also, headquarters has two laser printers, magnetic tape for processing payroll and accounts receivable, and digital audio tape (DAT) used for archiving and backup.

Each plant is serviced by a Token Ring LAN. Each LAN contains workstations, erasable optical disk, and laser printers. LANs for plants A and B are interconnected by a midrange computer in Warehouse AB located in Oregon. This midrange computer serves as a LAN bridge between plants A and B. It is also connected to Peoria via VSAT. It processes customer orders for the western half of the country and controls an inventory database stored on erasable optical disk. Several workstations and a laser printer are connected to the midrange via a bus topology. In the east, a similar arrangement exists with warehouse CD located in Virginia.

Required: Outline the computer architecture explained above. Use material in Chapters 13 and 14 as your guide.

SUGGESTED READING

Coffee, Peter. "'Interoperability' Is a Label for Many Ideas." *PC Week,* February 11, 1991.

Ellis, C. A., S. J. Gibbs, and G. L. Rein. "Groupware: Some Issues and Experiences." *Communications of the ACM,* Vol. 34, No. 1, January 1991.

Elms, Teresa. "The AS/400 Alternative." *Datacenter Manager,* November/December 1990.

Feldman, Steve. "Open Systems: Crawling Toward Connectivity." *Digital News,* October 1, 1990.

Green, Lee. "How to Rate Benchmarks." *InformationWEEK,* December 7, 1989.

Halvorsen, Jann-Marie. "What to Look for in Leasing." *Datamation,* March 15, 1990.

Ingram, Ray. "Do Your Applications Fit Your Platforms?" *Information Center,* February 1989.

Korzeniowski, Paul. "To Split or Not to Split? Tools Little Help Answering." *Software Magazine,* February 1991.

Krohn, Nico. "Magneto-Optical Drives Make Inroads." *Infoworld,* April 9, 1990.

Lazzaro, Joe. "Networking with Optical Disk Technology." *LAN Technology,* April 1990.

Osmundsen, Shelia. "Kodak Gives Wide Exposure to Outsourcing." *Digital News,* November 12, 1990.

Paulson, Ed. "Standard Benchmarks Fail to Dispel Controversy." *The HP Chronicle,* February, 1991.

Sayed, Husni, Andrew Patterson, and Deborah Cobb. "Mass Storage Technology." *Interact,* April 1990.

Serlin, Omri. "Toward an Equitable Benchmark." *Datamation,* February 1, 1989.

JOCS CASE: Designing a Computer Architecture

At this point in the JOCS case, the SWAT team has designed most of the output and input for the system, as well as outlined the majority of processes that will be performed. As the system is developed, the team anticipates that the end users will continue to identify new output for the system. If end users do identify new output, it's possible that more input data and processes may be required to produce the new output. Systems development tends to be an ongoing, almost circular, process rather than a set of linear steps.

The team hopes to develop a system that will allow end users to create some of their own output. As a result, not all output, input, and processes are fully defined at this time. The SWAT team has also identified the data that will be stored for the system, decided on a DBMS for data access and retrieval, designed the necessary controls, and drawn up a tentative design for the overall network that will eventually support a computerized JOCS application. The next step in the design process is to identify the specific software and computer hardware that will act as the technological platform to support JOCS. The SWAT team is having a meeting to discuss the most suitable hardware and software for JOCS.

"I think we already know what kind of system we want," says Tom Pearson. "We have been developing a model of the best system throughout this design process."

"You may know the best hardware and software," responds Carla Mills, "but I really don't know what's available."

"Carla, you're an accountant and we don't expect you to know about all the hardware and software options," says Tom. "But I read *Computerworld* and *PC-Magazine* every week. I think we should just be able to place the orders for our own system. In fact, with all the salespeople that hang around here for the systems in engineering and accounting, we could just ask them to give us quotes."

"Tom," interrupts Christine Meyers, "those magazines represent just a tiny bit of the available options for hardware and software. In fact, the articles seem to talk mostly about the vendors that advertise. What about all the rest of the options out there? Do you really feel comfortable ordering equipment based on their ads?"

"Come on, Christine, it's easy to keep current. You just have to keep your eyes open. I have drawn up a diagram for our system." Tom hands out a picture of his idea for the best hardware solution (Figure 14.6).

Everybody examines the diagram, and Cory Bassett observes, "It looks like you are proposing a set of file servers and diskless workstations to handle the JOCS applications for the accounting area and manufacturing department."

Tom agrees and says, "Yes, I don't think we will need to download anything to the workstations. We can have better controls if the users are not allowed to change anything on their individual workstations."

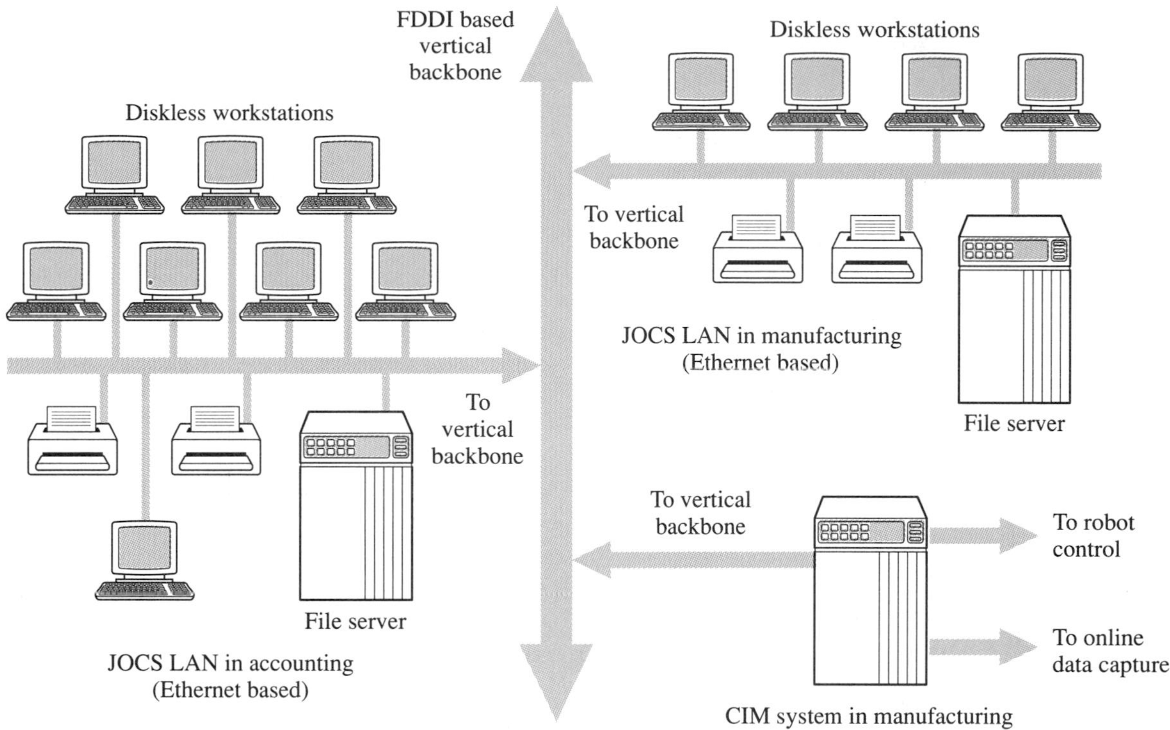

Figure 14.6 Hardware architecture for JOCS.

Christine groans, "Controls again. I thought one of our objectives was to allow the end users to create their own output. In fact, I thought we wanted them to become so comfortable with the DBMS that they designed both reports and queries. Jake, wasn't that one of our goals?"

"Yes, Christine," responds Jake, "We need a flexible, easy-to-use system, but I don't want the end users to have too much freedom to change existing data. Maybe we could have diskless workstations in manufacturing, and workstations with disks in accounting."

"Wait a second," says Cory, "you are showing two separate file servers for JOCS. I thought we had discussed having a single file server rather than having to maintain two copies of the database."

"We had decided on a single file server. But when we measured the distance between the manufacturing workstations and the accounting workstations, it was too far for a single local area network. Ethernet can't handle the distance. We have to have each workstation within about 2000 meters of each other. So I thought if we created two local area networks, it would solve our problem," responds Tom.

"But that creates another problem," interjects Carla. "How are we going to make sure that we have the same data on both file servers? Isn't there more of a chance that we will have problems with our database controls?"

"Well, we could have problems with duplication of data, or data not being the same on both systems," responds Tom. "But it's not a big deal. We can just make sure that we copy the data onto both systems each night."

"That sounds like a pretty contrived solution," says Christine. "I want the data to be the same on both systems all the time. Let's try to think of another way to do it."

"We could write programs to make updates on both file servers for each transaction that is applied in JOCS. The transactions could go through the vertical backbone and perform an update each time. No, that might cause some big delays in response time for online updates."

"I was reading an article last night about using radio signals to send data over short distances. We could install a radio to digital converter and transfer the data to the LAN immediately," adds Cory.

"This just proves my point," states Christine. "It is impossible to keep current on every little new piece of technology. Tom, in one of our first meetings, you suggested Oracle as our DBMS and Unix as our operating system. I know we have all been doing our design with those products in the backs of our minds. I think those are good suggestions, but I know there are other options out there and I want to know what they are. The vendors are supposed to be experts in the current technology, so let's use their expertise. I suggest we write a RFP, describe what we want the system to accomplish, and let the vendors respond with hardware and software alternatives. We've done our job. Let's let them do their jobs."

Tom protests, "This system is too small to go out for bid. We'll waste our time and the vendors' time."

"Christine has a good point," responds Jake. "The vendors know more about the available options than we do. We don't have to make this into a big production, but I would like to devise a RFP, give the vendors a couple of weeks to respond, and see what they come up with. We certainly have plenty to keep us busy during those two weeks."

Tom sighs, "I suppose you will even want to do machine benchmarks to evaluate the proposed systems."

"There are other methods of evaluation," says Cory. "Our system will not be processing a large number of transactions, so we don't really have to worry about a large-scale benchmark. We can check the transaction ratings through a fairly objective computer rating source, such as Datapro, and then talk to current users to find out how they are doing."

"We could get the end users involved during the evaluation process," Carla says excitedly. "We could find out how easy or hard it is to use the proposed DBMS by letting some of our key end users attempt to design a few reports and input screens. If this system is really for the end users, they should be involved in the evaluation process."

"OK, everyone, do we all agree?" asks Jake. "Do you think that a RFP is our best bet? Tom?"

"Well, I wouldn't mind seeing what the vendors can come up with," responds Tom. "Somebody grab a laptop and let's start writing."

The rest of the meeting is spent in producing a brief RFP for JOCS. The SWAT team decides to stay away from specifying the type of system that they want. They decide instead to describe what they want the system to do, and then let the vendors devise the specific hardware and software that will be appropriate. The team decides to use the documentation produced during the design process rather than write the RFP completely from scratch. They organize the RFP to accommodate their existing documentation easily.

The table of contents for the RFP is shown in Figure 14.7. As you can see,

Figure 14.7
Table of contents of JOCS request for proposal.

PEERLESS, INC.
JOB COSTING APPLICATION SYSTEM, RFP

CONTENTS

the team has stayed away from defining hardware and software requirements. Instead, they have provided a description of what they want to accomplish with their hardware and software.

The first two sections of the RFP introduce the scope of the new system, the evaluation process, and plot the significant dates. The SWAT team clearly states the necessary order and delivery dates in this section. Occasionally, a vendor will propose a system that is still under development; the SWAT team wants to avoid any misunderstandings about when they expect their system to be ready for delivery. In addition, this section also discusses the type of contract that Peerless expects to sign with a vendor.

Section 3 describes the current information systems environment so that the vendors will understand how important it is for the new system to fit within existing and planned future systems. Since Mary Stockland, CIO, wants to maintain a multivendor, open systems environment, Section 3 is used to explain this to the vendors. This section also includes a subsection discussing the qualifications of the current employees. Since a vendor usually has a set of choices available to propose, sometimes it is helpful to know which software packages are already part of a requestor's knowledge base. For example, the SWAT team is very familiar with the UNIX operating system. If a vendor has packages that are available for either UNIX or OS/2, the vendor would probably propose the UNIX-based package in order to fit best with the customer's specified skills. If, however, the customer wants to hire new people, or expand the skill base, the customer should provide that information to the vendor. A vendor can provide the best system if there is a free-and-easy exchange of information between customer and vendor.

JOCS is designed to interface directly with end users. It's critical for the vendor to be aware of the importance of the end-user interface to the success of this system so that a truly user-friendly interface can be recommended.

Section 4, Requirements and Features for JOCS Development Environment, defines exactly what Peerless wants the JOCS system to accomplish. The SWAT team wants to purchase a development environment, so they clearly tell the vendors that they want tools to assist the development process in the subsection discussing CASE tools.

A vendor must attempt to estimate the performance of the proposed system. The functional processing subsection discusses the design of the JOCS package. The SWAT team is planning on developing its own JOCS application software. The team is not looking to purchase an application software solution. This subsection, as a result, discusses the software that they are planning to develop. The team provides this information so that the vendors will understand what the system will be required to do in the future.

The last section of the RFP details the necessary requirements for service and support so that the vendor knows what will be expected after the sale. For example, there is a subsection describing training. The training requirements include both training at the time of installation and ongoing training as staff turnover occurs.

The SWAT team wants to describe each aspect of its planned system so that the vendors will not have any surprises after they turn in their proposals.

Part IV

Implementing and Maintaining the System

At this point, the systems professionals have completed the detailed systems design phase, producing a Detailed Systems Design Report, which is a documented deliverable that contains design specifications for output, input, processes, database, controls, and technology (telecommunication networks, computers, operating software). This documented deliverable serves as a "blueprint" for construction and implementation of the new system.

For some new systems, a major part of the construction and implementation effort involves developing application software. Therefore, Chapter 15 will introduce you to the software development life cycle (SWDLC) and its management. Chapter 16 will cover software design, the first phase of the SWDLC. Chapter 17 will present software coding, the second phase of the SWDLC. And software testing, the third and final SWDLC phase, will be treated in Chapter 18.

To implement all components of the new system requires other tasks besides software development. In Chapter 19, these additional tasks will be discussed and demonstrated. After the total new system is implemented, it will generally require maintenance over a long life span. Chapter 20, the final chapter of this text, will cover this very important phase of the systems life cycle.

Chapter 15
Software Development

WHAT WILL YOU LEARN IN THIS CHAPTER?

After studying this chapter, you should be able to:

1 Differentiate the sources of application software and discuss how to evaluate and select commercial software packages.
2 Define the software development life cycle (SWDLC) and briefly discuss its phases.
3 Discuss organizing for software development projects; define the program development team, chief programmer team, and egoless programming team; and analyze their advantages and disadvantages.
4 Define software productivity and present two ways to measure this productivity. Also explain management's impact on software productivity.
5 Discuss software quality and explain quality assurance and quality control.
6 Describe how the program evaluation and review technique (PERT) is used as a SWDLC game plan and project management technique.

INTRODUCTION

This chapter discusses software development, one of the largest and most challenging implementation tasks (see Figure 15.1). Software development ranges over two extremes, from a "spreadsheet for every application" syndrome to "reinventing the wheel" syndrome. The first syndrome occurs because some people believe that for every application there is a ready-made spreadsheet or some other commercial software package that will run the application. All one has to do is buy it and install it, and the application software problem is solved. At the other extreme, systems people develop a new computer program from scratch for every systems application without regard to what has already been developed in-house or what might be available from software vendors.

In many instances, after or near completion of the detailed systems design phase of SDLC, the systems project team may begin a search for commercial software packages that show promise of supporting the systems design specifications and running on the computer architecture. Commercial software packages are widely available for function-specific applications and well-defined business applications.

In systems designs dealing with unique or special user requirements, however, prewritten software packages may not support these user requirements

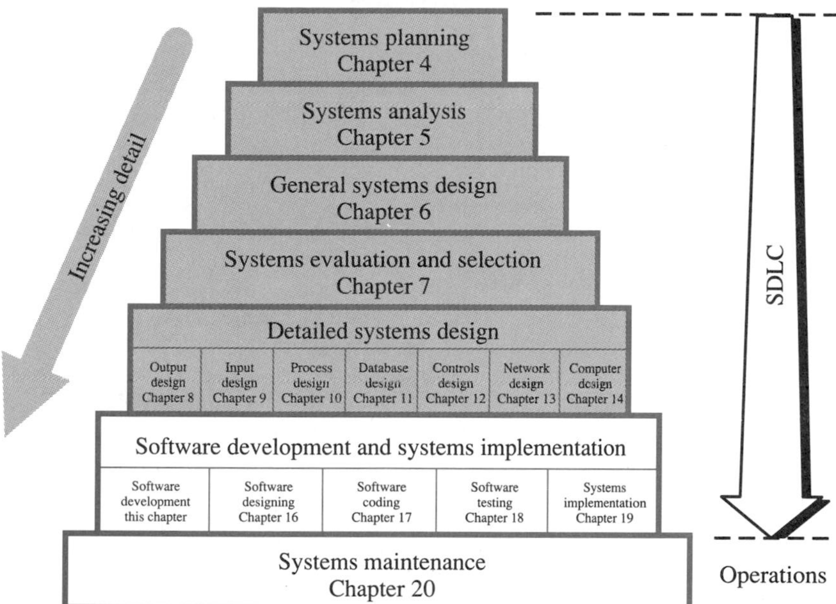

Figure 15.1
SDLC phases and their related chapters in this book. Chapter 15 deals with software sources, followed by myriad aspects of software development.

directly. Therefore, the software necessary to support such systems designs must be written from scratch to meet user requirements and systems design specifications.

After introducing you to the sources of software, we will assume that the detailed systems design calls for software that is developed in-house. Thus the remainder of this chapter discusses organizing for software development, measuring software development productivity, ensuring software quality, and preparing a game plan for software development. The accompanying case will provide you with insight as to how a systems project team moves from the detailed systems design phase to the systems implementation phase, that is, from design to construction.

Software Development at Tranquil Industries

Tranquil Industries is a small manufacturer of customized noise abatement devices for automobiles, airplanes, and other equipment. Early on, Jim Thompson, systems project leader, and his team members were charged with the responsibility for developing an accounting system and an executive information system (EIS) for the plant manager and her staff.

Jim's systems project team has just completed their Detailed Systems Design Report, which contains over 300 documents, bound and distributed to appropriate end users and management. Sections of the report that pertain to a specific end user or manager are highlighted so the end user can review only that which will be of special interest to him or her. For example, the controller is primarily interested in accounting output, input, process, and controls designs. The plant manager is chiefly interested in graphical output designs. The database administrator is interested in the database design. The CIO wants to know how the technology platform is configured and the results of the request for proposals and benchmarks.

At this point, all interested parties are thoroughly informed about the new systems design. Now, Jim is ready to begin software development, acquisition, and other implementation tasks. The EIS, for example, will have to be supported by a customized software program developed by an in-house programming team. Some preliminary work with several software vendors indicates that the accounting system will be supported by one of three modular accounting software packages that will be evaluated further to determine the best one. Each accounting software package contains the general ledger, accounts receivable, accounts payable, inventory, sales order, payroll, and job-order cost modules called for in the Detailed Systems Design Report.

This chapter and the following chapters examine what Jim and his fellow systems professionals must do to convert the Detailed Systems Design Report, which is the "blueprint" of the new system, to a constructed and implemented system ready for operations. The material in this and subsequent chapters does not pertain specifically to Tranquil Industries, however, but to all organizations that have developed a Detailed Systems Design Report containing all the material that fully and functionally describes the new system's output, input, process, database, controls, and technology platform. In fact, some concepts and techniques will be presented that will not necessarily relate to Tranquil's Detailed Systems Design Report, because the report does not call for such concepts and techniques. For example, we will discuss object-oriented design in Chapter 16, but Tranquil will develop the EIS using structured design, which we also discuss.

SOURCES OF APPLICATION SOFTWARE

There are two main sources of application software:

- Commercial software vendors
- Customized software developed in-house or by an independent programming contractor

From either source, the software must meet the design specifications outlined in the Detailed Systems Design Report, which evolved from user requirements. Thus the SDLC is applicable for development of any system, whether the software is acquired as a completed package or is developed from scratch.

Commercial Software Packages

Available commercial software packages cover a wide variety of business needs. Some packages are generic and multifunctional, allowing users to program the software to their particular needs. Examples are spreadsheets and DBMSs. Other packages are application-specific. Examples of this type are materials resource planning, general ledger, accounts receivable, accounts payable, payroll, job costing, and word processing. These packages automate basic business functions that do not vary significantly from one organization to another.

Advantages

Advantages of purchasing and implementing a commercial software package normally include the following:

Rapid Implementation The major advantage of purchased software is rapid implementation. The software is ready, tested, and documented. Depending on the size of the application, an organization should usually be able to implement a purchased package far more quickly than if it developed the same program in-house or had it developed by an independent contractor. This advantage has the potential to help solve the application backlog that plagues a number of organizations.

Cost Saving Total costs of a commercial software package should be lower than those of a similar customized program, because one package is sold to many organizations, thus spreading the cost of development over many users. It is doubtful that a customized program could be developed for less than the cost of an off-the-shelf commercial package.

Time and Cost Estimations The cost of the commercial package is known. Also, its implementation date is easily estimated. Customized development, on the other hand, is notoriously prone to exceed budget estimates in both time and cost.

Reliability The commercial software package was presumably tested rigorously before it was released to the general market. Through extensive usage by a number of organizations, any errors discovered have been detected and corrected. Although no program can ever be certified totally error-free, the chances of a commercial package having fewer errors than a customized program of equal size and complexity are greater.

Disadvantages

The disadvantages of commercial software packages include the following:

Poor Systems Design Match Although we have the common-sense belief that commercial software packages are quicker and easier to implement than in-house-developed software, this is not always the case. Because the commercial software package is written for many different organizations, not for any specific organization, the package will either have some functions not required or will not have functions that are required. In some cases, the package itself will have to be modified. If the vendor does not make the source code available for customization and does not provide customizing services, the systems design will have to be changed to fit the package. Either way, it may be better to develop the program in-house so that it meets the systems design specifications precisely.

Vendor Dependent The organization is dependent on the vendor for support. The organization is at risk if the vendor goes out of business and changes are needed in the package.

Indirect Costs from Crashing SDLC In some instances, management may want to bypass SDLC phases and go immediately to a commercial software package. The leading priority of the quick-fix strategy is to get the technical system up and running and to deal with specific user requirements, training, and to deal with systems and organizational problems later. Such strategies usually backfire. Often, the commercial software package doesn't work as intended, and the systems and organizational problems that existed before implementation of the package still exist. As a matter of fact, one of the prime reasons for applying the SDLC in the first place is to resolve systems and organizational problems and improve systems performance. Often, what is gained by not performing a complete SDLC is given back, and more, in increased costs of implementation, operations, and maintenance.

Preparing a Performance-Oriented Request for Proposal

In those cases where commercial software is applicable, the advantages in acquiring commercial software, however, far outweigh the disadvantages. Thus we will now turn our attention to issues related to acquisition of commercial software.

The first step in selecting the right vendor and commercial software package is to prepare a performance-oriented RFP similar to the one for computer architecture designs discussed in Chapter 14. The main evaluation factors include meeting detailed design specifications for output, input, processes, and database spelled out in the Detailed Systems Design Report. It should contain controls specified during detailed design. Furthermore, it should meet time and cost constraints. The use of a benchmark that simulates the new systems requirements should be applied on each vendor's package.

Rating the Packages

Each package should be rated. Part of the ratings may result from benchmarks. Other ratings can be gathered from a number of publications that are based on surveys of a large number of readers who use such packages.

- ***Operating Performance*** This performance rating is the result of benchmarks run on the packages measuring such things as transactions per second (TPS) and response time (RT).
- ***Documentation*** The documentation rating reflects the quantity and quality of both written and online procedures, including a quick-start guide, online tutorials, and help features. Poor organization, missing information, incomplete indices, and ambiguities lower the rating.
- ***Ease of Learning*** The rating for this performance factor depends on the user interface and the intuitive design of the package. The quality of documentation also influences this rating. A package must be learnable by the average user.
- ***Ease of Use*** This performance factor is in large part a function of the

package design. It determines how easy the average user finds the package to use once the basics have been mastered. Easy-to-follow menus and clear commands aid ease of use.

- ***Controls and Error Handling*** In Chapter 12 we presented the kinds of controls, especially input controls, that should be embedded in the program logic. These controls check input for range, reasonableness, and other values; access and compare values in identification (or validity) tables; and recompute check digits. All errors detected should generate clear error messages that explain the source of the problem. The package should also write errors to an error file.
- ***Support*** This performance factor is generally divided into two categories: policies and technical. Support policies include toll-free lines, warranties, money-back guarantees, and training. Technical support comes from available and knowledgeable technicians.

Selecting a Package

The ultimate objective is to determine which vendor's software package offers the greatest benefits at the least cost. To achieve this objective, total rating points and total cost are determined. A method to determine total rating points is presented in Figure 15.2. The relative weights are assigned to each general performance factor based on their relative importance. The base is 100. Next, based on the performance statistics gathered from vendor proposals, benchmarks, and published ratings, a rating is assigned to each performance factor (1 = poor and 10 = excellent). The weights are multiplied by the ratings. Each resulting score is summed to give total rating points for each vendor. In our example, we are evaluating Vendor A and Vendor B.

The cost for Vendor A's package is $22,700, and for Vendor B's package the cost is $27,690. Which package should be selected? The answer is determined by dividing the total rating points into the total cost to calculate the cost per rating point. These computations are summarized in Figure 15.3. Even though Vendor A's package had a lower rating, its cost per rating point of $37 makes it a better cost/benefit choice than that of Vendor B.

Figure 15.2 General performance rating.

General Performance Factors	Weight	Vendor A		Vendor B	
		Rating	Score	Rating	Score
Vendor assessment	10	6	60	8	80
Operating performance	20	7	140	8	160
Documentation	10	8	80	9	90
Ease of learning	20	7	140	6	120
Ease of use	10	5	50	6	60
Controls and error handling	20	4	80	6	120
Support	10	7	70	8	80
Total	100		620		710

Figure 15.3
Determining cost per rating point.

	Total Cost	Total Rating Points	Cost per Rating Point
Vendor A	$22,700	620	$37
Vendor B	27,690	710	39

Customized Software Programs

If the system under development cannot be supported by off-the-shelf software, one will have to be customized or built to fit the systems design. Original or customized software programs are developed by people who either work for the enterprise or for contractors. Our major emphasis throughout this book is on the in-house development of software, and that's what we present in this and the next three chapters.

What Is the Software Development Life Cycle?

Building programs follows a three-phase **SOFTWARE DEVELOPMENT LIFE CYCLE (SWDLC)**:

- Design
- Code
- Test

The following three chapters will treat each phase in order and provide sufficient material to prepare the SWDLC documented deliverables.

We spotlight the SWDLC and make it a component life cycle of the SDLC for several reasons. First, the SDLC encompasses the development of the entire system, which entails other components besides software. Second, in systems that do require the development of software based on the systems design created by the SDLC, the SWDLC is initiated. Third, when the SWDLC comes into play, it, like the broader-based SDLC, provides a reference set of expected phases necessary to develop the software. They contain or describe tasks and procedures that must be performed in each phase; deliverables produced by each phase; and metrics for setting schedules, estimating costs, and measuring productivity. Fourth, the SWDLC provides a standard operating procedure and framework that supports a structured, engineered approach to software development.

One recommended distribution of effort across the SWDLC phases is sometimes called the 40-20-40 rule. This rule emphasizes strong front-end design and back-end testing, whereas the mechanical programming (coding) task is deemphasized. The 40-20-40 rule is used only as a guideline. If the design is error-free, programming and testing should follow with little difficulty. Also, the more critical and larger a program is, the more effort will be devoted to the designing and testing phases.

Design The parts of the detailed systems design that will be converted to an application program are designed at a level that can be used by a programmer to write the code (i.e., write the program in a language such as C or COBOL). Some CASE packages will generate code from some detailed designs, thus eliminating the need for human coders. In any event, prime software design tools are:

- Structure charts
- Structured English
- Decision tables
- Decision trees
- Equations
- Data dictionaries
- Warnier–Orr diagrams
- Jackson diagrams

Code The coding phase (i.e., the writing of statements in a programming language), which we will assume is performed by programmers and not automatically generated by a CASE package, maps the design into program procedures (also called statements or instructions). To write code, the programmer must be highly skilled in using a programming language such as C or COBOL.

Test All modules of code, both separately and together, are subjected to a variety of tests to detect and remove errors. After thorough testing, the program is converted to operations. Although much testing is performed after the modules have been coded, some form of testing is actually conducted throughout the SWLDC.

ORGANIZING SOFTWARE DEVELOPMENT PROJECTS

A small, local-based system may be developed from planning to analysis and general design to detailed design to programming to conversion by one person called a programmer/analyst. If a customized program is needed, it is normally a very small program. The organization for such a systems project is fairly simple. When one moves to larger projects, however, requiring a variety of specialized skills, organizing such skills into workable project teams is a challenging management task.

The Importance of Good Coordination, Integration, and Communication

Passing the systems development baton to too many different players can result in a final system far different from what was originally intended. Continuity may be lost.

Developing systems in a haphazard fashion without clear documented deliverables, project control, and integration of skills may be like the party game of whispering a message into the ear of the person next to you and this person whispering supposedly the same message into the ear of the person next to him or her. More than likely, as the message is passed along, it becomes more and more distorted, often ending in a message far different from the original one. At a party, this game often results in hilarity. In systems development, it can end in a failed systems project.

If, however, the systems design requires a customized program to be developed in-house that is larger than a very small program, a great deal of organizing and work must still be done before one can achieve a successful ending.

The people who have performed most of the work to this point may believe that once they've completed the Detailed Systems Design Report, their work is completed. They hand off this documented deliverable to a programming team to write the program or programs. Because systems analysts and designers are similar to architects, we believe that they should be involved in the development of software and other implementation tasks just as architects are involved in the construction and successful completion of their design.

Like architects, systems professionals need to focus on the end result. Although architects do not lay brick, hang steel, or wire buildings, they know how each of these, as well as other construction jobs, should be performed. In the same way, systems analysts and designers may not actually code software programs, but they need to know how they should be coded and what the end result should be.

Organizational Approaches

In addition to the SWAT team approach, there are many different ways to organize a programming team. We will present three:

- Program development team
- Chief programmer team
- Egoless programming team

We will describe these three organizational approaches first. Then we will present Japan's software factory, an interesting and effective organizational concept.

Program Development Team

The **PROGRAM DEVELOPMENT TEAM**, shown in Figure 15.4, is managed by the team manager (or leader). Preferably, this person is the same person who manages the total systems project or someone from the systems project team who has been involved in the SDLC from its inception. Program development teams may be supported by librarians, editors, and program clerks to perform clerical, administrative, and documentation duties.

The skill levels on the team may range from those of trainees to those of highly skilled individuals. If a company follows the 40-20-40 program development rule, more highly skilled people will be assigned to designing and testing, as presented in Figure 15.5. The theory behind the 40-20-40 rule is: If design is complete, clear, and accurate, the coding function will be a simple mechanical process that can be performed by anyone familiar with the programming language syntax. (Normally, a trainee will, at the very least, be trained in the syntax.) This same idea supports the movement toward CASE technologies that can generate code based on precise designs, thus eliminating the need for hand coding.

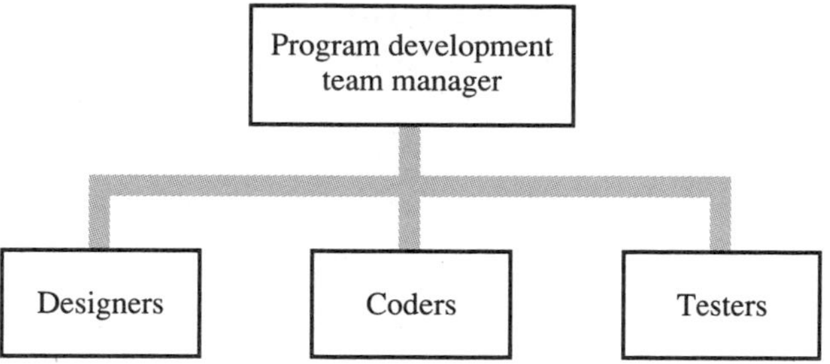

Figure 15.4 Program development team.

Chief Programmer Team

A **CHIEF PROGRAMMER TEAM** is built around a chief or super programmer who has a wealth of experience and programming knowledge. Such a person can communicate effectively with systems analysts and designers, users, and various technicians. He or she is a good manager. The chief programmer team is often compared to a surgical team in which the ultimate responsibility rests with the chief surgeon. That person is supported by a skilled, specialized staff, including a backup surgeon, other observing surgeons, an anesthetist, a chief nurse, assisting nurses, and other technicians.

The chief programmer is supported by a lead assistant, who is the main backup person on the project and who communicates with everyone else on the team, acting as a sounding board for the chief programmer's ideas. Depending on the size of the project, these two people are supported by some or all of the following personnel:

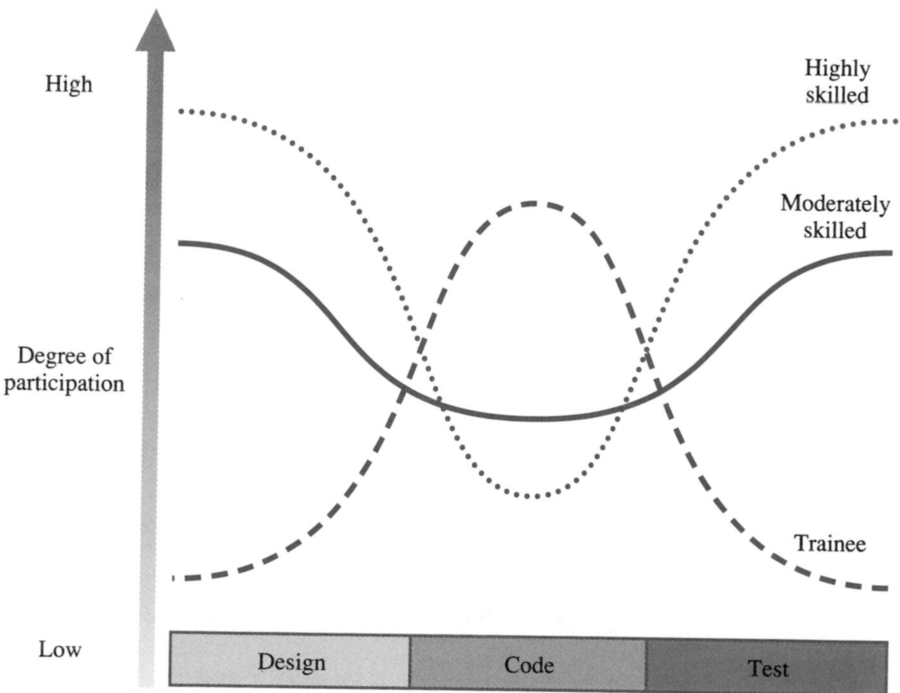

Figure 15.5 Skill-level participation in the SWDLC phases.

- ***Support Programmers*** Junior programmers are needed for large projects that cannot be handled by the chief programmer and lead assistant alone. Support programmers generally code lower-level modules.
- ***Librarian*** This person maintains the program production library, indexes files of compilations and tests, and keeps source code and object code libraries up-to-date.
- ***Administrator*** This person handles all nontechnical support details, such as budgets, personnel matters, and space allocation, and interacts with the rest of the organization's bureaucracy.
- ***Editor*** The editor is responsible for producing documentation, researching references, and overseeing all phases of documentation reproduction and distribution. In smaller projects, the editor may also perform librarian duties.
- ***Program Clerk*** This employee keeps track of all technical records for the programming team and provides whatever secretarial duties are needed by the programming team.

Team of Programming Peers

An **EGOLESS PROGRAMMING TEAM** is one without a boss or leader. The egoless team is composed entirely of peers who share software development responsibilities. The team manages its day-to-day affairs without direct supervision. Team members make decisions in a democratic fashion that avoids personal conflict. Each member of the team reviews and critiques each other's work. Egoless programming teams may be supported by librarians, editors, and program clerks to perform clerical, administrative, and documentation duties.

What Are the Differences Between Approaches?

The glaring difference between the program development team approach and the chief programmer team and egoless programming team approaches is that the program development team mirrors the 40-20-40 program development rule, whereas both the chief programmer and egoless programming teams emphasize the coding function. Because the program development team approach emphasizes designing and testing, most of our discussion in this book assumes this type of organization.

The number and size of communication paths between team members are different among team approaches. Most designing and coding errors in program development occur at interfaces; that is, Designer A communicating with Designer B, or Programmer A communicating with Programmer B.

The communication paths and interfaces of each organizational approach are presented in Figure 15.6. The program development team is composed of two designers, one coder, and two testers. The major communication paths and interfaces are between designers and the coder, and then between the coder and testers. Communication paths and interfaces are also between the designers and the testers. Because the team manager is not directly involved in the actual work, the communication path and interface to this person involves

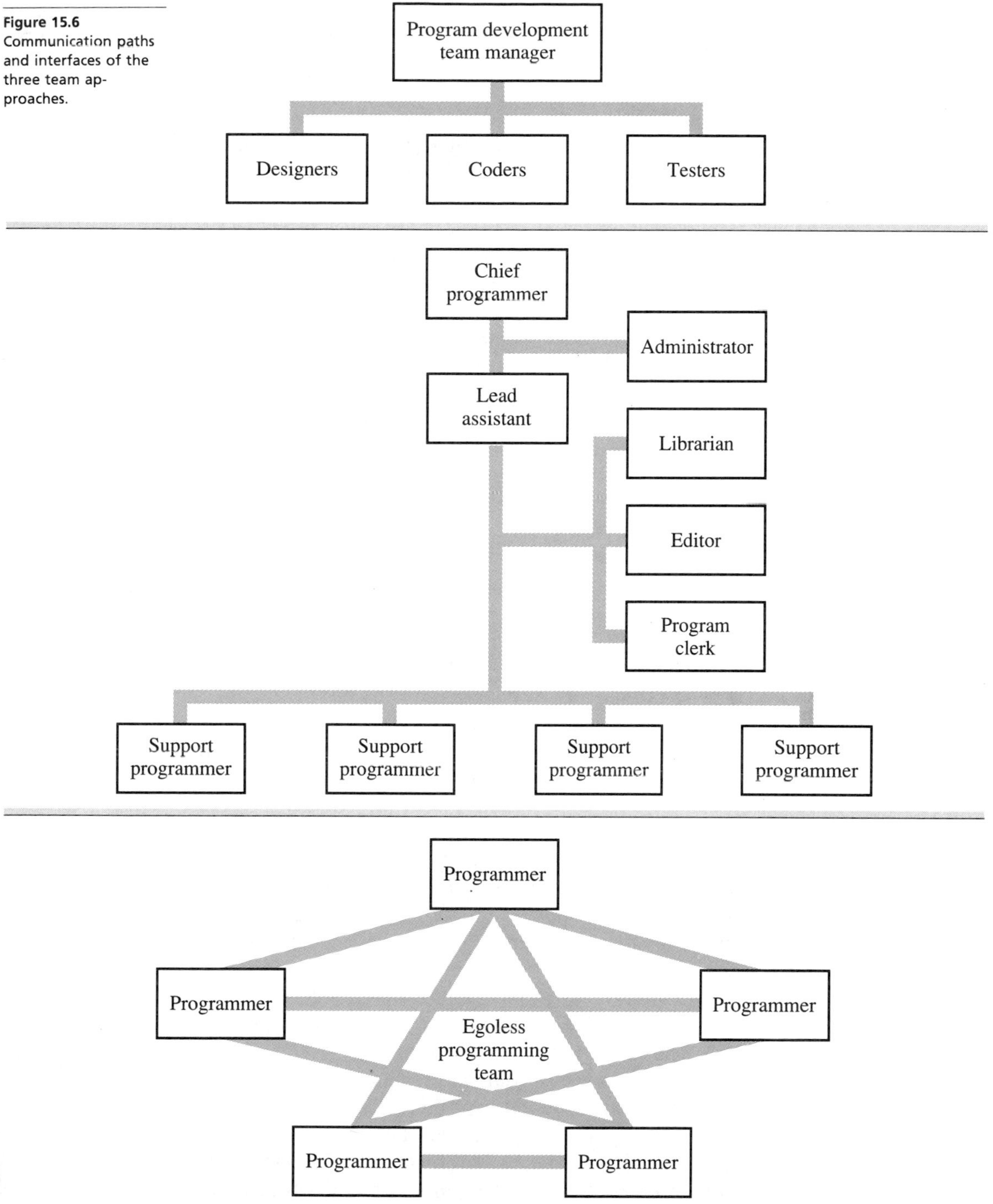

Figure 15.6
Communication paths and interfaces of the three team approaches.

summary and performance information. So the total communication paths and interfaces for the actual work being done is five, with one management interface.

In the chief programmer team, a team of five support programmers has five main communication paths and interfaces coordinated and channeled by the lead assistant and librarian. Compared to that of the egoless programming team, this arrangement reduces chances of errors and increases productivity. As a matter of fact, a number of authorities suggest that a chief programmer team is approximately twice as productive as the egoless programming team approach. Therefore, chief programmer teams are more likely than the egoless programming team to meet tight deadlines.

The number of communication paths and interfaces between n people in an egoless programming team is $n(n - 1)/2$. The egoless programming team in the figure is composed of five programmers. Therefore, there are $5(5 - 1)/2$, or ten communication paths and interfaces.

Normally, communication takes time and eats into productivity. The nature of any kind of development work is such that restart time is 30 or more minutes after each interruption. If a typical programmer, for example, devotes only eight 10-minute periods over a day to a programming task, the chances are little programming will get done. Although there are a lot of differing viewpoints and very few empirical data to support any particular viewpoint, it seems that the majority of authorities believe that the problem with an egoless programming team is that there's a lot of talk but very little action and productivity of high quality. As a general rule, therefore, where there are more than three programmers involved, a supervisor or leader is recommended. But the validity of the preceding statement depends on the members of the team. If all the members work well together like a finely tuned football team and a lot of communication is not needed, an egoless programming team may be very productive. Such a situation is, however, a rarity. That's why a serious football team, or any other team, hires a coach or manager.

What Is the Software Factory Concept?

In Japan, software factories are applying to software development the same principles of quality control and project management that enable Japan to dominate in the manufacturing sector.[1] Systems designers, coders, and testers are assigned workstations, which are interconnected via networks. These people are organized in a manner similar to that of factory workers. Although beyond the scope of this book, a study of just-in-time (JIT) manufacturing, kanbans, total quality control (TQC), and nonvalue-added cost reduction, would provide a model of Japanese manufacturing methods that are being implemented in Japanese software factories.

Although the project development team approach is organized differently from that of factory workers, the objectives of Japan's software factories are similar to the objectives espoused in this book, which are:

---

[1] Neil Gross, "New Software Isn't Safe from Japan," *Business Week,* February 11, 1991, p. 84.

- Application of an engineered approach to systems and software development
- Use of modeling tools and CASE technologies
- Installation of project management techniques
- Emphasis on maintainability, usability, reusability, reliability, and extendability (MURRE) design factors
- Pursuit of optimum systems and software development productivity

To meet a high demand for customized software, software factory personnel spend a great deal of effort and time in developing reusable code. Often, when a new customized application is needed, new code, if required, is combined with old code, thus reducing the total software development time.

The installation of software factories in Japan has been instrumental in doubling and tripling software development productivity from where it was just a few years ago without software factories. Late projects have been reduced from as much as 80 percent late projects to as little as 5 percent late projects. The average is about 10 percent late projects. The average lines of executable code (LOEC) delivered in the late 1970s and early 1980s was about 1200 assembly programming LOEC per programmer per month. A rule-of-thumb conversion from assembly to FORTRAN is about one-third, or about 400 (1200 ÷ 3) FORTRAN LOEC. Today, the average is about 3600 assembly programming LOEC (or 1200 FORTRAN LOEC) delivered per programmer per month, including reused code totaling about 45 to 50 percent of delivered LOEC. So the installation of the software factory concept along with the emphasis on reusability has increased software development productivity dramatically.

At the same time, software quality has improved from about 20 errors per thousand LOEC. Residual errors (i.e., errors found after the software is converted to operations) after rigorous testing (the subject of Chapter 18) is about 0.2 error per thousand LOEC. Some companies report 0.01 residual errors per thousand LOEC.

Toshiba uses software factories to develop enormously complex software with high productivity and reliability. Toshiba has been a world leader in real-time automated process control software for electric and nuclear power plants. Moreover, software developed by Toshiba's software factories is proving to require little maintenance, and the maintenance that is required is very easy to perform. The same can be said for Japan's other software factories.

MEASURING PRODUCTIVITY IN SOFTWARE DEVELOPMENT

Productivity can be defined by the following formula:

$$\text{Productivity} = \frac{\text{outputs produced}}{\text{inputs consumed}}$$

Thus productivity of software development can be improved by increasing outputs, decreasing inputs, or both. Inputs are relatively easy to measure, such as labor, workstations (e.g., CASE system), supplies, and various support facilities. The problem, however, lies in measuring outputs.

A number of alternative metric techniques have been advanced by various authorities. They attempt to measure software development output. The two metrics that seem to have the greatest following are:

- Lines of executable code (LOEC)
- Function points

By using metrics, it is possible to manage the software development process. It is also possible to measure the impact of change, such as a change to CASE technologies or from one generation of computer language to another generation. Furthermore, using metrics provides a perception that software development is more of a science (i.e., a product that is engineered) than an art. Early on, many programmers insisted that measuring their work was as impossible as measuring the work of William Faulkner or Pablo Picasso. But from the viewpoint of developing software following the SWDLC, which is project- and process-oriented, metrics are not only possible, but necessary. Whatever metric is used, the most important features of any metric system are reasonableness and consistency.

Counting Lines of Executable Code

Some systems professionals use source lines of code (SLOC) as their metric. A line of source code is any line of program that is not a comment or blank line, regardless of the number of statements or fragments of statements on that line, including executable and nonexecutable statements. A further refinement of the SLOC is to measure lines of executable code (LOEC) only. We use the **LOEC METRIC** in the following presentation.

Some authorities believe that LOEC is the best of all metrics, including function points, because it is simple and widely used. These authorities state that all-in-all no alternative metrics have demonstrated a clear superiority over LOEC. Indeed, several advantages induce organizations to continue to use the LOEC metric as their primary software output productivity measurement, such as:

- ***Easy to Define and Discuss with Clarity*** End users, managers, and systems professionals alike normally understand what a line of executable code is.
- ***Widely Quoted*** LOEC (or sometimes simply lines of code) is the metric frequently quoted by vendors of software development tools when referring to the productivity of their tools.
- ***Easy to Measure*** The lines of executable code are simply counted to determine the size of a program.
- ***Easy to Use for Estimates*** The approximate size of a program is determined from detailed systems design documentation. Then this number is used to estimate the time and cost of the software development project.

For example, a proposed program will contain 100 KLOEC, and if 2 KLOEC can be produced in one person-month, it will require 50 person-months to complete the project. If the inputs required to support one person-month total $9000, the software development project will cost $450,000.

What Are the Limitations of the Person-Month and Counting LOEC?

As Frederick P. Brooks, Jr., points out in his famous book *The Mythical Man-Month,* the person-month (or man-month) as a unit of measuring the size of a job is a myth. It implies that people and months are interchangeable. He further states that men and months are interchangeable only when a task can be partitioned among many workers with no communication among them. But as we know from our previous discussion on communication paths and interfaces, such is not the case. Therefore, the doubling of the size of the team (as Brooks points out) will not double its productivity, because of the exponential increase of communication paths and interfaces. In fact, productivity may not increase at all but actually decrease. For this reason, it is important to organize and manage a development team properly, and to minimize the number of communication paths and interface points.

Using a person-month as a benchmark or estimating parameter is not perfect. But if used in a reasonable and consistent manner, it can provide a way to help estimate the time and cost of projects, and it can be used as an aid in measuring and evaluating productivity. Proponents of the LOEC per person-month metric state emphatically that it is a more practical software productivity metric than any other currently available alternative.

In addition to inherent weaknesses in the person-month, there are also some drawbacks to the LOEC metric itself. The LOEC can favor misplaced productivity. For example, an assembly programmer may take four weeks and 1500 LOEC to write a program. A COBOL programmer may take two weeks and 500 LOEC to write the same application, and a C programmer may take one week and 300 LOEC. On the basis of the LOEC metric, the assembly programmer is the most productive. In real terms, however, the most productive programmer is the C programmer who finished in one week. Moreover, if we factor in the designing and testing phases, especially testing, the software development project may be performed much more effectively and efficiently with C and COBOL programming languages than with an assembly programming language. Furthermore, the LOEC metric measures the coding phase alone, which, of course, is another drawback because in a properly developed program, coding, as we have stated before, should require only about 20 percent of total software development effort. Finally, a programmer can easily inflate productivity by breaking instructions into more lines of code than is necessary.

What about quality of the program? Counting LOEC can establish a basis for comparing the speed of one programmer with that of another. This measure, however, does not indicate the quality of the work being done.

Using the Function Point Metric

The FUNCTION POINT METRIC is designed to overcome some of the deficiencies of the LOEC metric. Five functions are analyzed:

- Amount of input, such as forms and screens
- Amount of output, such as reports and screens
- Number of queries by end users
- Number of logical files accessed and used
- Number of interfaces to other applications

These five function types are what systems professionals count when using the function point metric.

Function points measure what the software development team will deliver to the end user. The function point metric therefore encompasses designing, coding, and testing, where the LOEC metric emphasizes only coding. Moreover, the function point metric measures both efficiency and effectiveness.

The function point metric gives an advantage in that it is a fairly uniform way to measure software productivity regardless of what programming language is used. It is also accepted by a number of authorities. Moreover, activities are in progress to provide better counting rules and extend its application. An example of this support is through the International Function Point User Group (IFPUG).

To count function points for a new system, the software development staff converts the detailed systems design into the five function point categories of:

- Input
- Output
- Inquiries
- Files
- Interfaces

For example, the staff counts input. Input may vary in size and complexity. Input that contains 60 data fields is obviously more complex than one that contains five fields. Similarly, input that requires access to ten files (or tables) is more complex than one that requires access to one file (or table).

A way to perform function point analysis is illustrated in Figure 15.7. The degree of complexity ranges from 1 on the low end to 10 as the most complex. The figure includes an analysis and total function points required to develop a particular software project. The number of functions is counted and weighted with a complexity factor. For example, twelve inputs are of fairly low complexity (2), three are of average complexity (5), and twelve are of fairly high complexity (8). All the other function points are counted, weighted, and calculated in the same manner. The total number of function points is 1272 for this particular software development project.

Figure 15.7
Function point analysis for a software development project.

Function Point	Degree of Complexity			Total
	Low	Average	High	
Input	12 × 2 = 24	3 × 5 = 15	12 × 8 = 96	135
Output	10 × 3 = 30	15 × 5 = 75	14 × 9 = 126	231
Inquery	10 × 3 = 30	16 × 6 = 96	17 × 8 = 136	262
File	9 × 4 = 36	20 × 7 = 140	10 × 10 = 100	276
Interface	12 × 4 = 48	20 × 6 = 120	20 × 10 = 200	368
		Total function points		1272

The development productivity rate, sometimes referred to as the delivery rate, is computed as follows:

$$\text{Development productivity rate} = \frac{\text{number of function points delivered}}{\text{number of person-months}}$$

The average development productivity rate for general applications software runs between 5 and 10. That is, one person can deliver about five to ten function points per month. If the new software development project requires 1272 function points and the average delivery rate of the team that will develop it is eight function points per person-month, the project will require about 159 person-months. If one person-month consumes $10,000 in resources (i.e., input), the project will cost about $1,590,000.

If the organization installs a CASE system, the delivery rate may jump to 15 or 20 function points per person-month. And if a CASE environment is combined with software development procedures that practice reusability of code and the organization can achieve 50 percent reuse of code, it may be possible to achieve a development productivity rate of more than 70 function points per person-month. So, with the adoption of CASE technology, the design approaches discussed in Chapter 6, and reusability of code, the cost of one person-month may increase initially by, let's say 20 percent, level off, and eventually decrease. For example, let's assume that the cost per person-month in our preceding example jumps to $12,000, but the development personnel can increase their delivery rate to 62 function points per person-month by using advanced techniques. The number of person-months to deliver 1272 function points is reduced to about 20.5 person-months (1272/62) at a cost of $246,000, or a cost savings of $1,344,000 ($1,590,000 − $246,000).

The preceding estimated time can be interpreted into elapsed months, adjusted according to the number of people assigned. For example, if five skilled and motivated people are assigned to the project, the total elapsed time to complete the project will be about four months (20.5/5).

This estimate ignores Brooks' mythical man-month, although his analysis is thorough, well presented, and in many instances, correct. But in the final analysis, the validity of using the person-month for estimating time and cost and productivity evaluation depends on the people doing the work. For example, in some cases, five people working together may actually produce a delivery rate far greater than five times what one person working alone could do. Maybe synergism takes hold and the whole is greater than the sum of its parts. In any event, a formula to measure relative delivery rates of different size teams is beyond the scope of this book. Even though we are aware that, in some instances, doubling the size of a team will not necessarily double the delivery rate of that team, we have nonetheless made that assumption.

Management's Influence on Productivity

Management of the software development project can be a positive or negative influence. Poor management can decrease software productivity more rapidly than any other factor. Here are some examples of management activities that frequently cause a loss in productivity.

Poor or Nonexistent Productivity Metrics Without a reasonable and consistent productivity metric, a manager has no way of estimating software development time and cost, or of measuring the productivity or delivery rate. In addition to serving as an estimating tool, a good metric can be used to identify increases or decreases in team productivity. To identify improvement, a manager must know what the team's current productivity rate is and compare it with an established baseline.

A productivity metric should measure both efficiency and effectiveness. Efficiency involves the resources consumed in development of a given application in a timely manner. Effectiveness applies to the quality of the finished program and its ability to meet user requirements.

Poor Planning and Control The project manager permits the project to evolve aimlessly without requiring target dates and deliverables. The application of project management techniques such as PERT will help alleviate poor planning and control.

Poor Mix of Skills To weight a software development project team with all coders will tend to deemphasize design and testing functions and emphasize coding. An ideal skills mix is about 40 percent designers, 20 percent coders, and 40 percent testers. Testers should be organized and managed as a separate, independent group.

Premature Coding Some project managers' philosophy may be summarized in this statement: "We'd better hurry up and start coding, because we're going to have a lot of debugging to do." Numerous examples can be cited in organizations in which failure to do the job right the first time cost far more than any benefits gained by ignoring the design phase.

Inequitable Rewards Outstanding team performers may receive a 5 percent raise, and sloppy, mediocre performers may receive a 4½ percent raise. Eventually, top performers become frustrated and leave. A great deal can be done by creative application of rewards, such as special performance bonuses, travel, and special seminars.

In some cases, using highly skilled and motivated team members will reduce the time required by a factor of three to five times less than using mediocre, uninspired team members. Moreover, additional productivity leverage can be achieved by providing the top performers with advanced tools such as CASE.

PRODUCING HIGH-QUALITY SOFTWARE

The ultimate goal of software development is to produce high-quality software at a high productivity rate that adds value to the enterprise. We already know about productivity or delivery rates, but what is software quality and how do we obtain it? The answer to this question has three dimensions.

First, from an end user's perspective, software quality is measured by performance factors, such as:

- Overall operating performance
- Ease of learning
- Controls and error handling
- Support from developers and maintainers

MURRE design factors and PDM strategic factors, discussed in Chapter 7, provide the other two dimensions—that of software design quality. MURRE factors include:

- Maintainability
- Usability
- Reusability
- Reliability
- Extendability

High-quality software supports PDM strategic factors by helping to:

- Increase productivity
- Augment product and service differentiation
- Improve management functions

Now, how do we obtain software quality? We do so by implementing strong quality assurance and quality control techniques.

What Is Quality Assurance?

QUALITY ASSURANCE (QA) ensures that the SWDLC used in developing a quality software product adheres to standards set forth for that product. On a broader scale, quality assurance includes the continuous monitoring of all systems and software development phases from systems planning to implementation. In addition, it includes corrections to the development process so that the quality of systems and software produced in the future will improve.

There is a cliché in systems that states: "There's never enough time to do it right, but plenty of time to do it over." The reality is that if the work is done right the first time, which is the essence of quality assurance, it will not contain errors and will not have to be redone. Occasionally, tight control of the development process will result in a later initial delivery date, but the delivery date of a well-managed project will also be the final delivery date of a completed program ready for operations. In the case where a project is not well managed and rushed, what is called a completed deliverable is actually still being developed, and real completion will come considerably later.

What Is Quality Control?

Whereas quality assurance focuses on the process of systems and software development, **QUALITY CONTROL** focuses on the product; that is, what is delivered. While the process may succeed, the product (i.e., the system and software) may be a failure. Because quality control evaluates the system and software after they are developed, no amount of quality control activity can improve quality. Quality must be built into the system and software while they are being built, not after development is complete. So, whereas quality assurance is an error-prevention technique, quality control is an error-removal technique. Quality assurance is never a substitute for quality control, or vice versa.

Creating a QA Group

Some organizations form an independent quality assurance (QA) group. Such a group may be composed of end-user representatives, systems analysts, systems designers, and skilled programmers, all of whom are independent from the developers. A QA group's tasks include:

- Establish standards for systems and software development such as following the SDLC and SWDLC.
- Evaluate documented deliverables.
- Perform systems and software design walkthroughs.
- Conduct code walkthroughs.
- Perform tests.

In what some authorities call a "cleanroom approach," development programmers not only do not test the code they have written, but are not even allowed to compile it. All testing is conducted by the QA group.

In a **SOFTWARE DESIGN WALKTHROUGH**, the QA group carefully reviews design specifications to detect design inconsistencies and errors before coding begins. In a **CODE WALKTHROUGH**, the QA group reviews the actual code to see if it matches design specifications and simulates how such code will be processed by the computer to discover coding errors. All of this is done before computer-based testing begins.

Some authorities state that walkthroughs can find 75 to 90 percent of design defects and coding errors prior to testing. Some authorities believe that it is the single most important quality improvement technique for software. Moreover, they state that organizations conducting independent walkthroughs have increased their productivity rate from 14 to 25 percent.

You may be asking yourself at this point: Where does quality control come into play? The quality control activity is normally performed after nearly all other testing procedures have been applied and the QA group and developers are fairly confident that the software is ready to be released for implementation and conversion to operations. Thus the quality control activity comes into play during systems and acceptance testing—especially acceptance testing. Both of these testing procedures are discussed in Chapter 18. Again, using manufacturing as our analogy, this approach coincides with quality control in manufacturing, whereby the product is evaluated at the end of the assembly line when production is complete and the product is ready to be transferred to finished goods for shipment to customers.

PLANNING THE SOFTWARE DEVELOPMENT LIFE CYCLE PROJECT

The SWDLC game plan's purpose is to enable the project manager to schedule and monitor all the tasks required to complete the SWDLC. A popular tool used to formulate a software development game plan is the program evaluation and review technique (PERT). Figure 15.8 demonstrates a PERT network of tasks required to develop a completed program. (We introduced a simplified version of PERT in Chapter 3.)

The objectives of PERT are to determine the sequence in which software development tasks must be performed and to estimate how long the project will take from start to completion given the task dependencies. The duration of the project will be the longest series of tasks that must be performed sequentially, which is the critical path of the project.

Four steps must be taken to set up a software development PERT network:

1 Identify all software development tasks that must be performed.

2 Estimate the time required to perform each task.

3 Determine the sequence of tasks.

4 Determine the critical path that dictates total software development time.

The result of the first step is a list of software development tasks shown in Figure 15.9. Each task is identified by number and description.

In the second step, expected times to complete tasks are estimated by experienced managers and staff personnel and the use of a productivity metric (e.g., function point metric). The goal in getting three estimates is to use them to calculate a single weighted average for each task. This weighted average is the expected time indicated by the following formula:

$$TE = \frac{O + 4M + P}{6}$$

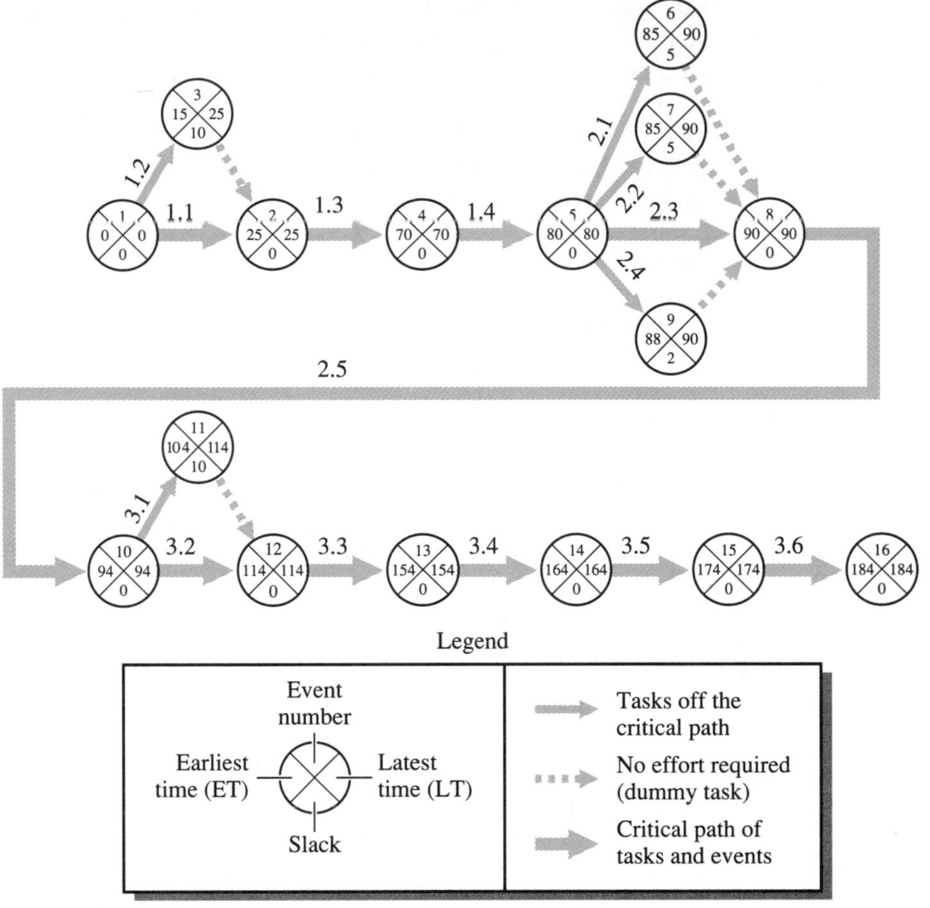

Figure 15.8
PERT network for a software development project.

Figure 15.9
Tasks for structured software development and their expected duration times.

Task	Description	Expected Time (in Person-days)
1.1	Transform data flow diagram to structure chart	25
1.2	Transform output, input, and data stores to data dictionaries	15
1.3	Prepare structured English for each module	45
1.4	Conduct design walkthrough	10
2.1	Program output modules	5
2.2	Program input modules	5
2.3	Program decision modules	10
2.4	Program computation modules	8
2.5	Conduct code walkthrough	4
3.1	Design test cases	10
3.2	Program drivers and stubs	20
3.3	Perform module testing	40
3.4	Perform integration testing	10
3.5	Perform systems testing	10
3.6	Perform acceptance testing	10

where TE = expected time
O = optimistic time estimate
M = most likely time estimate
P = pessimistic time estimate

This formula is based on the assumption that the time estimates approximate a beta distribution.

The third step in creating a software development PERT network is sequencing tasks. The sequence of tasks is determined by dependencies. Certain tasks must be completed before others can be started. In some instances, tasks can be performed simultaneously. In our software development PERT network, we see that Tasks 1.1 and 1.2 can be performed simultaneously; that is, transforming the data flow diagram into a structure chart and preparing data dictionaries from output, input, and data stores can be worked on at the same time. But Task 1.3, writing structured English for the modules, cannot be done until the structure chart of modules is complete. After these three major tasks of the software design phase have been completed, a software design walkthrough can be conducted to see if the software design meets the specifications of the detailed systems design.

After coding Tasks 2.1, 2.2, 2.3, and 2.4 are completed, a code walkthrough is conducted. If the modules are deemed acceptable by the walkthrough team, software testing can begin. In this phase, Task 3.1, Designing test cases, and Task 3.2, Preparing drivers and stubs, can occur at the same time. Each testing step—3.3, Module testing; 3.4, Integration testing; 3.5, Systems testing; and 3.6, Acceptance testing—must be performed in sequence. Assuming that the software passes all tests, the software development project is considered complete and the program itself is ready for implementation into the overall new system and converted from a developmental status to a full-time operable status.

The last step is to analyze the PERT network completed so far and compute the critical path. The project manager needs to know how early an event can begin and how late the event can begin without affecting the overall project schedule. The earliest time, designated as ET, is 0 for Event 1. For the other events, ET is the highest sum of the task duration and ET of any immediately preceding event. For example, four tasks immediately precede Event 5. The ET of Event 3 is 15 person-days (0 + 15); the ET of Event 2 is 25 person-days (0 + 25); the ET of Event 4 is 70 person-days (it takes 25 person-days to complete Event 2 and 45 person-days to complete Event 4); the ET of Event 5 is 80 (70 + 10). The remaining ETs for each event is computed in the same manner.

The latest event time, indicated by LT, is the latest time at which the event can begin without delaying the project. To calculate LT, it is necessary to work backward through the PERT network, starting from the right or last event on the network. The LT is the smallest difference between the LT of the event minus the task duration time. For example, the LTs for Events 6, 7, 8, and 9 are 90 person-days each. To compute the LT for Event 5, each expected time (or duration) is determined, as shown in Figure 15.9. Task 2.1 is 5 person-days,

2.2 is 5, 2.3 is 10, and 2.4 is 8. Subtracting the duration times from 90, we get 90 − 5 = 85, 90 − 5 = 85, 90 − 10 = 80, and 90 − 8 = 82, respectively. Thus the latest time we have to complete Event 5 is 80 person-days. For example, if we complete it in 85 person-days, we will not be able to complete Event 8 on time because we would be 5 person-days over schedule.

The critical path in our example is indicated by the broad lines. It was determined by connecting all events that contain zero slack time; that is, all events where ET and LT are equal. This critical path represents the set of tasks that the project manager must monitor closely. It identifies the events that must be begun and completed on time and that require no more than the estimated duration time. Any delay of tasks on the critical path results in an equivalent delay on the overall project. Other tasks that have slack times enable the project manager to allocate resources from these tasks to those on the critical path if the project schedule begins to slip.

Because PERT is available for personal computers, a project manager is advised to use it to improve control of projects and gain high productivity rates. Estimating, scheduling, and controlling interdependent project tasks will help in determining if they are being performed properly, on schedule, and within budget.

REVIEW OF CHAPTER LEARNING OBJECTIVES

The major goals of this chapter were to enable each student to achieve seven important learning objectives. We will now summarize the responses to these learning objectives.

Learning objective 1:
Differentiate the sources of application software and discuss how to evaluate and select commercial software packages.

Application software is either purchased ready-made from a commercial software vendor or developed by the company's information systems staff or an independent contractor. General software performance factors are:

- Operating performance
- Documentation
- Ease of learning
- Ease of use
- Controls and error handling
- Support

Each factor is weighted and rated to produce total rating points for the commercial packages being evaluated. These points are divided into the cost of each package to calculate a cost per rating point. The commercial package with the lowest cost per rating point is selected for installation.

Learning objective 2:
Define the software development life cycle (SWDLC) and briefly discuss its phases.

The software development life cycle (SWDLC) is the methodology followed to develop a software project; that is, an application program that runs part or all of the systems design. The SWDLC is composed of three phases:

- Design
- Code
- Test

The design phase converts systems design to the design of program modules—for example, the conversion of a data flow diagram to a structure chart. The code phase involves the actual writing of the programming language into program modules that represent the design. The purpose of the test phase is to detect and remove design and coding errors. Relative percentages of effort devoted to these phases are 40-20-40, respectively.

Learning objective 3:
Discuss organizing for software development projects; define the program development team, chief programmer team, and egoless programming team; and analyze their advantages and disadvantages.

Organizing people into a focused, highly motivated team working together in an integrated and coordinated fashion is paramount for the successful development of a software project. The program development team has a mix of designers, coders, and testers that reflect the 40-20-40 SWDLC rule in number. A chief programmer team is made up of a chief programmer, lead assistant, support programmers, librarian, administrator, editor, and program clerk. An egoless team is made up of a number of programmers who work together in a totally democratic fashion.

The program development team is tailor-made for the SWDLC because of its mix of designers, coders, and testers. The chief programmer team is tightly organized and coordinated. It is, therefore, well suited for coding, but it does not stress the design and test phases. The egoless programming team works without supervision. Like any team without a coordinating force, it may tend to drift and not reach its goal, or reach its goal far beyond a target date. It also emphasizes coding over designing and testing.

Learning objective 4:
Define software productivity and present two ways to measure this productivity. Also explain management's impact on software productivity.

The general definition of productivity is:

$$\text{Productivity} = \frac{\text{outputs produced}}{\text{inputs consumed}}$$

The definition of software productivity using the LOEC metric is LOEC per person-month. The definition of software productivity using the function point metric is:

$$\text{Development productivity rate} = \frac{\text{number of function points delivered}}{\text{number of person-months}}$$

Function points are composed of the number of inputs, outputs, inquiries, files, and interfaces, along with their degrees of complexity.

Poor management decreases software productivity. Poor management results from:

- Not having or using productivity metrics
- Poor planning and control
- Poor organizing
- Stressing coding over designing and testing
- Rewarding top performers the same (or close to the same) as mediocre performers

Learning objective 5:
Discuss software quality and explain quality assurance and quality control.

Software quality is made up of three dimensions:

- End-user dimension, which provides operating performance, ease of learning, controls and error handling, and support
- Design dimension, which achieves maintainability, usability, reusability, reliability, and extendability (MURRE) design factors
- Value-added dimension, which helps support the enterprise's productivity, differentiation, and management (PDM) strategic factors

Quality assurance (QA) is a process that designs quality into the system. Quality control is a process which ensures that quality of the system was achieved. A QA group ensures that both quality assurance and quality control processes are applied.

Learning objective 6:
Describe how the program evaluation and review technique (PERT) is used as a SWDLC game plan and project management technique.

The PERT technique helps the project manager:

- Identify all tasks that must be performed and the logical sequence in which they are to be performed
- Estimate the time required to perform each task
- Determine the critical path of the project

This results in a game plan for planning, scheduling, and controlling the software development project.

SOFTWARE DEVELOPMENT CHECKLIST

Following is a checklist on how to conduct software development. Its purpose is to remind you of major issues pertaining to software development. Specific phases of software development are covered in Chapters 16, 17, and 18.

1 Complete the Detailed Systems Design Report, and make sure that all interested parties are in agreement with it.

2 Use the Detailed Systems Design Report as a blueprint for systems construction, as a builder would use an architect's blueprint to construct a building.

3 Determine which applications require commercial software package support or customized software support.

4 If a commercial software package is indicated, prepare requests for proposals (RFPs) for commercial software vendors.

5 Evaluate and rate each commercial software vendor's package and select the one most appropriate.

6 If a customized software program is indicated, organize a programming team, and insist that they follow a software development life cycle (SWDLC).

7 Develop a workable metric for measuring software development productivity. Two workable metrics are the lines of executable code (LOEC) metric or the function point metric.

8 Set up a total quality assurance and control system to ensure a high level of software quality.

9 Use PERT to schedule, monitor, and control the SWDLC.

KEY TERMS

Chief programmer team
Code walkthrough
Egoless programming team
Function point metric
LOEC metric
Program development team
Quality assurance (QA)
Quality control
Software design walkthrough
Software development life cycle (SWDLC)

REVIEW QUESTIONS

15.1 What are the two major sources of application software?

15.2 Under what circumstances will a commercial software package be appropriate? Under what circumstances will customized software be appropriate?

15.3 List and briefly discuss the advantages and disadvantages of commercial software packages.

15.4 Why should a performance-oriented RFP be prepared for commercial software vendors? What should this RFP contain?

15.5 List and define the general performance factors used to evaluate commercial software.

15.6 Define the software development life cycle (SWDLC). Briefly describe each phase. What is the 40-20-40 rule?

15.7 Explain how large software development projects can cause problems of design coordination and communications.

15.8 Explain how excessive communication paths and interfaces among team members can cause a decrease in productivity. Give an example.

15.9 Explain why systems analysts and designers should be involved in software development.

15.10 Name three kinds of teams used to develop software. Which one is adapted to the 40-20-40 rule?

15.11 Explain why the people with the greatest skills should be assigned to designing and testing.

15.12 An egoless programming team is composed of 20 members. How many communication paths and interfaces are there between team members?

15.13 Explain why a chief programmer team may meet coding deadlines better than egoless programming teams.

15.14 Define productivity for any endeavor. What is the formula for productivity?

15.15 What are two popular metrics that are used to measure software development productivity?

15.16 Give the advantages and disadvantages of the LOEC metric.

15.17 Explain why Frederick P. Brooks, Jr., says that the person-month (man-month is his term) used as an estimating and productivity calculating parameter is a deceptive myth.

15.18 Give an example where five people working together on a project may actually be more productive than five people working alone. Define synergism and explain why, in some instances, the whole is greater than the sum of its parts.

15.19 List and give a brief example of the five function points.

15.20 Explain how the function point metric encompasses software designing, coding, and testing.

15.21 What are the advantages of the function point metric? What's the name of the organization that supports the function point metric?

15.22 What is the development productivity rate (also called the delivery rate)? How is it computed?

15.23 Explain management's influence on the productivity rate of a project team. Point out management activities that can frequently cause a loss in productivity.

15.24 Define software quality from a design viewpoint, from a value-added viewpoint, and from an end-user viewpoint.

15.25 What's the primary difference between quality assurance and quality control?

15.26 Discuss how a project team achieves good quality assurance.

15.27 What does quality control focus on?

15.28 Should a QA group be part of the development team? Explain why or why not.

15.29 Explain this statement: Quality assurance is never a substitute for quality control, and vice versa.

15.30 What is the "cleanroom approach?"

15.31 What are the key tasks performed by a QA group?

15.32 Explain how PERT is used to estimate, schedule, and manage a software development project.

CHAPTER-SPECIFIC PROBLEMS

These problems require exact responses based directly on concepts and techniques presented in the text.

15.33 You have narrowed your search for commercial inventory-management software to two packages: Fifo and Eoq. Out of a possible 100, you rate the general performance factors as follows: vendor assessment (20), operating performance (15), documentation (10), ease of learning (10), ease of use (10), controls and error handling (20), and support (15). After researching and benchmarking Fifo and Eoq, you rate each performance factor as follows:

Fifo	Eoq
5	6
7	9
9	5
7	4
6	7
4	9
8	9

Fifo costs $37,290 and Eoq costs $36,980.

Required: Compute a general performance factors score for each commercial package. On a cost/benefit basis, which package do you recommend?

15.34 It is estimated that a software project will require 300 KLOEC. 1.5 KLOEC can be produced per person-month. Inputs required to support one person-month amount to $8000.

Required How many person-months are needed to complete the project, and what is the estimated cost of the project? If 20 people are assigned to the project, how many elapsed months will the project require?

15.35 Software development costs for the Space Shuttle were $1.2 billion. They required 25,600,000 lines of code that took 22,096 person-years to develop. Software development costs for Citibank teller machine software was $13.2 million. Development required 780,000 lines of code that took 150 person-years.

Required: Compute the number of person-years required to develop one KLOEC and the cost per one KLOEC for both software projects. Which system required more effort in terms of time and cost per one KLOEC? Which system do you think required more testing? Why? Would sound estimating, scheduling, and monitoring techniques be applicable for these kinds of software development projects? Explain.

15.36 A proposed software project will include 60 inputs, 20 outputs, 70 inquiries, 10 files, and 100 interfaces, all with a degree of complexity of 5. The development productivity rate (or delivery rate) is 10 function points per person-month. It takes $5000 to support one person-month.

Required: Compute the number of function points and the estimated cost of the project. If 10 people are assigned to this project, how many elapsed months will it take to complete the project? If the company converts to a joint application development (JAD) CASE workstation approach, it is estimated that the delivery rate will increase to 40 function points per person-month at a cost of $8000 per person-month. If the company converts to JAD and CASE, how much will the project cost? If 10 people are assigned to the project, how many elapsed months will it take to complete the project?

15.37 A table of tasks and their expected duration times is presented below.

Phase	Task	Estimated Time (in weeks)
1	1.1	4.2
	1.2	4.0
	1.3	3.2
	1.4	3.0
	1.5	3.8
2	2.1	2.0
	2.2	1.2
	2.3	2.0
	2.4	3.8
3	3.1	2.0
	3.2	4.2
	3.3	3.8

Tasks in each phase must be performed in sequence, but the three phases can be performed simultaneously. There are 11 events, and the three phases converge at event 11, which represents completion of the project.

Required: Calculate the earliest time (ET), latest time (LT), and slack time. Draw a PERT network indicating event number, tasks, calculated times, and the critical path.

15.38 The following PERT network includes the event numbers and the ET of each task. The TE of each task is placed on the lines that represent that particular task.

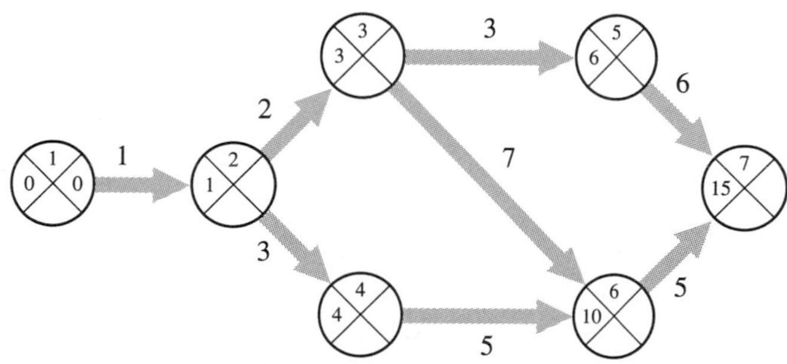

Required: Compute LT and slack time at each event and list the events in order that comprise the critical path. Slack time =

LT − ET; LT = latest allowable time for an event to occur; ET = earliest expected time for an event to occur.

THINK-TANK PROBLEMS

These problems call for a feasible approach rather than a precise solution. Although the problems are based on chapter material, extra reading and creativity may be required to develop workable solutions.

15.39 A public utility started a software development project in 1991. The new system was to combine charges for water and electricity service on a single monthly statement instead of mailing out separate bills for each. The original timetable called for the system to be implemented within one and a half years. So far the programming staff has spent 40 person-years and $2 million on the project. The project manager thinks that it will take half again that much work and another $1.5 million before the system will be operable.

Required: Analyze the situation above and discuss what you believe went wrong and how these kinds of delays could have been prevented. Explain the relevance of the SDLC, SWDLC, metrics, software development teams, productivity, quality control and quality assurance, and PERT or Gantt chart to this problem.

15.40 Dorothy Hamilton is heavy-equipment manager for Monarch Construction Company. Her major responsibility is to plan and coordinate the transfer of heavy equipment, such as bulldozers, cranes, pumps, pile drivers, and pavers, from one construction project to another in the most efficient manner possible. For example, the airport project manager estimates that that project will not need three bulldozers after March 1; the highway project manager can release one bulldozer March 1. The shopping center project manager will need four bulldozers March 15. Dorothy uses such information to make sure that project managers receive equipment as needed.

Dorothy also needs online, real-time equipment maintenance information. For example, bulldozer crankcase oil should be changed immediately after 50 hours of operation. Dorothy has established similar maintenance standards for all equipment. Both Dorothy and mechanics in the field are to receive the same maintenance information. As soon as maintenance tasks are completed, data describing such tasks are to be entered into the system to keep the maintenance status of each piece of equipment current. On a random basis, Dorothy also uses this information to check equipment to make sure that the mechanics are in compliance with her maintenance standards.

About a year ago, Dorothy submitted a Systems Service Request to the information systems (IS) group at Monarch. Clarence Moody, systems analyst/programmer, interviewed Dorothy to determine her requirements in more detail. She never heard from Clarence again until today. During the interim, Dorothy had become totally outraged with the lack of service from IS. Upon confronting Clarence in the company coffee shop, she told Clarence that she wanted to speak with him in her office. After both were in her office and the door was closed, the following is what she told Clarence:

> "Nearly a year ago, we met in this office to start the development of a computer-based equipment management (CBEM) system. I spelled out my requirements and you said you understood them. You also said that you could build the system in about three or four months at most—'no problem.' Well, there is a problem, a bad problem. Months have passed. I've lost my assistants who were doing manually what the CBEM system was supposed to do. I'm weeks behind in updating the manual system. Every project manager is on my back. They don't think I know what I'm doing.
>
> "I've tried to reach you dozens of times, and all I get is a silly message on your answering machine. I've also been by your office on numerous occasions, also with no luck. All I see are a bunch of people sitting at PCs playing stupid games. The IS area looks like an arcade. You people may know how to 'tickle keys' and the intricacies of DOS, but you apparently don't know how to develop systems for business.
>
> "Monarch has been in business for three years. It's my understanding that IS implemented a commercial accounting software package soon after we went into business, and it's working quite well. I also understand, however, that many systems requests have been made by users throughout Monarch for customized applications, but none have been implemented to date, including mine. No one seems to know what's going on. I've tried to contact Bill Mallory (Bill is the CIO) and he's never in his office. I met him in the coffee shop about two months ago and questioned him about my system. He told me he'd check into it. I haven't heard from him since.
>
> "I am sick and tired of the way things are done around here. Tomorrow morning, I have an appointment with Martin (Martin Handelman is Monarch's CEO). I think it's only fair to warn you that I'm going to get to the root of the problem and find out why you people in IS are not doing your job. Things are going to change dramatically around here, one way or the other. You can bet on it."

Required: You work for Sirius Systems Consultants, a firm comprised of first-rate systems people. Martin Handelman, CEO at Monarch, has called you and said, "We've got a real mess in our IS group, and we need some outside, independent help to straighten things out. Can you help us?" Two months have passed since your agreeing to take on the assignment at Monarch. Although the comments made by Dorothy were caustic, and some even uncalled for, her assessment of the IS group was fairly accurate.

Prepare an executive summary that outlines the key recommendations you are making to improve the IS group's performance and increase its credibility. (*Hint:* Although outside research may be helpful in deriving a feasible approach to Monarch's problem, a quick review of the preceding chapters should be sufficient. An executive summary report calls for a brief, but comprehensive set of recommendations. Normally, executives seek major points, not a lot of detail.)

15.41 "It seems there is never enough time to do a job right the first time, but there is always time to do it over."

Required: Discuss how the SWDLC overcomes the tendency of human behavior implied in the statement above. What kinds of errors are made if people rush into the coding phase of a software project? Compare the relative effort required to prevent errors with that needed to detect and correct them. Read one or two articles on Japan's software factories, and try to relate what you learned to the statement.

SUGGESTED READING

Abdel-Hamid, Tarek K., and Stuart E. Madnick. "On the Portability of Quantitative Estimation Models." *Information & Management,* 1987.

Arthur, L. J. *Measuring Programmer Productivity and Software Quality.* New York: Wiley-Interscience, 1985.

Benbasat, Izak, and Iris Vessey. "Programmer and Analyst Time/Cost Estimation." *MIS Quarterly,* June 1980.

Best, Laurence J. "Building Software Skyscrapers." *Datamation,* March 15, 1990.

Boehm, B. W. *Software Engineering Economics,* Englewood Cliffs, N.J.: Prentice-Hall, 1981.

Brooks, Frederick P., Jr. *The Mythical Man-Month.* Reading, Mass.: Addison-Wesley, 1975.

Carlyle, Ralph Emmett. "Fighting Corporate Amnesia." *Datamation,* February 1, 1989.

Fairley, Richard E. *Software Engineering Concepts.* New York: McGraw-Hill, 1985.

Fourteenth NASA Goddard Software Engineering Workshop.

Gray, Paul, William R. King, Ephraim R. McLean, and Hugh J. Watson. *Management of Information Systems.* Chicago: Dryden Press, 1989.

Gross, Neil. "Now Software Isn't Safe from Japan." *Business Week,* February 11, 1991.

Inman, W. H. *Information Engineering for the Practitioner: Putting Theory into Practice.* Englewood Cliffs, N.J.: Yourdon Press, A Prentice-Hall Company, 1988.

Jeffery, D. R. "A Software Development Productivity Model for MIS Environments." *Journal of Systems and Software,* 1987.

Jones, T. Capers. *Programming Productivity*. New York: McGraw-Hill, 1986.

Kaner, Cem. *Testing Computer Software*. Blue Ridge Summit, Pa.: TAB Professional and Reference Books, 1988.

Smith, L. Murphy. "Using the Microcomputer for Project Management." *Journal of Accounting and EDP*, Summer 1989.

JOCS CASE: Managing the Software Development Project

Jake Jacoby, SWAT team project manager for the job-order costing system (JOCS), started to decide during the systems analysis and design phases that JOCS would require a custom software solution. He began to make that decision when one of the future JOCS end users, a cost accountant, stated during a meeting:

> I'm not really sure what reports we will need from this system six months from now. We're working in a very competitive environment, and this system is supposed to help us control our costs and monitor our productivity. I can describe the reports that will help me do that right now, but I can't predict what will help me in the future. Instead of just designing reports, design me a system that *I* can use. I know how to use computers; make this system something that *I* can change when I need it to change. Let me make the reports that I will need down the road.

This was not the only end user who wanted to have the option to change the output from JOCS in the future. As the SWAT team began detailed systems design, many end users stepped forward and encouraged a flexible systems design. A supervisor in the manufacturing department said that he wanted

> a way to make estimates for each individual job. Each hydronautical lifter we make is unique. I can't put in a standard set of materials and labor and have those standards work for every lifter. I need a system that will allow me to change the requirements for each different machine. What if management decides to make a new product in the future? Will you have to design another JOCS to handle the new product line? Give me a flexible system. I'm willing to learn how to use one of those PC-things if I can get what I want from it. But I don't want to be a programmer or anything, so buy something I can learn in a couple of weeks.

Jake knew that prewritten software would not support these user requirements. Commercial software packages function best when there is a strict set of input/output defined for a system. JOCS is a job costing system, which is a very common accounting system, but the end users at Peerless, Inc. want a unique system; one that will support some defined input/output, but will also support end users defining their own data and information in the future. So Jake decided to develop the software for JOCS within the organization rather than purchasing the application from a software vendor.

This decision worries Jake. Jake has developed many computerized applications over the last ten years, and he is aware of the potential for problems when developing software. He knows that cost and time overruns are a common problem; that software frequently does not live up to the unreasonably high expectations of the user; and that defects can occur in even the most well-tested software packages. He also knows that some people doing research in software engineering think it is necessary to create an initial software package, and simply throw it away and start over, before a correct package can be produced. Jake is entering this phase of systems development fully aware of the potential pitfalls.

Jake thinks that the use of good managerial skills during software development will help him avoid most of the common problems. "Maybe good coordination and management will circumvent the need to create a 'throw-away' JOCS," Jake thinks. "We already have a strong SWAT team. I think we have done a good job of systems analysis and design, and we have the opportunity of defining our own development environment." Before he assigns any tasks to the SWAT team or starts to design the algorithms or codes even one line of the package, Jake develops a managerial approach to control the project.

Management usually consists of four functions: planning, controlling, organizing, and leading. Jake produces a software development managerial approach consisting of the following facets to incorporate each of those managerial functions:

1 ***Planning*** Jake develops a project estimation budget. Jake produces an initial estimation, in both time and dollars, for each part of the project. This budget will be used to mark the progress of each phase of the project.

2 ***Organizing*** Jake produces an initial Gantt chart to consolidate the components of the project and to portray the critical milestones. Jake also decides upon the organizational structure that will be used for the employees assigned to the project.

3 ***Controlling*** Jake chooses a method to measure the progress of the project accurately and continuously, including a way to evaluate the productivity of individual project participants. Jake also chooses a set time period that will be used to compare the actual results of software development against the budget to make sure that the project is proceeding according to the initial plan.

4 ***Leading*** Jake selects a type of communication for project participants. Jake creates a plan to keep personnel informed of critical decisions, develops a way to provide for interpersonal communication, and schedules appropriate training.

After outlining his managerial approach, Jake's next step is to make sure that the current SWAT team is sufficient to handle software development. Developing software is a labor-intensive task. The people working on a software development project are the most important element of the project. Luckily, the current SWAT team has been able to communicate very well during the prior phases of the systems development life cycle. Of the four people working on the project, Christine and Tom have a strong background in both software design and development. Cory is right out of college, but she has a good working knowledge of database management systems. She did program development while in school. Carla, however, does not have any prior experience developing software. She was chosen for the SWAT team because she understood the system requirements from an end-user perspective. Since Carla is simply "on loan" from the finance department to the JOCS project, this might be a good time for her to return to her normal job duties. But Jake is hesitant to give her up. Carla understands what the end users want from the system. She is able to judge whether or not a program is truly user friendly. Programmers tend to have a different opinion about the usability of software than users, and occasionally they make poor decisions about the structure of a user interface. Jake thinks that Carla could help the team design good software. He wants her to stay with the SWAT team and help design and test the software.

To be able to complete the project on time, Jake adds another person to the SWAT team. Jannis Court, a programmer with three years' experience, is added to the team to help develop the JOCS software. Jake adds only one new person to the SWAT team because research from other application projects has shown that large numbers of people frequently slow the development process because of communication difficulties.

Jake has established an organizational structure for the JOCS application that combines the best elements of both the program development team approach and the egoless programming team approach. Figure 15.10 provides an overview of the organizational placement of the personnel in the project. The solid dark lines on the figure represent direct reporting and management responsibility. The dotted lines show the potential for communication among the team members. Jake is managing the project, and each programmer/analyst reports directly to him. However, it's not necessary for one of the team members to go through the project manager to communicate with another of the team members.

Jake thinks communication is a critical component of a software development project. He also thinks it is important for all project participants to communicate freely with each other. However, using strictly oral communication sometimes pulls people away from their tasks and interferes with their work. To avoid this problem, Jake has installed an electronic mail system on the network used for JOCS software development.

Jake establishes a general account for the JOCS project participants to provide a method of disseminating information about the development effort. Each participant can easily send messages to another team member or to

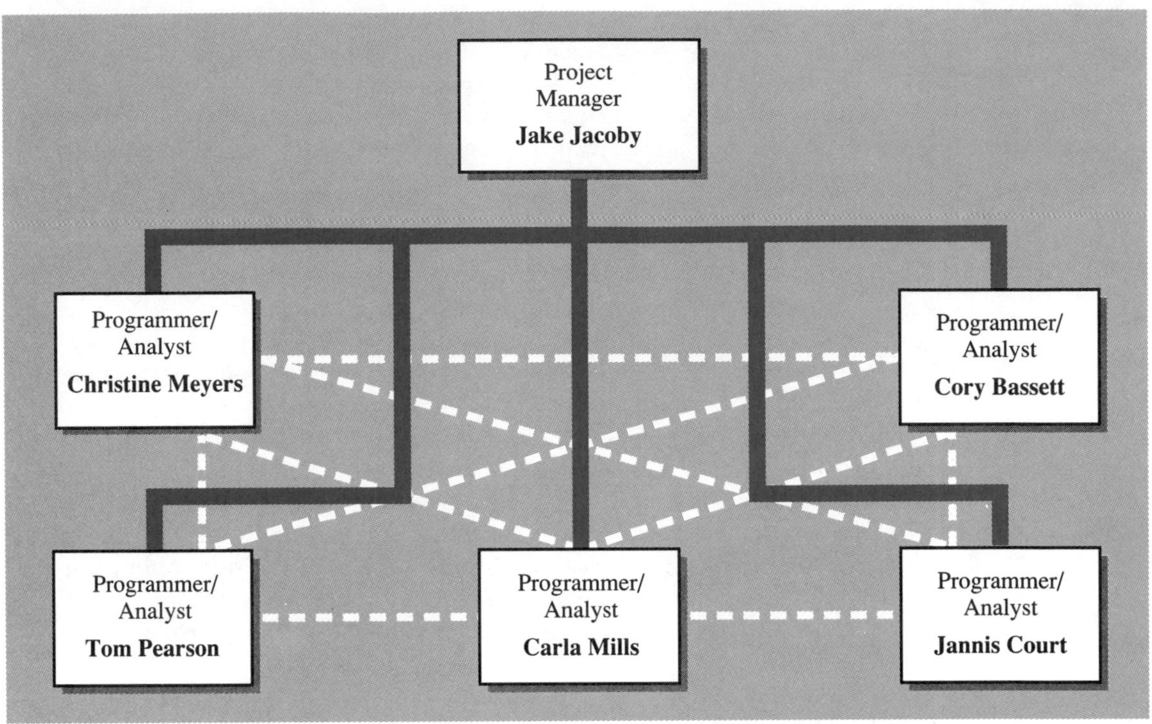

Figure 15.10
JOCS SWAT team organizational structure.

all members. Jake also uses the calendaring function of the electronic mail system to post required milestones by month and day. Using this system, each participant is always aware of current deadlines; each has an understanding of the entire system rather than just one specific program in the system. In addition to electronic communication, Jake sets up a biweekly meeting, with a predetermined length, to encourage group interaction.

Jake wants to provide a managerial environment that encourages each participant to communicate his or her ideas and needs, while at the same time it provides structure and control for the project. To achieve this goal, he works with his team to complete the next step in managing this project: Estimating the amount of time it will take to complete the JOCS project. These estimates are broken down and depicted graphically in the Gantt chart shown in Figure 15.11. Included in the estimation is the time required to train project participants on the technology of the database management system. Some of the tasks, such as developing a project plan, were finished while the chart was being prepared. They are designated as completed by a "C" on the Gantt chart.

The Gantt chart is used to mark the overall progress of the JOCS project. To keep track of individual tasks, Jake uses a PERT network. Figure 15.12 is an example of a PERT network for the tasks involved in structured software design. Each task is listed on the Gantt chart shown in Figure 15.11 and then is depicted in critical path format in Figure 15.12. Reading Figures 15.11 and

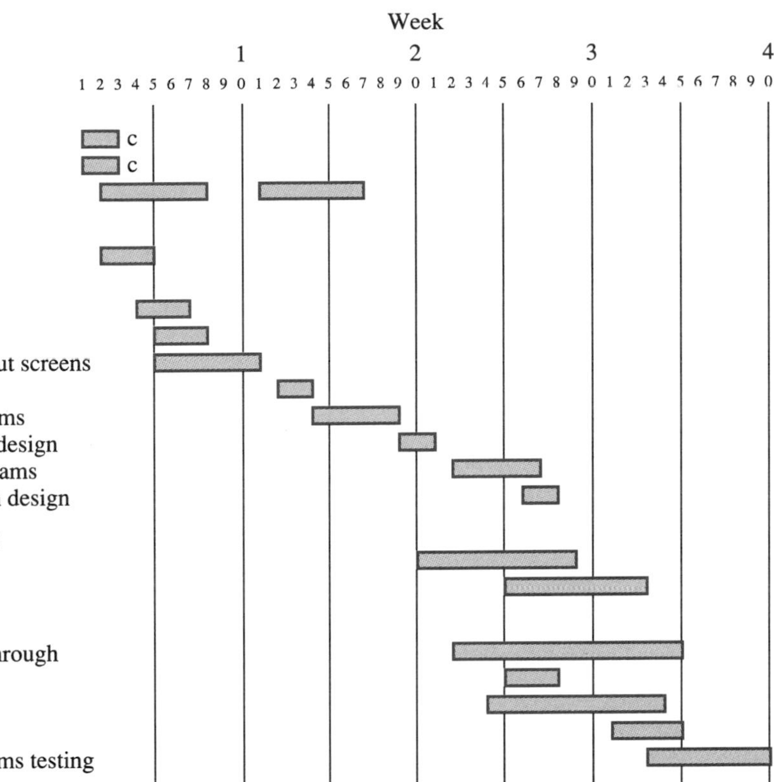

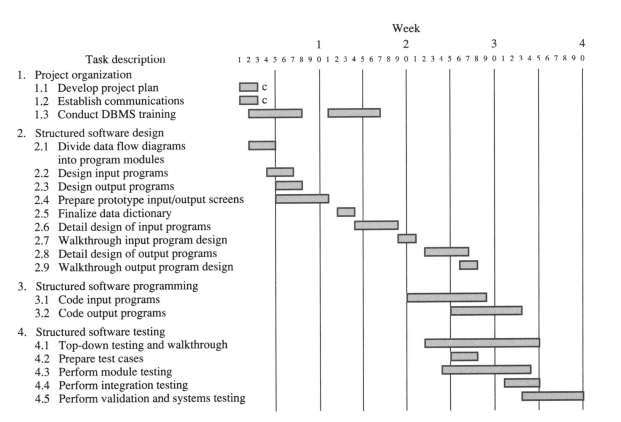

Figure 15.11
Gantt chart for JOCS development.

15.12, you should note that Tasks 2.2 (Design input programs) and 2.3 (Design output programs) are shown on the PERT network as tasks off the critical path that can be done in parallel. The goal of these tasks is Task 2.4 (Prepare prototype input/output screens).

Note that Task 2.2 is scheduled to be completed at the earliest time on Day 25 of this set of tasks, and at the latest time, must be completed by Day 35. This is necessary to complete Task 2.4 by, at the earliest, Day 50, or at the latest, Day 65. The SWAT team uses PERT networks both to organize and to control project progress.

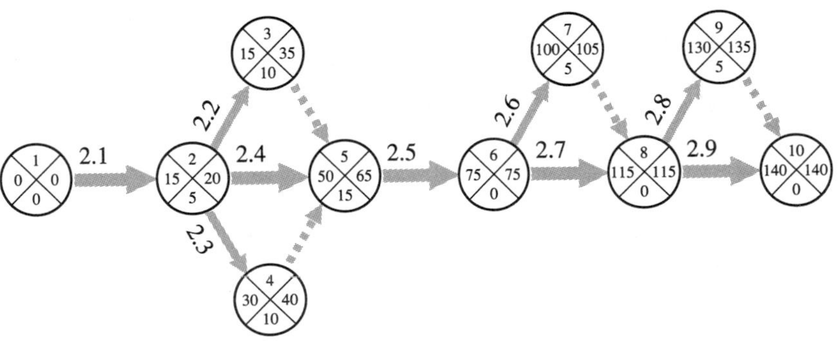

Figure 15.12
PERT network for JOCS structured software design.

As an ongoing managerial task, Jake must control the productivity of his team and measure the quality of their output. To accomplish this task, each team member knows that he or she is accountable for time and performance; and that management is also accountable for estimates and systems quality. SWAT team members feel comfortable with the estimates, because deadlines have been established by people who understand the complexity of the project. They are the same people who designed the requirements of the system.

The SWAT team has chosen to use function point analysis to evaluate the productivity of individual project members. This requires that each member keep track of the time spent on each facet of the software development effort. Figure 15.13 is a copy of the entry-screen time sheet that will be filled out each day by each project participant. Peerless has an automated time tracking system for its programming staff.

Jake now has his management approach in place. He has compiled a strong team of people to develop the software; he has created an initial plan of the time required to complete the project; he has organized and publicized the plan through the use of Gantt charts and PERT networks; he has decided on a method of exercising productivity and quality control; and finally, he has established an environment through meetings, electronic mail, and training to motivate and lead the team. All that's left is to cross his fingers, hope for the best, and develop the JOCS application.

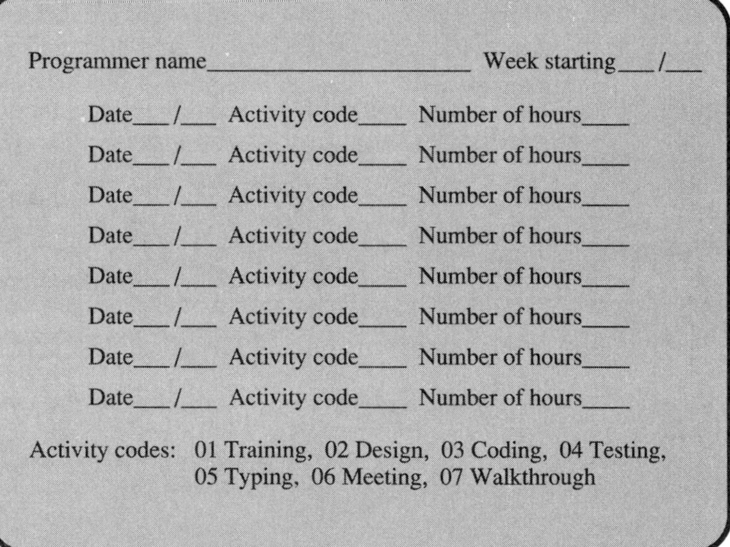
Programmer name______________________ Week starting___/___

Date___/___ Activity code____ Number of hours____

Date___/___ Activity code____ Number of hours____

Date___/___ Activity code____ Number of hours____

Date___/___ Activity code____ Number of hours____

Date___/___ Activity code____ Number of hours____

Date___/___ Activity code____ Number of hours____

Date___/___ Activity code____ Number of hours____

Date___/___ Activity code____ Number of hours____

Activity codes: 01 Training, 02 Design, 03 Coding, 04 Testing, 05 Typing, 06 Meeting, 07 Walkthrough

Figure 15.13
Time tracking system data-entry screen.

Chapter 16
Designing the Software

WHAT WILL YOU LEARN IN THIS CHAPTER?

After studying this chapter, you should be able to:

1 Explain reasons for performing the software design phase.
2 Describe the structured software design approach.
3 Describe the object-oriented software design approach.
4 Relate how to conduct a software design walkthrough.

INTRODUCTION

This chapter presents software design (see Figure 16.1). Software design is performed to make sure that we achieve an accurate systems design-to-software design transition. To ensure that this transition has been achieved, a software design walkthrough is conducted.

The preceding chapters of this book have been directed mainly to the design of a system, including its technology platform. Outputs and inputs have been designed in detail. Processes have been designed, using DFDs, STDs and, possibly, other modeling tools. If necessary, procedural models and equations for processing have been developed. Data dictionaries and logical data models have been created. Controls that must be embedded in the software have been specified. And, if required, information technology has been evaluated and acquired that will support the systems design. From the systems design specifications, especially those represented by DFDs and STDs, software designers have a clear vision of what they must do to convert these specifications to a software design that abides by such specifications.

This chapter presents two widely used software design approaches:

- Structured software design
- Object-oriented software design

Neither is confined to any particular computer programming language. Both approaches promise quality control, ease of maintenance, usability, reusability, reliability, and extendability. Today, many executives are trying to decide on which approach to use, or whether to implement both approaches. The accompanying case describes such situations.

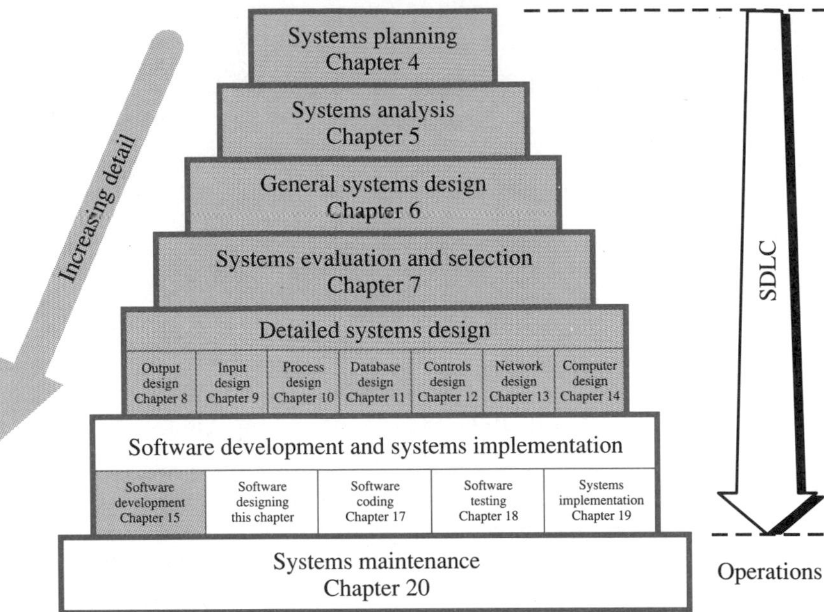

Figure 16.1
SDLC phase and their related chapters in this book. In Chapter 16 we deal with software design from two principal approaches: structured design and object-oriented design.

Deciding on the Software Design Approach at Tranquil Industries

Sylvia Porter is CIO at Tranquil Industries. She has called a meeting with Jim Thompson, systems project leader. Sylvia started the meeting by asking, "What's all this hoopla about object-oriented design and programming? Does it make sense to move our in-house development efforts from our conventional structured approach, which has served us well, to an object-oriented approach?"

"The answer to your first question is that it's not hoopla; the object-oriented approach is real," Jim responded.

"As far as your second question is concerned, I'm not quite ready to throw out all our nonobject-oriented tools," said Jim, "but I am ready to give the object-oriented approach a fair trial, maybe on the new executive information system."

"Refresh my memory about the essential difference between the structured approach and the object-oriented approach." stated Sylvia.

"Well, there are some significant technical differences, such as classes, encapsulation, inheritance, and so on. And, of course, the big promised payoff is reusability and extendability. But you have to have a large class library to reap the rewards of reusability, and . . ."

"Please bear with me, Jim," Sylvia interrupted, "I'm not familiar with some of those terms. What I'm really wanting to know is what is the conceptual difference between the two approaches."

"The conceptual difference is the way systems people—that is, systems designers and programmers—approach problems. Our traditional structured approach, as you well know, is more or less divided into two camps: the process-oriented people (also called procedure-oriented and algorithmic-oriented) and the data-oriented folks. The first camp thinks of processes to achieve a task and then builds data structures for those processes to use. The second camp thinks of how to structure (or model) the data and build processes around that structure. So we have process-centrics and data-centrics.

With the object-oriented approach, data structures and procedures (also called operations) are packaged together as objects. The objects possess repertoires of methods; that is, things they can do."

"That makes sense," said Sylvia. "I've read a few articles on the object-oriented approach, but I haven't really had time to dig into it to understand the details."

"You know with conventional procedural languages, such as COBOL or FORTRAN, we say, at least in generic terms, 'Write the value of this number on the printer.' The object-oriented programmers say, 'Write yourself.' In a conceptual way, object-oriented programmers are working with trained animals; that is, objects as trained animals. A user sends a message to an object to invoke its methods, such as 'print income statement.' "

"What about division of labor?" Sylvia asked. "Can we apply the concept of modularity to the object-oriented approach and have the programmers working on different objects simultaneously?"

"Oh, sure," Jim responded. "Modularity applies to the object-oriented approach as it does to the structured approach."

"Earlier, you mentioned inheritance. What does this term mean?"

"Inheritance provides a means for programmers to start with an existing object that is close to what is needed and use data attributes and operations in another application. Also, if needed, more code and data attributes can be added. Inheritance is a key object-oriented characteristic that promotes reusability. However, not all object-oriented programming languages feature inheritance."

"I like this idea of reusability," said Sylvia. "I've always thought that programmers spend too much time writing and rewriting the same kinds of programs because they're unable to reuse their previous work or take advantage of others' work."

"That's true," said Jim, "but object-oriented design does not guarantee reusability, nor does the structured approach prevent it. It's possible to design objects that are difficult to reuse. It's also possible to design software using the structured approach that possesses reusable modules."

"Well, from what you've said, I certainly don't believe the object-oriented approach is a fad. It seems to me that the systems world is really embracing it. What do you think we should be doing with it?"

"I think we should try it on our new EIS design," said Jim. "The object-oriented approach should work well here and provide us with a good test environment. If it works as well as I think it should, I also recommend that we try it on the LAN-based order-entry system that we've been planning."

"Sounds good to me," Sylvia said enthusiastically. "Let's do it. I'll give you all the support you need."

"It will require some research, because I'm not as knowledgeable about the object-oriented approach as I would like to be. Also, I want to select a workable modeling tool for the software design phase and also choose an object-oriented programming language. It's my understanding that Julie and Dana, our two top programmers, have already been experimenting with several object-oriented programming (OOP) languages. So we really have a head start."

"Great!" exclaimed Sylvia. "Keep me posted on your progress."

Jim immediately set out to learn more about the object-oriented approach and prepare for its application. He firmly believed that the new EIS would provide him with a nice, neat system that he could focus on and give the object-oriented approach a fair trial.

REASONS FOR PERFORMING THE SOFTWARE DESIGN PHASE

People inexperienced with information systems often regard software development as simply writing code in some language such as C, COBOL, FORTRAN, RPG, Pascal, or C++. The more time and effort devoted to software design, however, the better the software quality, the fewer problems will occur in later phases, and the fewer resources will be required for the new system's total life cycle.

Some people may declare that too much time is spent on software design. They ask: "Why don't we just start coding and worry about correcting errors later?" The objective, however, is not just getting a software program to work, but getting it to work right. Studies indicate that the cost of correcting an error is much higher after the program has been implemented than during the design and coding phases. Figure 16.2 demonstrates how the cost of correcting an error or design flaw increases at a geometric rate over the course of software development. Estimates from some authorities, including the Department of Defense, indicate that it costs between $30 and $50 to write one line of executable code (LOEC) and about $4000 per LOEC to maintain it. Information systems professionals must realize that companies cannot continue to operate competitively under such burdensome programming and maintenance costs. Furthermore, costs of losing user goodwill are almost impossible to measure,

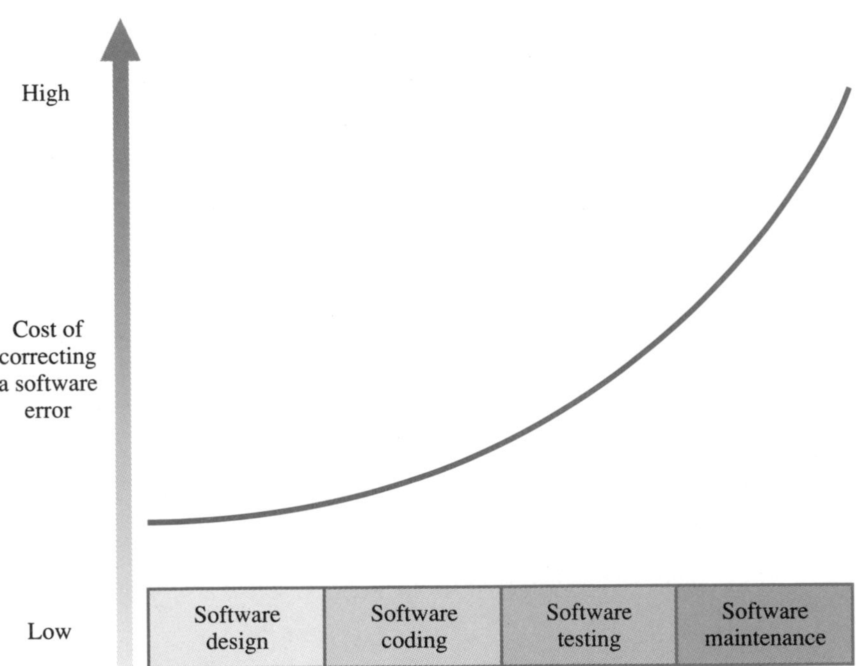

Figure 16.2 Relative cost of correcting a software error during software development and maintenance.

but they are sure to occur in various forms should most of the errors need to be corrected during operations of the new system.

The reason that the cost curve in Figure 16.2 behaves as it does is because of repeating previous phases that were done earlier. Obviously, discovering errors during maintenance usually causes a longer and more expensive correction process. If, for example, a user requirement error is discovered during maintenance, the entire SWDLC process may have to be repeated; that is, redesigning, recoding, and retesting to repair the system.

The Saga of Code King

Code King has been coding for years. On Code King's desk sit stacks of return messages from users. ("Why bother to call them back?" he growls. "They never know what they want. Anyway, I know their business better than they do!" Code King has a bumper sticker on his car that says, "We love to code and it shows." He thinks that analysis is paralysis, and that design is a resting place before code. Users should be shot, and documentation should be outlawed.

What we have done over the years is create a perfect, symbiotic relationship for the Code King. We reward firefighting and the ability to make quick fix after quick fix. We have become desensitized to the fact that systems are not documented, and that our code is unstructured. It's okay that testing consists of unit tests, when there's time. If there isn't time, "load and go"; besides, we can fix it later.

What happens, then, when a CASE tool is placed on Code King's desk? Can we expect a radical transformation from the love of coding to the love of effective systems? Recently, I decided to travel to Techtown, where Code King lives, to find answers to these questions.

When I arrived, Code King had his feet up on the keyboard of his dumb terminal. I asked him, "So how do you feel about doing design using the CASE tools that management has recently purchased?"

Code King quickly pulled his feet off the desk and clutched his mug, knuckles white. Fortunately, the silence in the room was broken by the ring of the phone. I figured it was an annoyed user calling about a quick fix that wasn't quickly fixed. But instead, it was his buddy, the systems programmer, ribbing Code King that his life was over.

"Pretty soon, you'll be a real arrr-TEEST dude, drawing all those pretty pictures" I heard over the earpiece. King winced. His buddy said that no CASE tools would ever be in his operations (God forbid). I had always thought that there was a special bond between programmers, but this guy was cruel. Was nothing sacred anymore?

Code King hung up, and we got back to our conversation. He said that "it" would arrive next Monday along with the PC XT (monochrome) to run it on. The dumb terminals would be removed that weekend when no one was around.

"What about training?" I asked. I know that the industry standard for CASE tool training was a minimum of 20 days. Code King abruptly said that there would be no training at this time. Management voted to postpone training until everyone had time to adjust to "it."

The drop-it-in approach to CASE is a popular paradigm today; it is also highly recommended by that friendly CASE tool salesperson on commission.

I asked Code King how was he going to use this analysis and design tool in his everyday work. He responded, "Your guess is as good as mine. But what really excites me is this announcement in *PC Daily* about the PC with a 586 chip. Can you imagine, C-quad-plus will compile a 3-Meg program in about 1.05 nanoseconds?" "Great," I thought, "wait till Grace Hopper feels the breeze."

King continued, "This will be great. Oh, but I bet you want to talk about doing those circle diagrams and box charts. Well, let's see." He pulled a purple book from his bookcase. (Tom DeMarco would probably be offended by the dust; I know I was. To me, the "purple" book will always be the analysis bible.) A cockroach fell from the binding; I realized how long it has been since Code King had opened a book on structured design.[1]

USING STRUCTURED SOFTWARE DESIGN

By the time the systems project is ready for **STRUCTURED SOFTWARE DESIGN**, a data dictionary and logical data model have been developed. Entity relationship diagrams (ERDs) describe relationships among data stores of the DFD. If required, STDs model time-dependent behavior. These STDs describe timing of execution and data access triggered by events.

Various modeling tools, such as DFDs, are used to describe systems functions. In the structured software design phase, the lowest-level DFD is converted to a structure chart with descriptions of programming language code. To get to this point, the DFDs have been recursively divided into subdiagrams until many small processes are left that are relatively easy to implement. The functional specifications that will be converted into programming language code may be expressed with decision tables, decision trees, structured English (or pseudocode), equations, or other techniques (e.g., Jackson's structured narrative).

Design Characteristics

Structured programs have these design characteristics (see Figure 16.3):

- Modules are arranged hierarchically. Structure charts, Jackson diagrams, and Warnier–Orr diagrams arrange modules hierarchically.
- CALL-based or PERFORM-based logic is used.
- Top-to-bottom design and control flow, and top-to-bottom or bottom-to-top coding are used.
- Tight repetitions or loops that span only one module are designed.
- Standard control constructs of sequence, selection, and repetition are employed.

---

[1] Adapted from Donna Wickes, "The King Lives!" *Software Magazine*, May 1991, pp. 127–128.

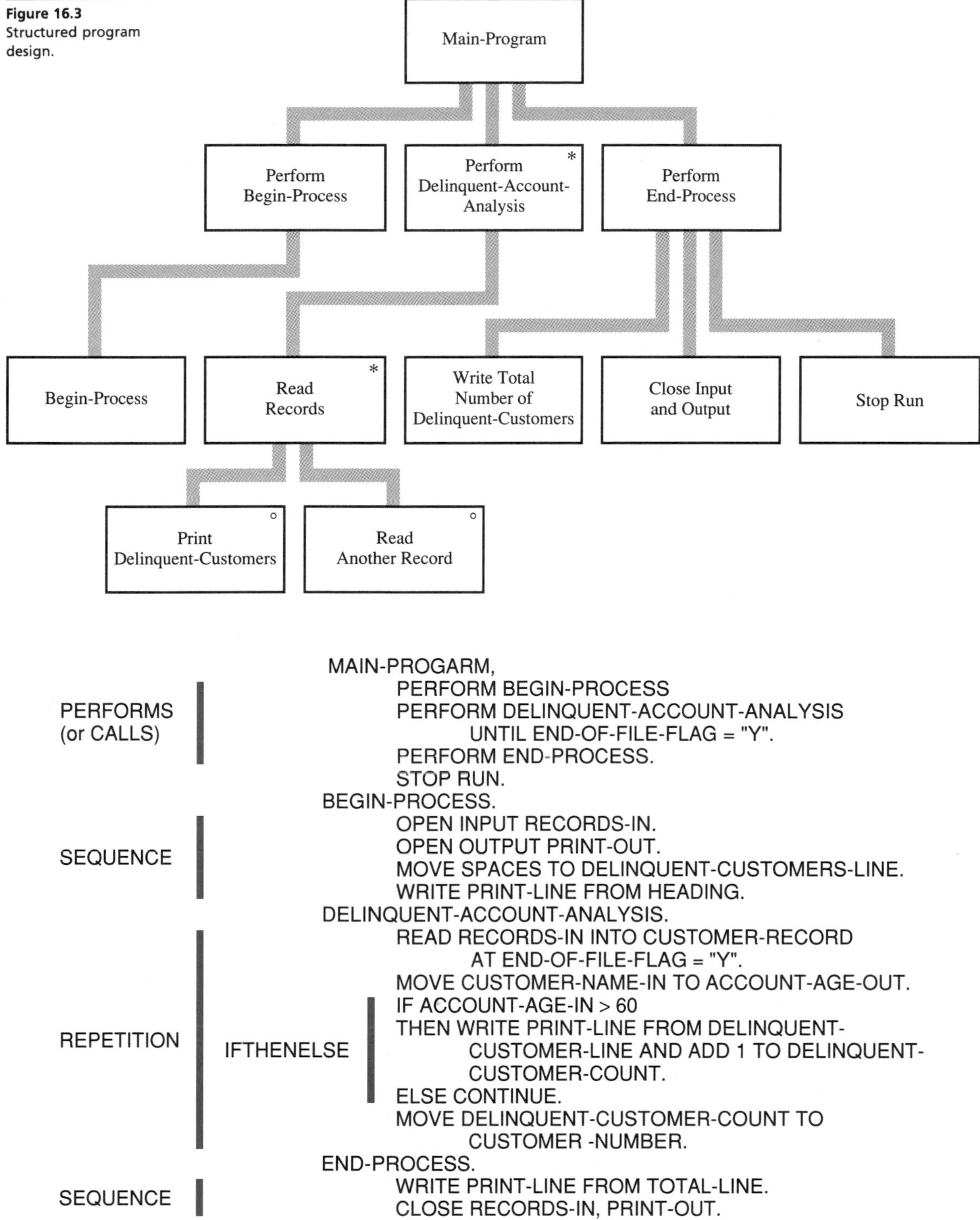

Figure 16.3
Structured program design.

A Jackson diagram is used (a structure chart could just as easily be used) for depicting the hierarchical design of modules that will be coded in the procedure division of a COBOL program. The identification, environment, and data divisions are not shown. The executive module is MAIN-PROGRAM, which calls the other modules of the procedure division. BEGIN-PROCESS prints headings. DELINQUENT-ACCOUNT-ANALYSIS reads in customer data, determines if the amount owed is over 60 days past due, and prints out the names of customers who are delinquent on their accounts. The repetition (or loop) is controlled by an end-of-file (EOF) flag. END-PROCESS prints the number of customers who are delinquent. STOP RUN in MAIN-PROGRAM ends the program.

Dividing the Design into Modules

MODULARITY is the partitioning of a software design into individual components, called modules or objects, to reduce its complexity. The goals of decomposing a software design into modules are to:

- Design, code, and test modules independently
- Revise and maintain modules easily after they are converted to operations

The technical objectives of modularity are to:

- Pass as few data as possible between modules
- Minimize the use of control data
- Effect module independence
- Focus modules on single functions

The attainment of these objectives depends on the level of coupling between modules and the degree of cohesion within modules.[2]

Coupling

COUPLING measures the degree of independence and interaction between modules. If a module is independent, it is not subject to modification by other modules. Interaction means the exchange and modification of variables between modules. If little interaction occurs between two modules, the modules are highly independent and considered loosely coupled. When a great deal of interaction occurs between modules, the modules are regarded as interdependent and tightly coupled. If modules are tightly coupled, it is almost impossible to maintain one module without making changes in other modules. On the other hand, some connection must exist between modules in a program or else they would not be part of the same program. A high-quality design means, however,

[2] Meilir Page-Jones, *The Practical Guide to Structured Systems Design,* 2nd ed. (Englewood Cliffs, N.J.: Yourdon Press, A Prentice-Hall Company, 1988) and A. Ziya Aktas, *Structured Analysis and Design of Information Systems* (Englewood Cliffs, N.J.: A Reston Book, Prentice-Hall, 1987).

that the modules are loosely coupled and are, therefore, easily maintained and reusable for another application.

Figure 16.4 demonstrates the coupling characteristics between Module A and Module B. An arrow from Module A to Module B indicates that Module A contains one or more CALLs to Module B. The term "CALL" (PERFORM paragraph-name in COBOL or the module name in a FORTRAN subroutine expression) refers to any mechanism used to invoke a module. The CALL or PERFORM statement is the simplest, most flexible and visible of the linkage mechanisms in most high-level languages. The side arrows annotate the parameters that are passed between modules. These arrows also show the direction in which data are passed. Module A passes Data item Y to Module B. Module B passes Data item Z back to Module A. Good coupling would be a pair of modules that communicate on two pieces of data from one to another, as shown in Figure 16.4. Weaker coupling would be another pair of modules that communicate many different pieces of data back and forth.

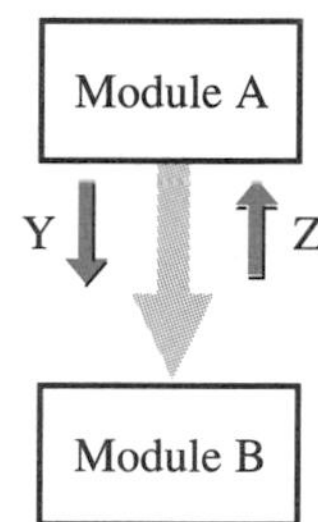

Figure 16.4
Module A calls Module B.

A more precise definition of coupling is illustrated in Figure 16.5. Because good coupling is such a strong feature of structured design, we will look at it more closely in terms of data, stamp, control, common, and content coupling. Data, stamp, and control coupling represent normal coupling. Common and content coupling represent abnormal coupling; they are to be avoided. A practical test of the quality of coupling is how easily each module can be coded by different programmers while maintaining a high level of independence from each other. Thus loose coupling facilitates division of labor between programmers.

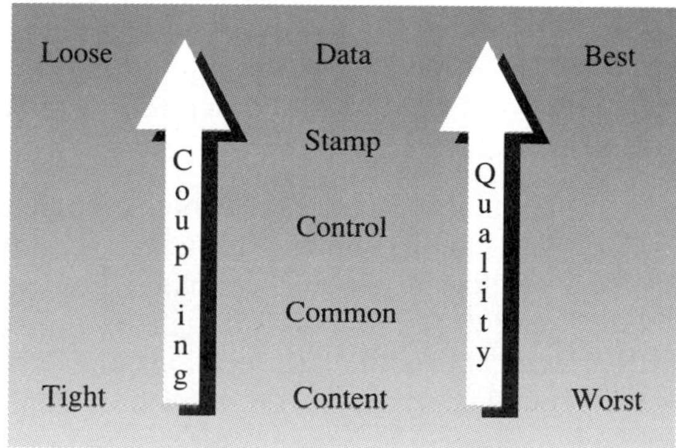

Figure 16.5
Types and levels of coupling.

Some authorities argue for an even higher level of coupling, in which Module A is coupled to Module B only in that A calls B's name, and no data are passed between them. That is, they are coupled in name only. The types of coupling we focus on in this chapter, however, are those listed in Figure 16.5.

Data Coupling Any two modules have data coupling if necessary data are communicated between them. Module A calls Module B, B returns to A, and data passed between them are by means of parameters presented with the CALL (or PERFORM) itself. For example, Module B is an algorithm that recomputes check digits of account numbers to determine their validity. A valid or invalid message is passed to Module A which reads customer transactions.

Stamp Coupling Two modules are said to be stamp coupled if they communicate a group of related data items such as a record consisting of various fields. Any change in either format or structure of a record or field in a record will have an effect on the modules that use that record. Therefore, stamp coupling tends to expose a module to more data items than it needs. The rule here is: Never pass records containing many fields to modules that need only one or two of those fields. For example, Module B reorders inventory based on data from Module A that controls access to inventory records. Rather than passing inventory description, price, cost, warehouse location, and other superfluous data, Module A passes only inventory number, quantity on hand, and vendor code, the only data items that Module B needs to perform its function.

Control Coupling Two modules are control coupled if one of the modules communicates data that control the internal logic of the other module. For example, Module B communicates to Module A an end-of-file (EOF) message. Module A then terminates its function and transfers control to another module.

Common Coupling Two or more modules are common coupled if they share data that are stored in a common area. An example of common coupling occurs when a programmer uses the REDEFINES clause of COBOL. The REDEFINES clause permits the storage of many different data items in the same physical storage location. It is one of the most complex clauses in COBOL. It can cause user report errors and maintenance problems. The reason for its use early on was to save physical memory storage, but that is no longer necessary, considering today's megabyte storage capacities. Thus the REDEFINES clause should not be used.

Content Coupling Two or more modules are content coupled if one refers to or alters the inside of another in any manner. This is the worst type of coupling. It is sometimes called pathological or sick coupling. For example, it alters a statement in another module, or it falls through into another module. The villainous ALTER statement used in early versions of COBOL is an excellent example of content coupling. It has been abolished from later versions of COBOL.

Cohesion

The second and complementary measure of module independence is **COHESION**, sometimes called binding. Cohesion measures the strength of the relationship between elements of code within a module. It is a way of describing the degree to which the module carries out a single, well-defined function. The term was borrowed from sociology, where it means the relatedness of humans within a group.

Figure 16.6 depicts the types and levels of cohesion. A system with highly cohesive modules will normally have low or loose coupling, both of which are design objectives of structured software design. If designers develop modules that go beyond communicational cohesion, they have crossed the boundary from well-designed, easily maintainable modules to poorly designed, less easily maintainable modules.

Figure 16.6 Types and levels of cohesion.

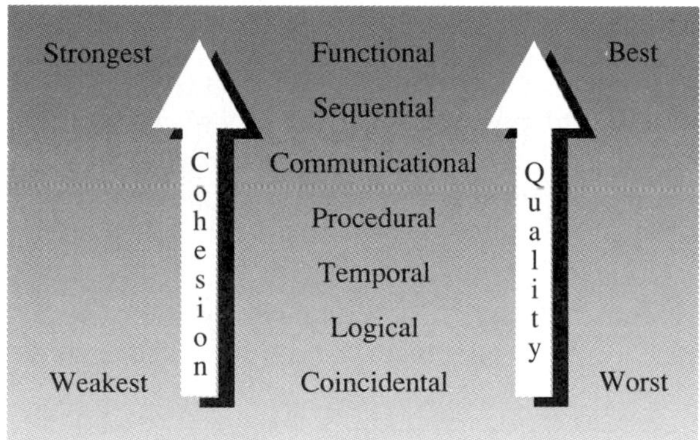

Functional Cohesion A functionally cohesive module contains procedures that all contribute to the execution of one and only one well-defined task. The objective of software design is for all modules to be functionally cohesive. Modules that read records, compute net pay, calculate a reorder point, or print a report are examples of modules that carry out one task to completion and are therefore functionally cohesive.

Sequential Cohesion A module has sequential cohesion if its procedures consist of a sequence of activities such that output of one activity is an input to the next activity of that module. Examples are procedures that read and edit data, create and store a record in a file, and accumulate data and print them.

Communicational Cohesion A module is said to have communicational cohesion if its procedures contribute to activities that use the same data for different purposes. For example, procedures may produce multiple output from the same stream of data. A transaction is input, and it is used to update a master file and to print a transaction report, or procedures can update or delete a record.

Procedural Cohesion A procedurally cohesive module is one whose procedures consist of different and possibly unrelated activities in which control flows from each activity to the next. A module, for example, may read an employee record, compute net pay, display absenteeism rate, print evaluation reports, allocate vacation time, and print paychecks. These procedures are related by order of execution rather than by any single task-specific function.

Temporal Cohesion This level of cohesion is time-related. Here, the strongest relationship between procedures is that they are all executed at the same time. A module that edits all the fields in a record in one pass is temporally cohesive. An initialization module that rewinds tape, sets a counter, opens files, moves spaces to output, and sets switch operations is another example of a temporally cohesive module.

Logical Cohesion A module has logical cohesion if its elements are not related by flow of data or by flow of control, but related only to tasks of the same general class of functions. Often, code is shared between functions in the same module. A logically cohesive module may edit records, make updates, and delete or create additions. Each of these functions should be divided into several, more function-specific modules.

Coincidental Cohesion In a coincidentally cohesive module, no meaningful relationship between the procedures of the module exists; they just happen to be together. Coincidental cohesion is the exact opposite of what we are trying to achieve in structured design; therefore, it should be avoided.

Guidelines

Underlying modular design is a set of highly interrelated guidelines to help development of a well-designed set of modules. These guidelines are presented next.

Factoring and Module Size Factoring, also called leveling, is decomposing or dividing one module into several modules. The objectives in factoring are to reduce module size and to improve coupling and cohesion properties of modules. If a question exists as to whether to divide a given module or not, divide it. It is better to develop too many modules than too few. But modules with fewer than five or six lines of executable code should be examined to see if they should be merged into their callers.

Generally, a module should comprise no more than 50 LOEC. Normally, a well-defined, one-function module will not exceed 15 to 25 LOEC and 5 to 10 lines of comments or narrative, which are nonexecutable.

Decision Splitting A decision split occurs if the two parts of a decision are separated and placed into two different modules. Decision splitting should be avoided if possible: for example, the decision "If Reorder-Point is equal to or less than Reorder-Point-Number, then Perform Reorder-Routine." This decision should remain intact; that is, the If and Then parts of the decision should not be split.

Fan-in/Fan-out Fan-in of a module is the number of modules that call it. Fan-out, or span of control, is the number of direct subordinates to a module. Both fan-in and fan-out are illustrated in Figure 16.7. Generally, fan-in modules should be as high as possible for modules that serve a common function, such as a print routine or some kind of utility. These kinds of modules are highly

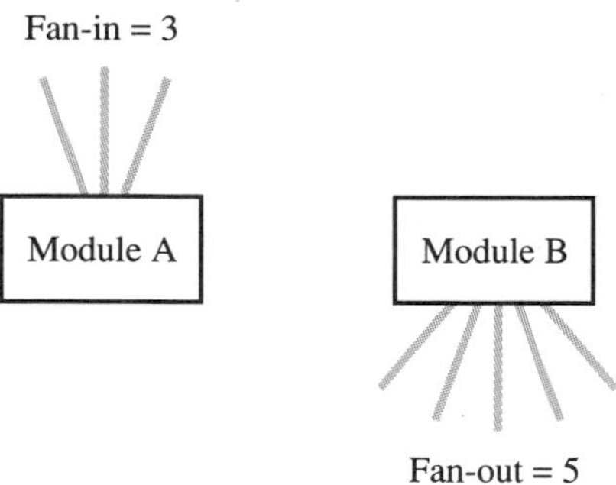

Figure 16.7
Fan-in and fan-out of modules.

reusable. Except for utility and reusable modules, if a module is called by more than three modules, it may contain two or more functions with different calling modules using different functions. If this situation exists, divide the invoked module into separate modules, each containing a specific function. Generally, a superior module should not call more than five subordinate modules.

Number of Modules There is an optimum number of modules, as demonstrated in Figure 16.8. To understand structured software design, it is useful to examine why modularity tends to reduce cost of software development and maintenancc.

A guideline measurement is to divide the total estimated LOEC by 50 (or less). For example, if a software development project is estimated to require 200 KLOEC, then about 4000 (200,000 LOEC/50 LOEC per module) modules will be close to the optimum number of modules for the project. Other authorities may increase or decrease the divisor by five or more LOEC, but 50 is a fairly common number. It is a guideline; it is not precise, but it gives a fairly good estimate.

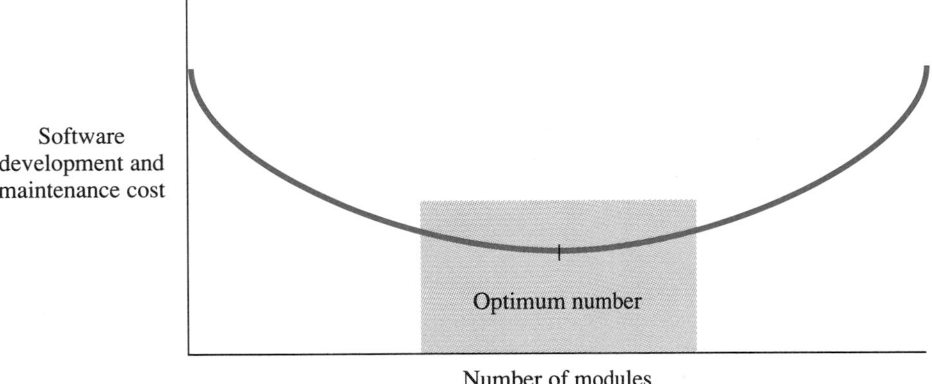

Figure 16.8
Optimum number of modules related to software development and maintenance cost.

Top-to-Bottom Design and Top-to-Bottom or Bottom-to-Top Coding

Although the structured approach for systems analysis and design is top-down, structured coding may follow either a top-down or bottom-up approach. The result of either approach is a hierarchically structured, modular program. In some situations, a mixture of top-down and bottom-up coding and testing is desirable. For example, it is beneficial to develop utility or common modules bottom-up, because they will be invoked by higher-level modules.

Using Standard Control Constructs to Build Structured Programs

The theory behind structured software design is that any program logic can be constructed from combinations of three control constructs:

- Sequence
- Selection
- Repetition

These constructs are illustrated with structured modeling tools in Figure 16.9. The key to these constructs is that each has a single entry and exit point.

Sequence

With the sequence control construct, program statements are executed one after another in the same order in which they appear in the source code. The commands for a sequence are always action verbs, such as MOVE, READ, WRITE, SUBTRACT, ADD, DISPLAY, COMPUTE, and the like. See the following examples:

```
MOVE ZEROS TO AMOUNT-OUT
COMPUTE GROSS-PAY = RATE TIMES HOURS-WORKED
SUBTRACT DEDUCTIONS FROM GROSS-PAY GIVING NET-PAY
```

Selection

A condition causes the system to make a transition and perform a specific action based on the value of the condition's variables. A set of statements is executed only if a stated condition applies, according to an IF-THEN-ELSE condition. For example:

```
IF
   HOURS WORKED GREATER THAN 40
THEN
   COMPUTE GROSS-PAY WITH OVERTIME RATE
ELSE
   COMPUTE GROSS-PAY WITH REGULAR PAY
ENDIF
```

Repetition

The repetition construct (also called looping or iteration) allows the program to execute one or more series of steps. There are two types of iteration: DO UNTIL and DO WHILE, or equivalent statements.

With DO UNTIL, a set of instructions is executed. Then a loop termination condition is tested. If the condition is true, the loop is terminated, and execution continues with the next sequential instruction. If the condition is false, the instruction set is executed again. Note that the DO UNTIL loop is always executed at least once. For example, if we wanted to print the names of 50 customers, the following commands would perform this task:

Figure 16.9 (opposite) Standard control constructs and modeling tools representing these constructs.

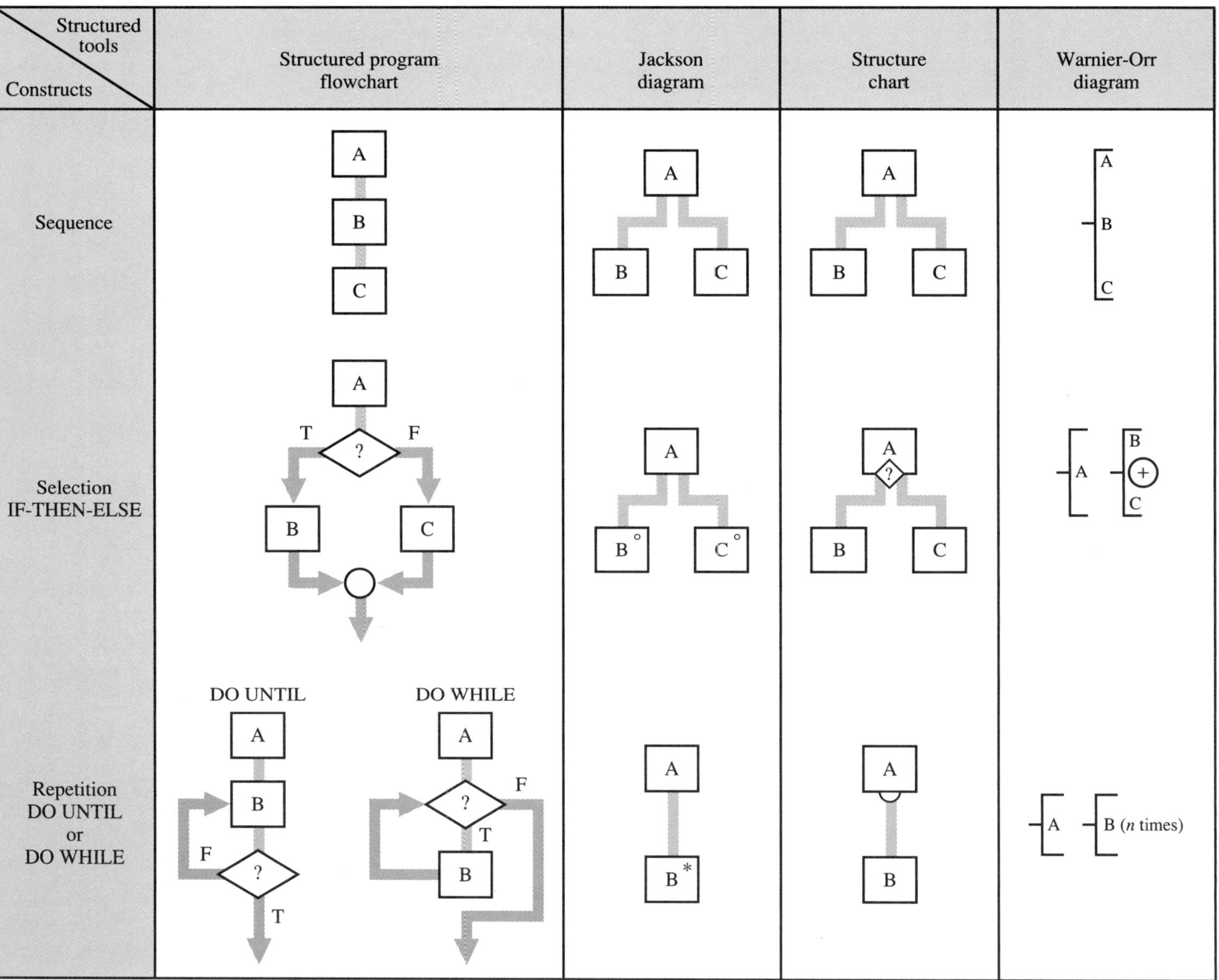
Structured tools
Constructs
Structured program flowchart
Jackson diagram
Structure chart
Warnier-Orr diagram
Sequence
Selection
IF-THEN-ELSE
Repetition
DO UNTIL
or
DO WHILE
DO UNTIL
DO WHILE
A
B
C
T
F
?
B°
C°
B*
B (n times)

```
PRINT HEADINGS
INITIALIZE COUNTER TO 1
DO UNTIL COUNTER EQUALS 51
  PRINT CUST-NAME
  INCREMENT ROW COUNTER BY 1
ENDDO
STOP
```

With DO WHILE, a termination condition is tested. If the condition is false, the loop is terminated and execution continues with the next sequential instruction. If the condition is true, a set of instructions is executed, and the condition is tested again. If the condition is initially false, the loop will not be executed at all. The following example shows a DO WHILE loop for reading a file of customer records and printing customer addresses:

```
PRINT HEADING
READ CUST-RECORDS
DO WHILE CUST-RECORDS REMAIN TO BE PROCESSED
   PRINT CUST-ADDRESS
   READ CUST-RECORDS
ENDDO
STOP
```

The DO WHILE loop requires that iterations continue as long as (while) a condition tests true. In this case, the condition being tested is whether or not the record just read is the last record in the file (i.e., end-of-file or EOF). When the condition tests false, the loop operation ends.

What Are Two Popular Modeling Tools Used for Designing Structured Software?

Two of the more popular modeling tools used in structured software design are:

- Structure charts
- Structured English

Structure charts and structured English are normally used to model the software design abstracted from systems design modeled by DFDs and STDs. In some quarters, Jackson and Warnier–Orr modeling tools are also popular. After reviewing structure charts and structured English, we will present an example of how they are used together to convert a detailed systems design into a structured software design, ready for the coding phase.

Structure Chart

The **STRUCTURE CHART** is a hierarchical diagram that defines the overall architecture of the software design by displaying the program modules and their interrelationships. Figure 16.10 shows a generalized structure chart. All structure charts follow this generalized form no matter how many modules they contain. A simple payroll example, shown in Figure 16.11 is used to demonstrate the application of the generalized structure chart.

Structured English

STRUCTURED ENGLISH is a subset of the English language. It serves as a specification tool that describes detailed input, process, and output of structure charts.

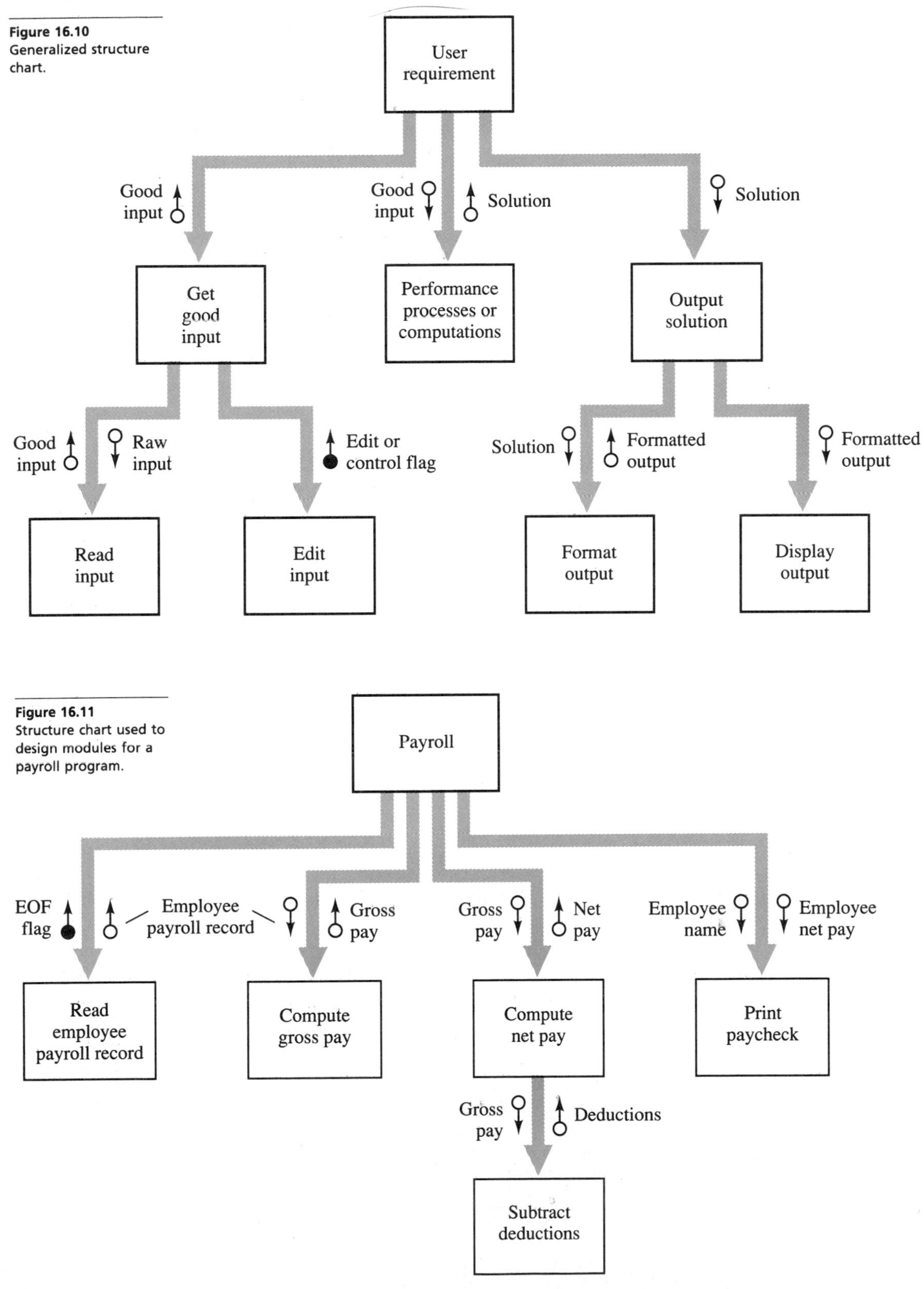

Figure 16.10
Generalized structure chart.

Figure 16.11
Structure chart used to design modules for a payroll program.

Structured English is a programming-like language (i.e., pseudocode) used to communicate to programmers and nonprogrammers alike. Its vocabulary is composed of verbs, terms from the data dictionary, and reserved words that denote logical processes. Depending on the nature and complexity of the application, decision tables, decision trees, and equations may be more appropriate than structured English for some applications, or they may be used to supplement structured English.

Rules for Writing Structured English Structured English follows the basic programming constructs of sequence, selection, and repetition. Statements are indented to denote logical hierarchy. Some structured English keywords are FOR, GET, READ, SET, ENTER, COPY, WRITE, IF, NOT, EQUAL, THEN, PERFORM, MOVE, ELSE, COMPUTE, ENDIF, REPEAT WHILE or DO WHILE, REPEAT UNTIL or DO UNTIL, ENDREPEAT or ENDDO, and EXIT. Keywords used for logic are AND, OR, GREATER THAN, and LESS THAN.

Structured English Example An example of how structured English is used to specify the processing of a customer order is shown in Figure16.12. Notice how this set of instructions not only communicates the process clearly but also can be easily converted to a program language such as C or COBOL.

Figure 16.12
Structured English used to specify order entry.

```
ORDER_ENTRY:
     FOR each customer PURCH_ORD
          GET CUSTOMER record
          IF CUS_NUM is valid
               SET INVOICE_HEADER record
               ENTER CUS_NAME, CUS_ADDR, DATE_ORD,
                    and PO_NUM in INVOICE_RECORD
               WRITE INVOICE_HEADER record
               GET DISCOUNT from table
          ELSE
               Display "Invalid Customer Number"
               QUIT ORDER_ENTRY
          ENDIF
          FOR each line item
               COPY ITEM_NUM and QTY_ORD on
                    INVOICE record
               GET ITEM_PRICE from price table
               SET ITEM_SUBTOTAL to ITEM_PRICE ×
                    QYT_ORD × (100-DISCOUNT)
               SET INVOICE_TOTAL to sum of ITEM_SUBTOTAL
               Write INVOICE_RECORD
          ENDFOR
          Prepare bill of lading
     ENDFOR
EXIT ORDER_ENTRY.
```

Transforming a Data Flow Diagram into a Structure Chart

The DFD illustrated in Figure 16.13 represents the systems design specifications for processing sales transactions. This DFD has been functionally decomposed to a detailed level, and thus it becomes the basis for structured software design using a structure chart.

Transforming a DFD such as the one displayed in Figure 16.13 into a structure chart involves **TRANSFORM ANALYSIS**, which entails an examination of the DFD to divide its processes into those that:

- Perform input and editing, such as validating a sales order
- Perform processing, such as checking inventory
- Generate output, such as preparing shipping documents

The resulting structure chart derived from dividing the DFD into input, processing, and output processes is shown in Figure 16.14. This structure chart represents the modules that will be coded.

Nearly all CASE systems facilitate transform analysis. They are especially useful in drawing structure charts and producing structured English. Some CASE systems are also able to derive structure charts directly from DFDs.

WHAT IS OBJECT-ORIENTED SOFTWARE DESIGN?

Chapter 6 introduced you to the object-oriented systems approach. In this chapter, we provide more detail, especially as it relates to preparing **OBJECT-ORIENTED SOFTWARE DESIGN**.

More About Objects

OBJECTS provide people or things with identity. Objects represent concrete entities from the application domain being designed. Some examples of objects are entities such as student, customer, systems analyst, inventory, account, and order entry. Objects are instances of one or more classes. Objects **ENCAPSULATE** data **ATTRIBUTES** (also called data structures or simply attributes) and **OPERATIONS** (also called procedures). Operations contain **METHODS** (i.e., program code) that operate on the attributes. Some authorities define objects as encapsulated data and code, or data and methods. The reason we use the terms "attributes" and "operations" is that these terms provide a way by which we can develop object-oriented software designs (i.e., the logical semantic framework) without thinking in terms of data formats and code syntax. In other words, the design model should be programming-language-independent. Once the software design is created, data formats will have to be described in detail, and program code (or methods) will have to be written to implement the design.

More About Classes

An object class describes a group of objects with similar attributes, common behavior, and common relationships to other objects. (Behavior is what an object does when it is sent a message, such as what a tiger does when given an

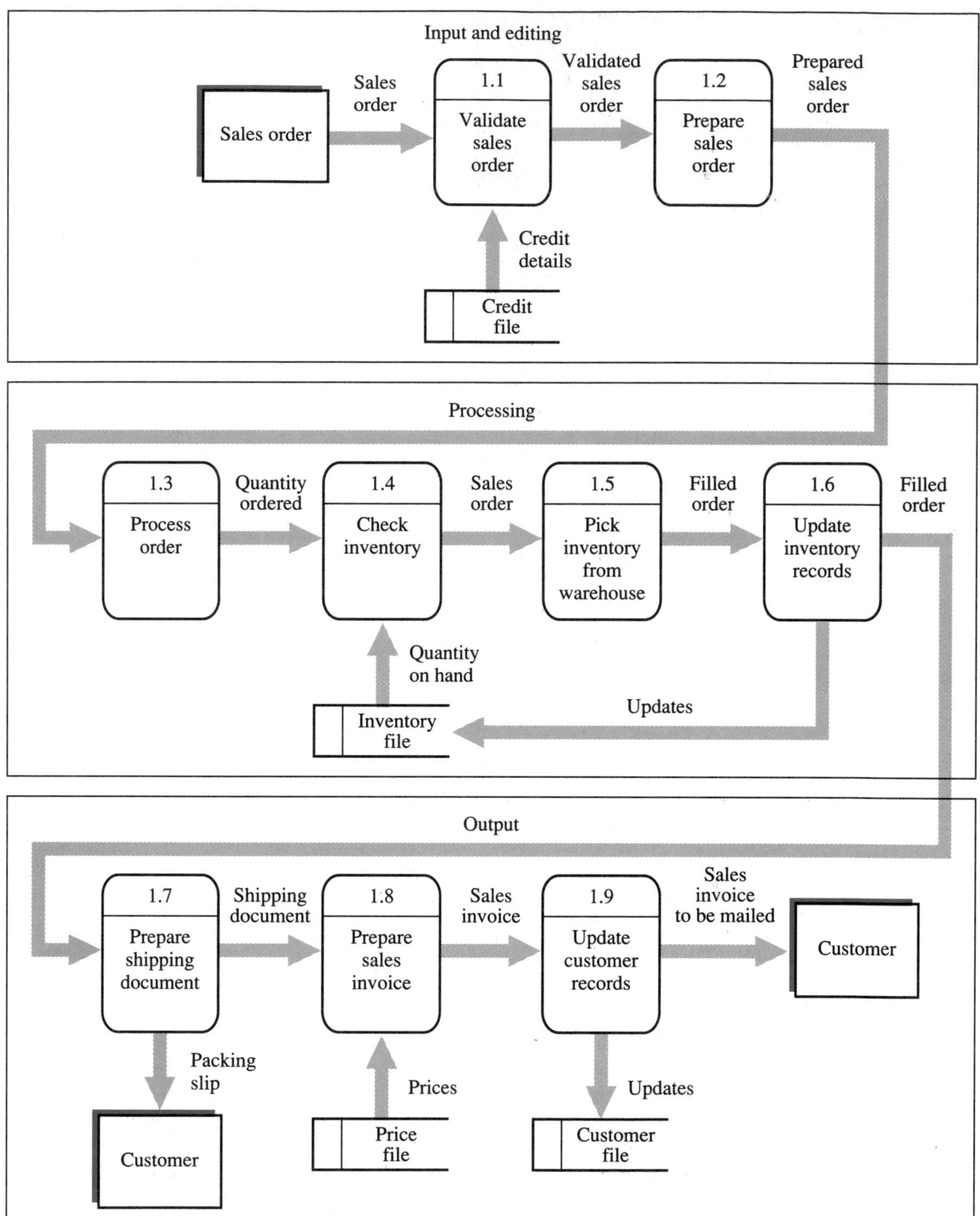

Figure 16.13 DFD design of a sales order processing system.

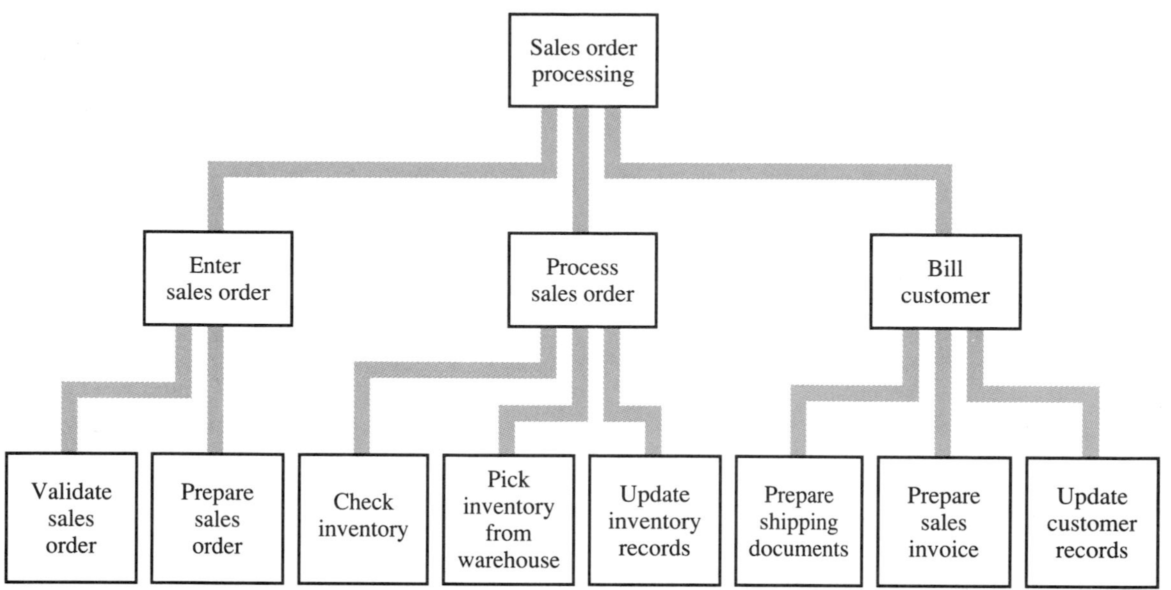

Figure 16.14
Structure chart for sales order processing derived from the data flow diagram.

instruction from a trainer.) The abbreviation **CLASS** is often used instead of object class.

A class may be declared to be a subclass of other classes known as superclasses. In that case, all objects belonging to the subclass also belong to the superclass. **INHERITANCE** is the capability to define subclasses of objects from an object class. Looking at it in terms of genetics, inheritance is the ability of a descendant to receive properties or characteristics from an ancestor. For example, in Figure 16.15, PROJECT MANAGER and SYSTEMS ANALYST are subclasses (descendants) of the superclass EMPLOYEE (ancestor) and they inherit all of its properties. Thus Employee-Number, Name, Age, and Salary, and code (not shown) that operates on these data attributes do not have to be redefined in the subclasses. But other attributes can be added to these objects, such as Project-Number.

A subclass may have many superclasses, thus multiple inheritance. Or, we could say that a descendant or child may have many ancestors. Just as a child inherits certain characteristics from its father and mother, an object can inherit characteristics from more than one ancestor. For example, a TEACHING ASSISTANT may be a subclass of both a STUDENT superclass and a TEACHER superclass, as portrayed in Figure 16.16.

Objects may gain and lose classes dynamically. An object representing a given person may be created as an instance of the STUDENT class. When that happens, all the attributes and operations defined in EMPLOYEE become applicable to the object, and the attributes and operations in STUDENT become inapplicable.

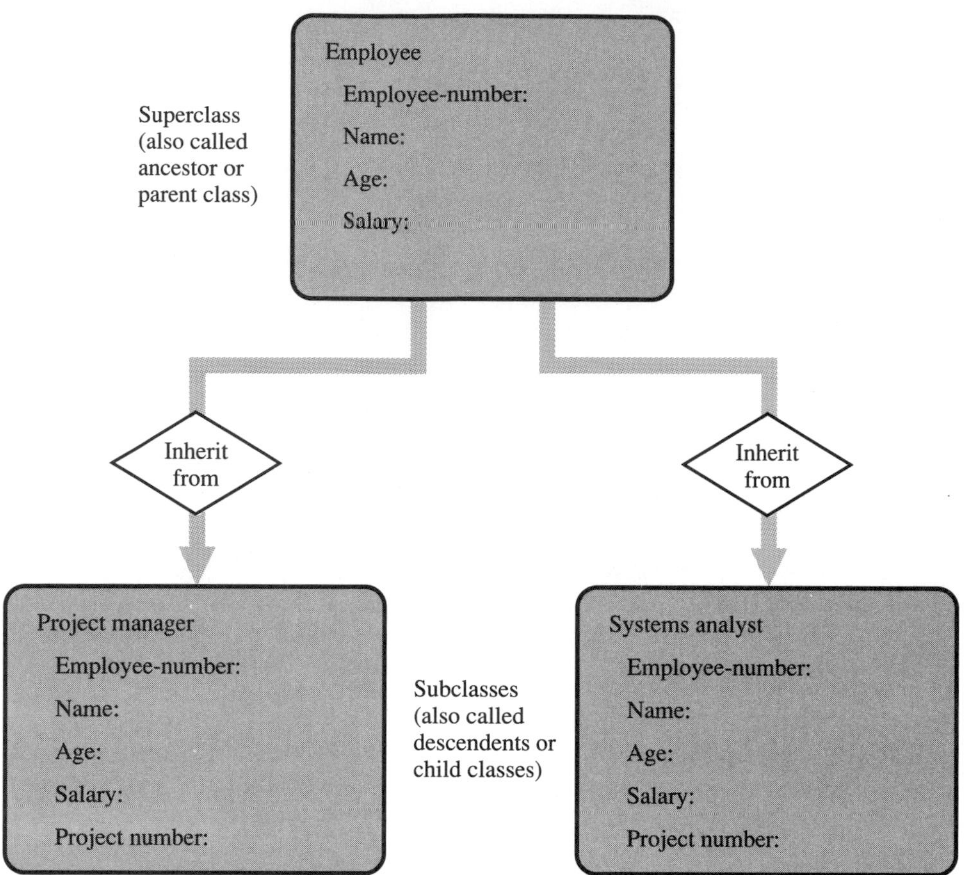

Figure 16.15 Subclasses inheriting properties from a superclass.

More About Relationships

Relationships describe the association between classes and objects. A class relationship indicates some sort of sharing or meaningful connection between classes. Class relationships enable inheritance. Relationships may be one-to-one, one-to-many, or many-to-many. In object-oriented design, this aspect of relationships is called **MULTIPLICITY**, which specifies how many instances of one class may relate to a single instance of an associated class. Generally, the software designer shouldn't worry about multiplicity early in software development. The first thing to do is to determine objects, classes, and relationships. Then, later, decide on multiplicity.

USING THE OBJECT-ORIENTED DESIGN APPROACH TO DESIGN SOFTWARE MODELS

Object-oriented software design results in a model that describes objects, classes, and their relationships to one another. The identification of classes and objects is the fundamental goal of object-oriented software designs. Systems

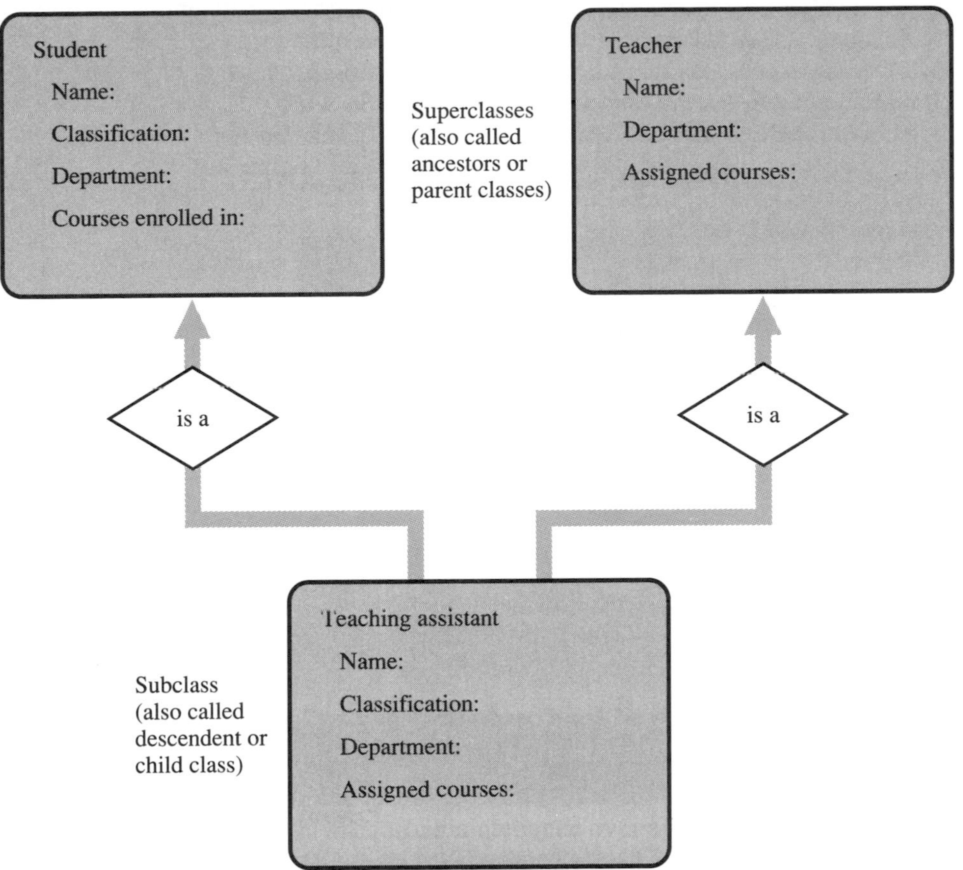

Figure 16.16
Subclass with a number of superclasses.

design models, such as DFDs, data dictionaries, and logical data models define the systems domain. Object-oriented software design identifies the classes and objects that will implement this domain. Generally, the best way to create object-oriented software design models is to build a hierarchy of classes in which subclasses inherit the attributes and operations from more generalized classes or superclasses.

Object-Oriented Modeling

We need a formalized way of expressing object-oriented designs that is logical and correct. Although several diagramming tools provide a way to do this, we offer the ERD, with some modifications, as a workable approach. (In fact, there is no standard ERD.) Although Chen's ERD version is commonly used, especially for data modeling, we use a variation of his version to model object-oriented designs. It provides a formal graphic notation for modeling objects, classes, and their relationships to one another. However, good software designs are produced by good designers, not from a modeling tool. An object-oriented modeling tool, such as a modified ERD, simply empowers the designer to concentrate on the creative aspects of design without worrying much about mechanics. Like structured software design, the object-oriented design is often abstracted from a DFD. In fact, each source or sink, process, or data store on a fully decomposed DFD may indicate a candidate object, or, possibly, several objects. Candidate classes may be derived from the DFD's data flows. The data dictionary and logical data model prepared during detailed systems design become the data attributes (or simply attributes) in an object, or, possibly, several objects. These models also indicate possible class and object relationships. A STD indicates the dynamic behavior associated with certain classes.

Describing an Object Using Modified ERD Notation

The modified entity symbol of the conventional ERD (i.e., the ERD used for enterprisewide modeling and data modeling) is a rounded rectangle, as depicted in Figure 16.17. This symbol is now called a class box.

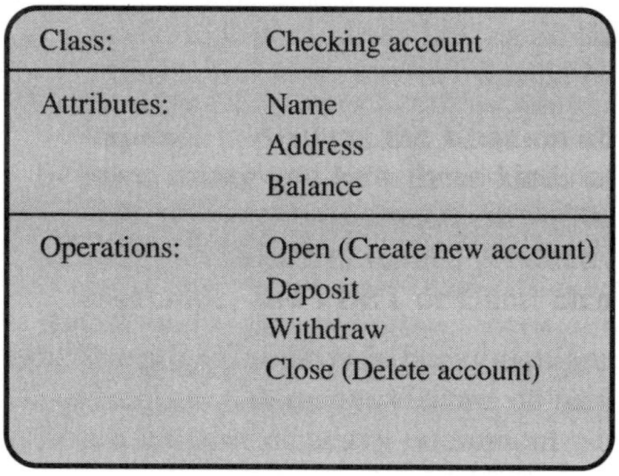

Figure 16.17
Class box.

Class:	Checking account
Attributes:	Name Address Balance
Operations:	Open (Create new account) Deposit Withdraw Close (Delete account)

What Is a Class Box?

The **CLASS BOX** has three regions. The regions, from top to bottom, contain class name, list of attributes, and list of operations. On the first object-oriented design model, attributes and operations may or may not be listed. It depends on the level of detail desired. Normally, a first design will include the class name. Then, as more definition is needed, attributes and operations are added to a second design and so on until the design process is completed. This approach to modeling is similar to functional decomposition of the structured design approach, where more and more detail is added as additional facts are gathered.

Listing Attributes

When attributes are added, they represent properties of individual objects, such as Name, Weight, Color, Amount, Model, or Style. For example, a CAR is an object; Color is an attribute of the CAR object. Name, Address, and Balance, in our banking example, are attributes of the CHECKING ACCOUNT object. Each attribute has a value for each object instance. For example, attribute Name may have a value of Mary Doakes, Address value of 24 Morningstar Lane, and Balance value of 1500.00. The class is CHECKING

ACCOUNT; the instance of an object (i.e., an individual object) is Mary Doakes. Different object instances may have the same or different values for a given attribute. For example, John Doakes is another object instance of the CHECKING ACCOUNT subclass, which has an Address value of 24 Morningstar Lane, the same Address value for the Mary Doakes object instance, and a Balance value of 500.00. Terry Parsons is another object instance of the CHECKING ACCOUNT subclass; it has an Address value of 214 Culver Street, and so on.

Listing Operations and Methods

Operations are listed in the lower third region of the class box. An operation is a function or transformation that may be applied to or by objects in a class. Open, Deposit, Withdraw, and Close are operations on subclass CHECKING ACCOUNT. All object instances in a class share the same operations.[3]

The same operation may apply to many different classes, which is the POLYMORPHIC characteristic of the object-oriented approach. This polymorphic characteristic means that the same operation takes on different forms in different classes.

A method (i.e., program code) is the implementation of an operation for a class. For example, the subclass CHECKING ACCOUNT has an operation, Deposit. Different methods are implemented to Deposit by teller or by electronic funds transfer. Both methods logically perform the same task, that is, depositing funds to checking accounts. Each method, however, is implemented by a different set of object-oriented programming code.

Modeling Relationships Among Objects and Classes

Relationships describe the associations between classes and objects. For example, Mary Doakes banks with First National Bank, as is shown in Figure 16.18. Relationships are inherently bidirectional. The name (or verb) of the relationship normally reads in a particular direction, but the relationship can be traversed in either direction. For example, ''banks with'' connects a CHECKING ACCOUNT subclass with the First National Bank. The inverse of ''banks with'' could be ''services'' or ''has'' that connects First National Bank with its CHECKING ACCOUNT subclass. Both directions are equally meaningful; both refer to the same underlying relationship. It is only the names of the relationships that establish a direction. Although relationships are modeled as bidirectional, they do not have to be implemented in both directions.[4]

A relationship between objects means that the objects can send messages to one another. Typically, messages are bidirectional. For example, a message may be GetMaryDoakesAccountBalance. It displays the amount of money presently in Mary Doakes' checking account.

---

[3] James Rumbaugh, Michael Blaha, William Premerlani, Frederick Eddy, and William Lorensen, *Object-Oriented Modeling and Design* (Englewood, Cliffs, N.J.: Prentice-Hall, 1991), p. 35.

[4] Ibid, p. 27.

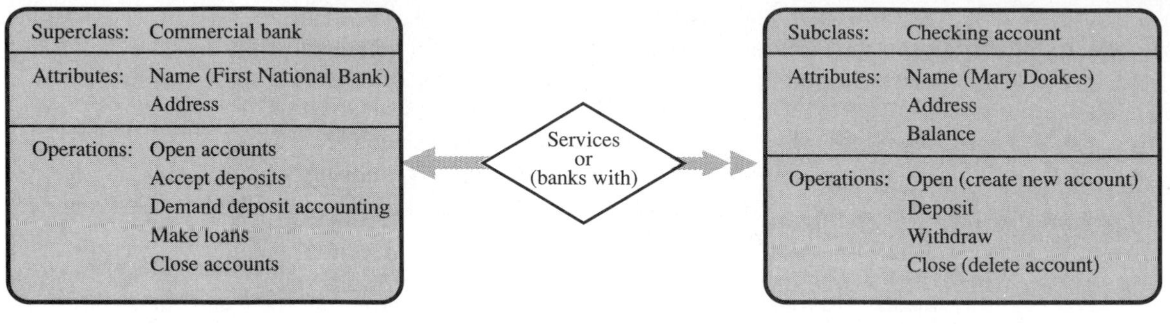

Figure 16.18
Object class relationship.

Modeling Inheritance

Any operation on any superclass can be applied to any instance of a subclass. Each subclass can not only inherit all the features of its superclass but can also add its own specific attributes and operations as well. This is a feature of inheritance called extension. For example, Figure 16.19 shows the superclass VEHICLE. The attributes and operations inherited from VEHICLE by subclasses CAR and TRUCK are shown parenthetically.

Besides design simplification, the chief advantage of inheritance is that it provides reusable code. For example, the code that implements the Add Model operation is written once and inherited wherever it is needed. Several OOP languages provide strong support for inheritance. In fact, inheritance is synonymous with code reuse as far as the OOP community is concerned.

A common task of object-oriented modeling is to group similar classes together and reuse common code. There are two kinds of code reuse: sharing newly written code within a project and reuse of previously written code on new projects. A completely modeled object-oriented design indicates where code reuse is possible to avoid duplicating programming effort. In some instances, code may also be available in class libraries from past object-oriented implementations, which can be reused as is or slightly modified to achieve the desired behavior. Some OOP languages offer extensive class libraries.

Incorporating MURRE Design Factors in Object-Oriented Software Designs

The MURRE design factors—maintainability, usability, reusability, reliability, and extendability (also sometimes called extensibility)—are just as applicable to object-oriented software design as they are to structured software design. Modularity, which supports the MURRE design factors, is applicable to object-oriented design. For example, highly cohesive classes are desirable. The least desirable form of cohesion is coincidental cohesion, in which entirely unrelated objects are thrown together in the same class. As with structured software design, the most desirable form of cohesion for object-oriented software design is functional cohesion, in which all the objects in a class work together to provide some well-bounded behavior. Not only will such function-specific sets of objects be easier to maintain, but their reliability can be tested more easily, their reusability will be enhanced, and extending them, if necessary, will be more efficient.

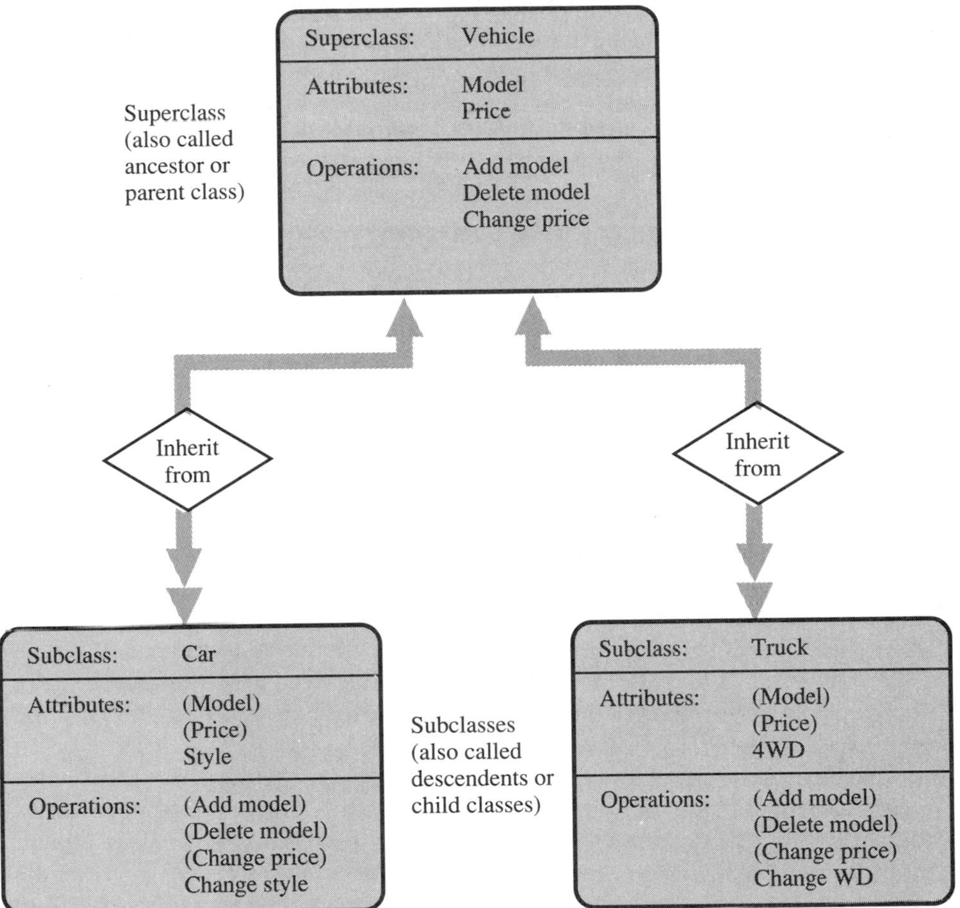

Figure 16.19
Model showing relationships and inheritance.

Modules for structured software design should be small and understandable. So, too, should sets of methods. Methods should use standard and meaningful variable names, avoid abbreviations, and use standard conditions and data types. A method should be understandable to someone other than the programmer of the method, and after a lapse of time, it should also be understandable by the programmer. As with structured code, object-oriented code should be documented. The documentation of a method describes its purpose, function, inputs, and outputs, as well as a description of the object. Internal comments within the method should describe major steps.

Encapsulation (also called information hiding) involves the separation of the external aspects of an object that are accessible to other objects from the internal attributes and operations of the object. This feature prevents something like content coupling from happening. Thus encapsulation prevents objects from becoming so tightly related (or coupled) that a small change in one object ripples throughout the object-oriented design. Good modular design (loose coupling between objects and tightly cohesive objects) supports encapsulation.

There is, however, some tension between encapsulation and inheritance. As we've just noted, encapsulation requires very loose coupling, but inheritance calls for tighter coupling. On the one hand, loosely coupled classes are desirable. On the other hand, inheritance calls for tightly coupled superclasses and their subclasses to exploit commonality and reusability.[5]

Encapsulation is violated when code associated with one class directly accesses the attributes and operations of another class. Many OOP languages declare attributes and operations as public or private. **PUBLIC** attributes may be read and public operations may be executed. **PRIVATE** attributes and operations cannot be read and executed. They are for internal use only by other methods of the same class and are hidden from other classes so other classes cannot use them.

The preceding guidelines, in addition to improving MURRE design factors, also improve inheritance of code. A common way to increase the chance of inheriting shared code is to factor out common code into a single method as a subroutine that is called by other method.

What Is the Difference Between Structured Software Design and Object-Oriented Software Design?

Modeling structured software designs and object-oriented software designs have some things in common. Both use similar modeling tools. A key difference, however, is that the structured approach designs a system around processes or data, whereas the object-oriented approach designs a system around objects.

In the structured software design approach, the structure chart shows processes (or functionality). In the object-oriented software design approach, the object-oriented model shows object classes and relationships between objects and classes. Changing user requirements with a structured process-based approach necessitates changes in processes rather than changes in objects. So, a change in a process-based design is relatively more difficult than a change in an object-based design, because a change in an object is done by adding or changing operations and supporting methods, leaving the object structure itself unchanged.

Structured software design normally is based on a systems scope or boundary, which may make it difficult to extend the design to a new scope or boundary. It may be much easier to extend an object-oriented design merely by adding objects.

Object-oriented software design proponents believe that this design approach is more flexible to change and more extendable. They also say that objects are more easily understood by users, because the objects represent concepts and things in the real world. Users, therefore, do not have to comprehend attributes and operations; in fact, these are usually hidden from the end user. All users have to do is understand the object's behavior and what messages to send to invoke the required behavior. For example, behavior is in-

[5] Grady Booch, *Object Oriented Design* (Redwood City, Calif.: Benjamin/Cummings, 1991), p. 124.

voked by messages, such as ChangeModel, GetStudent, AssignGrade, Change-Pay, DeleteCustomer, and GetBirthdate. In other words, a message tells an object what to do. How the object implements the message is of no concern to the user as long as the results are correct.

At the center of the choice of structured software design approach versus the object-oriented software design approach is reusability. The ideal goal of reusability is to have in a class library all the objects available to support design applications. This class library is documented and publicized to let others know that it exists. Because of the newness of the object-oriented software design approach, we are, however, a long way from widely employing large libraries of reusable objects, especially for business applications. Also, bear in mind that well-designed modules following the structured software design approach also provide opportunities for reusing such modules.

PERFORMING A SOFTWARE DESIGN WALKTHROUGH

Software designs based on the structured approach or the object-oriented approach should be subjected to a **SOFTWARE DESIGN WALKTHROUGH**. A software design that contains design flaws or errors, inconsistencies, and ambiguities is dysfunctional and worse than no design at all. Therefore, after the software design is completed, we need to conduct a software design walkthrough to ensure that the design is correct and that it meets the Detailed Systems Design Report specifications.

What Is a Software Design Walkthrough?

There are two major variables in any kind of walkthrough. First, is the degree of formality or structure of the walkthrough. If it is strictly formal with an independent quality assurance or peer group performing the walkthrough (or review), such an organization is sometimes referred to as a structured walkthrough. In this book, we assume highly structured (i.e., formal) walkthroughs. The second major variable is timing. We can perform structured walkthroughs at any time during the SDLC or SWDLC. For example, walkthroughs generally occur after the systems analysis, design, and evaluation phases.[6] Some companies also perform structured walkthroughs both before and after all phases of the SDLC and SWDLC. In the series of chapters on software development in this book, we focus on performing structured walkthroughs after designing and coding software. A structured walkthrough of software program code is often referred to as white box testing; it is discussed in Chapter 18.

A software design walkthrough conducted by the quality assurance group or an independent peer group simulates the way the software is supposed to work. The simulation shows how different parts of the software design interact.

---

[6] Edward Yourdon, *Structured Walkthroughs,* 4th ed. (Englewood Cliffs, N.J.: Yourdon Press, A Prentice-Hall Company, 1989), p. 18.

It can expose ambiguities, redundancies, or missed systems design specifications.

A software design walkthrough is not conducted to "catch people" but, instead, to uncover the unexpected. Both the point at which errors are introduced and the point at which they are discovered are especially critical in software development. Two-thirds of software errors emanate from software design rather than software coding.

The obvious benefit of a software design walkthrough is the early discovery of software design errors so that each error may be corrected prior to the next step in the software development process.

How Is a Software Design Walkthrough Conducted?

An ideal software design walkthrough is administered by a meeting manager or moderator. Software design documents (e.g., structure chart, structured English, object-oriented models such as a modified ERD, decision tables, etc.) are scrutinized by reviewers. They may be software engineers (i.e., those programmers who understand designing and testing procedures as well as coding), user representatives, and systems analysts or designers.

Reviewers must have clear-cut instructions and procedures. They must read the design documentation in advance of the walkthrough meeting and challenge or question it in the meeting. At the meeting, neither the meeting moderator nor the recorder comments on software design. The moderator runs the meeting. Doing so requires several tasks, including establishing meeting rules, setting meeting time and place, recognizing reviewers and software designers, stopping interruptions, keeping the walkthrough on track, and preparing a summary report. The recorder writes all significant comments on flip chart sheets that everyone in the meeting room can easily see. These comments are incorporated into the results. Results of a software design walkthrough are captured in a **SOFTWARE DESIGN WALKTHROUGH REPORT**, as illustrated in Figure 16.20.

At the end of the software design walkthrough, the walkthrough team must decide whether to:

- Accept the design without further modification
- Reject the design because of major errors
- Accept the design conditionally

In other words, a decision is made to release the design for coding or send it back for redesign. Once the decision is made, all members of the software design walkthrough team complete a sign-off, indicating their participation in the walkthrough and their concurrence with the walkthrough team's findings. After some reasonable time elapses, the walkthrough team will conduct a follow-up walkthrough to make sure all errors have been corrected, and none have been ignored.

A software design should not be considered complete until it is approved by the software design walkthrough team. Several walkthrough meetings may be held before such approval is finally made.

Figure 16.20
Software Design Walkthrough Report.

SOFTWARE DESIGN WALKTHROUGH REPORT

Walkthrough Identification:

Project: EIS — Walkthrough number: W104

Date: MMDDYY — Location: Room 302

Time: 9:00 AM

Software Identification:

Material reviewed: Detailed design modules for EIS
Software designer: Herb Miller

Material Reviewed:

1. Detailed design of Modules A, B, and C
2. Structured English for modules

Walkthrough Team:

Name	*Signature*
1. Anne Graham, Moderator	1. ____________
2. Jim Kincaid, Recorder	2. ____________
3. Julie Byars, Software Engineer	3. ____________
4. Dana Sellers, Software Engineer	4. ____________
5. Felix Garcia, User Representative	5. ____________
6. Patsy Matusz, Systems Analyst	6. ____________

Software Design Appraisal:

Accepted: as is ________ with minor modification ________
Not accepted: major revision ________ minor revision ___X___

Errors List:

1. The formula for the economic order quantity is shown as $Q = \sqrt{4ysp/c}$; it should be $Q = \sqrt{2ys/pc}$
2. The structured English statement:
   "If division 2 sales are greater than 50,000 units, then print an exception report."
   It should be:
   "If division 2 sales are less than 30,000 units or greater than 50,000 units, then print an exception report."

Suggestions:

1. Object B could be improved by dividing it into two modules. One module would compute the sales forecast; the other would compute the economic order quantity.
2. Object A, which inputs and edits inventory data, is not needed. Object Z is in the class libraries of OOP languages that we are reviewing.

REVIEW OF CHAPTER LEARNING OBJECTIVES

The major goals of this chapter were to enable each student to achieve four important learning objectives. We will now summarize the responses to these learning objectives.

Learning objective 1:
Explain reasons for performing the software design phase.

The more time and effort is spent on software design, the better the software quality becomes. If the software design is correct, the likelihood that a high-quality software program will be developed increases. If a software design is not prepared, errors may not manifest themselves until the program is converted to operations. Correcting errors during operations is much more expensive than creating a correct software design in the first place.

Learning objective 2:
Describe the structured software design approach.

The essential elements of the structured approach are:

- Modularity
- Top-to-bottom control
- Sequence, selection, and repetition control constructs

Modules should be highly independent and function-specific. Such modules are said to be loosely coupled and highly cohesive. Superior modules invoke subordinate modules via a mechanism such as a CALL or PERFORM.

Modules should not contain more than 50 LOEC. Splitting a decision function between modules should be avoided. Normally, a subordinate module should not be called by more than three superior modules, and a superior module should not call or invoke more than five subordinate modules. If these design guidelines are followed, the resulting design will generally be made up of an optimum number of modules.

All of the tools discussed in this book can be used to aid structured software designers. Two popular modeling tools are structure charts and structured English. The structure chart is transformed from a data flow diagram. It shows each module that comprises the software design. Structured English gives the specifications followed by the coder. In some instances, decision tables, decision trees, and equations for special algorithms are used to supplement structured English.

Learning objective 3:
Describe the object-oriented software design approach.

The object-oriented software design model can be abstracted from DFDs, STDs, data dictionaries, and logical data models. The resulting object-oriented software design is composed of objects, classes, and relationships.

A variation of the ERD serves as an excellent object-oriented modeling tool. The chief symbol is the class box that contains three regions:

- The class name
- List of attributes
- List of operations

The ERD shows relationships of class boxes.

Learning objective 4:
Relate how to conduct a software design walkthrough.

Typically, a software design walkthrough team is composed of members from the quality assurance group. Ideally, the software design walkthrough should be organized by a moderator and recorded by a recorder, both of whom do not review the design documentation. The design documentation is reviewed by the other members of the walkthrough team prior to the software design walkthrough meeting. Designers are given a change to respond to and clarify their design documentation. After the meeting, a report is sent to all attendees. All errors are corrected by the designers and verified by the walkthrough reviewers at the next meeting. Coding of a software design is not permitted until the software design is approved by the software design walkthrough team.

SOFTWARE DESIGN CHECKLIST

Following is a checklist on how to perform software design. Its purpose is to remind you of major aspects of software design.

1 Recognize the importance of software design and make a major commitment to this SWDLC phase.

2 Understand user requirements and the systems design fully as specified in the Detailed Systems Design Report and modeled by modeling tools, such as DFDs, STDs, and the like. For example, a detailed DFD shows what has to be done. A STD describes when it is done. Structure charts, structured English, decision trees, decision tables, equations, and modified ERDs model, at the software design level, show how it is done.

3 Abstract the software design from systems design, using one or a combination of modeling tools, such as structure chart, structured English, Jackson diagram, Warnier–Orr diagram, modified ERD, decision tables, decision trees, and equations.

4 Don't try to design everything precisely the first time. More often than not, your models will require revisions (iterations). Each iteration will probably require more detail. Both structured design and object-oriented design involve an incremental, iterative process. When you think that the design is complete at one level of abstraction, add more detail and flesh

out the design further at a finer level of detail. You will possibly discover new modules or new attributes and operations that must be added to classes. Maybe new classes will be identified. Maybe new modules will be required. It may be necessary to revise coupling between modules. It may be necessary to revise relationships between objects. Iterating three, four, or more times is not uncommon in software design for either the structured approach or the object-oriented approach. Of course, because iterations are costly and time-consuming, it is always the aim to come as close as you can to being right the first time.

5 Identify all modules or objects that will play a role in the new systems application.

6 Specify the functions or operations performed by each module or object and the data attributes acted upon.

7 Indicate relationships between modules or objects following the objectives of modularity, that is, loose coupling and tight cohesion.

8 Subject completed design models to a structured software design walkthrough to make sure that the software design meets systems design specifications and can be used correctly by programmers.

KEY TERMS

Attributes
Class
Class box
Cohesion
Coupling
Encapsulate
Inheritance
Methods
Modularity
Multiplicity
Object-oriented software design
Objects
Operations
Polymorphic
Private
Public
Software design walkthrough
Software Design Walkthrough Report
Structure chart
Structured English
Structured software design
Transform analysis

REVIEW QUESTIONS

16.1 Explain the benefits of devoting at least 40 percent of the software development effort to design.

16.2 What are the two factors that measure the degree of modularity?

16.3 Define coupling and cohesion.

16.4 Explain why tightly coupled software modules are difficult to code, test, and maintain.

16.5 What are two common commands used to invoke modules?

16.6 List and describe modular design guidelines.

16.7 The prime objective of structured design is modularity. But can a software design have too many modules? Explain.

16.8 Explain why large, spaghetti-like programs are difficult to code, test, and maintain.

16.9 What role does a structure chart play in structured software design? What role does structured English play? Why are decision tables, decision trees, and equations sometimes needed as supplementary design tools?

16.10 What are the rules for writing structured English?

16.11 If a data flow diagram is used as a systems design tool, why is it transformed into a structure chart?

16.12 Define objects, classes, and relationships.

16.13 Explain how inheritance supports reusability.

16.14 Describe the contents of a fully defined class box.

16.15 Explain how to achieve MURRE design factors using the object-oriented approach.

16.16 Explain encapsulation and its relationship to public and private features of object-oriented designs.

16.17 Explain the similarities between the structured approach and the object-oriented approach. Explain how the two approaches differ.

16.18 Explain why it is more difficult and costly to remove an error during the maintenance phase than it is during the design phase.

16.19 What is the purpose of a software design walkthrough? Explain how a walkthrough should be conducted.

16.20 Where do the majority of the software errors come from?

16.21 Explain the function of moderator, recorder, and reviewers.

16.22 Describe the elements that make up a Software Design Walkthrough Report.

16.23 When is a software design deemed complete and ready for coding?

CHAPTER-SPECIFIC PROBLEMS

These problems require exact responses based directly on concepts and techniques presented in the text.

16.24 Part of a program executes a loop. An EOF condition is tested, and if the condition is false, the loop is terminated. Within the loop, if the

SALES-CODE is equal to the TRANS-CODE, selling price is computed, INV-UPDATE is performed, and another record is read. If the SALES-CODE is not equal to the TRANS-CODE, "No-Sale" is displayed, and another record is read.

Required: Write structured English for this process.

16.25 Review the following structure chart.

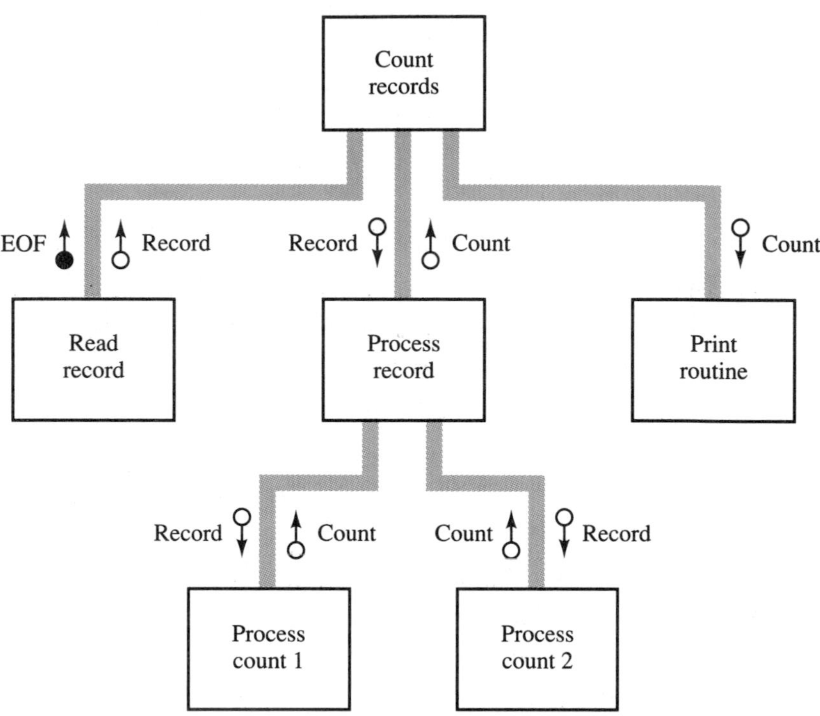

Required: Write structured English (or pseudocode) that specifies this structure chart in more detail.

16.26 Study the following data flow diagram:

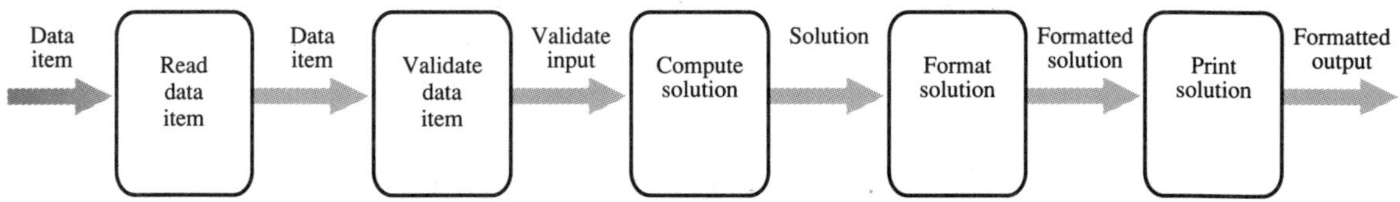

Required: The data flow diagram will consist of nine modules when it is transformed to a structure chart. The top module is the name of the central transform. Three intermediate modules read data, process a solution, and write. The bottom modules read from an input file, validate data, and produce an EOF flag, format a solution, and write to an output file. Draw a structure chart of these modules with the flow of data among them properly designated.

16.27 Review the following instructions for their readability and instructional hierarchy.

```
ORDER-ENTRY
FOR ALL ORDERS
ENDIF
IF CUSTOMER-NUMBER IS VALID
GET CUSTOMER-RECORD
ENDFOR
ELSE DISPLAY ''INVALID-RECORD''
CALCULATE ORDER-DETAILS
EXIT ORDER-ENTRY
WRITE ORDER-RECORD
```

Required: Using indentation, improve readability and clarify instructional hierarchy for this set of instructions.

16.28 A payroll software application at level 0.0 calls modules to read payroll data (1.0), process payroll record (2.0), and print payroll checks (3.0). Other processes include: read time tickets (1.2), read payroll file (1.3), compute gross pay (2.1), compute straight time wages (2.2.1), compute overtime (2.2.2), print payroll register (3.1), and print paychecks (3.2).

Required: Draw a structure chart for the system described above.

16.29 The following table contains a list of modules and their procedures.

Module	Procedures	Coupling Type
A	This module transfers a record with twelve fields to a module that requires three of these fields	
B	This module alters statements in another module	
C	This module contains three REDEFINES clauses	
D	This module sends the appropriate INTEREST RATE to a loan calculation module	
E	This module sends an EOF message to another module	

Required: Complete the table by entering appropriate coupling types.

16.30 The following table contains a list of modules and their procedures.

Module	Procedures	Cohesion Type
A	Gets transactions, edits, and returns them	
B	Using employee number, determines employee name and other employee data, and returns these data items	
C	Gets data, goes to process, matches codes, runs utility, updates master, and processes audit routine	
D	Updates weekly billing schedule or invoice, depending on switch	
E	Accesses transaction, edits, gets master record, matches codes	
F	Initializes, reads, edits, writes, and closes file in that order	
G	Computes gross pay	

Required: Complete the table by entering appropriate cohesion types.

16.31 Module A contains two functions. It is called by five modules.

Required: Construct a structure chart that shows a better calling arrangement.

16.32 Module A calls Module C, and Module B calls Module C, as demonstrated below.

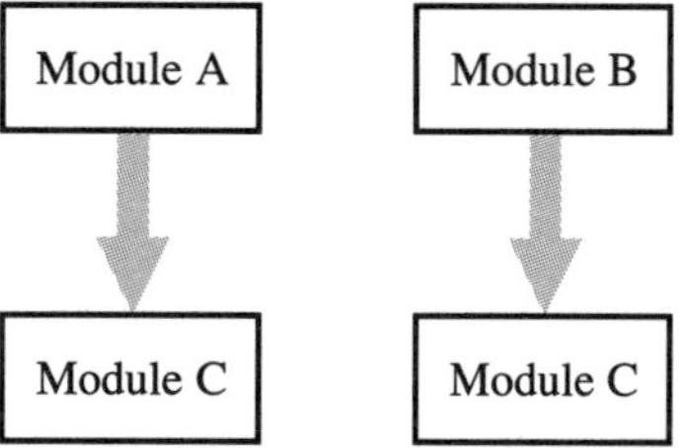

Required: Draw a structure chart that demonstrates a better calling arrangement.

16.33 You are designing a software application to verify customer orders. A customer number accesses a credit record. The customer either has or does not have a record. These data are returned to the verify order module. Then an inquiry is made to an authority module that calculates a self-checking digit, checks credit limit, and generates a credit authorization number. A response from the authority module is returned to the verify order module. The verify order module transmits the response to a present response module, which formats the response and transmits the response.

Required: Design a structure chart for this application, labeling each module and the passing of data between modules. Design your structure chart according to sound coupling and cohesion principles.

16.34 Study the following structured flowchart:

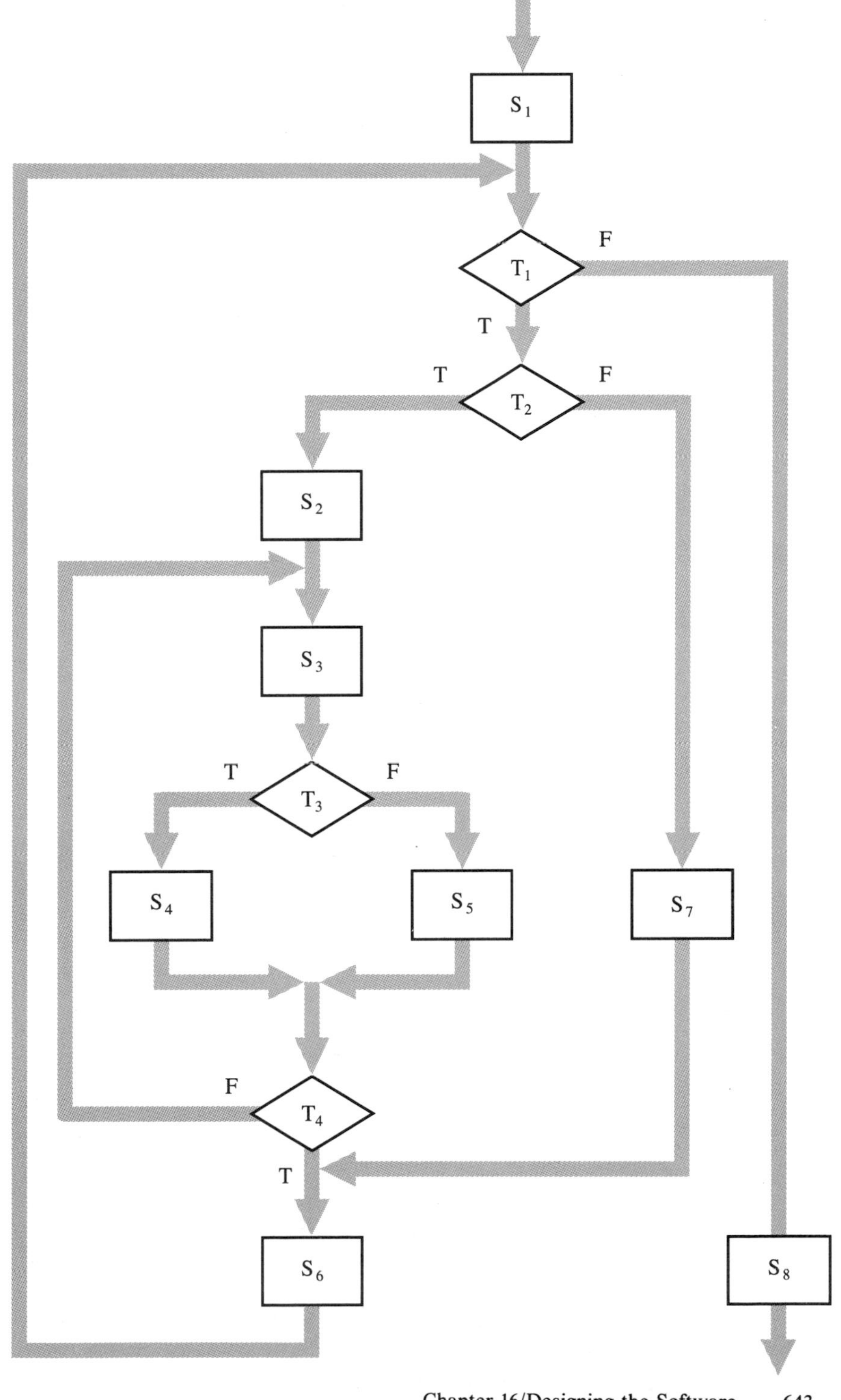

Required: Write the structured English equivalent.

16.35 Analyze the following structured flowchart:

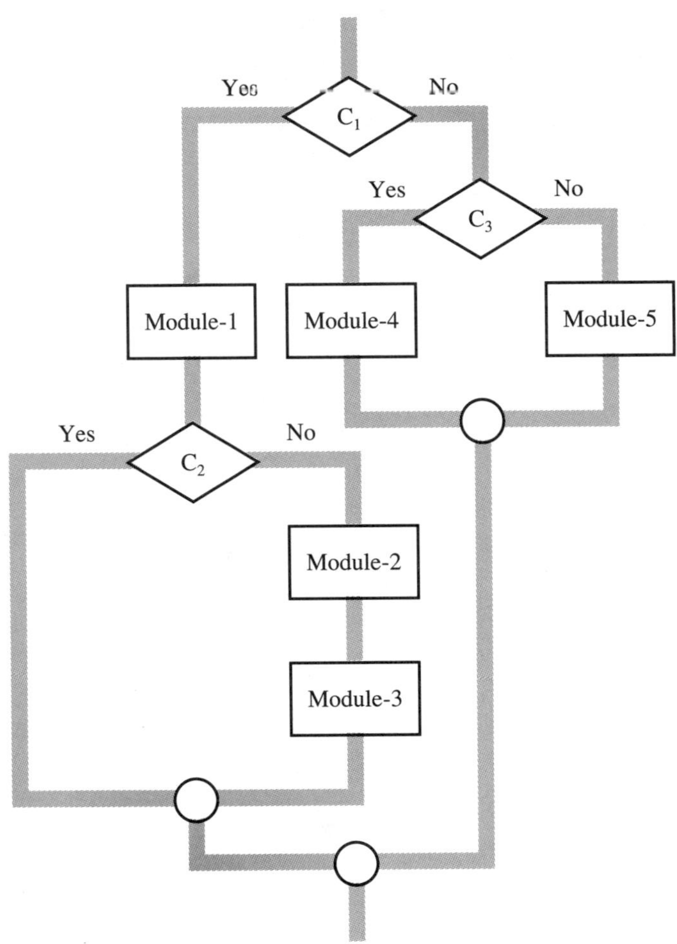

Required: Write the structured English equivalent.

16.36 Analyze the structured flowchart on the opposite page.

Required: Write the structured English equivalent. Also, develop a decision table for the decision logic.

16.37 Study the following statement:

```
IF C1
      PERFORM MODULE-1
      IF C2
            NEXT SENTENCE
      ELSE
            PERFORM MODULE-2
            IF C-3
```

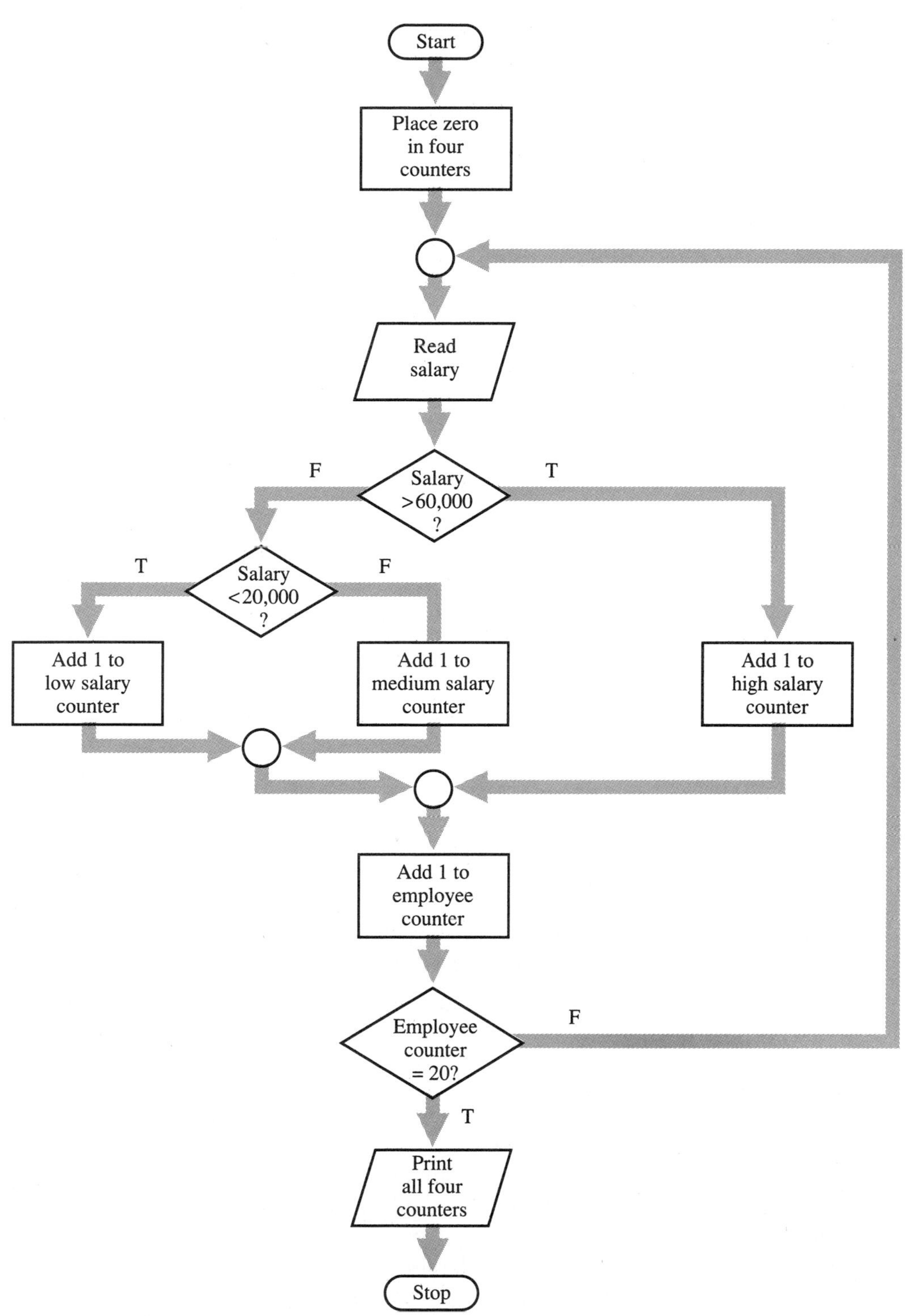
Start
Place zero
in four
counters
Read
salary
F
Salary
>60,000
?
T
T
Salary
<20,000
?
F
Add 1 to
low salary
counter
Add 1 to
medium salary
counter
Add 1 to
high salary
counter
Add 1 to
employee
counter
Employee
counter
= 20?
F
T
Print
all four
counters
Stop

```
                PERFORM MODULE-3
                PERFORM MODULE-4
           ELSE
                NEXT SENTENCE
ELSE
     NEXT SENTENCE
ENDIF
```

Required: Draw a structured flowchart (not a structure chart) of the preceding logic.

16.38 Study the following statements:

```
IF C1
     PERFORM C2-TEST
     PERFORM MODULE-1
ELSE
     IF C2
          PERFORM MODULE-2
     ELSE
          PERFORM MODULE-3
ENDIF
```

Required: Draw a structured flowchart (not a structure chart) of the preceding logic.

16.39 MANAGER and STUDENT/WORKER are subclasses of EMPLOYEE, and EMPLOYEE and STUDENT are subclasses of PEOPLE.

Required: Draw a modified ERD representing the preceding object-oriented relationships.

16.40 VEHICLE is a superclass composed of the attributes Model, Weight, and Manufacturer; and the operations Sell, Ship, Cancel, and Return. VEHICLE has two subclasses: CAR and TRUCK. There is a one-to-many relationship between VEHICLE and CAR and TRUCK. CAR's attributes are Model, Weight, and Color; its operations are Sell, Cancel, and Return. TRUCK's attributes are Model, Weight, and Axle; its operations are Sell, Cancel, and Return.

Required: Using a modified ERD, model the class above. Show attributes and operations inherited from the superclass parenthetically.

THINK-TANK PROBLEMS

These problems call for a feasible approach rather than a precise solution. Although the problems are based on chapter material, extra reading and creativity may be required to develop workable solutions.

16.41 Following are familiar objects:

- Students, courses, university, professors, books, classrooms, grades.

- Table, freezer, cabinet, kitchen, sink, refrigerator, light switch, ice, cheese, smoke alarm.
- Wheel, car, brake, engine, brake light, model, color, price, battery, muffler.

Required: Prepare an object-oriented ERD showing objects, object classes, and relationships. Also, give examples of attributes and operations.

16.42 You have just been hired as a systems analyst by Oldstyle Clothiers, which owns a chain of men's clothing stores in the southwest. Your first meeting with Oldstyle's two programmers, Jane and Bill, has not gotten off to a good start. Jane said, "I don't know what you mean by this software designing-engineering-modeling stuff." "Yeah," Bill chimed in. "This building of a formal, testable model of a software program before writing code—what a radical idea! What a waste of time! Why fool around with a model when we could be coding? Models are for mechanical or civil engineers, not for programmers. Programming is an art, not something that can be engineered."

Required: Prepare a response to Bill and Jane.

16.43 As an experiment, estimate the average cost of a two-day software design walkthrough based on the average salary of the participants plus a 20 percent overhead charge (i.e., 0.20 × total average salary). Then calculate the number of design errors that will be found in the walkthrough and savings that are likely to result. The software design that will be tested using a walkthrough is a program similar to three others already implemented. An average of 200 errors were found in these programs after they had been converted to operations. This is why management requires a software design walkthrough on this new software design. The previous software programs did not go through a formal software design phase, and, therefore, were not subjected to a software design walkthrough.

Required: Demonstrate the effectiveness of management's demand for a structured software design walkthrough. Make any assumptions you deem necessary to derive an adequate presentation.

SUGGESTED READING

Aktas, A. Ziya. *Structured Analysis and Design of Information Systems*. Englewood Cliffs, N.J.: A Reston Book, Prentice-Hall, 1987.

Booch, Grady. *Object Oriented Design*. Redwood City, Calif.: Benjamin/Cummings, 1991.

Chen, P. P. S. "The Entity-Relationship Model—Toward a Unified View of Data." *ACM Transactions on Database System 1,* March 1976.

Cox, Brad J. *Object Oriented Programming*. Reading, Mass.: Addison-Wesley, 1986.

Darst, Donald. "Balancing Productivity and Quality." *Datamation,* September 15, 1990.

DeMarco, Tom. *Structured Analysis and Systems Specification*. Englewood Cliffs, N.J.: Prentice-Hall, 1979.

Glass, Brett. "Object-Oriented Programming: An Executive Overview." *Infoworld,* February 11, 1991.

Hill, Tom. "Object Oriented Programming Using COBOL." *Interact,* April 1990.

Kim, Won. "A New Database for New Times." *Datamation,* January 15, 1990.

McNur, Barbara Canning. "Improving Large Application Development." *I/S Analyzer,* March 1988.

Martin, James, and Carma McClure. *Diagramming Techniques for Analysts and Programmers*. Englewood Cliffs, N.J.: Prentice-Hall, 1985.

Meyer, Bertrand. *Object-Oriented Software Construction*. Englewood Cliffs, N.J.: Prentice-Hall, 1988.

Orr, K. T. *Structured Systems Development*. Englewood Cliffs, N.J.: Yourdon Press, A Prentice-Hall Company, 1977.

Page-Jones, Meilir. *The Practical Guide to Structured Systems Design,* 2nd ed. Englewood Cliffs, N.J.: Yourdon Press, A Prentice-Hall Company, 1988.

Perreault, Bob, and Katie Rotzell. "Object Database Management Systems." *Interact,* April 1990.

Peters, L. J. *Software Design: Methods and Techniques*. Englewood Cliffs, N.J.: Yourdon Press, A Prentice-Hall Company, 1981.

Pressman, Roger S. *Software Engineering: A Practitioner's Approach,* 2nd ed. New York: McGraw-Hill, 1987.

Rumbaugh, James, Michael Blaha, William Premerlani, Frederick Eddy, and William Lorensen. *Object-Oriented Modeling and Design*. Englewood Cliffs, N.J.: Prentice-Hall, 1991.

Stevens, Wayne P. *Using Structured Design*. New York: John Wiley, 1981.

Stevens, W., G. Myers, and L. Constantine. "Structured Design." *IBM Systems Journal,* Vol. 13, No. 3, 1974.

Teague, Lavette C., Jr., and Christopher W. Pidgeon. *Structured Analysis Methods for Computer Information Systems*. Chicago: Science Research Associates, 1985.

Tse, T. H., and L. Pong. "Towards a Formal Foundation for DeMarco Data Flow Diagrams." *Computer Journal,* February 1989.

Yourdon, Edward. *Structured Walkthroughs*, 4th ed. Englewood Cliffs, N.J.: Yourdon Press, A Prentice-Hall Company, 1989.

Wickes, Donna. "The King Lives!" *Software Magazine,* May 1991.

JOCS CASE: Designing Software

The managerial approach Jake established for the JOCS application in Chapter 15 will be used for all phases of the software development life cycle (SWDLC). In Chapter 15, Jake finished the initial project cost and time estimation, which comprised the first phase of the SWDLC. The second phase involves the actual design of the necessary software. Each of the programs required for the JOCS application must be designed and walked through to test validity individually.

Christine Meyers has been made responsible for the software design for JOCS. She assigns specific detailed design components of the system to other members of the SWAT team. She has assigned one detailed design component to Jannis Court. The component that Jannis is currently working on is the online capture of direct labor data. As discussed in a previous chapter, Peerless has installed a computer-integrated manufacturing system (CIM) to capture direct material and labor data. The CIM system has special terminals to input the time started and ended for a given job. A manufacturing employee performs the following steps before beginning work on a job:

1 Passes the bar code wand over her ID card to capture her employee ID number.

2 Passes the bar code wand over the job traveler document that accompanies each job to capture the job number.

3 Passes the bar code wand over the operation number of the manufacturing operation sign that she is about to begin. This step will capture both the operation number, such as welding or assembly, and the time that she is starting to work.

When finished, the employee again passes the bar code wand over her ID card, the job traveler document, and the operation number sign, to capture the employee ID number, job number, operation number, and the time she finished working on that specific operation. The data flow diagram devised during systems analysis for this component of JOCS is provided in Figure 16.21.

Jannis uses the data flow diagram to create the tentative design depicted in the Jackson diagram shown in Figure 16.22. While developing the diagram, Jannis realizes that some issues must be discussed before the design can continue. The first issue concerns the extent of online interface for data capture.

"Should the interface simply collect data from the CIM system and store it on a secondary storage device?" Jannis wonders. "Or instead of that, should the interface provide feedback to the CIM system so that the employee using the bar code wand can be informed of mistakes?" She continues to think about the design by coming up with possible employee mistakes. "I wonder

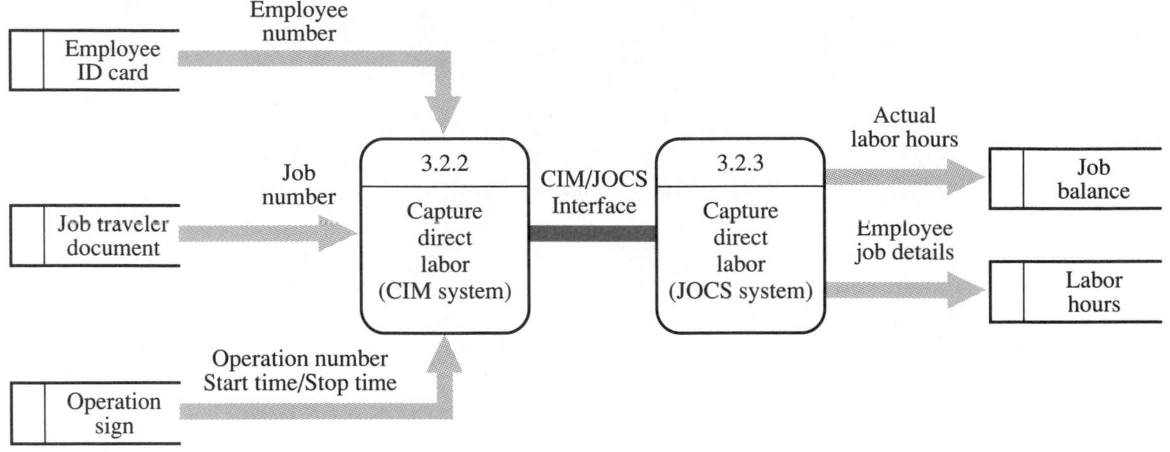

Figure 16.21
Data flow diagram for JOCS direct labor data capture.

what should happen if an employee tried to perform an operation for a job when the job didn't require that operation. The system could provide immediate feedback to the employee stating that an incorrect operation was being performed. I could generate a message telling the employee that he or she entered an invalid operation." This line of thinking opens up different errors that could be detected before the data are entered into the system.

Jannis thinks about other errors: "Maybe an employee would try to perform an operation for a job that has already been reported as completed. I mean, what if their supervisor had made a mistake and accidentally assigned an operation that was already done? Or maybe the operation wasn't really

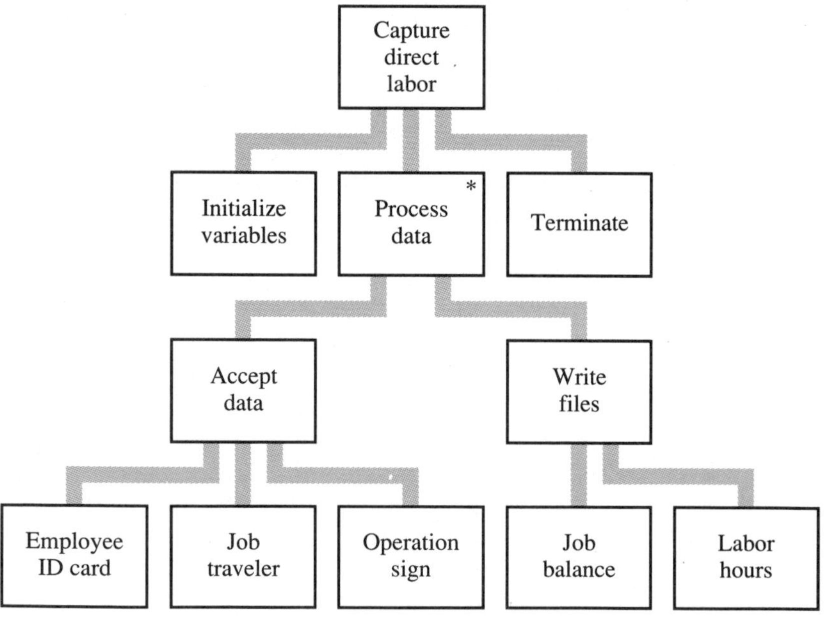

Figure 16.22
Jackson diagram—initial JOCS direct labor data capture.

done, but they had already spent their elapsed amount of time and the system thought the operation was done. I bet the supervisor would want to know about this situation before the employee started working on the operation."

This kind of online error checking will require a more extended interface between the CIM system and the JOCS application. If a more extended interface is desired, Jannis will have another issue to settle. This issue concerns the methods of settling potential discrepancies. Jannis thinks about ways to let the employee work on the operation or not work on the operation: "If the errors do happen, should I let an employee override the system and perform the operation? Maybe I should make each employee get his or her supervisor to enter a special supervisor code to approve the operation." Jannis is worried about these questions. She thinks, "These questions are pretty important. I'd better talk about it with the manufacturing supervisor. I don't think I should just decide this stuff and then tell the manufacturing people that this is the way we are going to do it. In fact, I am going to have to know more about how the CIM system will interface with JOCS to make sure that I can even do all this error checking."

She sends Jake an E-mail message outlining her problems. Jake sends back a message telling Jannis to contact Tom Pearson about the CIM/JOCS interface to find out if any error checking can be done between the two systems. Jannis then sends Tom an E-mail message. Tom sends back a message saying that the speed between the two systems will allow her to do online data error checking, and he gives her detailed information about the data transfer rates. Jannis sends Jake another E-mail message saying that she needs to talk to the manufacturing people to settle the design for this component of the system. Jake agrees, through an E-mail message, that this is an issue that must be discussed with the users of the system before the design of this component can be completed.

Jake arranges a meeting for Jannis with the supervisor of manufacturing to discuss these questions. This meeting, between Jannis and the manufacturing personnel, helps clarify the functions of the system under development for each person involved in the project. The meeting also helps to establish a strong working relationship between Jannis and the people who will use her software. Jake wants the designers of the software and the users to work together to define the exact software requirements.

After meeting with the personnel from the manufacturing department, Jannis is able to complete the design for the direct labor online capture component. Figure 16.23 is a Jackson diagram for this component of the JOCS application.

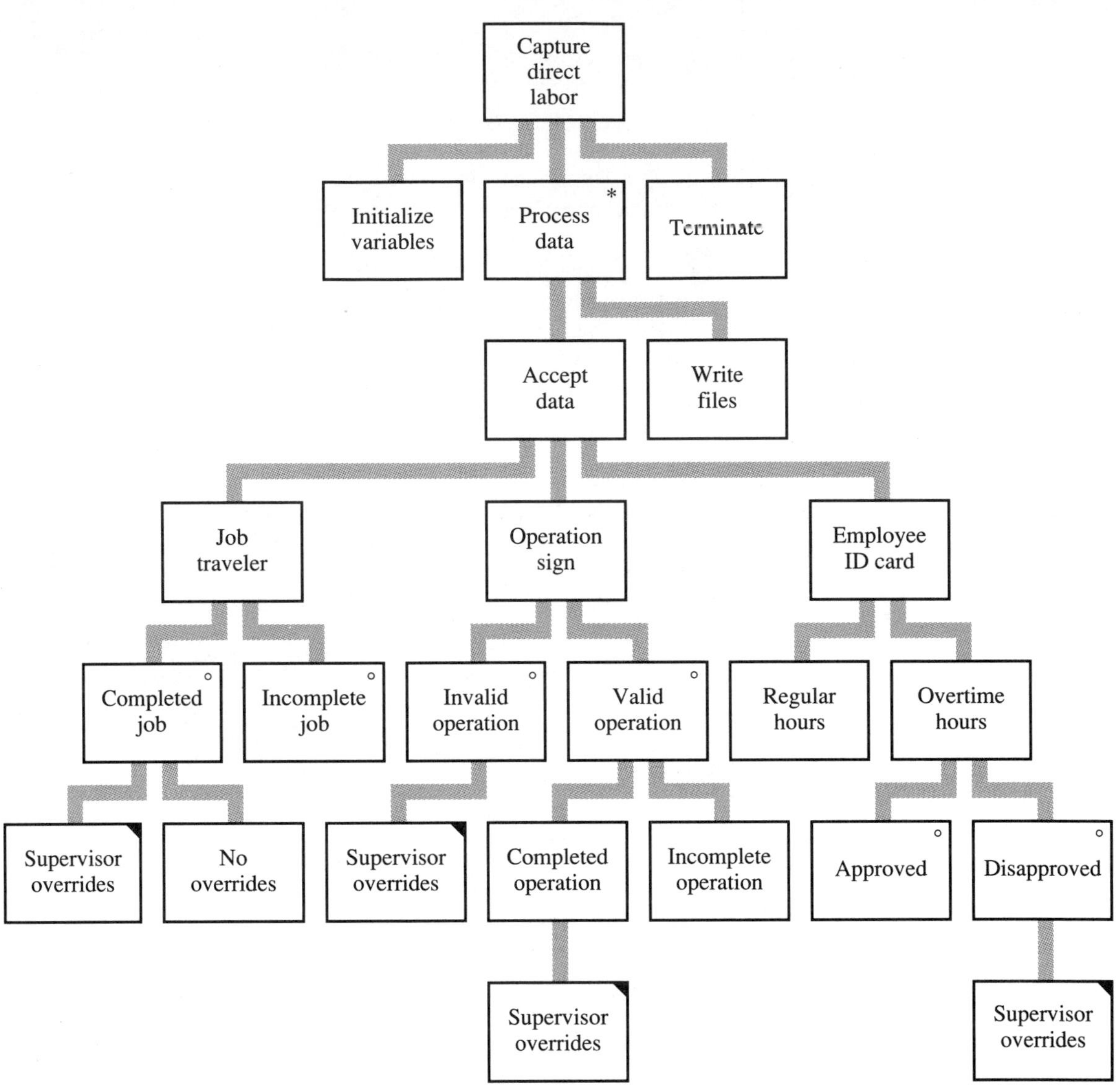

Figure 16.23
Jackson diagram—final JOCS direct labor data capture.

Chapter 17
Coding the Software

WHAT WILL YOU LEARN IN THIS CHAPTER?

After studying this chapter, you should be able to:

1 Compare fourth-generation languages (4GLs) with third-generation languages (3GLs).
2 Describe object-oriented programming (OOP) languages, review five important OOP languages, and briefly discuss OOP tools.
3 Explain how special-purpose language tools are used to enable end users to communicate with computer systems.
4 Match appropriate languages to software design applications.
5 Discuss the selection of languages relative to issues such as level of usage in the business world, expressiveness, convenience, portability, maintainability, and extendability.
6 Prepare software documentation, operations documentation, and user documentation.

INTRODUCTION

The objective of software coding is to convert software design applications to well-documented software programs (see Figure 17.1). The choice of computer programming language and tools that will be used to code software design applications is a critical decision for systems professionals. A poor choice can cause a host of problems, whereas a proper choice can dramatically improve development productivity, testing, and maintenance.

Once these decisions are made, the syntactical (i.e., grammatical) aspect of the language itself becomes a mechanical, and in some cases, an automatic process if one is using CASE automatic code generators. The purpose of this chapter, however, is not to treat syntax or the actual coding of language instructions, but to provide material that will aid in making the preceding critical decisions.

Coding Decisions at Tranquil Industries

"The windows environment is really working well," said Julie. "The windows-based development toolkits do a good job of supporting rapid application development."

"I couldn't agree more," Dana said. "We were able to develop the executive information system in about five days. Before these development tools, I believe it would have taken us several weeks, maybe months."

Figure 17.1
SDLC phases and their related chapters in this book. Chapter 17 has as its goal a review of software languages and a description of documentation.

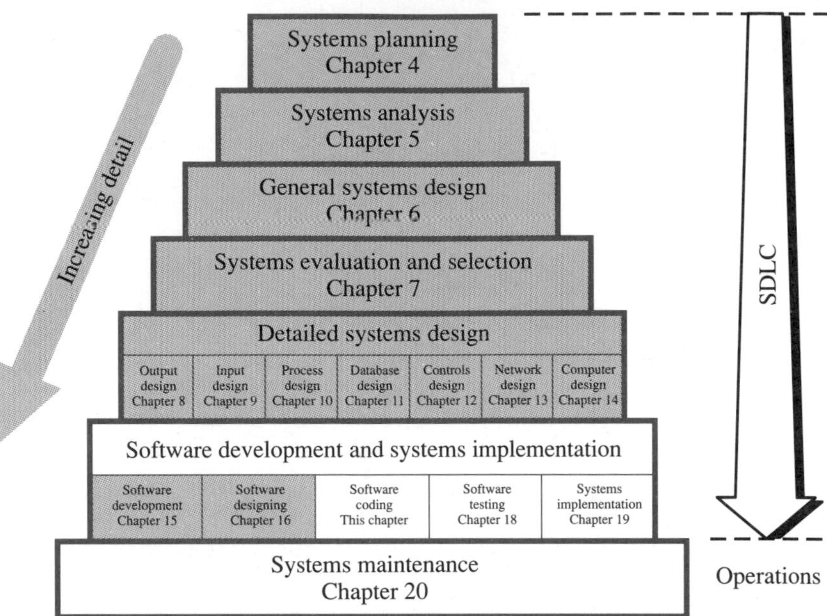

"The object-oriented tool set enabled the executives to 'see' what they wanted. They may not understand the intricacies of languages, but they sure do understand objects and how they want those objects to behave," added Julie.

"What about our new object-oriented programming language class library?" asked Jim. "Has it met your expectations?"

"Yes it has," Julie responded. "I'm especially pleased with the objects, such as dialogue boxes, icons, and menus, that helped us rapidly develop user interfaces that the executives intuitively understood. They can readily access information without assistance."

"Also, we built some objects of our own and included them in the class library for future systems design applications," Dana added.

"How's the LAN-based order-entry system coming along?" asked Jim.

"We completed the software design last week. All of our modules are completely specified with structure charts, structured English, decision tables, and equations," answered Julie. "And our structured software design walkthrough was successful. We plan to code it in COBOL."

"I thought you were going to code it in the new OOP language," said Jim, somewhat surprised.

"We're the only ones at Tranquil who can code in the OOP language, and we're presently working on a truck-dispatching system for Fred in transportation," Julie answered.

"The other programmers are excellent COBOL coders with years of experience, and they're reluctant to change to another language. They're very enthusiastic about trying out our new CASE workbench for COBOL," Dana said.

"How about using object-oriented COBOL?" Jim asked.

"We don't believe object-oriented COBOL is mature enough at this time to tackle, but we are looking at it," said Julie.

"Well, you guys are the programming experts, but it seems to me that we should consider a fourth-generation language for the new LAN-based order-entry system," said Jim.

"We plan to use a 4GL for the truck-dispatching system," said Julie.

"Our hopes are that the new COBOL CASE development toolkit will substantially increase productivity of our COBOL programmers," said Dana.

"I was reading an interesting article by Capers Jones, 'Using Function Points to Evaluate CASE Tools, Methodologies, Staff Experience, and Languages,' in the January/February 1991 issue of *CASE Trends,*" said Jim. "He observed that an inexperienced programmer using unstructured methods, ordinary pen-and-pencil, and low-level languages, can produce five function points per month. However, an experienced programmer, using structured methods, CASE tools, and a high-level language, can produce 100 function points per month. If we can come close to such a dramatic improvement in development productivity at Tranquil, I will consider our CASE development toolkits to be a success."

"Will COBOL give you this kind of increase in productivity?" asked Jim.

"At this time, I really don't know," said Dana. "But we're going to put it to the test using the CASE toolkit."

"Don't the CASE tools automatically generate code?" asked Tom Flanders, Jim Thompson's assistant.

"Lots of people have the notion that CASE means 'computer automatic software engineering' instead of 'computer-aided software engineering,' " Dana replied. "It's not magic. The tool set generates the code interfaces and the data structures, including procedure calls and parameter passing, but we still have to write the procedural code or the pseudocode, such as structured English."

"Will the new LAN-based order-entry system provide users with online inquiry capabilities?" asked Jim.

"Yes, we are going to embed some structured query language code into the COBOL program to facilitate online transaction processing and support queries from the relational database management system," said Julie.

"It's really just a matter of using the best language tools for the job, isn't it?" asked Jim.

"That's true," said Julie, "and that's why our home toolkits have saws, hammers, screwdrivers, pliers, and wrenches, instead of just one tool. When your only tool is a hammer, every problem looks like a nail."

"Well, it seems to me that you all have everything pretty well planned as far as coding of the software designs are concerned," said Jim. "Keep me posted."

"We will," said Dana.

WHAT ARE THE DIFFERENCES BETWEEN FOURTH-GENERATION LANGUAGES AND THIRD-GENERATION LANGUAGES?

The following subsections help to compare **FOURTH-GENERATION LANGUAGES (4GLs)** with **THIRD-GENERATION LANGUAGES (3GLs)**. Although some 3GLs can be used in object-oriented programming (OOP), we will revisit that subject when we devote another main section to OOP languages.

Advantages of 4GLs

The attractiveness of 4GLs centers on the fact that they use fewer lines of code to achieve what would require many lines in a 3GL. 4GLs are also designed to offer the potential for end-user programming of their own applications. More specifics on 4GL advantages are discussed in the following subsections.

Development Methodology

A 4GL development methodology does not usually resemble a 3GL development methodology. The development of 3GL applications is closely tied to the software development life cycle (SWDLC) and structured, top-down analysis and design. Some sage advice springing from the 3GL development methodology is:

- "Do it right the first time."
- "Don't start coding until user requirements are fully defined and the application is designed in detail."
- "The later you have to make changes to code, the more costly it is."

On the other hand, 4GL developers start with the assumption that users do *not* know precisely what their requirements are and exactly what they want. So the 4GL developer starts with a prototype of some kind, although it may be based on a great deal of guesswork. The idea is to give the user something no matter how different it may be from the final application. This approach gives the user something tangible that he or she can work with and evaluate. It starts the analysis and thinking processes required to develop any kind of successful system.

Then, the 4GL developer begins to develop more and more prototypes, working closely with the user. In this way, design comes almost last in the development process. With 3GL applications, the approach is to use design specifications before coding begins, but the 4GL programmer starts writing code to determine design.

Increased Productivity

One instruction in a 4GL is equivalent to a page or more of 3GL code. Moreover, many 4GLs contain specific functions, such as statistical and mathematical models. Screen painting and prototyping capabilities help define user requirements quickly and thus shorten the SWDLC. In fact, it might be more meaningful to call 4GLs fourth-generation design aids or software tools rather than languages.

Improved Service

Many users are unhappy with the time it takes to meet their system service requests. A large number of these requests are for local-based systems and ad hoc applications that can be easily and efficiently met with 4GLs. Applications like scheduling court cases or loan portfolio analysis or union grievance tracking systems are easily supported by 4GLs.

User Participation

Generally, end-user participation improves the final application that is implemented. Although computer literacy may differ among organizations, many end users are computer-literate enough to be taught 4GLs. Such users are generally excited and motivated to know that they can develop and maintain personal applications on their own or with minimal assistance from a systems professional. Although 4GLs may liberate end users from professional programmers in some situations, they do not free end users from professional programming and documenting standards.

Often, ad hoc requests need to be handled almost immediately for them to be of any value to the user. Putting such requests in a 3GL development queue may delay their development and installation far past the time that the user needs them, unless a 3GL programmer/analyst is supported by a CASE tool set. If so, 3GL development time is shortened significantly.

Advantages of 3GLs

We will now discuss the advantages of third-generation languages (3GLs) stemming from the following characteristics:

- Conciseness
- Machine efficiency
- Functionality
- Compatibility
- Coding productivity
- Testing and maintenance

We will use COBOL as our model 3GL because it is pervasive in business.

Conciseness

Most business applications contain a great deal of input and output, and most of COBOL's wordiness comes from describing and documenting all this input and output. COBOL requires input and output to be described in meticulous detail. (What the 4GL advertisements do not tell the reader is that the same amount of meticulous detail about input and output data must be—had better be—defined by somebody somewhere. The somewhere is normally a data dictionary, which is accessible to the 4GL for its data description.)

Machine Efficiency

Typically, 4GLs consume more resources and produce slower response times than do equivalent 3GL programs. Increased run-time overhead is caused primarily by the interpretive nature of 4GLs. These languages translate source code into machine language at execution time. Because any given program will be executed many times, this means that the overhead of interpretation is multiplied. This also means poor response time for end users and throughput.

Functionality
Fourth-generation languages do not contain the robustness and full capabilities of 3GLs. Any business application can be coded in a 3GL. On the other hand, 4GLs are normally not suitable for complex, high volume applications.

In fact, some disasters have occurred because 4GLs were misused in the development of complex, high volume systems. A classic 4GL disaster was New Jersey's vehicle registration system, which was coded in a well known 4GL. The system could not handle the volume of transactions and was disabled, costing the state tens of millions of dollars in cost overruns.[1] A number of authorities believe that the New Jersey case is merely a large-scale example of similar experiences that occur all too often among systems people who misuse 4GLs.[2] In many instances, the corrections of such problems involve recoding the failed applications in COBOL.

Compatibility
When vendors of computer architectures implement upgrades, this may cause the 4GL applications to be incompatible with the computer operating environment. This situation forces users to wait for matching upgrades from 4GL vendors to regain compatibility. In some cases, these upgrades may never come.

Coding Productivity
Hand coding of 3GLs such as COBOL is without doubt a tedious process, but CASE automatic code generators are available that automatically generate COBOL code to run well-designed applications. Structure charts, structured English, and decision tables are converted directly to COBOL code without the need for hand coding.

Major software vendors recognize COBOL's endurance and therefore continue to offer COBOL CASE development tools. Although such vendors are too numerous to mention here, Microsoft Corporation, a giant software vendor, serves as an excellent example. Microsoft's PC-based COBOL Professional Development System provides a fully integrated programmer's workbench, including such software development tools as compilers, editors, diagnostic routines, source-code formatters (for readability), and graphic user interface (GUI). These COBOL programming and development tools produce programs that run in various environments, such as DOS, OS/2, UNIX, MVS, and VMS. COBOL CASE workbenches also produce screen designs that help facilitate prototyping and joint application development (JAD). They also provide mixed languages that can be integrated with COBOL, such as structured query language (SQL). The CASE automatic code generator completely nullifies the purported coding efficiency advantage of 4GLs.

Moreover, a CASE central repository is used to store a module library, which provides the names, descriptions, and access procedures for reusable

---

[1] D. Kill, "Anatomy of a 4GL Failure," *Computer Decisions,* February 11, 1986, pp. 58–65.

[2] Al Lee, "The Problem with 4GLs," *Computer Programming Management* (Boston: Auerbach Publishers, 1989).

modules that have already been coded and tested. These modules may be standard across industries, such as certain statistical routines, or they may apply to particular functions within the company, such as order-entry processing. When needed as part of a new program, they are simply called from the module library by their names, such as CALL ORDER-ENTRY.

Testing and Maintenance

The concept of hidden logic—that is, macro 4GL instructions generating hundreds or thousands of machine instructions—makes testing and maintenance difficult and frustrating. Making a minor change in one area of the program may cause an apparently unrelated malfunction in another area, and subsequent efforts to correct the error can result in more confusion.

The code of 3GLs is openly visible to testers and maintainers. If structured design is followed and the code is well documented, a C or COBOL program can be easily read and understood. COBOL is heralded as a self-documenting language.

If an organization chooses an obscure or unpopular language, either a 3GL or 4GL, it will have difficulty finding enough programmers to develop, test, and maintain programs. The use of a standard, universal language, such as C or COBOL, makes the maintenance task a lot easier. A large number of people are available who know how to code software in C or COBOL. Indeed, C and COBOL are considered standard languages.

Summary Comments About 3GLs and 4GLs

Because 3GLs and 4GLs are presently the dominant language generations used for application software, we will now provide you with a summary comparison. Figure 17.2 summarizes some of the key features of 3GLs and 4GLs. The 3GL COBOL will probably remain the dominant language for transaction-intensive business applications at least for the next decade, if not well into the twenty-first century. Ada, C, and Pascal have many proponents, and one can expect these languages to grow in use. For example, Ada is mandated by the Department of Defense for new software development.

OBJECT-ORIENTED PROGRAMMING LANGUAGES

The Panacea

Marilyn Ageloff, CIO, is trying to coordinate 16 systems projects, all of which are behind schedule. She's spent nearly all morning dealing with irate users. She just got back from a meeting with the steering committee trying to explain why a $1.5 million investment in a new network and computer architecture has not increased productivity as was originally estimated.

Don Melanson, one of Marilyn's best programmers, pops into her office and begins to tell her that he has found the perfect solution to their problems.

"What we need to do is convert to object-oriented language tools," says Don. "They will give us objects, encapsulation, classes, inheritance, polymorphism, extendability, and they will . . ."

"Excuse me Don, but I've had a bad morning. Would you please fast-forward to the real benefits of what you're talking about? For example, how will object technology help us with our backlog?"

"We can clean up our backlog in a jiffy. We can write programs ten times faster. See, you can build programs out of existing objects instead of coding them from scratch. And if you need a new object, you can subclass it by inheriting from an existing object. Today, you don't build a house and build your own windows. You buy prefabricated windows and doors. And . . ."

"So, everything we've learned over the years about the structure-oriented approach is down the tubes?"

"Yeah," said Don smugly. "It's a paradigm shift, a moving away from today's structured programming methods to a more natural way of dealing with the computer. A whole new ballgame."

"Sounds a little farfetched to me, but I'm willing to give object technology a chance. How about your making a paradigm shift and using the object-oriented approach on Jim's portfolio analysis project? We're certainly not going to shift to OOP in a single swoop. There's too much existing expertise in the structured paradigm."

"Sounds good to me," said Don. "I'll keep you abreast of my progress."

"Fine," Marilyn responded. "You'll have to excuse me. I've got a meeting with the auditors to develop a new disaster recovery plan."

Figure 17.2
Summary comparison of 3GLs and 4GLs.

	Language Generation	
Features	**3GL**	**4GL**
Conciseness	No	Misleading
Machine efficiency	Yes	No
Compiled	Yes	Some quasi-compiled
Functionality	Complete	Limited
Compatibility	Yes	Can be a problem
Portability	Yes	Some
Batch and real-time	Yes	Some
Standard language	Yes	No
Vendor independence	Yes	No
End-user coding	Normally not	For some ad hoc applications
Coding efficiency	May be superior	Yes
Prototyping	Yes*	Yes
Built in controls	Yes	Difficult
Supports SQL	Some	Some
Easy to document	Yes	No
Availability of programmers	Yes	Some
Ease of maintenance	Yes if structured No if unstructured	Normally not

* Some CASE COBOL systems provide screen painting and report-generating utilities.

Some people who have devoted themselves to mastering the structure-oriented approach to systems and software development get this message from object-oriented enthusiasts: "Everything you've learned and practiced is wrong." This message is not only pompous, but it's also untrue. While it's true that the object-oriented approach differs in some respects from the structure-oriented approach, both, if followed correctly, will ensure successful systems development.

Indeed, the object-oriented approach doesn't provide anything stunningly new. It does, however, offer a good set of OOP language tools for achieving sound modularization and reusability. So the message to the traditionalists should be: "Everything you know is worthwhile and correct. Here are some language tools to help you do things a little better in some situations."

OBJECT-ORIENTED PROGRAMMING (OOP) LANGUAGES support most of the object-oriented design concepts described in this book. An OOP language with an extensive class library can dramatically decrease SWDLC time and cost. Most OOP languages, however, are in a state of flux.[3] For example, C++ is periodically coming out with a new revision.

The Object Management Group (OMG) is the standard-setting body for object-oriented software design and programming. Its Object Management Architecture Guide defines a framework that vendors can use to develop an object-oriented environment. This environment consists of an OOP language, such as C++, a graphic user interface (GUI), and an object-oriented database. The goal of OMG is to define specifications that developers can use to build software design applications. They will run under any compliant object-oriented environment and across a wide variety of computer platforms.

Whereas 4GLs and especially 3GLs implement structure-oriented software designs, OOP languages implement various features of object-oriented software designs discussed in Chapter 16. A wide array of OOP languages serve as implementers. Specific OOP languages, however, vary in their support of object-oriented software design concepts. Some languages, such as C++ and Eiffel, support multiple inheritance. Many others do not.

Types of OOP Languages

There are two types of OOP languages:

- Pure
- Hybrid

Pure OOP Languages

PURE OOP LANGUAGES build objects from scratch. They contain more object-oriented capabilities than do hybrids. Probably the best example of a pure OOP language is Smalltalk, developed by Alan Kay at Xerox Corporation. Because everything related to pure OOP languages such as Smalltalk is an object, it provides a useful study in complete OOP implementation. The Smalltalk learning curve is, however, steep for programmers who are well grounded in struc-

[3] Tim Chase, "Everything You Wanted to Know About Object Oriented Programming But Were Afraid to Ask," *Interact*, February 1991, p. 20.

tured programming, and the learning curve is literally straight up for programmers who have been writing nonstructured programs and who are unfamiliar with object-oriented design concepts.

Hybrid OOP Languages

Probably the more common application of the OOP approach is via **HYBRID OOP LANGUAGES**. Hybrid OOP languages graft object-oriented capabilities onto familiar 3GLs, such as C, COBOL, or Pascal. The most popular, and generally purported to be the de facto standard hybrid OOP language, is C++, based on C.

The main advantage of hybrid OOP languages is to enable programmers already familiar with a base 3GL, such as Ada, C, COBOL, or Pascal, to migrate to the hybrid OOP languages gradually. Whereas Smalltalk and Eiffel appeal to the object-oriented purists, C++, Object Pascal, and Object COBOL appeal to the traditionalists. To be able to claim mastery of the OOP approach, however, designers and programmers must be able to understand the major OOP design characteristics, discussed in Chapters 6 and 16, which are:

- Reusability
- Extendability
- Messages
- Objects and object classes
- Encapsulation
- Inheritance
- Polymorphism

Once these design characteristics are mastered, the syntax of the specific language should be more easily mastered. At that point, systems professionals can consider themselves to be fully knowledgeable about the OOP approach.[4]

Smalltalk

Smalltalk was the first popular pure OOP language. Its early success caused the development of other OOP languages. Smalltalk originated much of today's object-oriented concepts and terminology.

Smalltalk is built around two simple concepts:

- Everything is treated as an object.
- Objects communicate by passing messages.[5]

Smalltalk is excellent for reusability and extendability.

---

[4] Grady Booch, *Object Oriented Design* (Redwood City, Calif.: Benjamin/Cummings, 1991), p. 474.

[5] Ibid., p. 475.

Smalltalk is an interpreted-based language that is connected to a graphically based environment that totally integrates windows, editing, debugging, compiling, and graphics. It therefore provides the designer/programmer with a highly interactive development environment which avoids the edit-compile-link cycle delays of traditional compiler-based languages. It possesses an extensive class library, which can contribute to rapid prototyping of an application.

Eiffel

Eiffel is a pure OOP language developed by Bertrand Meyer. Programs consist of collections of class declarations that include methods. Multiple inheritance is strongly supported, and a small class library is provided. Eiffel is supported by excellent development tools. The Eiffel compiler translates source programs into C. It has good facilities for encapsulation, and is an excellent business-oriented OOP language.

C++

C++ is a hybrid OOP language developed by AT&T Bell Lab researcher Bjarne Stroustrup as an object-oriented variant of C. Some people call C++ "C with classes." In general, the biggest drawback of C++ may be its relative complexity. That is, C is typically the language of choice for many professional software engineers, and C++ possesses similar intricacies and robustness.

With C++, some entities are treated as objects, and some are not. C++ was developed not only to add object-oriented capabilities but also to correct some of C's weaknesses. Because it is a superset of C, C++ guarantees upward compatibility with the existing base of C programs.

C++ does not offer a standard class library, although several class libraries have been developed by various C++ vendors. These different class libraries may be incompatible. C++, however, provides a good method of specifying access to attributes and operations of a class. Inheritance allows developers to reuse existing, pretested objects to create their own customized (or application-specific) objects. Moreover, multiple inheritance permits developers to inherit the properties of more than one object.

C++ is a nonproprietary language supported by major computer vendors. C++ is presently the dominant OOP language for general use.

Object Pascal

Object Pascal is a hybrid OOP language created by developers from Apple Computer in conjunction with Niklaus Wirth, the designer of Pascal. Because Object Pascal is more English-like and adds fewer new commands in its OOP implementation, it is normally considered a friendlier language for learning OOP than C++. Like Smalltalk, Object Pascal has an extensive and rich class library.[6] Designers/programmers normally start by using the most obvious classes in a library. As they become accustomed to using these obvious classes, they move incrementally to the use of more sophisticated classes. Eventually, they may even add classes of their own.

---

[6] Ibid., p. 286.

Object Pascal allows single polymorphism and single inheritance. Unlike Smalltalk and many other OOP languages, all attributes in Object Pascal are unencapsulated. And unlike C++, all methods are public in Object Pascal.[7]

Like C++, Object Pascal provides compatibility with existing Pascal programs. These programs can have object-oriented constructs added to them. This feature creates a shorter learning curve to object expertise for those who already know Pascal.

Object COBOL

Most mundane business systems use tried-and-true COBOL. There is no way that billions of lines of COBOL code will be discarded. This is the main reason for Object COBOL to exist.[8]

Object COBOL is a hybrid OOP language that will provide a gateway to an array of services that are currently not readily accessible from traditional COBOL, such as the ability to assemble objects that have already been coded and tested. By being able to assemble precoded, pretested objects rather than coding all of the functionality from scratch, Object COBOL will cut designing, coding, and testing time by a large percentage. Moreover, the adoption of Object COBOL will be easier for traditional COBOL programmers, and also for those who feel more comfortable with English-like commands.

OOP Tools

Language and development tool vendors provide a wide array of object-oriented development toolkits. For example, Smalltalk is adapted to work with windows and windowing standards.[9] Borland's Turbo C++ offers an excellent C++ development environment.

With groups like the Object Management Group (OMG) producing and publishing standards for OOP tools, windows-based OOP development systems are becoming standardized. Standard mouse-driven point-and-click graphic user interfaces (GUIs) include class libraries such as the following:

- Dialogue boxes
- Menus
- Buttons
- List boxes (plus object and class browsers)
- File and processing editors
- Compilers
- Interactive testers and debuggers
- Interactive code generators

---

[7] Ibid., p. 290.

[8] Roger J. Dyckman, "COOL/3000: The Object Is the Subject," *Interact,* February 1991.

[9] Don Crabb, "OOP Tools Ease Windows Developers' Pains," *Infoworld,* February 25, 1991, p. 52.

Microsoft's Application Factory provides an integrated development system that includes:

- Tools for creating and manipulating objects, including object-library browsing and object-assembling features
- Visual programming environment for constructing application and interface objects
- A shell for supporting both object-oriented and structure-oriented programming

This CASE workbench contains a central repository that stores all the details related to a given software project.

Objectworks, a CASE environment for C++, provides the following features:

- Incremental compilation and object assembling
- Access to code by class
- Graphic representation of single and multiple inheritance
- Object inspection that provides inspection and alteration of objects at run time
- Interactive testing and debugging tools

Many OOP tools help speed development time by providing a set of tools that enable designers/programmers to escape the edit-link-compile cycle. Additionally, some development tool vendors blend into their packages expert systems and hypertext and hypermedia (or multimedia) features.

Armed with mouse-driven point-and-click GUI development toolkits, programmers get part of their work done for them in the form of predetermined and pretested objects. Particularly helpful are objects for building the user interface, such as dialogue boxes, menus, icons, buttons, pointers, scroll bars, and so on. If these objects are readily available, programmers can complete their projects faster because they don't have to write as much code. Also, with fewer lines of code written, there are fewer places for bugs to appear, making it easier to produce reliable programs.

SPECIAL-PURPOSE LANGUAGE TOOLS

The programming languages described above are generalized, multipurpose languages normally used for big projects. Although they can also be used for special-purpose applications, we want to devote this section to a group of language tools (some people would refer to them as development tools or design aids) that are designed for particular (i.e., ad hoc) purposes. Three examples of **SPECIAL-PURPOSE LANGUAGE TOOLS** are:

- Interactive user-oriented language tools
- Database management system (DBMS) query language
- Hypertext and multimedia (or hypermedia) language tools

The accompanying case provides a real-world setting in which these special-purpose language tools are applied.

Designing a Decision Support System at Delta Manufacturing Company

Delta Manufacturing Company is a midsized manufacturer of hand tools and plumbing fixtures. Derek Wright is director of systems development at Delta. A new replacement information system was implemented four months ago. It has met with great success. It has fulfilled the business transaction and information processing needs of most end users. However, Derek has a meeting with three top executives at Delta to discuss the development of a decision support system (DSS) for them.

Clyde Williams, plant manager, congratulated Derek and his staff on the success of Delta's new information system. "Derek," said Clyde, "We are extremely pleased with what you've accomplished. The reason for this meeting is to explore the feasibility of developing a decision support system for us that we could use without the need to go to you or your staff every time we need a report. In other words, something we could do on our own."

"What I envision," said Margaret Benfeld, marketing manager, "is a workstation at my fingertips where I can access the database for sales information and information about trends and the competition. I'd also like to make some what-if analysis. But I want to do it myself."

"The same setup for me," said Jeff Trainor, chief financial officer. "I would like to access the new database management system from time to time from my own workstation."

"Derek, are we wanting something that is not feasible?" asked Clyde.

"No, not at all," Derek responded. "My staff and I can develop a decision support system for you all. We will need to set up a series of interviews to clarify what your requirements are. Also, let me stress this point: You will have to commit to training, because even with all the advancements we've made in information technology, programming languages, and development tools, users still must devote some amount of time and effort to learning how to work with the system."

"We're ready to spend as much time and effort as is needed to learn how to use the system," said Clyde. Jeff and Margaret nodded in agreement.

"Very good," said Derek. "But first we need to set up our interviews with you all."

"How about a JAD session like the one we did for the new system?" asked Margaret. "Those early JAD sessions really set a good foundation and got us off to an excellent start."

"Sounds good to me," Derek responded enthusiastically (Derek had been the major champion of joint application development at Delta). "How about setting up a miniJAD session next Wednesday? Are all of you free at that time?" Everyone agreed that they were and could meet next Wednesday.

It required three months of design and redesign before the decision support system was implemented at Delta. The general architectural design is illustrated in Figure 17.3. The top part of the figure includes components that were already in place.

Figure 17.3
Decision support system (DSS).

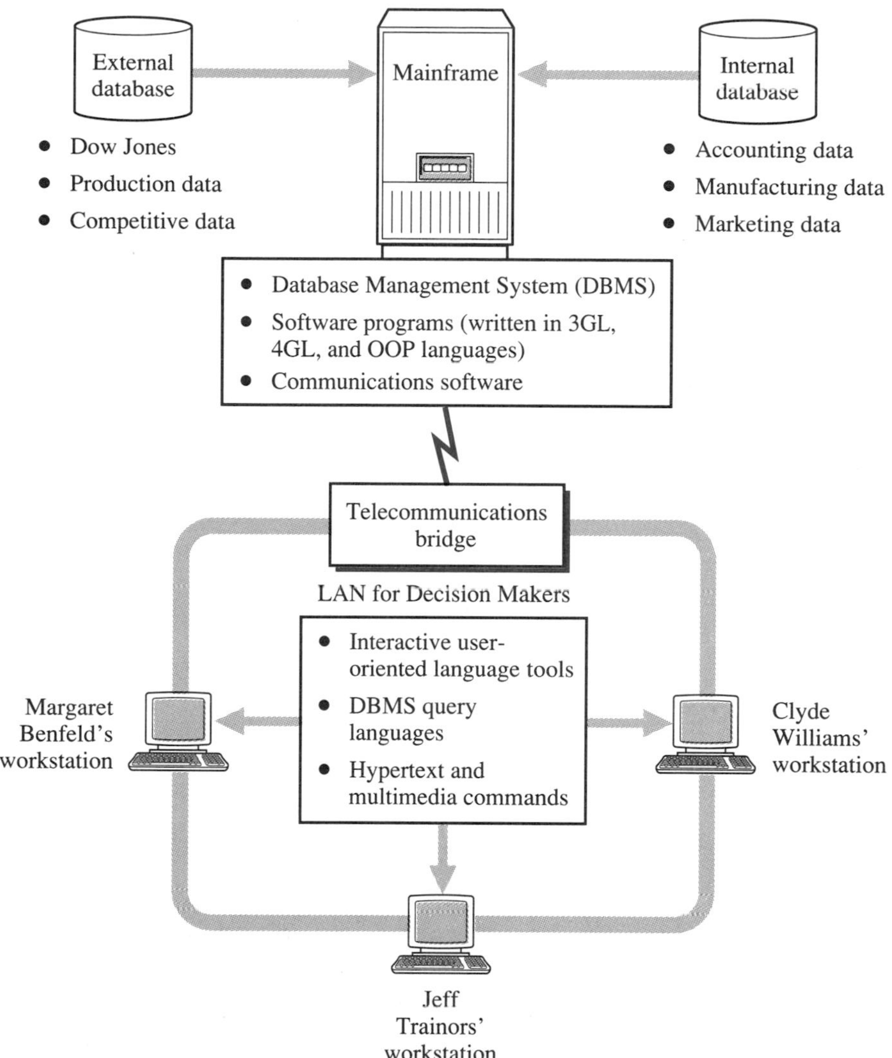

The LAN represents the new DSS and its connection to Delta's information system. The external database has been added to provide data about the environment in which Delta operates. The mainframe, in addition to performing transaction processing tasks and serving other users, also acts as a database server and communications controller for the DSS LAN. What follows is material that describes the main features of this new DSS.

Interactive User-Oriented Language Tools

INTERACTIVE USER-ORIENTED LANGUAGE TOOLS permit users to interact with the system in the form of:

- Interactive dialogue language tools
- Spreadsheet commands for what-if analysis

Using Interactive Dialogue Language Tools

The following **INTERACTIVE DIALOGUE LANGUAGE TOOLS** allow Delta's executives to carry on a dialogue with the system.

Commands This form of dialogue with the system uses one of two types of languages:

- Syntactical
- Natural

For syntactical commands, Delta's executives learned a special vocabulary to run the DSS. Such commands as COPY, RETRIEVE, PRINT, EDIT, and MOVE are followed by one or more arguments, such as FILEA, FILEB, and FILEC. Syntactical command-driven systems offer the greatest interactive flexibility, but they require the greatest amount of learning and are the easiest to forget. To overcome this problem, the syntactical commands were compiled in a command dictionary for Delta's executives.

The aim of the natural language interface is to reduce the amount of special syntax that executives have to learn. A natural language command, for example, is: "Display division B's sales for the first quarter."

Menus A menu-driven DSS permits executives to select an array of items. To a great extent menus guide the interaction.

Figure 17.4 depicts a typical DSS menu interface, in this case making use of touch screen technology. Note that it provides a straightforward means for the busy executive with little computer experience to retrieve both company-spc-cific information and information concerning the firm's competitive environment. An electronic mail utility is also available, so the executive may request additional information from key personnel.

Icons These involve providing pictorial representations of different procedures. A file cabinet indicates a procedure to save a file, or a picture of a bar chart represents a procedure to display the information on a bar chart.

Question and Answer With this approach, the DSS asks a series of questions, the executives provide the answers, and at the end of the Q/A session, the system makes a recommendation. A simple example follows:

DSS: What is your objective?
USER: To acquire firms for growth.
DSS: What are the names of companies you are investigating?
USER: Inputs ticker symbols of candidate companies.
DSS: What are their credit ratings, high–low stock prices of 1992 and 1993, price/earnings ratio, earnings per share since 1991, current assets, current liabilities, long-term liabilities, and capital?
USER: Inputs these variables via a data-entry electronic form.

The DSS combines these variables with extensive data stored in the external database and processes them via investment acquisition models. The executives are then given acquisition scenarios and recommendations.

Figure 17.4
Menu-driven DSS interface.

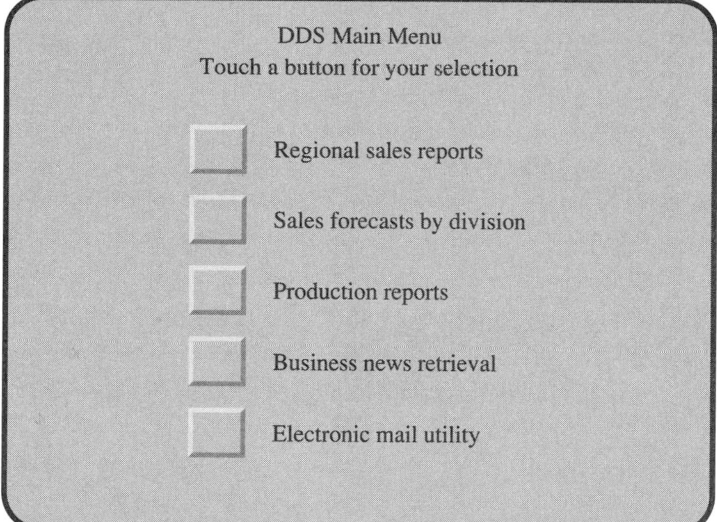

Dialogue Boxes These boxes, displayed on the screen, contain options asking for different information. After the requested information is provided, the user chooses a button and touches it to carry out the command or select an option.

The electronic mail utility we mentioned earlier is an ideal application for a dialogue box. Figure 17.5 shows how the electronic mail screen might look. The executive has just typed a message to one of her marketing managers concerning a sales report. A pop-up dialogue box appears on the screen, allowing her to send the message or cancel it, and offering the option of a return

Figure 17.5
Pop-up dialogue box as part of a decision support system.

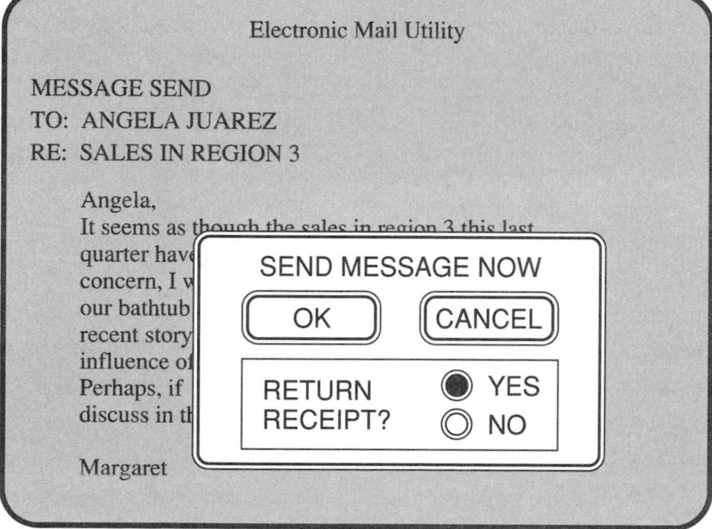

Figure 17.6
Initial what-if screen (in thousands).

	1	2	3	4	5
Sales	100.00	125.00	156.25	195.31	244.14
Variable costs	75.00	100.00	125.00	150.00	175.00
Contribution margin	25.00	25.00	31.25	45.31	69.14
Profit volume margin	0.25	0.20	0.20	0.23	0.28

Sales = 100, Previous sales * 1.25
Variable costs = 75, 100, 125, 150, 175
Contribution margin = Sales – Variable costs
Profit volume ratio = Contribution margin / Sales

receipt when the message is read. The options of the dialogue box are easily selected with a mouse pointer or by touching a touch-sensitive screen.

Using Spreadsheet Commands for What-If Analysis

One of the most powerful features of Delta's DSS design is its ability to provide what-if analysis for its executives. Assume, for example, that they want to project cost-volume-profit (CVP) data for five periods. The marketing department has estimated that sales for the first period for a particular product will be $100,000, and they will increase by 25 percent each period. The cost accounting department projects the variable costs at $75,000 for the first period, $100,000 for the second period, $125,000 for the third period, $150,000 for the fourth period, and $175,000 for the fifth period. With these data, a what-if screen is set up to reveal the results of changing conditions. The initial what-if screen and its results are displayed in Figure 17.6.

At the bottom of the initial what-if screen are the models and data that are used to generate the values at the top of the screen. This initial screen becomes the base case. When a new definition is given to a variable, this definition replaces the one in the base case. The base case, however, is still stored in memory and can be recalled at any time.

In many what-if analyses, executives may want to change a variable in one or more periods from its value in the base case. For example, the what-if statement:

```
Sales = Prior * 1.10
```

tells the software to use the definition in the base case and increase it by 10 percent.

Another powerful feature of the what-if feature is goal-seek. Through goal-seek, executives find out the value a particular variable has to have to achieve a desired level of performance. For example, executives might type the goal as:

Figure 17.7
Goal-seek screen (in thousands).

	1	2	3	4	5
Sales	100.00	125.00	156.25	195.31	244.14
Variable costs	75.00	80.00	91.25	110.31	139.14
Contribution margin	25.00	45.00	65.00	85.00	105.00
Profit volume margin	0.25	0.36	0.41	0.44	0.43

Sales = 100, Previous sales * 1.25
Variable costs = 75, 100, 125, 150, 175
Contribution margin = Sales – Variable costs
Profit volume ratio = Contribution margin / Sales

Goal-seek:
Goal: Contibution margin = 25, Previous + 20
Adjust: Variable costs

```
Goal: Contribution Margin = 25, Previous + 20
```

This goal states that the contribution margin is to start at $25,000 and to increase by $20,000 each period.

Once the goal is established, the system asks for the variable to be adjusted. To find out how executives reach the goal by changing variable costs, they type:

```
Adjust: Variable Costs
```

As soon as the executives press the Return key, the DSS solves the model and displays the results as shown in Figure 17.7.

The results of the goal-seek screen indicate to executives that they must reduce variable costs by those amounts shown for each period if actual sales equal those projected in the base case, and if Delta is to realize a contribution margin increase of $20,000 each period beginning with a contribution margin of $25,000 in period 1.

Using Relational Database Management System Languages

Two notable DBMS (or RDBMS) query languages that were chosen for Delta's DSS are:

- Structured query language (SQL)
- Query-by-example (QBE)

Structured Query Language

The structured query language (SQL) is a set of commands that enables Delta's executives to define, manipulate, and control data in Delta's relational DBMS.

SQL can be used for mathematical operations and subqueries. For example, Clyde wants to find all employees whose salary is greater than the average of all employees in the company. The SQL command that Clyde writes is

```
SELECT   Employee_Name
FROM     Employee_Table
WHERE    Salary > (SELECT AVG(Salary)
                   FROM Employee_Table)
```

SQL can also be used to insert, update, and delete data in the database. For example, to insert a new customer named Tricor, the necessary SQL commands may look like this:

```
INSERT INTO  Customer_Table (Customer_Number,
               Customer_Name, Street, City,
               Amount_Owed)
VALUES       (19978, Tricor, 49 Comstock,
               Reno, 0.0);
```

The executives are not the only ones who can take advantage of SQL. As the preceding examples indicate, other users also require access to the relational DBMS to keep it updated and to make certain ad hoc queries. Therefore, Jeff's staff embedded SQL in COBOL, the information system's main transaction-processing language. This embedded SQL significantly reduced the time and effort required to design and code the mainline 3GL programs. It also enhanced query-and-control features of the information system.

Query-By-Example

Query-by-example (QBE) is one of the most popular visual interfaces to relational DBMSs. QBE specifies an example of the query or update that is needed. The notion of querying by example is that instead of describing the commands to be followed in obtaining the desired information, the executive gives an example of what is required.

Figure 17.8 illustrates an application of QBE. The attribute value P. identifies those attributes to be printed on a report or displayed on a screen. For example, assume the query: ''I want to find all valves ordered alphabetically by description in warehouse A with quantity on hand greater than 30 and a price of more than $12.00.'' This query is at the left of the figure; the results of the QBE query are shown at the right.

Figure 17.8
QBE query and its response.

QBE Query

Inventory	Inventory Item	Description	Quantity on hand	Price	Warehouse
P.	Valves	Order by	>30	>12.00	A

QBE Response

Description	Quantity on hand	Price
2″ Ball	61	14.50
3″ Check	37	18.90
6″ Seat	58	64.90

Figure 17.9
QBE insert, delete, and update procedures.

QBE Query

Inventory	Inventory Item	Description	Quantity on hand	Price	Warehouse
I.	Wrenches	24" Box	= 14	9.00	B
D.	Saws	Crosscut			C
U.	Valves	2" Ball		14.50 * 1.20	A

Like SQL, QBE can also perform procedures such as updates, arithmetic operations, relation joins, inserts, and deletes. For example, suppose that Margaret wants to insert a new inventory item, which includes fourteen 24-inch box wrenches priced at $9.00 and stored in warehouse B. Also, crosscut saws that are located in warehouse C represent an inventory item that is to be deleted from inventory. Finally, the price of 2-inch Ball valves located in warehouse A is to be increased by 20 percent. The current price of the 2-inch valve is $14.50. The table that handles these QBE procedures is displayed in Figure 17.9.

Hypertext and Multimedia Tools

HYPERTEXT is a method of organizing and associating textual information in a nonsequential hierarchical manner. When the concept of hypertext is expanded to include linking together not only textual information but also graphic, icon, video, sound, programs, and other forms of pertinent information, the concept is expanded to become **MULTIMEDIA**, also referred to as **HYPERMEDIA**. These tools are used to help manage an ever-increasing flood of information at Delta, and to help executives assimilate important information without being drowned by it.

A typical application of hypertext and multimedia tools is the development of a production performance report for Clyde Williams. He browses through chunks of information, and then selects, links together, and maps these chunks into a multimedia report.

The first screen presented to Clyde is a hypertext retrieval menu, shown at the top of Figure 17.10. At the bottom is the result of choice 3, which is another menu giving basic plant production reports: direct material cost summary, direct labor cost summary, manufacturing overhead cost summary, general inspector's reports, and the choice to go to another topic. Clyde's choice is 4, the general inspector's reports.

The result of the last choice by Clyde is a screen, displayed in Figure 17.11, used to link together chunks of information in a multimedia (or hypermedia) workspace. The textual material contains reams of reports dealing with the general inspector's evaluation of many production operations, including producing, receiving, storing, and shipping.

Figure 17.10
Hypertext menus.

Hypertext Retrieval Menu

15 performance reports, 12 cost summaries, and 14 graphics are available for plant production.

Do you want to:
1 BROWSE this complete set of text and graphics?
2 BROADEN this information space?
3 NARROW this information space?
Type 1, 2, or 3 and press RETURN to indicate your choice
ENTER CHOICE: 3

When appropriate, type either HELP, RESTART, or STOP and press RETURN.

Basic Plant Production Information Menu

Do you want:
1 Direct material cost summary?
2 Direct labor cost summary?
3 Manufacturing overhead cost summary?
4 General inspector's reports?

Type 1, 2, 3, or 4
and press RETURN

ENTER CHOICE : 4

When appropriate, type either HELP, RESTART, or STOP and press RETURN.

First, Clyde wishes to browse some of these reports. This is like electronically thumbing through a stack of papers. He selects BROWSE and clicks the mouse to enable him to do this. Finding a specific chunk or fragment of text that he wants, he moves the cursor to TYPE and clicks the mouse. The fragment of text he selects is automatically copied in the multimedia workspace. Then he browses through the graphics space and chooses a spider web chart to show four performance-evaluation elements of manufacturing (i.e., working on the shop floor): preventive maintenance, housekeeping, raw materials quality, and tool control. (These elements were rated by the general inspector.) A rating scale is also included from the icon space to clarify the meaning of the spider

Figure 17.11 (opposite) Use of hypertext and multimedia to associate information chunks.

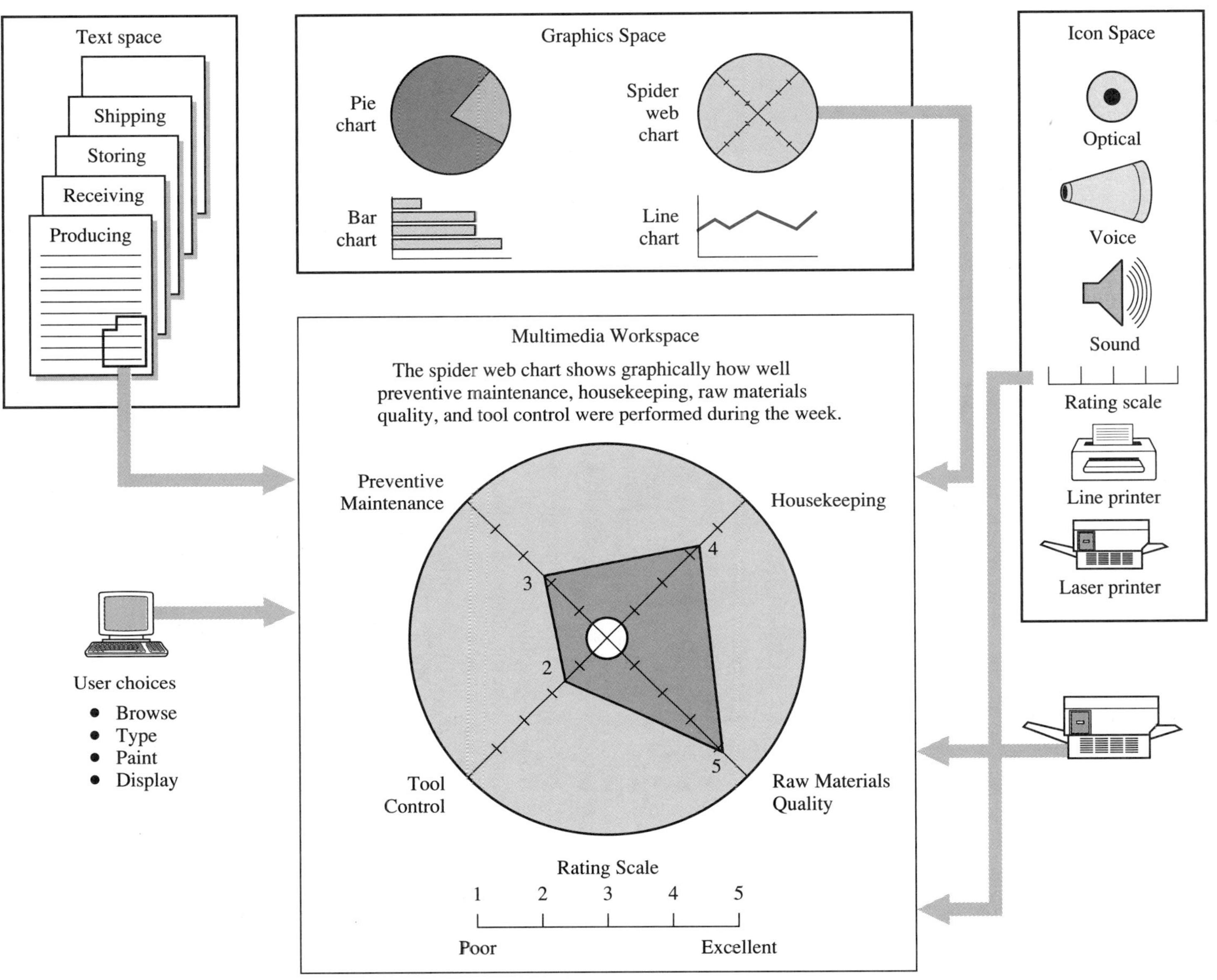
Text space
Shipping
Storing
Receiving
Producing
Graphics Space
Pie chart
Spider web chart
Bar chart
Line chart
Icon Space
Optical
Voice
Sound
Rating scale
Line printer
Laser printer
Multimedia Workspace
The spider web chart shows graphically how well preventive maintenance, housekeeping, raw materials quality, and tool control were performed during the week.
Preventive Maintenance
Housekeeping
Tool Control
Raw Materials Quality
3
4
2
5
Rating Scale
1 2 3 4 5
Poor
Excellent
User choices
• Browse
• Type
• Paint
• Display

web chart. Both the spider web chart and rating scale are automatically painted on the multimedia workspace screen by selecting the appropriate graph and icon, moving the cursor to PAINT, and clicking the mouse. And finally, Clyde wants the multimedia workspace with its text fragment, spider web graph, and rating scale printed on a laser printer. This is done by selecting a laser printer icon, moving the cursor to DISPLAY, and clicking the mouse. A hard copy of the multimedia workspace is printed automatically via a laser printer.

Graphic user interface (GUI) multimedia development tools enable developers to incorporate audio, graphics, text, and animation in their applications. A real advantage to such tools is that they allow nonprogrammers to create applications easily. Hardware manufacturers have released multimedia PCs that support all the multimedia features. These hardware platforms contain a CD-ROM drive, as well as audio and animation support.

SELECTING THE APPROPRIATE LANGUAGES

Probably the only perfect languages and tools for users are natural spoken languages; that is, users simply tell the computer what they want and the computer responds immediately and precisely. A great deal of research is under way to achieve this ultimate user/system interface. Whether this ideal will ever be achieved is open to a great deal of speculation. In the meantime, we must select the appropriate language currently available for the software design.

Matching Languages with Software Design Applications

To see a summary guide in helping choose the right language or mix of languages and special-purpose language tools, refer to Figure 17.12. When choosing which languages and tools to use for coding specific software design applications, several questions should be considered:

- Does the present computer system have an assembler, compiler, or translator for a particular language?
- Are there programmers in the organization who are proficient in the language?
- Are there large numbers of programmers throughout the business community who are skilled in the language?
- If the program is to be run on different computer architectures, is the language it is coded in portable?
- Do many CASE vendors support the language?
- If special-purpose language tools are to be used, are users willing to devote sufficient time and effort to be trained in their use?

Possibly the most important question, however, is: What is the nature of the design application? Answers to this question follow.

Software Design Application	Language			
	3GL	4GL	OOP Language	Special-Purpose Language Tools
Application where machine efficiency is paramount	Excellent	Fair to poor, depending on application	Good	NA
I/O-bound	Excellent	Poor	Excellent	NA
Process-bound	Varies, FORTRAN: Excellent COBOL: Fair to poor	Fair	Fair to excellent	NA
Decision support system (DSS)	Fair to excellent	Excellent	Excellent	Excellent
Executive information system (EIS)	Fair to excellent	Excellent	Excellent	Excellent
Prototypes	Fair to excellent	Excellent	Excellent	NA
Ad hoc	Excellent if embedded SQL is used	Excellent	Excellent	Excellent
Voice interaction	NA	NA	NA	Fair and evolving
Supports structured design	Yes	Not very well	NA	NA
Supports object-oriented design	Ada, C, COBOL, and Pascal can be modified to implement some OO design concepts.	No	Smalltalk, Eiffel, C++, Object Pascal, and Object COBOL are designed specifically for OO design concepts.	NA

NA = not applicable.

Figure 17.12 Summary of software design applications and languages that support them.

- If the application is process-bound, requiring a large number of complex calculations, a language such as Ada, APL, C, FORTRAN, Pascal, or PL/1 is suggested.
- If the application is input/output-bound, requiring extensive file manipulation, such as processing a large payroll or accounts receivable file, COBOL or RPG is the likely choice.
- If the application involves data manipulation for management analysis, quick reports, or prototyping, a 4GL should be considered.

- For very special information processing needs, interactive user-oriented language tools, DBMS query languages, and hypertext and multimedia (hypermedia) tools should be considered.
- To implement a structured, modular design approach, a host of 3GLs are recommended, such as Ada, C, COBOL, and Pascal.
- To implement object-oriented software designs, a large number of OOP languages are available, such as Smalltalk, Eiffel, C++, Object Pascal, and Object COBOL.

C++ appeals to the scientific and engineering community because of its C heritage. Smalltalk, Eiffel, Object Pascal, and especially Object COBOL appeal to the business community. Indeed, members of the American National Standards Institute (ANSI) COBOL committee have merged with the Conference on Data Systems Languages (CODASYL) COBOL committee and the Object Oriented COBOL Task Group (OOCTG). Their objectives are to support, develop, and standardize Object COBOL.

Additional Issues for Selecting Languages

Other real issues in choosing programming languages include the following:

Level of Usage in the Business World If a language is widely used in the business world, it is likely to have ongoing software development support from major CASE vendors and a large number of people who are conversant in it. It is also likely to be easily portable from one technology platform to another, because most, if not all, computer vendors support it. Assume the following situation: The entire systems staff of ABC Company, including systems analysts, systems designers, and programmers, went in together and bought the winning ticket of a $100 million lottery. If they don't return to work, will other systems professionals be able to review the work in progress, understand the language(s) being used, and continue the work in progress with minimal loss of time? If the answer is no, it's time for a change to systems development methods and languages that are standard in the business world and are more expressive.

Expressiveness The language's ability to represent the software design fully in terms that both programmers and nonprogrammers can understand is termed expressiveness. The expressiveness of COBOL is one of the reasons that this 3GL is so widely popular in the business community.

Convenience The level of convenience is based on features such as:

- Ease of use
- Ease of learning
- Development productivity

3GLs with extensive CASE development tools, 4GLs, OOP languages with rich class libraries, and special-purpose language tools provide convenience.

Portability Portability is the ability to move languages to different environments. It is based on an accepted language definition, consistency and completeness of the language features, and a standard set of library functions.[10]

Maintainability After a software program has been coded in a particular language, tested, and converted to operations, its maintainability becomes a very important issue. Thus its ease of change to adapt to the dynamics of the business it serves is critical. However, maintainability is more a function of design and documentation than it is of the language unless the language is obscure and maintenance programmers are unavailable. In such a case, the maintainability, to a large extent, becomes a function of the language. If sound structured design principles are employed, and the design is coded in COBOL, the resulting software program should be relatively easy to maintain. If, on the other hand, the same software design is coded in an obscure language, maintenance will become a problem because not many people will be conversant in it.

Extendability The language's ability to increase the scope of the original software design without recoding the program is termed extendability. OOP languages are noted for their extendability. 3GL programs, if coded following the structure-oriented software design approach, are also easily extended.

SOFTWARE DOCUMENTATION

SOFTWARE DOCUMENTATION describes the programs designed and coded to support the system. Software documentation is composed of:

- Internal documentation
- External documentation

Internal Software Documentation

INTERNAL SOFTWARE DOCUMENTATION is embedded in the program's code. Because of this feature, a number of languages are considered "self-documenting" languages. For example, COBOL is considered self-documenting because the program coded in COBOL reads like simple English. But for a program to read like simple English, the systems analyst must specify meaningful and standard names. For example, a field in a record that contains unique numbers that identify customers may be called CUSTOMER-NUMBER. This name is meaningful; it may also be the standard because any program using it would have to spell it the same way. For another example, review the following COBOL program segment:

```
P1.     IF HOURS > 40 PERFORM P2
        ELSE PERFORM P3.
```

[10] Michael Beckman and Dmitry Lenkov, "The C++ Language," *Interact,* May 1991, p. 42.

```
P2.     COMPUTE RP = R * 40.
        COMPUTE OT = 1.5 * R * (H - 40).
        COMPUTE GP = RP + OT.
P3.     COMPUTE RPGP = R * H.
```

The preceding segment will execute because it is written properly as far as the syntax of COBOL is concerned. However, semantically, it is virtually meaningless to a nonprogrammer or another programmer who has to maintain it. Yet, with the application of meaningful and standard names, the segment can be converted to one that can be understood by programmers and nonprogrammers alike, such as

```
COMPUTE-GROSSPAY.
  IF HOURS-WORKED > 40
       PERFORM GROSSPAY-WITH-OVERTIME
  ELSE
       PERFORM GROSSPAY-WITHOUT-OVERTIME.
GROSSPAY-WITH-OVERTIME.
  COMPUTE REGULAR-PAY = HOURLY-RATE * 40.
  COMPUTE OVERTIME-PAY = 1.5 * HOURLY-RATE *
       (HOURS-WORKED - 40).
  COMPUTE GROSSPAY = REGULAR-PAY + OVERTIME-PAY.
GROSSPAY-WITHOUT-OVERTIME.
  COMPUTE REGULAR-PAY-GROSSPAY = HOURLY-RATE *
       HOURS-WORKED.
```

A programmer can also increase the internal documentation of the program by inserting comments and explanations bordered by asterisks into the program. For example, the following may be inserted just before a program segment that calculates economic order quantities for inventory items.

```
* * * * * * * * * * * * * * * * * * *
*                                   *
*        THE FOLLOWING MODULE       *
*         CALCULATES ECONOMIC       *
*     ORDER QUANTITIES (EOQ) FOR    *
*          INVENTORY ITEMS.         *
*                                   *
*                                   *
* * * * * * * * * * * * * * * * * * *
```

As we have seen with structured English, it is a good idea to indent the code in the same manner, and never to write more than one sentence (i.e., program statement) on the same line. These rules increase readability of the code. For example, view the following code segment:

```
IF EMPLOYEE-NO = RETIREMENT-NO PERFORM RETIREMENT-UPDATE ELSE
DISPLAY "NO UPDATE." PERFORM READ-ROUTINE.
```

For better readability, change the segment above as follows:

```
IF EMPLOYEE-NO = RETIREMENT-NO
     PERFORM RETIREMENT-UPDATE
ELSE
     DISPLAY "NO UPDATE."
PERFORM READ-ROUTINE.
```

Figure 17.13
Title page for program manual.

Page 1 of 1

Title Page

Program Name:	CREATE_SALE
Program Number:	P7212
Program Purpose:	This program reconciles sales transactions with sales master records, and generates reports.
Programmer:	Gerald Dwayne Orland
User Department:	Sales 415
Date Written:	September 17, 1992
Project Manager:	Helen T. Berg
Controls:	Transaction serial numbers must agree with those contained in file CONFIG.DAT at end of run.
Input:	(1) MASTER_FILE File label: SALESMASTR.DAT Data Records: MASTER_REC
	(2) TRANSACTION_FILE Source: SALE_TRANS Data Records: TRANS_REC
Output:	(1) SALES REPORT (2) EXCEPTION REPORT
Distribution of Output:	Copy 1 to Sales Copy 2 to Warehouse Manager Exception Report to Marketing Manager

Paper-Based External Software Documentation

There are numerous forms that can be used in documenting a software program. Each company normally establishes its own **PAPER-BASED EXTERNAL SOFTWARE DOCUMENTATION** standards with respect to items that make up this documentation. These items are usually compiled into what is called a **PROGRAM MANUAL.** Some of the more important items found in this program manual are:

- Title page
- Various modeling tools that describe the software design, such as structured program flowcharts, structure charts, Jackson diagrams, Warnier-Orr diagrams, decision tables and trees, structured English, and equations
- Description of input

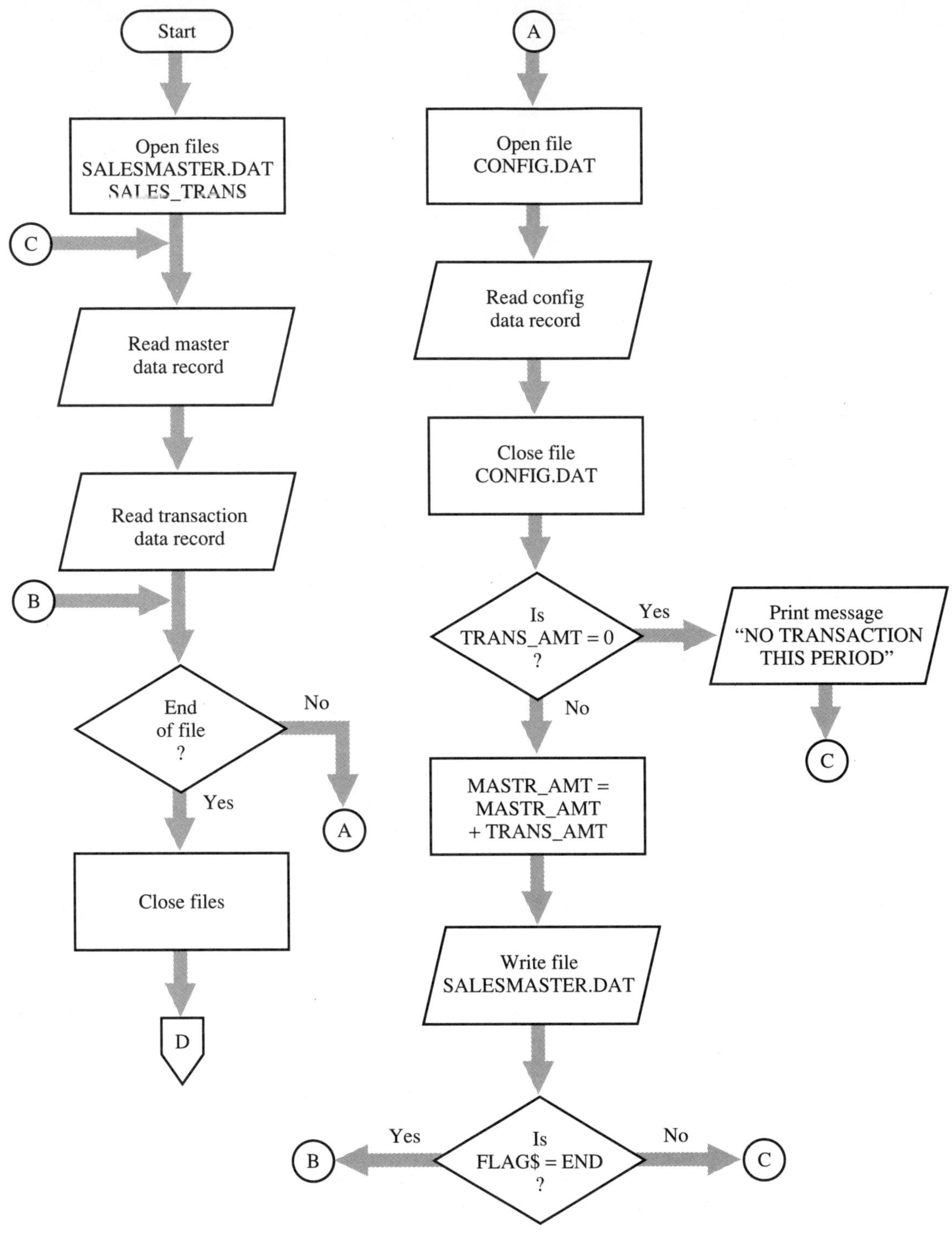

Figure 17.14 Structured flowchart that documents the program's logic.

Figure 17.15
Decision table used in conjunction with the structured program flowchart to document the program's logic.

		RULES	
		1	2
IF (Conditions)	TRANS_AMT = 0	Y	N
	TRANS_AMT ≠ 0	N	Y
THEN (Actions)	PRINT MESSAGE: "NO TRANSACTION THIS PERIOD"	X	
	MASTR_AMT = MASTR_AMT + TRANS_AMT		X

Figure 17.16
Record layout worksheet.

RECORD LAYOUT WORKSHEET

Program Name: CREATE_SALE   Program Number: P7212
Record Name: TRANS_REC   File Name: SALE_TRANS

Field					
From	To	Size	Char.	Field Name	Remarks
1	6	6	N	CUSTOMER_NUMBER	9(6)
7	21	15	A	LAST_NAME	
22	31	10	A	FIRST_NAME	
32	33	1	A	MIDDLE_INITIAL	
34	36	2	N	DEPT_CODE	10=SHOES; 20=SHIRTS; 30=OTHER
37	38	1	N	SALES_CODE	1=WHOLESALE; 2=RETAIL
39	58	20	AN	ITEM_DESCRIPTION	
59	67	9	AN	ITEM_IDENTIFIER	99AAA99VA
68	73	6	N	UNIT_PRICE	9(4)V99
74	88	15	A	SALES_PERSON	

CHAR:
A—Alpha   N—Numeric   S—Signed Numeric   AN—Alphanumeric

[x] New   Replaces Pages

Approved: ________   Date: 7/15/92   [ ] Revision   ________ thru ________

Figure 17.17
Sample output of the program.

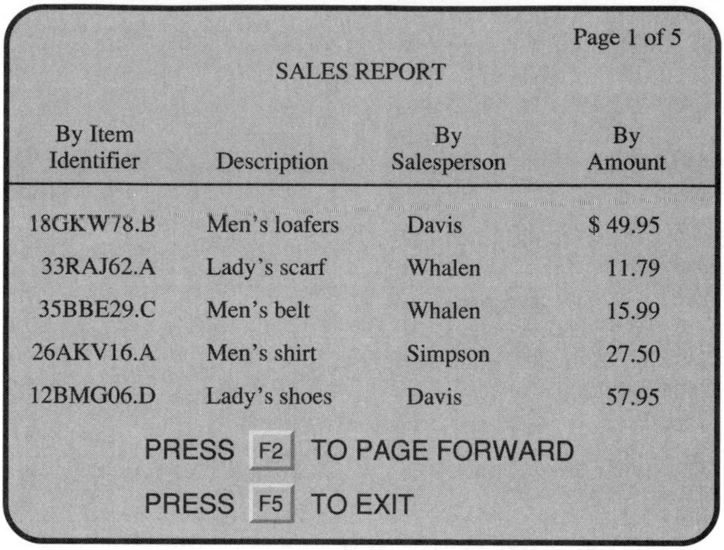

By Item Identifier	Description	By Salesperson	By Amount
18GKW78.B	Men's loafers	Davis	$ 49.95
33RAJ62.A	Lady's scarf	Whalen	11.79
35BBE29.C	Men's belt	Whalen	15.99
26AKV16.A	Men's shirt	Simpson	27.50
12BMG06.D	Lady's shoes	Davis	57.95

Page 1 of 5

SALES REPORT

PRESS F2 TO PAGE FORWARD

PRESS F5 TO EXIT

- Description of output
- Copy of the source code
- Program change sheet

After the software program is tested, the next chapter's subject and test cases used to test the program are also included.

Figure 17.13 shows an example of a title page. It identifies the program and provides other information about the program for easy reference.

The structured program flowchart (other modeling tools could have been used), illustrated in Figure 17.14, represents the logical design of the program. It shows the order of the steps that must be taken by the computer to convert input to output. Figure 17.15 is the decision table used with the structured program flowchart. This decision table describes the logic in the decision "Is Trans_Amt = 0?" in the structured program flowchart.

Figure 17.16 depicts a record layout sheet used to document transaction records contained in the transaction file. It includes such information as the type of characters that are contained within the record fields. A similar layout can be used to document the master records in the master file.

The report shown in Figure 17.17 is an example of the output produced by the software program. The output layout worksheet used in its original design is portrayed in Figure 17.18.

Program Change Sheet

A **PROGRAM CHANGE SHEET** is used to record changes made to the program after it is converted to operations. It is illustrated in Figure 17.19. Normally, a company will use a **MAINTENANCE WORK ORDER (WO) FORM** to initiate and authorize a change to the program after it has been converted to operations. A maintenance work order (WO) form is described in Chapter 20.

Figure 17.18 (opposite)
Output layout worksheet.

TERMINAL SCREEN DISPLAY LAYOUT FORM

APPLICATION CREATE_SALE

☐ INPUT

☒ OUTPUT SALES REPORT

SCREEN NUMBER 3 SEQUENCE 1

COLUMN

Row	Screen content
01	
02	PAGE 1 OF 5
03	
04	SALES REPORT
05	
06	BY ITEM BY BY
07	IDENTIFIER DESCRIPTION SALESPERSON AMOUNT
08	---------- ----------- ----------- -------
09	
10	99AAA99.A A————A A————A $9999.99
11	
12	9999.99
13	
14	
15	
16	
17	
18	99AAA99.A A————A A————A 9999.99
19	----
20	PRESS ¦ F2 ¦ TO PAGE FORWARD
21	----
22	----
23	PRESS ¦ F5 ¦ TO EXIT
24	----

PROGRAM CHANGE SHEET

To: ______________________ Date: ______________

Description of Program Change Requested:

Date Desired: ____________________

Requested By:
Name: ____________________ Title: ______________ Phone: ______________
Department: ____________________

SPACE BELOW FOR PROCESSING USE ONLY

Program Name: ____________________ Program Number: ______________

Change Approved By: ________________________ Date: ______________

Assigned To: ____________________________

Change Reviewed By: ______________ Approved By: ______________

Date: ______________ Date: ______________

NEW DOCUMENTATION CHECKLIST:

- ☐ Change title page, if necessary
- ☐ Change software design, if necessary
- ☐ Change description of input, if necessary
- ☐ Change descripiton of output, if necessary
- ☐ Create new source code listing
- ☐ Change test cases, if necessary
- ☐ Change operations documentation, if necessary
- ☐ Change user documentation, if necessary

Figure 17.19
Program change sheet.

OPERATIONS DOCUMENTATION

Big processing jobs are normally handled by a special group of people called computer operators who work with mainframes. Operators are distinct from end users who typically sit at terminals and workstations connected to a network to perform their tasks.

OPERATIONS DOCUMENTATION is generally in the form of a **RUN MANUAL** kept near the computer console. The run manual includes the following information:

- Identification of the job (application) and when it is run
- Identification of input media (e.g., disk, tape)
- Form numbers for special output forms or number of parts to be used
- Instructions for aligning forms on the printer
- Hardware devices required (e.g., disk, printer, tape drives)
- Expected processing time
- Special instructions should the program abnormally terminate
- Program messages and required operator actions
- Controls
- Output distribution

In large, centralized mainframe-based systems, the run manual may even include the office and home telephone numbers of the maintenance programmers and analysts who will be on call to handle problems 24 hours a day. A

Figure 17.20
Typical run sheet for providing instructions to computer operators.

Program Name: CREATE_SALE
Job Name: CSREV2
Frequency: Daily
Type: Transactions

Input:
1. Sales master file
2. Sales transaction file

Output:
1. Sales report printed on two-part forms
2. Exception report printed on stock paper

Special Instructions:
1. Access sales report generator module from program librarian by CSREV2 and password.
2. In the event of a job abort, the restart procedures are outlined on the attached page.

Programmed Messages:
1. Mount sales report in printer A, align forms, and enter S1 on the console.
2. Mount single-part paper stock on printer B and enter R1 on the console.

Controls
The beginning and ending transaction serial numbers contained in file CONFIG.DAT at the end of job run must agree with the transmittal form.

Output Distribution:
1. Burst and decollate the sales report. Give copy 1 to sales and copy 2 to the warehouse manager.
2. Give the exception report to the marketing manager.
3. Return a copy of the transmittal form to the accounting control clerk.

typical **RUN SHEET** contained in the run manual is displayed in Figure 17.20. Generally, operators do not run a job without consulting a run sheet. A large system may have several thousand jobs and hundreds being added and deleted.

USER DOCUMENTATION

Users generally interact with the system via a screen and keyboard, mouse, or stylus pen. To design **USER DOCUMENTATION** for this interface, it is helpful to classify these users first.

Classifying Users

Some systems professionals classify users along a sophistication scale, such as:

- Parrot
- Novice
- Intermediate
- Expert
- Master[11]

Ben Schneiderman has classified users as:

- Novices
- Occasional users
- Frequent light users
- Frequent power users[12]

The user classification that is used in this text is based on Horton's categories, which are:[13]

- ***Novice Users*** These users have no syntactic knowledge about computers and software. Also, **NOVICE USERS** typically know little about their assigned tasks. Generally, they are extremely nervous about making embarrassing mistakes. They have trouble in differentiating between what is critical and what is trivial. Novices are reluctant to ask for help, lacking the vocabulary to express their questions.
- ***Occasional Users*** These users knew how to work with the system at one time, but, because of infrequent use, **OCCASIONAL USERS** have forgotten essential commands and procedures. They will accept only a limited amount of

[11] M. L. Schneider, "Information Hiding in Complex Displays," *Directions in Human/Computer Interaction* (Norwood, N.J.: Ablex Publishing, 1982), pp. 137–148.

[12] Ben Schneiderman, *Designing the User Interface: Strategies for Effective Human-Computer Interactions* (Reading, Mass.: Addison-Wesley, 1987).

[13] William K. Horton, *Designing and Writing Online Documentation: Help Files to Hypertext* (New York: John Wiley, 1990), pp. 34–35.

initial training and are intolerant of paper-based software documentation (e.g., various procedure manuals). They are impatient with precise and formal computer terms.

- ***Transfer Users*** They already know how to work with systems; **TRANSFER USERS** are simply trying to transfer what they already know to a new system.
- ***Expert Users*** These people understand how to work with the new system as well as most other systems. **EXPERT USERS** are intolerant of any procedures or instructions that waste their time. Their chief demand is for fast response time. They like shortcuts, macro commands, and abbreviations.

Designing Online Documentation for Users

Generally users require a combination of:

- Tutorials
- Messages
- Menus
- Icons
- Help features
- Shortcuts
- Online reference manuals

ONLINE DOCUMENTATION can contain all of these features to guide and instruct all users interactively.

Tutorials

Many commercially available software packages provide **TUTORIALS**, or online lessons, to guide the novice or occasional user in the use of the software. Although separate from the working software program itself, these tutorials always resemble the actual package in most respects.

The success of online tutorials as training tools is dependent to a great extent on the amount of realism built into the lesson. When designing tutorials, then, it is important to provide users as much access to the program's features as possible. Many poorly designed tutorials prohibit users from seeing the results of a mistake. Instead, these tutorials display a message, such as "You pressed the wrong key. Press this one instead." A well-designed tutorial permits users to make a mistake, see the results of that mistake, and finally correct the mistake with the aid of messages and prompts from the system.

Messages

A means of alerting users to vital information concerning the system is through onscreen **MESSAGES**. Specific, concise, and courteous messages have been shown to be most effective.

Figure 17.21
Two alternative designs for an error message.

MISCELLANEOUS ERROR 216

INVALID!

An example of a poorly designed message

FILE "DISCOUNT.EXT" NOT FOUND
ON DRIVE A.

PLEASE CHECK FILENAME SPELLING
OR CHANGE DRIVE DESIGNATION

A well-designed error message

When designing messages, it should be kept in mind that the onscreen message is not permanent. It typically disappears from the screen before users act on it. For the message to be effective, it must be brief enough that users can remember all its key points until they have acted on them fully.

The style of messages should be consistent throughout all online documentation. Figure 17.21 illustrates a poorly designed message, hostile to the user, which provides no helpful information. Also shown is a well-designed error message, specifying exactly what went wrong, and courteously advising the user on how to correct the problem.

Menus

Frequently, users can learn a lot about what a system does and how it operates simply by studying the system's command **MENUS**. A menu allows novice or occasional users to operate a sophisticated system successfully by taking advantage of the human mind's strength in recognition rather than its comparative weakness in recall.

In designing menus, keep the user's task easy by making each menu option self-explanatory. One way of doing this on larger menus is by providing descriptions of each menu choice right on the menu itself. For smaller menus, where space is at a premium, one allows the user to jump directly to a help display.

The entire purpose of a menu is to eliminate the complexities of memorizing and entering a myriad of keyboard commands, and simple menus allow users to be more productive with fewer errors made. It is better to phrase the menu choices as action verbs, such as "Realign Text" rather than "Text Justification."

Finally, designers must avoid distracting users with menu choices that are not available to them or which are not appropriate to the user's task. For instance, most large systems assign various privilege levels to users for security reasons. While the high-privilege-level user will have access to most, if not all, system commands, other users are restricted to a smaller set of specific commands. It is better for the low-privilege-level user if the designer leaves unavailable menu choices off the menu entirely, or deemphasizes their appearance by displaying them dimly.

Icons

Many users, particularly novices, benefit from the use of graphic **ICONS** rather than words in online documentation. Some systems even go so far as to let users design their own icons, thereby maximizing their ability to recognize choices on a menu quickly. The human mind can recognize a small picture, such as an icon, almost instantaneously, while a text command must first be read. Hence the time required to select program options is reduced.

Help Features

Users at all experience levels can be classified as impatient. Novices want to be productive on the system as quickly as possible, and they prefer learning by trial and error. Expert users are quickly frustrated when they are forced to stop their work and look up information in a manual. Online **HELP FEATURES** better serve the needs of users at all experience levels.

Help features range from a simple command summary screen, which simultaneously describes all program commands, to context-sensitive and diagnostic help which depends on where users are in the system and what they are trying to do at the time help is requested. For instance, if you were trying to execute a particular command and you requested help, a context-sensitive help feature would provide assistance on that command, and possibly related commands, only. Figure 17.22 shows the context-sensitive help feature found in Microsoft Windows 3.0.

Shortcuts

The expert user has learned the system. This user does not require the prompting provided by menus, and finds menu selection of commands to be slow and inefficient. It's important to give these people **SHORTCUTS** that allow them to access commands while bypassing most or all menu interfaces.

An ideal shortcut is one that assigns a unique keyboard input for each available menu choice. The printing of a sales report might require wading through three levels of menus before the command to print can be issued by the user. An experienced person will appreciate being able to bypass all this with the simple key sequence "SHIFT P," for example. Much time is saved when commands can be issued from the keyboard, allowing the hands to remain in position rather than having to move to a mouse-based pointing device.

Finally, let the expert user chain a sequence of commands on one command line rather than entering each command individually. An extension of this is the macro, sometimes called a script file, which allows a user to record a series of keystrokes for a repetitive operation, then play the recording back over and over every time that operation is performed. Such a feature has the benefit of reducing keyboard input from perhaps hundreds of keystrokes down to just two or three.

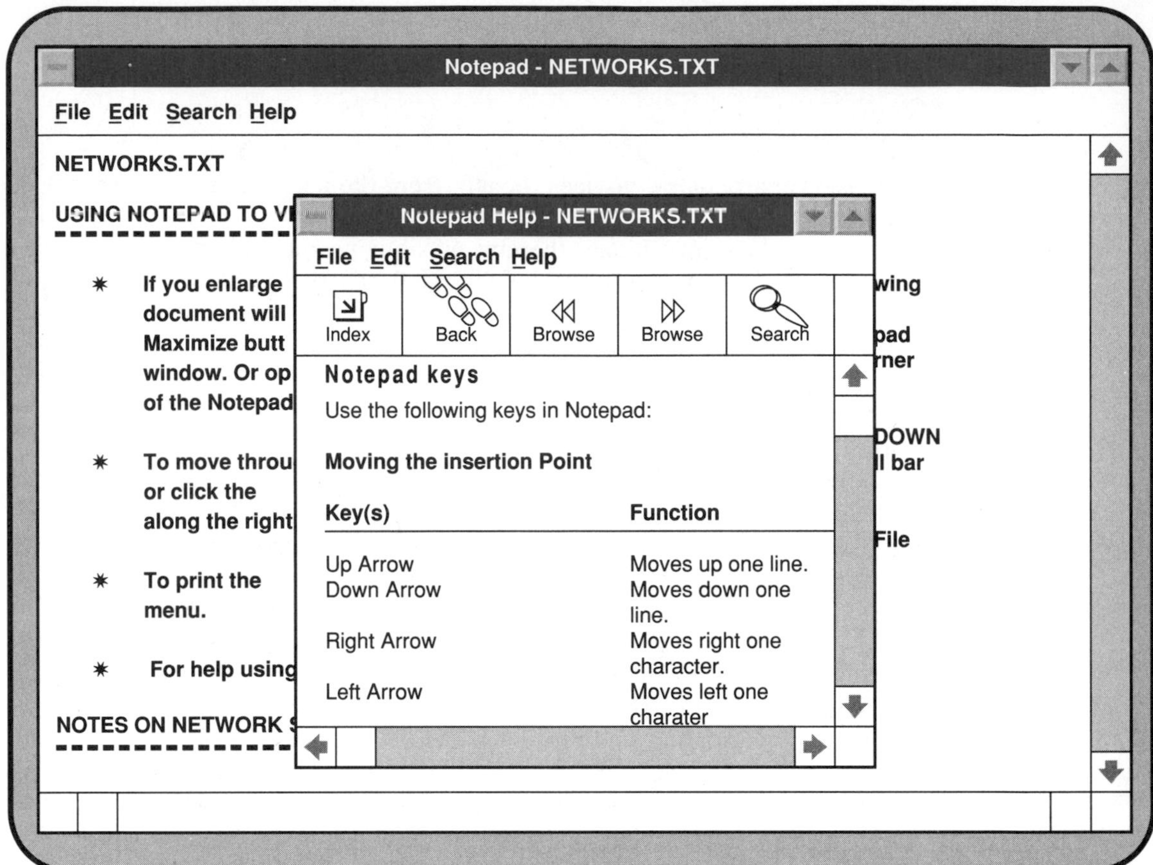

Figure 17.22
Context-sensitive help feature found in Microsoft Windows 3.0.

Online Reference Manual

While the **ONLINE REFERENCE MANUAL** will typically provide all the information of its paper counterpart, it has been found that users will use such a manual only for quick reference, not for lengthy reading or study. Therefore, while the online manual should duplicate the logical organization of the paper manual, additional information retrieval features are necessary.

When creating such a reference manual, one must make the online version behave like its paper cousin, with a table of contents, index, glossary, and headings. Figure 17.23 shows one possible implementation of an online reference manual, made to look like a book on the screen. In addition, designers include menus for accessing information search commands and provide links between related topics which allow the user to jump directly to those topics without performing an additional search. For good results, designers will always use the major tasks performed by the program as the headings and topic names of the manual.

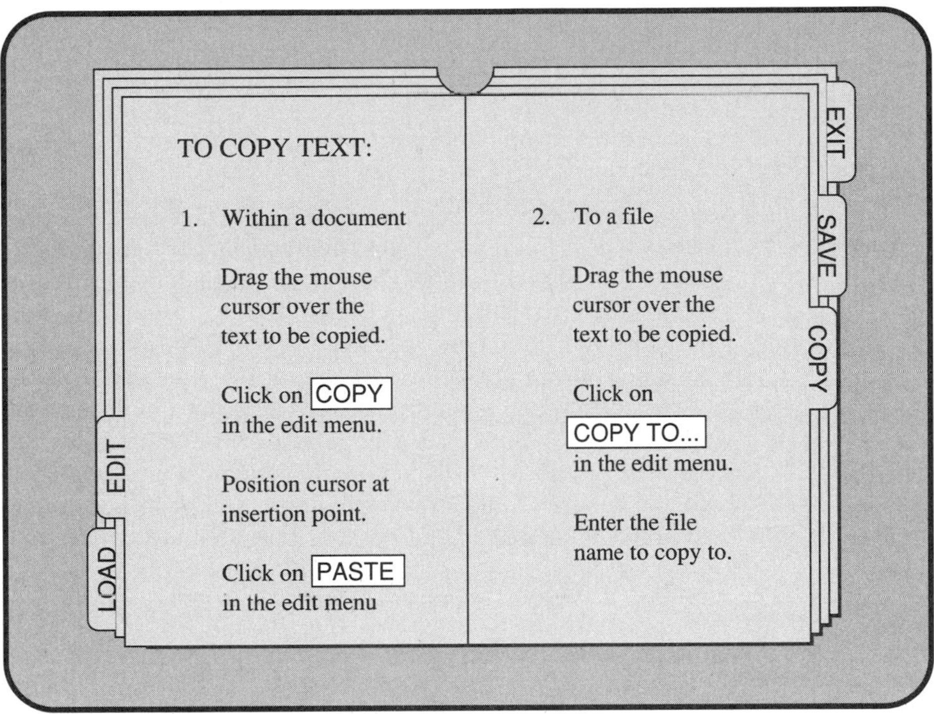

Figure 17.23 Example of an online reference manual made to look like a book.

REVIEW OF CHAPTER LEARNING OBJECTIVES

The main goals of this chapter were to enable each student to achieve six important learning objectives. Summary responses to these learning objectives follow.

Learning objective 1:
Compare fourth-generation languages (4GLs) with third-generation languages (3GLs).

Fourth-generation languages help users define their requirements by providing what can be done and then allowing users to react. Fourth-generation languages contain sophisticated functions that perform a host of functions, such as statistical or financial analysis. Users can often code their own ad hoc applications.

Third-generation languages are superior for transaction-intensive, I/O-bound software design applications. For some applications, 3GLs are as concise as 4GLs. Because 3GLs are compiled, they provide superior machine efficiency. Third-generation languages such as COBOL are portable between and compatible with most computer platforms. With the advent of CASE code generators for 3GLs, especially COBOL, their coding productivity may be superior to 4GLs. Once a 3GL is coded, it is normally easier to test and maintain, if properly designed and documented, than a 4GL.

Learning objective 2:
Describe object-oriented programming (OOP) languages, review five important OOP languages, and briefly discuss OOP tools.

Implementation of the methods (i.e., code that executes objects' operations) is done by using OOP languages. OOP languages vary in their support of object-oriented design concepts. There is no single language that supports every object-oriented design concept. Therefore, the selection of an OOP language is based on the same concept as the selection of other languages. The language must be appropriate to the software design application. For example, if a software design requires multiple inheritance, the OOP language chosen must support multiple inheritance. Choosing an OOP language with an extensive and mature class library will increase development productivity, as will the application of an OOP menu-driven point-and-click windows-based tool set. Five noted OOP languages that deserve consideration are:

- Smalltalk
- Eiffel
- C++
- Object Pascal
- Object COBOL

Learning objective 3:
Explain how special-purpose language tools are used to enable end users to communicate with computer systems.

Special-purpose language tools enable end users who have little computer and programming knowledge to use the system on their own for their local, ad hoc applications. Special-purpose languages are:

- Interactive user-oriented language tools
- DBMS query languages
- Hypertext and multimedia (or hypermedia)

Learning objective 4:
Match appropriate languages to software design applications.

If the software design application is developed using the structure-oriented approach, the language selected should support this approach. Examples of candidate languages are Ada, C, COBOL, and Pascal. If the software design application is developed using the object-oriented approach, an object-oriented programming (OOP) language should be selected. Example candidate languages are Smalltalk, Eiffel, C++, Object Pascal, and Object COBOL.

For big transaction processing applications, a 3GL is normally recommended. For scientific and engineering applications requiring a great deal of internal processing, a mathcmatical-oriented language such as FORTRAN is suggested. For decision support systems (DSS), executive information systems (EIS), and ad hoc applications, a combination of 4GLs, OOP languages, and special-purpose languages are generally appropriate. However, using a 3GL such as COBOL supported by a COBOL CASE tool set may also be appropriate.

Learning objective 5:
Discuss the selection of languages relative to issues such as level of usage in the business world, expressiveness, convenience, portability, maintainability, and extendability.

The level of usage of a language in the business community means that the language will be widely supported by:

- Computer hardware manufacturers, CASE vendors, and a large number of programmers skilled in its application
- Groups that will promulgate standards for it and revise it to meet current needs

The language coded in the form of a program is the embodiment of the software design. Therefore, it should express this design so that even nonprogrammers can understand it.

Clearly, the more convenient a language is to work with, the better. For example, the availability of extensive and rich class libraries means that many objects need not be developed by the programmer. The software development tools that are available for browsing the class library, editing source code, compiling, debugging, and so on, can enhance convenience and increase productivity.

A number of systems projects may be ideally suited for a particular language with many dialects from the viewpoint of its functionality. However, it may not be wise to choose such a language because of its absence of standards, which increases problems with portability.

After the software program is converted to operations, the people who designed, coded, and tested the program may not be employed by the enterprise it serves. People come and go. Therefore, the language the program is coded in must be totally clear to those who will maintain the program over its life. In business, it is not unusual for a program to be in operation for 10 to 20 years, maybe more. To be without total support from vendors and programmers over the program's life can place a real maintenance burden on the enterprise.

Normally, most software will require extension. The language must therefore be amenable to the techniques of modularity. Modularity supports both extendability and reusability.

Learning objective 6:
Describe software documentation, operations documentation, and user documentation.

Software documentation includes both internal and external documentation. Internal documentation is an integral part of the program's source code. It includes formatting and indenting code for readability, the use of standard and meaningful data names, and a liberal use of comments. External software documentation is usually paper-based. It includes a title page, various modeling tools that represent the software design and logic, description of input and output, program change sheet, and test cases.

Operations documentation is normally in the form of a run manual, which contains run sheets. Run sheets include program identification and when it is run, input media, forms, hardware used, various instructions, controls, and output distribution.

User documentation provides online documentation for end users, such as novice, occasional, transfer, and expert users. Online documentation for the first three users usually contains a mix of:

- Tutorials
- Messages
- Menus
- Icons
- Help features

For expert users, documentation typically includes:

- Shortcuts
- Online reference manuals

SOFTWARE CODING CHECKLIST

The software coding phase converts the software design into a software program using a computer programming language or a combination of languages and tools. The final software program is fully documented. Following is a checklist on how to make sure that the appropriate language(s) is selected, the software coding phase is performed properly, and the final software program is thoroughly documented.

1 Do not begin software coding prematurely. It is imperative to complete the software design application first and subject it to a structured software design walkthrough.

2 Select a language (or a mix of languages) appropriate for the software design application. Such a language should be widely used throughout the business community. It should also be expressive, convenient, portable,

maintainable, and extendable. The language chosen should be one that will enjoy long-term support from vendors and the business community.

3 For software documentation, prepare a title page; include models of the software program based on modeling tools, description of input and output, and a program change sheet.

4 For operations documentation, prepare a run sheet for the run manual.

5 For user documentation, first classify users and determine their documentation needs. Then create a mix of tutorials, messages, menus, icons, and help features to meet these needs. For the expert user, provide command shortcuts and detailed online reference manuals.

KEY TERMS

Expert users
Fourth-generation languages (4GLs)
Help features
Hybrid OOP languages
Hypertext
Icons
Interactive dialogue language tools
Interactive user-oriented language tools
Internal software documentation
Maintenance work order (WO) form
Menus
Messages
Multimedia (hypermedia)
Novice users
Object-oriented programming (OOP) languages
Occasional users
Online documentation
Online reference manual
Operations documentation
Paper-based external software documentation
Program change sheet
Program manual
Pure OOP languages
Run manual
Run sheet
Shortcuts
Software documentation
Special-purpose language tools
Third-generation languages (3GLs)
Transfer users
Tutorials
User documentation

REVIEW QUESTIONS

17.1 Explain why a 4GL may not be more concise than a 3GL.

17.2 Explain why a 3GL is more machine-efficient than a 4GL.

17.3 Explain why a 3GL is more robust and functional than a 4GL.

17.4 Explain why 3GLs, especially COBOL, are portable across computer architectures. Explain why 4GLs may not be.

17.5 Explain how a 3GL may be coded even faster than a 4GL.

17.6 Explain how a CASE system can increase 3GL development productivity.

17.7 What's a pure OOP language? Name one. What's a hybrid OOP language? Name one. What are they good for?

17.8 Can a 3GL be used for coding an object-oriented design? Explain your answer.

17.9 Is there an OOP language that supports and implements all object-oriented design concepts? Explain your answer.

17.10 What is the key component of an OOP language development tool set that increases productivity and reusability?

17.11 Name three special-purpose language tools. Explain their purpose and describe how they are used.

17.12 Besides a language's power and functionality relative to the software design application, name and briefly describe six other language selection issues.

17.13 Describe software documentation and its essential elements.

17.14 What are the key contents of a run manual? For whom is a run manual prepared?

17.15 List and briefly describe the four categories of end users. What kind of documentation should be prepared for each category?

17.16 Describe the characteristics of a well-designed on-screen message.

17.17 What are some reasons why the designer of a menu might want to deemphasize certain menu selections?

17.18 What is a macro, and how does it benefit the expert user?

CHAPTER-SPECIFIC PROBLEMS

These problems require exact responses based directly on concepts and techniques presented in the text.

17.19 Insert an F for false or T for true beside each phrase relative to 3GLs.

______________ Developed in a step-by-step algorithmic manner.

______________ Very concise.

______________ Easy to learn.

______________ Difficult to document.

______________ Machine inefficient.

______________ Proprietary.

______________ A great deal of user participation.

______________ Difficult to maintain.

17.20 Indicate whether the following phrases are T (true) or F (false) as they pertain to 4GLs.

_______________ Long development cycles.

_______________ Very wordy.

_______________ Procedural.

_______________ Difficult to learn.

_______________ Excellent for prototyping.

_______________ Little user participation.

_______________ Easy to maintain.

17.21 Check those terms that apply to OOP languages.

_______________ Hybrid

_______________ Pure

_______________ GO TO

_______________ REDEFINES

_______________ Reusability

_______________ ISDN

_______________ Class libraries

_______________ Inheritance

_______________ C++

_______________ FORTRAN

_______________ Smalltalk

_______________ Modem

_______________ Methods

_______________ Extendability

_______________ VSAT

17.22 A STUDENT relation is composed of the following attributes: S_Name, SSN, S_Address, Major, and GPA.

Required: Design SQL commands that:

1 Give the names and addresses of all students whose major is systems.
2 Give the names of all students in alphabetical order whose major is accounting, who are female, and have a grade point average (GPA) of greater than 3.0.

17.23 You are designing a query-by-example (QBE) interface for an inventory control clerk. The name of your table is INVENTORY. The INVENTORY table includes three attributes:

- Description
- Quantity on hand
- Selling price

Required: Draw the table and include the following three QBE entries: (1) Print all shovels with a selling price greater than $30.00, (2) Delete snow shovels, and (3) Increase the selling price of square-pointed shovels by 15 percent.

17.24 You are sitting at a workstation with access to the hypertext and multimedia design features. You have browsed through a voluminous sales report and found a key paragraph that describes sales performance of department B. You want to retrieve this text fragment and copy it in your multimedia workspace. To give a graphic presentation of department B's sales performance, you choose a pie chart from a graphics space that also includes bar charts and line charts. You choose a laser printer to generate a hard copy of your completed multimedia workspace containing your selected text fragment and pie chart.

Required: Draw a schematic of the preceding hypertext and multimedia application and explain how it works.

17.25 You are a programmer creating a menu interface for an information system used by a law firm. This system is mainframe-based, and a number of different privilege levels are assigned to its various users. These levels range from the very limited access provided for the secretary to the full access provided for the system operator.

Required: List and describe some techniques for deemphasizing certain selection choices on your menu, such that different users have access only to pertinent commands for their privilege levels.

THINK-TANK PROBLEMS

These problems call for a feasible approach rather than a precise solution. Although the problems are based on chapter material, extra reading and creativity may be required to develop workable solutions.

17.26 In the past, COBOL has been the recipient of widely publicized negative comments. Some of the top computer scientists and software engineers heap invective on what they call a "verbose dinosaur of a language." Many "experts" have predicted its death for years. Proponents of 4GLs and proponents of other 3GLs (e.g., PL/1) have been trying to replace COBOL without making a dent. Today, proponents of COBOL say that it is stronger than ever. Indeed, the fact remains that COBOL is the most widely known and used business computer language. As one authority said: "These other languages, especially 4GLs, don't meet industrial-strength demands for ruggedness in the face of the real world. Do you want to use 4GL source code or an obscure 3GL developed by someone else and originally written for a different computer platform, which uses a different function library? I hope you like detective stories. By contrast, look at COBOL or Ada, cast-iron languages that govern what coders code and how they code it. In fact, I just saw a research report and as I went down the list of things, such as programming development tools and maintaining programs coded by someone, I realized I rediscovered COBOL. The more things change, the more they stay the same."

Required: Explain why experts have predicted the death of COBOL, but COBOL is stronger today than it ever has been. Explain why COBOL is referred to as a "rugged language that can meet industrial-strength demands."

17.27 The company you work for has a history of coding software with obscure nonstandard languages and failing to develop documentation. The new CIO recently stated: "We will develop new systems that will be correct, standard, and maintainable. The Achilles heel of any system is documentation. Without clear, current, and correct documentation, we don't even have a system."

Required: You have been selected by the CIO to investigate and report to her how standard, well-documented application software can be developed in the future. Prepare this report. Refer to Chapters 15 and 16 to prepare your report.

17.28 A compiler translates all source code in a program to machine language instructions before any of the coded instructions are executed. Execution of the program will generally occur many times without the need to recompile. An interpreter (or translator), in contrast, accepts source code instructions one by one and immediately generates and executes the corresponding machine language, without waiting to see what the next coded instruction may be. A compiler is similar to an English edition translation of *War and Peace*. An interpreter is somewhat analogous to the simultaneous sentence-by-sentence translating done at the United Nations.

Required: Which method, compilation or translation, is the most machine efficient? Which method do you recommend for business data processing applications? Which method would work well for random, interactive, and real-time applications? Explain your recommendations.

17.29 Following are some applications that have been designed and now require coding:

______________ Large accounting system with extensive accounts payable and accounts receivable files.

______________ Simulation application with multiple equations that requires source code that is excellent for machine efficiency.

______________ Scientific application that simulates the colliding of planets. This application involves thousands of complex mathematical computations. Coding efficiency is more important than machine efficiency.

______________ User must manage a large volume of information and convert it to a multiple presentation.

______________ User needs to access the relational DBMS on an ad hoc basis.

______________ User needs to carry on a dialogue with the system.

______________ Market tracking system for the marketing manager.

______________ An application that will contain a large number of reusable objects.

Required: Choose a language that you believe to be the most appropriate for the above applications. Explain the reason for your choices.

17.30 You are a member of a five-person programming team. This team has just completed the code, written in C++, for a major program module as part of a large systems project.

Required: Prepare an outline of the complete external program documentation package to be submitted to your firm's CIO. This outline should list all paper-based software and operations documents to be included in the documentation package.

17.31 Describe five poorly designed messages you have seen while using microcomputer, minicomputer, or mainframe software programs. Briefly describe the context of the program in which each message appears, and discuss reasons why you feel each message is poorly designed. How could these messages be improved?

SUGGESTED READING

Alderson, Rusty. "Standards: Good, Bad, or Ugly?" *Interact,* May 1991.

Beckman, Michael, and Dimtry Lenkov, "The C++ Language." *Interact,* May 1991.

Booch, Grady. *Object Oriented Design*. Redwood City, Calif.: Benjamin/Cummings, 1991.

Chase, Tim. "Everything You Wanted to Know About Object Oriented Programming But Were Afraid to Ask." *Interact,* February 1991.

Coffee, Peter. "COBOL Evolves as Mission-Critical Tool." *PC Week,* January 21, 1991.

Coffee, Peter. "Heavy-Duty Applications Need Cast-Iron Tools." *PC Week,* May 28, 1990.

Cox, Brad J. *Object Oriented Programming.* Reading, Mass.: Addison-Wesley, 1986.

Crabb, Don. "OOP Tools Ease Windows Developers' Pains." *Infoworld,* February 25, 1991.

Dern, Daniel. "Wanted: Productivity Improvements." *Digital News,* January 21, 1991.

Dyckman, Roger J. "COOL/3000: The Object is the Subject." *Interact,* February 1991.

Eskow, Dennis. "Programmers Find Smooth Road to PCs." *PC Week,* October 22, 1990.

Gray, Paul. *Guide to IFPS.* 2nd ed. New York: McGraw-Hill, 1987.

Hill, Tom. "Object Oriented Programming Using COBOL." *Interact,* April 1990.

Horton, William K. *Designing and Writing Online Documentation: Help Files to Hypertext.* New York: John Wiley, 1990.

Kill, D. "Anatomy of a 4GL Failure." *Computer Decisions,* February 11, 1986.

Lee, Al. "The Problem with 4GLs." *Computer Programming Management.* Boston: Auerbach Publishers, 1989.

McIntyre, Scott C., and Lexis F. Higgins. "Object-Oriented Systems Analysis and Design Methodology and Application." *Journal of Management Information Systems,* Summer 1988.

Meyer, Bertrand. *Object-Oriented Software Construction.* Englewood Cliffs, N.J.: Prentice-Hall, 1988.

Parsaye, Kamran, Mark Chignell, Setrag Khoshafian, and Harry Wong. *Intelligent Databases: Object-Oriented, Deductive Hypermedia Technologies.* New York: John Wiley, 1989.

Perreault, Bob, and Katie Rotzell. "Object Database Management Systems." *Interact,* April 1990.

Pinson, Lewis J., and Richard S. Wiener. *Applications of Object-Oriented Programming.* Reading, Mass.: Addison-Wesley, 1990.

Rumbaugh, James, Michael Blaha, William Premerlani, Frederick Eddy, and William Lorensen. *Object-Oriented Modeling and Design.* Englewood Cliffs, N.J.: Prentice-Hall, 1991.

Stroustrup, Bjarne. *The C++ Programming Language.* Reading, Mass.: Addison-Wesley, 1987.

Taylor, David. "How to Explain This Stuff to Your Boss." *Object Magazine,* July/August 1991.

Vesely, Eric Garrigue. *A Guide to Structured, Portable, Maintainable, and Efficient Program Design*. Englewood Cliffs, N.J.: Prentice-Hall, 1989.

Winblad, Ann L., Samuel D. Edwards, and David R. King. *Object-Oriented Software*. Reading, Mass.: Addison-Wesley, 1990.

JOCS CASE: Coding Software

"I have only a few minutes to discuss this issue," says Kyle Bartwell, president of Peerless, Inc., talking to Mary Stockland, Peerless's chief information officer. "I'm glad that you were free for a quick meeting." Kyle ushers Mary into his office and closes the door. "You understand that we have had a few problems in the past with our computer systems, and I want to be sure we are on the right track this time around."

"Is there a specific area that you want to discuss, Kyle?" responds Mary. She gropes for a reason for this meeting by saying, "We are currently developing one system to handle our job cost accounting and estimation, as I outlined in the MIS steering committee, and we are beginning to analyze the feasibility of developing a financial decision support system."

"Yes, I know about JOCS and FIDS. The cost accounting people and manufacturing people are really looking forward to having up-to-date information." Kyle looks at the picture over Mary's head, and continues, "I'm just wondering about our personnel allocation to these projects."

Still trying to understand the reason for this meeting, Mary says: "Are you concerned about anyone in particular? Are you wondering about the number of people working on these projects?"

"Well, now that you mention it, personnel costs have increased over 400% in your department during the last six months. The finance department is still picking up the costs for a person loaned to your department."

Mary freezes for a moment, then says in a very precise voice, "I started these projects with strong support from the MIS steering committee. I understood that you were willing to provide the resources necessary to develop these systems. I need to know if this understanding has changed."

Kyle, a little affronted by her response, says, "I *am* providing the resources. I just want to be sure that we are using our resources in the best possible way."

"I can assure you that I have complete faith in my staff and their decisions. I am working closely with my project manager, Jake Jacoby, to control our development efforts." Mary's voice has become more formal as the meeting has progressed. She is bombarded with thoughts at this point and is trying to decide how to continue this meeting. She's thinking, "I knew it. Management just can't devote the resources necessary to completing a project. This happened at my last job. You just start making real progress and somebody decides that you are getting too big a slice of the pie." She looks directly at

Kyle and asks, "Is there a problem with the resources I am using for these projects?"

"No. No, I am not making myself clear," responds Kyle. "Let me put this another way. I was walking through the lunchroom yesterday, and I heard a couple of your people discussing the JOCS development. They mentioned a computer language I remember from college called COBOL."

Mary has absolutely no idea what to say at this point. She is wondering if Kyle is concerned about her employees discussing their job with people outside their department. Her mind is racing trying to figure out what the problem is.

Kyle continues by saying, "I went to college quite a few years ago. Even then people were saying that FORTRAN was a better language than COBOL. My daughter in junior high is writing programs in C. Her teacher says that COBOL should have died years ago. Last night, I read an article in *Business Week* which said that COBOL was a third-generation language. The article went on to say that all businesses should be using a fourth- or fifth-generation language. It said that third-generation languages required more people and more time to finish any business-related project."

Mary finally understands the problem. She tactfully ignores the statement from Kyle's daughter's teacher and asks, "Did the article give a definition of third-, fourth-, and fifth-generation languages?"

"Well, it was a little vague. However, from what I understand, a fourth-generation language allows programmers to develop applications ten times faster than a third-generation language."

"I've heard that ten times faster rule of thumb a few times," sighed Mary. "I often wonder if it means ten times faster than ten years ago or ten times faster than right now. Ten years ago we had a much different computing environment available to business programmers."

"I don't think that's important," retorted Kyle. "I just want to be sure we are developing our systems with the best and most current technology that is available today."

Mary decides that it is time to be blunt about her systems development effort. "Kyle, you hired me to install computer systems that support the business functions at Peerless. That's what I'm doing. I'm using the best possible tools to complete that task."

"But COBOL? How can you say that COBOL is the best possible tool available? The language has to be over thirty years old."

Mary laughs. "I could say that I'm over thirty years old, and I'm the best possible CIO available, but I guess that the parallel doesn't quite hold true. Let me outline our development technology for you. For the JOCS project, we are using a series of high-powered microcomputers linked together by a local area network. That's just hardware and a cabling system designed to make the machines communicate with each other; that isn't the really important part of our system."

"From what I read," responded Kyle, "That sounds like the most important part. Looking at your expenditures, it appears to be one of the most expensive parts."

"That's because the cost of software is deceptive. You pay for hardware as an initial cost, but the price of software must be added up over the whole life of an application. Software costs includes initial development, initial enhancements, and then ongoing enhancements and maintenance. Add to that figure initial and ongoing training, and you will have a better idea of the true cost of software."

"All right. Computer applications in general look pretty expensive to me," sighs Kyle. "However, these systems support our business, and I have already accepted the costs. Let's move on. You mentioned your development technology. What exactly are you doing?"

"Let me tell you about the important part of our development approach. The backbone of our development is a system, called UNIX, that allows my programmer/analysts to move easily among a variety of applications. A programmer can be coding a program, and then receive mail communication from another programmer. UNIX lets the programmer briefly, and quickly, stop what he or she is doing and look at the mail. The programmer can respond immediately or simply file the mail message for a later time. The staff can work with very few interruptions and very little time waiting for the computer to access another piece of software. They can then continue coding without missing a beat."

"The development team performed all systems analysis and design with the use of a software product called CASE, which stands for computer-aided software engineering. As the project was designed, all data about it were kept in a central repository of design documentation. Any changes made to the design were immediately reflected in the documentation stored in the central respository."

Mary continues by discussing their database technology. "We are using a database management system that integrates with our CASE product. As an analyst identifies new data requirements, these items are immediately added to our data dictionary."

"Screen designs, such as a screen to look at the material variances for a specific job, and report designs, such as a report to list material variances for all jobs, were completed through the CASE product. Our CASE product creates COBOL libraries for screens and reports that programmers then integrate into their program with a COBOL command called Copy."

"Does this mean that screens and reports can be changed without modifying the COBOL program?" asks Kyle.

"Yes. If, in the future, we have new personnel in cost accounting who want a report structured in a different manner, our analysts can simply change the report design and then recompile the applicable programs."

"So where does the COBOL part come in?" inquires Kyle.

"We are using COBOL to write the processing portion of our programs. Input, such as screen data and database data, comes from our CASE COBOL generator. Output, such as screen designs and report formats, also comes from our CASE COBOL generator. Our programmers take the code produced from our CASE product and integrate it into working programs that transform the input into the output."

"Before you even ask, Kyle, let me explain why we chose COBOL," added Mary before Kyle could ask his next question. "We needed a language that allowed us the flexibility to create complex programs. We had to consolidate the computer-integrated system in manufacturing with our new job costing system. This required some unique programs that could not be created without the help of a procedural language. COBOL is the only language that is truly business-oriented, handling files and records of data easily, but still allowing programmers the freedom to create unusual applications."

"But I don't want my people in cost accounting and manufacturing to take the time to learn a detailed language like COBOL. One semester in college was plenty for me," says Kyle. "Yet I know that these people are anticipating writing their own computer reports."

"I agree. I don't want our end users to become programmers, either. But I do want them to be able to create their own ad hoc applications when necessary. That's why Jake Jacoby is setting up training classes for the end users. He will establish ongoing classes training end users in the use of the database query language and report generator."

"I have another meeting that started five minutes ago," says Kyle. "It sounds like you have our development under control. The next time a vendor comes in telling me about new products, or I read an article discussing so-called obsolete technology, I'll talk it over with you. Maybe you can teach me how to use that electronic mail package so you and I can work a little more efficiently, too."

Chapter 18
Testing the Software

WHAT WILL YOU LEARN IN THIS CHAPTER?

After studying this chapter, you should be able to:

1 Describe a software testing scenario and give its purpose.
2 Define two basic ways to design test cases.
3 Discuss what should be done when errors are detected.
4 Present a way to help measure the reliability of test cases.
5 Name and briefly discuss the four software testing stages.
6 Differentiate between top-down integration testing and bottom-up integration testing.

INTRODUCTION

The purpose of this chapter is to present ways to test software, the final phase of software development (see Figure 18.1). Many CASE systems provide test utilities and debugging tools that support the techniques presented in this chapter.

The purpose of software testing is reliability, which requires error detection and removal. Software testing cannot turn a poorly designed software program into a good one, but it can help determine the level of reliability before the software is released for use. Reliability generally increases with the amount of testing; it is closely linked with project schedule and cost. Since a project manager must control schedule and cost, reliability is intimately tied into how well the total systems project is managed. If the software design phase was hurried and not conducted properly, design changes during the testing phase will tend to decrease reliability as a result of these changes. If a large number of design changes appear to be necessary, it may be advisable to change the schedule and go back to the design phase of the SWDLC and start the design-code-test process again.

Usually there is a trade-off between time and cost of completing the program and its reliability. An unreliable program will cause the project team to gain a bad reputation and lose credibility. On the other side of the scales is the problem of being late. Missing a completion date can also cause similar negative results. More often than not, the project manager is caught in the middle of a tug-of-war between these conflicting interests. Project managers, aided by systems team members and users, often must revise schedules and budgets and reliability goals in order to achieve a happy balance among conflicting interests.

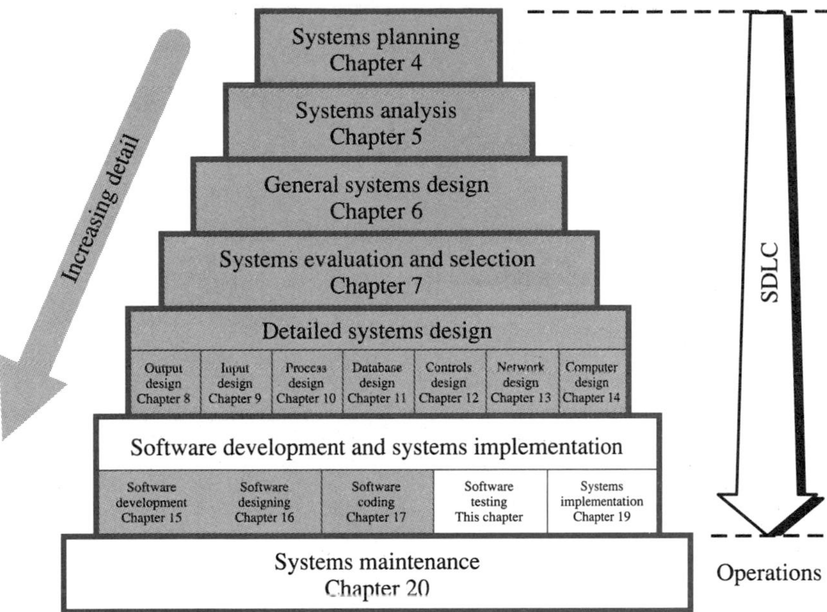

Figure 18.1
SDLC phases and their related chapters in this book. In Chapter 18 we look at software testing and the handling of error detection and correction.

An authority stated, "Anyone who believes that his or her program will run correctly the first time is either a fool, an optimist, or a novice programmer." Regardless of whether software is developed in-house or acquired from a vendor, software testing provides a documented basis for ensuring that the program will perform as required. The following material discusses how one goes about performing this critical process.

WHAT IS SOFTWARE TESTING?

Software testing is a process that should follow a pattern and well-defined plan. This testing process is often performed by an independent quality assurance group, also called a test team, to help ensure a high-quality, highly reliable software package by finding and correcting errors.

Preparing Test Cases

Testing is conducted in a manner shown in the chart in Figure 18.2. The target software is subjected to test cases and strategies, which generate test results. **TEST CASES** are procedures that examine the software and produce results that will lead to acceptance, modification, or rejection of the software. Test cases should be recorded and their results well documented. The quality and efficiency of testing is largely dependent on the development and application of test cases. The following is a list of areas in which test cases are usually developed:

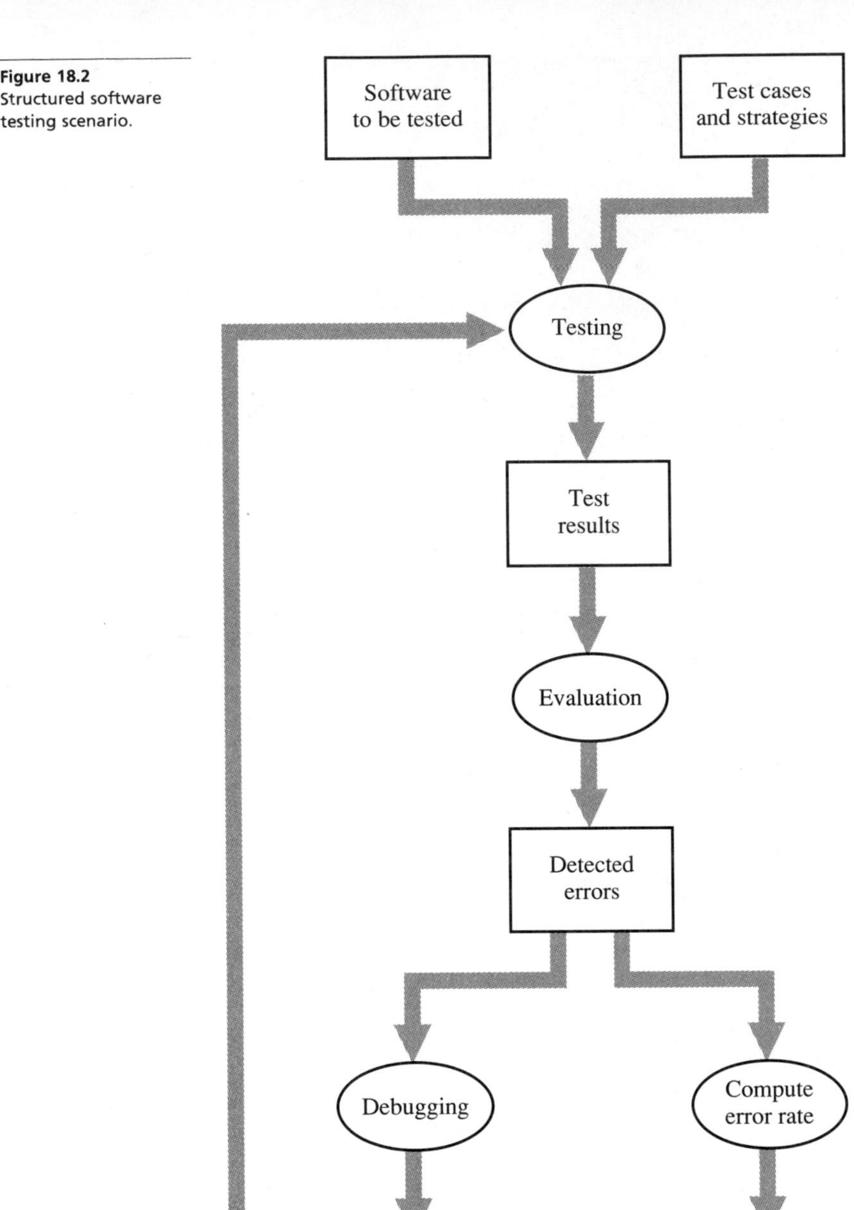

Figure 18.2
Structured software testing scenario.

- ***Field*** Test the attributes of individual fields.
- ***Record*** Test the entry, storage, retrieval, and processing of records.
- ***File*** Test the opening, retrieval, use, and closing of files.
- ***Data Entry*** Test for inaccurate, incomplete, or obsolete data. Validate functions necessary to enter the proper data according to design specifications.
- ***Controls*** Check all embedded controls discussed in Chapter 12 for ensuring the accurate, complete, and authorized processing of transactions.
- ***Program Flow*** Determine that the software performs sequence, selection, and repetition constructs properly.

These test results are evaluated. Detected errors are debugged, and corrections to the software are made. If corrections cannot be made, the software is rejected.

What Are Software Errors and Their Relationship to Software Reliability?
Error rates are computed to predict software reliability. Quantitatively, **SOFTWARE RELIABILITY** can be expressed in errors per KLOEC delivered. Following are the types of errors which testers are likely to encounter.

FATAL ERRORS are of three types:

1 Crash, in which the program terminates abnormally.

2 Logic, in which the program does not perform a function properly; for instance, executing a wrong branch or opening a wrong file.

3 Hang, in which the program or a portion of it appears to loop (i.e., repeat) indefinitely.

SERIOUS ERRORS produce incorrect output. **MINOR ERRORS** cause user dissatisfaction about the program's results, such as a misaligned column of figures; they are not serious, but irritating.

The cause of all software program errors is something called a **BUG**. A bug is an unexpected defect, flaw, fault, or imperfection. If fatal or serious errors are discovered, software reliability is questioned, and redesign and recoding are indicated, or, possibly, outright rejection occurs. If, on the other hand, minor errors are easily resolved, and the software modules seem to be functioning properly, one of two conclusions can be derived:

1 Software reliability is acceptable.

2 Test cases are inadequate to detect fatal and serious errors.

Using an error-seeding model, discussed later in this chapter, can help ascertain the effectiveness of test cases.

What Is Debugging?

DEBUGGING is the removal of bugs. It occurs because of successful testing. Test cases detect errors; debugging removes bugs and corrects errors. In many cases, an error is a symptom of an underlying cause not evidenced in the error. The debugging process tries to match symptom with cause, thereby leading to total error correction. For complete debugging, it is always imperative to find the root cause of an error.

In software design and walkthrough, the question, "Are we building the right software?" was answered. In software testing, the question, "Are we building the software right?" is answered. Debugging will identify poor coding procedures and control and uncover incorrect, unauthorized, ineffective, inefficient, and nonstandard code.

What Is the Objective of Testing?

The objective of testing is reliable software. Testing, however, can never demonstrate that software is totally reliable; it can only show the presence of errors, not prove their absence. It is always possible that undetected errors may exist even after the most comprehensive and rigorous testing is performed.

All the systems professionals' work up until the testing phase has been constructive. When testing starts, there is an attempt to destroy or break the system. At least to the coders, testing appears to be destructive. However, errors do exist, and they must be found before the user discovers them during operations, which is obviously the worst time to uncover errors.

Software Testing at Capital Industries

"I just don't like the idea of testing," said Maurice Singh, newly hired junior programmer at Capital Industries. "It is tedious, never-ending, and just plain no fun."

"Is it good or bad to find a bug?" asked Jane Cappola, project leader.

"I don't know," Maurice responded. "I guess it depends on who finds it."

"Precisely," said Jane. "To me, it is *good* to find bugs. In some groups, finding a bug is an embarrassment, as though good programmers never make mistakes and only incompetents produce bugs. Actually, finding a bug means you saved the cost and embarrassment of having the end user find your bug for you. Indeed, finding bugs during software development is much less costly and much less embarrassing."

"I never thought of it that way," said Maurice. "I can certainly see your point. Do we have a software testing methodology in place?"

"Yes," said Jane. "We have white box testing, or some people call it glass box testing. Then we have black box testing."

"I'm sorry," said Maurice with a quizzical look. "My university courses ignored the subject of testing. What are white box and black box testing?"

"In simple terms, white box testing looks at the code inside the modules—that is, the 'boxes'—to make sure the code is put together properly. On the other hand, black box testing makes sure that the modules produce the correct output. Some people call black box testing specification-based testing or function-based testing."

"So, you're testing the code to confirm its structure and to see if there are any coding oddities or untried paths with white box testing. Then, with black box testing, you're determining if the program does what it's supposed to do," said Maurice.

"That's it," said Jane. "Also, because we believe in testing at Capital, we actually think a lot about testing during the design phase and therefore perform 'defensive programming.' We believe that test planning during design is a very healthy thing to do."

"Sounds like a lot of work to me," said Maurice.

"It is," said Jane. "But we test a little bit at a time. We assign modules to different programmers. For example, you may be responsible for four or five modules. These modules will involve no more than one or two days of coding effort. Then, coding is followed by one or two days of testing. We integrate all the modules from different programmers, and test again. When we're satisfied with the results, we assign more modules, then code and test them. Next, they're integrated with the previous modules. We continue this process until all modules of the software design have been coded, tested, and integrated. Taken in small steps like this, testing is almost fun. Our motto at Capital is: 'To walk a long journey, take lots of small steps.'"[1]

DESIGNING TEST CASES

Any engineered product, including software, can be tested in two ways:

1 Knowing the internal workings of a product and testing to see if they have been adequately exercised (white box testing).

2 Knowing functions that the product is supposed to perform and testing the product to see if it performs the functions properly (black box testing).

White Box Testing

WHITE BOX TESTING is based on the direct examination of the internal logical structure of the software. It uses knowledge of the program structure to develop efficient and effective tests of the program's functionality. Logical paths through the software are tested by providing test cases that exercise specific sets of SEQUENCE, IF-THEN-ELSE, DO WHILE, and DO UNTIL constructs.

One could assume that white box testing can result in a totally correct program. But the logical paths through a program can become overwhelming even in fairly simple programs. Therefore, exhaustive white box testing is impracticable. A limited number of important or high-risk paths can be selected and exercised thoroughly, using the white box method.

Specific commands and their testing ramifications are:

- ***SELECT*** This command relates a file to an input/output device. It can be checked to see if the program processes proper and authorized files.
- ***OPEN/CLOSE*** This type of command makes a file available or unavailable for processing. Multiple OPEN/CLOSE commands in a program could mean a file is being made available for unauthorized processing.

[1] Marc Rettig, "Testing Made Palatable," *Communications of the ACM*, Vol. 34, No. 5, May 1991, pp. 25–29.

- ***COPY REPLACING*** This command changes the definition of data items copied into a program from a source library. It should be examined to see that the changes give the right results.
- ***IF*** This can be the major conditional statement used in a program. It could be used to execute an unauthorized or erroneous section of code when a certain condition is true or false.
- ***PERFORM UNTIL and PERFORM WHILE*** The UNTIL and WHILE statements permit a loop to be executed. The loop should be examined to make sure that it is activated the proper number of times.
- ***CALL*** This command is used to call a subprogram or module. The module called may be the wrong one, or it may be unauthorized.

Black Box Testing

BLACK BOX TESTING demonstrates that software functions are operational, that output is correctly produced from input, and that databases are properly accessed and updated. It requires knowledge of the user requirements to conduct such tests. Black box testing thus does not directly examine the syntax and internal logical structure of the software, and is therefore not an alternative to white box testing.

Black box test cases consist of sets of input conditions, either intentionally valid or invalid, that fully exercise all functional requirements of a program. Generally, these tests will uncover a different class of errors from white box methods. Both white box and black box testing uncover errors that occur during coding, but only black box testing is focused on uncovering errors that occur in implementing user requirements and systems design specifications.

Unlike white box testing, which is performed early in the testing process, black box testing tends to be conducted during later stages of testing. Black box test cases are therefore more appropriate at the integration, systems, and acceptance testing levels and are normally not used at the module level.

Testing Equivalence Classes

Testing **EQUIVALENCE CLASSES** is a key part of black box testing. Two input values are in the same equivalence class if they are handled by the program in the same way. For example, suppose that acceptable input for a data field is a number between 1 and 50. It is a waste of effort to test 45, 38, 12, and 6, for example, because these numbers reside in the same equivalence class. If 38 works, for example, there is a high probability that all the others will work. It is more efficient to list just one number between 1 and 50, test the end points 1 and 50 (bounds testing), then move to other equivalence classes such as 0, a negative number, or a number greater than 50.

There are two types of equivalence classes: valid and invalid. In the foregoing example, the numbers 1 through 50 represent a valid equivalence class. Any numbers outside this range are invalid equivalence classes, as are nonnumeric characters. Full testing requires the application of both valid and invalid test cases.

Using Equivalence Classes to Build Test Cases

Using equivalence classes seeks to define a test case that discovers classes of errors, thereby reducing the total number of test cases that must be designed. Typically, an input condition is a specific numeric value, range of values, a set of related values, or a yes-or-no condition. Some specific examples of equivalence class test cases follow:

- Check to see if control totals are prepared properly. For example, if 100 test records are processed, the number of transactions processed should read 100.
- Try to process a sensitive transaction without proper authorization (e.g., change of customer's credit limit) and see if the system rejects it.
- Make numeric, alphabetic, and special character checks. For example, if all the characters in a customer number are supposed to be numeric, input an alphabetic character in this field. A properly working control will detect this mistake before processing is performed.
- Input a field with a negative sign to see if it is handled as a negative value. In some systems, without proper control, the negative sign is converted to a positive sign.
- Divide an amount by zero.
- Perform validity checks on key data fields. For example, input an invalid code or try to process one department number as another department number.
- Make range and reasonableness checks. If no employee can work more than 60 hours per week, process a time card with more than 60 hours worked.
- Check for proper transaction sequence. Where transactions are supposed to be in sequence, shuffle the order of several test transactions so they are out of sequence.
- Include an account number with a predetermined check digit and see if it is processed properly.
- Use units of measure different from those allowed, such as feet for pounds.
- Input several fields with incomplete or missing data.
- Insert characters in fields to cause an overflow condition.
- Try to read from or write to a wrong file.

The test group should devise an orderly way to record cases. Figure 18.3 presents a **TEST CASE MATRIX** that records and documents test objectives, expected results from the test, the test cases conducted, and the actual results from the tests.

Test Objective	Expected Results: Reject	Expected Results: Display Error Message	Expected Results: Automatically Compute Correct Amount	Test Case Design	Actual Results
To determine if program computes check digit and rejects transposed account number	X	X		Input transposed account numbers	Rejected and displayed error message
To determine if department numbers are checked for validity	X	X		Input invalid department numbers	Rejected and displayed error message
To verify accuracy of overtime pay computations			X	Pay an hourly employee for 15 hours overtime	Overtime pay was computed at 1.5 times regular rate
To determine if program processes a credit memo accurately with missing general-ledger code		X	X	Input a credit memo with all data except general-ledger code	Credit memo was processed accurately and missing code error message was displayed

Figure 18.3
Test case matrix.

Reporting and Resolving Errors

As soon as an error is discovered, an **ERROR REPORT** should be filled out. Such a report is illustrated in Figure 18.4. A suggestion doesn't necessarily mean that anything is wrong, but is typically an idea submitted by the tester on how something can be improved. A coding error is unintended by the coder. For example, the coder may write IF A > B PERFORM SOMETHING, when it should be IF A = B PERFORM SOMETHING. Design errors are typically user interface errors, such as incorrect translation of input or output design into program code. Documentation error means that the code and documentation don't match. Either the documentation is in error, or the code is in error, or both. A query means that the program does something that the tester doesn't understand, and therefore needs clarification from the coder.

Attachments include such things as a disk containing test data, printouts, a memory dump, or a memo describing in detail what tests were performed. Also, if the tester cannot reproduce the error, a description of what might have prompted the error should be included. In some instances, the tester may have uncovered an error, tried to reproduce it, and failed. The tester is unsure how the error was triggered. To help the one who is assigned the responsibility to fix the error, the tester should write down everything about what occurred just before the error was triggered. Even good guesses should be noted. If any or all of these items could be helpful to the person assigned to resolve the error, they should be attached to the Error Report.

Figure 18.4
Error Report.

Error Report Number: ____________

Program Name: ____________________

Report Type (1–5) ______ Severity (1–3) ______ Attachments (Y/N) ______

1-Suggestion	1-Minor	If yes, describe:
2-Design error	2-Serious	
3-Coding error	3-Fatal	____________
4-Documentation error		
5-Query		____________
		____________

Can the error be reproduced? (Y/N) ______

Error and how to reproduce it:

____

____

Suggested fix: ______________________________

Name of tester: ______________ Date ____/____/____

- -

To be filled out by coding team

Assigned to: ______________ Date ____/____/____

Resolution code (1–6) ______

1-Fixed	4-Disagree with suggestion
2-Can't reproduce it	5-Withdrawn by tester
3-Can't be fixed	6-Works according to specifications

- -

Resolution certification
Resolved by: __________ (Coder) __________ (Tester) Date ____/____/____

Project manager approval: ______________ Date ____/____/____

Resolution code indicates what the assigned programmer did with the reported error. Resolution of the reported error includes one of the following:

1 It is fixed as recommended.

2 The error cannot be reproduced.

3 The error cannot be fixed.

4 The coder disagrees with the tester's suggestion.

5 The reported error has been withdrawn by the tester.

6 The code about which the error is reported works in accordance with the design specifications.

The tester and coder must reach an agreement for final resolution. They both must sign and date the report to verify resolution. The project manager must also sign off to certify final resolution.

After a program error has been fixed, the test that uncovered the error in the first place should be repeated. This is a **REGRESSION TEST**. This retesting may discover new program errors that were masked by errors found in the first test. Added variations on the initial test, to make sure that the fix works are part of the regression test.

Later, after the software is converted to operations, it will normally have to be maintained. Changes to the software require that regression testing be conducted to ensure that the change does not disturb other modules of the program; such disturbances are referred to as a ripple effect. Test cases prepared during the testing phase of the SWDLC are saved in a test case library to be used to perform regression testing during the maintenance phase of the software's life cycle. In a CASE-based system, the test case library is stored in the central repository.

Testing the Test Cases

In theory, testers can find the number of errors in a coded program by a method called error seeding or bebugging. A program is randomly seeded with a number of known artificial errors that represent the kinds of errors typically encountered. Such errors are unknown to the testers. The program is run with the test cases. The probability of finding i real errors in a total population of I unknown errors can be related to the probability of finding j seeded errors from J errors embedded in the code.

SEEDING (or prebugging or bebugging) a program motivates testers to find errors. If they know there are errors in the program that are being monitored, they will build strong test cases to find them. In doing so, they will also find the real errors.

The error seeding method assumes that program reliability is related to the number of errors removed from it. After both real and seeded errors are detected during a test run, the number of remaining real errors is approximated by the formula

$$\frac{\text{remaining number of real errors}}{\text{remaining number of seeded errors}} = \frac{\text{number of real errors detected}}{\text{number of seeded errors detected}}$$

The proportion of errors not detected helps determine the quality of the test cases and the general testing process, which in turn helps estimate software reliability. Thus, if 100 errors are seeded into the program, and testers find 40 of these along with 200 real errors, the odds are that another 300 real errors remain in the program:

$$\frac{\text{remaining number of real errors}}{60} = \frac{200}{40}$$

$$\text{remaining number of real errors} = 300$$

For the seeding method to give a fair approximation of remaining number of real errors; that is, the level of program reliability, the seeded errors must be similar to the real errors.

DEVELOPING A SOFTWARE TESTING STRATEGY

A software testing strategy integrates software test case approaches into a well-planned series of the following stages:

- Module testing
- Integration testing
- Systems testing
- Acceptance testing

These stages and their progression are illustrated in Figure 18.5.

Testing Individual Software Modules

MODULE TESTING is the process of testing the smallest units in the total software program before they are put together to form a whole program. After source-level program code has been written, reviewed, and verified for correct syntax, module test case design begins.

Module testing is the purest form of white box testing. The goals of module testing are to:

- Execute each command in a module.
- Follow each logic path in the module.
- Recompute each computational command.
- Test the module with each possible set of input data.

It is, however, impossible to achieve full coverage of the module's structure because the number of tests required would be overwhelming.

Module testing is simplified and much less costly when high-cohesion modules are designed. When only one function is handled by a module, the number of test cases is reduced and errors can be more easily discovered and corrected.

Which Areas Present the Greatest Risk?

In a program with 50 or more branches, there are more possible test cases than there are grains of sand on Malibu Beach. In a program with n branches, there are 2^n different possible paths through the program, and 2^n power gets big very quickly. While all combinations of branches cannot be tested, each branch must, at least, be tested once.

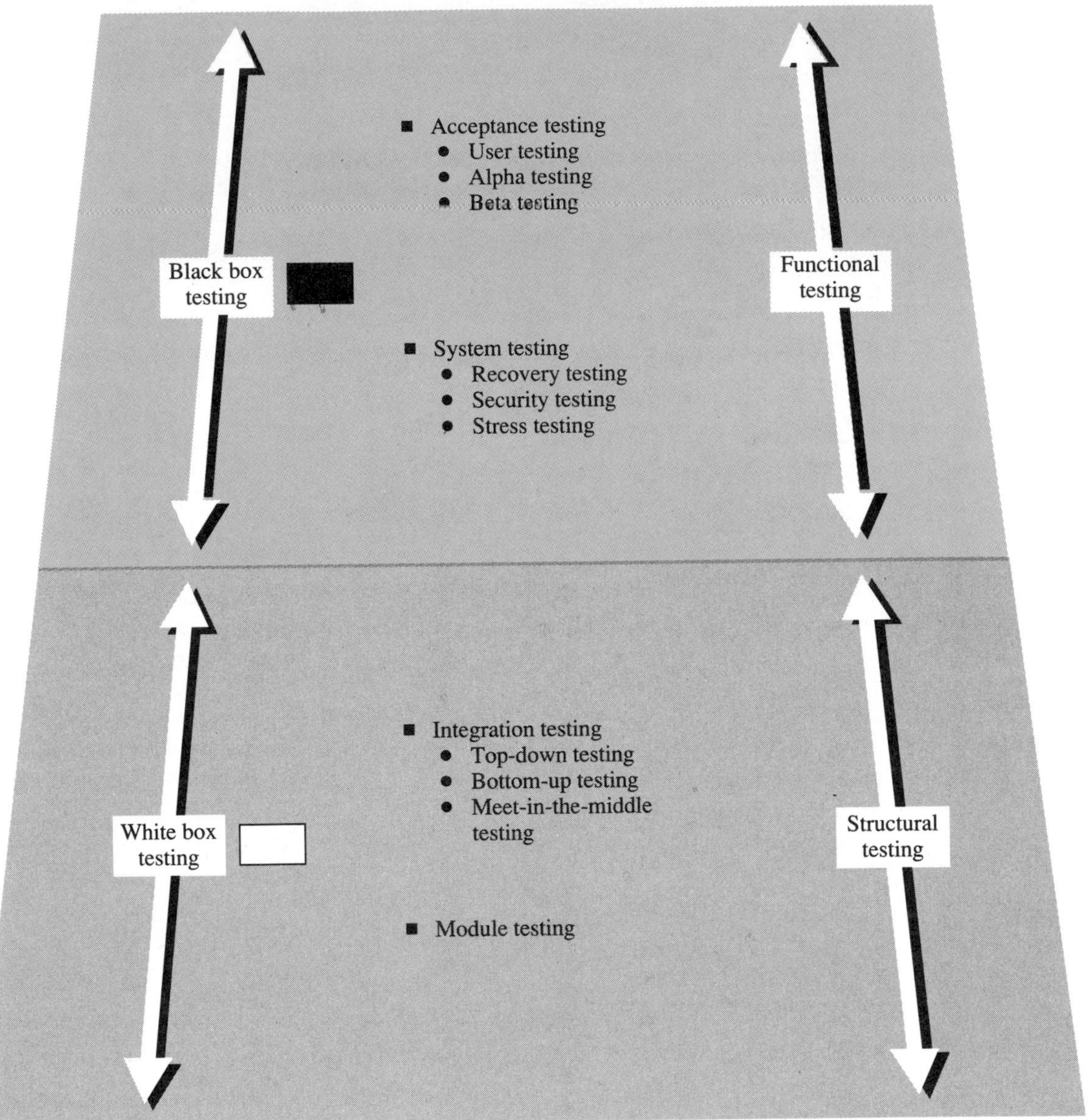

Figure 18.5
Software testing stages.

Because we cannot test all possibilities, it makes sense to concentrate on the areas that present the greatest risk. Software modules should be plotted on a risk grid based on their probability of error and level of impact. See Figure 18.6.

From the module risk rating grid, we see that Modules B, C, and E have the highest risk. These modules, therefore, need to receive most of the testing effort, because they have both the highest probability of an error occurring and the strongest impact on the program.

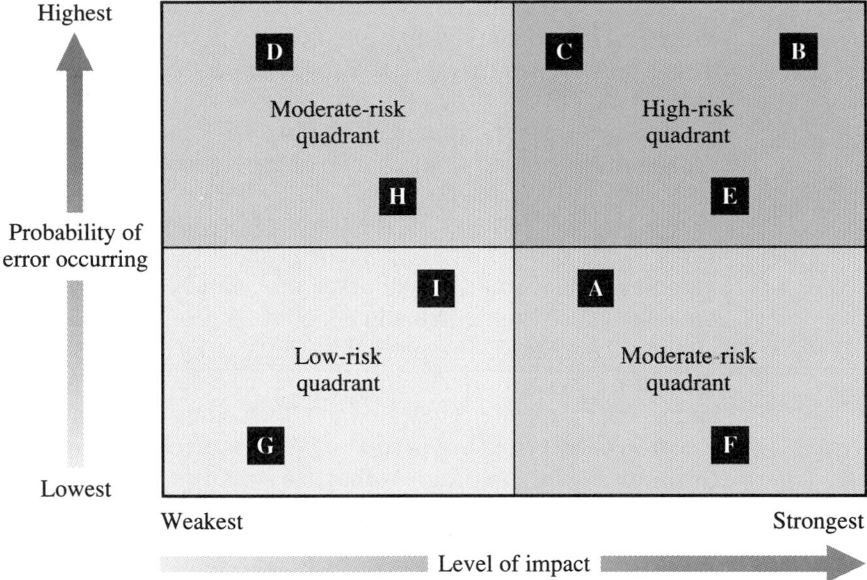

Figure 18.6 Module risk rating grid.

Impact depends on what the program does. The strongest impact relates to errors that can cause the worst consequences.

- If a module executes incorrectly, what might happen?
- Could bills be paid more than once?
- Could checks be printed with extra zeros in the amount field?
- Could a command cause a machine to drill holes in wrong locations in a truck chassis?
- Could a cost report contain incorrect figures that would cause management to underbid on contracts?
- Are incorrect results updating and corrupting the database?

Yes answers to these questions would presumably indicate strong impact on the system. Certainly, an error that destroys a customer accounts receivable file is worse than one that only misnumbers the pages of a report.

Probability of error occurring is an estimate of the likelihood that the module will fail, resulting in the impact. Some factors that contribute to high probability of impact occurring include:

- The size and complexity of a module
- How many other modules it interacts with
- The uniqueness of the new system
- The experience level of the coders

After the modules have been rated by relative risk, levels of testing effort can be more efficiently and effectively allocated. The rating of the modules from a risk perspective should also influence the order in which the modules are written. As a rule of thumb, the highest-risk modules should be produced first. From a practical standpoint, if the high-risk modules are delivered late in the schedule, they stand little chance of receiving the testing effort due them.

Which Modules Account for the Greatest Number of Errors?

Error density is generally present in most programs. Twenty percent of the modules may account for 80 percent or more of the errors. The probability of the existence of more errors in a module is proportional to the number of errors already detected in that module. That's why some managers will not spend more time trying to make error-prone modules error-free, but will scrap them and rewrite the code. They also give the coding assignment to a different coder. Because early errors are predictors of later errors, the appropriate strategy is to throw away work in progress that has already shown a tendency toward errors.

Testing Integrated Software Modules

INTEGRATION TESTING is performed by combining modules in steps. While module testing concentrates on specific modules, integration testing is performed on a hierarchy of modules, especially on interfaces between modules.

If modules work alone, why won't they work when combined? Because integrating modules may develop the following interface errors:

- Data can be lost across interfaces.
- A function may not perform as expected when combined with another function.
- One module can have an adverse affect on another.

Integration testing is a systematic way to build the program while performing tests to detect interfacing errors.

Integrating modules follows an incremental approach. In it, testing starts with a single module that is subjected to appropriate test cases. Once the testing of this module produces satisfactory results, a second module is introduced, and more test cases are applied. The process continues until all modules are eventually integrated into a complete program. Because of the incremental nature of this approach, it can be assumed that if errors occur when a new module is introduced, those errors are caused by the new module, or some aspect of introducing it. The source of errors is therefore localized, and detecting and correcting errors is made easier.

Using Stubs and Drivers

Because a module is not a stand-alone program, stub and driver modules must be developed for each module test. As illustrated in Figure 18.7, a test module that invokes and transmits data requires STUB MODULES to model this relationship. The stub takes the place of a called module that hasn't been coded yet. The

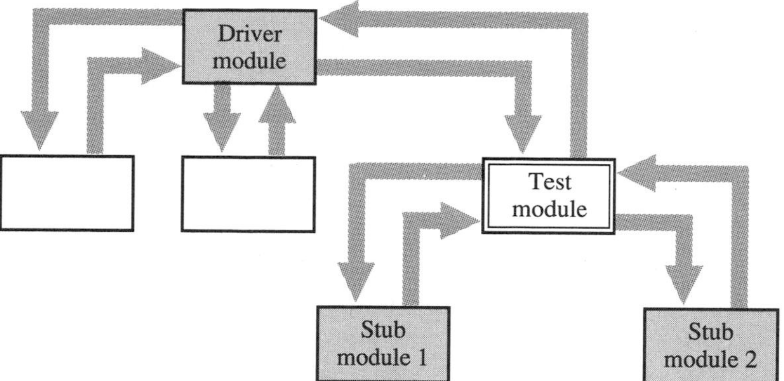

Figure 18.7
Role of stub and driver modules in integration testing.

DRIVER MODULES call the module under test and pass it test data. Stubs and drivers link modules to enable them to run in an environment close to the real one of the future.

Stubs and drivers are often seen as throw-away code. Although they aren't in the final version of the program under development, they can be reused to retest the program whenever it changes, as in regression testing. A comprehensive library of drivers and stubs represents a very powerful testing tool.

Ways to Test Integrated Modules

Two basic means of performing integration testing are:

- Top-down
- Bottom-up

Top-Down Integration

TOP-DOWN INTEGRATION TESTING relies on the incremental delivery of high-level modules and stubs as bases for testing. Modules are integrated by moving downward through the control hierarchy, beginning with the main executive module, as indicated in a typical structure chart. As modules are written at each level, stubs are replaced by their corresponding real modules until, eventually, the bottom real modules are written, and the total program can be tested as an integrated whole.

Figure 18.8 shows an example of top-down integration testing in which the control module, Module A, is tested first. This test requires stubs corresponding to Modules B, C, and D. When Modules B, C, and D are written, they are tested with Module A and stubs for Modules E, F, and G. Finally, when Modules E, F, and G are written, the entire program is tested as an integrated package.

A proper test of an invoking (i.e., calling) module requires that the stub module check the input parameters and return reasonable values for the output parameters. The stub must simulate the lower levels of the program closely for top-down testing to be effective.

Figure 18.8
Top-down integration testing.

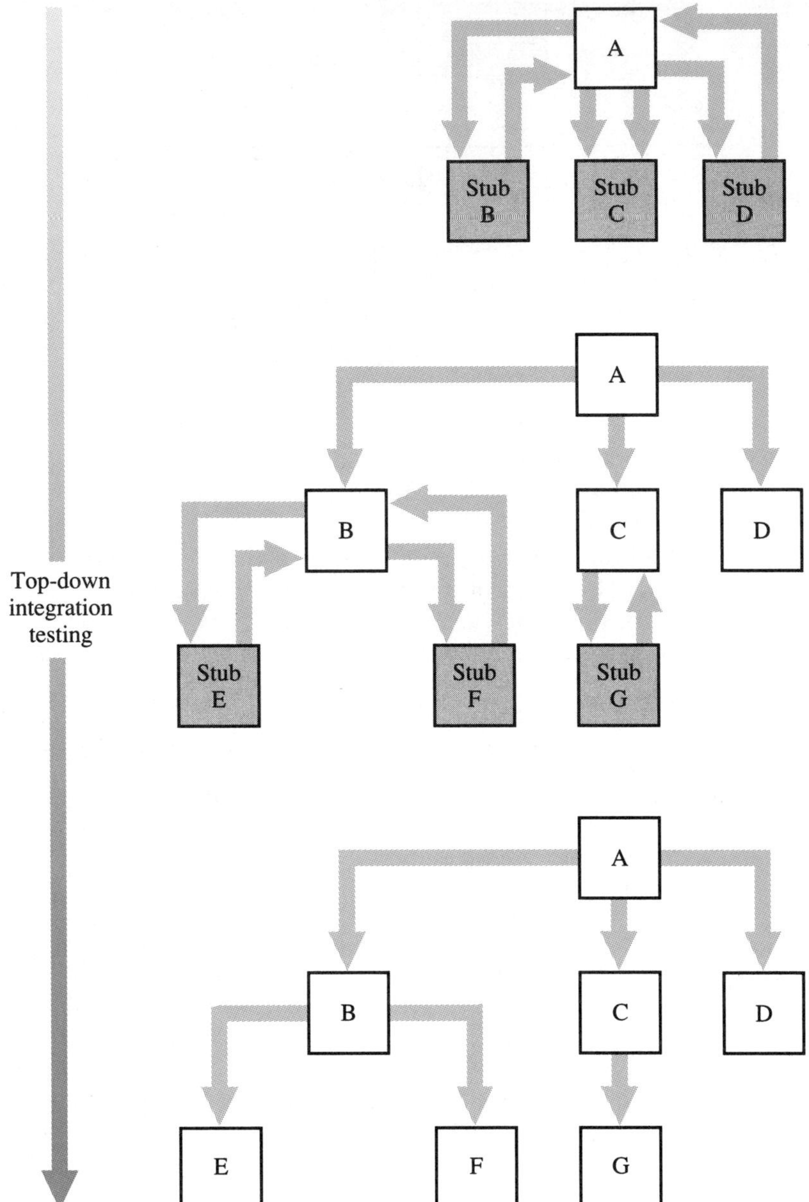

Bottom-Up Integration

Figure 18.9 illustrates **BOTTOM-UP INTEGRATION TESTING**, in which the lowest level modules are coded and tested first with appropriate drivers. This form of testing relies on the delivery of low-level modules and drivers as a base for integration testing.

In our example, Modules E and F need Driver B; Module G needs Driver C; and Module D needs Driver A. At the next level, Modules B and C are fully

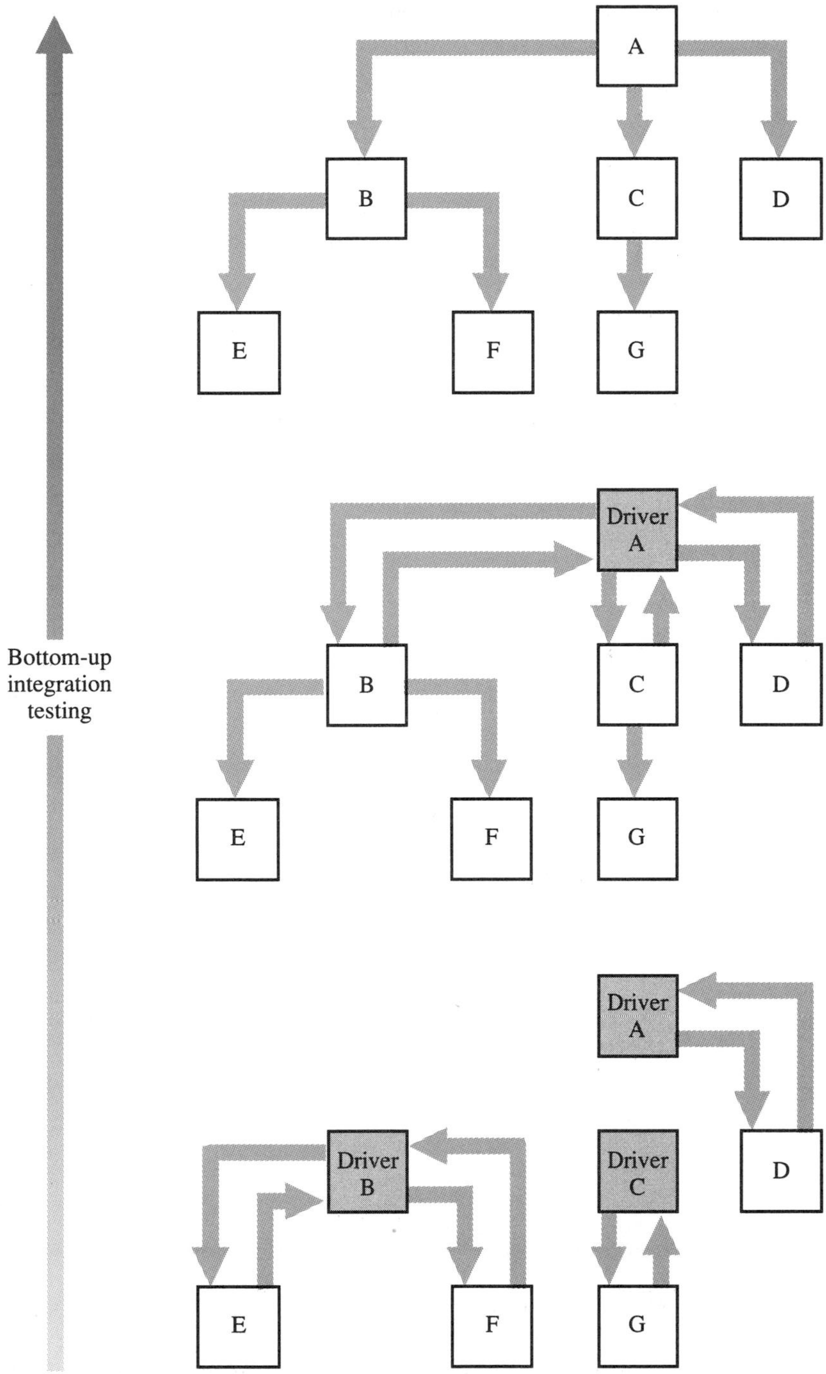

Figure 18.9
Bottom-up integration testing.

A
B
C
D
E
F
G

Driver A
B
C
D
E
F
G

Bottom-up integration testing

Driver A
Driver B
Driver C
D
E
F
G

coded, integrated, and tested with Driver A. Then, Driver A is fully coded and the modules are fully integrated and tested as a whole.

Advantages and Disadvantages of Top-Down and Bottom-Up Approaches

Generally, the advantages of the top-down approach are the disadvantages of the bottom-up approach, and vice versa. Both approaches code a little, then test a little.

One of the major factors to consider when choosing between top-down and bottom-up integration testing is the cost of preparing stubs and drivers. Generally, driver modules are easier to develop, making bottom-up testing less costly.

With the top-down approach major modules are coded and tested first. As a result, major errors or design flaws are discovered early in the process, thereby saving much wasted effort in redesigning and rewriting code. Furthermore, developing a working, albeit limited, program early on provides a strong psychological boost to all involved in systems development. The major disadvantage of the top-down approach is the need for stubs. If the systems application is complex, it is difficult, if not impossible, to write stub modules that adequately simulate the application and generate realistic output.

The bottom-up approach involves testing modules at lower levels in the structured hierarchy first, and then working up the hierarchy of modules until the final control module is tested. Drivers provide lower-level modules with appropriate input, and generally, test cases are easier to develop than for top-down testing.

A major disadvantage of the bottom-up approach is, however, that no working program can be demonstrated until the last module is tested. Therefore, systems design errors that may exist will not be detected until the last module is tested, requiring the entire program to be rewritten and retested. Such a situation reinforces the need for strong design and design walkthroughs.

In summary, if the control and upper-level modules are complex and critical to the software's success, the top-down approach is recommended. On the other hand, if lower-level modules are more critical, the bottom-up approach should be used. In general, a meet-in-the-middle approach that combines top-down for upper levels of the software structure, coupled with bottom-up for subordinate levels, may be the best compromise.

What Is Systems Testing?

SYSTEMS TESTING is the process of testing the integrated software in the context of the total system that it supports. Tests conducted at this level include the following:

- ***Recovery Testing*** A system test forces the software to fail in various ways and verifies that complete recovery is properly performed.
- ***Security Testing*** Test cases are conducted to verify that proper controls have been designed and installed in the system to protect it from a number of risks. The system testers (sometimes called a Tiger Team) do anything to penetrate the system and perform harmful acts.

- ***Stress Testing*** This type of testing, similar in concept to security testing, executes a system in a manner that demands resources in abnormal quantity, frequency, or volume. For example, testers may generate 20 interrupts per second or generate thousands of instructions to try to break the program. Stress testing validates the program's ability to handle large volumes of transactions in a timely manner.

What Is Acceptance Testing?

ACCEPTANCE TESTING evaluates the new system to determine if it meets user requirements under operating conditions. When the software and documentation are deemed stable by the test group, it's time to release the software and documentation to a user test group to get user feedback. A user test group should be composed of all or a sample of people who will work with the system under development. All or a large part of their jobs will depend on having this system work correctly. Two acceptance testing procedures can be used:

- Alpha testing
- Beta testing

Acceptance testing is the last chance to test and rethink things before the software is converted from development to operations.

Alpha Testing

ALPHA TESTING is conducted in a natural setting (i.e., in a real-world operating environment) with systems professionals in attendance, usually as observers, recording errors and usage problems. Two techniques that are especially applicable to alpha testing are the following:

Usability Labs The designs of USABILITY LABS differ, but the center theme is constant: Get a representative sample of the people who will eventually use the software to do the testing, and place them in a controlled and structured test environment. Sample size may range from only one user for small, local-based applications to 10 or more for large, global-based applications.

A usability lab may be a room with one PC, or a large room divided and equipped with one-way mirrors, television cameras, and several PCs or workstations. Users are in the test side of the lab and observers (e.g., systems professionals) are positioned behind one-way mirrors to watch users work with the system. One television camera focuses on the computer screen and keyboard; a second focuses on the user; and a third, mounted overhead, records the users' interaction with the user documentation. The purpose of a usability lab is not so much to determine if the software works, from a technical viewpoint, but to determine how well users think that it works. For example, observation may show a confused user trying to fill in a form or follow unclear procedures. After reviewing a videotape of the testing session, observers find out precisely why users are confused. For example, the reason for the confusion may be the form design or unclear procedures.

Usability Factors Checklist Recall the MURRE design factors first explained in Chapter 7 and alluded to in other chapters (e.g., Chapter 15). To this point maintainability has been evaluated by design and code walkthroughs. Usability has been tested throughout development but not necessarily under working conditions. For example, screen designs were evaluated, and specific test transactions have been run. The level of reusability may have been extraordinary. For example, the system under development may have required 100 KLOEC, but 80 percent of the code may have been written already for other applications, and this 80 percent can be used for the new system. So only 20 KLOEC had to be written from scratch. The level of reliability may have been rated high because the software passed module and integration testing and recovery, stress, and security testing without a hitch. The level of extendability may be high because of excellent modular design or the use of object-oriented programming methods. That is, the new system will be able to adapt to and grow with changing user requirements easily.

Even though all the MURRE design factors are rated excellent, the evaluation of the usability factor, however, has not been completed. A complete and final assessment of this design factor has to be performed by the user test group under conditions that are representative of how users will work with the system after it is released to operations.

Since the user is the final arbiter of software quality, a technique that can be applied to help evaluate and finalize the usability design factor is to have the user test group complete a **USABILITY FACTORS CHECKLIST** (see Figure 18.10). This checklist can be expanded and tailored. It can also be used along with usability labs. The usability factors checklist gives users the opportunity to evaluate the quality of the new software before it is converted from development to operations.

In addition to completing a usability factors checklist, the user test group must also explain to systems professionals why factors were rated low if that is the case. For example, why does a user find the program to be rated low for ease of use? Is it because of poor documentation and lack of instructions? The user should answer such questions and, if possible, provide suggestions as to how any or all usability factors can be improved.

Figure 18.10
Software usability factors worksheet.

USABILITY FACTORS

	1 = low to 5 = high				
A. Ease of use	1	2	3	4	5
B. User friendliness	1	2	3	4	5
C. Understandability	1	2	3	4	5
D. Level of confidence	1	2	3	4	5
E. Conformance to requirements	1	2	3	4	5
F. Conformance to response time	1	2	3	4	5
G. Comfort level	1	2	3	4	5

Please circle the number applicable to each quality factor.

Beta Testing

BETA TESTING is similar to alpha testing, except that no systems professionals are present (e.g., in a usability lab) during user acceptance testing. In addition to employing a usability factors checklist, the user test group records all problems, real or imagined, encountered during beta testing and reports them to the systems people periodically. Modifications are made to make the system ready to be released for full implementation and operations.

Releasing the Software

How much testing is enough? Actually, one never completes testing. The testing phase itself in the SWDLC is simply stopped when software passes acceptance tests.

After all the preceding tests have been conducted, and the software has successfully passed such tests, it is probably reasonable to deem the software acceptable for release from development status to operations. Releasing the software represents the end point of the SWDLC. The software is accepted for implementation along with other components comprising the total system. Completion of the implementation phase represents the end point of the SDLC.

Any project manager would like to certify the software to be totally error- and trouble-free. Such a certification, however, is usually impossible. The project manager knows what has been tested, but there is no way of knowing what hasn't been tested.

So, to a great extent, releasing the new software to final implementation and operations is a judgment call. Using the preceding testing procedures will reinforce and give a strong base on which to make this judgment. For example, the project manager may make the following statement: ''Judging from the testing procedures conducted and documented herein, I believe Software A to be 95 to 99 percent error-free and its maintainability, usability, reusability, reliability, and extendability design factors to be rated excellent. I hereby recommend its release for systems implementation and operations.''

REVIEW OF CHAPTER LEARNING OBJECTIVES

The major goals of this chapter were to enable each student to achieve six important learning objectives. We will now summarize the responses to these learning objectives.

Learning objective 1:
Describe a software testing scenario and give its purpose.

A software testing scenario involves:

- Developing test cases and testing strategies
- Running the tests
- Reporting errors detected

- Correcting errors
- Predicting the level of software reliability

Testing is a vital phase in software development. Testing software and removing errors before it goes into operations will help to reduce the number of errors that may crop up during operations—the worst time for errors to occur.

Learning objective 2:
Define two basic ways to design test cases.

Two basic ways to design test cases are:

- White box testing
- Black box testing

White box testing focuses on the interworkings of the code. Black box testing, on the other hand, tests functionality of the software. All test cases should be recorded in a test case matrix along with expected results and actual results.

Learning objective 3:
Discuss what should be done when errors are detected.

If errors are detected, they should be recorded in an Error Report and communicated to the person assigned to correct and resolve errors. The report types are:

- Suggestion
- Design error
- Coding error
- Documentation error
- Query

Errors may be minor, serious, or fatal. The final resolution is one of the following:

- The error is fixed
- The error cannot be reproduced
- The error can't be fixed
- The coder disagrees with the tester's suggestion
- The error or suggestion is withdrawn
- The software works in accordance with design specifications

The coder assigned to resolve errors, the tester, and the project manager sign off on the Error Report upon resolution.

Learning objective 4:
Present a way to help measure the reliability of test cases.

Seeding errors into the program that is to be tested helps to measure the reliability of the test cases. If the seeded errors represent the type of real errors, and the test cases uncover all the seeded errors, one can infer that all of the real errors have also been detected.

Learning objective 5:
Name and briefly discuss the four software testing stages.

The four software testing stages are:

- Module
- Integration
- Systems
- Acceptance

Module testing is the process of testing the smallest components in the total program before they are put together to form a software whole. Integration testing is the process of testing the joined modules to see if the software plays together as a whole. Systems testing is the process of testing the integrated software in the context of the total system that it supports. Acceptance testing is the process of allowing both users and operators to test the program to see if it meets their requirements before the program is released and converted to operations.

Learning objective 6:
Differentiate between top-down integration testing and bottom-up integration testing.

Top-down integration testing starts with the coded superior module and stubs to test the top-most level of the structured hierarchy first. Then the next-level modules are coded and tested in a similar fashion until the bottom-level modules are coded and tested in an integrated whole. For bottom-up integration testing, this process is reversed. The bottom-level modules are coded first and drivers are prepared to test them. Then the next level of modules is coded and tested in the same manner until the final top module is coded and tested in an integrated program.

SOFTWARE TESTING CHECKLIST

The following is a checklist on how to conduct software testing, the last phase of the SWDLC.

1 Establish a software testing plan using a structured software testing scenario as your basic guide.

2 Develop white box and black box test cases, and use a test case matrix to record test cases. Where you deem it necessary, seed errors to test the effectiveness of test cases.

3 Prepare an Error Report to document all discovered errors and their resolution.

4 Create a software testing strategy that breaks testing into stages such as module, integration, systems, and acceptance testing.

5 Determine level of reliability. If the tests indicate high reliability, release the software for operations.

KEY TERMS

Acceptance testing
Alpha testing
Beta testing
Black box testing
Bottom-up integration testing
Bug
Debugging
Driver modules
Equivalence classes
Error Report
Fatal errors
Integration testing
Minor errors
Module testing
Regression test
Seeding
Serious errors
Software reliability
Stub modules
Systems testing
Test case matrix
Test cases
Top-down integration testing
Usability factors checklist
Usability labs
White box testing

REVIEW QUESTIONS

18.1 Does the software testing phase ensure that a program is totally error-free? Explain your answer.

18.2 Outline a software testing scenario.

18.3 Explain the purpose of white box testing.

18.4 Explain the purpose of black box testing.

18.5 Briefly describe the reasons to test the following commands: SELECT, OPEN/CLOSE, COPY REPLACING, IF, PERFORM UNTIL or PERFORM WHILE, and CALL.

18.6 Define an equivalence class. Give two examples of equivalence class tests.

18.7 What elements should a test case matrix include?

18.8 What elements should an Error Report contain?

18.9 Name error report types and their degree of severity.

18.10 List six ways to resolve error report types.

18.11 If a tester is unsure of what triggered an error, what should he or she do?

18.12 Define a regression test and explain its purpose.

18.13 What's the purpose of a seeding model? Explain how it works and its relationship to helping estimate software reliability.

18.14 Define risk-driven testing. What is its purpose?

18.15 Define error density.

18.16 Differentiate between module testing and integration testing.

18.17 Define a stub module and describe how it works. Define a driver module and describe how it works.

18.18 What are the three components of systems testing? Briefly describe the purpose of each.

18.19 What is the purpose of acceptance testing? Who are the two key players in acceptance testing?

18.20 Define alpha and beta tests.

CHAPTER-SPECIFIC PROBLEMS

These problems require exact responses based directly on concepts and techniques presented in the text.

18.21 A coder writes the following instruction:

```
IF
   A = B
THEN
   SET B = 10
   SET B = 20
```

Required: Explain what this instruction does. Recode the instruction to set B equal to 10 if A is equal to B, else set B equal to 20.

18.22 A coder writes the following instruction:

```
COMPUTE A = B/C
```

Required: Develop four test cases for this instruction, and state the expected result for each test case.

18.23 Review the following instruction:

```
IF
   A < B AND C = 3
THEN
   DO SOMETHING
```

Required: Develop four test cases for this instruction, and describe the expected result of each test case.

18.24 A program is supposed to accept any number between 1 and 99.

Required: Design four equivalence class tests, one that is valid and three that are invalid.

THINK-TANK PROBLEMS

These problems call for a feasible approach rather than a precise solution. Although the problems are based on chapter material, extra reading and creativity may be required to develop workable solutions.

18.25 Chapter 12 presented a number of controls that should be designed into programs. Some of these controls are:

- Control totals
- Error reports
- Transaction log
- Sequence checks
- Missing data
- Range check
- Reasonableness check
- Check digits
- File labels
- Validity (or identification table)

Required: Prepare a test case for each one of these controls to make sure that the program is executing them properly. Derive any numbers or make any assumptions you deem necessary to develop a reasonable approach to this problem.

18.26 Study the following structure chart:

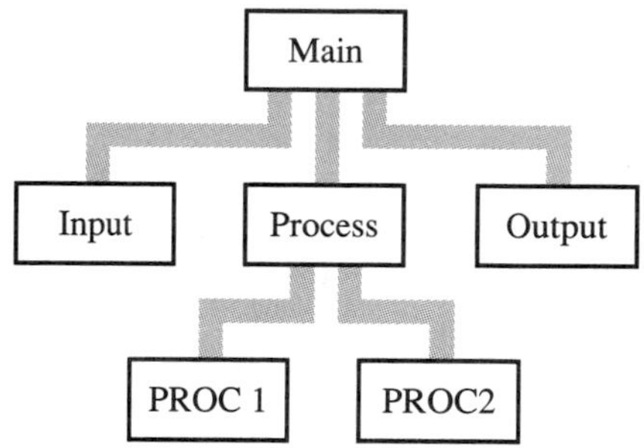

Required: In structured English form, or using a brief narrative of your own, specify how to perform top-down integration testing for the structure chart above. Do the same for a bottom-up integration testing.

18.27 Assume that the following set of modules is to be tested:

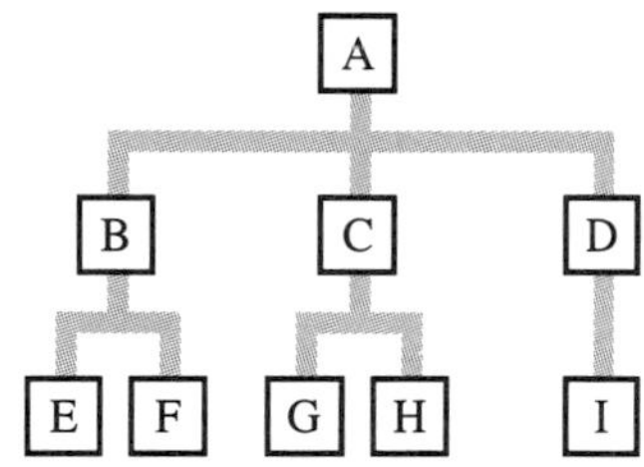

Required: Demonstrate how you would go about performing bottom-up and top-down integration testing for the foregoing set of modules. There is a third kind of integration testing. What is it called? Demonstrate how it works. Discuss the problems with bottom-up testing. Explain how top-down testing offers several advantages over bottom-up testing. State the problem with top-down testing.

18.28 In the Maltese Drapery Company, software programs are seldom tested before their conversion to operations. Users often encounter errors in programs, but often ignore them. For example, a program may crash, and users may not be able to isolate the activity that caused the crash. Their attitude is that it is quicker to try a slightly different method to get the current job done than it is to find and report the error. You are a systems consultant brought in by Maltese to assess the current situation and make recommendations to improve software reliability. Some sample comments heard from users are:

"My experience is that it is largely useless to report errors unless I can supply the exact cause of the error and reproduce the error."

"I haven't reported this error, because recovery from it is usually faster than the time needed to report and have the programmer correct it."

"I don't generally report an error, because the systems people will either not look at it, chalk it up as a one-time event or user mistake, or take forever to fix it."

Required: Prepare a report to top management on your assessment of the present situation concerning software reliability at Maltese and your recommendations to increase software reliability.

SUGGESTED READING

Beizer, B. *Software System Testing and Quality Assurance*. New York: Van Nostrand Reinhold, 1984.

Burch, John G. "Basic Program Development and Testing." *Journal of Accounting and EDP,* Summer 1989.

Crosby, P. B. *Quality Is Free: The Art of Making Quality Certain*. New York: McGraw-Hill, 1979.

Desmond, John. "The Friendly Master." *Software Magazine,* February 1988.

Durant, Jerry E. "Applying Systematic Testing to Application Development Audits." *Internal Auditor,* February 1991.

Kaner, Cem. *Testing Computer Software,* Blue Ridge, Pa. TAB Professional and Reference Books, 1988.

Keyes, Jessica. "Gather a Baseline to Assess CASE Impact." *Software Magazine,* August 1990.

Martin, James, and Carma McClure. *Structured Techniques for Computing*. Englewood Cliffs, N.J.: Prentice-Hall, 1985.

Meyers, G. *The Art of Software Testing*. New York: John Wiley, 1979.

Miller, Barton P., Lars Fredriksen, and Brian So. "Study of the Reliability of UNIX Utilities." *Communications of the ACM,* Vol. 33, No. 12, December 1990.

Musa, John D., Anthony Iannino, and Kazuhira Okumoto. *Software Reliability*. New York: McGraw-Hill, 1990.

Perry, William E. *A Standard for Testing Application Software*. Boston: Auerbach Publishers, 1990.

Pressman, Roger S. *Software Engineering: A Practitioner's Approach,* 2nd ed. New York: McGraw-Hill, 1987.

Rettig, Marc. "Testing Made Palatable." *Communications of the ACM,* Vol. 34, No. 5, May 1991.

Sivula, Chris. "Back to the Lab." *Datamation,* August 15, 1990.

Stahl, Bob. "The Ins and Outs of Software Testing." *Computerworld,* October 24, 1988.

JOCS CASE: Testing Software

Jake Jacoby has implemented an ongoing, almost cyclical software testing strategy for the JOCS project. Instead of testing only at the end of the cycle, he has decided that testing should be performed throughout the SWDLC. Figure 18.11 is a graphic representation of the software testing strategy that Jake has devised. Just as a programmer uses a DFD to understand a system, Jake felt that devising a graphic model would make his testing strategy understandable to the rest of the SWAT team.

The individual circles represent the four distinct stages in software testing. For example, circle 1 represents a circular process composed of two tasks: general purpose design and the accompanying structured walkthrough of that design. These two tasks are part of the first circle because the design and walkthrough are done repetitively, or in a circular fashion, until the programmer is confident that an acceptable design has been achieved.

The second stage in the process, Circle 2, includes the circular tasks of module design, module walkthrough, and top-down testing. Jake thinks that software is best developed with the incremental policy: Design a little, code a little, test a little. Design a little more, code a little more, test a little more. He prefers this policy to that of designing all the software, then coding all the software, then crossing one's fingers and testing all the software. The second circle on Figure 18.11 represents Jake's incremental policy.

Circles 1 and 2 represent the white box testing stages for a given program. The tasks that compose these circles will be completed by the programmer who is responsible for the design, coding, and initial testing for that program. The programmer will have the opportunity, while performing these tasks, to test the overall and detailed logic of the program. Jake thinks that a programmer should have responsibility for, and pride in, a program. Finishing the tasks of the first two circles should satisfy the very human need for a feeling of job completion.

The next two stages are performed by personnel other than the original programmer. Circle 3 is composed of three tasks: stress testing, recovery testing, and security testing. Jake plans to have these tasks done by a separate person within the SWAT team. This person will be responsible for testing the validity of programs by running the programs with a series of inputs and checking what outputs are created. This is black box testing. As shown by the arrows on the chart, the tasks in Circle 2 always feed the tasks in Circle 3. However, the dashed line shows that sometimes the tasks done in Circle 3 will require that the program be returned to Circle 2. For example, if the tester finds problems with a program during Circle 3 testing, the tester will either fix the problem or return the program to the programmer, who will again go through the tasks shown in Circle 2.

Circle 4 represents the testing performed by the user: user and beta testing. User testing will be conducted by users with a SWAT team member

Figure 18.11
JOCS software testing strategy.

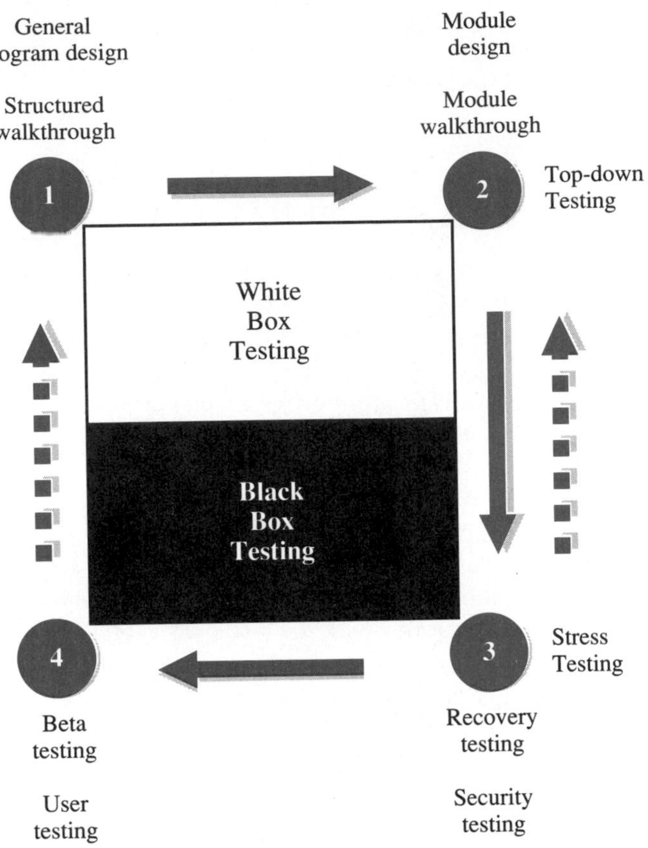

standing by to observe any programming problems. This testing will be done to simulate a live environment, but without entering real data. Beta testing, on the other hand, will be performed by the users in their own department environment. During beta testing, users will execute the programs with accurate data as if the software were fully functioning. In Figure 18.11, there is a dashed line between Circles 1 and 4. If users discover problems during beta testing, these problems will probably cause a programmer to redesign the program. It is possible that users may discover new system requirements during beta testing. These new requirements may be tackled at that time, or they may be delayed until systems implementation is complete.

As an example of Jake's software testing strategy, let's follow the development of the direct labor-hours data capture program created by Jannis Court.

Circle 1: Jannis translated the systems analysis DFD into a Jackson diagram in Chapter 16. Using the testing strategy established by Jake, Jannis performed a structured walkthrough on the design with another team member.

A few problems were discovered during the walkthrough, and it was necessary for Jannis to redo the general program design. Another walkthrough of the design approved her changes.

Circle 2: Jannis takes the Jackson diagram created in Chapter 16 and converts it into structured English. An example of the structured English for her program is shown in Figure 18.12. Jannis walks through the structured English with another member of the SWAT team and gains the team's approval of her design.

Jannis now types in the main modules of the program with stubs representing the detailed modules and she conducts top-down testing to identify syntax and logic errors in the main modules.

Figure 18.12
Overall structured English: JOCS direct labor data capture program.

```
Begin Capture Direct Labor Hours


Clear employee id, job number, operation number
Set data input complete flag = "no"
Do process data module until data input complete flag = "yes"
   Get job number
   If job status = "complete"
      Do completed job module
   Else
      Get operation number
      If valid operation for job status = "no"
         Do invalid operation module
      Else
         If operation status = "complete"
            Do completed operation module
         Else
            Get employee id
            If hours > weekly total hours
               Do overtime hours module
            Else
               Set process transaction flag = "yes"
            Endif
         Endif
      Endif
   Endif
   If process transaction flag = "yes"
      Get current date and time
      Do write files module
      Set data input flag complete flag = "yes"
   Endif
Enddo
   Do terminate module


End Direct Labor Hours Capture
```

After she has assured herself through online testing with sample data that the overall logic is correct, detailed modules will be designed and coded with the structured walkthrough technique described above. If, however, the overall logic were identified as incorrect during either the structured walkthroughs, or during top-down testing, Jannis would return to the design phase of the cycle in this circle and redesign the program. The detailed modules are typed in and tested one at a time for syntax and logic errors.

Circle 3: After Jannis has completed top-down testing on the program, the program is turned over in its entirety to Tom Pearson for black box testing. In Jake's software testing strategy, the team member responsible for designing and coding a program does not perform exhaustive data tests on that same program.

Tom Pearson has been assigned the task of completed program testing for the project, because he has experience in programming and analysis of functions, and he has shown an aptitude for identifying and repairing defects in other people's programs.

Tom uses three different sets of data to stress-test the programs: *Normal data,* data that are within the expected bounds of the program; *abnormal data,* data that stretch the anticipated range of values; and *incorrect data,* data that contain invalid values. For example, Tom tested Jannis's program using a job number that was incomplete, a job number that was complete, a job number that didn't exist in the system, and a job number that contained alphabetic values as well as numeric values.

He also ran a variety of tests to find out what happened to the program if it aborted during processing, such as if the power failed in the middle of processing, and he checked the security provisions for program access.

Circle 4: After Tom finishes testing Jannis's program, he calls in a user to run the program. Tom observes any problems the user experiences, such as data input errors, interface difficulties, and misunderstandings due to unclear error messages.

The final task of testing, beta testing, will be delayed until the rest of the system is ready for it. When this occurs, Jannis's program will be released for final testing approval.

Jake wants to monitor the entire development process. As discussed in Chapter 15, he measures the amount of time the SWAT team members work on the project to track productivity. In addition to measuring productivity, Jake also measures the quality of his staff. To accomplish this, Jake instructs Tom to keep track of the defects he finds within a program, and also identifies what is causing a defect. Tom may also actually fix any defects if time permits. Tom's information is fed back to the programmer, who may fix a defect (if Tom has not already repaired it). Programmers are kept aware of the defects discovered in their programs so that they can use this information to produce new programs without the same types of defects.

Chapter 19

Implementing the System

WHAT WILL YOU LEARN IN THIS CHAPTER?

After studying this chapter, you should be able to:

1 Prepare a systems implementation plan using PERT.
2 Discuss how to prepare a site for the technology platform.
3 Describe ways to train people to work with the new system.
4 Describe the documentation that must be prepared.
5 Explain four methods of systems conversion.
6 Discuss a postimplementation review.

INTRODUCTION

Systems implementation is the last phase in the systems development life cycle (SDLC). To gain a perspective of what we have achieved to this point, refer to Figure 19.1.

SYSTEMS IMPLEMENTATION involves integrating all systems design components, including the software, and converting the total system to operations. Systems implementation involves a two-step process:

1 Planning

2 Executing

Planning determines what systems implementation tasks are to be performed and when. Executing actually performs the systems implementation tasks.

CREATING A SYSTEMS IMPLEMENTATION PLAN

The **SYSTEMS IMPLEMENTATION PLAN** is a detailed formulation and graphic representation of how systems implementation will be achieved. Generally, the systems implementation plan is prepared several weeks or months in advance, depending on scope and complexity of the project. The **SYSTEMS IMPLEMENTATION REPORT** contains the implementation plan and the results of tasks performed according to this plan. For a very simple project, such as installing a software package for a microcomputer, actual implementation may require an hour or less. In this case, the Systems Implementation Report may be a brief memo containing user instructions.

For a major systems project built from scratch, the implementation plan-

Figure 19.1
SDLC phases and their related chapters in this book. In Chapter 19 the emphasis is on system implementation, including planning and postimplementation review.

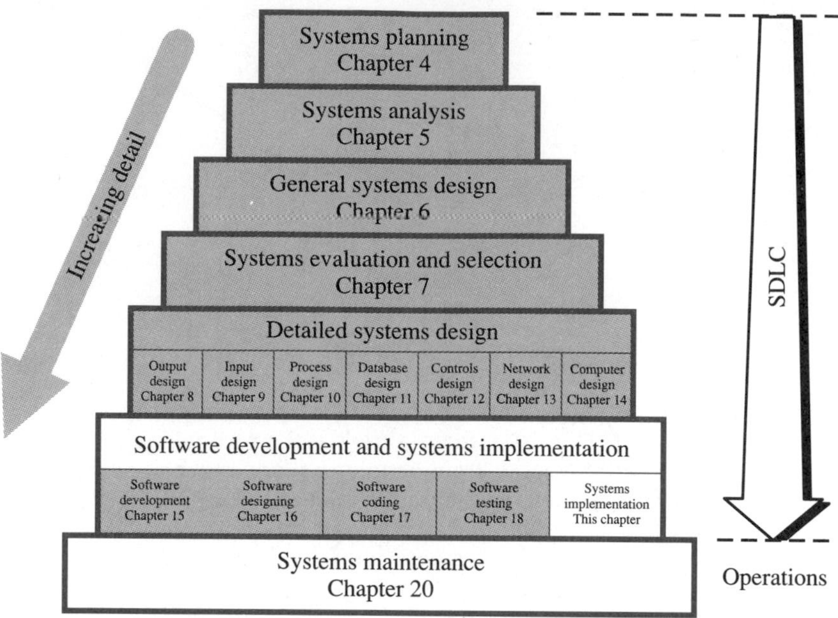

ning lead time is measured in months. It entails the coordination and scheduling of a number of activities and tasks performed by an implementation team made up of the following:

- Systems professionals who designed the system
- Managers and various staff
- Vendor representatives
- Primary users
- Coders
- Technicians

A typical big systems project implementation plan is illustrated in Figure 19.2. At this point, technology has already been evaluated and selected. Purchase orders are released to acquire the selected technology. Site preparation begins immediately, so it is ready to receive the technology platform from the vendor(s).

A programming team is formed, and coders begin review of the design specifications to ensure clear understanding before coding and testing starts. Training begins immediately for all personnel and users. It continues until it is completed. When the equipment arrives, its site is ready for installation.

Programs are subjected to various test cases to ensure a high level of program reliability. Test cases are also created and applied to input, output,

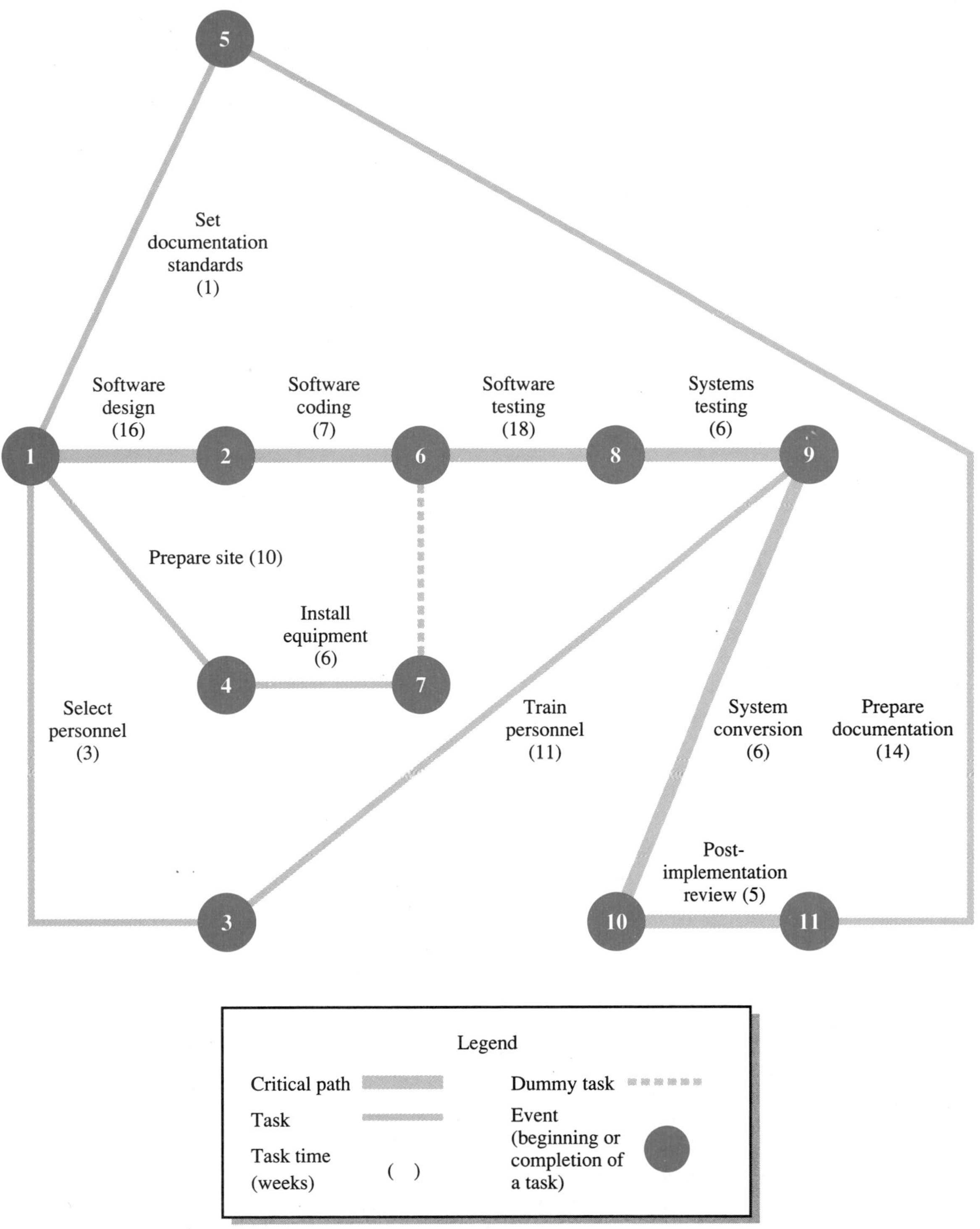

Figure 19.2 Systems implementation plan represented by a PERT chart.

database, and controls to make sure that the total system is working as expected.

If test results are successful, the system goes through a conversion process. Usually, a postimplementation review is performed to make sure the system is successfully converted and is operating as planned using live data. Fine-tuning may be appropriate to enhance the new system's performance.

During this time, users or their representatives have been running an acceptance test. If everything at this stage is satisfactory—smooth runs, correct results, clear and comprehensive documentation, trained personnel, and clear operating procedures—the system moves from development status to operational status.

Because software development is normally such a significant part of systems implementation, we devoted the three preceding chapters to it. Now we will turn our attention to other key parts of systems implementation:

- Site preparation
- Personnel training
- Documentation preparation
- Systems and file conversion
- Postimplementation review

PREPARING THE SITE

If the technology platform is a microcomputer, little site preparation is required, except for establishing environmental, physical, and database controls presented earlier in the controls chapter. If a mainframe or large network is to be installed, a new building or extensive remodeling of present facilities may be required. Review Chapter 12 to determine how these facilities should be prepared.

Adequate space should be provided for each piece of equipment and furniture. Appropriate wiring, cables, supplies, ventilation, and air conditioning should be installed to ensure a clean, workable environment. Antistatic comfort mats and carpets should be laid. Sufficient shelving, stands, tables, and disk and tape storage trays and cabinets should be made available. Human factors require acoustic privacy panels, printer enclosures and acoustic printer cushions, and ergonomically designed furniture and workstations.

The site should be relatively inaccessible to the public and unauthorized employees. Two doors creating a mantrap controlled by biometric access control devices will provide maximum security. The site, if not a standalone structure, should be above the ground floor and toward the center of the building.

A properly prepared site ensures quick and easy installation when the technology arrives. Furthermore, on-site technology tests can begin with a high degree of assurance that if errors are detected or failures occur, they will not be caused by site inadequacies.

Before a vendor delivers the equipment, a diagnosis of the equipment should be run at the vendor's site to simulate several weeks of continuous operation. This burn-in test checks for shorts in the electrical system, switches, connectors, and various components.

While burn-in tests are performed, site preparation for the equipment should be near completion. Further, this site should meet environmental standards to ensure valid software and systems testing once the equipment arrives from the vendor.

TRAINING PERSONNEL

No system can work satisfactorily unless the users and others who interact with the system are properly trained. Without an appropriate level of expertise and acceptance by personnel, the system will probably fail. Personnel training not only increases people's expertise, but it also facilitates people's acceptance of the new system—a very important factor in a system's ultimate success.

Generally, three groups will need to be trained:

1 Technical personnel who will operate and maintain the system.

2 Various workers and supervisors who interact directly with the system to perform their tasks and make decisions.

3 General managers.

If the company has extended its information system beyond corporate walls, as with an electronic data interchange system, a fourth group who will need to be trained are those who do not work for the company but who will interact with the system, such as customers and suppliers. All of these groups are conveniently placed under the umbrella term "users."

Training should begin early enough for it to end about the same time the system becomes operational. Training gives users self-confidence and streamlines and minimizes disruption during early stages of systems operation.

Ideally, training should be performed before the new computer system and network are installed, assuming that the system requires new technology. Although this approach is not always feasible, as much preinstallation training as possible should be done. Otherwise, the system will sit idle while users are being trained.

Training Programs

Training ranges in scope from a brief tutorial for one user on how to perform a new but simple task to training most, if not all, of the users throughout the organization for a major new system. Training programs that support extensive training agendas include:

- In-house training
- Vendor-supplied training
- Outside training services

Any one or a combination of these training programs can be used.

In-House Training

The advantage of training users on-site is that the instruction can be tailored to the specific needs of that system and its users. The disadvantage is that users are in their own environment and might be subjected to telephone calls and emergencies that disrupt the training sessions.

Some companies use information centers and a training staff to help in user training. Information center staff can develop an introductory course and equipment or application demonstration, provide individualized training, help users apply a particular software package to a specific business problem, and arrange workshops. The goal of the information center is to encourage users to expand and explore the information system's benefits and services and to show how users can solve their own problems. In some cases, highly motivated users are trained as superusers who serve as training resources to help train other users in their respective departments.

Vendor-Supplied Training

Often the best source of training on how to use computer equipment, networks, or database management systems is the vendor supplying them. Most vendors offer extensive training programs as part of their service. In some cases, the vendor will charge for this service; in other cases, training is free. Most vendor training is hands-on, so the trainees actually use the system in the presence of trainers. Moreover, the system that trainees are using is devoted exclusively for training purposes. Therefore, there is not a rush to complete training so the system can convert to full operations as the case may be if the users were being trained on the system acquired by the company. Sending users to off-site short courses providing in-depth training may be preferable to in-house training. Also, users can be trâined before their hardware and software are installed. When they are installed, the users will be able to start working with them immediately.

Outside Training Services

If the vendor cannot provide the necessary training, or the company does not have an ongoing need or funds to support an in-house training program, management should consider outside training services, such as universities, professional societies, training institutes (e.g., MIS Training Institute), and training centers.

Business schools often provide a variety of courses in programming languages, software applications, and computer use. Although these courses may be helpful and informative, many users may not be able to wait an entire semester before mastering a vital skill. However, the continuing education division of a university typically offers short courses on popular software packages. It can also be contracted to provide tailored courses.

Both professional societies and training institutes offer a wide variety of seminars dealing with specific topics. These seminars are well suited for busy users, because they take only a few hours or days and generally provide hands-on training.

Training centers are established to train users for a specific vendor's product line. Authorized training centers (i.e., those authorized by the vendor) fulfill a need for effective, uniform training. Many authorized training centers offer regularly scheduled classes, customized curriculum and course materials, and corporate group rates.

Training Techniques and Aids

To perform the actual training, different training techniques and aids are used. These include:

- Teleconferencing
- Interactive training software
- Instructor-based training
- On-the-job training
- Procedures manual
- Textbooks

Teleconferencing

Teleconferencing, also called videoconferencing, can be used for many purposes, including training. As a training aid, teleconferencing permits users and trainers who are geographically separated to display and view text, data, graphics, and images and to interact with one another in real-time while sitting at their workstations or in a teleconferencing station. A subset of full-blown teleconferencing is one-way video and two-way audio. Images and voices of the trainers are transmitted to users who can be heard but not seen, as in some call-in TV shows. Users call in with questions for the trainers.

Everyone involved in the training session can meet at once and answer questions without spending excessive time away from the job. Normally, it is impossible to get a large number of people together at one location and one time. For those who can attend, money spent on airfares, hotel rooms, and meals can be saved.

This training aid allows trainers to reach many users at one time. It is particularly useful when trainers are presenting an overview of the system. It is also useful in large organizations where many people perform the same tasks. Teleconferencing, however, will not replace the need for tutorial or one-on-one instruction.

Interactive Training Software

Interactive training software permits a dialogue between users and the computer. Such training software is available in many forms:

Computer-Based Training (CBT) Computer-based training (CBT) software uses a microcomputer to guide users through a series of effective, easy-to-learn lessons. Most CBT programs allow users to review lessons, but include error trapping features that prevent users from proceeding to the next lesson or question unless correct answers are given for the current assignment.

Audio-Based Training Usually audio-based training software comes with a workbook. To use this kind of training program, the user needs to have access to a cassette player and a computer. Like CBT programs, the most effective audio programs enable students to use the actual software application for which they are being trained.

Video-Based Training Video-based training requires access to a TV, VCR, and computer. Normally, a workbook is provided. This training program requires more physical space and equipment. Moreover, it demands more eye movement by the user. Such visual Ping-Ponging makes it awkward watching the video and trying to use the computer simultaneously.

Video-Optical Disk Video-optical disk (also called interactive video) training programs, using CD-ROM, deliver prerecorded material onto a monitor that is connected to a computer. The user can watch the lessons and complete the lessons using the same screen. This eliminates excessive eye movement required by video-based training programs. Video-optical disk training programs are, however, relatively expensive, because they require special monitors to support toggling between video and programs.

All of the preceding interactive training programs are effective if users take advantage of them and use them conscientiously. Permitting users to work alone and at their own pace, however, makes it difficult for the trainer to monitor users' progress. If a user has a unique need, interactive software is not customized. Therefore, if users are not using the interactive training software, or their needs require customization, another training technique should be used alone or in conjunction with interactive training software. Such a training technique is instructor-based training.

Instructor-Based Training

Instructor-based training permits the instructor to monitor users' progress and customize the lessons to fit users' specific needs. Having an instructor available can also help reduce the intimidation felt by some users of being left alone with a computer. Normally, only a human instructor can help users overcome their fear of computers and reduce confusion and frustration.

As the term implies, this training technique is more personal and, consequently, fairly expensive. By using this technique in conjunction with other training techniques, instructor-based training can eliminate any remaining void that prevents a satisfactory understanding of the system. In systems in which certain tasks are highly complex or particularly vital to successful operations, instructor-based training may be the only feasible technique.

On-the-Job Training

Perhaps the most widely used approach to training operating personnel is simply to put them to work. Usually, the individual is assigned simple tasks and given specific instruction on what is to be done and how it is to be done. As these initial tasks are mastered, additional tasks are assigned. The learning curve in this approach can be quite lengthy, and, in many cases, what appears to be immediate results or production can be very deceptive. Moreover, if a particular operation is highly complex and difficult to master, the individual designated to carry it out may become frustrated and request a transfer.

Procedures Manual

The procedures manual (also called a user manual) assists users who do not know how to operate the system. Instructions tell users how to run different applications, how to save documents on secondary storage media, how to update files, how to use the printer, and so on. For example, Figure 19.3 shows

Figure 19.3 Instructions on how to fill in a purchase contract screen.

Purchase Contract Screen Design

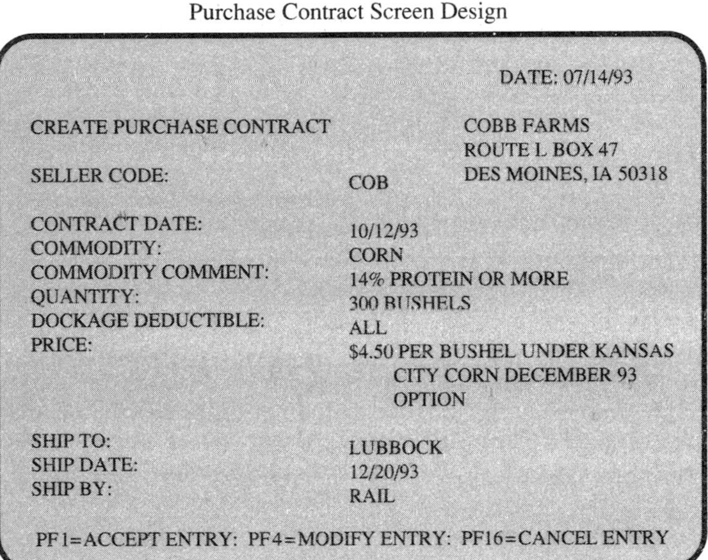

DATE: 07/14/93

CREATE PURCHASE CONTRACT — COBB FARMS, ROUTE L BOX 47, DES MOINES, IA 50318

SELLER CODE:	COB
CONTRACT DATE:	10/12/93
COMMODITY:	CORN
COMMODITY COMMENT:	14% PROTEIN OR MORE
QUANTITY:	300 BUSHELS
DOCKAGE DEDUCTIBLE:	ALL
PRICE:	$4.50 PER BUSHEL UNDER KANSAS CITY CORN DECEMBER 93 OPTION
SHIP TO:	LUBBOCK
SHIP DATE:	12/20/93
SHIP BY:	RAIL

PF1=ACCEPT ENTRY: PF4=MODIFY ENTRY: PF16=CANCEL ENTRY

Process Procedures

Get CRPURCNT Program

Do until operator indicates end of program
by pressing PF16.

Edit data field

Seller code:	must be valid code from SELLERS file
Contract date:	must be MM/DD/YY format
Commondity:	must be valide code from COMDTY file
Commondity comment:	no restrictions
Quantity:	must be positive value, no number greater than 999,999 and units must be bushels, pounds, or tons
Dockage Deductible:	must select all, all over 1%, or none
Price:	must be positive value not greater than 999,999.99 per bushel, pound, or ton
Ship to:	must enter a non-blank value
Ship date:	must be MM/DD/YY
Ship by:	must select rail, truck, or water

End-Do

how instructions are put together to demonstrate how a purchase contract screen is to be filled in. The screen at the top of the page illustrates how the screen looks after it has been filled in. The procedures at the bottom tell users exactly how data entry fields should be filled in. With minimal initial tutoring, users should be able to perform this data entry task with very little trouble and further training.

Textbooks

The traditional textbook is an excellent training aid for various software and hardware applications. Reading a book, however, is often time-consuming, and without an instructor available to answer questions, the user may become frustrated very quickly. A book, therefore, is best used as a supplement to other training techniques.

Probably the best approach generally is to use a combination of the foregoing training techniques. For example, interactive training techniques and books used in conjunction with instructor-based training in a classroom setting can provide a strong learning base. If time is short and in-depth training is required quickly, any self-directed training technique using books or audio cassettes is normally not recommended. An instructor-based training technique with one-on-one tutorials is probably the most effective training technique for this situation.

PREPARING THE DOCUMENTATION

DOCUMENTATION is the written material (although some portions may be in video or audio form) that describes how a system operates. It includes what programs do and procedures that users follow. Proper systems documentation is a subject with which some people agree on but then proceed to neglect completely. This circumstance is surprising, since almost everyone benefits from documentation, including managers and supervisors, users, trainees, operators, systems professionals, maintenance programmers, and auditors. Documentation is used for the following purposes:

- Training
- Instructing
- Communicating
- Establishing performance standards
- Maintaining the system
- Historical reference

CASE systems central repositories store the documentation for easy reference and updating. Changes in business operations and user requirements necessitate systems and program modifications. Before changes can be made, however, the person making the changes must first understand what the system

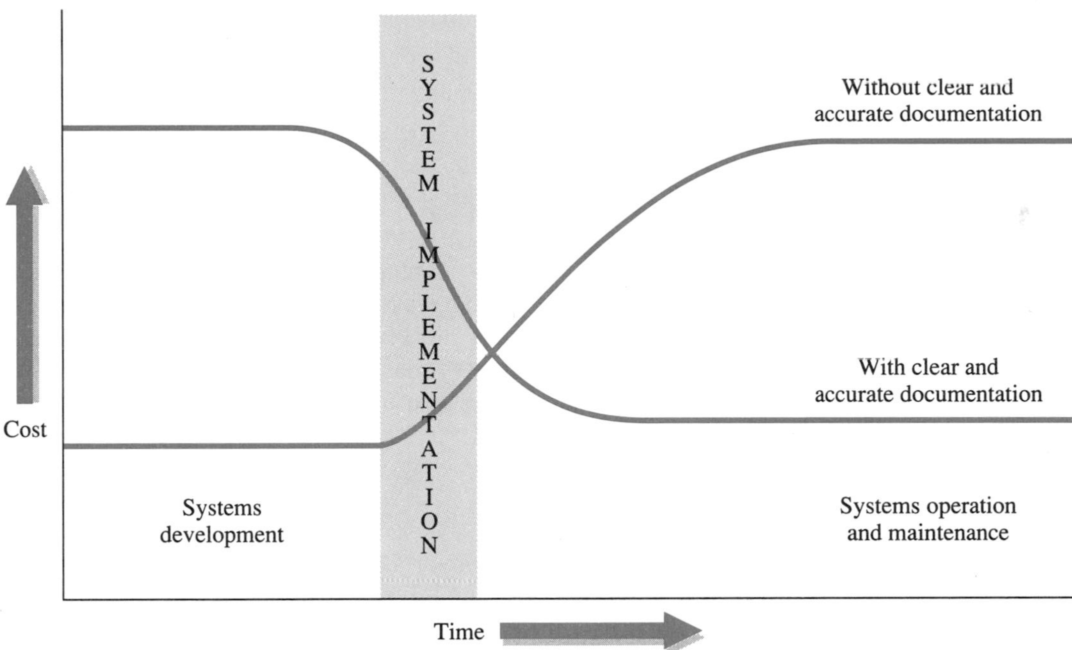

Figure 19.4
Systems cost over the life of the system.

is supposed to do. Programs that were written several months or years earlier must be well documented, because it is easy to forget details in just a few days. Additionally, more than one project has been left virtually unmaintainable and inoperable because personnel who built the system left the company for other employment without leaving adequate documentation behind.

Analysis of the total cost of the new system over its life indicates that development costs are less than ongoing operation and maintenance costs, as indicated in Figure 19.4. Good system design will help reduce the need for maintenance and its costs. But when maintenance is required it will be performed easier and quicker if documentation is available. The four main areas of documentation are:

- User documentation
- Systems documentation
- Software documentation
- Operations documentation

User Documentation

In Chapter 17, we presented **USER DOCUMENTATION**, which is a group of procedures that tell users how to work with the system and perform their tasks. User documentation may be online or contained in a procedures manual.

Documentation tells users how to fill in source documents and data entry screens; how to establish controls and generate reports; and how to check the validity of output. Expected results should always be included in the documentation. If the task is to fill in a form, an example of a filled-in form should be provided to the user for a guide. User documentation should reflect accurately what users learned during training.

Systems Documentation

The documented deliverables generated by systems professionals and the design elements contained in the CASE central repository serve as **SYSTEMS DOCUMENTATION**. As the system was being developed, documented deliverables served as communication devices to keep every participant informed about how the system was taking shape and the system's progress. This systems documentation stored in a CASE central repository can keep all participants, even those who worked with the systems project less than full time, linked together and focused on the system. If the systems project were interrupted along the way, or if key people were replaced, systems documentation made it possible for the work to continue.

Later, after the system is converted, it is subjected to a postimplementation review. The collection of documented deliverables is used as a source of review facts for this process. Systems documentation provides management with an accurate gauge to review and evaluate the new system. Furthermore, if changes are to be made to the system, the systems documentation should be consulted. If systems documentation is unavailable, the work of the systems analyst or the maintenance programmer assigned to make the changes becomes far more difficult.

Software Documentation

SOFTWARE DOCUMENTATION, covered in Chapter 17, facilitates maintenance and enhancement of the program. Maintenance programmers assume responsibility for the program after it is implemented. Because it has been written by someone else, to modify it, maintenance programmers must understand clearly the functions and logic of the software. Although a thorough examination of the code may provide the maintenance programmer with an understanding of the logic, this is a tedious and time-consuming way to find out what the program should do or is doing. Even if the programmer who coded the software is still employed by the company, it is difficult to get back into the logic and explain what the program is supposed to do even if only a few months have elapsed.

Operations Documentation

Computer operators don't have time to read a lot of documentation. They need **OPERATIONS DOCUMENTATION** that contains forms and diagrams with key words and numbers. They don't want to know what a program does or anything about its interworkings. (For control reasons, the company also doesn't want computer operators to understand a program or to be able to change it.) Basically, a computer operator wants to know what data files the program uses and what kind of paper to put in the printer. Operations documentation may, in some

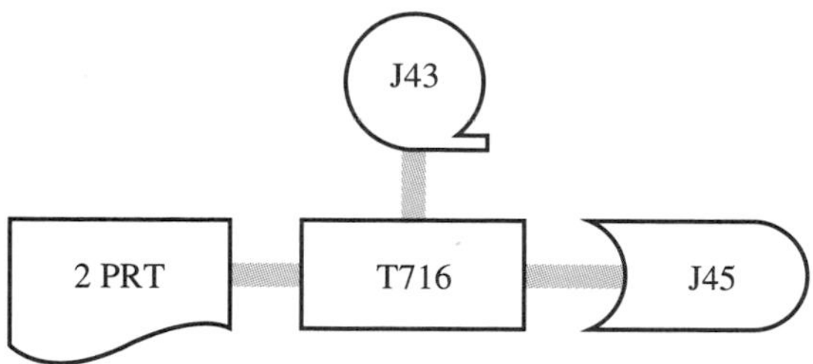

Figure 19.5
Operator documentation.

applications, be as simple as that presented in Figure 19.5. It shows a job, which requires the mounting of two files, tape file J43 and disk file J45. The job produces two-part paper on a printer. This diagram simply states job and files according to installation standards.

CONVERTING TO THE NEW SYSTEM

Conversion is the process of changing from the old system to the new system. The degree of difficulty and complexity in converting from the old to the new system depends on a number of factors. If the new system is a canned software package that will run on its present computer, the conversion will be relatively simple. If conversion entails new customized software, new database, new computer equipment and control software, new networks, and a drastic change in procedures, conversion becomes quite involved and challenging.

In the following material, we first present four methods of total systems conversion. Although file conversion is obviously a part of systems conversion, we include a separate section on this vital task following the section on systems conversion methods.

Methods for Converting Systems

Four different systems conversion methods are available:

- Direct conversion
- Parallel conversion
- Phase-in conversion
- Pilot conversion

A graphic representation of these four conversion methods is shown in Figure 19.6.

Direct Conversion

A **DIRECT CONVERSION** is the implementation of the new system and the immediate discontinuance of the old system, sometimes called the cold turkey approach.

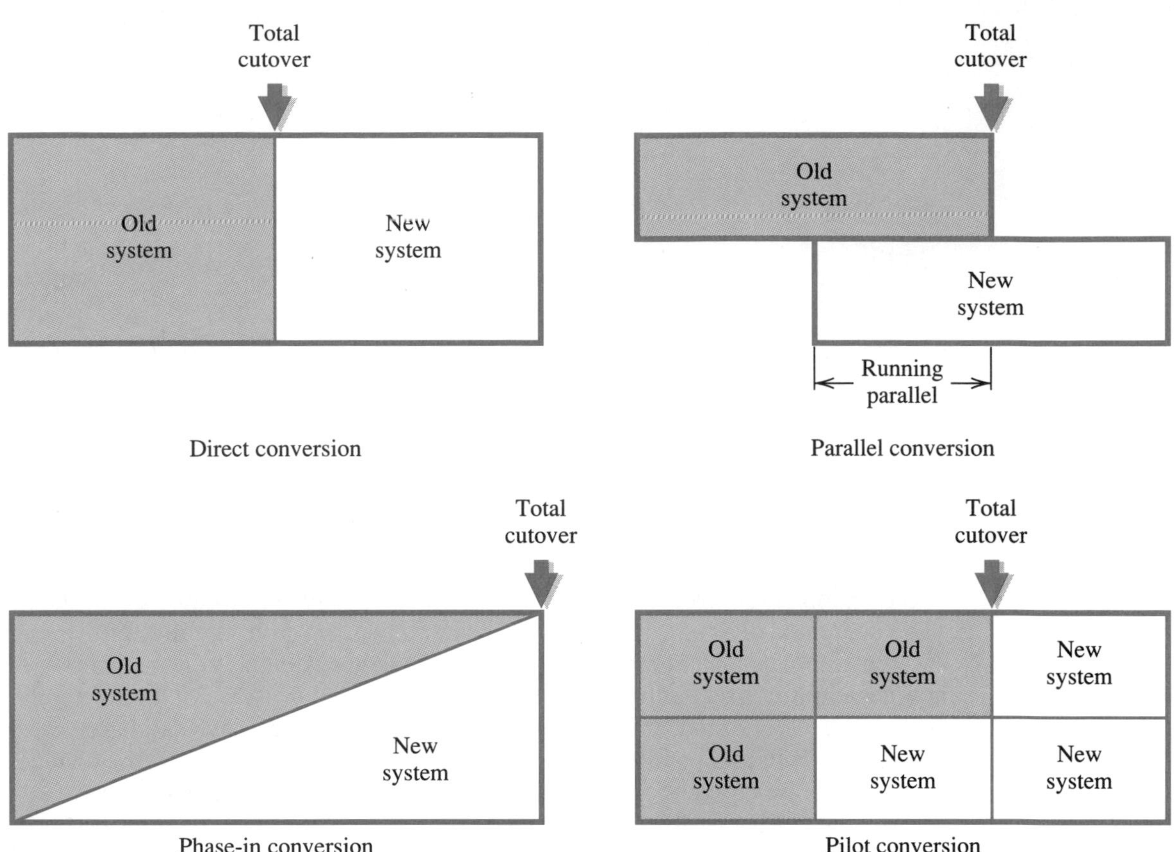

Figure 19.6
Graphic representation of systems conversion methods.

Once the conversion is made, there is no way of going back to the old system. This conversion approach is meaningful when:

- The system is not replacing any other system.
- The old system is judged absolutely without value.
- The new system is either very small or simple or both.
- The design of the new system is drastically different from that of the old system, and comparisons between systems would be meaningless.

The primary advantage of this approach is that it is relatively inexpensive; the primary disadvantage is that it involves a high risk of failure. When direct conversion is to be used, the testing and training activities discussed earlier take on even greater importance.

A modification to direct conversion is a trial conversion, which is like a dress rehearsal. Every user does his or her scheduled part. Instead of going into actual production with the new system, files are copied and jobs are run. The output, the updated files, and other components are studied in detail to make sure everything is working correctly. After a successful dress rehearsal, the next cutover is real.

Parallel Conversion

PARALLEL CONVERSION is an approach wherein both the old and the new system operate simultaneously for some period of time. It is the opposite of direct conversion. In a parallel conversion mode, the outputs from each system are compared, and differences are reconciled. The advantage of this approach is that it provides a high degree of protection to the organization from a failure in the new system. The obvious disadvantages to this approach are the costs associated with duplicating facilities and personnel to maintain the dual systems. But because of the many difficulties experienced by organizations in the past when a new system was implemented, this approach to conversion has gained widespread popularity. When the conversion process of a system includes parallel operations, the systems development people should plan for periodic reviews with operating personnel and users concerning the performance of the new system. They must designate a reasonable date for its acceptance and for discontinuance of the old system.

Because of faulty training and testing activities, conversion projects are often burdened with additional tasks of training, testing, rewriting procedures and documentation, changing files, attempting to retrofit controls, and making major computer architecture adjustments. If this is the case, parallel conversion is really the only sensible approach to use. Other situations may exist, however, when different production methods, decision rules, accounting procedures, and inventory control models are to be used in the new system. In these situations, parallel conversion makes little sense. It will not work. In this case, and in all cases, we recommend stringent training and testing procedures before the conversion process begins.

If parallel conversion is used, a few points should be kept in mind. First, a target date should be set to indicate when parallel operation will cease and the new system will operate on its own. If possible, the target date should be set at the end of the longest processing cycle (e.g., at the end of the fiscal period and after year-end closings). Second, if a discrepancy occurs between the old and new system, it should be verified that the inputs to both systems were the same. If the inputs were the same, the new program should be reviewed to make sure that it is processing the transactions properly. In some instances, the old system may be the one that is not processing correctly.

Phase-In Conversion

With the PHASE-IN CONVERSION method, the new system is implemented over time, gradually replacing the old. It avoids the riskiness of direct conversion and provides ample time for users to assimilate to the changes.

To use the phase-in method, the system must be segmented. For example, the new data collection activities are implemented, and an interface mechanism with the old system is developed. This interface allows the old system to operate with the new input data. Later the new database access, storage, and retrieval activities are implemented. Once again, an interface mechanism with the old system is developed. Another segment of the new system is installed until the entire system is implemented. Each time a new segment is added, an interface with the old system must be developed. The advantages to this approach are that the rate of change in a given organization can be minimized, and data processing resources can be acquired gradually over an extended period of time. The disadvantages to this approach include the costs incurred to develop temporary interfaces with old systems, limited applicability, and a demoralizing atmosphere in the organization because people never have the sense of completing a system.

Pilot Conversion

With the **PILOT CONVERSION** method, only part of the organization tries out the new system. Whereas the phase-in method segments the system, the pilot method segments the organization. One plant or branch office, for example, may serve as the guinea pig for experimentation or an alpha or beta testing site for a working version of the new system. Before the new system is implemented throughout the organization, the pilot system must prove itself at the test site.

This conversion method is less risky than the direct method and less expensive than the parallel method. Any error can be localized and corrected before further implementation is attempted.

When a new system involves new procedures and drastic changes in equipment and software, the pilot method is often preferred. In addition to serving as a test site, the pilot system can also be used to train users throughout the organization in a "live" environment before the system is implemented at their own location.

METHODS FOR CONVERTING EXISTING DATA FILES

The success of systems conversion depends to a great degree on how well the systems professional prepares for the creation and conversion of data files required for the new system. By converting a file, we mean that an existing file must be modified in at least one of three places:

1 In the format of the file

2 In the content of the file

3 In the storage medium where the file is located

It is quite likely in a systems conversion that some files can experience all three aspects of conversion simultaneously.

Two basic methods can be used to perform file conversion:

1 Total file conversion

2 Gradual file conversion

Total file conversion can be used in conjunction with any of the four preceding systems conversion methods. Gradual file conversion is especially amenable to parallel and phase-in methods. In some instances, it will work for the pilot method. Generally, gradual file conversion is not applicable for direct systems conversion.

Total File Conversion

If the new system files and the old system files are on computer-readable media, a simple program can be written to convert files from the old to the new format. Generally, however, converting from one computer system to another involves tasks that cannot be done automatically. New file design almost always includes additional record fields, new coding structures, and new ways of relating data items (e.g., relational files). Therefore, some nonprogrammable file conversion is generally necessary.

If file conversion consists of converting from a manual file to a computer file, the task of file conversion is even more arduous. Converting from manual records to computer records requires a great deal of keying. Ten thousand records containing 360 characters per record requires the entry of 3,600,000 characters of data. At a traditional keystroke entry rate of 1200 clean keystrokes per hour, this equates to 3000 keystroke-hours or more than 90 operator-weeks of effort. This computation is for data entry only and does not include other required tasks, such as determining causes of missing fields and correction of data errors.

Often, during the conversion of files it is necessary to construct elaborate control procedures to ensure the integrity of the data available for use after the conversion. Using the following classification of files, note the kind of control procedures used during conversion:

- ***Master Files*** These are the key files in the database. Usually at least one master file is created or converted in every systems conversion. When an existing master file must be converted, the analyst should arrange for a series of hash and control totals to be matched between all the fields in the old file and all the same fields in the converted file. Special file backup procedures should be implemented for each separate processing step. This precaution prevents having to restart the conversion from the beginning in the event an error is discovered in the conversion logic at a later date.

- ***Transaction Files*** These files are usually created by the processing of an individual subsystem within the information system. Consequently, they must be checked thoroughly during systems testing. The transaction files that are generated in areas of the information system other than the new subsystem, however, may have to be converted if the master files they update change in format or medium.

- ***Index Files*** These files contain the keys or addresses that link various master files. New index files must be created whenever their related master files have undergone a conversion.

- ***Table Files*** These files can also be created and converted during systems conversion. Table files may also be created to support software testing. Thus they will be ready for conversion. The same considerations required of master files are applicable here.

- ***Backup Files*** The purpose of backup files is to provide security for the database in the event of a processing error or a disaster in the data center. Therefore, when a file is converted or created, a backup file must be created. The backup procedures for the converted file more than likely will be the same as the procedures that existed for the original file. The file media, however, may differ. For example, a magnetic disk master file may be backed up on a cartridge tape.

Gradual File Conversion

Some companies convert their data files gradually. Records are converted only when they show some transaction activity. Old records that do not show activity are never converted. This method works this way:

1 A transaction is received and is entered into the system.

2 The program searches the new master file (e.g., inventory file or accounts receivable file) for the appropriate record to be updated by the transaction. If the record has already been converted, the record update is completed.

3 If the record is not found in the new master file, the old master file is accessed for the proper record, and it is added to the new master file and updated.

4 If the transaction is for a new record, that is, a record not found on either the old or new file (e.g., a new customer), a new record is prepared and added to the new master file.

Normally at the end of a 30-day period, most of the active records will be converted to the new master file. Some companies will discover that a sizable portion of their old master file was composed of inactive records. Therefore, using gradual file conversion triggered by transactions can significantly decrease the file conversion effort and file space that would have been required if the total file conversion method was used.

EVALUATING THE NEW SYSTEM AFTER IMPLEMENTATION

Although the goal of developing quality information systems is to produce systems that are within budget, on time, and that meet user requirements, that goal is not always accomplished. A process is needed to analyze what went right and wrong with successful as well as unsuccessful projects. This analysis is performed by a **POSTIMPLEMENTATION REVIEW** (also called postimplementation audit or evaluation). A postimplementation review is an organized search for ways of improving the efficiency and effectiveness of the new system, and to provide information that will be helpful in the development of future systems.

Normally, the postimplementation review is conducted two to six months after systems conversion. This period allows for at least two monthly cycles of operating the system. At this point, the memories of all the people involved are fairly fresh and reliable, and the system has had time to settle into a somewhat normal operating pattern.

A postimplementation review may be conducted by a team composed of user representatives, internal auditors, and systems professionals. In some companies, however, an external consultant or independent auditor is brought in to conduct the review to help increase objectivity and reduce any political ramifications that may exist between internal groups.

The following four postimplementation review areas merit the most attention:

- Systems factors
- Systems design components
- Accuracy of estimates
- Level of support

The role they play in the postimplementation review process is shown in Figure 19.7.

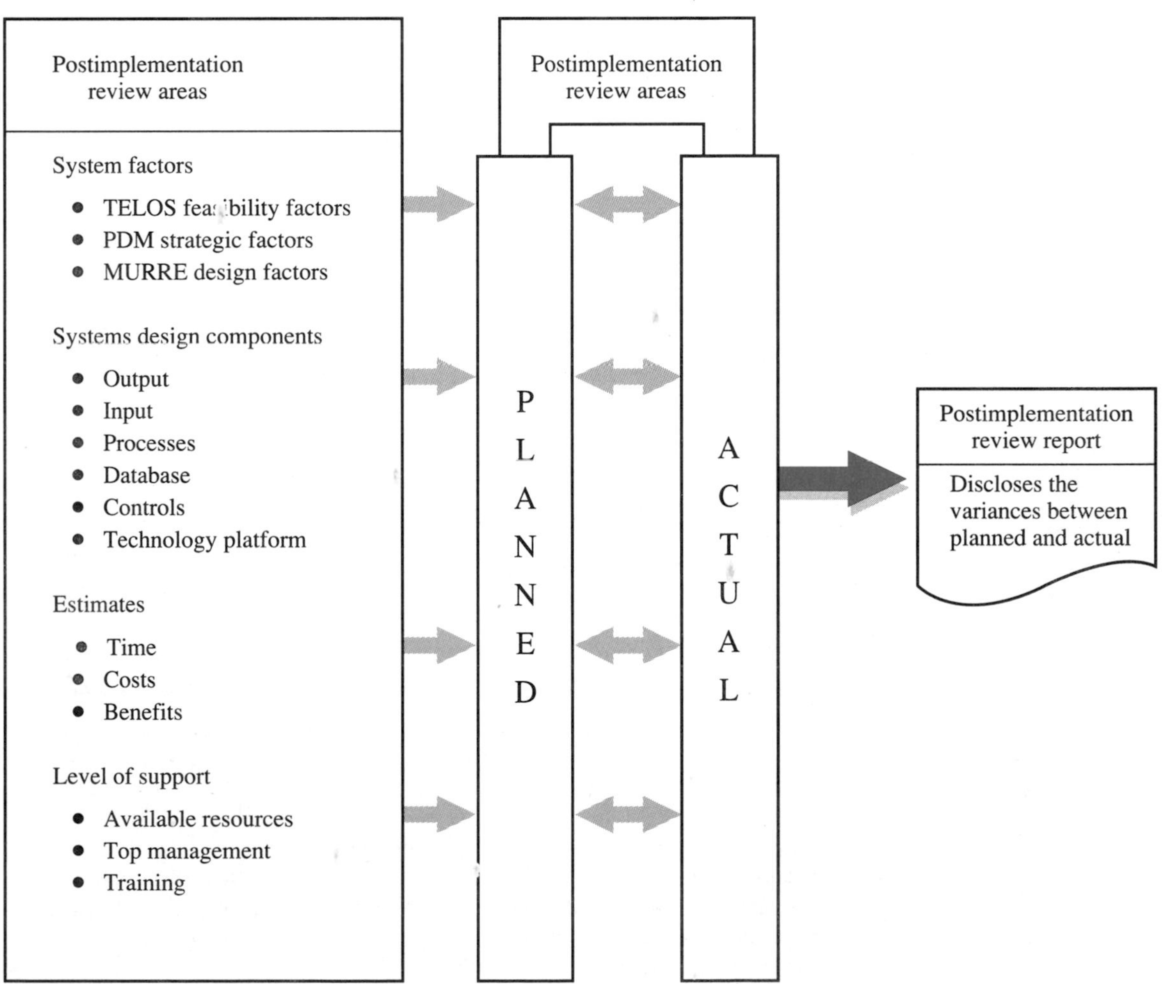

Figure 19.7 Postimplementation review process.

The first review area includes the systems factors, which are:

- The technical, economic, legal, operational, and schedule (TELOS) feasibility factors
- The productivity, differentiation, and management (PDM) strategic factors
- The maintainability, usability, reusability, reliability, and extendability (MURRE) design factors

The second review area deals with:

- Output
- Input
- Processes
- Database
- Controls
- Technology platform

The third review area investigates:

- Time
- Costs
- Benefits

The fourth and final review area pertains to:

- Available resources
- Top management
- Training

Postimplementation Review at Vulcan Industries

Vulcan Industries has been operating its new information system for about four months. Recently, it hired SystemPro, an independent systems consulting firm, to perform a postimplementation review of the new system.

The postimplementation review team included Noel Stevenson, the review team leader; Sam Bennison; and Trudy Carson. All members have a rich background in systems development and consulting.

The prime components used by the SystemPro postimplementation review team to document its work are checklists. Each postimplementation review area has its own checklist. The checklists contain key review questions, a rating scale, and the review area's final score. These checklists serve as key documents presented to the steering committee and other interested parties at Vulcan. Postimplementation review results, for example, are particularly helpful to systems professionals for performance evaluation. Also, the results provide information for estimating time, cost, and problem areas of future systems development, and for correcting mistakes.

A large number of questions directed to both users and systems professionals is the key tool used by the SystemPro postimplementation review team. Besides interviewing and questioning various people throughout Vulcan, the review team also conducted a number of observations and tests on their own. For example, Noel Stevenson ran some test transactions to determine the effectiveness of the software program's embedded controls. He also simulated the destruction of customer files to see if they could be recreated using current database backup procedures. So SystemPro's postimplementation review method consists of interviewing people throughout the organization and asking them questions. Then, where appropriate, the postimplementation review team makes independent observations and tests to make sure that the system is in compliance with answers given by interviewees. Such a review method provides a solid base on which to assign scores to postimplementation review areas.

Reviewing Systems Factors

Early in the systems planning phase, ratings were assigned to TELOS feasibility and PDM strategic factors for the proposed systems project. Again, during the systems evaluation and selection phase, these factors were rated again along with MURRE design factors for general systems design. During postimplementation review, it is determined how well systems factors of the new system that is now in operation stack up against the systems factors predicted for it while it was under development.

TELOS Feasibility Factors

During systems planning and again during systems evaluation and selection, feasibility of the systems project was rated in accordance with TELOS feasibility factors. The TELOS ratings originally assigned to the system should be compared with how well the implemented system fared.

PDM Strategic Factors

The reason for developing the new system in the first place was because of an opportunity to augment the productivity, differentiation, and management (PDM) strategic factors. The original PDM ratings are now compared with the PDM ratings that are assigned to the system after its implementation. The ratings should be reasonably close. Specifically, management wants to know if the new system has given the company a competitive advantage.

MURRE Design Factors

By now, it should be relatively easy to ascertain how maintainable the new system is. Perhaps the maintenance programmer has already had to make a change based on a change in business operations. Did the maintenance programmer understand the program and was the change made easily? Usability and reusability should be well established by now, because the system has been in actual use. Also, reliability should be well tested at this point. The degree of extendability, however, may not yet be well established, because the system has not been in operation long enough.

The systems factors review results compiled by the SystemPro postimplementation review team appear in Figure 19.8. Questions asked by the review

Figure 19.8
Systems factors review results at Vulcan Industries.

SYSTEMS FACTORS REVIEW RATING CHECKLIST

TELOS Feasibility Factors: 8.5

- Is technology supporting the systems design as originally estimated?
- Were adequate funds available for acquisition of computing resources?
- Are adequate funds available for operating and maintaining the system?
- Is the system in compliance with laws and regulations?
- Do people possess the necessary skills to operate, maintain, and use the system?
- Was the overall schedule of the systems project followed?

PDM Strategic Factors: 8.0

- Have productivity gains been produced by the new system?
- Has the system contributed to differentiating the company's product such as style, quality, price, and uniqueness?
- Has the system contributed to differentiating the company's service to its customers by providing product and tracking information and improving response to orders and reducing delivery time?
- Does the system produce information that augments management planning, controlling, and decision making?

MURRE Design Factors: 7.5

- Is documentation comprehensive, clear, and current?
- Is a change management system (CMS) in place and working?
- Are user requirements being met?
- Are software modules reusable?
- Is the system error-free?
- Is the system flexible and adaptive?

Total	24.0
Final score (24.0/3)	8.0

Rating scale:

0	5	10
Poor	Fair	Excellent

team to derive reasonable ratings are included under each systems factors category.

Reviewing Systems Design Components

The systems design components are reviewed to see how well they were designed and implemented. They are discussed next.

Output

The information system is reviewed and evaluated in terms of the information it provides. A system that merely functions properly may not be a system that is appropriate for users who must perform tasks and make decisions. Output produced by the system must be usable, accessible, timely, relevant, and accurate. Therefore, the output design component merits special attention by the postimplementation review team.

Technical output review pertains more to the form of the output versus substance. This review entails checking for proper headings, edited amounts (e.g., leading zero suppression, debit/credit notation, dollar signs), correct page number sequence, clear end-of-report indicators, and correct dates (e.g., date the report was prepared and the current date).

Input

Review of the input design component involves determining if paper and electronic forms and data-entry screens meet design guidelines and specifications. Users are also tested to determine if they are completing forms properly. Presumably, they were taught how to fill in forms during training. Postimplementation review is performed to determine if they were properly trained.

If input is handled by a point-of-sale device, a simple random sample of products is selected and passed by the reader to determine correctness of price and description. If certain products do not contain a bar code, a keyboard must be available to enter the data manually. If input entered from a keyboard is displayed on a VDT, proper layout on the screen is important. Any screens that are cluttered and contain unnecessary or confusing instructions or captions should be identified and corrected.

Processes

To a great extent, reviewing the other design components is tantamount to reviewing processes. For example, if output is appropriate, the software may be considered to be performing correctly. Recall that software was subjected to a number of tests during the early stages of implementation. But software can be producing appropriate output and still not be operating satisfactorily from the viewpoint of the operator. For example, a program that produces perfectly good output may terminate on occasion for no apparent reason. From the viewpoint of the person who has to operate the system, such software is clearly of poor quality and should be corrected.

Other aspects of the processes design component that warrants close review are procedures and documentation. Procedures must be easy to follow and complete in order for various personnel to perform their appointed tasks. Poorly written procedures are frustrating to understand and follow, and in some instances can cause a task to be executed improperly. If such procedures exist, they will have to be rewritten. Any irregularities discovered in documentation require prompt action to correct them.

Database

To a large extent, the review of output subjects the database to simultaneous review. Additional review, however, should be made to make sure that the database meets all demands placed on it, especially from the viewpoint of controls and general performance. Controls review includes the following tasks:

- Creating a new record after the last record
- Creating a record for a nonexistent division or department
- Trying to read from or write to a file with the wrong header label
- Attempting to process past an end-of-file indicator
- Trying to create and enter a record that is incomplete

Files should be checked for completeness. Predetermined control totals should be compared with the totals produced from the new files. For instance, predetermined control totals might be checked against the number of records in a file or the total amount of a specific amount field, such as AMOUNT-OWED in an accounts receivable master file. File description and layouts should be compared with the ENVIRONMENT and DATA DIVISION in COBOL programs for agreement.

In a query-intensive environment, tests of SQL joins are critical. Two-table, three-table, and in some instances, four-table joins are common. All these joins should be tested with data that reflect live transactions.

The quality of the query optimizer is the single most important component of an SQL-based DBMS. Therefore, it should be tested to ascertain its ability to handle joins efficiently. Test cases should be devised to test the optimizer's sensitivity to join conditions in the WHERE clause, such as

```
SELECT  NAME
FROM    EMPLOYEE
WHERE   EMPLOYEE.EMP-NO = DEPARTMENT.EMP-NO
```

Then test by transposing the WHERE condition as follows:

```
DEPARTMENT.EMP-NO = EMPLOYEE.EMP-NO
```

The results may be different from the first query because optimizers can sometimes be tricked into using the wrong table indexes.

Controls

The purpose of controls review is to ensure that controls are in place and are working as intended. This is called compliance testing. The three phases involved in compliance testing are:

1 Study and observe controls.

2 Conduct the actual test of compliance.

3 Evaluate how effectively the controls meet these compliance tests.

Test transactions, for example, help to ensure that processing controls, such as range and reasonableness checks and access controls, are in place and working correctly.

Reviewers may be able to satisfy themselves that a number of controls are working merely by observation. For example, they can observe a fire-suppression system and inspection tags. Reviewers may even try out particular control techniques to see if they work (e.g., various access controls). In other cases, the only way to tell that particular controls work is to set up a test situation and see what happens. For example, a disaster simulation may be run on a surprise basis. To run disaster simulations, the reviewers seal certain master files and tell the computer center manager the system is down. This simulated situation requires data processing personnel to bring the system to a current status by using cycled and backup files, other backup facilities, and contingency procedures. If they fail, reasons for failure are ascertained and swift corrective action is taken. Also, with executive approval, reviewers pull the power switch to the computer center to test recovery procedures. Sometimes, professional penetrators or tiger teams are hired to attempt to obtain access to the computer center and database. Unexpected fire drills are performed to see if standard operating procedures are followed.

Technology Platform

The technology platform for the new system is also reviewed, including peripherals, workstations, processors, and networks. The primary target is a comparison of current performance with design specifications. The outcome of the review indicates any differences between expectations and realized results. It also points to any necessary modifications to be made.

Some of the more popular technology platform review tools are:

- Job accounting system
- Hardware monitor
- Software monitor

These tools act as stethoscopes or probes that check how well the computer architecture is operating.

Job Accounting System This system can be used to test design efficiency, help in capacity planning, and project growth patterns. IBM's system management facility (SMF) is an example. It shows who used the system and how long they used it, along with the data files accessed. SMF indicates the amount of available space on direct-access storage devices and gives basic error statistics for magnetic tape files. Other vendors provide similar software packages.

Hardware Monitor A hardware monitor consists of sensors connected to the processor's circuitry. It measures processor active, processor wait, disk seek, disk data transfer, disk mount, tape active, tape rewind, and internal memory timings and utilization. The sensors are, in turn, connected to a small computer that records and displays various signals.

With sufficient utilization statistics, properly evaluated, the total computer architecture budget may be reduced by improving overall efficiency through changing the processor, adding channels, or reconfiguring the storage hierarchy. The broad purpose of using a hardware monitor is to match the power of the computer architecture to the demands of the information system. Except for terminal response time for individual terminals and lines, the hardware monitor, however, is not equipped to measure systems application performance. This area is measured by a software monitor.

Software Monitor To evaluate overall systems performance, a software monitor can be used by itself or in conjunction with a hardware monitor and a job accounting system. A software monitor is a program that resides in the computer architecture under review. Software components commonly measured are operating systems, support software, and application programs. Measurement of the operating system identifies inefficient sections of code. The existence of such sections could compel management to ask the vendor for improvement, or if this is not possible, other commercially available software might be considered as a replacement.

Support software falls in the gray area between application programs and the operating system. Although handled in the system much like application software, support software is written by the vendor and appears to the ordinary user as part of the operating system. Should the software monitor indicate that the support software is not operating in an optimal manner, the vendor should change it.

Application software is measured to determine resource utilization and code efficiency. These measurements record and report the amount of time a program used each resource, such as internal memory, disk, and tape. For example, a program may request eight tape drives, even though it would normally need only four. If there is need to limit tape drives, the reviewer should ask the maintenance programmer to change the program.

A software monitor can isolate heavy paging areas in virtual storage systems. Heavy paging can occur when application software accesses routines repeatedly, causing a thrashing-of-pages condition to occur. With analysis and program rewrites, this software inefficiency may be reduced. Also, by caching frequently used data and programs in semiconductor disk (also called solid-state disk), the I/O gap between the processor and auxiliary storage can be narrowed, thus improving response time and throughput.

The systems design components review checklist displayed in Figure 19.9 represents findings of the SystemPro postimplementation review team. The checklist contains questions asked about systems design components, each component's rating, and the final score for this review area.

Reviewing Estimates of Time, Costs, and Benefits

As we have just seen, the postimplementation review is a formal process to determine how well the system is working, how it has been accepted, and whether adjustments or redesigns are needed. Another important reason for performing a postimplementation review is to help refine the ability to estimate project time, costs, and benefits. By comparing actual time, costs, and benefits with those which were estimated at the beginning of the project, people learn where mistakes were made and how to avoid them in future projects.

Figure 19.9
Systems design components review results at Vulcan Industries.

SYSTEMS DESIGN COMPONENTS REVIEW RATING CHECKLIST

Output Design Component: 9.0

- Is output secure from unauthorized users?
- Is output accurate, timely, and relevant?
- Can users get access to information when they need it?
- Does the output fit users cognitive styles? That is, are "big picture" users provided with graphics, and are "detail" users provided with numbers?
- Are reports appropriately edited and identified?

Input Design Component: 9.5

- Are there sufficient input controls to verify and validate data entry?
- Do users know how to fill in forms?
- Are electronic forms and direct-entry devices used instead of source documents?

Processes Design Component: 9.0

- Are adequate procedures available and documented?
- Is software reliable?
- Does software documentation accurately reflect software design and functionality?
- Are appropriate accounting and statistical models being used?
- Is conversion of input to output performed in a timely manner?

Database Design Component: 7.5

- Is the database backed up and secure from loss?
- Are files properly labeled?
- Do these files match the programs that process them?
- Is a data dictionary available?
- Is the database designed to respond quickly and accurately to ad hoc queries?
- Is the database designed to process long job streams (e.g., payroll and billings) in the most efficient manner?

Controls Design Component: 9.0

- Does the company have a strong control policy?
- Is a disaster recovery plan in place?
- Did the disaster recovery plan pass simulated disaster tests?
- Does the system contain strong access controls?
- Did these access controls meet certain compliance tests?
- Is the system safeguarded against destructive software such as viruses?

Technology Platform Design Component: 7.0

- Does the computer configuration meet system and user processing requirements?
- Is the technology optimally configured?
- Is response time acceptable?

Total	51.0
Final score (51.0/6)	8.5

Rating scale:

0	5	10
Poor	Fair	Excellent

Analysis of Time

Target dates based on PERT, Gantt, or similar techniques were used to estimate time to complete various phases or tasks. Also, the number of person-months (or some other unit of time) required to produce so many lines of executable code (LOEC) or function points was used as a way to estimate the time to develop software. The actual time required to complete tasks and the project is compared to estimated time to determine the estimation error. If the error is large, new ways must be found to provide more accurate time estimates.

Analysis of Costs/Benefits

A critical decision was made after the general systems design phase as to which alternative design should be selected for detailed design and implementation. The key quantitative technique used to make this decision was cost/benefit analysis. Costs and benefits as presently determined are compared to the costs and benefits estimated during this early phase of the SDLC. Present calculations should be reasonably close to early estimates.

In many organizations, user departments are charged for development and ongoing costs of the new system. Therefore, such users are very sensitive about the accuracy of cost estimates. Generally, if actual costs are within 5 to 10 percent of estimated costs, users are satisfied.

The estimates review checklist presented in Figure 19.10 with attendant questions, represents the level of estimates accuracy. The checklist contains each estimate made early on in the systems project and the final score that

Figure 19.10 Estimates review results at Vulcan Industries.

ESTIMATES
REVIEW RATING CHECKLIST

Time: 6.5

- Were PERT, Gantt, or other similar project planning and control techniques used?
- Were metrics, such as lines of executable (LOEC) or function points, used to estimate the time and cost of the project and to monitor productivity?
- Was the estimation error acceptable to management?

Cost/Benefit Analysis: 8.5

- Were actual costs in line with estimated costs?
- Are users satisfied with costs allocated to their departments?
- Are users receiving benefits that were originally promised?

Total	15.0
Final score (15.0/2)	7.5

Rating scale:

0	5	10
Poor	Fair	Excellent

signifies how close these estimates are to what actually occurred. Such information is valuable for making future estimates.

Reviewing the Level of Support Provided

To a great extent, the success of a systems project has a lot to do with the level of support the systems professionals received during its development.

Available Resources

At the beginning of a systems project, sufficient resources are normally promised or committed by management to develop and implement the new system. If the original design had to be modified because of a cutback in resources, then more than likely, original user requirements also had to be modified. Adequacy of resources should be ascertained and fully disclosed to all affected parties.

Support from Top Management

In addition to the commitment of adequate resources to a systems project, top management should also provide moral and political support. If top management is actively involved in a project and supports it, you may find that fewer problems will occur; participants will be more cooperative; and red tape will be cut.

Training

Various training techniques are applied to train all users based on their particular needs. All users should have received adequate training, and the system should be deemed easy to understand and use as a result of the training.

The level-of-support checklist shown in Figure 19.11 discloses how well the systems project was backed throughout the SDLC. The checklist includes each level-of-support category, its rating, and the final score for this postimplementation review area. Questions asked by the SystemPro team to derive reasonable ratings are included under each category.

REVIEW OF CHAPTER LEARNING OBJECTIVES

The major goals of this chapter were to enable each student to achieve six important learning objectives. We will now summarize the responses to these learning objectives.

Learning objective 1:
Prepare a systems implementation plan using PERT.

If the system to be implemented is a global-based system, PERT or an equivalent technique is required to prepare a systems implementation plan. Systems professionals must take the following five steps to establish a PERT-based implementation plan.

1 Identify all implementation tasks that must be performed.

2 Determine the sequence of tasks.

Figure 19.11
Level-of-support review results at Vulcan Industries.

LEVEL-OF-SUPPORT
REVIEW RATING CHECKLIST

Available Resources: 9.0

- Were funds, equipment, supplies, and skilled systems professionals available throughout the systems development life cycle?
- Did the systems project suffer any delays due to inadequacy of such resources?
- Did the systems project have to be scaled down or redesigned because of inadequate resources?

Top Management Support: 7.0

- Did senior-level managers participate in systems planning?
- Did the systems project and the project team receive enthusiastic support from top management throughout the SDLC?

Training: 8.0

- Are users competent in interacting with and working with the system?
- Was a combination of in-house and vendor-supplied training programs used?
- Were new training techniques, such as teleconferencing and interactive training software, used?
- Was instructor-based training used for complex tasks?
- Were documentation and user procedures utilized during training sessions to ensure their appropriateness and clarity?

Total	24.0
Final score (24.0/3)	8.0

Rating scale:

0	5	10
Poor	Fair	Excellent

3 Estimate the time required to perform each task.

4 Prepare a time-scaled network of tasks.

5 Determine the critical path in the network.

Learning objective 2:
Discuss how to prepare a site for the technology platform.

All technology platforms, whether for a microcomputer or for a cooperative-based computer architecture, should be housed in an environmentally safe, well-controlled site. This site should be prepared and ready to receive the technology platform when it arrives.

Learning objective 3:
Describe ways to train people to work with the new system.

A prime implementation task is to gain a proper level of expertise necessary for people to work with the system. Generally, people have to be trained to gain this expertise and to acclimate them to the new system. Training programs include:

- In-house training
- Vendor-supplied training
- Outside training services

Training techniques and aids consist of:

- Teleconferencing
- Interactive training software
- Instructor-based training
- On-the-job training
- Procedures manual
- Textbooks

Learning objective 4:
Describe the documentation that must be prepared.

To use, operate, and maintain the system, documentation must be prepared. User procedures, including written material, video, and screen messages, aid the user to work with the system. Systems documentation aids in conversion and postimplementation review. Program documentation is necessary for software maintenance. Operations documentation tells operators how to work with the system.

Learning objective 5:
Explain four methods of systems conversion.

Systems conversion can be done by using one of four methods:

- Direct
- Parallel
- Phase-in
- Pilot

Direct conversion is the simultaneous installation of the new system and discontinuance of the old system. Parallel conversion permits both the old and new system to operate jointly for some extended period. Phase-in conversion is the gradual installation of the new system and elimination of the old system.

Pilot conversion installs a replica or working version of the total system in some segment of the organization. Once the pilot system is deemed successful, all segments receive the total new system.

Learning objective 6:
Discuss a postimplementation review.

A postimplementation review is conducted several months after the new system has been in operation to determine how successful it is. Postimplementation reviewers determine how well the system is living up to its expectations. The postimplementation review process investigates four areas:

- Systems factors
- Systems design components
- Accuracy of estimates
- Level of support

What was presupposed or planned is compared with what has actually happened. Variances between planned and actual are disclosed in a postimplementation review report.

SYSTEMS IMPLEMENTATION CHECKLIST

Following is a checklist on how to conduct systems implementation. Its purpose is to summarize the chapter and to remind you of how systems implementation tasks are performed.

1 Prepare a systems implementation plan using PERT or a similar technique. Organize systems implementation personnel.

2 Make ready the new system's site beforehand.

3 Train personnel using one or a combination of in-house, vendor-supplied, or outside training services.

4 Employ one or a combination of teleconferencing, interactive training software, instructor-based training, on-the-job training, procedures manual, and books as training techniques and aids.

5 Develop user procedures, systems, software, and operations documentation.

6 Conduct systems conversion using one or a combination of the following methods: direct, parallel, phase-in, and pilot.

7 Employ an independent postimplementation review team to assess the efficiency and effectiveness of the new system after it has been in operation for two to six months. Review and evaluate systems factors, systems design components, accuracy of estimates, and level of support. Report

the results to the steering committee and other interested parties. Use this postimplementation review information to take corrective action if called for and to help guide future systems development.

KEY TERMS

Direct conversion

Documentation

Operations documentation

Parallel conversion

Phase-in conversion

Pilot conversion

Postimplementation review

Software documentation

Systems documentation

Systems implementation

Systems implementation plan

Systems Implementation Report

User documentation

REVIEW QUESTIONS

19.1 What are the two main parts of a Systems Implementation Report?

19.2 Explain the use of PERT or Gantt charts in the implementation process.

19.3 Discuss the role of site preparation. Describe site preparation required for a new computer architecture.

19.4 What is a burn-in test? Why is it performed?

19.5 List the three groups of people that will have to be trained for the new system.

19.6 List and briefly discuss training programs.

19.7 List and describe training techniques and aids.

19.8 Name the types and purposes of documentation.

19.9 List the four methods of systems conversion. Discuss the advantages and disadvantages of each method.

19.10 Discuss the two basic methods used for file conversion.

19.11 List and briefly describe the kinds of files converted.

19.12 Explain how gradual file conversion works.

19.13 Define postimplementation review. Why is it performed?

19.14 List and briefly discuss the postimplementation review areas.

19.15 Essentially what does the postimplementation review report disclose? What can be learned from this report?

CHAPTER-SPECIFIC PROBLEMS

These problems require exact responses based directly on concepts and techniques presented in the text.

19.16 You have just completed detailed systems design for Wilmington Sporting Goods. You have determined that the following tasks must be performed to effect systems implementation:

Task	Description
1,2	Order technology
1,3	Review specifications
1,4	Prepare site
2,5	Install technology
3,6	Write programs
4,9	Train personnel
5,7	Test technology
6,8	Test programs
7,9	Dummy
8,9	Test input, output, database, and controls
9,10	Convert system
10,11	Postimplementation review

Required: Sketch a PERT chart of the foregoing tasks. Prepare site, train personnel, convert system, and perform postimplementation review are tasks on the critical path.

19.17 For each of the following situations, select the most appropriate training program:

1 Training is to be conducted in the trainee's environment.

2 The enterprise cannot provide the necessary training, but it is close to a university that provides a strong continuing education program in the systems field.

3 The trainees are to learn specifics about a vendor's new computer architecture.

19.18 In the following situations, select the most appropriate training technique or aid.

1 The trainee needs to carry on an audio dialogue with the training technique.

2 Trainees are geographically dispersed and cannot leave their jobs,

but all must interact with each other and be trained at the same time.

3 Each trainee requires personalized training.

19.19 In the following situations, select the most appropriate conversion method.

1 A computer-based accounts receivable system is to be converted from manual to computer-based operation. The system project team failed to perform tests, and consequently, users are not sure whether the new system will produce accurate results.

2 A lottery system is to be converted in a state where one had not existed before.

3 A new bank demand deposit system will eventually serve 60 branches.

4 A new executive information system has been thoroughly tested. The old EIS was of little value.

5 One of the plants of Hercules Manufacturing wants to serve as a site for the new just-in-time performance reporting system.

6 The corporate culture in which the new system is being implemented is slow to change its ways. The new system can be segmented.

THINK-TANK PROBLEMS

These problems call for a feasible approach rather than a precise solution. Although the problems are based on chapter material, extra reading and creativity may be required to develop workable solutions.

19.20 The manual accounts receivable system of Calico Pet Supply is being converted to a computer-based system. The present manual accounts receivable system has the following characteristics:

1 Each customer with a nonzero accounts receivable balance has a folder containing a copy of all unpaid invoices and credit notes issued.

2 When a payment and accompanying remittance is received from a customer, the remittance is matched to an unpaid invoice, and both documents are placed in a current closed file, which, in turn, is purged every six months.

3 A permanent closed file, composed of purged current closed file documents, is maintained for a period of seven years.

4 At month's end, the balance of each customer's account is classified

according to the age of the balance outstanding. This is done by a clerk tallying and dating folder amounts.

The frequency of access to the three files varies considerably. The folder file is accessed frequently. The current closed file is accessed periodically, normally at the request of the credit manager or a customer. The permanent closed file is accessed infrequently. The new system to be implemented will contain three files:

1 A customer master file, containing information pertinent to each customer.

2 An open item file, corresponding to the "folder" file of the manual system.

3 A closed item file, assembling a combined current closed and permanent closed file.

Assume that you were given the responsibility of converting the old system to the new system. Please answer the following questions:

1 What conversion method would you recommend for this case? Justify your answer.

2 What special clerical procedures would you need to establish to validate the correctness of the new system's operation against the old system?

3 How would you handle file conversion? Explain your answer.

19.21 You have plotted actual cost and benefit figures against those estimated early in the systems project as follows:

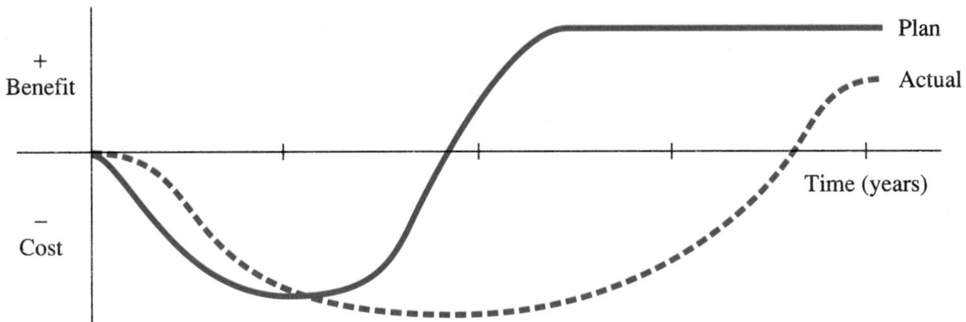

Required: What does this graph say about meeting cost and benefit expectations?

19.22 Terminal response time is the time that the user must wait to begin a transaction after completing the previous one. Definition of a good response time depends on the application and user. Typically, data-entry applications are the most demanding, because little thought on the part of the user is required. Therefore, any delay between the completion of one transaction and the beginning of the next is considered a significant interruption in work flow. Skilled data-entry operators complain bitterly about terminal response time greater than 2.5 seconds. You have been using a hardware monitor as part of your postimplementation review. The following graph displays an average online response time during one month of monitoring operations.

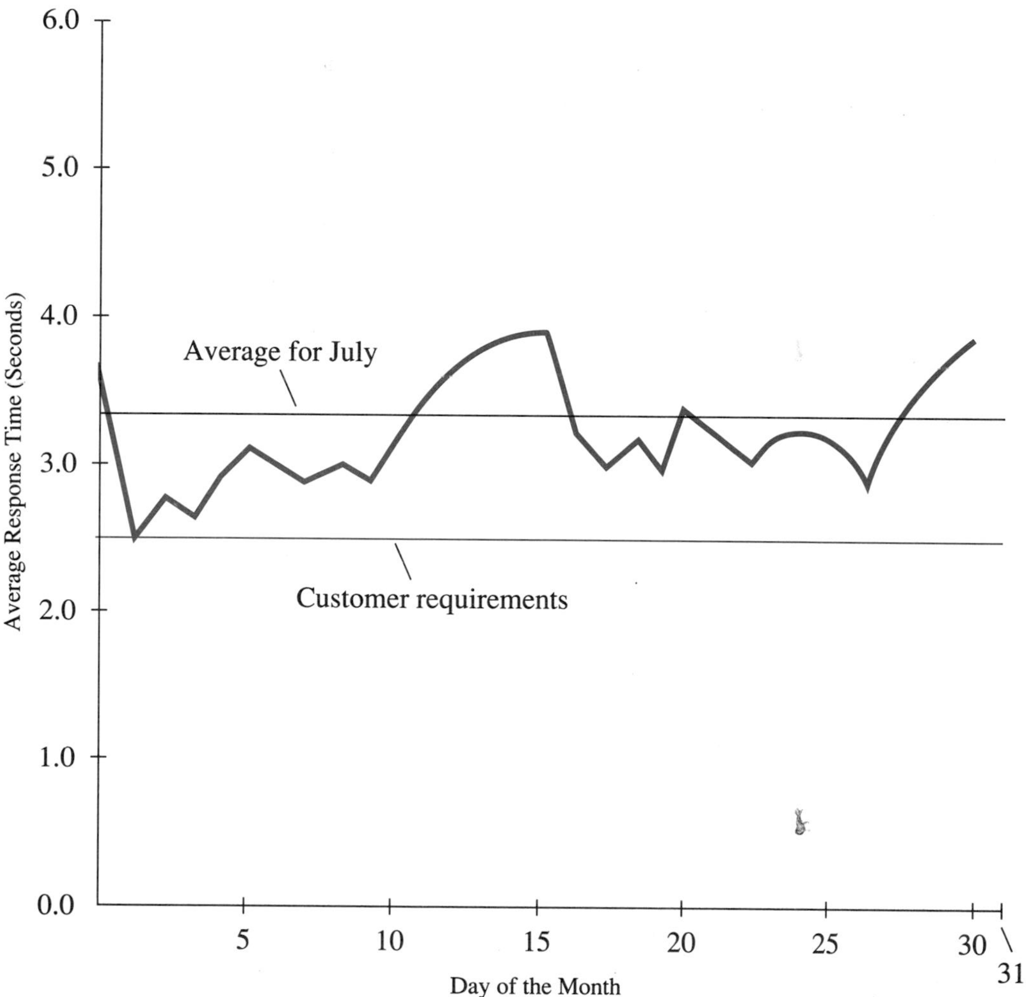

Required: Basing it on the information displayed in the preceding graphic, prepare a response to be included in your postimplementation review report. If the system you are reviewing is designed to support mostly data-entry operators, is the system adequate? What might be some feasible changes in the system that would make the system more adequate? Make specific recommendations. Reviewing detailed design chapters on telecommunication networks and computer systems might be helpful in formulating your response.

19.23 The percentage of systems-availability time is closely related to terminal response time. A poor percentage of availability and poor terminal response time, however, may be symptoms of different problems. The percentage of systems availability is considered an important measure of systems performance.

The percentage of systems availability is calculated with the following equation:

$$\text{Percentage availability} = \frac{(\text{total scheduled availability} - \text{downtime}) \times 100}{\text{total scheduled availability}}$$

Total scheduled availability is the total number of hours the system is scheduled to operate within a given time period. The time period may be a day, week, month, or year. Downtime is the number of hours the system is unavailable for use during the same time period.

You have monitored a system for a month immediately following systems conversion. The results of this monitoring are presented in the following graph.

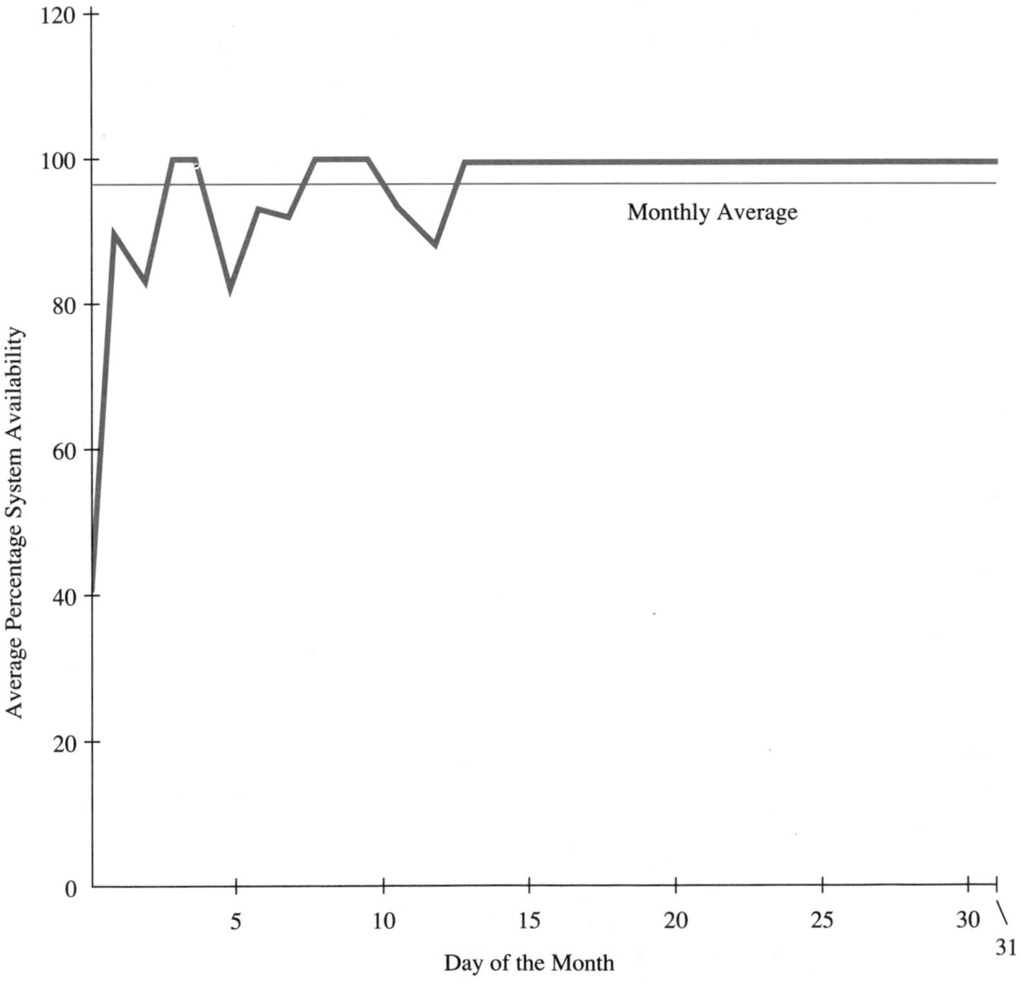

Required: What assessment can you make of the system's availability from the foregoing graphic? Make any assumptions that you deem appropriate and explain the reason or reasons behind the fluctuations in thc first half of the month. Do you plan to continue monitoring the system next month? Why? Why not?

19.24 The new order-entry system has been in operation for four months after direct conversion. During this time, orders have been reported late 35 percent of the time because of system malfunctions and user incompetency. Reruns are common, and the order-entry software has suffered 14 abnormal terminations. On the first three abnormal terminations, operations took over two days to restart the system.

Required: Prepare a critique of this system's implementation. What should have been done to prevent these problems from occurring? Do you believe direct conversion to be appropriate? Which conversion method do you recommend? What testing and training procedures are appropriate? Justify all of your answers.

SUGGESTED READING

Auerbach Information Management Series. Pennsauken, N.J.: Auerbach Publishers, 1980.

Burch, John, and Gary Grudnitski. *Information Systems: Theory and Practice,* 5th ed. New York: John Wiley, 1989.

Carr, Houston H. *Managing End User Computing*. Englewood Cliffs, N.J.: Prentice-Hall, 1988.

Chandler, John S., and H. Peter Holzer. *Management Information Systems*. New York: Basil Blackwell, 1988.

Cohen, Isabelle. "Computer Training Programs: What's Available and How to Select Them." *Computers in Accounting,* August 1988.

Eckols, Steve. *How to Design and Develop Business Systems*. Fresno, Calif.: Mike Murach & Associates, 1983.

Emery, James C. *Management Information Systems: The Critical Strategic Resource*. New York: Oxford University Press, 1987.

Inmon, W. H. *Information Engineering for the Practitioner*. Englewood Cliffs, N.J.: Yourdon Press, A Prentice-Hall Company, 1988.

Leeson, Marjorie. *Systems Analysis and Design,* 2nd ed. Chicago: Science Research Associates, 1985.

Long, Larry E. *Design and Strategy for Information Systems: MIS Long-Range Planning*. Englewood Cliffs, N.J.: Prentice-Hall, 1982.

Martin, Merle P. "The Day-One Systems Changeover Tactic." *Journal of Systems Management,* October 1989.

Methlie, Leif B. *Information Systems Design: Concepts and Methods*. New York: Columbia University Press, 1988.

Walsh, Robert. "The Postimplementation Audit." *EDPACS,* August 1989.

JOCS CASE: Implementing the System

Status meetings for the JOCS software development project are scheduled for the third Wednesday of each month. During status meetings, Jake Jacoby, the JOCS project manager, and the rest of the SWAT team meet to evaluate the progress of the JOCS software development project. At the last status meeting, Jake noted that most of the project is proceeding as originally scheduled. However, there has been some time slippage in the design and development of the online data capture system. As discussed in Chapters 16 and 18, Carla Mills is working on the application programs to capture the online data from the CIM (computer-integrated manufacturing) system, while Tom Pearson is responsible for the interface between the CIM system and the JOCS local area network. During the last status meeting, Carla wanted to know when the interface between the CIM system and JOCS would be completed so that she could begin program testing. Tom was just starting to work on the interface. He said that he would get back to her through electronic mail with a firm interface delivery date.

The SWAT team meets as infrequently as possible so that each team member is able to complete his or her work with as few interruptions as possible. Most of the time, the team members communicate through electronic mail (rather than telephone conversations and meetings) when there is anything important to discuss. The next status meeting will be coming up in a few days. Jake is preparing to discuss a systems implementation schedule at that meeting. Carla is still concerned about the delivery date for the CIM/JOCS interface. She wants some answers before the next meeting. What follows is some of the electronic mail sent to discuss these issues:

Date: September 12, 1991 09:13 A.M.
From: Carla Mills
To: Tom Pearson
Copies:
Attach:
Subject: Online Capture Delay

What's going on with the interface between the CIM system and JOCS? I want to test the first part of my online data capture program this week, but I can't test my program until the interface is finished. What's the delay? When do you think it will be ready?

Date: September 12, 1991 09:30 A.M.
From: Tom Pearson
To: Jake Jacoby
Copies: Carla Mills
Attach:
Subject: Reply to: Online Capture Delay

I have been unable to install the interface between the CIM system and JOCS due to delivery problems with our vendors. The hardware connection between the two systems was due for delivery over a month ago. The vendor, Connect-Plus, kept promising me that the hardware portion of the bridge would be available any day, and now he has stopped returning my calls.

The software bridge vendor, Interface International, has also been giving me excuses about its missed delivery date. Yesterday, I finally spoke with the technical product manager. She said that the product is still in testing. She was pretty upset about the whole thing and said that their salesman sold the product with the understanding that installation was to be at the end of October rather than the beginning of September. She guarantees shipment of a fully tested product at the end of October, but she says that her group is currently in beta testing.

I am sending over a copy of the log that I have kept of my conversations with both vendors. I hate to say this, but I need some help pushing these people to deliver their products.

Carla is ready to begin testing her online data capture program, but she is completely dead in the water without the CIM/JOCS interface.

Date: September 12, 1991 01:02 P.M.
From: Jake Jacoby
To: Tom Pearson
Copies:
Attach: Hardware Vendor Listing
Subject: Vendors Can Be a Pain

I finally contacted someone in management at Connect-Plus. Those people are difficult to pin down, but here is what I found out. The product we purchased is available and ready for shipping. However, the product we want is still under development and won't be ready for another six months. The salesman we were working with didn't understand exactly what we wanted and sold us the wrong product. He obviously didn't want to admit his mistake and has been covering it up hoping that the product we wanted would be ready by the time we were ready. Anyway, the upshot is that I canceled our order with Connect-Plus, and we have to find a new vendor for our hardware connection as soon as possible.

I have attached a list of hardware vendors for you to contact. This is your number one priority project right now, Tom. Find a new bridge as quickly as you can. As soon as you have a delivery date, send me a message so that I can update our schedule.

By the way, I also spoke with the technical product manager at Interface International. She really seems to know what she's doing, and I would like to buy their product. If our negotiations work out, we will be a beta test site for their bridge software. I will let you know early tomorrow if everything works with Interface, or if we need to find a new software vendor.

Date: September 12, 1991 01:08 P.M.
From: Jake Jacoby
To: Carla Mills
Copies: Tom Pearson
Attach:
Subject: Online Capture Delay

We will have to delay testing of your online data capture programs until we complete the interface between the CIM system and JOCS LAN. Tom is working on the interface problems and will keep you posted on the time schedule. He should have been able to give you the new delivery schedule by the end of the week.

Since you will have to stop working on the capture programs, please contact Christine and help her with the variance reporting programs.

Date: September 13, 1991 02:25 P.M.
From: Jake Jacoby
To: Mary Stockland
Copies:
Attach:
Subject: Telecommunications Support

We have discussed the implications of our growing telecommunications network many times, but now it's time to get serious about this problem area. I think we need an additional person qualified to support our network. Tom is doing a pretty good job, but we should have a backup in case he isn't available at a given time.

Tom is scheduled for a two-week training class beginning the week of September 23. I am sending him to a vendor-sponsored class. The vendor offered a competitive price and will be able to explain all the necessary technical details of its system during the class. Tom is really looking forward to being away from the office and his sometimes tedious testing responsibilities so that he can concentrate on the course material. I will encourage Tom to make contacts during the class; then he will be able to call other class members for assistance during our implementation. I would really like to send two people to this class. Tom and the other telecommunications person could help each other during class and perform the installation of the system together. We could get a half-price discount on the additional person if we register within the next week. What do you think?

Date: September 13, 1991 02:45 P.M.
From: Mary Stockland
To: Jake Jacoby
Copies:
Attach:
Subject: Telecommunications Support

Jake, you must have been reading my mind. Since our last discussion, I have been negotiating with the VP of engineering to share a telecommunications support position with our department. We are going to share David Martinez between engineering and MIS for telecommunications installation and support. David has five years' electrical engineering experience. He is familiar with the Ethernet system used in the engineering department. In addition, he worked with UNIX systems while in college and originally configured the UNIX-based minicomputer located in engineering. He really enjoys working with computer systems, and he is excited about this new position.

Please register David immediately for the local area network class. I will contact him and tell him about the class as well as our status meeting next week.

Date: September 13, 1991 03:24 P.M.
From: Jake Jacoby
To: Mary Stockland
Copies:
Attach:
Subject: JOCS Implementation Ideas

David Martinez is now registered in local area network training starting September 23. I have also had my secretary take care of his flight and lodging reservations.

I am in the process of drawing up our JOCS implementation schedule. I want to bounce a few ideas off you.

User Training: We have to train users in two different areas: DBMS syntax and use and JOCS detailed usage. We will train users on JOCS using our in-house staff. However, I would prefer that the users have outside training of some kind on the DBMS. Vendor training is exorbitantly expensive. Do you have any ideas?

Conversion: JOCS is a fully integrated, DBMS-based package replacing a manual cost accounting system. This fact makes it difficult for us to convert gradually to the new system. It would be almost impossible to install single pieces of the new system, because each piece relies heavily on data from other pieces of the system. For example, it would be impossible to install the online data capture module without also installing the inventory module and job accounting module. In addition to this problem, the manual system does not provide all the information that will be available through JOCS. Currently, totals about a job, such as labor and materials, are available two to four weeks after they have been applied to the job. The current totals are frequently incorrect. Using the current data to validate the new system would be time-consuming and potentially misleading. As a result, it is virtually impossible to perform a gradual cutover to the new system, and it would be completely impossible to run parallel. I suggest we take the risk and perform a direct conversion to the new system. What do you think?

Date: September 13, 1991 04:10 P.M.
From: Mary Stockland
To: Jake Jacoby
Copies:
Attach:
Subject: Reply to: JOCS Implementation Ideas

User Training: How about contacting the local university and/or community college and checking out their classes? We are using a common DBMS, so they might offer classes in this area. Check to see if they offer any short seminars through their continuing education departments. You might also check the local technical business school. Contact one of the local computer distributors and see if they would be willing to sponsor a class for us. If these suggestions don't work, I know a consultant who would be willing to hold a class for us. The only problem is that we would have to use our own hardware facilities and I would prefer that the training occur off-site.

Conversion: Sounds like you have really thought this through. If you think direct conversion is our best option, let's do it.

Figure 19.12
Gantt chart for JOCS implementation (starting at week 24 of software development).

1. Conduct technical training
   - 1.1 Communications
   - 1.2 DBMS
   - 1.3 JOCS system concepts
2. Install network
   - 2.1 Delivery of peripherals terminals and storage
   - 2.2 LAN installation
   - 2.3 Unix - LAN installation
   - 2.4 Ethernet - Unix - LAN installation
   - 2.5 Storage installation
   - 2.6 Terminal installation
   - 2.7 Integrated testing
3. Write documentation
   - 3.1 Operator documentation
   - 3.2 User documentation
4. Perform systems conversion
5. Conduct postimplementation review

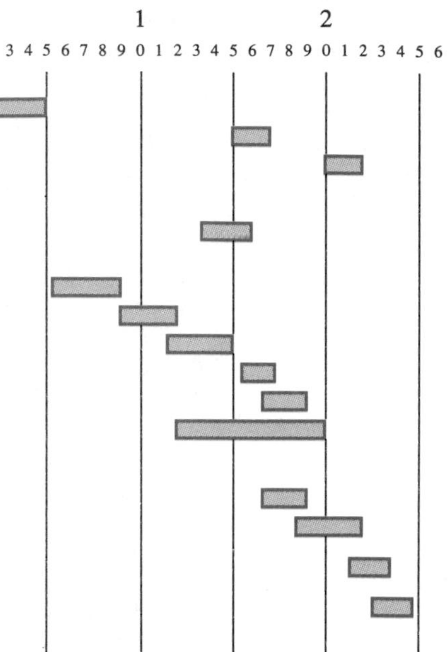

Date: September 15, 1991 03:25 P.M.
From: Jake Jacoby
To: Mary Stockland
Copies: SWAT team group
Attach: Gantt Chart for JOCS Implementation
Subject: Implementation Schedule

We are currently beginning week 24 of the JOCS development cycle. JOCS development is approximately three weeks behind the original schedule. We have had a few delays from problems with our hardware and software bridge vendors and more than five unanticipated user changes to report and screen design. However, even with these delays, I think we can recover enough time during implementation to finish the project close to our original delivery date.

The subject of our status meeting next Wednesday is an implementation schedule for JOCS. I have attached a Gantt chart (see Figure 19.12) that we will use as a framework for our implementation plan. This chart begins at week 24 of our original schedule (see Figure 15.12).

The tasks shown in this chart will overlap the tasks shown in our original schedule. To help us implement this overlapping schedule, I have added another person to the SWAT team. We are temporarily borrowing David Martinez from the engineering department to help us with our telecommunications installation and testing. David and Tom will work together, after they return from training, to finish the installation of our telecommunication network.

See you in conference room 308 at 10:00 A.M. on 09-18-91 for our status meeting.

Chapter 20

Maintaining the System

WHAT WILL YOU LEARN IN THIS CHAPTER?

After studying this chapter, you should be able to:

1 Discuss systems maintenance and define the various types of systems maintenance.
2 Outline the steps of the systems maintenance process.
3 List the procedures for improving systems maintainability.
4 Name and discuss the CASE tools that aid systems maintenance.
5 Discuss managing systems maintenance.
6 Describe a change management system (CMS) and state its purpose.

INTRODUCTION

At this point, the new system has been developed and converted to operations (see Figure 20.1). The SDLC is complete, with the system converted to operations. Now the system is in the maintenance phase of its life cycle. Although both hardware and software composing a system must be maintained during operations, most of the material in this chapter is directed toward the maintenance of software.

MAINTAINING THE SYSTEM

All information systems are subject to change. **SYSTEMS MAINTENANCE** is the activity that makes these changes.

Systems maintenance begins as soon as the new system becomes operational and lasts its lifetime, as shown in Figure 20.2. The dashed lines showing systems maintenance indicate that the systems maintenance phase of the systems life cycle exceeds time, effort, and cost of all systems development life cycle (SDLC) phases combined.

Normally, most of the information compiled during the postimplementation review is used to perform initial maintenance. Periodic reviews, audits, and user requests will continue to be the principal sources for performing systems maintenance throughout the systems life span.

After early systems maintenance is performed following the postimplementation review, the cost and effort of systems maintenance should decline. But after some months or years, more and more change requests will be made, requiring an increasing amount of effort and cost.

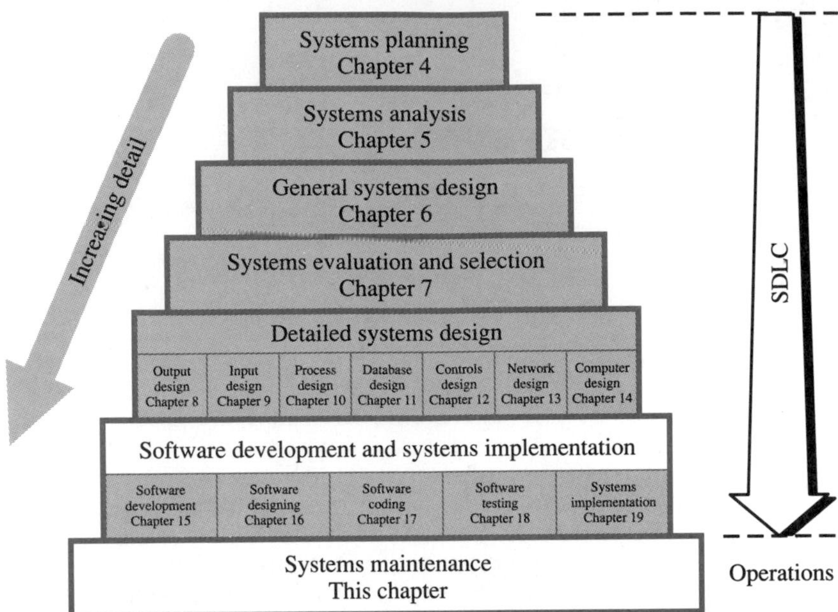

Figure 20.1
Maintenance phase relative to the SDLC.

At some point, the system is more troublesome and costly to maintain than it's worth. When a system becomes a significant problem to users or new opportunities are available, the system is replaced by a new system.

The cost of software maintenance has increased steadily during the past 25 years. Some organizations spend 80 percent or more of their systems budget on software maintenance. This means that backlogs of new applications are getting longer. New user requirements lag behind by as much as two or more years. It is, indeed, not difficult to visualize organizations becoming so maintenance-bound that all of the systems budget is spent on maintenance, and no new systems are developed.

Types of Systems Maintenance

Systems maintenance can be categorized into the following four types:

- Corrective maintenance
- Adaptive maintenance
- Perfective maintenance
- Preventive maintenance

Corrective Maintenance

CORRECTIVE MAINTENANCE is the less noble and more burdensome part of systems maintenance, because it corrects design, coding, and implementation errors that should never have occurred. The need for corrective systems maintenance can often be traced back to poor application of the SDLC and SWDLC.

Commonly, corrective maintenance involves an urgent or emergency condition that calls for immediate attention. The ability to diagnose and remedy the error or malfunction rapidly is of considerable value to the organization.

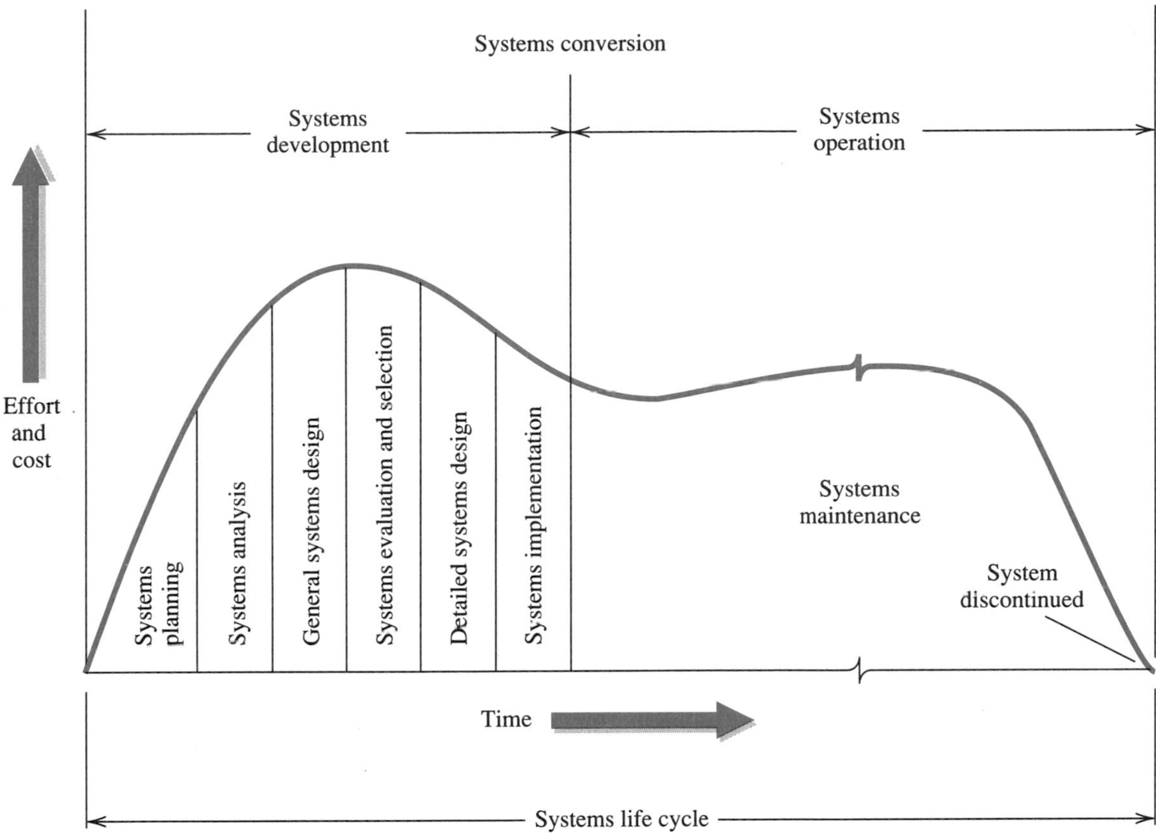

Figure 20.2
Total systems life cycle.

Adaptive Maintenance

ADAPTIVE MAINTENANCE is performed to satisfy changes in the processing or data environment and meet new user requirements. The environment in which the system operates is dynamic; therefore, the system must continue to respond to changing user requirements. For example, a new tax law may require a change in the calculation of net pay. Or a new report is required, or a new accounting depreciation method must be installed before the end of the fiscal period.

Generally, adaptive maintenance is good and inevitable. Too much of it, however, may mean that phases of the SDLC and SWDLC were not thoroughly and properly performed.

Perfective Maintenance

PERFECTIVE MAINTENANCE enhances performance or maintainability. It also allows the system to meet user requirements unrecognized earlier. When making substantial changes to any module, the maintainer also exploits the opportunity to upgrade the code, to remove outdated branches, correct sloppiness, and improve documentation. For example, this maintenance activity may take the form of reengineering or restructuring software, rewriting documentation, altering report formats and content, defining more efficient processing logic, and improving equipment operating efficiency.

Preventive Maintenance

PREVENTIVE MAINTENANCE consists of periodic inspection and review of the system to uncover and anticipate problems. As maintenance personnel work with a system, they often find defects (not really errors) that signal potential problems. While not requiring immediate attention, these defects, if not corrected in their minor stages, could significantly affect either the functioning of the system or the ability to maintain it in the near future. A motto for preventive maintenance could be: "Pay me now, or pay me later."

The Systems Maintenance Life Cycle

A number of authorities recommend that software maintenance be performed in a **SOFTWARE MAINTENANCE LIFE CYCLE (SMLC)**.[1] Essentially, the SMLC includes the following phases:

1 ***Maintenance Request*** A systems maintenance request is documented and submitted by a user. This maintenance request is used to prepare a maintenance work order (WO), explained later in this chapter.

2 ***Transform the Maintenance Request to a Change*** The maintainer has a description of the existing system as well as one of the desired system. Transforming the request to a change involves finding the differences between the two systems and eliminating the proper differences.

3 ***Specify the Change*** The change will involve one or all of the existing code, data, procedures, or hardware. The new code, data, and procedures are sometimes referred to as a patch. A change in hardware may involve replacing or repairing a part or reconfiguring the equipment.

4 ***Develop the Change*** The change or patch in software programs, unless very simple, is designed and coded in a fashion similar to the designing and coding phases of the SWDLC.

5 ***Test the Change*** Test cases used in the testing phase of the SWDLC stored in the CASE system's central repository can be used for testing the software after it has been changed. The purpose of testing helps to validate and verify that the right change has been made and made correctly.

   Regression or revalidation testing tries to confirm that the functions of the system which were to have been left unchanged still perform as they were intended to perform before the required change to the system was made. Thus regression testing focuses on the functional integrity of the system.

6 ***Train Users and Run an Acceptance Test*** If the change is relatively simple, this step may be skipped. If, on the other hand, it introduces new ways in which users perform their tasks, these users will have to be trained. After they are trained, acceptance tests should be conducted using alpha testing as described in Chapter 18.

---

[1] Ned Chapin, "Software Maintenance Life Cycle," *Proceedings of the Conference on Software Maintenance 1988* (Washington, D.C.: Computer Society Press of the IEEE, 1988), pp. 6–13.

7 ***Convert and Release to Operations*** Once the newly changed system has passed testing successfully, it is ready to be converted and released back to operations.

8 ***Update the Documentation*** All the documentation pertaining to the maintenance activities should be updated to reflect the change and the new procedure for the system.

9 ***Conduct a Postmaintenance Review*** After the system has been operating for a few weeks (or maybe a few days) after maintenance has been performed, the maintainer should conduct a postmaintenance review to determine if the change continues to meet user expectations.

Maintaining the Software

Unlike plant machinery or computer equipment, software doesn't wear out, because software programs do not contain moving parts. But even though it doesn't wear out in a physical sense, software does require change.

Application software may be structured or unstructured, and documented or undocumented. Some software that is unstructured and undocumented may be on the verge of being unmaintainable. In fact, one of the major reasons why systems maintenance takes such a large bite out of the systems budget is because of excessive effort spent on trying to maintain poorly structured and documented software.

In other instances, software programs that are unstructured and undocumented are also unmaintainable. Should a change in operations require them to be changed, they will have to be scrapped and a new program developed, thus wasting all the resources that were spent to develop the original, unmaintainable programs, not counting disruption of business operations when this apocalyptic day arrives.

Maintaining the Hardware

Although most of this chapter is devoted to software maintenance, hardware maintenance is an important part of systems maintenance. Hardware maintenance is chiefly preventive maintenance entailing the repairing, replacing, or adding of parts and components to restore or keep the hardware in working order. In many ways, computer equipment is similar to a car; when it is not maintained properly, it can break down and cause problems. Cars are serviced periodically to prevent major mechanical problems. In the same manner, the hardware components of an information system should be checked and serviced periodically.

Clearly, the most common source of new hardware maintenance is the vendor that supplied it. Typically, a mainframe or minicomputer vendor makes available a maintenance support group that provides maintenance and service for a standard price. If the hardware is a microcomputer, the dealer will generally provide maintenance as a chargeable service. The buyer may pay a lower price for microcomputers purchased from mail-order companies, but may lose the convenience of local service. Maintenance is also available from specialized companies known as third-party vendors.

Mean time to repair (MTTR) is an important measurement used to evaluate equipment maintenance efficiency. Mean time between failures (MTBF) is a measurement used to evaluate equipment maintenance effectiveness.

PROCEDURES FOR MAINTAINING THE SYSTEM

SYSTEMS MAINTAINABILITY is the capacity of the maintainer to perform corrective, adaptive, perfective, or preventive maintenance. The more maintainable the system, the less effort and cost must be spent for systems maintenance.

Designing Systems for Easy Maintenance

Systems maintainability is increased if the system is designed to make changes easier. This aspect encompasses the following procedures:

- ***SDLC and SWDLC*** Professional application of the SDLC and SWDLC and their supporting modeling tools and techniques are the best overall things one can do to increase systems maintainability.
- ***Standard Data Definitions*** The trend toward relational database management systems underpins the push for standard data definitions and data normalization. Many organizations have redundant and inconsistent data definitions. These inconsistent data definitions are found in procedure manuals, source program documentation, and data files. They add to the problem of maintenance. A glossary or data dictionary of terms for data elements and other items in the system should be provided. For example, all data elements should have a standard name, description, size, source, location, and maintenance responsibility designation. It is also important to use the name precisely in each application. (To a computer, CUST-NAME is not the same as CUSTOMER-NAME, for instance.)
- ***Standard Programming Languages*** The use of a standard programming language, such as C or COBOL, makes the systems maintenance task easier. If C or COBOL software contains complete and clear internal documentation, even a novice maintenance programmer or user can understand what it is doing. Moreover, C and COBOL are universal languages generally known to vast numbers of people. Therefore, maintenance programmer turnover has less impact on the company's ability to maintain old C and COBOL programs.
- ***Modular Design*** As with the maintenance of home appliances, in which a repair person can determine which module is causing trouble and quickly replace it, the maintenance programmer can change modules of a program much easier than he or she can deal with the total program.
- ***Reusable Modules*** Common modules of reusable code can be accessed by all the applications requiring them.
- ***Standard Documentation*** Standard system, user, software, and operations documentation are needed so that all the information required to operate and maintain a particular application will be available.

- ***Central Control*** All programs, documentation, and test data should be installed in a central repository of a CASE system. Later, we present a change management system (CMS) that does this and more.

Three Approaches for Organizing the Systems Maintenance

No clear-cut answer exists as to the best way to organize for systems maintenance.[2] We will, however, explore three:

1 Separate systems development and systems maintenance

2 Combine systems development and systems maintenance

3 Position systems professionals who have responsibility for both systems development and systems maintenance within the enterprise's functional areas

Separate Approach

Traditionally, information systems have been organized into two distinct groups:

- Development
- Maintenance

This SEPARATE APPROACH of duties provides a natural way for one group to force the other group to perform its work properly. For example, a maintenance programmer would be against accepting a new program for operations unless it had been thoroughly tested during development. Maintenance separated from development forces better documentation to be prepared, formalizes the conversion of the new system from a development status to an operations status, and formalizes change procedures. Also, senior maintenance programmers may be promoted to development project leaders, because they have a good knowledge of documentation requirements, standards, and operations. Maintenance is also a good training ground for junior programmers.

Combined Approach

The COMBINED APPROACH brings together both development and maintenance personnel into one major group of the information system. If both development and maintenance are performed in the same group, the user departments will have one point of contact with the information system personnel who can effect change. User departments often do not know if a request for work will be classified as development or maintenance, because large revisions or system improvements are often treated as development. Furthermore, the analysts, designers, and coders who originally developed the systems can best assess the full impact of changes. Some systems are so critical and complex that mainte-

[2] Ned Chapin, "The Job of Software Maintenance," *Proceedings of the Conference on Software Maintenance 1987* (Washington, D.C.: Computer Society Press of the IEEE, 1987), pp. 4–12.

nance must be handled by only the most capable people and, in many instances, the most capable people are those who developed the system in the first place.

Functional Approach

The FUNCTIONAL APPROACH is a variation of the combined approach. The functional approach, however, removes systems professionals from information systems and assigns them to business functions for both development and maintenance. Organizing according to functions puts systems professionals much closer to end users, both in terms of physical proximity and knowledge of the users' jobs and requirements.[3]

Essentially, by employing the functional approach, systems professionals are actually acting as members of the user departments, and consequently, they have a vested interest in the success and feasibility of system applications. Moreover, systems professionals become more business-literate and users become more systems-literate in a combined approach. The payoff is understanding, communication, and mutuality of interest.

USING CASE TOOLS TO MAINTAIN THE SYSTEM

Five special CASE tools that aid systems maintenance of old systems and help break the new development backlog jam are:

- Forward engineering
- Reverse engineering
- Reengineering
- Restructuring
- Maintenance expert systems

Forward Engineering

Figure 20.3 depicts the FORWARD ENGINEERING process. This process should look very familiar because, to this point, forward engineering has been the major subject of this book. Forward engineering represents the way new systems should be developed. If they are developed this way, the systems maintenance burden is reduced considerably.

Reverse Engineering

Figure 20.4 shows REVERSE ENGINEERING, which is forward engineering performed backwards. Reverse engineering is a process of examining and learning more about old, legacy systems by creating their design. The entire system is read, including source code, screens, reports, data definitions, and job control language. The results are abstractions of design specifications in the form of models such as data flow diagrams and structure charts. Now, systems profes-

[3] Christine B. Tayntor, "Maintenance Magic," *System Builder*, October/November 1990, pp. 26–29.

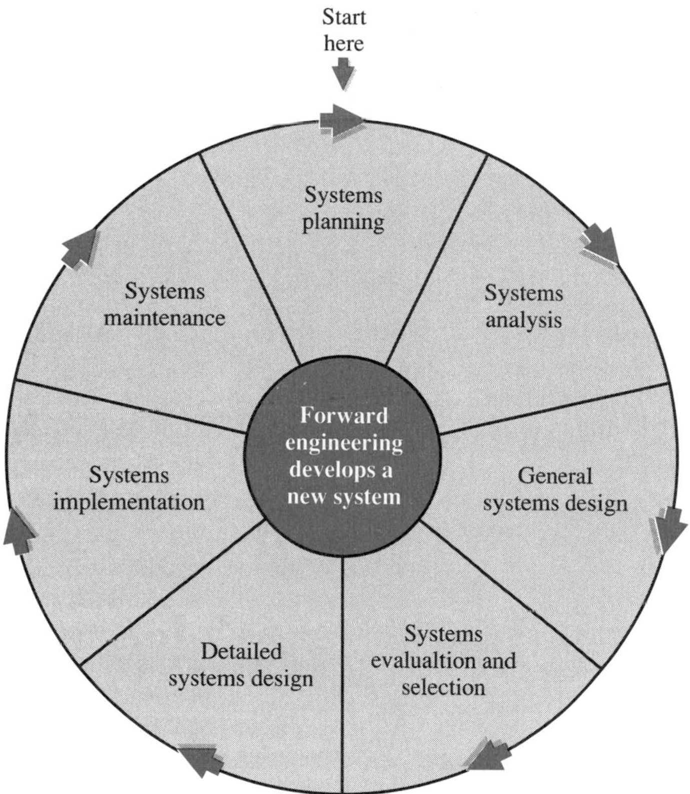

Figure 20.3
Forward engineering.

sionals have a clear view of the old system to analyze and evaluate its design quality and functionality. After the old system is fully evaluated, a decision may be made to scrap it.

Some software was developed 10 to 25 years ago. Unfortunately, much of this software was poorly designed, structured, and documented. Should we reengineer or restructure it to increase its maintainability, or should we practice software euthanasia? Sometimes scrapping the existing software and developing new software from scratch is more effective and efficient. However, reverse engineering provides sufficient design-level understanding to reengineer or restructure the software should management decide not to scrap the old system.

Reengineering the System

Some old software may be worth reengineering. **REENGINEERING** generally includes some form of reverse engineering to gain a better understanding of the existing software, followed by forward engineering to redesign and change its form and functionality. Unnormalized data are normalized. Data names are standardized. Ambiguities, redundancies, anomalies, and unused code are eliminated. Inefficient code is recoded to be made efficient. Unstructured code is structured. Documentation is prepared. The system is then reimplemented with new form and functionality.

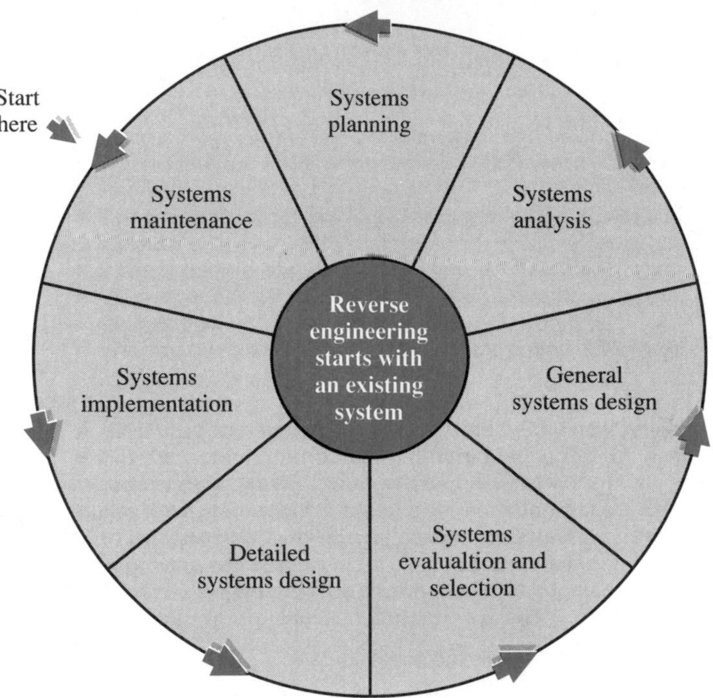

Figure 20.4
Reverse engineering.

Restructuring the System

Recall that software doesn't wear out physically. If its functionality essentially meets the needs of the user, the software is functionally sound. But old software that is functionally sound is often structurally unsound, which makes it difficult to understand and thus difficult to maintain.

The **RESTRUCTURING** process converts unstructured spaghetti-like code into fully structured documented code. The restructuring of software is one of the simplest methods available to decrease the cost of maintenance.

Using a Maintenance Expert System

A **MAINTENANCE EXPERT SYSTEM** extends the capabilities of systems maintenance personnel. Although not commonplace, a maintenance expert system offers many positive features:

- It upgrades the skills of the maintenance staff.
- It enables the company to retain the experience and knowledge of skilled experts who may be about to retire or leave the company.
- It is an effective method for training junior maintenance personnel.

A skilled systems maintenance person is not always available to respond in a timely manner. The maintenance expert system emulates the logical reasoning process that human maintenance experts use to solve problems. Members of the systems maintenance staff can sit at a computer and hold a consultation about a problem with the maintenance expert system in much the same way as they would confer with a human expert. In operation, the maintenance expert system asks about the circumstances of the problem and stores the responses in its database. The computer then displays suggestions for taking action.

Maintenance expert systems are designed to provide explanations for their decisions and recommendations. Thus inexperienced personnel can increase their knowledge by engaging in what-if exercises at the computer for training purposes. This way, they will gain experience that otherwise might take years to acquire.

Here are the advantages of maintenance expert systems:

- They help convert reactive maintenance to proactive preventive maintenance, thus improving the mean time between failures (MTBF).
- They reduce mean time to repair (MTTR) by correctly diagnosing the problem the first time.
- They can be applied by users and operators in some situations to provide practical advice and to handle troubleshooting and maintenance of common problems.
- They help reduce the systems maintenance burden on senior and skilled maintenance personnel.

MANAGING THE SYSTEMS MAINTENANCE

The challenge of managing systems maintenance is the same as that of managing any other endeavor. That's the challenge of managing people. The first priority of managing maintenance, therefore, is to put together a group of competent and motivated maintainers, and supply them with the tools and resources to perform scheduled and unscheduled systems maintenance.

Scheduled systems maintenance may be managed by a calendar or Gantt chart. Unscheduled maintenance is typically initiated by users and operators. In any case, management should establish a way to initiate, record, and evaluate maintenance activities. Presumably with an evaluation of maintenance activities, the manager can eventually optimize the overall systems maintenance program.

Defining Scheduled Systems Maintenance Activities

Scheduled systems maintenance activities, which are primarily for hardware, are triggered by a date, which falls into one of three categories:

- Daily
- Weekly
- Occasional

Daily maintenance activities are concerned primarily with user administration, such as adding and deleting users; assigning elementary levels of security such as passwords; and allotting disk space.

Weekly maintenance involves disk maintenance, backups, and server configurations. Disks are reformatted and files are backed up and stored off-site.

Occasional maintenance activities are performed on an ad hoc basis. These activities include adjusting system parameters; conducting printer maintenance; and making additions and upgrades.

Upgrades and additions are simpler if network software and applications reside only on servers. Only the network drivers and log-on command should reside on the workstation. Although this will increase traffic on the network, the advantages of control, standardization, single software versions, and single location justify centralizing network software and applications. This makes scheduled maintenance much easier.

Initiating and Recording Unscheduled Systems Maintenance Activities

What triggers unscheduled systems maintenance? The postimplementation review, discussed in Chapter 19, generally initiates systems maintenance early in a new system's operating life. During this postimplementation review, errors may be detected and ways to enhance software may be discovered. Moreover, managers should perform periodic audits to ensure that existing goals, policies, and procedures are being executed properly. But ongoing systems maintenance needs a formal way to initiate a maintenance request. This can be achieved by using a **MAINTENANCE WORK ORDER (WO) FORM**.

The WO, as depicted in Figure 20.5, is the document used to initiate and record systems maintenance work. Management is particularly interested in the:

- Work requested
- Work performed
- Estimated time versus actual time
- Maintenance code
- Maintenance cost

The description of work requested should be as clear as possible, explaining what needs to be changed. The person who makes the request should enter his or her name and title. A number is used to identify each maintenance WO form uniquely. Date of the request is also entered. Priority indicates how quickly systems maintenance should begin. If it's an emergency, maintenance work begins immediately. Emergency priority generally means that the system has stopped operating. If it's urgent, maintenance work for this priority interrupts the maintenance schedule and begins on the next available daily schedule. Urgent priority normally implies that a problem exists that has a probability of stopping operations in the very near future. If the WO request is routine, this work is placed on the next available weekly schedule. Routine priority typically means that a defective condition has been identified. This condition will most likely not stop operations or cause damage if corrected during the next week to four weeks.

Figure 20.5
Maintenance work order form.

MAINTENANCE WORK ORDER

Work required: Add vendor performance code to inventory record.

Number: [| | | | |]

Request Date: MM | DD | YY

Name: Tom Barkley   Title: Purchasing Agent

Priority
- [ ] Emergency
- [ ] Urgent
- [ ] Routine

Work performed: Added vendor performance code. Tested and documented change.

Estimated time: HH:MM

Name: Maria Gomez   Title: Maintenance Programmer

Actual time: HH:MM

Change approved: Harry Feldman   Date: MM/DD/YY

Maintenance code: [| |]

Labor cost (actual time * labor rate) $ [| | | | | | . |]

Material or parts cost   [| | | | | | . |]

Total maintenance cost $ [| | | | | | . |]

The maintenance supervisor, along with the maintenance programmer or technician, should enter an estimate of the time required to perform the maintenance work in terms of hours and minutes. After the maintenance work is completed, actual time required is entered and compared with the estimated time. This comparison is used for control and future scheduling purposes.

The maintenance code is an abbreviated description of the systems maintenance work. It indicates the types of maintenance as:

- Corrective
- Adaptive
- Perfective
- Preventive

It designates the priority of the request as:

- Emergency
- Urgent
- Routine

It classifies the type of task involved as:

- Scrap
- Reengineer
- Restructure
- Replace
- Repair
- Modify

Scrapping, reengineering, and restructuring were explained earlier. Replacing means that a software module is deleted and replaced by a new module. Repairing means that some element of the system is restored to a sound state. Modifying involves minor changes. For example, an instruction may be inserted on a data-entry screen to make an electronic form more understandable. The maintenance code can be structured as shown at the top of the opposite page. For example, a code 116 indicates corrective-type maintenance of emergency priority that requires a modify activity.

Work performed should be explained to give additional meaning to the maintenance code. In some instances, the actual work performed may entail more effort and time than was originally thought by the person making the request. The person who performs the maintenance work should enter his or her name and title. When the work is completed, the supervisor should sign his or her name on the completion approved line and enter the date of signature. The signature will verify that the supervisor has inspected the completed work

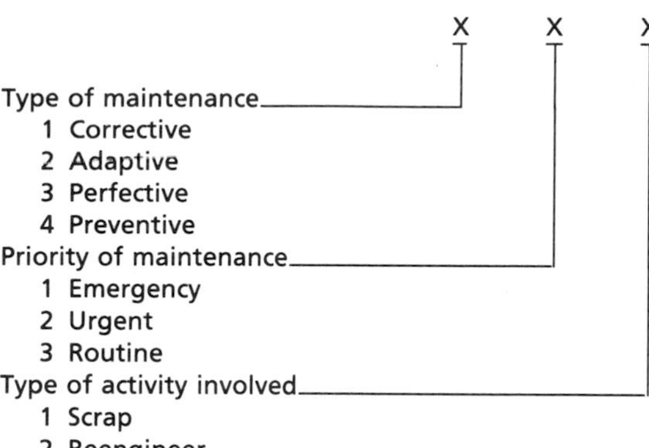

2 Reengineer
3 Restructure
4 Replace
5 Repair
6 Modify

and that the work has been done in accordance with maintenance policies of that particular company and that it was of proper quality.

The primary purposes of the maintenance WO form, besides triggering maintenance work, are to:

- Provide a means for screening and authorizing work
- Generate maintenance cost data
- Provide feedback information on repetitive failures for analysis
- Disclose impact on operations
- Serve as input for planning, scheduling, and controlling maintenance work

Determining the reason for maintenance request is very important to management. Good recordkeeping of maintenance WO forms over time and analysis of these WOs may indicate:

- Lack of an adequate preventive maintenance program
- Faulty previous maintenance
- Incorrect equipment evaluation and selection
- Defective systems and software program designs
- Errors in implementation
- Operator or user abuse or error

Learning the reasons for maintenance in detail is an important way to help improve a company's systems maintenance program.

Using a Help-Desk Software System

Software for a help-desk system is similar to what would be used in the technical support or customer-service department of a software or hardware vendor.

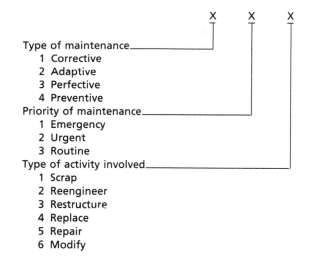

The help-desk system enables requests for maintenance to be logged into a priority queue. The system generates a maintenance WO form. It also provides information on previous similar maintenance problems and their solutions. This feature eliminates the potential problem of maintenance personnel continually figuring out how to solve recurring problems when they could search a database for the problem's solution. It also helps when maintenance personnel leave the company or go on vacation.[4]

Applying a help-desk system follows these steps:

1 An end user encounters a problem and calls the help desk.

2 The help-desk coordinator answers the call and logs the problem into the help-desk system.

3 A maintenance WO form is printed out and given to a maintenance technician.

4 The maintenance technician searches the help-desk system's database for any solutions to similar problems.

5 After solving the problem, the maintenance technician enters the solution in the database for future reference.

6 The maintenance WO form is completed and entered into the database.

Evaluating Systems Maintenance Activities

To work toward the goal of a maintenance optimization program (discussed in the next section), the maintenance manager can use the maintenance WO form as a key document for each of the following activities:

- Compute a variety of maintenance cost analyses.
- Measure the number of failures per program.
- Calculate total person-hours spent on each maintenance type.
- Compute the proportion of emergency, urgent, and routine maintenance.
- Derive the average number of changes made per program, per language, and per maintenance type.
- Develop a profile of those making most of the maintenance requests and common problems encountered.
- Calculate the average turnaround time per maintenance WO. This calculation entails the MTTR.
- Build a MTBF profile on all applications. From this profile determine those applications that are requiring the largest portion of the maintenance budget, and why. It is often discovered that 20 percent of the applications cause 80 percent or more of the problems.

---

[4] Eric Rayl, "SupportMagic Tops Help-Desk Software," *PC Week,* February 4, 1991, p. 63.

Optimizing the Systems Maintenance Program

The goal of a well-managed systems maintenance program is to achieve the lowest cost of the sum of two quantities:

- Maintenance cost, including labor, material, parts, and computer time
- Operation loss (cost) caused by downtime, inefficiency, or production of incorrect results

As Figure 20.6 points out, the lowest combined cost occurs where the level of systems maintenance is optimized.

As systems maintenance effort is intelligently increased, the operation loss (cost) decreases until the lowest combined cost is achieved. At this point, the goal of a well-managed maintenance program is achieved. Systems maintenance effort required beyond this point increases total cost and changes systems maintenance from a necessary and optimizing function to a necessary evil.

How much effort should be devoted to preventive maintenance as part of the maintenance optimization program? Preventive maintenance tasks will increase maintenance costs when first initiated, until the beneficial effect of these tasks have time to take effect. Then total costs should begin to decrease.

Figure 20.6
Graph showing an optimized systems maintenance program.

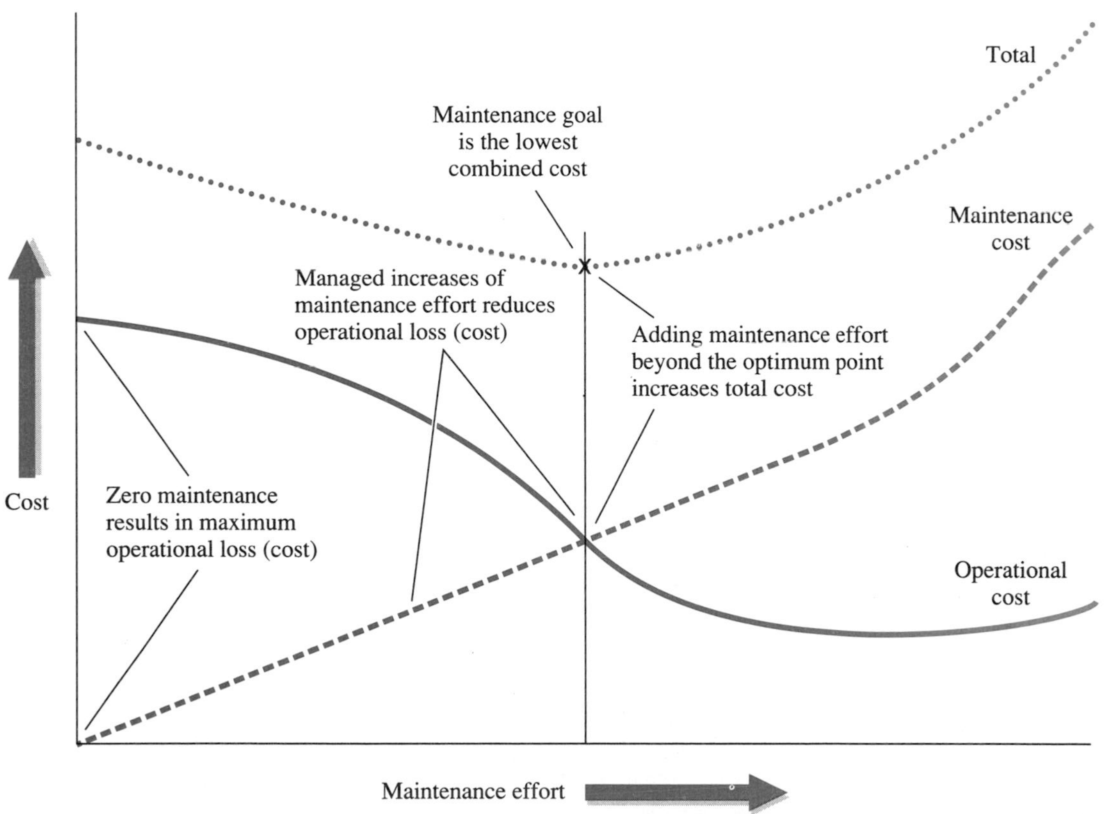

For some equipment items of very small dollar value, possibly the best course of action is to not perform any preventive maintenance. Rather than perform preventive maintenance, it is more cost-effective to allow the item to run until breakdown and then either repair or replace it. At the other extreme, equipment and software with relatively large dollar value that must be operated at the lowest achievable downtime because of the criticality of the application they support must have an aggressive total systems maintenance optimization program.

DEVELOPING A CHANGE MANAGEMENT SYSTEM

A comprehensive **CHANGE MANAGEMENT SYSTEM (CMS)** can help reduce the confusion and complexity of developing new systems and maintaining existing systems. A CMS, which may be part of a CASE system or a stand-alone system, can also facilitate many of the systems maintenance procedures discussed in the preceding material.

A general model of a CMS is portrayed in Figure 20.7. A CMS:

- Restricts access to production source and object code
- Reduces errors and design defects from being introduced into production

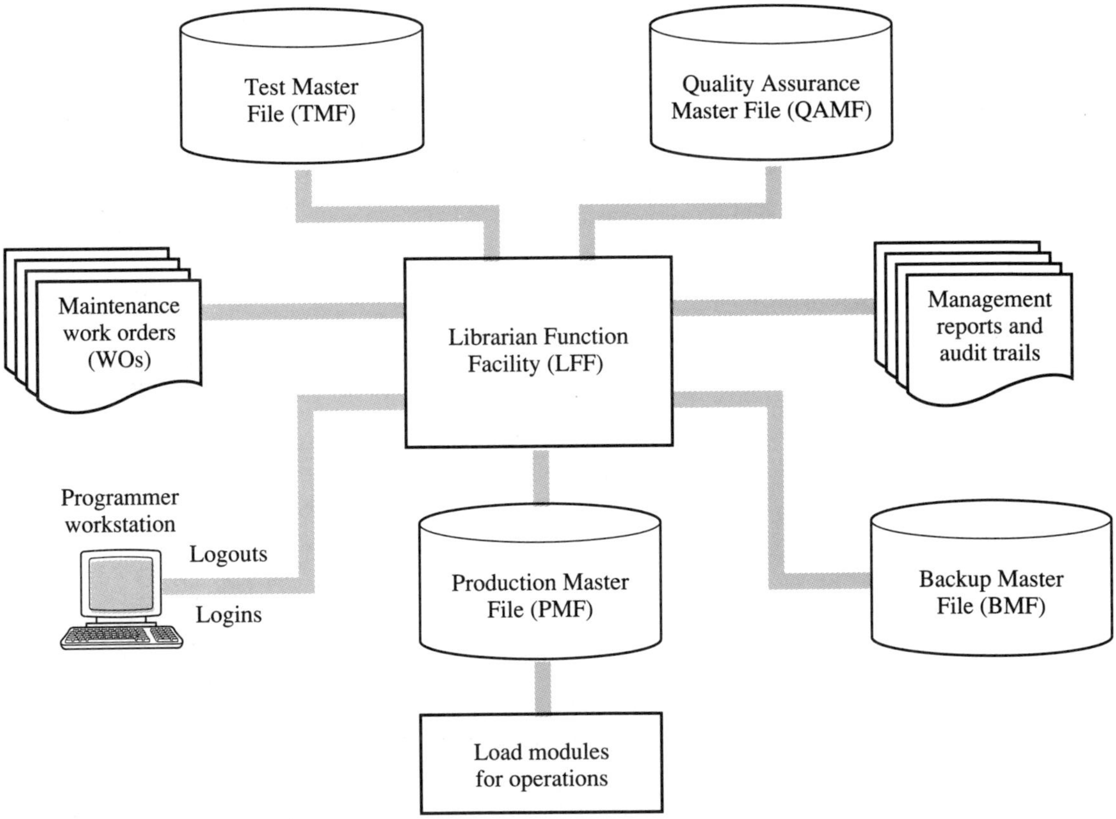

Figure 20.7 Change management system (CMS).

- Prevents the existence of more than one version of source and object code programs in the production master file (PMF)
- Improves quality and reliability of software
- Increases security and better overall control of software development and systems maintenance activities
- Increases software productivity

What Are the Components of a CMS?

Components of a CMS include:

- The librarian function facility (LFF)
- The maintenance work order (WO)
- Programmer workstations
- The test master file (TMF)
- The quality assurance master file (QAMF)
- The production master filc (PMF)
- The backup master file (BMF)
- Management reports and audit trails

Librarian Function Facility

The LFF is the heart of the CMS. It works in a manner similar to any librarian function. It is a software package that centralizes, tracks, controls, and automates changes to programs against an approved maintenance WO. It also controls the implementation of newly developed or acquired programs. Figure 20.8 depicts the program promotion and release hierarchy. If a program that is already in operations has to be changed, it is logged into the test master file (TMF). If a new program is to be placed into production, it must first enter the TMF. No changes are allowed in the production master file (PMF).

The LFF controls linkage between source and object code, and automatically loads modules online for execution thereby assuming synchronization of the two codes. Comprehensive management reports and audit trails are available via screen and hardcopy for history, status, tracking, and performance information. All master files are backed up to safeguard the system from disasters. Programmers have online access to the CMS to augment change productivity. Programmers access privileges are controlled by passwords or biometric control devices.

Programmer Workstations

The CMS acts as a single point of control, and approved maintenance WOs initiate logouts and logins that enable programmers to do their work at their workstations. Logouts are done against an approved and assigned WO. (WOs are assigned by the CMS supervisor.) When working against a specific WO,

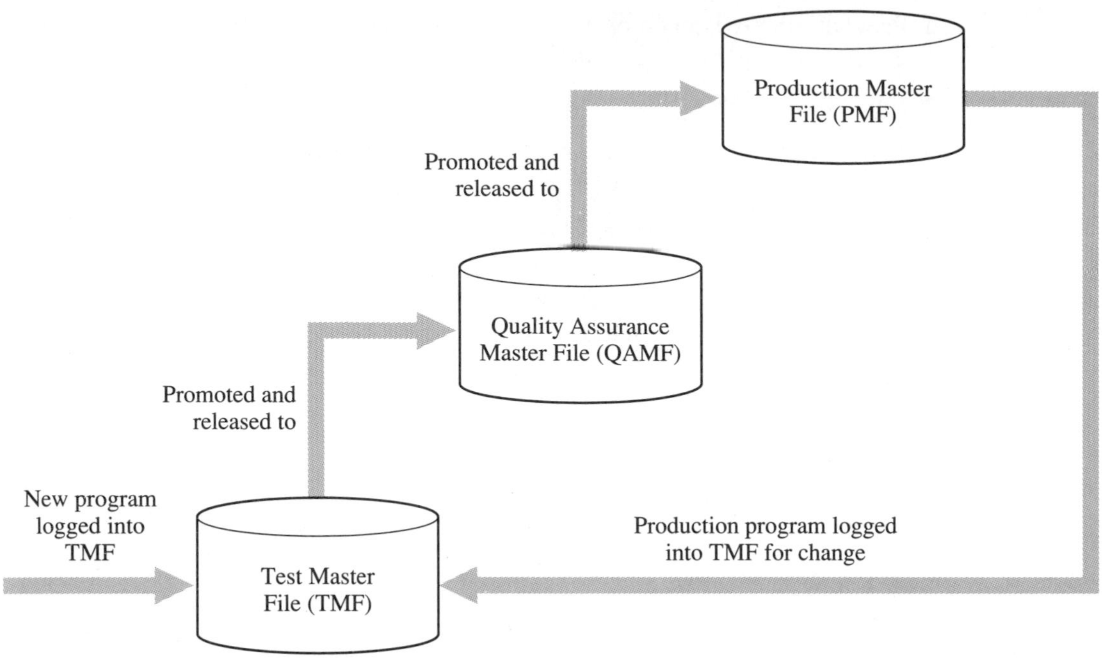

Figure 20.8
Program promotion and release hierarchy.

programmers are permitted to logout as many modules as necessary to complete the assigned unit of work. Program modules outstanding to other WOs are noted during the logout process.

Logins are performed to promote and release a changed and tested program to the quality assurance master file (QAMF), and then to the production master file (PMF). Careful testing and walkthrough procedures are conducted before any program or program modules are promoted and released to the PMF.

Middle-of-the night production problems generally require a maintenance programmer to be onsite or, if at home, to come back to the office to correct the problem. On-call programmers with workstations installed at their homes and connected to the CMS can increase the efficiency of maintenance work. Although this setup doesn't eliminate the dreaded phone call, it may reduce the need for a trip to the office.

Test Master File

To change a production program, a WO must be opened and the program demoted from the PMF to the TMF. This changes the program from production status to test status. As long as the test status is maintained, the program can be changed.

When a program first enters the TMF, it is automatically assigned a name and level 1, or level 0 in some systems. Each time it is changed, the level is increased by 1. To change a program, the programmer must use both the name and correct level number. The level number serves to:

- Protect against the use of old source code listings in previous versions
- Protect against duplicate updating
- Provide a history of program changes and programmer activity

A new software program entering the TMF is subjected to module testing, module integration testing, systems testing, and acceptance testing, covered in Chapter 18. If it passes these tests, it is promoted and released to the QAMF. A production program that is demoted to the TMF has already been subjected to these tests, but after it is changed, it is subjected to regression testing. The purpose of regression testing is to make sure that the change is working as intended, and that it didn't cause any adverse side effects.

Quality Assurance Master File

Some CMSs employ an independent quality assurance (QA) group that has access to the QAMF. The QA group performs software design walkthroughs and various other tests. The QA group also ensures that new software is developed in accordance with company standards before it is promoted and released to the PMF.

Production Master File

Once a program enters the PMF, it is locked into production status, and it cannot be changed. With proper authorization, a program can be copied and logged into the TMF with a new name, and the copied version can be changed. This protective feature helps to ensure that production software programs will not be changed inadvertently. It also helps to prevent malicious or unauthorized changes.

When management no longer wants a particular program in the PMF, the status is changed from enable to disable. This change in status does not delete the program from the PMF, but flags it for deletion. Only authorized managers, possibly requiring dual or triple custody, can delete the disabled production program from the PMF. This feature supports program control, continuity, and housekeeping.

Backup Master File

In the event that any master file is destroyed, the CMS permits information systems personnel to recover from a BMF any files or specific modules that may have been lost. Usually, a copy of the BMF is maintained locally, and another copy is stored off-site in a secure location.

Management Reports and Audit Trails

The reporting features help managers to develop an optimized CMS and auditors to attest to the integrity of the CMS. As discussed earlier in this chapter, management reports help to evaluate and optimize systems maintenance activities. Audit trails help to ensure the integrity of the system.

What Risks Does the CMS Avoid?

A CMS helps the information system avoid the following risks:

- Lack of an accurate inventory of software programs and other information system resources
- Incomplete history of program changes
- Duplicated software program modules
- Unauthorized software program changes
- Lack of clear, comprehensive, and current documentation
- Poor software quality and reliability

REVIEW OF CHAPTER LEARNING OBJECTIVES

The major goals of this chapter were to enable each student to achieve six important learning objectives. We will now summarize the responses to these learning objectives.

Learning objective 1:
Discuss systems maintenance and define the various types of systems maintenance.

After a system has been implemented, it must be maintained until it is no longer economical to do so, at which time the existing system is eliminated and a new one is developed. Systems maintenance entails both software and hardware.

There are four types of systems maintenance:

- Corrective
- Adaptive
- Perfective
- Preventive

Corrective maintenance rectifies design, coding, or implementation errors. Adaptive maintenance adjusts to changing environmental conditions. Perfective maintenance tends toward making the system perfect. Preventive maintenance anticipates and forestalls potential problems.

Learning objective 2:
Outline the steps of the systems maintenance process.

The systems maintenance process (i.e., the SMLC) involves the following nine steps:

1 Understand the maintenance request.

2 Transform the maintenance request to a change.

3 Specify the change.

4 Develop the change.

5 Test the change.

6 Train users and run an acceptance test.

7 Convert and release to operations.

8 Update the documentation.

9 Conduct a postmaintenance review.

Learning objective 3:
List the procedures for improving systems maintainability.

All systems should be developed with maintainability in mind. The more maintainable a system, the less cost and time must be spent for maintenance. Designing for systems maintainability includes the following procedures:

- Apply the SDLC and SWDLC.
- Specify standard data definitions.
- Use standard programming languages.
- Design well-structured modules.
- Employ reusable modules.
- Prepare clear, current, and comprehensive documentation.
- Install software, documentation, and test cases in a CASE system's central repository or CMS.

System maintenance personnel may be organized using the:

- Separate approach
- Combined approach
- Functional approach

Learning objective 4:
Name and discuss the CASE tools that aid systems maintenance.

Special CASE tools that aid systems maintenance and reduce the development backlog are:

- Forward engineering
- Reverse engineering
- Reengineering
- Restructuring
- Maintenance expert system

Forward engineering develops a new system right the first time. Reverse engineering starts with a poorly designed system and abstracts high-level design specifications from it for study and evaluation. Reengineering uses a combination of reverse and forward engineering to refurbish an existing system by changing its form and functionality. Restructuring converts spaghetti-like code to structured code, but does not change functionality of the code. A maintenance expert system guides junior maintainers.

Learning objective 5:
Discuss managing systems maintenance.

The prime objective of managing systems maintenance is to hire or train highly motivated and skilled maintainers. Their maintenance activities should be scheduled and accounted for. Unscheduled systems maintenance should be triggered by a maintenance WO. Further, a comprehensive maintenance work order form will provide management with sufficient data to make a number of measurements to evaluate how well the systems maintenance program is progressing. The ultimate objective of management is to achieve an optimized systems maintenance program.

Learning objective 6:
Describe a change management system (CMS) and state its purpose.

One of the most vulnerable activities in the information system is installing new systems and changing existing systems (sometimes called legacy systems). During the change process, strict controls must be employed to ensure that program changes are properly requested, approved, assigned, coded, tested, documented, and released to production. A major resource that facilitates these controls is a change management system (CMS). A CMS provides management with a wealth of information to enable management to evaluate and optimize systems maintenance activities.

SYSTEMS MAINTENANCE CHECKLIST

Following is a checklist on how to perform systems maintenance. Its purpose is to remind you of how key systems maintenance methods are applied.

1 Understand that systems maintenance may extend for years after implementation and become by far the most costly phase of the systems life cycle.

2 Comprehend the four types of systems maintenance: corrective, adaptive, perfective, and preventive.

3 Establish a systems maintenance life cycle (SMLC).

4 Install methods that will ensure the development of maintainable systems in the first place.

5 Organize the systems maintenance, using one of three approaches: separate, combined, or functional.

6 For maintenance of legacy systems, employ special CASE tools.

7 Install a sound means of managing the systems maintenance. Managing systems maintenance should include a formal way to initiate a maintenance request, such as a maintenance WO form supported by a help-desk system.

8 Implement a change management system (CMS).

KEY TERMS

Adaptive maintenance
Change management system (CMS)
Combined approach
Corrective maintenance
Forward engineering
Functional approach
Maintenance expert system
Maintenance work order (WO) form
Perfective maintenance
Preventive maintenance
Reengineering
Restructuring
Reverse engineering
Separate approach
Software maintenance life cycle (SMLC)
Systems maintainability
Systems maintenance

REVIEW QUESTIONS

20.1 What's the longest phase of the systems life cycle? Generally, how much of the total systems budget is allocated to systems maintenance?

20.2 What are the two principal sources for performing systems maintenance?

20.3 List and briefly describe the four types of systems maintenance.

20.4 Explain why corrective maintenance may be referred to as bad maintenance. What generally causes the need for corrective maintenance?

20.5 Give an example of adaptive maintenance. Is adaptive maintenance necessary and worthwhile? What is the cause of performing too much adaptive maintenance?

20.6 What's the purpose of perfective maintenance?

20.7 What's the purpose of preventive maintenance?

20.8 List and briefly describe the activities of the systems maintenance process.

20.9 A telephone company, following a maintenance change, introduced an error in its billing program for long-distance calls. When the error was discovered and corrected, the company was unable to recover $35 million in unbilled revenue. Explain how this problem could have been avoided.

20.10 What's the chief type of systems maintenance used in hardware maintenance?

20.11 Explain why MTTR measures maintenance efficiency, and MTBF measures maintenance effectiveness.

20.12 List steps for, and briefly describe, how designing to achieve a high level of maintainability is achieved.

20.13 Briefly discuss three ways to organize the systems maintenance process.

20.14 Differentiate between forward engineering and reverse engineering.

20.15 Describe the reengineering process. What is its purpose?

20.16 Describe the restructuring process. What is its purpose?

20.17 What's the purpose of a maintenance expert system? List its advantages.

20.18 How is scheduled systems maintenance typically managed?

20.19 Who normally initiates unscheduled systems maintenance?

20.20 Explain why the postimplementation review usually triggers some of the first maintenance performed.

20.21 List and explain the purpose of each element in the maintenance WO form.

20.22 Describe how the maintenance WO form helps optimize systems maintenance.

20.23 Describe CMS and its components.

20.24 What is the purpose of a CMS?

CHAPTER-SPECIFIC PROBLEMS

These problems require exact responses based directly on concepts and techniques presented in the text.

20.25 Your company uses the coding structure for unscheduled maintenance service shown on the opposite page:

Assign a code based on this maintenance coding structure to the following systems maintenance situations:

1 An order-entry system has been operational for 18 months. During this period, the company's product line has been expanded, two new warehouses have been added, and three sales offices have been opened in Canada. Additionally, users have indicated that the report showing backorders would be of greater value if it were to reach their desk before noon rather than by 4:00 P.M. Key users of the new system will be available Monday for analysis of the new system.

```
                                     X    X    X
Type of maintenance__________________|    |    |
   1 Corrective                           |    |
   2 Adaptive                             |    |
   3 Perfective                           |    |
   4 Preventive                           |    |
Priority of maintenance___________________|    |
   1 Emergency                                 |
   2 Urgent                                    |
   3 Routine                                   |
Type of activity involved______________________|
   1 Scrap
   2 Reengineer
   3 Restructure
   4 Replace
   5 Repair
   6 Modify
```

2 You, as the maintenance analyst for the payroll system, have just received a memorandum from the vice president of public affairs. She has requested information on the year-to-date number of minority hires. It is indicated further in the memorandum that henceforth this information will be required on a quarterly basis.

3 As you leave the tennis court, there is a message requesting you to call operations immediately. It seems the sales analysis program, which had been running smoothly for the last six months, has just abnormally terminated.

4 Accounting has just received its VGA monitors. The accounting folks want them installed to replace their monochrome monitors.

5 An inventory application was developed 15 years ago without using the structured approach. Documentation is sketchy and out-of-date. Users are satisfied with the application's functionality, but they have requested the description field be expanded. Your supervisor wants you to restructure and fully document the old software.

6 Jeff Travis, a data-entry operator, has called to tell you his keyboard has locked up. He further complains that it never has operated correctly, with keys sticking, and now it won't operate at all.

7 The capstan in the tape drive has frozen and will not turn, thus making the reading of the payroll master file impossible. It's 1:30 P.M. and the payroll must be processed by 4:00 P.M.

8 Based on design specifications generated from reverse engineering, it is decided that the existing software be discontinued and new software developed.

9 An old accounts receivable software package is to be structured and

documented. Also, some code is to be changed to gain efficiencies and to modify the packages functionality.

20.26 Given the following table:

```
01   CITIES
    05 FILLER     PIC X(10)       'BIGCITY'
    05 FILLER     PIC X(10)       'LITTLECITY'
    05 FILLER     PIC X(10)       'DREADVILLE'
    05 FILLER     PIC X(10)       'ROSYVILLE'
01   CITY-TABLE REDEFINES CITIES
    05 CITY  OCCURS 4 TIMES   PIC X(10)
```

what city name will be MOVEd to LINE-OUT (not shown) by the following pair of statements?

```
MOVE 2 TO CITY-SUBSCRIPT
MOVE CITY(CITY-SUBSCRIPT) TO LINE-OUT
```

This exercise contains a portion of a COBOL program. Assume that you know very little about COBOL. Can you understand this code anyway? If you were a junior maintenance programmer, could you readily understand the code and easily make any necessary changes?

20.27 Given the following program code:

```
IF C1
   PERFORM P1
   IF C2
      NEXT SENTENCE
   ELSE
      PERFORM P2
      IF C3
         PERFORM P3
         PERFORM P4
```

Reverse engineer this code in the form of a structured program flowchart. Comment on using names such as C1, C2, and so on. Are these names of any value to you in understanding what kind of application the program is performing? What would make this program more understandable besides the structured program flowchart?

20.28 A software program written 20 years ago is unstructured and without documentation. You have pored over meaningless source code and interviewed an operator. You have concluded that the program processes an accounts receivable file with a header and trailer. This file contains N records, and each record contains payment or invoice or both.

Required: According to your notes, reverse engineer the program in the form of a Warnier–Orr and Jackson diagram.

THINK-TANK PROBLEMS

These problems call for a feasible approach rather than a precise solution. Although the problems are based on chapter material, extra reading and creativity may be required to develop workable solutions.

20.29 A manager at Nept Company has hired you to help manage the changes made to programs. He outlines the following problems:

- ***Lack of an Accurate Program Inventory*** No one in information systems knows precisely what programs have been approved for use, what programs are actually in use, where the programs are located, and which programs are scheduled for changes.
- ***Incomplete History of Program Changes*** No one has a complete history of who requested or authorized changes, what programs have been changed, who changed them, or why.
- ***Program Abends*** A programmer spends hours reviewing the program listing trying to find out what went wrong, only to discover that the source code the programmer is looking at is not the version that was originally used to create the load module currently in use. The real source code is sitting in someone's private file, or worse, it's missing altogether.
- ***Duplicated Program Modules*** A number of programs can use the same modules for standard processes. Often modules that perform these standard processes are duplicated rather than being developed once and reused. Obviously, this duplication wastes both personnel and computing resources. Similarly, modules that are shared may be changed to accommodate one program only to disrupt another program's operation.
- ***Unauthorized Changes*** For whatever reasons, some people may make unauthorized changes to production programs. Such changes may be innocent. Others may be fraudulent, destructive, or the work of the notorious "midnight programmer."
- ***Lack of Documentation and Testing*** Pressures to fix a program may constrain effective program change documentation and may encourage using a program before it has been adequately tested.
- ***Inability to Back Out of a Change*** What happens if the revised code fails? Will the programmer be able to retrieve the earlier working version? Reconstructing the original file may be impossible.

Required: Explain to the manager how a CMS may help solve each problem.

20.30 Marsha Colvin, one of your more experienced and effective maintenance programmers, has requested a meeting with you to discuss the possibility of transferring to the systems development area.

"I'd like to move to the development area because I'm tired of cleaning up someone else's mess. And I would like to get over to development because there is a greater chance for exposure to top management and, consequently, I might be able to progress a bit faster," she begins.

"My progress here has been quite good, and I have nothing against you or the people in this area. It just seems that I might be able to do something more important—something I could call my own—over in development. I mean, that's where the action is, and that's where they're creating all the new systems that are really going to contribute to the company," Marsha continued. "Besides, development works with new equipment and new languages. It is not very exciting to be stuck here in maintenance and have to work with second generation and outmoded equipment."

Required: How would you respond to Marsha to convince her of the value of the systems maintenance function to the company, and the important role she can play within that function? Assume that you were put in charge of the information system. What kind of systems maintenance program would you recommend? Give reasons.

SUGGESTED READING

Carlyle, Ralph Emmett. "Fighting Corporate Amnesia," *Datamation,* February 1, 1989.

Chapin, Ned. "Changes in Change Control." *Proceedings of the Conference on Software Maintenance 1989*. Washington, D.C.: Computer Society Press of the IEEE, 1989.

Chapin, Ned. "The Job of Software Maintenance." *Proceedings of the Conference on Software Maintenance 1987*. Washington, D.C.: Computer Society Press of the IEEE, 1987.

Chapin, Ned. "Software Maintenance Life Cycle." *Proceedings of the Conference on Software Maintenance 1988*. Washington, D.C.: Computer Society Press of the IEEE, 1988.

Corder, A. S. *Maintenance Management Techniques*. New York: McGraw-Hill, 1976.

Dallimonti, Renzo. "Smarter Maintenance with Expert Systems." *Plant Engineering,* June 18, 1987.

Foster, Gerald D., and Hien Van Tran. "Maintenance and Money." *Information Strategy: The Executive's Journal,* Spring 1990.

Hanna, Mary Alice. "Defining the 'R' Words for Automated Maintenance." *Software Magazine,* May 1990.

Heintzelman, J. *Complete Handbook of Maintenance Management.* Englewood Cliffs, N.J.: Prentice-Hall, 1976.

Herbaty, Frank. *Cost-Effective Maintenance Management.* Park Ridge, N.J.: Noyes Publications, 1983.

Higgins, David A. "Structured Maintenance: New Tools for Old Problems." *Computerworld,* June 15, 1981.

Longstreet, David H. (Ed.). *Software Maintenance and Computers.* Los Alimitos, Calif.: Computer Society Press of the IEEE, 1990.

Marek, Bill. *CA Librarian Change Control Facility: Source Management the 1990's and Beyond.* Phoenix, Ariz.: Computer Associates International, Inc., March 1990.

Martin, James. "Restructuring Code Is a Sound Investment in the Future." *PC Week,* May 7, 1990.

Martin, James. "Reverse-Engineering Gives Old Systems New Lease on Life." *PC Week,* April 16, 1990.

Parikh, Girish (Ed.). *Techniques of Program and System Maintenance.* Boston: Little, Brown, 1982.

Parikh, Girish, and Nicholas Zvegintzov. *Tutorial on Software Maintenance.* New York: Institute of Electrical and Electronics Engineers, Inc., 1983.

Pressman, Roger S. *Software Engineering: A Practitioner's Approach,* 2nd ed. New York: McGraw-Hill, 1987.

Rayl, Eric. "SupportMagic Tops Help-Desk Software." *PC Week,* February 4, 1991.

Tayntor, Christine B. "Maintenance Magic." *System Builder,* October/November 1990.

Walters, Roger E. *The Business Systems Development Process: A Management Perspective.* New York: Quorum Books, 1987.

Weinberg, Gerald M., and Dennis P. Geller. *Computer Information Systems.* Boston: Little, Brown, 1985.

JOCS CASE: Maintaining the System

"I don't know about the rest of you, but I'm still tired," said Carla Mills. "It's been two months since we implemented JOCS, and I'm still trying to catch up on my errands. Nobody told me that systems people work 80-hour weeks when they're installing a new system. This was like working during tax season when I was in public accounting."

"Yes, but after the new system is installed, systems people get the joy of working 20-hour weeks until the next crunch hits," said Christine Meyers. She laughed. "Accountants tend to work steady hours all the time."

"I like the roller coaster aspect of my job," said Cory Bassett. "One of the reasons I went into this profession is the sense of accomplishment you get when a job is finished and done. I like the project orientation of systems work."

"Just a minute," exclaimed Jake Jacoby. "Who says this project is done? We finished the systems development part of the cycle. Now we have to work on systems operation. I'm planning for the JOCS package to have a long and healthy life. According to the company's strategic plan, JOCS should live at least a good seven years."

"Oh, come on, Jake," sighed Tom Pearson, "we all know that maintenance work doesn't take much effort. How about if we assign a new, entry-level person the maintenance tasks so that we can move on to developing the financial decision support system? I have some ideas about the new system I would like to talk about today."

"Is that what you guys thought this meeting was for?" asked Jake. "Discussing the new FIDS package?"

The JOCS, wanting-to-be-ex-SWAT team members looked at each other and nodded. A few vague comments such as "JOCS is working; it's time to move on"; "We've been looking at the new FIDS system"; and "I thought that the new FIDS project was why I went to the expert systems class," passed among the group.

Jake shook his head and said, "JOCS isn't finished and done yet. If this system is going to be in operation for seven years, we have to support it for seven years. The purpose of this meeting is to organize a long-term maintenance plan for the package."

Jake continues: "The time, effort, and planning we put into JOCS brought it in on time and within our cost estimates. I know we had a few problems, and it wasn't easy. Everybody on this team virtually sweated blood to finish this project. So I assumed that you'd all be interested in making sure that JOCS continued to function effectively over the long haul."

"Of course, we want JOCS to live, Jake," said Christine in a sweet tone. "But we did such a good job, that I hope we won't have any errors to correct in the future."

"Christine," replied Jake, "don't try to appease me. You've been working in systems long enough to know that that statement is never true. There are always defects to correct. But even if we never have to perform any corrective maintenance on the JOCS software, we still will have to add enhancements and modifications to the package. No piece of software can satisfy business requirements for seven years without a substantial amount of adaptive and perfective maintenance."

Cory jumped into the discussion with, "What you're saying, Jake, is that our system has to be evolutionary rather than static."

"Exactly," agreed Jake. "We aren't in a static business. New products, new personnel, new governmental requirements, new management, new competitors; any one of those factors will require us to help JOCS evolve. We all know software doesn't evolve naturally. Somebody has to make it change. I have a feeling that this isn't a popular issue with my team, but let me outline for you what I have done regarding system maintenance, and what we have to discuss today."

Jake hands out a brief worksheet (Figure 20.9) that sketches the maintenance areas that he wants to discuss.

"We'll start with hardware maintenance because, for us, that's truly the easy part. The initial hardware warranty was 90 days for all components. That time expired during the implementation phase of system development. I've contracted with the two hardware vendors to provide ongoing hardware support for the minicomputer, the file servers, and the bridge. For a monthly fee, the vendors will provide all parts and labor necessary to support the systems. In addition, the vendors perform preventive maintenance functions one morning each month. This is an expensive alternative, but I think this service will help us avoid potential hardware problems and will save us money

Figure 20.9
Outline of JOCS maintenance worksheet.

JOCS ONGOING SYSTEMS MAINTENANCE

I. Hardware Maintenance
   A. Vendor-supplied maintenance agreement
      1. LAN file servers
      2. Minicomputer
      3. Bridge hardware
   B. On-call maintenance agreement
      1. Individual microcomputers
      2. Terminals
II. Software Maintenance
   A. Vendor-supplied maintenance agreement
      1. Operating system
      2. DBMS
      3. Bridge software
   B. To be identified
      1. JOCS application software
      2. Data dictionary
III. Documentation—To Be Identified
IV. User Training—To Be Identified

in the long run. I calculated the costs for a hardware maintenance agreement for our microcomputers and workstations, and figured that we could buy two new micros each year for the price of a maintenance agreement. The insurance of a maintenance agreement just doesn't pay off for those components, so we will work with on-call support for them."

"I have also arranged with each vendor to supply ongoing software support for our operating systems, DBMS, and bridge software. This support will include automatic upgrades to the new releases, new manuals, and unlimited telephone support. We are a beta test site for the bridge software, so I hope we will receive excellent support from that vendor."

"Now we come to a more difficult area. We have to develop a method to handle JOCS maintenance. This includes corrections, enhancements, modifications, and anything we decide to add in the future to prevent software problems. Now, Tom, before you say anything, I think that an original JOCS team member should be responsible for these tasks. We know the software, the specifications, and the documentation. We should support this system."

"I wasn't going to say anything," said Tom. "I just don't think maintenance is all that difficult. Anybody should be able to understand the system."

"Tom," Jake responded sharply, "we have over 25,000 lines of COBOL code, a DBMS interface, an interface between the computer-integrated manufacturing system and JOCS, and are using some pretty specialized terminals. Could you walk in the door of this company and understand that kind of system?"

"I have a suggestion, Jake," offered Christine. "How about if we take turns maintaining the software? You could assign an original JOCS SWAT team member to the project for six months, and also assign a person unfamiliar with JOCS. The two could work together to maintain the system. After six months, switch the JOCS SWAT team member to a new project so that he or she will remain challenged by working with a new system. You could cycle the new person out on a different schedule."

"I like that idea," said Tom. "The biggest problem with maintenance is that boredom sets in after awhile. I don't mean to belittle maintenance. It's just that it can be ungodly dull at times. If we use Christine's suggestions, nobody will become permanently buried on a maintenance project."

"You know," Tom continued, "I just don't see why we should have two people working on maintenance, though. One should be plenty."

Carla, barely able to contain her comments, looked at Tom and said, "You must be kidding, Tom. With all the turnover we have in accounting, one of those people will be busy just conducting training classes."

"Yes," said Cory attempting to avoid an argument, "I think we should talk about training, too. We have to arrange an ongoing training program for both the DBMS report generator and query language syntax, as well as JOCS usage. Maybe the people assigned to maintenance could use their first few weeks to create and organize an ongoing training program."

"That sounds like a good idea," said Jake. "After we create a maintenance team, it could establish training and documentation procedures as two of its first tasks."

"What a relief," exclaimed Jannis Court. "I was so quiet because I couldn't figure out what you had in mind for the maintenance team during the first six months of maintenance time. I mean, the system is installed and working, and I can't imagine that we will have any enhancements these first few months."

"I think you may be right, Jannis," said Jake. "We have spent the last two months shaking down the JOCS system and correcting the obvious defects. I want the maintenance team to provide a method of ongoing documentation, problem identification and repair, and training during these first relatively easy months."

"Well," said Jannis, "I would like to volunteer to work on that project. I have some ideas about how to do those functions, and I would love to develop the initial procedures. Ever since I came back from my expert systems class, I have been thinking about creating an expert system to perform maintenance functions. I would like to take an expert systems shell and see if my theories will work. I think I could design an expert system to help us identify program defects, as well as locate program placement of potential enhancements. For example, one of my objectives is to be able to input the type of enhancement that we want to do, say something like adding a new type of transaction, and the expert system will tell us the likely area to be modified in the system."

"That sounds really interesting," said Cory. "I heard about that in college, but we never had time to finish up a real expert systems project. Jake, if you don't have a new hire available at this time, I would like to work with Jannis on this project."

"OK, Cory! Well," said Jake, "it looks like we have our first maintenance shift in place. I am impressed with the maintenance expert systems concept, Jannis. Work up the costs, and send me some E-mail detailing what you would need to do a prototype for the project. I think we could find the funds to support the development of a good maintenance system." He laughed and said, "That was an odd-sounding sentence. I guess we have come full circle at this point. I mean, here we are developing our maintenance system. Maybe systems developers are always developing systems, no matter if we call those systems old or new."

Glossary

Acceptance Testing Means for determining a system's completeness and readiness for implementation. It is the last test before the new information system is installed for operation. Two acceptance testing procedures are alpha and beta testing.

Access Controls Procedures and physical mechanisms that prevent unauthorized users and applications from using the system, while permitting authorized users and applications to use specified system resources.

Adaptive Maintenance Maintenance performed in response to changes in the system or business environment.

Alpha Testing Type of acceptance testing conducted by users with assistance from, and monitoring by, systems professionals.

Attributes (or Attribute) In the relational model, the term means one column of a table; in traditional file terminology, the term means a field. In general, it is a named property of an entity or a file.

Audit Trail A comprehensive record in magnetic or hardcopy form of all systems activity, including who logged on when and where, accessing what and for how long. The audit trail records complete data on all transactions. The audit trail also tracks errors and unsuccessful attempts to violate access privileges.

Automatic Code Generator A software development tool used to produce code without the need for hand coding.

Automatic Teller Machines (ATMs) Devices for transaction processing that allow individuals to bank electronically.

Back-End Phases Development and implementation phases that involve the functional design and implementation of the selected conceptual design derived during the front-end phases. The back-end phases include detailed systems design and systems implementation. The systems maintenance phase is often referred to as a back-end phase of the total systems life cycle.

Bar Codes Zebra-striped vertical marks on bars used to represent data. An example of a bar code is the universal product code (UPC) used by most grocery and other retail stores.

Bar Graphs Block-like graphs used to compare quantities. There are two types of bar graphs: the horizontal bar graph and the vertical bar graph. In each graph quantities are presented by bars placed on a grid. The length or height of the bars represent amounts determined by the scale of the grid.

Basic Accounting Equations Expressions of variables used to compute income for the period and financial position at the end of the period.

Batch Controls Control procedures and devices that are applicable to batch processing, such as field, record, and file checks and control of reports and sensitive documents.

Batch Processing A set up of computing resources that collect transactions over a period of time and process these transactions as a group or batch. In computer operations, the processing of a group of related transactions or other items at planned intervals. Large cyclic processing needs, such as payroll and accounts receivable, benefit from this individual processing of application programs. In any information system, batches are normally processed overnight or during slow periods.

Batch System Output Controls Pertain to the safeguarding of documents, processing, printing, transportation, and eventual scheduled destruction of output in a batch processing system.

Benchmark Tests Standardized procedures and transactions that are run on candidate computer architectures for the purpose of evaluating and selecting the one that performs best.

Beta Testing Type of acceptance testing performed by users without assistance from or monitoring by systems professionals.

Biometric Controls Mechanisms that control access to the system based on the physiological and behavioral characteristics of people, such as hand geometry, retina, fingerprint, body weight, dynamic signature, voice, and keystroke.

Black Box Testing The software program (or any system being tested) is viewed as a black box. The tester inputs test transactions and observes output, but does not know how it works internally. Black box tests the software's functionality.

Block Code A specified sequential block of numbers reserved for each classification to be coded, distinguishing the classifications to which each item belongs.

Bottom-Up Design A process of starting with the basic objects and assembling them until the final product is completed.

Bottom-Up Integration Testing The lowest-level modules are tested first. Test drivers are written to call them and pass them test data. If this level of testing is successful, the next-higher-level modules are written along with their drivers and so on until all modules are tested together.

Bridges Devices that connect different local area networks (LANs) at the data-link layer.

Budget and Performance Evaluation Equation Method used to measure actual amount spent against the amount budgeted, and to indicate a budget variance, either favorable or unfavorable.

Bug A fault or flaw in the software code that causes errors to occur.

Business Plans A detailed formulation of the enterprise's strategic business goals and methods by which these goals are to be achieved.

Captions The headings, titles, or designations accompanying a field.

Central Repository A database of a computer-aided systems and software engineering (CASE) system that stores and updates all elements of a system under development.

Change Management System (CMS) A formal system that identifies, authorizes, evaluates, tracks, and coordinates required changes to information system resources.

Check Digit A digit generated from the code or account number to which it is assigned to help ensure that the code or account number is recorded correctly in subsequent processing.

Chief Information Officer (CIO) A person who manages the information system and ensures that it supports the enterprise's business plan and strategic goals.

Chief Programmer Team A group aimed at developing high-quality software efficiently through a team effort led by a chief programmer.

Class Box A rectangular shape, borrowed from the entity relationship diagram (ERD), that names a class object, lists its attributes, and describes its operations.

Classes (or Class) A set of objects that share a common structure and a common behavior.

Coaxial Cable A type of transmission cable composed of one wire, called a conductor, surrounded by a stranded shield that acts as a ground.

Code Controls Countermeasures such as check digits used to make sure codes are recorded correctly.

Code Walkthrough A peer-group evaluation of software structure and interworkings, sometimes referred to as white box testing.

Codes Systems of symbols (letters, bars, colors, icons, numbers, words) used to classify and represent assigned meanings.

Cohesion A measure of the degree by which each module carries out a single, problem-related, and well-understood function.

Cold Sites Backup sites that provide an environment for quick installation of computing facilities.

Color Codes Use of colors to set items or groups apart from each other.

Combined Approach A method that brings together development and maintenance personnel into one group.

Company-Owned Backup Facility A remote computing facility that is a mirror image of the main operating facility.

Comparative Reports Output that shows likenesses and differences based on horizontal, vertical, and counterbalanced analysis.

Computer-Aided Systems and Software Engineering (CASE) A product that helps automate the systems development life cycle, software development life cycle, and the systems maintenance phase.

Concatenated Linked together.

Concurrency Controls Database controls that prevent different users from accessing the same data simultaneously.

Context Level Diagram The highest-level model of the system described by a data flow diagram. The system is portrayed as a single process with sources and sink. The model sets the boundaries of the system.

Controls Procedural and physical mechanisms executed by one or more components to prevent, detect, or correct abuses, errors, or irregularities that affect the reliability and integrity of the system.

Cooperative-Based Architecture Configuration of computer technology distributed throughout the enterprise to serve all users in an optimal manner.

Corrective Controls Procedures that rectify a mishap or enable a system to recover from some failure.

Corrective Maintenance Maintenance performed in response to failure to identify or fulfill user requirements or to correct errors and design flaws.

Cost/Benefit Analysis Determination and comparison of costs and tangible and intangible benefits of a project to ascertain its level of economic desirability.

Cost-Volume-Profit Equation Measures the interrelations among cost of the product, volume of sales, and profit of the product. One of the most pervasive equations used in management reporting.

Counterbalance Report A report that discloses all sides of a situation under consideration. Generally a counterbalance report gives worst-case and best-case scenarios, and an intermediate scenario called a likely-case scenario or moderate-case scenario.

Coupling The measure of relative interdependence among two or more modules.

Data Dictionary File that details all of the attributes for a system and defines their relevant characteristics.

Data-Entry Screen Form, mechanism, or layout on a computer screen that facilitates data input.

Data Field A single data element that is treated as a unit within a record.

Data Flow A symbol of the data flow diagram that depicts the flow of data throughout the system.

Data Flow Diagram (DFD) A modeling tool that shows the sources and destinations of data, identifies and names the processes performed, and identifies data stores accessed by processes of an existing system or a proposed system.

Data Store A symbol of a data flow diagram that shows where data of the system are stored.

Database A set of interrelated records, placed in a series of files, that are linked together through the use of a database management system. The DBMS is organized to facilitate user interaction, especially the retrieval of information in response to queries posed by users.

Database Controls Measures that provide authorization, concurrency, encryption, and backup and recovery controls for safeguarding and maintaining integrity of the database.

Database Design The process of determining the content and arrangement of data needed to support various systems designs.

Database Management System (DBMS) Software that enables users with different application needs to create, access, modify, and maintain data in a database.

Database Management System (DBMS) Approach A database design approach that employs a DBMS.

Data-Oriented Approach A subapproach of the structure-oriented design approach that concentrates on decisions made by users. It is used when the processes of a system, as well as the input and output, are relatively undefined. Often an entity relationship diagram is used as a modeling tool. Users are interviewed. Depending on their responses, certain nouns become entities, certain verbs indicate relationships between entities, and certain adjectives or other modifiers indicate data attributes.

Debugging The removal of bugs from the software.

Decision Table A presentation in tabular form of a set of conditions and their corresponding actions.

Decision Tree A graphic representation of a selection control structure in which the combination of conditions is shown as sequences of decisions.

Destructive and Fraudulent Software Programs that possess the potential to cause damage to the information system or enable persons to steal assets.

Detailed (functional) Systems Design A process by which all of the systems components are designed precisely, both physically and logically.

Detailed Systems Design Report A documented deliverable that contains precise design specifications for output, input, process, database, controls, and technology. It is sometimes called the systems design blueprint.

Detective Controls Monitoring procedures and countermeasures that discover threats to the system as they are occurring.

Differentiation A strategic factor used by businesses to distinguish their products or services from those of their competitors.

Digitizer Device that can be moved over a form to convert the picture to digital data, which can be stored in a computer, displayed on a screen, or printed out on paper.

Direct-Access Storage Devices (DASDs) Secondary storage devices that are connected directly to the computer, such as magnetic and optical disk drives.

Direct Conversion Implementation of the new system and the immediate discontinuance of the old system, sometimes termed cold turkey or crash conversion.

Direct Entry In transaction processing, data entry is made into computer-processible form at the time the transaction is made.

Disaster Recovery Plan A plan that consists of four subplans: prevention plan, which establishes controls to try to anticipate and keep a disaster from happening in the first place; contention plan, which formulates how to deal with a disaster if one occurs; contingency plan, which provides a method to keep the business running as far as computer information system applications are concerned; and recovery plan, which formulates how to bring back the normal condition of the computer information system after a disaster has occurred.

Disk-Mirroring The writing of data on duplicated storage systems simultaneously at the same site.

Document Image Processing (DIP) A system composed of scanners, optical storage media, server, workstations, and various output devices, used to input, process, output, and control paper documents electronically.

Documentation Material that describes how the system operates.

Documented Deliverables Compiled results of work conducted during phases of the systems development life cycle. They indicate that milestones have been reached, provide a clear trail of the SDLC, and show progress of the total project.

Domain A pool from which the values for an attribute must be chosen.

Double-Key Public Key Cryptosystem Protective system using two mathematically complementary keys, the public one for encryption and the private one for decryption.

Downsizing The process of designing a computer architecture that is smaller and less expensive than a traditional mainframe-based system. Downsizing is achieved by offloading applications to minicomputers, microcomputers, and local area networks that might otherwise be handled by mainframes.

Driver Modules Groups of special test code written to call lower-level, fully developed modules and pass them test data.

Economic Feasibility Likelihood that sufficient funds will be available to support the estimated cost of the proposed systems project.

Economic Order Quantity Equation Method of measuring the optimum quantity of items to purchase or manufacture.

Egoless Programming Team A closely knit peer group of programmers who allow colleagues to check their code. The object is not to identify those who have made errors, but rather to increase the likelihood of overall project effectiveness and timeliness, to the credit of the whole team. Sometimes referred to as an adaptive programming team.

Electronic Data Interchange (EDI) Telecommunications hardware and software that enable trading partners, such as vendors, customers, the buying and selling company, banks, and transporters to be linked together as a meta-enterprise to conduct business and process transactions electronically.

Electronic Forms Forms displayed on screens that are filled in by keying operations. Electronic forms replace paper forms.

Elementary Process/Entity Matrices A modeling tool offered by some CASE systems that defines the effects of elementary processes on entity types.

Elementary Processes Processes that are defined at their most basic level.

Encapsulate The process of combining data attributes with the operations. The term "operations" is used at the software design level because operations, like structured English, specify the functions of the design so that both users and programmers can understand the design. During coding, programmers will convert operations into actual object-oriented program code, which is often referred to as methods. At this point data attributes are encapsulated with methods, which are dedicated to manipulating the data attributes. One of the most important tenets of object-oriented programming is that the programmer should think in terms of data attributes and methods (code) *together*.

Encryption A process that scrambles a message so that it cannot be read by an unauthorized user. Encryption devices consist of either software or hardware that safeguards data in files or during transfer between nodes in a network. Encryption keys are used to encode and decode the message. The two most popular encryption systems are the single-key data encryption cryptosystem and the double-key public key cryptosystem.

Enterprisewide Model A very high-level systems design that shows all the major entities of an enterprise and how the entities are related.

Entities (or Entity) A set of persons, places, things, or objects, all of which have a common name, a common definition, and a common set of properties or attributes.

Entity Relationship Diagram (ERD) A modeling tool that shows entities and relationships that may exist between these entities.

Equations Expressions that help specify processes.

Equivalence Classes A group of tests that will give the same results. If one test in an equivalence class catches an error, then the others should also catch the same error.

Error Report Contains all discovered errors and their final disposition.

Exception Report A report that is generated when a process or activity exceeds or falls below predetermined limits or quotas.

Executive Information System (EIS) A locally based system that serves a few strategic-level managers.

Expected Value Equation Measures the average value that would result from observations of a distribution.

Expert Users People who completely understand a system and use it regularly. Also called power users.

Fatal Errors Catastrophic errors that will cause the system to terminate abnormally, perform functions improperly, or perform endless loops.

Feasibility Factors Technical, economic, legal, operational, and schedule factors contributing to the possibility that a system will succeed.

Fiber Optic Cable Transmission cable made of any filament or fiber (usually glass fiber) of dielectric materials that are used to transmit laser- or light-emitting diode (LED)-generated light signals.

Field Checks Procedures that determine the correctness of fields within records.

File Checks Procedures that determine the correctness of the files to be processed.

Filter Report A report that minimizes the information detail needed at specific levels in a user hierarchy. Generally, an executive needs less detail than a lower-level manager.

First Normal Form (1NF) Data containing no repeating groups and attributes; each attribute is named, and the primary key is not null.

Foreign Corrupt Practices Act of 1977 An act passed by the U.S. Congress that requires (among many other stipulations) all companies registered with the Securities and Exchange Commission (SEC) to establish and maintain a sound system of accounting controls.

Foreign Key (FK) An attribute, or collection of attributes, in a relation whose value is required either to match the value of the primary key in another relation or to be null.

Forward Engineering Development of systems using methodologies, modeling tools, structured and modular approaches, and automated technologies.

Fourth-Generation Languages (4GLs) A term generally used to label an array of software productivity tools. As used in this book, 4GLs are proprietary, nonprocedural languages developed to replace third-generation languages (3GLs), especially COBOL.

Front-End Phases The early phases of the systems development life cycle, which include systems planning, systems analysis, general (conceptual) systems design, and systems evaluation and selection. These phases set the stage for detailed systems design and systems implementation, which are referred to as the back-end phases of the SDLC.

Front-End Processor (FEP) A computer designed for communications control of the information system. The FEP is usually located between a mainframe and the various nodes.

Function Point Metric Measure developed by A. J. Albrecht of IBM, to assess productivity of software development by counting five functions delivered to users: amount of input, amount of output, number of queries, number of logical data files the system uses, and number of interfaces to other applications. To develop a function point index each function point is assigned a numerical complexity level.

Functional Approach A method that locates systems professionals within the operating areas of the business enterprise.

Gantt Chart A schedule of a systems project from the first phase to the last. Within a Gantt chart, the horizontal axis depicts time, and the vertical axis lists phases. The chart is used to compare planned performance against actual performance to show whether the systems project is ahead of, behind, or on schedule.

Gateways Devices that interconnect two otherwise incompatible networks (usually wide area networks). Gateways perform protocol-conversion operations across numerous communication layers.

General (conceptual) Systems Design A phase in the systems development life cycle in which alternative broad designs of systems are presented and recommended for further evaluation.

General Systems Design Report A documented deliverable that presents the findings and general systems design alternatives at the end of the general (conceptual) systems design phase of the systems development life cycle.

Gentralized Information Systems Model A preliminary and broad design of the technology platform, made up of a telecommunication network and computer architecture, that supports the enterprisewide model, prepared during systems planning to provide a broad computer information systems perspective.

Global-Based System An information system that pervades the entire organization and interconnects group- and local-based systems.

Grandfather-Father-Son Recovery Plan A three-file backup system in which the son is the current file onsite, the father is in the local vault or library, and the grandfather is stored in a safe facility offsite.

Graph A diagram or chart displayed as a series of one or more points, lines, line segments, curves, or areas that represent one variable in comparison with one or more other variables.

Group Code Hierarchically structured codes that are arranged so that the interpretation of each succeeding symbol depends on the value of the preceding symbol.

Group-Based System An information system that serves users who are affiliated with a special group, segment, or department in an organization.

Help Features Information in online documentation that can be called by users to accomplish an end.

Hierarchical Models Data models of entities in which the relationships between entities are viewed as an upside-down tree structure.

Hierarchical Reports Output which condenses, aggregates, and levels information to enable managers at all levels to receive the information that meets their specific requirements without having to sort through irrelevant detail.

Horizontal Bar Graph Graphic which compares different items during the same time frame.

Horizontal Report Comparison of amounts of the current time frame with amounts of previous periods.

Hot Sites Secure computing facilities that provide clients with ready access to complete backup processing.

Hybrid OOP Languages A number of popular third-generation languages that have been extended to include object-oriented capabilities.

Hypertext A method of organizing and associating textual material in a nonsequential hierarchical manner.

Iceberg Effect The results of systems development in which more than 80 percent of the systems budget is being spent on systems maintenance and less than 20 percent is being spent on development of new systems.

Icon Menu A list or assortment of options that a computer can perform, represented by symbols.

Icons Emblems or symbols displayed on the screen to stand for objects, menu options, commands, or applications.

Image Scanners Devices that identify typewritten characters or images to convert them to digital data to be stored and processed by computers.

Information Engineering Life Cycle (IELC) A type of systems development methodology used by some systems professionals. The IELC is similar to the systems development life cycle presented in the book.

Information Engineering Methodology (IEM) A term used by some practitioners to describe a systems development methodology similar to the systems development life cycle presented in the book, especially regarding its emphasis on the systems planning phase.

Inheritance A mechanism that permits classes to share attributes and operations according to a specific relationship. Inheritance defines a hierarchy of classes in which a subclass inherits from one or more superclasses. A subclass usually augments or adds new attributes and operations (or methods).

Input Controls Physical and procedural devices and checks used to verify the validity, accuracy, and completeness of data entered into the system for processing.

Input Design The process of conceiving and devising methods and devices to enter data into the information system for storage and processing.

Input Identification Controls Methods used to authenticate transactions.

Input Validation Controls Procedures used to discover errors in data before the data are processed. Validation controls are performed at the field, record, and file levels.

Inputs Data that are entered into the system via input devices or retrieved from a database and processed to produce output.

Insourcing A practice by which an enterprise develops and manages all aspects of its computer information system and provides information system services to all users throughout the enterprise. In many situations, the computer information system is a business within a business that also provides information system services to businesses outside the enterprise in which it is based.

Intangible Benefits These are financial contributions generated by the information system that are difficult to estimate at an actual value. They generally arise out of, or are identified by means of, one's perception and experience. Intangible benefits reduce expenses or generate revenue, or both.

Integrated Services Digital Network (ISDN) All-digital telecommunications service capabilities offered by telephone companies.

Integration Testing This type of testing is performed by bringing together modules in steps and concentrating on and checking the interfaces between modules.

Interactive Dialogue Language Tools An array of language tools that enables novice, occasional, and transfer users to interact with the system easily.

Interactive User-Oriented Language Tools These are language tools that are easily understood by novice, occasional, and transfer users.

Internal Revenue Service (IRS) A federal agency that requires certain tax, accounting, and other records to be kept a certain way for a period of time.

Internal Software Documentation Specific information about the program which is embedded in the program code.

Interoperable A term borrowed from the military referring to situations where military groups of the North Atlantic Treaty Organization (NATO) joined together cooperatively to share key resources rather than having each military group maintain separate stocks of common resources. Interoperable computer architecture designs distribute information system services to all users throughout the enterprise in an optimal sharing and cooperative manner.

Interviewing An exchange of information, planned with a specific purpose, and centered on asking and answering questions to gain study facts for analysis.

Iteration Repeating steps; an important feature of the structured approach that permits systems professionals to go back to earlier phases and tasks to make incremental improvements and respond to refined or changed definitions of user requirements.

Jackson Diagram A hierarchical diagram used primarily for software design to indicate each module and the sequence, selection, and repetition constructs. It is named after its developer, Michael Jackson.

Joint Application Development (JAD) A technique used to involve users in systems development.

Lease A contract by which the lessor (i.e., owner of the computer architecture) conveys the computer architecture to a lessee (i.e., user) for a specified term and a specified rent.

Legal Feasibility The likelihood that the proposed system will be in compliance with regulations and laws.

Line Graphs Graphs used to compare how several items vary over time.

Local Area Network (LAN) A network backbone that is either contained inside a single building or site or situated within a compact area whose longest dimension is not more than several miles.

Local-Based System A system usually designed for one or a few people for a parochial, mission-critical application, such as an executive information system.

LOEC Metric Measures productivity of software development by counting the lines of executable code.

Logic Bomb A piece of destructive and fraudulent software that performs its destructive act when triggered by an event of some kind.

Magnetic Ink Character Recognition (MICR) Method of machine-reading characters made of ink containing magnetized particles.

Maintenance Expert System The capabilities and knowledge of human maintenance experts.

Maintenance Work Order (WO) Form A document used to initiate and record maintenance work.

Management People who plan, control, and make decisions for the enterprise. This term is used to represent one of the strategic factors that the information system supports.

Manifolding A process of designing a form in which many copies are created with one writing.

Matrix Method to display information arranged and related by rows, columns, and cells.

Mean Time Between Failures (MTBF) The quantitative measure of reliability.

Mean Time to Repair (MTTR) The quantitative measure of maintainability.

Menu A list of options displayed on a screen, ready for selection by the user.

Messages (pertinent to object-oriented design) Invocations of objects and classes with directions to behave in a certain way.

Messages (pertinent to online documentation) Communications in text or symbols that are displayed on the screen as user instructions or guides.

Methods The implementation of operations for a specific object. Methods are the actual object-oriented programming code that manipulate attributes (also called data attributes).

Microwaves Data transmission is performed via high-frequency waves that travel in straight lines through the air. The waves (or signals) are relayed via strategically placed microwave antennas.

Minor Errors Defects that cause annoyances but are not serious or fatal.

Mobile Data Center A backup system installed in a large vehicle to be transported to a site requiring computing facilities.

Modeling Tools Methods used to create diagrammatic descriptions of systems and software designs on paper or screen for review and evaluation by both systems professionals and users.

Modem MOdulator/DEModulator device used to convert serial digital data from a transmitting terminal to a signal suitable for transmission over a telephone channel, or to reconvert the transmitted signal to serial digital data for acceptance by a receiving terminal.

Modularity A principle of both structured- and object-oriented design that supports maintainability, reusability, reliability, and extendability.

Modularization The segmentation and structuring of software into modules that can be coded, tested, and documented as a unit. Building a large program from loosely coupled and tightly cohesive modules is one of the precepts of the structured approach.

Module Testing Subjecting a single software module to white box and black box testing.

Monitoring Reports These outputs are generated by the system's observing certain activities and processes and reacting to predetermined conditions. Such reports include variance and exception reports.

Monthly Progress Report A summary of the weekly progress reports prepared by individual systems project team members or systems project leaders.

Mouse Device that can be moved on a desktop to move a cursor on the screen.

Multimedia (Hypermedia) The linking together of text, graphics, video, and audio to produce output with a high degree of usability.

Multiplexer (Mux) A device used for division of a transmission facility into two or more subchannels, either by splitting the frequency band into narrower bands (frequency division) or by allotting a common channel to several different transmitting devices one at a time (time division).

Multiplicity Measure showing how many instances of one class may relate to a single instance of an associated class.

MURRE Design Factors Five design quality factors represented by the MURRE acronym: maintainability, usability, resuability, reliability, and extendability.

Natural Language Interface A means by which people can communicate with the computer in their conversational language.

Nested Menus A set of selection and processing alternatives in which a selection from one set of choices leads to a subsequent decision about other alternatives included within the one chosen.

Net Present Value (NPV) Method Discounting all expected future net cash inflows to the present, using some predetermined cost of capital rate.

Network Interface Card (NIC) The network element that determines cable access method by making the physical connection between a workstation and other terminal devices and the network cable. The NIC determines the topology of the network.

Network Models Data models in which any entity in a network may be related to any other entity. Complex network models can be used to show all possible relationships among entities in a database.

Network Operating System A software product that determines the network's functionality.

Network Servers Special-purpose computers that control peripheral devices such as printers and storage devices and assist workstations in accessing these devices.

Network Topology The physical and logical relationship of nodes in a network. Networks are typically of either a star, bus, or ring topology, or some combination.

Normal Forms Methodology for grouping attributes to prevent insertion, deletion, and update anomalies.

Normalization A formal approach to logical data modeling which groups data attributes that have been converted into fields together to form stable records.

Novice Users People with little experience and systems knowledge who are using the system for the first time.

Null A special value meaning "not applicable" or "unknown."

Object Something that you can do things to or have behave in a certain way by sending messages. One element of a class.

Object Class Library A centralized and documented collection of object classes available for design.

Object-Oriented Design (also Object-Oriented Software Design) At the general systems level, a process of identifying object classes, identifying relationships, identifying attributes, identifying inheritance, and building a class hierarchy. At the software design level, a process of defining in detail each object's attributes and operations.

Object-Oriented Design Approach A means by which objects encapsulated with data and operations are assembled together to form a system.

Object-Oriented Programming (OOP) The development of reusable objects that encapsulate data and procedures in a single package or module and behave according to messages.

Object-Oriented Programming (OOP) Languages Programming languages that support all or some of the object-oriented design concepts.

Observing To watch carefully, especially with attention to details or behavior, to gather study facts.

Occasional Users People who have learned a system once but who use it so infrequently that few details are likely to be remembered.

Online Documentation Information communicated by the computer to users to guide users in how to work with the system via the screen rather than paper documents.

Online Reference Manual A detailed source of on-screen information to which users are referred. A representation of a paper-based reference manual, containing a table of contents, index, glossary, and headings. Topics within the online reference manual can be accessed easily via menus and commands.

Online Transaction Processing (OLTP) A process in which transactions are captured and entered into the system as soon as they occur and make an immediate impact on the system.

Operational Feasibility The likelihood that the proposed system will be functional according to personnel and resources available when the system is implemented.

Operations Functions that are applied to objects in a class. Operations describe what the object does at the software design level in terms that users and object-oriented programmers alike can understand. During the coding phase, operations are converted to methods, which represent the object-oriented programming code.

Operations Documentation Information and substantive material that guides operators in working with the computer system and performing their tasks. The term "operations" refers to operating a computer system. It is not related to the term "operations" used in object-oriented design.

Optical Character Recognition (OCR) Direct-entry technology based on special characters, letters, and numbers that can be read by a device that converts these data into digital data, which can be processed by computers.

Optical Mark Recognition (OMR) Direct-entry device that senses the presence or absence of a mark and converts signals into digital data which are processed by the computer.

Output Information produced for users by the information system. The substance of output should be accurate, relevant, and timely. The form of output must agree with the user's cognitive style, such as graphs for big picture people and numbers for detail people.

Outsourcing A practice by which an enterprise turns over its computer information system to an outside vendor that then supplies information system services to the enterprise for a fee.

Paper Forms Documents that contain captions and fields in which data are entered. Source-document paper forms are used for data entry. Special printer paper forms such as invoices or payroll checks are used for output.

Paper-Based External Software Documentation Information about the software program contained in a program manual.

Parallel Activities A feature of the structured approach by which more than one person can work on and develop different parts of the system under development simultaneously and thus speed up systems and software development.

Parallel Conversion In converting from an old information system to a new one, the old and new systems are operated side by side until the new one has shown its reliability and everyone has confidence in it.

PDM Strategic Factors The three critical success factors represented by the PDM acronym: productivity, differentiation, and management. The purpose of information systems is to support and enhance these strategic factors.

Perfective Maintenance Support performed to enhance the quality of the system

Phase-In Conversion Implementing the new system over a period of time, gradually replacing the old system.

Picturegraph Columns of small signs or icons that are used instead of bars as one finds in a bar graph. Each sign or icon represents a certain quantity of the item illustrated.

Pie Chart A circle that has been segmented into portions, each of which represents a certain percentage of the whole.

Pilot Conversion In converting from an old information system to a new one, only a part of the enterprise tries out the new system. The new system must prove itself at the test site before it is implemented throughout the enterprise.

Point-Of-Sale (POS) Devices that capture data at the point where a sale is transacted and transmit the data directly to the computer for immediate processing.

Polymorphic (or Polymorphism) Giving a message or name that is shared up and down a class hierarchy, with each class in the hierarchy implementing the message in a way appropriate to itself.

Postimplementation Review Audit of the new system soon after its implementation to see how well it supports systems factors, how well its systems design components are meeting user requirements, number and degree of variances from estimates, and level of support provided the new system.

Premises Distribution System (PDS) A structured cabling methodology for connecting nodes in a local area and for linking these local networks to wide area networks.

Present Value Index (PVI) Method Placing all competing investment alternatives on a comparable basis for the purpose of ranking them. The PVI used in this ranking is a ratio that compares the present value of net cash inflows to the present value of the net investment.

Preventive Controls Countermeasures to forestall anticipated threats.

Preventive Maintenance A periodic inspection and review of the system to uncover and anticipate problems and deal with them before they negatively impact the system.

Primary Key (PK) One or more attributes in a table that make that table different from the others in the table. In a traditional file system, a primary key is one or more fields in a record that make that record different from the others in the file.

Private An interface that declares an object or class unaccessible by any other classes or objects. (Contrast with the term "public.")

Process An action that converts input to output. Dimensions of processes include time, modeling tools, and technology platform.

Process Action Diagram A modeling tool offered by some CASE systems that shows elementary processes and detail steps within processes for a specific application.

Process Design Involves specifying when and how processing activities should be performed to support user requirements and convert input to output.

Process-Oriented Approach A subapproach of the structure-oriented design approach that concentrates on the input, output, and processing of a fairly well defined system to determine the data needs of the system. A data flow diagram can be used effectively to model processes.

Process Specification Tools Devices that enable description of elementary-level processes. Popular process specification tools are structured English, decision tables, decision trees, and equations.

Productivity Having the ability to produce high-quality results, utilities, and profits from the use of resources effectively and efficiently.

Program Change Sheet A record of changes made to a program after it is converted to operations.

Program Development Team A traditional hierarchical organization of designers, coders, and testers under the supervision of a program development team manager.

Program Evaluation and Review Technique (PERT) Chart A commonly used project planning, scheduling, and controlling technique. It represents the tasks required to complete a project. It explicitly establishes sequential dependencies and relationships between tasks. A PERT diagram consists of both tasks and events.

Program Manual Paper-based external software documentation that contains a title page, software design, description of input and output, copy of source code, a program change sheet, and test cases.

Programmer/Analyst A systems professional who is responsible for performing all phases of the systems development life cycle.

Project Director A systems professional who is in charge of all systems projects under development.

Project Manager (or Project Leader) A systems professional who supervises a systems project team and who is normally in charge of one systems project.

Project Schedule Report A document that identifies the proposed systems project, the resources required to complete it, and a breakdown of its schedule using Gantt charts and PERT charts.

Protocols Formal sets of conventions governing the formatting and relative timing of message exchange between two communicating systems. Protocols make possible the transmission of computer communications over telecommunication channels.

Prototyping Building a model of the actual system that will look much like the final system based on the concept that people can describe what they do not like about an existing system easier than they can describe what they think they would like in an imaginary system.

Public An interface that declares an object or class accessible by other objects or classes. (Contrast with the term "private.")

Public Output Information from the system that is distributed to the public, such as stockholders, investment analysts, and various governmental agencies.

Pull-Down Menus Submenus that drop down from a main menu option.

Purchase An acquisition method in which the enterprise buys the computer architecture from a vendor and takes title to it. Contrast with lease.

Pure OOP Languages Programming languages developed strictly for the object-oriented approach.

Quality Assurance A process that occurs simultaneously with development to assure that high-quality systems and software are developed. Focuses on the methods used to develop systems and software, walkthroughs, and testing procedures.

Quality Control An after-the-fact evaluation of a completed system to determine if it operates satisfactorily. Generally, quality control evaluation occurs during systems and acceptance testing.

Query-By-Example (QBE) A database query language that permits users to fill in skeleton forms or tables to express queries or commands.

Rapid Application Development (RAD) A methodology that combines joint application development, specialists with advanced tools (SWAT) teams, computer-aided systems and software engineering tools, and prototyping to develop major parts of a system quickly.

Real-Time Processing Entering data in the computer and processing it as soon as a transaction or event occurs that produces such data. Data are processed fast enough so that the results can be used immediately.

Real-Time System Output Controls Physical and procedural devices and techniques that safeguard online real-time output and ensure its accuracy. These controls include telecommunication, terminal, and floppy disk controls.

Reciprocal Agreement A contract signed by two companies agreeing to provide backup computing facilities to each other in the event one suffers failure or disaster.

Record Checks Procedures used to determine if the proper records (data fields) are being processed.

Reengineering A redesign of a system to improve its form and quality and change its functionality.

Regression Test A procedure performed to make sure that a change in software does what it is supposed to do. An error is found and it is fixed. Then a test is repeated that exposed the problem in the first place. Regression testing also executes a standard series of tests to make sure that the change didn't disturb anything else.

Relational Database Management Systems (RDBMSs) Complex software packages that support the relational database model.

Relational Model A data model in which data are presented as two-dimensional tables and relationships between these tables.

Relations In the relational model, a mathematical term used to describe two-dimensional tables.

Relationships (or Relationship) A term used to describe associations between entities, relations, tables, or objects.

Remote Secure Storage Facilities Buildings or other enclosures that provide safe places to store data and documentation far from the main site for added protection.

Request for Proposal (RFP) A document that solicits proposals from vendors.

Resource Requirements Matrix A list of resources and amounts of funds required to support the new information system.

Responsibility Report Document that provides information specifically related to users' areas of responsibility and what they have control over.

Restricted Output Information generated by the system that is disseminated only to users within the enterprise.

Restructuring Conversion of nonstructured software to fully documented, structured software without changing its functionality.

Reverse Engineering A process of examining and learning more about the existing system by creating its design.

Run Manual A main part of operations documentation that instructs computer operators how to run jobs.

Run Sheet A key document in a run manual that tells the computer operator how to set up and run a particular program or job.

Salami Technique Unauthorized software code that processes data in such a way that it steals from the enterprise a little at a time.

Sampling The process of selecting representative items from a larger whole to gain information about the whole.

Satellite (or Satellites) A transmission medium launched into space that receives uplinks (signals) from earth stations, amplifies the signals, then retransmits them to earth on different frequencies, called downlinks.

Scatter Graphs (Scatterplots or Scatter Diagrams) Plots of observations relative to the x and y axes. For example, cost may be shown on the vertical or y axis and volume may be shown on the horizontal or x axis. The plotted points show the behavior of cost as it relates to volume. When a line is fitted to the plotted points, it is generally known as a regression line.

Schedule Feasibility The likelihood that the system will be implemented and ready for operations on its estimated completion date.

Second Normal Form (2NF) A logical data model is in second normal form when it is already in first normal form and all nonkey attributes are fully functionally dependent on the primary key.

Sectorgraphs Graphs that show how total amounts are divided. Two popular sectographs are pie charts and layer graphs.

Securities and Exchange Commission (SEC) A federal agency that specifies disclosure procedures for financial and operating information, and occasionally announces accounting changes. All companies that are listed on the stock exchanges must abide by many SEC regulations, most of which deal with reporting requirements.

Seeding A controlled process of prebugging or bebugging the software to determine the effectiveness of test cases.

Separate Approach A method that divides systems professionals into the development group and the maintenance group.

Sequence Code A serial structure that represents a one-for-one consecutive assignment of numbers to such items as payroll checks, account numbers, inventory items, purchase orders, and employees. Any list of items is simply numbered consecutively, usually starting with one.

Serious Errors Defects that produce incorrect output.

Shingled Menus Overlapping menus that enable the user to perform multiple tasks or to refer to several parts of text or images at the same time. The front menu is the most recent, or current menu.

Shortcuts Commands that permit expert users, and possibly transfer users, to perform tasks more directly and quickly than ordinary procedures.

Single-Key Data Encryption Standard (DES) Cryptosystem A secret key is used to encrypt and decrypt the message. It is possessed by the sender and receiver.

Sinks (Destinations) Symbols of the data flow diagram that indicate the net outputs of a system.

Smart Electronic Forms Show users how to fill in the form displayed on the screen, provide online messages and instructions, perform calculations, and transmit data for additional processing.

Software Coding The second phase of the software development life cycle, which converts the software design into a program by writing statements in a programming language.

Software Design The first phase of the software development life cycle, which converts systems design into a set of process specifications following the structure-oriented approach or the object-oriented approach.

Software Design Walkthrough A formal and structured review process used to verify that the software design meets the systems design specifications.

Software Design Walkthrough Report A document that results from a structured software design walkthrough. Includes identification information, walkthrough team names and signatures, software design appraisal, error list, and suggestions.

Software Development Life Cycle (SWDLC) A three-phase methodology for developing software required to support the detailed systems design. The three SWDLC phases are designing, coding, and testing.

Software Documentation Material that describes the software in detail.

Software Maintenance Life Cycle (SMLC) A prescribed methodology for performing software maintenance. Its phases include enter maintenance request; transform maintenance request to a change; specify the change; develop the change; test the change; train users and run an acceptance test; convert and release to operations; update the documentation; and conduct a postmaintenance review.

Software Reliability A quantitative measurement expressed in errors per thousand source lines of executable code delivered.

Software Testing The application of test cases and procedures to discover and eliminate errors and increase the form, quality, and reliability of the software.

Sound Cues Various beeps, chirps, tones, voice messages used to guide users through a series of tasks.

Source Documents Original paper forms that contain transaction entry data. Data are first recorded on a source document such as a sales order form and then converted to computer-readable form.

Sources (or Origins) Symbols of the data flow diagram that indicate the net inputs of the system.

Specialists With Advanced Tools (SWAT) Teams Small groups of highly trained and motivated systems professionals who have the latest systems development resources at their disposal (e.g., CASE tools). The purpose of SWAT teams is to develop high quality systems quickly and successfully.

Special-Purpose Language Tools Languages that serve specific purposes, usually for a small group of users; they include interactive user-oriented tools, DBMS query languages, and hypertext and multimedia design aids.

Spoofing A deception in which a sender of a message masquerades as the authorized sender.

Standard Control Constructs (or Control Structures) The essential elements of structured programs, which are sequence, selection, and repetition.

State Transition Diagram (STD) A modeling tool that represents a collection of finite states and state transitions that cause a process, object, or entity to change their states.

Statistical Equation Method used to measure the mean, median, and mode of data in addition to other measurements.

Steering Committee A group of executives who decide which systems projects proposals will be developed, backlogged, or scrapped. The group also provides budgets, oversees systems development, and resolves conflicts between users and systems professionals.

Straight-Line Equation Method used to fit a line to data with Y as the dependent variable; a as the fixed element; b as the degree of variability or slope of the line; and X as the independent variable.

Strategic Factors Include productivity, differentiation, and management factors that support the strategic goals of the enterprise.

Strategic Information Systems Plan (SISP) A formulation of a program of action that describes how the proposed information system will support the strategic goals of the enterprise.

Structure Chart A treelike diagram that displays software modules hierarchically and identifies the data that passes between them. Each box in the structure chart describes the module's function.

Structured Approach A disciplined, engineered approach to development of systems and software. The features of the structured approach are modeling tools, modularization, top-down decomposition, iteration, parallel activities, and systems development automation.

Structured English A narrative notation or linguistic modeling tool used to describe the function of software modules unambiguously so that both users and coders can understand the functions. Structured English follows the control constructs of sequence, selection, and repetition.

Structured Program Flowchart A modeling tool describing the standard control constructs sequence, selection, and repetition of a software program.

Structured Query Language (SQL) A de facto standard query language used in relational database management systems.

Structured Text A narrative modeling tool that is similar to structured English. It is used to define module functions designated by a Jackson diagram. The structured narrative can then be translated easily into a programming language.

Structure-Oriented Design (Structured Design Approach or Structured Software Design) A design process based on the methodologies, modeling tools, and techniques of the structured approach.

Stub Modules Lower-level modules used to take the place of a module that has not yet been fully coded. Stub modules are used in top-down testing to check the calling function of fully coded higher-level modules.

Study Facts Evidential matter upon which the systems design is based. Study facts come from three sources: existing system, internal sources, and external sources.

Subclasses Classes that inherit from one or more superclasses.

Superclass The class from which another class inherits.

Systems Analysis The second phase in the systems development life cycle that is conducted to gather and analyze study facts about the system under development.

Systems Analysis Report A documented deliverable that discloses the findings of systems analysis.

Systems Design Walkthrough A structured peer group review of the systems design under development.

Systems Development Automation The use of computers and software to develop systems and software.

Systems Development Life Cycle (SDLC) A set of prescribed phases and tasks for the development of an information system. Phases of the SDLC include systems planning, systems analysis, general (conceptual) systems design, systems evaluation and selection, detailed (functional) systems design, and systems implementation.

Systems Documentation Material that describes the functional design features of the system.

Systems Evaluation and Selection The fourth phase of the systems development life cycle, which entails a process by which systems value, costs, and benefits of alternative conceptual systems designs are compared, and one is chosen for detailed systems design.

Systems Evaluation and Selection Report A documented deliverable that discloses the findings of the systems evaluation and selection phase.

Systems Implementation This is the sixth and final phase of the systems development life cycle. It entails planning for and conducting conversion of the system from development to operations.

Systems Implementation Plan A detailed formulation, usually specified by a PERT diagram or Gantt chart, that shows a schedule of implementation tasks that are to be performed to complete the systems implementation phase.

Systems Implementation Report A documented deliverable of the systems implementation phase that includes two major segments: implementation plan and description of implementation tasks with their final results.

Systems Maintainability The capacity of the maintainer to perform corrective, adaptive, perfective, and preventive maintenance.

Systems Maintenance A phase in the systems life cycle in which the information system is monitored and evaluated, and, if necessary, modified.

Systems Plan Report A formulation of resources and applications that guides the development of an information system that supports the enterprise's business plan and puts into action strategic factors.

Systems Planning A process that derives and establishes a framework for satisfying an enterprise's information systems requirement. It is congruent with and supportive of the enterprise's business plan and strategic goals.

Systems Professionals People who play a significant role in the development and maintenance of information systems. Some examples are systems analysts, systems designers, systems auditors, programmers, database administrators, computer technicians, and telecommunication experts.

Systems Project Proposal Form A document used during systems planning to record requested systems projects and the TELOS feasibility factors score and PDM strategic factors score of each requested systems project.

Systems Project Proposals Systems project requests or suggestions put forward by planners to be considered and scored on their feasibility factors and contribution to strategic factors.

Systems Project Proposals Portfolio Statement A document that lists approved and tentatively approved systems project proposals.

Systems Scope A designation of the boundaries of a system under development.

Systems Testing A process that checks the integrated software in the context of the total system with which it is to work. Systems testing includes recovery, security, and stress testing.

Systems Value Measure of a system's TELOS feasibility factors, PDM strategic factors, and MURRE design factors.

T-Span Services Digital private-line services offered by telephone companies that support transmission rates beyond those available in standard dial-up or leased telephone lines.

Table (as used in output design) Columns of data with subject titles arranged on a grid.

Tables (as used in relational database design) Another name for relations.

Tangible Benefits Financial contributions generated by the information system that are capable of being appraised at an actual or approximate value. Tangible benefits reduce expenses or generate revenue, or both.

Technical Feasibility The likelihood that sufficient technology will be available to support the proposed system.

Technology Platform A configuration of hardware, software, and telecommunication devices that supports the systems design.

Technology Platform Controls Safeguards that protect the technology platform from abuse and destruction. These controls include mainframe data center controls, fire suppression systems, uninterruptible power supplies, PC controls, and disaster recovery plan.

Televaulting The process of transmitting data to backup sites electronically. Typically, the computer information system is connected to a secure remote site by a high-speed transmission medium, such as T1 service. Televaulting is also called electronic vaulting.

TELOS Feasibility Factors Five feasibility factors represented by the TELOS acronym: technical, economic, legal, operational, and schedule feasibility factors.

Test Case Matrix An orderly method of recording test cases and their objectives, expected results, and actual results.

Test Cases Procedures that examine the software and produce results that will lead to acceptance, modification, or rejection of the software.

Third-Generation Languages (3GLs) High-level compiler-based languages that focus on system application procedures.

Third Normal Form (3NF) A logical data model is in third normal form when it is already in second normal form and any partially transitive dependencies have been removed.

Tickler File A file that serves as a reminder and is indexed to bring matters to the attention of certain people in a timely manner.

Tiled Menus A number of menus displayed on the screen in a side-by-side or patchwork manner.

Top-Down Decomposition The process of breaking a system down into smaller and smaller parts that are easy to understand.

Top-Down Design A process of starting with the most general features of a system and designing them into finer and finer detail until the system is finally designed at its elementary level.

Top-Down Integration Testing A software testing practice by which a program's uppermost modules are coded first along with lower-level stub modules, continuing this process from least detailed to most detailed modules.

Top-Secret Output Extremely sensitive information that can be seen by only a few authorized users.

Touch Menus A number of options displayed on the screen that can be activated by touching the selected option.

Traditional File System Approach This is an approach to data storage and access that uses separate data files for each different application system. In a traditional file system, data are created, updated, and accessed by individual programs, typically written in a third-generation language (3GL).

Transaction Equation Method used to describe how business transactions are processed.

Transfer Users People who already know how to operate one computer system and are trying to learn another similar system.

Transform Analysis An examination of the detailed level data flow diagram to divide the processes into those that perform input and editing, those that perform processing such as calculations, and those that format and generate output. The result of transform analysis is usually a structure chart.

Transient Electromagnetic Pulse Emanations Standard (TEMPEST) Federal government source suppression security guidelines that limit signal emanations from computers.

Transmission Media Devices that carry signals, including twisted pair cable, coaxial cable, fiber optic cable, microwaves, and satellites.

Transmission Modes Types of data transmission that correspond to the three types of circuits available: simplex, half duplex, and full duplex.

Trojan Horse A set of software codes that is embedded in the authorized software package, put there to perform destructive or fraudulent acts.

Tutorials Sessions used in online documentation as lessons to instruct, train, and guide novice or occasional users.

Twisted Pair Cable A type of transmission cable which gets its name from the fact that its two wires are twisted together. Twisted pair cable may be shielded or unshielded. This was the original type of cable used by the telephone industry, and it is still widely used today.

Uninterruptible Power Supply (UPS) A facility that consists of a rectifier/charger, battery, and inverter that prevents power surges and other power transients from harming the computer system. A UPS also protects the information system from power failure.

Usability Factors Checklist A form that catalogs and provides weights to a number of factors that indicate the level of usability of a system from the viewpoint of users.

Usability Lab A place where a sampling of users actually interacts with the new software before it is released to operations. The activities of the users are monitored by systems professionals to ascertain the usability of the software.

User Documentation Information and procedures that guide users in working with the system.

Users People who employ and interact with the information system to perform their tasks. Referred to sometimes as end users, they are differentiated from systems professionals who develop, manage, operate, and maintain the information system. Users fall into the following types: novice, occasional, transfer, and expert.

V Curve This is a time vs. resources diagram shaped like the letter V which indicates that users are involved in systems development much more during the front-end phases, less during detailed systems design, and more during implementation.

Variance Report A document which compares standard performance or amounts with actual performance or amounts at the end of an activity or process. The deviations between standards and actuals are called variances.

Vertical Bar Graph Measures the same item at different periods of time.

Vertical Report Uses percentages to show the relationship of the different items to a total within a single report.

Very Small Aperture Terminals (VSAT) Small, portable telecommunication devices that send and receive satellite signals.

Virus A self-replicating segment of destructive software code that damages a host program.

Voice Recognition Direct-entry technology that acknowledges speech and converts this speech into digital data for computer processing.

Warnier–Orr Diagram (WOD) A bracket diagram that decomposes the software design from left to right. A Warnier–Orr diagram is fundamentally a series of brackets used with other symbols to portray a software design, named after its developers, Jean-Dominique Warnier and Kenneth Orr.

Weekly Progress Report A document that tracks the degree of progress that has been achieved and the level of expenditures occurring during the week for a specific systems project.

White Box Testing Procedures that involve a structured walkthrough of the interworkings of software code.

Wide Area Network (WAN) A network that uses microwave relays, satellites, and public digital services to reach users over long distances.

Wireless LANs A special form of local area networks in which nodes transmit data via radio signals.

Worm A segment of software code that replicates and spreads throughout the system, but does not require a host program.

X.25 Packet-Switching Networks Standard governing the interface between data communication devices and terminals for packet-mode transmission on public data networks.

Zoning A forms design technique in which the form is divided into specific regions where data will be input.

Index